Sustainable Agriculture in the Era of Climate Change

Rajib Roychowdhury • Shuvasish Choudhury
Mirza Hasanuzzaman • Sangeeta Srivastava
Editors

Sustainable Agriculture in the Era of Climate Change

Springer

Editors
Rajib Roychowdhury
Department of Vegetables and Field Crops,
Institute of Plant Sciences
Agricultural Research Organization
(ARO) – The Volcani Center
Rishon Lezion, Israel

Shuvasish Choudhury
Plant Stress Biology & Metabolomics
Laboratory, Central Instrumentation
Laboratory (CIL)
Assam (Central) University
Silchar, Assam, India

Mirza Hasanuzzaman
Department of Agronomy, Faculty of
Agriculture
Sher-e-Bangla Agricultural University
Dhaka, Bangladesh

Sangeeta Srivastava
Division of Crop Improvement
ICAR-Indian Institute of Sugarcane
Research
Lucknow, Uttar Pradesh, India

ISBN 978-3-030-45668-9 ISBN 978-3-030-45669-6 (eBook)
https://doi.org/10.1007/978-3-030-45669-6

This Springer imprint is published by the registered company Springer Nature Switzerland AG
The registered company address is: Gewerbestrasse 11, 6330 Cham, Switzerland

Preface

Since ancient times, human beings have been dependent on agricultural production, and previously the production systems were more natural and sustainable. Presently, climate change and global population increase jointly challenge sustainable agriculture and crop productivity. One of the imposing tasks for crop researchers globally is to diminish the negative effects of climate change on crop biology, especially in relation to yield and nutritive values of the harvested products of cereals, legumes, vegetables and fruits. This is of special significance in view of the impending climate change, with complex consequences for economically profitable and both ecologically and environmentally sound global agriculture. The challenge at the hands of the crop scientist in such a scenario is to promote competitive and multifunctional agriculture, leading to the overall crop improvement in a sustainable agricultural system. Hence, crop improvement is an urgent need for us to feed the ever-increasing world population. As a result of continuous climatic changes in the form of environmental hazards, abiotic stresses (water scarcity, salinity, high and low temperature, heavy metals and metalloids, oxidants, and others), high disease incidents, global warming, etc., natural and cultivated habitats of crop plants are continuously disturbed. Crop productivity and biology are seriously affected by such changing patterns. Beyond such kind of hazardous barriers, this book is aimed at crop improvement through sustainable agricultural processes.

In this book *Sustainable Agriculture in the Era of Climate Change*, Chap. 1 emphasizes the landraces of the crops which serve as the genetic stock having the many useful alleles/genes and can bridge between their genetic progenitors and modern cultivars and to utilize them in the modern breeding program. Chapter 2 is focusing on how to improve the grain damage of cereals due to pre-harvest sprouting and late maturity for gearing up production. Chapters 3 and 4 are dealing with the macro- and micronutrients for better crop development and physiological productivity. Chapter 5 describes the plant-mediated mitigation strategies of air pollution in the environment caused by the urbanization of lands. Chapters 6 and 7 are depicting the drought stress responses of crop plant and mitigation strategies with special aid from the involvement of silicon. Chapters 8, 9, and 10 are providing scientific background of crop plants that cope with salinity, high temperature, and irradiation stress, respectively. As a part of abiotic and biotic stress response pathways, reactive oxygen species (ROS) are formed and needed to activate the crop's anti-oxidant mediated mitigation pathways, which are described in Chaps. 11 and 12. Chapter 13 is providing heavy metal–based tolerance of crops. More specifically, arsenic becomes a serious threat to contaminate the crops and human health. Chapters 14 and 15 are describing the arsenic stress response of plants and their signaling pathways to mitigate the serious issues caused by such metalloid. Chapter 16 is providing the idea of heme oxygenase enzyme activity for its role to mitigate the stress conditions. In organic agriculture, plant growth-promoting rhizobacteria (PGPR) are found to be very useful for crop's stress tolerance and it is described in Chaps. 17. 18 is focusing on different tissue cultural methods to propagate the crop *in vitro*. Chapter 19 is showing the involvement of different state-of-the-art omics technologies that can be utilized in crop stress response research. Chapter 20 is describing microbial influences for crop's abiotic stress response. Chapter 21 is briefing molecular mechanism of plant–pathogen interaction for disease resistance of crop plants. Chapter 22 is focusing on the role of small RNA or microRNA (miRNA) for crop's abiotic and biotic stress tolerance. Chapter 23 is dealing with the transgenic improvement of crops and Chap. 24 is describing the utilization of very new gene-editing tool CRISPR/CAS for crop improvement. Different bioinformatics tools and their utilization for crop improvement research have been detailed in Chap. 25. Chapter 26 is emphasizing on nano-biotechnological utility in crop's stress tolerance. Chapter 27 is describing how to improve crop's agrobiodiversity with special reference to the underutilized cereals "millets" which can be an alternative source of carbohydrate and nutrition for the global population. Thus, *Sustainable Agriculture in the Era of Climate Change* covers a wide range of topics under present-day environmental challenges, agronomy and agriculture processes, and biotechnological approaches.

We are grateful to all the authors/contributors who gave their valuable time to write the scientific chapters within the deadline for the possible success of this book. We shall be highly thankful to the readers for pointing out the errors and omissions which, in spite of all care, might have crept in. All suggestions for further improvement of this edited volume will be highly appreciated and accepted. We, along with all the contributors, apologize to those researchers for the parts of their work that could not be cited in this edited volume/chapters due to the space limitation.

We are highly thankful to Kenneth Teng (Publishing Editor, Springer Nature, New York), Jacco Flipsen (Vice President, Springer, New York), Eric Schmitt (Managing Director, Springer International Publishing AG, New York), Nicholas DiBenedetto (Editorial Assistant, Springer Nature, New York), Saveetha Balasundaram (Production Editor for Springer Nature, SPi Global, Chennai, India), Mario Gabriele (Senior Project Manager, SPi Global, Chennai, India), Anthony L Dunlap (Project coordinator, Springer Natute, New York), and other members of the editorial staff for their approval of this book project, prompt and timely responses during the acquisition, production coordination and support, formatting, proof checking, and editorial corrections in the manuscripts to make this volume complete and publishable.

In this book *Sustainable Agriculture in the Era of Climate Change*, the diverse chapters are contributed by experienced, highly dignified, and internationally reputed scientists, researchers, and academicians from around the world who have worked on the challenging problems of achieving crop improvement in the variable environments. Every chapter has been written in such a way that it deals with the theoretical as well as applied aspects of its specific theme. Hope such hot topics will be helpful to formulate the future keys toward the crop improvement in the present challenging and changing climate. This book has been written for a wide range of readers with easy-to-understand tables and simplified diagrams.

We believe this book will be very useful for scientists, researchers, and students working in the fields of agriculture, plant science, environmental biology, and biotechnology.

Rishon Lezion, Israel — Rajib Roychowdhury
Silchar, India — Shuvasish Choudhury
Dhaka, Bangladesh — Mirza Hasanuzzaman
Lucknow, India — Sangeeta Srivastava

Contents

About the Editors

Rajib Roychowdhury is presently working as Postdoctoral Visiting Scientist at the Institute of Plant Sciences of Agricultural Research Organization (ARO) – Volcani Center, Ministry of Agriculture and Rural Development, Israel. His current research is focusing on Genetics and Breeding of Berry fruits (strawberry, blueberry, raspberry) to improve yield and quality under Mediterranean climate. In the postdoctoral tenure, Dr. Roychowdhury extensively worked on Winter Cereal Genetics and Breeding, especially to improve wheat in Israeli semi-arid environments (a part of fertile crescent of wheat domestication) using diverse genetic stocks like modern cultivars, landraces, durum, spelt, dicoccum, emmer, etc. Dr. Roychowdhury received his Ph.D. in Plant Biotechnology from Visva-Bharati Central University (India) for research on previously unreported rice landraces for their agro-morphology, grain quality, and abiotic stress tolerance. Later, he also did short-term postdoctoral research on "Metalloid stress tolerance in rice" under a CSIR project (Government of India) from Assam Central University, Silchar, India. During master's and post-master's tenure, he worked a lot on mutation breeding in Carnation, which was a pioneering work on this floricultural crop in South-East Asia. Dr. Roychowdhury was elected as a Fellow Member of the Linnaean Society of London, UK. His biography was published in the Pearl Edition of Marquis Who'sWho of the World, USA. In 2019, he got the travel award by International Wheat Initiative (Germany) for presenting his research in the 1st International Wheat Congress

(Saskatoon, Canada) and to participate in the pre-conference workshop on 'Un-manned Aerial Vehicle (UAV) based phenotyping for wheat improvement'. Dr. Roychowdhury has written several impacted research papers, review articles, invited chapters, and books in both international and nationally reputed journals and publishers. Till date, his publications got a total of 1400 citations as per GoogleScholar with the h-Index 15. Dr. Roychowdhury serves as Editorial Board member in 'South African Journal of Botany' (Elsevier), and reviewed manuscripts for 'Genetic Resources and Crop Evolution' (Springer), 'Phytoparasitica' (Springer), 'Ecotoxicology and Environmental Safety' (Elsevier), and 'Journal of Crop Improvement' (Taylor & Francis).

Shuvasish Choudhury is an Assistant Professor at Assam University, Silchar, India, since 2009. He obtained his M.Sc., M.Phil., and Ph.D. in Life Sciences from Assam University, Silchar, and Post Doctorate from Oklahoma State University, USA. He has published more than 40 research and review articles including book chapters in journals and volumes of national and international repute. Dr. Choudhury has been awarded Young Botanist Award by Indian Botanical Society in 2010 and thereafter with FAST TRACK Young Scientist Award and Overseas Associateship by SERB and Department of Biotechnology (DBT), Government of India respectively. Dr. Choudhury has undertaken several research projects in the multidisciplinary areas of plant sciences, funded by agencies of Government of India such as University Grants Commission (UGC), Department of Biotechnology (DBT), Science and Engineering Research Board (SERB), and Council of Scientific and Industrial Research (CSIR). He is also enlisted reviewer for several journals published by Springer, Elsevier, American Chemical Society, etc. His research interest involves understanding the molecular physiological and metabolomic basis of abiotic stress perception and tolerance in crops, especially metalloid and drought stress. Besides this, Dr. Choudhury's research also involves identification of bioactive marker compounds in medicinal plants used by local and indigenous people of Northeast India. Dr. Choudhury teaches biochemistry, plant systems biology and analytical instrumentation.

Mirza Hasanuzzaman is Professor of Agronomy at Sher-e-Bangla Agricultural University in Dhaka, Bangladesh. He received his Ph.D. in "Plant Stress Physiology and Antioxidant Metabolism" from the United Graduate School of Agricultural Sciences, Ehime University, Japan, as a recipient of a scholarship from the Japanese government (MONBUKAGAKUSHO). Later, Prof. Hasanuzzaman completed his postdoctoral research at the Center of Molecular Biosciences, University of the Ryukyus, Okinawa, Japan, as a recipient of the Japan Society for the Promotion of Science (JSPS) postdoctoral fellowship. Subsequently, he received the Australian Government's Endeavour Research Fellowship for postdoctoral research as an Adjunct Senior Researcher at the Tasmanian Institute of Agriculture, University of Tasmania, Australia. Prof. Hasanuzzaman has supervised 20 M.S. students. His current work is focused on the physiological and molecular mechanisms of environmental stress tolerance. Prof. Hasanuzzaman has published over 120 research publications in peer-reviewed journals. He has edited 12 books and written 45 book chapters on important aspects of plant physiology, plant stress responses, and environmental problems in relation to agricultural plants. According to Scopus®, Prof. Hasanuzzaman's publications have received roughly 4400 citations with an h-index of 33. He is an editor and reviewer for more than 50 peer-reviewed international journals and was a recipient of the "Publons Peer Review Award 2017, 2018, and 2019." He has been honored by different authorities for his outstanding performance in different fields like research and education, and has received the World Academy of Science Young Scientist Award (2014). He has attended and presented 25 papers at international conferences in many different countries (USA, UK, Germany, Australia, Japan, Austria, Sweden, Russia, Indonesia, etc.). Prof. Hasanuzzaman is an active member of 40 professional societies and is currently the Acting Research and Publication Secretary of the Bangladesh JSPS Alumni Association. He is also a fellow of The Linnean Society of London.

Sangeeta Srivastava is presently working as Principal Scientist at ICAR-Indian Institute of Sugarcane Research (IISR), Lucknow, India. She earned her M.Sc. (Botany) with 1st position from Lucknow University in 1985 and Ph.D. from CSIR-CIMAP, Lucknow, in 1989 as CSIR-NET JRF/SRF and secured the first rank in ARS examination for Agricultural Scientist in India. Since her Ph.D. and post-doctoral work in Germany and USA, Dr. Srivastava has made noteworthy contributions to realize genetic enhancement and diversity of sugarcane and sugar beet through classical and molecular breeding interventions. Through her dedicated efforts, a first-time Indian resource of 26,451 sugarcane ESTs and molecular diagnostic assays to detect incipient infection of pathogens of sugarcane diseases have been developed. She has more than 175 publications from internationally and nationally reputed publishers/journals including book compilation, chapters, popular articles, etc. Dr. Srivastava is a recipient of many prestigious national awards including ICAR-Young Scientist award, DBT-CREST Award, ISCA Platinum Jubilee award, Prof. YS Murty Gold Medal, Woman Botanist Gold Medal, SAB Award of excellence, Best Woman Scientist award, and fellowships of NABS, UPAAS, IBS, ISGPB, SAB, and SSRP. She has guided Ph.D. and M.Sc. theses and dissertations of many UG/PG students. She is the Editor in Chief of *Indian Journal of Sugarcane Technology* and has served as editor of *Journal of Environmental Biology*, *SugarTech* (Springer), *Plant Cell Biotechnology and Molecular Biology*, and *Indian Journal of Fundamental and Applied Life Sciences*, and reviewer of several national/international journals.

Contributors

Malay Kumar Adak Plant Physiology and Plant Molecular Biology Research Unit, Department of Botany, University of Kalyani, Kalyani, India

Sinchan Adhikari Department of Botany, University of Kalyani, Kalyani, Nadia, West Bengal, India

Muhammad Afzaal School of Environment and Life Sciences, Salford, UK

Shakeel Ahmad Department of Agronomy, Bahauddin Zakariya University, Multan, Pakistan

Mukhtar Ahmed Department of Agricultural Research for Northern Sweden, Swedish University of Agricultural Sciences, Umeå, Sweden

Department of Agronomy, Pir Mehr Ali Shah Arid Agriculture University, Rawalpindi, Pakistan

Afroz Alam Department of Bioscience and Biotechnology, Banasthali Vidyapith, Banasthali, Rajasthan, India

Aqsa Ali Department of Botany, Government College University, Faisalabad, Pakistan

Qasim Ali Department of Botany, Government College University, Faisalabad, Pakistan

Shafaqat Ali Department of Environment Sciences, Government College University, Faisalabad, Pakistan

Muhammad Azeem Department of Botany, Government College University, Faisalabad, Pakistan

Rajib Bandopadhyay Department of Botany (UGC-Centre for Advanced Study), The University of Burdwan, Burdwan, West Bengal, India

Chiranjib Banerjee Department of Environmental Science & Engineering, Indian Institute of Technology (ISM), Dhanbad, Jharkhand, India

Dina Barman Department of Botany, Gauhati University, Guwahati, India

Kaushik Bhattacharjee Division of Life Sciences, Institute of Advanced Study in Science and Technology, Guwahati, India

Bharati Bist Department of Botany, Rizvi College of Arts, Science and Commerce, Mumbai, Maharashtra, India

Koushik Biswas Department of Biotechnology, Visva-Bharati University, Bolpur, West Bengal, India

Kiranmai Chadipiralla Department of Biotechnology, Vikrama Simhapuri University, Nellore, Andhra Pradesh, India

Priyanka Chakraborty Department of Botany (UGC-Centre for Advanced Study), The University of Burdwan, Burdwan, West Bengal, India

Chunoti Changwal Division of Plant Physiology, ICAR-Indian Agricultural Research Institute, New Delhi, India

Sudeshna Das Department of Plant Physiology, College of Basic Sciences & Humanities, G. B. Pant University of Agriculture & Technology, Pantnagar, Uttarakhand, India

Arnab Kumar De Plant Physiology and Plant Molecular Biology Research Unit, Department of Botany, University of Kalyani, Kalyani, India

Debabrata Dolui Plant Physiology and Plant Molecular Biology Research Unit, Department of Botany, University of Kalyani, Kalyani, India

Pachipala Gayathri Department of Biotechnology, Vikrama Simhapuri University, Nellore, Andhra Pradesh, India

Arijit Ghosh Plant Physiology and Plant Molecular Biology Research Unit, Department of Botany, University of Kalyani, Kalyani, India

Parthadeb Ghosh Department of Botany, University of Kalyani, Kalyani, Nadia, West Bengal, India

Surendra Kumar Ghritlahre ICAR-National Rice Research Institute, RRLRRS, Kamrup, Assam, India

Sandeep Ghughe Institute of Plant Sciences, Agricultural Research Organization (ARO) – The Volcani Center, Rishon LeZion, Israel

Rachna Gupta Department of Biotechnology, Visva-Bharati, Santiniketan, Bolpur, West Bengal, India

Lisha Gurung Department of Botany, Gauhati University, Guwahati, Assam, India

Noman Habib Department of Botany, Government College University, Faisalabad, Pakistan

Alkesh Hada Department of Biotechnology, National College (Autonomous), Tirruchirappalli, India

Muhammad Zulqurnain Haider Department of Botany, Government College University, Faisalabad, Pakistan

Urmi Halder Department of Botany (UGC-Centre of Advanced Study), The University of Burdwan, Golapbag, Burdwan, West Bengal, India

Mirza Hasanuzzaman Department of Agronomy, Faculty of Agriculture, Sher-e-Bangla Agricultural University, Dhaka, Bangladesh

Naeem Iqbal Department of Botany, Government College University, Faisalabad, Pakistan

Muhammad Tariq Javed Department of Botany, Government College University, Faisalabad, Pakistan

Ambika Joshi Department of Botany, Jai Hind College, Mumbai, Maharashtra, India

Nitesh Joshi Department of Botany, Rizvi College of Arts, Science and Commerce, Mumbai, Maharashtra, India

Ashutosh Kabiraj Department of Botany (UGC-Centre of Advanced Study), The University of Burdwan, Golapbag, Burdwan, West Bengal, India

Jyotirmay Kalita Department of Botany, Gauhati University, Guwahati, Assam, India

Imran Khan Department of Agronomy, University of Agriculture, Faisalabad, Pakistan

Zeba Khan Department of Botany, Aligarh Muslim University, Aligarh, Uttar Pradesh, India

Ajay Kumar Institute of Post-Harvest, Agricultural Research Organization (ARO) – The Volcani Center, Rishon LeZion, Israel

Anil Kumar Department of Entomology Nematology and Chemistry Units, Institute of Plant Protections, Agricultural Research Organization (ARO) – Volcani Center, Rishon LeZion, Israel

Narendra Kumar Forest Ecology and Climate Change Division, Forest Research Institute, Dehradun, Uttarakhand, India

Navin Kumar Plant Ecology and Climate Change Science, National Botanical Research Institute, Lucknow, India

Roshan Kumar Department of Human Genetics & Molecular Medicine, Central University of Punjab, Bathinda, Punjab, India

Anubhab Laha Department of Botany, Chandernagore College, Chandernagore, West Bengal, India

Department of Botany (UGC-Centre for Advanced Study), The University of Burdwan, Golapbag, Burdwan, West Bengal, India

Lipika Lahkar Department of Botany, Gauhati University, Guwahati, Assam, India

Moitri Let Department of Botany (UGC-Centre of Advanced Study), The University of Burdwan, Golapbag, Burdwan, West Bengal, India

Faisal Mahmood Department of Environment Sciences, Government College University, Faisalabad, Pakistan

Krishnendu Majhi Department of Botany (UGC-Centre of Advanced Study), The University of Burdwan, Golapbag, Burdwan, West Bengal, India

Amjad Malik Faculty of Education, University College of the North, Thompson, MB, Canada

Biology Department, Brandon University, Brandon, MB, Canada

Parth Malik School of Nano Sciences, Central University of Gujarat, Gandhinagar, Gujarat, India

Shekhar Mallick Plant Ecology and Climate Change Science, National Botanical Research Institute, Lucknow, India

Arpan Modi Institute of Plant Sciences, Agricultural Research Organization (ARO) – The Volcani Center, Rishon LeZion, Israel

Narayanan Nair Anjali Jawaharlal Nehru Tropical Botanic Garden and Research Institute (JNTBGRI), Thiruvananthapuram, Kerala, India

Shyam Narain Pandey Department of Botany, University of Lucknow, Lucknow, Uttar Pradesh, India

Anindya Sundar Panja Post-Graduate Department of Biotechnology and Biochemistry, Oriental Institute of Science and Technology, Burdwan, West Bengal, India

D. A. Patil Post-Graduate Department of Botany, S.S.V.P. Sanstha's L. K. Dr. P. R. Ghogrey Science College, Dhule, Maharashtra, India

Nisha Patwa Soft Wheat Quality Laboratory, United States Department of Agriculture, Wooster, OH, USA

Shilpi Paul Science and Engineering Research Board, New Delhi, India

Bryan W. Penning Soft Wheat Quality Laboratory, United States Department of Agriculture, Wooster, OH, USA

Amit Kumar Pradhan Department of Botany, Gauhati University, Guwahati, Assam, India

Avinash Chandra Rai Institute of Plant Sciences, Agricultural Research Organization (ARO) – The Volcani Center, Bet-Dagan, Israel

Krishna Kumar Rai Department of Botany, Institute of Sciences, Banaras Hindu University, Varanasi, Uttar Pradesh, India

Vemula Rajani Department of Biotechnology, Vikrama Simhapuri University, Nellore, Andhra Pradesh, India

Pratibha Rawat Department of Plant Physiology, College of Basic Sciences & Humanities, G. B. Pant University of Agriculture & Technology, Pantnagar, Uttarakhand, India

Muhammad Ali Raza College of Agronomy, Sichuan Agricultural University, Chengdu, China

Pichili Vijaya Bhaskar Reddy Department of Life Science and Bioinformatics, Assam University, Diphu Campus, Diphu, Assam, India

Rajib Roychowdhury Department of Vegetables and Field Crops, Institute of Plant Sciences, Agricultural Research Organization (ARO) – The Volcani Center, Rishon Lezion, Israel

Swarnendu Roy Plant Biochemistry Laboratory, Department of Botany, University of North Bengal, Darjeeling, West Bengal, India

Kalluvettankuzhy Krishnannair Sabu Jawaharlal Nehru Tropical Botanic Garden and Research Institute (JNTBGRI), Thiruvananthapuram, Kerala, India

Farah Saeed Department of Botany, Government College University, Faisalabad, Pakistan

Indraneel Saha Plant Physiology and Plant Molecular Biology Research Unit, Department of Botany, University of Kalyani, Kalyani, India

Soumen Saha Department of Sericulture, Raiganj University, Raiganj, West Bengal, India

Bipul Sarkar Plant Physiology and Plant Molecular Biology Research Unit, Department of Botany, University of Kalyani, Kalyani, India

Buddhadev Sarkar Plant Biochemistry Laboratory, Department of Botany, University of North Bengal, Darjeeling, West Bengal, India

Durre Shahwar Department of Botany, Aligarh Muslim University, Aligarh, Uttar Pradesh, India

Deepti Shankhdhar Department of Plant Physiology, College of Basic Sciences & Humanities, G. B. Pant University of Agriculture & Technology, Pantnagar, Uttarakhand, India

Shailesh Chandra Shankhdhar Department of Plant Physiology, College of Basic Sciences & Humanities, G. B. Pant University of Agriculture & Technology, Pantnagar, Uttarakhand, India

Iti Sharma Birla Institute of Technology and Science (BITS), Pilani, Rajasthan, India

Satish Kant Sharma Forest Ecology and Climate Change Division, Forest Research Institute, Dehradun, Uttarakhand, India

Nilofer Sheikh Department of Botany, University of Science and Technology, Meghalaya, Baridua, India

Avinash Singh Department of Biochemistry and Molecular Biology Section, Medical University of South Carolina, Charleston, SC, USA

Hukum Singh Forest Ecology and Climate Change Division, Forest Research Institute, Dehradun, Uttarakhand, India

Prashant Kumar Singh Institute of Plant Sciences, Agricultural Research Organization (ARO) – The Volcani Center, Rishon LeZion, Israel

Department of Biotechnology, Pachhunga University College, Mizoram Central University, Aizawl, Mizoram, India

Toolika Singh Department of Biochemistry and Molecular Biology Section, Medical University of South Carolina, Charleston, SC, USA

Sudhakar Srivastava Plant Stress Biology Laboratory, Institute of Environment and Sustainable Development, Banaras Hindu University, Varanasi, India

Bhaben Tanti Department of Botany, Gauhati University, Guwahati, Assam, India

Avijit Tarafdar Department of Botany, University of Kalyani, Kalyani, Nadia, West Bengal, India

Legumes Pathology, ICRISAT, Hyderabad, India

Khushbu Verma Faculty of Agriculture & Veterinary Science, Jayoti Vidyapeeth Women's University, Jaipur, Rajasthan, India

Wenjing Wang State Key Laboratory of Cotton Biology, Henan Key Laboratory of Plant Stress Biology, School of Life Science, Henan University, Kaifeng, Henan, China

Youcai Xiong MOE (Ministry of Education) Engineering Research Center for Dryland Agriculture and Ecological Conservation, School of Life Sciences, Lanzhou University, Lanzhou, Gansu, China

Deepanker Yadav Department of Botany, Guru Ghasidas Vishwavidyalaya, Bilaspur, Chhattisgarh, India

Vaishali Yadav Plant Stress Biology Laboratory, Institute of Environment and Sustainable Development, Banaras Hindu University, Varanasi, India

Chapter 1
Stress Management in Crops by Utilizing Landraces: Genetics and Plant Breeding Perspective

Amit Kumar Pradhan, Jyotirmay Kalita, Lipika Lahkar, Lisha Gurung, Surendra Kumar Ghritlahre, and Bhaben Tanti

Abstract Native germplasm is often the untapped resource that is supposed to hold the key to unlocking nature's repository of stress overcoming genes of crop plants. With an increasing global population, crop plants are expected to yield more under a variety of environmental conditions, hence exposing them to severe stress conditions. Locally available landraces of various crop plants that are often ignored in the presence of high-yielding developed varieties, however, have survived environmental turmoil over the ages and thus are the urgent need of the hour in order to couple yield or another desirable trait with stress tolerance. To bring these landraces to use, various mechanisms by which they tolerate environmental stress need to be understood. The use of conventional breeding techniques such as hybridization, selective backcrossing, gene pyramiding, etc. along with modern techniques such as mutagenesis, use of genetic markers, development of transgenic crops, etc. enables the use of trait diversity of landraces to counter environmental stresses while maintaining crop productivity. This chapter deals with various methods for identification of causal genetic factors responsible for stress tolerance in crop landraces and the possible techniques for developing tolerant varieties in the face of increasing frequency and severity of biotic and abiotic stresses.

Keywords Stress management · Landraces · Genetics · Plant breeding · Domestication

A. K. Pradhan · J. Kalita · L. Lahkar · L. Gurung · B. Tanti (✉)
Department of Botany, Gauhati University, Guwahati, Assam, India
e-mail: btanti@gauhati.ac.in

S. K. Ghritlahre
ICAR-National Rice Research Institute, RRLRRS, Kamrup, Assam, India

R. Roychowdhury et al. (eds.), *Sustainable Agriculture in the Era of Climate Change*, https://doi.org/10.1007/978-3-030-45669-6_1

1 Introduction

The process of crop domestication is closely associated with the various agricultural practices, human preferences, and the changing environment (Meyer et al. 2012; Smýkal et al. 2018). With the change in climate or environment and changing patterns in the use of natural resources, plants have been subjected to a variety of stresses. This led to the process of development and evolution of different adaptive strategies that permit plant cells to sense environmental stimuli and to activate responses that allow avoidance or survival of the environmental stresses encountered. The interaction of genotype and environment, which affects both agricultural practices and crop yield, is a critical point of research during the development of stress-tolerant plants under different field regimes (Mohammed 2009; Yumurtaci 2015). In the process of domestication, plants emerged in variable types, some as wild types, landraces, or the traditional domesticated genotypes. The crops plants have been globally dispersed with the spread of humans or evolution and due to various other environmental factors such as wind, water, or different natural disaster creating a geographical barrier. The crop plants thereby varied across the world based on the different process of cultivation and in accordance served to be the landrace to a particular area. The residues of the ancient crops undergo a very slow process of domestication like that of an evolutionary process, making the crop tolerant or highly diverse (Tanno and Willcox 2006; Smýkal et al. 2018).

Stress can be defined, in the widest biological sense, as any factor that may produce an adverse effect in individual organisms, populations, or communities. It is also defined as the overpowering pressure that affects the normal functions of individual life or the circumstances in which plants are interrupted from fully expressing their genetic potential for overall growth, development, and reproduction (Rhodes and Nadolska-Orczyk 2001). As per the agricultural viewpoint, stress includes the external factors that limit crop productivity or destroys biomass (Grime 1979). Stress can be biotic or abiotic in nature. Biotic stresses emerge through interactions between organisms, whereas abiotic stresses are those that depend on the interaction between organisms and the physical environment. Abiotic stresses, which include the various external factors of disturbance, are often interrelated, and either individually or in combination, they hamper the physiological, biochemical, and molecular processes that adversely affect plant growth and productivity. The combined effect of the stress factors is based on its nature of interactions with its effect on the crop plants ranging from germination stage to grain filling. Crop plants in response to stress undergo a series of modifications and process of adaptations that includes both morphological and genetic changes such as genome rearrangement, induction of various tolerance genes, etc., as per its ability to tolerate the adverse effect (Tester and Bacic 2005; Pandey et al. 2017). Plants to withstand against the stress effect induce a series of regulatory mechanisms which are either constitutively expressed or induced (Roy et al. 2014; Pradhan et al. 2019). Thereby, understanding the genetic regulatory mechanism of the landraces against the stress effect and its ability to adapt to such adverse conditions serves as a source to identify or understand the process of tolerance against stress.

Breeding approach to enhance the crops' ability to adapt adverse stress conditions though due to various effects of cultivation bottleneck that has a negative impact on crop production must be a continuous process with an added focus on the different stress factors in addition to its yield (Bertoldo et al. 2014). Application of constant monitoring of the individuals using the modern noninvasive method as well as sequencing of the genotype helps in the proper screening of the crops for its application in breeding and development of cultivar with high adaptability. In the majority of the cases toward the increase of the agronomic value of cultivated crops, its genome structure, the type of stress factors exposed, and the variable environmental conditions help in the proper selection of the suitable strategies (Yumurtaci 2015). So there is a need to build effective crop screening methods and use the prevalent crop biodiversity on the basis of present technical advancements for the detailed evaluations of the different type of landraces which will enhance and improve the efficiency of breeding and crop improvement.

With advances in fields like physiology, molecular biology, and genetics, our understanding of crops' response to stress and the basis of varietal differences in tolerance have greatly improved. In this chapter, the biotic and abiotic stress management of crops by utilizing landraces and their wild relatives through plant breeding and genetic approaches are addressed.

2 Crop Landraces and Their Significance

Landraces include a group of progressive cultivated plant populations with a distinct location of origin and specific distinguishing characters, with high genetic diversity and ability to adapt variable local weather conditions. The process of landrace origin includes the set of seed selection practices or the crop domestication by farmers and the process of field management as per the different agricultural practices by farmers which vary with the area (Fuller 2007; Casañas et al. 2017). As reported by Dwivedi et al. (2016), plant landraces encompass mainly the domesticated plant species with heterogeneous local adaptations which provide the genetic elements required to meet the growing challenges of farming crops in stressful environments. These local heterogeneous genotypes with wide genetic diversity show changing phenotype with its yield ranging from low to moderate level but with high nutritional quality. Thereby, landraces with high ability to cope with stress factors include the traditional cultivars with high nutritional content and yield ability under low input system of agriculture (Zeven 1998; Azeez et al. 2018). A landrace which serves as the repository of various mechanisms of tolerance completely varies from that of the new selectively improved cultivar or variety developed by breeder taking into account a specific trait. Landraces with high genetic diversity and different traits responsible for adaptation to adverse conditions thereby serve as the source of efficient crop development using plant breeding techniques contributing mainly toward the efficient traits of nutrient uptake, utilization, and genes of tolerance against stress conditions. An unambiguous evaluation of the landrace with the

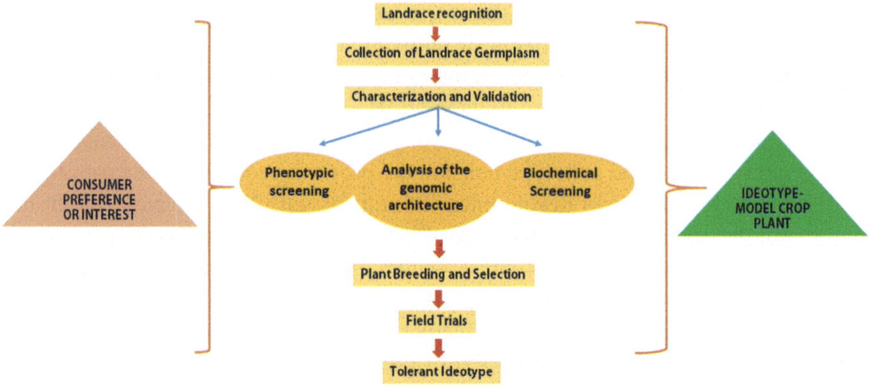

Fig. 1.1 Schematic representation of the process of landrace utilization in crop breeding programs

improved techniques of identification (Fig. 1.1) may uncover the different patterns of diversity present, which will expedite in identifying alleles/gene loci for increasing growth, yield, and adaptation to stress, thus increasing the productivity and stability of staple crops in vulnerable environments.

Landraces are normally low yielding and less productive in comparison to the developed commercial varieties, but with the advancement in modern research in the recent years, they now serve as the repository source of genetic variability in the search for genes or traits responsible for tolerance or resistance to biotic and abiotic factors for the proper establishment of agriculture in that specific locality (Casañas et al. 2017). However, the intense use of developed varieties or transgenic plants led to decrease in use, management, and conservation of local landraces, thereby causing less presence of the agronomic or genetic data of the landraces which appear to be of great importance with changing the environment. Zeven (1998) proposed the importance and role that landraces played in the development of the crop worldwide and also its role in the improvement of crops and agricultural production which served to have been in existence since the origin of agriculture itself. Since the time of origin of agriculture, landraces have been subjected to various natural genetic modifications which include both the abiotic and biotic interventions, leading to the development of a mechanism of adaptation and tolerance (Bansal et al. 2014). Crop landraces are thereby the principal focus for the development of the agricultural system from century's immemorial (Umakanth et al. 2017). The traditional process of farmers sowing, harvesting, and storage of some proportion of seeds for sowing in the subsequent season highly enriched the genetic pools promoting intraspecific diversity (Frankel et al. 1998). This traditional concept of crop cycle for conservation and maintenance of crop landrace with the beginning of plant breeding leading to generation of generally higher-yielding cultivar resistant to specific stress subsequently replaced the traditional process leading to decrease in diversity and gene pool, thereby causing downfall to agricultural production in the present condition (Frankel and Hawkes 1975; Casañas et al. 2017).

Hawkes (1983) remarked landraces as crops associated to one specific geographical area in contrast to that of cultivars, which are breed and trialed in diverse locations seeking its ability of tolerance to that of the trait opted for. Thus, landraces of a particular location are highly specific to that particular location and with most of the landrace nomenclature taken as per the name of the location likewise the Kent Wild White Clover from Kent, UK. However, various ecological factors often led to the introduction of informal landrace varieties and migration of traditional landraces from native region to new regions.

The occurrence and maintenance of landrace germplasm include various processes like that of seed exchange or replacement among farmers in a geographical area (Louette and Smale 1996). The process of continuous cultivation by an individual farmer or discontinuous but collective cultivation of landraces also forms a significant process of maintenance. Therefore, open cultivation system with routine local or the remote introduction of landrace germplasm serves as the major process for the maintenance and occurrence of genetic diversity among landraces belonging to a particular location.

3 Genetic Mechanisms of Plant Against Different Stresses

Plants respond to various stresses with alterations in different gene expressions. Genes that are induced by the effect of biotic or abiotic stresses often impart stress tolerance to the plants. Genes induced by stress include functional genes or regulatory genes. Stress response genes also induce the activation of antioxidant response pathways and protection against damage from oxidative injury. Salinity, drought, heat, cold, heavy metals, radiation, and submergence are various abiotic factors responsible for stress in plants, whereas pathogens also induce biotic stress in plants. Genes responsible for stress tolerance have been identified in many plants for a variety of stresses, and overexpression of such genes in transformed plants has been shown to result in increased tolerance.

3.1 *Salinity*

Salinity induces both osmotic and ionic stress in plants. The salt overly sensitive (SOS) pathway is an important mechanism maintaining ion homeostasis during salt stress. The SOS signaling pathway comprises of three important genes, *SOS1*, *SOS2*, and *SOS3* (Ji et al. 2013). Maintenance of ion homeostasis is a key response to salt stress, where Na^+ extrusion is the primary process, providing tolerance to the plant against salt. Elevation in Na^+ levels disrupts enzymatic functions and is toxic to plant cells. During salt stress, a Ca^+ spike activates the cascade of SOS signal transduction in the root cell cytoplasm. *SOS3* is responsible for activation of a primary Ca^+ sensor, and with the binding of Ca^+ and SOS3, the SOS2 gets activated.

In shoots, however, SCaBP8 (SOS3-like calcium-binding protein 8) is the dominant activator of SOS2. Activation of SOS2 recruits it to the plasma membrane, thereby activating the downstream target SOS1. SOS1 is a Na^+/H^+ antiporter, and its activation leads to the extrusion of Na^+. *SOS1* mutants are found to be highly sensitive to salt (Ji et al. 2013). High-affinity potassium transporter (*HKT*) genes are also involved in ion exclusion. *HKT2* has been found to increase salt tolerance through Na^+ exclusion (Roy et al. 2014). Tolerance to salinity is also achieved by sequestration of Na^+ to vacuoles by vacuolar Na^+/H^+ antiporters such as OsNHX1, OsNHX2, OsNHX3, and OsNHX4 in rice. Activities of Na^+/H^+ antiporters increase during salinity, and the increase is found to be more in case of tolerant plants. Increased expression of *OsNHX1* leads to increased tolerance due to Na^+ compartmentalization in vacuoles and also increases tolerance in transgenic rice plants. Upregulation of *OsHKT1;1*, *OsHAK10*, and *OsHAK16* leads to increased Na^+ accumulation in old leaves and is an important stress adaptive feature in rice (Reddy et al. 2017).

High salinity induces the biosynthesis of abscisic acid (ABA) hormone that leads to the closure of guard cells and helps plants to overcome osmotic stress occurring due to salinity. Jasmonic acid (JA) is important in the ABA-dependent regulation of salinity response genes. Homologues of the JASMONATE ZIM DOMAIN (JAZ), *OsTIFY1, OsTIFY6, OsTIFY9, OsTIFY10,* and *OsTIFY11* have been identified in rice. Several protein kinases (MAPK, RLK, etc.) are also parts of the ABA-dependent pathways of regulation of gene expression besides transcription factors such as *ONAC022* and microRNAs. Transcription factors such as DREB (dehydration-responsive element-binding protein) are important in ABA-independent pathways. Overexpression of *OsDREB1A*, *OsDREB1F*, and *OsDREB2A* has improved salt tolerance in transgenic rice. Genes such as *PDH45*, *OsCPK12*, etc. regulate the accumulation of ROS during salt stress and improve tolerance to salinity (Reddy et al. 2017).

3.2 Drought

Drought induces osmotic stress in plants. Expressions of late embryogenesis abundant (LEA) proteins involved in protection against desiccation are upregulated during drought. Overexpression of some LEA class genes has led to increased tolerance to dehydration. A gene coding for galactinol synthase (*GolS*), which is involved in the biosynthesis of raffinose family oligosaccharide, has enhanced drought tolerance in transgenic *Arabidopsis*. Gene expressing raffinose synthase is also upregulated during drought. Raffinose functions as an osmoprotectant and alleviates cellular damage during drought stress (Shinozaki and Yamaguchi-Shinozaki 2007). Methionine sulfoxide reductase (MSR) proteins encoding gene *MsrB2* induces tolerance to drought in terms of reduction of oxidative injury by ROS. *CaMsrB2* from *Capsicum annuum* overexpressed in transgenic rice has improved rice tolerance to drought (Nahar et al. 2016).

DRE/CRT (dehydration responsive element/C-repeat) regulons cooperatively with ABRE (ABA-responsive element) regulate the expression of *RD29A* gene encoding LEA-like proteins. *RD29B*, also encoding for LEA-like proteins, is regulated by ABRE elements. Drought induced *DREB* genes to regulate the expression of several drought tolerance genes. Overexpression of *DREB1A* in transgenic rice increases drought tolerance. *DREB2* genes are also induced by drought stress. However, overexpression of *DREB2* does not result in increased drought tolerance in transgenic plants, which suggests that DREB2 protein requires posttranslational activation (Nakashima and Yamaguchi-Shinozaki 2006). Two basic leucine zipper (*bZIP*) transcription factors AREB/ABF activate ABA-dependent gene expression by binding to ABRE. Overexpression of *ABF3* or *AREB2/ABF4* leads to ABA hypersensitivity and reduces transpiration, enhancing drought tolerability of transgenic *Arabidopsis* plants. Transcription factors *AtMYC2* and *AtMYB2* cooperatively activate the *RD22*, a drought-inducible gene, following the accumulation of endogenous ABA. Overexpression of both *AtMYC2* and *AtMYB2* enhanced osmotic stress tolerance in transgenic plants (Shinozaki and Yamaguchi-Shinozaki 2007).

3.3 Heat

Thermo-tolerance in plants is measured in terms of cell membrane thermostability (CMS). QTL and co-segregation analyses revealed several heat shock proteins to be the genetic cause of thermo-tolerance in many cereals. Thermo-tolerance is not controlled by a single gene in cereals (Maestri et al. 2002). Heat stress results in the expression of heat shock genes (*HSG*) that encode heat shock proteins (HSP). Heat shock factors (HSF) bind to the specific binding sites of Heat Shock Elements (HSE) in the promoters of HSGs, inducing the expression of HSPs on heat treatment. HSPs act as chaperones and protect intracellular proteins against denaturation, maintaining their structural stability through protein folding and keeping their functions intact. HSPs are categorized on the basis of their molecular mass, viz., HSP100, HSP90, HSP70, HSP60, and HSP20 (sHSP). Small heat shock proteins (sHSPs) show the highest diversity among all the HSPs (Hasanuzzaman et al. 2013). All sHSPs have a conserved 90-amino acid carboxyl-terminal domain, the α crystallin domain (ACD). ACD distinguishes sHSPs from other heat-induced proteins. Tolerance to heat stress is induced in sHSP-overexpressing plants (Sun et al. 2002). *HSP70* and *HSP101* are involved in heat stress tolerance in *Arabidopsis*. Three HSP101 members, *Tahsp101a*, *Tahsp101b*, and *Tahsp101c*, have also been cloned in wheat (Maestri et al. 2002). *Hsp17.7* and *Hsp100* have been found to confer thermo-tolerance in transgenic plants overexpressing these genes. *NPK1*-related transcripts are significantly elevated by heat. Constitutive overexpression of H_2O_2-responsive *ANP1/NPK1* was found to increase protection against heat stress in tobacco. Overexpression of *APX1*, coding for antioxidant ascorbate peroxidase, was also found to confer moderate heat tolerance in barley (Hasanuzzaman et al. 2013).

DREB2A in its active form during high-temperature stress regulates the expression of heat shock-related genes (Sakuma et al. 2006). Heat induces the expression of many HSFs. Plants have multiple copies of HSF genes. Heat stress regulatory proteins have conferred thermo-tolerance in transformed plants. Arabidopsis *HSFA1a* and *HSFA1b* control early response of many genes to heat. *HSFA1* has been proposed to be the master regulator of heat shock response in tomato (Hasanuzzaman et al. 2013).

3.4 Cold

A number of functional and regulatory genes in plants respond to cold stress. Desaturation of fatty acids is important for membrane functioning during cold stress. Glycerol-3-phosphate acyltransferase (*GPAT*) gene from *Cucurbita maxima* and *Arabidopsis thaliana*, which induces desaturation of phosphatidyl glycerol fatty acid, results in an increase in unsaturated fatty acids in transgenic tobacco plants and reduces cold sensitivity. Mutants of *FAB1* (involved in fatty acid biosynthesis), *fad5*, and *fad6* or triple mutants of genes *fad3-2*, *fad7-2,* and *fad8* (genes involved in fatty acid desaturation) lead to increase in saturated membranes and show decrease in chlorophyll content and photosynthetic efficiency and growth retardation during cold stress, suggesting their significance in low-temperature tolerance. LEA proteins functioning against cellular damage and anti-aggregation of enzymes under freezing stress increase plant cold tolerance. *Arabidopsis* genes *LOS4* and *AtNUP160* responsible for the export of RNA from the nucleus to the cytoplasm are also crucial for chilling and freezing tolerance in plants (Sanghera et al. 2011).

Transcriptional regulation during cold stress is mediated by *ICE1* [inducer of C-repeat binding factor (*CBF*) expression 1]. *ICE1* induces expression of *CBF3*, which in turn regulates transcription during cold stress. CBFs regulate genes involved in membrane transport and hormone metabolism, phosphoinositide metabolism, osmolyte biosynthesis, ROS detoxification, and signaling (Chinnusamy et al. 2007). *DREB1A/CBF3*, *DREB1B/CBF1,* and *DREB1C/CBF2* regulons are found to be involved in cold stress-responsive gene expression. The products of cold-induced *DREB1/CBF* genes regulate the expression of many stress-inducible genes. Overexpression of these genes in transgenic *Arabidopsis* led to increased tolerance to freezing. *ICE1* gene was found to regulate the expression of *DREB1A* promoter without affecting the other *DREB1/CBF* genes. A homologue of *DREB1/CBF*, *LeCBF1* from tomato was found to induce freezing tolerance in transgenic *Arabidopsis*. Novel DREB1/CBF transcription factor, ZmDREB1A, from maize was found to regulate cold-responsive gene expression (Nakashima and Yamaguchi-Shinozaki 2006).

3.5 Heavy Metal

Heavy metals lead to the onset of oxidative injury in plants (Roychowdhury and Tah 2011h; Basu et al. 2012; Roychowdhury et al. 2018, 2019). In *Arabidopsis thaliana*, cadmium uptake induces expression of *Atcys-3A*, which is involved in cysteine biosynthesis under heavy metal stress. Cysteine is the precursor in glutathione biosynthesis, which in turn is required for phytochelatin production. Phytochelatins bind to heavy metals and increase plant tolerance against heavy metal stress. Transformed *Arabidopsis* plants overexpressing *Atcys-3A* have shown increased tolerance to cadmium (Domínguez-Solís et al. 2001). *CABPR1* overexpression in tobacco also showed increased tolerance to heavy metal stress (Sarowar et al. 2005). In rice, *OsPIP2;6* has been demonstrated to play a role in As^{3+} efflux, thereby increasing plant tolerance against arsenic. Transgenic *Arabidopsis thaliana* overexpressing *PvACR3* from *Pteris vittata*, involved in vacuole sequestration of arsenic, have been shown to have increased tolerance to arsenic (Kalita et al. 2018).

3.6 Radiation

Transcript levels of *PyroA*, *Ubq3,* and *MEB5.2* were found to be increased by a low dose of UV-B radiation. *PyroA*, involved in the biosynthesis of pyridoxine, is important in protection of cellular structures against singlet oxygen. *Ubq3* encodes ubiquitin, associated with protein degradation. *MEB5.2* is a novel gene with unknown function (Brosche et al. 2002). Plant flavonoids are important secondary metabolites that protect against UV-B exposure, due to its absorbance in this wavelength region and elevation of its levels in epidermal cell layers. Flavonoids are also found to inhibit oxidative stress. Flavanone 3-hydroxylase (F3H) is a key enzyme in the flavonoid biosynthetic pathway, which is encoded by *RsF3H* in *Reamuria soongorica*. *RsF3H* gene expression and enzyme activity increase rapidly under stress (Liu et al. 2013).

3.7 Submergence

A QTL for submergence response in rice, Sub1, has been identified near the centromere of chromosome 9 of rice. *Sub1A*, *Sub1B,* and *Sub1C* are three genes identified in the Sub1 locus, of which *Sub1B* and *Sub1C* are present in all genotypes. *Sub1A* which is the variant gene has two alleles, of which *Sub1A-1* is specific for submergence tolerance and *Sub1A-2* is specific for intolerance. *Sub1A-1* overexpression has been found to confer submergence tolerance in transgenic rice (Xu et al. 2006).

3.8 Biotic Stress

Expressions of various regulatory genes determine tolerance or susceptibility to biotic stress caused by pathogen attack (Mamgain et al. 2013). ABA-independent dehydration-responsive *DREB2A* signaling pathways were found to crosstalk with *adr* signaling pathways, associated with disease resistance (Agarwal et al. 2006). Transgenic tobacco plants constitutively expressing *OsDREB1B* were found to induce pathogenesis-related (PR) gene expression. The promoter of *OsDREB1B* contains several disease-responsive cis-elements, and transgenic tobacco overexpressing *OsDREB1B* has shown reduced disease manifestations and delayed systemic infections with induced expression of PR genes such as *PR1b*, *PR2*, *PR-3*, *PR5*, and *CHIN50* (Gutha and Reddy 2008). Plants overexpressing *OsEREBP1* showed the reduced impact of the fungus *Magnaporthe grisea* where transcriptome analysis revealed high expression of transcription regulators belonging to the NAC and WRKY families (Jisha et al. 2015). Overexpression of *MBF1a* transcriptional coactivator gene has been found to induce resistance to fungal disease in *Arabidopsis* (Kim et al. 2007). Rice 14-3-3 family genes *GF14b*, *GF14c*, *GF14e*, and *GF14f* are differentially regulated in the interactions of rice with fungal pathogen *Magnaporthe grisea* and bacterial pathogen *Xanthomonas oryzae* pv. *oryzae*. 14-3-3 proteins act as scaffoldings for the assemblage of large signaling complexes and are potential factors in disease resistance and tolerance of stress. Response of 14-3-3s in defense against pathogens has also been reported in many other plants such as soybean, cotton, and tomato (Chen et al. 2006). Rice gene *Osmyb4* which encodes an Myb transcription factor leads to the upregulation of several genes with known functions in resistance against pathogens. Myb is effective in induction of systemic acquired resistance (SAR) (Vannini et al. 2006). Overexpression of *SlAREB1*, a member of the AREB/ABF subfamily of bZIP transcriptional factors, from *Solanum lycopersicum* has been found to enhance the expression of PR proteins (Orellana et al. 2010). NAC transcriptional factors are important in plant pathogen interactions. NAC proteins activate PR genes, induce hypersensitive response (HR), and result in cell death at the infection site. Some NAC proteins, however, also increase the susceptibility of plants against pathogens. *ATAF2* overexpression was found to increase susceptibility toward *Fusarium oxysporum* by repressing PR genes. *ATAF1* increases resistance against *Blumeria graminis* f. sp. *graminis* (Bgh) but reduces resistance against *Pseudomonas syringae*, *Botrytis cinerea*, and *Alternaria brassicicola* (Puranik et al. 2012). Overexpression of *OsNAC6* in transgenic rice has led to increased tolerance against blast disease (Nakashima et al. 2007).

4 Crop Domestication and Its Impact on Genetic Diversity

Landraces with its origin and cultivation from the beginning of agriculture of about 12,000 years ago led to its domestication, thereby resulting into the conversion of wild crop plants to cultivated forms, e.g., rice landrace domestication from wild ancestor *Oryza rufipogon*. Worldwide, humans develop a special interest to a particular group of crop plants varying in location based on their taste or quality production, cultivation practices, and agricultural environments, thereby causing random selections which further led to genetic manipulation and erosion among the different landraces or crop species. Domestication indicates the approach of selective sweep on the standing as well as new variation in the genetic constituents of the landraces due to mutation or introgression (Smýkal et al. 2018). The domesticated species and its progenitor vary from each other mainly based on physiological and genetic makeup termed as domestication syndrome. Henceforth, the domestication syndrome causing loss of genetic diversity among landraces does include not only the extinction of species but the loss in allele or gene level during the process of selection trials (Flint-Garcia 2013; Karmakar et al. 2012; Roychowdhury et al. 2013; Ganie et al. 2014, 2016; Anumalla et al. 2015).

Modern cultivar developed after selection or trials, taking into account a specific trait, appears to be less tolerant to the different factors of stress in comparison to their wild relatives or the available landraces mainly because of the erosion of useful genes in the course of selection for high yield (Reif et al. 2005). In the present scenario of depleting environmental conditions and variable climatic change, modern cultivars have failed to meet the limit of tolerance, thereby developing a matter of concern among the farmers or researchers. Wild crop plants or landraces in the course of evolution without any human interference develop or adapt to different changes or stress which therefore serves as the major source to cope with the environmental stress impacts.

In the earlier period, plants breeder or researcher focused mainly on the yield and production of the crop plant, giving less importance to stress factors both abiotic and biotic. But in the present age, with an alarming increase of different abiotic and biotic factors, modern crops fail to serve with its inadequate genetic content present to cope with the different stress factors. The natural disasters due to various biotic stress-causing factors like potato blight in Ireland during the 1840s, coffee rust in Brazil, and maize leaf blight in the USA (Rogers 2004) changed the conception of the breeders toward the tolerance against stress. In the present time, along with breeding against biotic stresses, i.e., the different disease causing pathogens, various steps have also been taken forward to aggravate the abiotic factors to improve the growth and yield of crop plants, taking into account the capacity and capability of landraces to adapt against adverse climatic conditions. Different management practices using both genetic and breeding approach against both biotic and abiotic stresses have been discussed below.

5 Stress Management Through Various Genetic and Breeding Approaches in Crops

Biotic stresses are resultant of the interaction of plants with other living organisms, which include pathogenic fungi, bacteria, viruses, or pests and nematodes resulting in considerable yield reduction, while abiotic stresses arise from the interaction of plants with the changing physical environment. Conventional breeding, as well as modern advanced techniques, may be used to overcome or manage the different biotic and abiotic stress undergone by the crop plants using the landrace stability.

5.1 Conventional Methods Used in Breeding Approaches for Stress Management

These methods bear traditional significance and have been used for long time in the past in developing disease-resistant varieties of crops. Sometimes, exotic varieties are also introduced in some areas in order to overcome biotic stress. Some of these methods are discussed here.

5.1.1 Development of New Cultivars Through Crop Hybridization

This technique is used to incorporate genes for disease and insect resistance in a variety from other crops varieties belonging to the same or different genera. Introgression lines developed through hybridization of *Brassica juncea* or *Brassica napus* with *Erucastrum cardaminoides*, *Erucastrum abyssinicum,* and *Diplotaxis tenuisiliqua* showed much higher resistance to Sclerotinia rot fungus as compared to cultivars obtained from India, China, and Australia (Garg et al. 2010). Leaf spot black sigatoka caused by fungal pathogen *Mycosphaerella fijiensis* Morelet results in heavy losses in the production of banana and plantain. Hybrid progenies obtained from crossing susceptible plantains with wild resistant banana “Calcutta 4” have been found to contain partially resistant genotypes (Ortiz and Vuylsteke 1994). Identification of resistant wild varieties of crops is the important first step in the process of development of hybrids.

When subjected to abiotic stresses, such as photoinhibition, high temperature, and drought, the hybrid wheat line 1–12 was found to show better tolerance compared to its parents Jing-411 and Xiaoyan-54 (Yang et al. 2006). Managed abiotic stress screening during breeding can yield resultant genotypes with adequate stress tolerance in a target environment. Breeding with stress screening for drought and low N has resulted in the development of maize lines tolerant to these stresses (Bänziger et al. 2006).

5.1.2 Backcrossing

Brown plant hopper-resistant rice plants were developed using a backcross breeding program involving 203 accessions from different parts of the world as donors. Backcross populations involving *indica* donors were found to produce more resistant varieties compared to *japonica* donors (Ali et al. 2006). Backcross breeding has been used for the transfer of the *cryIA* gene of transgenic cotton imparting insect's resistance in susceptible cultivars (Zhang et al. 2000). Backcrossing is particularly important in order to avoid incompatibility or sterility while developing resistant varieties when the source of the resistance gene is any related species. Backcrossing has yielded important results while breeding plants for nematode resistance (Boerma and Hussey 1992).

Backcrossing methods have also been applied in developing varieties tolerant to abiotic stresses. In a large-scale project using over 160 donor cultivars, and 3 recurrent ones, rice cultivars tolerant to drought were developed (Lafitte et al. 2006). Using backcross breeding, promising rice lines have been developed, showing tolerance to a range of stresses such as salinity, zinc deficiency and submergence, and resistance to brown leafhoppers, which depict the tremendous potential of the rice gene pool diversity in overcoming abiotic and biotic stresses (Ali et al. 2006).

5.1.3 Gene Pyramiding

Gene pyramiding involves incorporation of multiple genes of resistance from different genotypes in a single genotype by the process of repeated crosses and helps to build horizontal resistance in resultant plants. Resistance genes *rym4*, *rym5*, *rym9*, and *rym11* against barley yellow mosaic virus have been incorporated in barley cultivars using gene pyramiding method (Werner et al. 2005). Pyramiding has been used to incorporate resistance genes for downy and powdery mildew in *Vitis vinifera* to obtain disease-free genotypes (Eibach et al. 2007). Resistance genes *xa5*, *xa13*, and *Xa21* of rice against bacterial blight pathogen *Xanthomonas oryzae* were pyramided in the rice cultivar PR106. This resulted in increased host resistance under a combination of resistance genes against several races of the pathogen (Singh et al. 2001). Genetic markers come handy in the process of gene pyramiding by identifying desired gene combinations. Pyramiding in order to incorporate the three loci for soybean mosaic virus resistance, viz., *Rsv1*, *Rsv3*, and *Rsv4*, in a single genotype has been achieved (Shi et al. 2009). The incorporation of several resistant loci into a single line is expected to impart broad-spectrum resistance to resultant crop genotypes.

5.1.4 Introducing Exotic Lines

During the 1970s, Southern corn leaf blight epidemic, an exotic Texas cytoplasm, was introduced to develop hybrid maize seed in the affected regions. The hybrids contained a mitochondrial sterility gene and were resistant to the pathogen *Cochliobolus heterostrophus* and high yielding (Ullstrup 1972). Exotic maize lines were found to show higher resistance to European pod borer *Ostrinia nubilalis* (Hübner), and introduction of such varieties could be a solution to improving maize crop in affected regions (Sullivan et al. 1974). Transfer of greenbug (*Schizaphis graminum*) resistance genes from Argentinian rye to wheat was only found to be suitable via the exotic "Gaucho" *Triticale* variety as chromosome pairing was rare in rye-wheat intergeneric hybrids.

5.2 Application of Modern Techniques in Stress Management

Owing to various limitations of the conventional methods of breeding, such as the requirement of more time, labor, uncertainty of results, etc., various modern methods have been developed as discussed below.

5.2.1 Mutation Breeding

Mutation breeding involves the treatment of plant material with physical (UV, X-ray, etc.) or chemical mutagens (EMS, colchicines, etc.) and screening the mutant plants for desirable traits and homozygosity while growing them for several generations. Mutants can be used as direct mutant variety such as barley cv. "Diamant" and rice cv. "Calrose" 76, or they can be used for crossbreeding, as in linseed mutants "M1722" and "M1589" (Maluszynski et al. 1995). Mutagen benzothiadiazole (BTH) has been found to develop systemic acquired resistance (SAR) in *Arabidopsis thaliana* against turnip crinkle virus, *Pseudomonas syringae* pv tomato, and *Peronospora parasitica* (Lawton et al. 1996). Mutation breeding methods were used to develop variation in disease and pest resistance and plant stature in tropical tuber crops cassava and yam by the process of in vitro mutagenesis and then by somatic cell manipulation (Novak and Brunner 1992). Two aromatic *indica* rice varieties "RD6" and "RD15" have been developed by the process of gamma irradiation from the "KDML 105" variety. Moreover, mutant varieties have been developed in many crops such as maize, wheat, tomato, barley, soybean, carnation, brinjal, mung bean, etc. (Roychowdhury 2011; Roychowdhury and Tah 2011a, b, c, d, e, f, g, h, 2013; Roychowdhury et al. 2011a, b, c, d, 2012a, b; Oladosu et al. 2016).

5.2.2 Marker-Assisted Selection (MAS) and QTL Mapping

DNA markers are a very important tool in plant breeding approaches as they help to identify plants containing specific DNA regions of importance which code for desired traits (Roychowdhury 2014; Roychowdhury et al. 2014). They are the most widely used types of markers predominantly due to their abundance. These markers are selectively neutral because of being usually located in coding regions. Widely used DNA markers are restriction fragment length polymorphisms (RFLP), random amplified polymorphic DNA (RAPD), and simple sequence repeats (SSR) or amplified fragment length polymorphisms (AFLP) (Collard et al. 2005). QTL mapping is an important approach required when traits are governed by more genes than one, located at particular loci within the chromosomes. QTL mapping is based on the principle that genes for specific traits segregate at the recombination events of chromosomes during meiosis. With the use of recombination frequency among progenies, the distance between markers is detected, and thereby, positions of QTLs within the genome are calculated (Collard et al. 2005). MAS can be used for screening germplasms for resistant genes (Simon-Mateo and Garcia 2011). Gene for head smut resistance in sorghum was detected using RFLP and RAPD markers (Oh et al. 1994). Ten QTLs for yellow rust resistance was detected in wheat using QTL mapping, which included five major QTLs, *QYr.sgi-4A.1*, *QYr.sgi- 2B.1a*, *QYr.sgi- 2B.1*, *Lr34/Yr18*, and *QYr.sgi- 4A.2* (Chu et al. 2009).

5.2.3 Transgenic Approach

Foreign gene transfer from related or unrelated species through vectors such as *Agrobacterium tumefaciens*, gene gun, electroporation, etc. into target organisms so as to modify its genetic makeup, and introduce desirable traits, is another method of crop improvement for stress tolerance. *A. tumefaciens*-mediated gene transfer is a commonly preferred method for the purpose. Cry protein gene derived from *Bacillus thuringiensis* (Bt) has been transferred to different crops for providing resistance against chewing insects. The Cry protein disrupts the insects' guts once consumed, resulting in its death. Bt crops such as cotton, eggplant, etc. have been developed and have been proven to be commercially successful, with increasing growing areas worldwide (Romeis et al. 2006). An *Rxo1* gene for resistance against bacterial blight was transferred from maize to rice resulting in resistance against *Xanthomonas oryzae* (Zhao et al. 2005). USA and Argentina were using a total area of 30.3 and 10 million hectares of cultivated land for transgenic crops in 2000 (James 2003).

5.2.4 Marker-Assisted Backcrossing (MABC)

Marker-assisted backcrossing is another convenient way of incorporating tolerance genes or QTLs into popular cultivars of crops. The widely cultivated Swarna variety of rice was incorporated with the rice submergence tolerance QTL *Sub1* in three backcrosses using polymorphic markers (Neeraja et al. 2007). Several MABC procedures have been utilized by ICRISAT in developing terminal drought tolerance in pearl millet (*Pennisetum glaucum* (L.) R. Br.) (Liu et al. 2000). *Sub1* QTL was also introgressed into the Bangladeshi mega rice variety using backcrossing assisted by SSR markers to incorporate submergence tolerance trait (Iftekharuddaula et al. 2011).

5.2.5 Marker-Assisted Recurrent Selection (MARS)

MARS is used in the situation when many genes are involved in giving rise to a particular trait, in order to increase the frequency of desired alleles. Genetic gain accounting for increased grain yield of maize up to 51 kg ha^{-1} $year^{-1}$ under drought stressed conditions was achieved in sub-Saharan African regions using MARS (Beyene et al. 2016).

6 Conclusion

During the process of crop domestication and various revolution like that of the green revolution to meet the growing demand of food based on the preference of the humans and the changing environmental conditions, different wild plants lose the natural habitat system and led to the effect of cultivation bottlenecks which later serve to be an inefficient domesticated crop. Therefore, with the evolving changes in environmental conditions and pathogenic races, untapped landraces of different crops, which were left out undomesticated during the rapid expansion of agriculture by means of high-yielding varieties and use of fertilizers and pesticides, have once again emerged as the repository source of high genetic diversity and stability. With the growing world population, and climate change pushing agricultural productivity to the edge, this genetic resource of landraces and wild varieties of crops need to be screened for tolerance or resistance traits using advanced high-throughput technologies and various procedures mentioned in the preceding sections. Stress management in the face of today's hurdles has to be an integrated approach, involving new technologies for selection and breeding using local landraces, as well as spreading of social awareness regarding the acceptance of transgenic crops that are less demanding for the environment.

References

Agarwal PK, Agarwal P, Reddy MK, Sopory SK (2006) Role of DREB transcription factors in abiotic and biotic stress tolerance in plants. Plant Cell Rep 25(12):1263–1274

Ali AJ, Xu JL, Ismail AM, Fu BY, Vijaykumar CH, Gao YM, Domingo J, Maghirang R, Yu SB, Gregorio G, Yanaghihara S (2006) Hidden diversity for abiotic and biotic stress tolerances in the primary gene pool of rice revealed by a large backcross breeding program. Field Crop Res 97(1):66–76

Anumalla M, Roychowdhury R, Geda CK, Mazid M, Rathoure AK (2015) Utilization of plant genetic resources and diversity analysis tools for sustainable crop improvement with special emphasis on rice. Int J Adv Res 3(3):1155–1175

Azeez MA, Adubi AO, Durodola FA (2018) Landraces and crop genetic improvement. In: Rediscovery of landraces as a resource for the future 2018 Sep 12. IntechOpen

Bansal KC, Lenka SK, Mondal TK (2014) Genomic resources for breeding crops with enhanced abiotic stress tolerance. Plant Breed 133(1):1–11

Bänziger M, Setimela PS, Hodson D, Vivek B (2006) Breeding for improved abiotic stress tolerance in maize adapted to southern Africa. Agric Water Manag 80(1–3):212–224

Basu A, Roychowdhury R, Bhattacharyya SS, Tah J (2012) Estimation of major heavy metals (Fe, Cu and Zn) in the fruit part of *Cucumis sativus* L. World J Sci Technol 2(7):01–03

Bertoldo JG, Coimbra JLM, Guidolin AF, Andrade LRBD, Nodari RO (2014) Agronomic potential of genebank landrace elite accessions for common bean genetic breeding. Sci Agric 71(2):120–125

Beyene Y, Semagn K, Crossa J, Mugo S, Atlin GN, Tarekegne A, Meisel B, Sehabiague P, Vivek BS, Oikeh S, Alvarado G (2016) Improving maize grain yield under drought stress and non-stress environments in sub-Saharan Africa using marker-assisted recurrent selection. Crop Sci 56(1):344–353

Boerma HR, Hussey RS (1992) Breeding plants for resistance to nematodes. J Nematol 24(2):242

Brosche M, Gittins JR, Sävenstrand H, Strid A (2002) Gene expression under environmental stresses – molecular marker analysis. In: Jain SM (eds) Molecular Techniques in Crop Improvement. Kluwer Academic Publishers, pp 371–408, https://doi.org/10.1007/978-94-017-2356-5_14

Casañas F, Simó J, Casals J, Prohens J (2017) Toward an evolved concept of landrace. Front Plant Sci 8:145

Chen F, Li Q, Sun L, He Z (2006) The rice 14-3-3 gene family and its involvement in responses to biotic and abiotic stress. DNA Res 13(2):53–63

Chinnusamy V, Zhu J, Zhu JK (2007) Cold stress regulation of gene expression in plants. Trends Plant Sci 12(10):444–451

Chu CG, Friesen TL, Xu SS, Faris JD, Kolmer JA (2009) Identification of novel QTLs for seedling and adult plant leaf rust resistance in a wheat doubled haploid population. Theor Appl Genet 119:263–269

Collard BC, Jahufer MZZ, Brouwer JB, Pang ECK (2005) An introduction to markers, quantitative trait loci (QTL) mapping and marker-assisted selection for crop improvement: the basic concepts. Euphytica 142(1–2).169–196

Domínguez-Solís JR, Gutiérrez-Alcalá G, Romero LC, Gotor C (2001) The cytosolic O-acetylserine (thiol)lyase gene is regulated by heavy metals and can function in cadmium tolerance. J Biol Chem 276(12):9297–9302

Dwivedi SL, Ceccarelli S, Blair MW, Upadhyaya HD, Are AK, Ortiz R (2016) Landrace germplasm for improving yield and abiotic stress adaptation. Trends Plant Sci 21(1):31–42

Eibach R, Zyprian E, Welter L, Topfer R (2007) The use of molecular markers for pyramiding resistance genes in grapevine breeding. Vitis-Geilweilerhof 46(3):120

Flint-Garcia SA (2013) Genetics and consequences of crop domestication. J Agric Food Chem 61(35):8267–8276

Frankel OH, Hawkes JG (1975) Crop genetic resources for today and tomorrow. Cambridge University Press, Cambridge

Frankel OH, Brown AHD, Burdon JJ (1998) The conservation of plant biodiversity, 2nd edn. Cambridge University Press, Cambridge, pp 56–78
Fuller DQ (2007) Contrasting patterns in crop domestication and domestication rates: recent archaeobotanical insights from the Old World. Ann Bot 100(5):903–924
Ganie SA, Karmakar J, Roychowdhury R, Mondal TK, Dey N (2014) Assessment of genetic diversity in salt-tolerant rice and its wild relatives for ten SSR loci and one allele mining primer of *salT* gene located on 1st chromosome. Plant Syst Evol 300(7):1741–1747
Ganie SA, Karmakar J, Roychowdhury R, Mondal TK, Dey N (2016) An exploratory study on allelic diversity among rice and its wild species as well as relatives with simple sequence repeat and inter simple sequence repeat markers. Indian J Biotechnol 15(3):357–362
Garg H, Atri C, Sandhu PS, Kaur B, Renton M, Banga SK, Singh H, Singh C, Barbetti MJ, Banga SS (2010) High level of resistance to *Sclerotinia sclerotiorum* in introgression lines derived from hybridization between wild crucifers and the crop *Brassica* species *B. napus* and *B. juncea*. Field Crops Res 117(1):51–58
Grime JP (1979) Plant strategies and vegetation processes. Wiley, Chichester, 222
Gutha LR, Reddy AR (2008) Rice *DREB1B* promoter shows distinct stress-specific responses, and the overexpression of cDNA in tobacco confers improved abiotic and biotic stress tolerance. Plant Mol Biol 68(6):533
Hasanuzzaman M, Nahar K, Alam M, Roychowdhury R, Fujita M (2013) Physiological, biochemical, and molecular mechanisms of heat stress tolerance in plants. Int J Mol Sci 14(5):9643–9684
Hawkes JG (1983) The diversity of crop plants. Harvard University Press, Cambridge, MA, 102 pp 184. ISBN: 06742186X
Iftekharuddaula KM, Newaz MA, Salam MA, Ahmed HU, Mahbub MA, Septiningsih EM, Collard BC, Sanchez DL, Pamplona AM, Mackill DJ (2011) Rapid and high-precision marker assisted backcrossing to introgress the SUB1 QTL into BR11, the rainfed lowland rice mega variety of Bangladesh. Euphytica 178(1):83–97
James C (2003) Global review of commercialized transgenic crops. Curr Sci 84(3):303–309
Ji H, Pardo JM, Batelli G, Van Oosten MJ, Bressan RA, Li X (2013) The Salt Overly Sensitive (SOS) pathway: established and emerging roles. Mol Plant 6(2):275–286
Jisha V, Dampanaboina L, Vadassery J, Mithöfer A, Kappara S, Ramanan R (2015) Overexpression of an AP2/ERF type transcription factor *OsEREBP1* confers biotic and abiotic stress tolerance in rice. PLoS One 10(6):e0127831
Kalita J, Pradhan AK, Shandilya ZM, Tanti B (2018) Arsenic stress responses and tolerance in rice: physiological, cellular and molecular approaches. Rice Sci 25(5):235–249
Karmakar J, Roychowdhury R, Kar RK, Deb D, Dey N (2012) Profiling of selected indigenous rice (*Oryza sativa* L.) landraces of Rarh Bengal in relation to osmotic stress tolerance. Physiol Mol Biol Plants 18(2):125–132
Kim MJ, Lim GH, Kim ES, Ko CB, Yang KY, Jeong JA, Lee MC, Kim CS (2007) Abiotic and biotic stress tolerance in *Arabidopsis* overexpressing the Multiprotein Bridging Factor 1a (*MBF1a*) transcriptional coactivator gene. Biochem Biophys Res Commun 354(2):440–446
Lafitte HR, Li ZK, Vijayakumar CH, Gao YM, Shi Y, Xu JL, Fu BY, Yu SB, Ali AJ, Domingo J, Maghirang R (2006) Improvement of rice drought tolerance through backcross breeding: evaluation of donors and selection in drought nurseries. Field Crop Res 97(1):77–86
Lawton KA, Friedrich L, Hunt M, Weymann K, Delaney T, Kessmann H, Staub T, Ryals J (1996) Benzothiadiazole induces disease resistance in *Arabidopsis* by activation of the systemic acquired resistance signal transduction pathway. Plant J 10:71–82
Liu H, Qi X, Sharma A, Kolesnikova-Allen MA, Bidinger FR, Witcombe JR (2000) Marker-assisted backcrossing to improve terminal drought tolerance in pearl millet. Molecular Approaches for the Genetic Improvement of Cereals for Stable Production in Water limited Environments: CYMMYT, Batan, pp 114–119
Liu M, Li X, Liu Y, Cao B (2013) Regulation of flavanone 3-hydroxylase gene involved in the flavonoid biosynthesis pathway in response to UV-B radiation and drought stress in the desert plant, *Reaumuria soongorica*. Plant Physiol Biochem 73:161–167

Louette D, Smale M (1996) Genetic diversity and maize seed management in a traditional Mexican community: implications for in situ conservation of maize, Natural Resources Group, Paper 96-03. International Centre for Maize and Wheat Improvement (CIMMYT), p 22
Maestri E, Klueva N, Perrotta C, Gulli M, Nguyen HT, Marmiroli N (2002) Molecular genetics of heat tolerance and heat shock proteins in cereals. Plant Mol Biol 48(5–6):667–681
Maluszynski M, Ahloowalia BS, Sigurbjörnsson B (1995) Application of *in vivo and in vitro* mutation techniques for crop improvement. Euphytica 85(1–3):303–315
Mamgain A, Roychowdhury R, Tah J (2013) *Alternaria* pathogenicity and its strategic controls. Res J Biol 1:1–9
Meyer RS, Du Val AE, Jensen HR (2012) Patterns and processes in crop domestication: an historical review and quantitative analysis of 203 global food crops. New Phytol 196:29–48
Mohammed MI (2009) Genotype x environment interaction in bread wheat in Northern Sudan using AMMI analysis. Am Eurasian J Agric Environ Sci 6:427–433
Nahar S, Kalita J, Sahoo L, Tanti B (2016) Morph – physiological and molecular effects of drought stress in rice. Ann Plant Sci 5(9):1409–1416
Nakashima K, Yamaguchi-Shinozaki K (2006) Regulons involved in osmotic stress-responsive and cold stress-responsive gene expression in plants. Physiol Plant 126(1):62–71
Nakashima K, Tran LSP, Van Nguyen D, Fujita M, Maruyama K, Todaka D, Ito Y, Hayashi N, Shinozaki K, Yamaguchi-Shinozaki K (2007) Functional analysis of a NAC-type transcription factor *OsNAC6* involved in abiotic and biotic stress-responsive gene expression in rice. Plant J 51(4):617–630
Neeraja CN, Maghirang-Rodriguez R, Pamplona A, Heuer S, Collard BC, Septiningsih EM, Vergara G, Sanchez D, Xu K, Ismail AM, Mackill DJ (2007) A marker-assisted backcross approach for developing submergence-tolerant rice cultivars. Theor Appl Genet 115(6):767–776
Novak FJ, Brunner H (1992) Plant breeding: induced mutation technology for crop improvement. IAEA Bull 4:25–33
Oh BJ, Frederiksen RA, Magill CW (1994) Identification of molecular markers linked to head smut resistance gene (*Shs*) in sorghum by RFLP and RAPD analyses. Phytopathology 84:830–833
Oladosu Y, Rafii MY, Abdullah N, Hussin G, Ramli A, Rahim HA, Miah G, Usman M (2016) Principle and application of plant mutagenesis in crop improvement: a review. Biotechnol Biotechnol Equip 30(1):1–16
Orellana S, Yanez M, Espinoza A, Verdugo I, Gonzalez E, Ruiz-lara SIMÓN, Casaretto JA (2010) The transcription factor *SlAREB1* confers drought, salt stress tolerance and regulates biotic and abiotic stress-related genes in tomato. Plant Cell Environ 33(12):2191–2208
Ortiz R, Vuylsteke D (1994) Inheritance of black sigatoka disease resistance in plantain-banana (*Musa* spp.) hybrids. Theor Appl Genet 89(2–3):146–152
Pandey P, Irulappan V, Bagavathiannan MV, Senthil-Kumar M (2017) Impact of combined abiotic and biotic stresses on plant growth and avenues for crop improvement by exploiting physio-morphological traits. Front Plant Sci 8:537
Pradhan AK, Shandilya ZM, Lahkar L, Hasnu S, Kalita J, Borgohain D, Tanti B (2019) Comparative metabolomics approach towards understanding chemical variation in rice under abiotic stress. In: Advances in rice research for abiotic stress tolerance. Woodhead Publishing, pp 537–550
Puranik S, Sahu PP, Srivastava PS, Prasad M (2012) NAC proteins: regulation and role in stress tolerance. Trends Plant Sci 17(6):369–381
Reddy INBL, Kim BK, Yoon IS, Kim KH, Kwon TR (2017) Salt tolerance in rice: focus on mechanisms and approaches. Rice Sci 24(3):123–144
Reif JC, Zhang P, Dreisigacker S, Warburton ML, Van Ginkel M, Hoisington D, Bohn M, Melchinger AE (2005) Wheat genetic diversity trends during domestication and breeding. Theor Appl Genet 110:859–864
Rhodes D, Nadolska-Orczyk A (2001) Plant stress physiology. eLS. https://doi.org/10.1038/npg.els.0001297
Rogers DL (2004) Genetic erosion: no longer just an agricultural issue. Nativ Plant J 5(2):113–122

Romeis J, Meissle M, Bigler F (2006) Transgenic crops expressing *Bacillus thuringiensis* toxins and biological control. Nat Biotechnol 24(1):63
Roy SJ, Negrao S, Tester M (2014) Salt resistant crop plants. Curr Opin Biotechnol 26:115–124
Roychowdhury R (2011) Effect of chemical mutagens on carnation (*Dianthus caryophyllus* L.): a mutation breeding approach. LAP Lambert Academic Publishing
Roychowdhury R (2014) Crop improvement in the era of climate change. IK International Publishing House, New Delhi, p 496
Roychowdhury R, Tah J (2011a) Assessment of chemical mutagenic effects in mutation breeding programme for M_1 generation of Carnation (*Dianthus caryophyllus*). Res Plant Biol 1(4):23–32
Roychowdhury R, Tah J (2011b) Chemical mutagenic action on seed germination and related agro-metrical traits in M_1 Dianthus generation. Curr Bot 2(8):19–23
Roychowdhury R, Tah J (2011c) Mutation breeding in *Dianthus caryophyllus* for economic traits. Electron J Plant Breed 2(2):282–286
Roychowdhury R, Tah J (2011d) Evaluation of genetic parameters for agro-metrical characters in carnation genotypes. Afr Crop Sci J 19(3):183–188
Roychowdhury R, Tah J (2011e) Genetic variability for different quantitative traits in *Dianthus caryophyllus* L. during mutation breeding. Int J Sci Nat 2(4):778–781
Roychowdhury R, Tah J (2011f) Germination behaviors in M_2 generation of *Dianthus* after chemical mutagenesis. Int J Adv Sci Tech Res 2(1):448–454
Roychowdhury R, Tah J (2011g) Genetic variability study for yield and associated quantitative characters in mutant genotypes of *Dianthus caryophyllus* L. Int J Biosci 1(5):38–44
Roychowdhury R, Tah J (2011h) Differential response by different parts of *Solanum melongena* L. for heavy metal accumulation. Plant Sci Feed 1(6):80–83
Roychowdhury R, Tah J (2013) Mutagenesis – a potential approach for crop improvement. In: Hakeem KR, Ahmad P, Ozturk M (eds) Crop improvement – new approaches and modern techniques. Springer, Boston, pp 149–187
Roychowdhury R, Sultana P, Tah J (2011a) Morphological architecture of foliar stomata in M_2 Carnation (*Dianthus caryophyllus* L.) genotypes using Scanning Electron Microscopy (SEM). Electron J Plant Breed 2(4):583–588
Roychowdhury R, Bandopadhyay A, Dalal T, Tah J (2011b) Biometrical analysis for some agro-economic characters in M_1 generation of *Dianthus caryophyllus*. Plant Arch 11(2):989–994
Roychowdhury R, Tah J, Dalal T, Bandyopadhyay A (2011c) Selection response and correlation studies for metrical traits in mutant Carnation (*Dianthus caryophyllus* L.) genotypes. Cont J Agric Sci 5(3):06–14
Roychowdhury R, Roy S, Tah J (2011d) Estimation of heritable components of variation and character selection in eggplant (*Solanum melongena* L.) for mutation breeding programme. Cont J Biol Sci 4(2):31–36
Roychowdhury R, Alam MJ, Bishnu S, Dalal T, Tah J (2012a) Comparative study for effects of chemical mutagenesis on seed germination, survivability and pollen sterility in M_1 and M_2 generations of Dianthus. Plant Breed Seed Sci 65(1):29–38
Roychowdhury R, Datta S, Gupta P, Tah J (2012b) Analysis of genetic parameters on mutant populations of mungbean (*Vigna radiata* L.) after ethyl methane sulphonate treatment. Not Sci Biol 4(1):137–143
Roychowdhury R, Karmakar J, Karmakar J, Adak MK, Dey N (2013) Physio-biochemical and microsatellite based profiling of lowland rice (*Oryza sativa* L.) landraces for osmotic stress tolerance. American J Plant Sci 16;4(12):52
Roychowdhury R, Taoutaou A, Hakeem KR, Gawwad MR, Tah J (2014) Molecular marker-assisted technologies for crop improvement. In: Roychowdhury R (ed) Crop improvement in the era of climate change. IK International Publishing House, New Delhi, pp 241–258
Roychowdhury R, Khan MH, Choudhury S (2018) Arsenic in rice: an overview on stress implications, tolerance and mitigation strategies. In: Hasanuzzaman M, Nahar K, Fujita M (eds) Plants under metal and metalloid stress. Springer, Singapore, pp 401–415

Roychowdhury R, Khan MH, Choudhury S (2019) Physiological and molecular responses for metalloid stress in rice – a comprehensive overview. In: Hasanuzzaman M, Fujita M, Nahar K, Biswas J (eds) Advances in rice research for abiotic stress tolerance. Woodhead Publishing, Elsevier, pp 341–369

Sakuma Y, Maruyama K, Qin F, Osakabe Y, Shinozaki K, Yamaguchi-Shinozaki K (2006) Dual function of an Arabidopsis transcription factor DREB2A in water-stress-responsive and heat-stress-responsive gene expression. Proc Natl Acad Sci 103(49):18822–18827

Sanghera GS, Wani SH, Hussain W, Singh NB (2011) Engineering cold stress tolerance in crop plants. Curr Genomics 12(1):30

Sarowar S, Kim YJ, Kim EN, Kim KD, Hwang BK, Islam R, Shin JS (2005) Overexpression of a pepper basic pathogenesis-related protein 1 gene in tobacco plants enhances resistance to heavy metal and pathogen stresses. Plant Cell Rep 24(4):216–224

Shi A, Chen P, Li D, Zheng C, Zhang B, Hou A (2009) Pyramiding multiple genes for resistance to soybean mosaic virus in soybean using molecular markers. Mol Breed 23(1):113

Shinozaki K, Yamaguchi-Shinozaki K (2007) Gene networks involved in drought stress response and tolerance. J Exp Bot 58(2):221–227

Simon-Mateo C, Garcia JA (2011) Antiviral strategies in plants based on RNA silencing. Biochim Biophys Acta 1809(11–12):722–731

Singh S, Sidhu JS, Huang N, Vikal Y, Li Z, Brar DS, Dhaliwal HS, Khush GS (2001) Pyramiding three bacterial blight resistance genes (*xa5, xa13* and *Xa21*) using marker-assisted selection into indica rice cultivar PR106. Theor Appl Genet 102(6–7):1011–1015

Smýkal P, Nelson M, Berger J, von Wettberg E (2018) The impact of genetic changes during crop domestication. Agronomy 8(7):119

Sullivan SL, Gracen VE, Ortega A (1974) Resistance of exotic maize varieties to the European corn borer *Ostrinia nubilalis* (Hübner). Environ Entomol 3(4):718–720

Sun W, Van Montagu M, Verbruggen N (2002) Small heat shock proteins and stress tolerance in plants. Biochimica Biophys. Acta (BBA) – Gene Struct Expr 1577(1):1–9

Tanno K, Willcox G (2006) How fast was wild wheat domesticated? Science 311(5769):1886–1886

Tester M, Bacic A (2005) Abiotic stress tolerance in grasses from model plants to crop plants. Plant Physiol 137:791–793

Ullstrup AJ (1972) The impact of the southern corn leaf blight epidemics of 1970–71. Annu Rev Phytopathol 10:37–50

Umakanth B, Vishalakshi B, Sathish Kumar P, Rama Devi SJS, Bhadana VP, Senguttuvel P, Madhav MS (2017) Diverse rice landraces of North-East India enables the identification of novel genetic resources for *Magnaporthe* resistance. Front Plant Sci 8:1500

Vannini C, Iriti M, Bracale M, Locatelli F, Faoro F, Croce P, Pirona R, Di Maro A, Coraggio I, Genga A (2006) The ectopic expression of the rice *Osmyb4* gene in *Arabidopsis* increases tolerance to abiotic, environmental and biotic stresses. Physiol Mol Plant Pathol 69(1–3):26–42

Werner K, Friedt W, Ordon F (2005) Strategies for pyramiding resistance genes against the barley yellow mosaic virus complex (BaMMV, BaYMV, BaYMV-2). Mol Breed 16(1):45–55

Xu K, Xu X, Fukao T, Canlas P, Maghirang-Rodriguez R, Heuer S, Ismail AM, Bailey-Serres J, Ronald PC, Mackill DJ (2006) *Sub1A* is an ethylene-response-factor-like gene that confers submergence tolerance to rice. Nature 442(7103):705

Yang X, Chen X, Ge Q, Li B, Tong Y, Zhang A, Li Z, Kuang T, Lu C (2006) Tolerance of photosynthesis to photo inhibition, high temperature and drought stress in flag leaves of wheat: a comparison between a hybridization line and its parents grown under field conditions. Plant Sci 171(3):389–397

Yumurtaci A (2015) Utilization of wild relatives of wheat, barley, maize and oat in developing abiotic and biotic stress tolerant new varieties. Emir J Food Agric 27:01–23

Zeven AC (1998) Landraces: a review of definitions and classifications. Euphytica 104:127–139

Zhang BH, Guo TL, Wang QL (2000) Inheritance and segregation of exogenous genes in transgenic cotton. J Genet 79:71–75

Zhao BY, Lin XH, Poland J, Trick H, Leach J, Hulbert S (2005) From the cover: a maize resistance gene functions against bacterial streak disease in rice. Proc Natl Acad Sci 102:15383–15388

Chapter 2
Environmental Impact on Cereal Crop Grain Damage from Pre-harvest Sprouting and Late Maturity Alpha-Amylase

Nisha Patwa and Bryan W. Penning

Abstract Pre-harvest sprouting (PHS) and late maturity alpha-amylase (LMA) result from increased alpha-amylase (AA) activity in wheat, rice, barley, rye, and maize grain prior to harvest leading to undesirable degradation of seed starch. The starch degradation can reduce grain quality and thus crop values by up to 30% and in severe cases make them unfit for human consumption and usable only as animal feed. Specific environmental conditions of temperature and rain occurring at or near seed maturity lead to the onset of PHS or LMA. While environmental effects cannot be controlled in the field, genetic resistance/tolerance is possible. Genetic control of PHS and LMA is complicated by the numerous factors involved in modulating AA which ultimately leads to starch degradation. This includes multiple enzymatic pathways, multiple plant hormones, a complex signaling pathway, several physical factors, and potential protein inhibitors that have been found to mitigate AA. Here we review how AA activity is measured; what environmental, physical, chemical, and genetic factors influence it; and perspective on how PHS and LMA can be best mitigated with changes in climate favoring an increase in their occurrence in cereal crops, which form the bulk of calories consumed worldwide.

Keywords Pre-harvest sprouting · Late maturity alpha-amylase · Seed dormancy · Alpha-amylase activity · Falling number · Quantitative trait loci · Plant hormones

Abbreviations

AA	Alpha-amylase
AACC	American Association of Cereal Chemists
ABA	Abscisic acid
BR	Brassinosteroid

N. Patwa · B. W. Penning (✉)
Soft Wheat Quality Laboratory, United States Department of Agriculture, Wooster, OH, USA
e-mail: bryan.penning@usda.gov

R. Roychowdhury et al. (eds.), *Sustainable Agriculture in the Era of Climate Change*, https://doi.org/10.1007/978-3-030-45669-6_2

ELISA	Enzyme-linked immunosorbent assay
FN	Falling number
GA	Gibberellin
JA	Jasmonic acid
LMA	Late maturity alpha-amylase
miRNA	microRNA
PHS	Pre-harvest sprouting
QTL	Quantitative trait loci
SNP	Single-nucleotide polymorphism

1 Introduction

Pre-harvest sprouting (PHS) is the germination of the seed while they are still on the spike with breakage of dormancy after maturity but before harvest. PHS occurs in the world's most grown cereal crops, ranked in order by production in 2017: maize, wheat, rice, barley, and sorghum (http://www.fao.org/faostat/en/#compare, 2019) as well as rye. PHS is reported in most regions of the world including Japan, China, India, the United States, Canada, Australia, North Africa, and throughout Europe (Wan et al. 2006; Biddulph et al. 2007; Nakamura 2018; Benech-Arnold and Rodriguez 2018). Damage occurs when cold and/or rainy conditions persist reducing yield, quality, and end-use applications (Paterson et al. 1989; Clarke et al. 2005; Barnard and Smith 2009; Gualano and Benech-Arnold 2009; Huang et al. 2012; Zhang et al. 2014; Lee et al. 2017; Benech-Arnold and Rodriguez 2018). In China, PHS has reduced the yield of rice, a staple food, by 10–50% in the Yangtze River Valley and Southwest China (Longxing et al. 2007; Lee et al. 2016). Sorghum as a native African crop is being used to produce ethyl alcohol for biofuels, beer, food, and animal fodder, but PHS can raise the cyanide concentration to levels too lethal for consumption limiting its use (Ikediobi et al. 1988; Benech-Arnold and Rodriguez 2018). PHS not only reduces the yield and end-use quality of barley and sorghum but also adversely affects malting quality important for brewers (Rodriguez et al. 2001; Li et al. 2004; Benech-Arnold and Rodriguez 2018). Late maturity alpha-amylase (LMA) activity occurs in winter wheat of the Pacific Northwest United States when large temperature changes occur close to harvest time (Brown et al. 2018). PHS and LMA are identified at the point of sale by a low falling number (FN) negatively affecting the grade, marketability, and price per bushel of wheat by reducing test weight and the functional properties of grain (Colombo et al. 2008; Simsek et al. 2014). In severe cases, reduced test weight and functional properties lead to downgrading of grain from human use to animal feed lowering the price per bushel up to 30% (http://www.ers.usda.gov). In 2015, soft winter wheat had the lowest FN recorded since widespread reporting began in the Northeastern United States and led to reduced test weight and functional properties that downgraded wheat quality from an average US grade 2 to US grade 3, and losses of crops were estimated at 10.6 million metric tons (http://www.uswheat.org). When PHS lowers FN, it adversely affects the baking quality of flour leading to bread loaves with large holes, sticky crumb, and dark-colored crusts or sticky noodles and pasta (Simsek et al. 2014; Olaerts and Courtin 2018). PHS also contributes to cereal crops becoming prone to black-point, red smudge, and *Fusarium* damage that ultimately leads to poor nutritive value (Wang et al. 2003; Clarke et al. 2005).

As our global climate changes, the conditions leading to PHS and LMA and subsequent reduction in seed quality may occur with more regularity and in more places as demand for higher quality and quantity of food increases over the next three decades leading to potential food production deficits in cereals (Maity and Pramanik 2013; Ray et al. 2013). Although the biochemical basis for the physiological effects is known, less is known about the genes involved in PHS. Quantitative trait locus (QTL) studies have linked PHS/LMA resistance to nearly all wheat, rye, and barley chromosomes (Ullrich et al. 2008; Masojć and Milczarski 2009; Gao et al. 2013). Many observed QTL and multiple biochemical pathways suggest a complex network of genes are involved in PHS and LMA. Because QTL regions can span thousands of genes, the causal genes often are not identified (Cabral et al. 2014). While some genes have been discovered such as those affecting dormancy through abscisic acid (ABA), gibberellin (GA), and a few signaling or other regulatory genes, much work is left to find durable and consistent resistance without impacting yield, quality, and germination rates (Henry et al. 1992; Yanagisawa et al. 2005; Chono et al. 2006; Tuttle et al. 2015; Martinez et al. 2016; Guo et al. 2018; Nakamura 2018). Considering an uncertain climate that could lead to crop loss, genes for PHS and LMA resistance must be found and deployed to meet global food needs without increasing land and/or water usage.

2 Mechanism and Measurement of Pre-harvest Sprouting and Late Maturity Alpha-Amylase

As seeds germinate, they release alpha-amylase (AA) which starts the process of starch digestion. During germination, starch provides carbohydrates for conversion to energy prior to seedlings being able to photosynthesize. AA activity in the seed is controlled by two genes, *α-Amy1* and *α-Amy2* (Laethauwer et al. 2013). When PHS and LMA occur, the process of germination proceeds prematurely, while cereal crop seeds like barley, wheat, rice, and maize are still on the spikelet or ear often before harvest or sometimes shortly after and are positively correlated with AA activity (Baun et al. 1970; MacGregor et al. 1972; DePauw and McCaig 1991; Thevenot et al. 1992; Lunn et al. 2002; Laethauwer et al. 2013). AA is a glycoside hydrolase family 13 endo-amylase that cleaves the alpha 1–4 linkage of the glycosidic bond of repeating glucose units of starch (van der Maarel et al. 2002; Lombard et al. 2014). This cleavage breaks starch molecules into smaller units. In addition, electron micrographs have shown a reduction in the protein matrix around starch molecules and partial hydrolyzation of starch granules during PHS (Simsek et al. 2014). The entire process leads to flour paste with reduced viscosity which significantly alters the properties of dough made with flour from sprouted seeds making noodle dough soft and sticky and baked bread and breadcrumb less firm (Bean et al. 1974; Morad and Rubenthaler 1983; Simsek et al. 2014). There are several often used measures of PHS, some pertinent for LMA and some specific by plant species and region.

A colorimetric enzyme assay is used to measure AA activity from crude extractions of whole ground seeds using American Association of Cereal Chemists (AACC) method 22-02.01 (AACC International 2010a) for both PHS and LMA. As the concentration of released AA increases, absorbance at 400 nm increases for a certain volume of extract. This procedure requires a certain extraction time and gives an estimate of activity over time, making consistent time measurement critical for all samples, usually by stopping the reaction after 20 min and measuring within 2 h as the color indicator, *p*-nitrophenol, is not stable. At extreme low amounts of extracted AA, little *p*-nitrophenol is released causing a greater error in measurement. If too much enzyme is released, the *p*-nitrophenol becomes too concentrated to be accurately read by spectrophotometer and must be diluted. Complete conversion of *p*-nitrophenol is possible for very high AA amounts leading to under-representation of activity.

A second AACC method, 56-81.03 (AACC International 2010b), was developed to measure FN also called the Hagberg number. This method measures ground seed paste viscosity and is used by industry as the standard for PHS/LMA damage to seeds and to determine suitability for use in baking. The procedure uses set amounts of ground seed or flour based on moisture content mixed with water, shaken, and heated to form a paste. The time it takes a plunger to fall to the bottom of the tube in seconds is the FN. Greater than 300 s indicates sound grain with little starch damage from AA. As FN becomes progressively <300 s, the paste becomes progressively thinner due to increased starch damage from high AA. Thus, FN has an inverse relationship with starch damage and AA activity. The *p*-nitrophenol indicator assay was tested versus a cross-linked dye-labeled starch assay available commercially and compared to FN with both methods correlating well to FN (Henry 1989). However, the correlation between AA activity and FN has been shown to be nonlinear, complicating the substitution of one technique for the other (Yu et al. 2015).

A rapid measure of damage, unique to PHS, is visual sprouting damage (Fig. 2.1). This measure relies on the absence or appearance of root/shoot tips and to what degree they have emerged. Visual sprouting is more loosely correlated with AA (positively) and FN (negatively) but can give an instant indication at harvest in the field of grain that is sprouted and will not be saleable. It has also become a high-throughput visual screen for PHS resistance in several studies (Paterson et al. 1989; Kulwal et al. 2012), mostly for presence/absence of resistance or category-based statistical genetic analyses, while AA activity or FN remains the preferred method for numerically based statistical genetic analyses despite the low throughput of these methods (Brown et al. 2018).

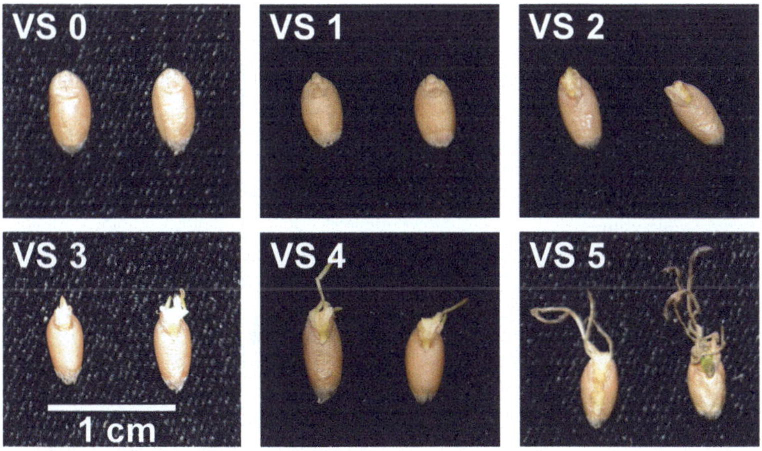

Fig. 2.1 *Visual assessment of sprouting damage*. Images of pre-harvest sprouting damage of soft winter wheat grain with degree of visual sprouting (VS) on a scale of 0 to 5, based on amount of sprouting observed. (Picture courtesy: C. Hampton)

An enzyme-linked immunosorbent assay (ELISA) test was designed for rapid detection of AA in Australian wheat in the field (Verity et al. 1999). For many Australian varieties of wheat, both the standard and rapid ELISAs show a strong correlation with FN in both field and artificially wetted samples (Verity et al. 1999; Skerritt and Heywood 2000). However, recent testing with wheat lines from the Pacific Northwest in the United States shows that the Australian wheat ELISA did not detect AA in breeding lines used in that region of the world (Kiszonas and Morris 2018). A human AA ELISA kit, AMY2A, was found to detect wheat AA for the United States Pacific Northwest (Kiszonas and Morris 2018). While fast and accurate, ELISA kits may be of limited scope and must be tested for proper function based on geographic region and species before wide-scale use.

3 Environmental Conditions That Impact PHS/LMA and Replicating Them by Artificial Means

Although PHS and LMA traits are controlled at the genetic level, they are heavily influenced by environmental factors depicted in Fig. 2.2 (Fenner 1991; Penfield and MacGregor 2017). In one study from 9.9% to 64.5% of seed dormancy could be attributed to non-genetic variation from weather conditions 14 days prior to physiological maturity (Strand 1989). Continued rain and high humidity along with high temperature contribute to PHS (Reddy et al. 1985; Springthorpe and Penfield 2015; Penfield and MacGregor 2017). Rain increases moisture on the spike and, with a conducive temperature around the time of seed maturity, provides suitable conditions for PHS prior to harvesting grain (Lunn et al. 2002). In 2000–2001, the grade

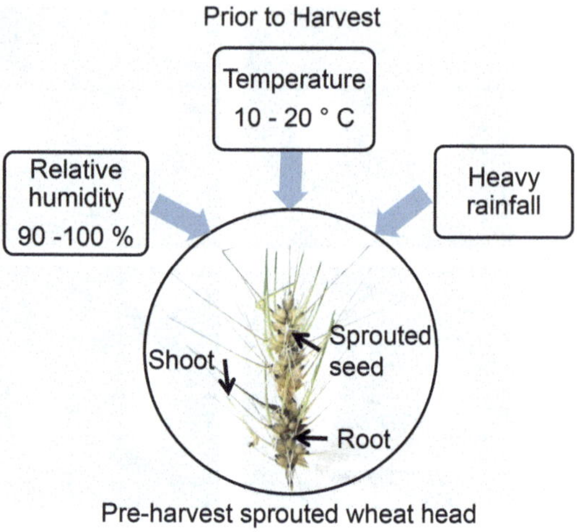

Fig. 2.2 *Environmental conditions affecting pre-harvest sprouting.* Image of a heavily sprouted soft winter wheat spike with many roots and shoots extended out from germinating seeds. Rain, humidity, and temperature can all contribute to sprouting before harvest. (Picture courtesy: N. Patwa)

of Western Red Spring wheat was downgraded and accounted for >$100 million loss due to PHS from heavy continuous rain in Canada (Clarke et al. 2005). LMA is influenced by large, fast changes in temperature rather than rain (Brown et al. 2018). Temperatures <10 °C induce higher and sustained dormancy during seed development, while the same temperature breaks dormancy in freshly harvested seeds (Reddy et al. 1985). A direct correlation was found between temperature and germination index in the barley cultivar Quilmes Palomar 12 days after physiological maturity (Rodriguez et al. 2001). Correlation between temperature and PHS was observed by Gualano and Benech-Arnold (2009) in four barley cultivars. Water, soil nitrogen availability, and day length were also correlated. PHS tolerance was enhanced in the hard white spring wheat variety Cunderdin, a non-dormant genotype, by drought and high temperature to levels similar to DM2001, a dormant genotype (Biddulph et al. 2005). Thomason et al. (2009) correlated dry high-temperature conditions with PHS tolerance before grain maturity and the inverse correlation of low temperature with high dormancy afterward in wheat.

While it is desirable to study PHS and LMA under field conditions, weather variance from year to year can make these studies difficult or not possible if conducive conditions do not happen at the right time. In the past few decades, numerous greenhouse, growth chamber, and other artificial rain and temperature conditions have been developed to study PHS and LMA under more controlled conditions. Most have a “rain” event of 20–135 mm of water either sprayed or misted on the plant spikes over 1–3 h followed by incubation at 95–100% relative humidity with water misting to keep high relative humidity for 2 days to 2 weeks at temperatures from 15 °C to 20 °C (DePauw and McCaig 1983, 1991; Edwards et al. 1989; Nielsen et al. 1984; King 1993; Trethowan 1995; Shorter et al. 2005). Artificial and natural PHS measurements are related. A moderate correlation was observed in sprouting damage of 444 breeding lines between simulated PHS conditions at the University of Sydney and field tests in Toluca Valley, Mexico, with rainfall during maturity (Trethowan et al. 1996). A study of natural versus artificial weathering in New

Zealand for nine cultivars of wheat over 2 years found visible sprouting between weathering types to be correlated but variable, FN was moderately correlated the first year but not correlated the next year (Shorter et al. 2005). Three cultivars with very late maturity showed much greater differences between naturally and artificially induced PHS suggesting maturity as a component of PHS resistance (Shorter et al. 2005). Overhead irrigation in the field to simulate rainfall has been shown to increase sprouting damage and reduce FN but was less severe and somewhat genotype dependent compared to natural weathering of plants left in the field (Humphreys and Noll 2002). A study of grain characteristics, water retention, and PHS under artificially controlled rain for 50 wheat varieties revealed no correlation between grain characteristics and water retention but found a limited correlation between water retention and PHS (King and Richards 1984). A link between PHS, high AA and protease activity, and end-product milling quality was demonstrated using a rain simulator (Edwards et al. 1989). A positive correlation between PHS and AA activity and stronger response in white winter wheat as opposed to red wheat was observed using simulated rain conditions. A study on hard winter wheat found high temperature before physiological maturity increased PHS after an artificial rain treatment, while high temperatures after physiological maturity reduced PHS (Nielsen et al. 1984). Increased PHS under simulated rainfall conditions was observed in awned versus awnless wheat (King and Richards 1984; Ji et al. 2018). Yanagisawa et al. (2005) developed PHS-resistant line Kitakei 1802 using artificial rain conditions, but the line has low yield and inferior quality.

4 Physical, Chemical, and Genetic Factors That Influence PHS and LMA

While PHS and LMA occur through environmental changes, namely, temperature (PHS and LMA) and rain or humidity (PHS), many physical, chemical, and genetic factors influence the onset and severity of PHS and LMA. Physical attributes such as seed coat color, presence/absence of awns, and epicuticular waxes have long been associated with differences in PHS severity. Plant hormones such as ABA or GA and the activity of AA affect both PHS and LMA. These factors are involved in several seed-related agronomic traits such as germination efficiency, dormancy, and barley malting quality. Numerous QTL and genome-wide association studies have been performed and many small effect locations found on nearly every chromosome of several species suggesting a very complex trait. Several genes have been isolated and those discussed are collected in Table 2.1. Considering the known interactions and/or genes in several pathways of seed germination, there are many possible avenues leading to damage from PHS and LMA and just as many potential ways to improve tolerance through allele selection. Genes from many of these processes are collectively called dormancy genes as they are related to lack of or slow seed germination or dormant seeds (Martinez et al. 2016; Nakamura 2018).

Table 2.1 Genes, hormones, or regulators affecting PHS and LMA

Functions	Factors[a]	Plant	PHS resistance	References
Signaling genes	MKK3	Barley Wheat	Positive	Nakamura et al. (2016) and Torada et al. (2016)
	MFT, TaPHS1	Wheat	Positive	Nakamura et al. (2011) and Liu et al. (2013)
	R-gene (red-testa pigmentation)	Wheat	Positive	Himi et al. (2011), Rodriguez et al. (2015), Kato et al. (2017), and Benech-Arnold and Rodriguez (2018)
Quantitative trait locus	Qphs.ocs-3A.1	Wheat	Positive	Mori et al. (2005) and Nakamura et al. (2011)
Epigenetic modification	ARGNAUTE gene	Wheat	Positive	Singh et al. (2013) and Nonogaki (2014)
Post-transcriptional regulation	Micro RNA9678 (miR9678)	Wheat	Positive	Guo et al. (2018)
Enzyme	α-amylase (AA)	Wheat	Negative	Laethauwer et al. (2013)
Regulatory protein	Thioredoxin h gene (Trx h)	Wheat Barley	Negative	Li et al. (2009)
Plant hormone	Abscisic acid (ABA)	All	Positive	Fang and Chu (2008) and Nonogaki and Nonogaki (2017)
	Gibberellins	All	Negative	Rodriguez et al. (2012)
	Auxin	All	Positive	Ramaih et al. (2003)
	Brassinosteroid (BR)	All	Negative	Hu and Yu (2014) and Shu et al. (2016)
	Jasmonic acid (JA)	All	Negative	Jacobsen et al. (2013)
Stimulus	Blue light	All	Positive	Barrero et al. (2014)

[a]Factors include genes, plant hormones, enzymes, proteins, or light stimulus

Many studies over the decades have associated red seed coat color with increased seed dormancy and PHS/LMA tolerance (Gfeller and Svejda 1960; DePauw and McCaig 1983; Himi et al. 2002; Bassoi and Flintham 2005; Lin et al. 2016, Brown et al. 2018; Ji et al. 2018). While red seed coat often appears to enhance PHS/LMA tolerance and often used by breeders as a convenient visual trait to improve PHS resistance, many other factors that contribute to PHS/LMA tolerance may be more effective (DePauw and McCaig 1983; Himi et al. 2002; Gao et al. 2013; Lin et al. 2016). In the absence of PHS conditions, AA activity was not significantly different between red and white wheat varieties (Yu et al. 2015). However, under conditions conducive to PHS, either naturally occurring in the field or by artificial means, red wheat varieties had significantly lower AA activity and higher FN than white varieties (Brown et al. 2018; Ji et al. 2018). The presence of awns has been positively correlated with uptake of water on the ear and increase in PHS in wheat and with a QTL for PHS in rice, but it remains unclear if this effect occurs under natural field conditions or if it is the awn itself or the lemma or glume of awned varieties that lead to the correlation (King and Richards 1984; Lee et al. 2017; Ji et al. 2018).

Epicuticular waxes which can repel water also provide a physical means of reducing PHS but would have little effect on LMA which occurs by temperature fluctuations rather than moisture. Glaucous barley with smooth, wild-type waxes reduces in-ear sprouting under artificial rainfall conditions compared to non-glaucous, glossy mutants through differences in water uptake by the ear (King and von Wettstein-Knowles 2000).

Plant hormone signaling plays a large role in germination and in PHS. ABA and GA are well-known as the primary hormones affecting seed dormancy and thus germination or premature germination on the spike but recently other hormones such as jasmonic acid isoleucine (JA-Ile), brassinosteroid (BR), and aleurone auxin/indole-3-acetic acid (IAA) have also been implicated in dormancy and PHS. Two 9-cis-epoxycarotenoid dioxygenase genes (NCEDs) involved in the production of ABA have been shown to affect the sprouting of barley seeds. *HvNCED2* is most highly expressed in the immature embryo when ABA is at its highest level and has a cyclical expression pattern repeated in multiple years of field grown plants regardless of conditions. In contrast, a closely related gene *HvNCED1* was expressed in both the immature embryo and the embryo of mature seeds well past grain filling. During artificial water imbibition tests, *HvNCED1* had increased expression, and field studies showed higher expression of *HvNCED1* in mature seed embryos under PHS conditions, while *HvNCED2* did not show a change in expression pattern based on weather or water imbibition. An ABA 8′-hydroxylase gene, *HvCYP707A1*, involved in the degradation of ABA expressed after seed maturation, showed some increase in expression during PHS conducive field conditions, but overall ABA remained higher in conducive PHS field conditions compared to dry conditions (Chono et al. 2006). The mutant *ENHANCED RESPONSE TO* ABA8 (*ERA8*) in the soft white spring wheat cultivar Zak was observed to have increased dormancy and PHS resistance and was hypersensitive to ABA without changes in the concentration of endogenous ABA but less sensitive to GA than wild type (Martinez et al. 2016). MicroRNAs (miRNAs) play an important role in biotic and abiotic stress response (Zhang and Wang 2015; Patwa et al. 2019; Patwa and Basak 2018). A miRNA, miR9678, in wheat, when bound by ABA, was found to activate its expression and led to a reduction in GA and increased PHS resistance (Guo et al. 2018). Exogenous application of gibberellin A_3 (GA_3) in rice was observed to break dormancy, while application of paclobutrazol, a GA inhibitor, delayed dormancy in sorghum (Steinbach et al. 1997; Hu et al. 2017). Thus, sensing of PHS conditions through increased ABA could be transmitted in the embryo by reducing GA for increased PHS resistance. Both JA-Ile and IAA levels increased in *ERA8* mutants, but the mechanism by which they might act is still unclear (Martinez et al. 2016). Suppression of auxin response factor 10 by miR160 led to increased seed germination, while application of IAA showed the opposite effect (Ramaih et al. 2003; Liu et al. 2007, 2013). Reduced sensitivity to IAA as seeds approach germination was also shown to be correlated with repression of the genes *auxin response factor* and *related to ubiquitin1* (Liu et al. 2013; Shu et al. 2016). Methyl jasmonate was observed to reduce ABA levels by reducing expression of the ABA-synthesizing gene, *TaNCED1*, and increasing expression of the ABA-catabolizing gene, *TaABA8O*

OH-1, in wheat (Jacobsen et al. 2013). In *Arabidopsis*, BR promotes germination by inhibition of ABA through the mother of flowering time (MFT)-mediated pathway and through *BRASSINOSTEROID INSENSITIVE2* (*BIN2*) and *ABSCISIC ACID INSENTITIVE5* (*ABI5*) (Xi and Yu 2010; Hu and Yu 2014; Shu et al. 2016). Increased amounts of BR in wheat seeds breaking dormancy after dry storage were correlated with suppression of the gene *BIN2* and increased expression of genes *de-etiolated2*, *dwarf4*, and *brassinosteroid signaling kinase* (Chitnis et al. 2014).

Other factors have been found that impact dormancy and PHS through signaling pathways or affecting the activity of AA. In barley, alanine transaminase (AlaAT) which interconverts glutamate to alanine was found to dramatically increase dormancy, but the mechanism is unclear (Nakamura 2018). Signaling genes such as the *mitogen-activated protein kinase kinase* (*MKK3*) of barley and wheat, as well as *MFT* of wheat, known as *TaPHS1*, have also been identified as major sources of seed dormancy and PHS tolerance (Nakamura et al. 2016; Nakamura 2018). The ABA signal is transduced through *MKK3* in *Arabidopsis* to prevent germination, and some *MKK3* alleles have been reported to be hypersensitive to ABA (Lu et al. 2002; Hwa and Yang 2008). A barley *bifunctional alpha-amylase/subtilisin inhibitor* (*BASI*) was found to reduce the activity of AA for both barley and wheat. While wheat and rye were found to have similar *BASI* genes to barley, none of the cultivars tested had the same significant reduction in activity (Henry et al. 1992). Structural analysis of the interaction of BASI with wheat ALPHA-AMYLASE 2 (AMY2) found a fully hydrated calcium ion-coordinated interaction between the proteins at three AMY2 catalytic amino acid residues. A total of 48 amino acid residues from the 2 proteins interact forming a BASI/AMY2 complex that is enzymatically inactive (Nielsen et al. 2004). In 2002, Kaneko et al. reported increased expression of *RAmy1A*, an AA gene, in rice endosperm in response to increased expression of *OsGA3ox2* gene encoding 3β-hydroxylase, an enzyme catalyzing GA synthesis.

Other factors can impact quality by influencing AA activity and FN. In the later stages of wheat grain development, a sudden large temperature shock produces high levels of AA without the need for water imbibition and without premature seed germination termed LMA (Mares and Mrva 2008). Blue light transduced through the CRYPTOCHROME (CRY1) blue light receptor can increase dormancy in barley by inducing the ABA-synthesizing gene *9-cis-epoxycarotenoid dioxygenase* and reducing expression of the ABA-catabolizing gene *8′-hydroxylase* while red and far-red light have no effect (Gubler et al. 2008; Barrero et al. 2014). Storage of field harvested wheat grain at different temperatures from 5 °C to 35 °C for 5 months resulted in significantly increased FN as temperature increased under experimental conditions suggesting storage and temperature may play a role in AA activity although moisture content between 10% and 13% had no effect (Ji and Baik 2016). In barley, many high malting quality varieties have a tendency toward PHS as seed dormancy is often detrimental to malting quality (Li et al. 2003). However, in bread baking, a lower FN and higher AA activity could be useful in reducing the need for baking improver containing amylases and xylanases. Grain high in AA due to increased expression of *alpha-amylase 3* as a proxy for LMA grain without improver had the same loaf volumes and browning as sound grain with baking improver and significantly more loaf volume and browning than sound grain without baking improver (Ral et al. 2016).

The number of QTL involved in PHS and LMA is predicted to be large as this is a complex trait with many sources of difference leading to the condition. In rye, 16 QTL for difference in AA activity were observed across two mapping populations covering all seven chromosomes. Nine of those QTL corresponded to PHS QTL indicating a close but not complete linkage between PHS and AA activity (Masojć and Milczarski 2009). Eight PHS QTL regions covering six of seven chromosomes in one barley mapping population were observed of which six QTL also showed a difference in AA activity and all corresponded to known locations for dormancy (Ullrich et al. 2008). A study using 21,000 single-nucleotide polymorphisms (SNPs) across 21 rice accessions found 39 candidate PHS resistance loci covering eight of twelve chromosomes (Lee et al. 2017). The 39 candidate loci were genotyped by SNPs and phenotyped for PHS resistance and germination index in 144 rice accessions yielding 7 locations overlapped with previous QTL studies and 6 new significant locations for PHS resistance (Lee et al. 2017). Gao et al. (2013) reviewed numerous QTL studies in wheat populations where dormancy and/or PHS QTL were found on 17 of 21 chromosomes missing only the D genomes of chromosomes 1 and 7 and the A and B genomes of chromosome 6. The wide range in location and number of QTL is expected. However, it makes breeding for the trait very difficult and requires further information on the action of the QTL in PHS/LMA tolerance as demonstrated for wheat and barley signaling genes found from QTL in Nakamura (2018). QTL analysis can provide insight into potential trait improvement especially when multiple traits are tested. A relationship between red grain color and PHS has been long established. One study found four QTL linked for grain color and PHS while grain color had eight independent QTL and PHS seven. Also, the study found the major QTL for grain color was only a minor QTL for PHS suggesting that, while some PHS tolerance can be obtained with red wheat, it is possible for strong PHS tolerance in white wheat (Kumar et al. 2009). Further studies identified and cloned genes in three QTL regions, one of which, Qphs.ocs-3A.1, contained the gene *mother of flowering time*, *TaMFT* also called *TaPHS1*, in wheat which was observed to confer PHS tolerance (Nakamura et al. 2011; Liu et al. 2013; Shao et al. 2018). A similar tactic will be necessary to find PHS/LMA resistance without affecting dormancy to allow for deep planting or overwintering without suffering winter kill. Many studies have linked AA activity, dormancy, and PHS to the same location for some but not all loci (Li et al. 2003; Ullrich et al. 2008; Masojć and Milczarski 2009; Gao et al. 2013; Lee et al. 2017). Comparative mapping in cereal species has shown many locations for PHS tolerance and dormancy overlap (Gale et al. 2002). Thus, much of what is learned in one grass species is likely useful in other grass species.

5 Other PHS Genes or QTL Regions and Their Potential Effect

While not meant to be an exhaustive list, Table 2.1 summarizes the major locations or potential genes that can impact PHS from published journal articles. Here we briefly describe the genes not covered elsewhere in the chapter. There are numerous germination-related QTL, often involved in dormancy or heading date including *Sdr1, Sdr6, Hd8, qDT-SGC3.1, qSD-1, qSD-3, qSD1, qDEG1, QPhs.ocs-3A.1, QPhs.ocs-4A.1, and QPhs.ocs-4B.1* (Miura et al. 2002; Takeuchi et al. 2003; Gu et al. 2004, 2006; Mori et al. 2005; Jiang et al. 2011; Li et al. 2011; Marzougui et al. 2012). The *R-1* gene encodes an *R2R3 MYB transcription factor* regulating anthocyanin synthesis and increasing red pigmentation in seeds often associated with PHS resistance in wheat (Himi et al. 2011; Rodriguez et al. 2015; Kato et al. 2017; Benech-Arnold and Rodriguez 2018). The expression of some ARGONAUTE genes involved in DNA methylation such as *AGO4_9, AGO1003*, and *AGO802B* are reported to be negatively correlated with PHS (Singh and Singh 2012; Singh et al. 2013; Nonogaki 2014). The thioredoxin h (Trx h) protein reduces the disulfide bonds of proteins including seed proteins such as AA and pullulanase in wheat and barley. Reducing *Trx h* expression can delay PHS (Zhou et al. 2006; Li et al. 2009; Guo et al. 2011). The plant hormone ABA is a known regulator of germination and can cause PHS. Several mutants in ABA synthesis, *Viviparous 5 (vp5), Viviparous 7 (vp7), Viviparous14 (vp14), Viviparous10/Viviparous13 (vp10/vp13), and Viviparous15 (vp15)* (maize) and *PHS1–4 (phs1-psh4)* and *Osaba1* (rice) have been observed (Fang and Chu 2008). A mutation in the rice gene *Dormancy1–2* (*qSD1–2*), a GA-synthesizing gene, resulted in increased dormancy (Ye et al. 2015). The ABA-catabolizing genes *TaABA8′ OH1* and *TaABA8′ OH2* have an inverse relationship with the gibberellin-synthesizing gene *TaGA3ox2* in PHS-tolerant varieties of wheat (Kashiwakura et al. 2016). A simplified diagram of the interactions of the known factors and genes is shown in Fig. 2.3.

6 Future Perspectives on PHS and LMA with a Changing Climate

A global challenge is to meet a doubling of food demand by 2050 without increasing usage of arable land or water resources (Ray et al. 2013). This is further confounded by changes in climate creating challenges to maintain current production quality due to more intense weather conditions that negatively impact seed quality traits including PHS and LMA (Maity and Pramanik 2013). Together, these two global challenges could lead to food production deficits in cereals. There are many genetic avenues to alter the response of seeds to germination in the cereals prior to

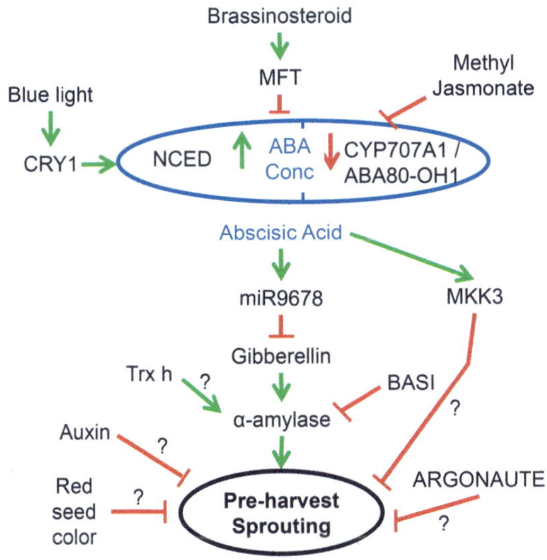

Fig. 2.3 *The complex signaling pathway to PHS*. The many factors and signals to germination or PHS are partially known. Several signal mechanisms interact with the synthesis/degradation of abscisic acid (ABA) through the *NCED* gene, part of the ABA synthesis pathway, and the *CYP707A1/ABA80 OH-1* gene that catabolizes ABA (gray circle). These include blue light transduced through the CRY1 receptor that increases ABA synthesis and reduces ABA degradation. Brassinosteroid reduces ABA through the MFT-mediated pathway. Methyl jasmonate reduces ABA synthesis. ABA induces the expression of the *MKK3* gene reducing PHS by an unclear pathway. ABA binds miR9678 which reduces gibberellin. Gibberellin increases the release of alpha-amylase (AA) which breaks down starch molecules and begins pre-harvest sprouting. Higher thioredoxin h (Trx h) levels reduce more disulfide bonds of AA and increase PHS. *Bifunctional alpha-amylase/subtilisin inhibitor* (*BASI*) reduces AA activity by forming a *BASI/AMY2* complex that is enzymatically inactive. *ARGONAUTE* and red seed color reduce PHS through unknown mechanisms. The blue circle represents ABA concentration. Up/angled arrowhead (green) represents induction and down/flat arrowhead (red) represents repression. Question marks indicate the exact interaction is not known

harvest mostly tied to seed dormancy, but they have come at the cost of slow germination and an undesirable penalty to yield (Yanagisawa et al. 2005; Martinez et al. 2016; Guo et al. 2018; Nakamura 2018). A delicate tightrope must be walked between delayed germination with low yield and pre-harvest germination with low quality. To perform this act, as much knowledge about the process and genes involved must be gained to breed new varieties of cereal crops adapted to high yield and less susceptibility to PHS and LMA.

Acknowledgment We would like to thank Cindy Hampton, USDA-ARS Soft Wheat Quality Lab, for taking images of visually sprouted seeds.

References

AACC International (2010a) Approved methods of analysis, 11th edn. Method 22-02.01. Measurement of alpha-amylase in plant and microbial materials using the ceralpha method. AACC International, St. Paul, MN, U.S.A. https://doi.org/10.1094/AACCIntMethod-22-02.01

AACC International (2010b) Approved methods of analysis, 11th edn. Method 56-81.03. Determination of falling number. AACC International, St. Paul, MN, U.S.A. https://doi.org/10.1094/AACCIntMethod-56-81.03

Barnard A, Smith MF (2009) The effect of rainfall and temperature on the preharvest sprouting tolerance of winter wheat in the dryland production areas of the Free State Province. Field Crop Res 112:158–164

Barrero JM, Downie AB, Xu Q et al (2014) A role for barley CRYPTOCHROME1 in light regulation of grain dormancy and germination. Plant Cell 26:1094–1104

Bassoi MC, Flintham J (2005) Relationship between grain colour and preharvest sprouting-resistance in wheat. Pesq Agropec Bras, Brasilia 40:981–988

Baun LC, Palmiano EP, Perez CM et al (1970) Enzymes of starch metabolism in the developing rice grain. Plant Physiol 46:429–434

Bean MM, Keagy PM, Fullington JG et al (1974) Dried Japanese noodles. I. Properties of laboratory-prepared noodle doughs from sound and damaged wheat flours. Cereal Chem 51:416–426

Benech-Arnold RL, Rodriguez MV (2018) Pre-harvest sprouting and grain dormancy in *Sorghum bicolor*: what have we learned? Front Plant Sci 9:1–8

Biddulph TB, Mares DJ, Plummer JA et al (2005) Drought and high temperature increases preharvest sprouting tolerance in a genotype without grain dormancy. Euphytica 143:277–283

Biddulph TB, Plummer JA, Setter TL et al (2007) Influence of high temperature and terminal moisture stress on dormancy in wheat (*Triticum aestivum* L.). Field Crop Res 103:139–153

Brown LK, Wiersma AT, Olson EL (2018) Preharvest sprouting and α-amylase activity in soft winter wheat. J Cereal Sci 79:311–318

Cabral AL, Jordan MC, McCartney CA et al (2014) Identification of candidate genes, regions and markers for pre-harvest sprouting resistance in wheat (*Triticum aestivum* L.). BMC Plant Biol 14:340

Chitnis VR, Gao F, Yao Z et al (2014) After-ripening induced transcriptional changes of hormonal genes in wheat seeds: the cases of brassinosteroids, ethylene, cytokinin and salicylic acid. PLoS One 9(1):e87543

Chono M, Honda I, Shinoda S et al (2006) Field studies on the regulation of abscisic acid content and germinability during grain development of barley: molecular and chemical analysis of pre-harvest sprouting. J Exp Bot 57:2421–2434

Clarke FR, Clarke JM, DePauw RM et al (2005) Strategic approach to mitigating weather induced defects of wheat quality. Euphytica 143:285–290

Colombo A, Pérez GT, Ribotta PD, León AE (2008) A comparative study of physiochemical tests for quality prediction of Argentine wheat flours used as corrector flours and for cookie production. J Cereal Sci 48:775–780

DePauw RM, McCaig TN (1983) Recombining dormancy and white seed color in a spring wheat cross. Can J Plant Sci 63:581–589

DePauw RM, McCaig TN (1991) Components of variation, heritabilities and correlations for indices of sprouting tolerance and seed dormancy in *Triticum* spp. Euphytica 52:221–229

Edwards RA, Ross AS, Mares DJ, Ellison FW, Tomlinson JD (1989) Enzymes from rain-damaged and laboratory-germinated wheat IEffects on product quality. J Cereal Sci 10:157–167

Fang J, Chu C (2008) Abscisic acid and the pre-harvest sprouting in cereals. Plant Signal Behav 3(12):1046–1048

Fenner M (1991) The effects of the parent environment on seed germinability. Seed Sci Res 1:75–84

Gale MD, Flintham JE, Devos KM (2002) Cereal comparative genetics and preharvest sprouting. Euphytica 126:21–25
Gao X, Hu CH, Li HZ et al (2013) Factors affecting pre-harvest sprouting resistance in wheat (*Triticum aestivum* L.): a review. J Anim Plant Sci 23:556–565
Gfeller F, Svejda F (1960) Inheritance of post-harvest seed dormancy and kernel colour in spring wheat lines. Can J Plant Sci 40:1–6
Gu XY, Kianian SF, Foley ME (2004) Multiple loci and epistases control genetic variation for seed dormancy in weedy rice (*Oryza sativa*). Genet 166:1503–1516
Gu X, Kianian S, Foley M (2006) Isolation of three dormancy QTLs as Mendelian factors in rice. Heredity 96:93–99
Gualano NA, Benech-Arnold RL (2009) Predicting pre-harvest sprouting susceptibility in barley: looking for "sensitivity windows" to temperature throughout grain filling in various commercial cultivars. Field Crop Res 114:35–44
Gubler F, Hughes T, Waterhouse P et al (2008) Regulation of dormancy in barley by blue light and after-ripening: effects on abscisic acid and gibberellin metabolism. Plant Physiol 147:886–896
Guo H, Zhang H, Li Y et al (2011) Identification of changes in wheat (*Triticum aestivum* L.) seeds proteome in response to anti–*trx s* gene. PLoS One 6(7):1–10
Guo G, Liu X, Sun F et al (2018) Wheat miR9678 affects seed germination by generating phased siRNAs and modulating abscisic acid/gibberellin signaling. Plant Cell 30:796–814
Henry RJ (1989) Rapid α-amylase assays for assessment of pre-harvest sprouting damage in wheat. J Sci Food Agric 49:15–23
Henry RJ, Battershell VG, Brennan PS, Oono K (1992) Control of wheat α-amylase using inhibitors from cereals. J Sci Food Agric 58:281–284
Himi E, Mares DJ, Yanagisawa A, Noda K (2002) Effect of grain colour gene (R) on grain dormancy and sensitivity of the embryo to abscisic acid (ABA) in wheat. J Exp Bot 53:1569–1574
Himi E, Maekawa M, Miura H et al (2011) Development of PCR markers for *Tamyb10* related to *R-1*, red grain color gene in wheat. Theor Appl Genet 122:1561–1576
Hu Y, Yu D (2014) BRASSINOSTEROID INSENSITIVE2 interacts with ABSCISIC ACID INSENSITIVE5 to mediate the antagonism of brassinosteroids to abscisic acid during seed germination in Arabidopsis. Plant Cell 26:4394–4408
Hu Q, Lin C, Guan Y et al (2017) Inhibitory effect of eugenol on seed germination and pre-harvest sprouting of hybrid rice (*Oryza sativa* L.). Sci Rep 7:1–9
Huang T, Qua B, Li H-P et al (2012) A maize *viviparous 1* gene increases seed dormancy and preharvest sprouting tolerance in transgenic wheat. J Cereal Sci 55:166–173
Humphreys DG, Noll J (2002) Methods for characterization of preharvest sprouting resistance in a wheat breeding program. Euphytica 126:61–65
http://www.ers.usda.gov (3/9/2019) United States Department of Agriculture, Economic Research Service
http://www.fao.org/faostat/en/#compare (3/9/2019) Food and Agriculture Organization of the United Nations
http://www.uswheat.org (3/9/2019) U.S. Wheat Associates
Hwa CM, Yang XC (2008) The AtMKK3 pathway mediates ABA and salt signaling in Arabidopsis. Acta Physiol Plant 30:277–286
Ikediobi CO, Olugboji O, Okoh PN (1988) Cyanide profile of component parts of sorghum (*Sorghum bicolor* L. Moench) sprouts. Food Chem 27:167–175
Jacobsen JV, Barrero JM, Hughes T et al (2013) Roles for blue light, jasmonate and nitric oxide in the regulation of dormancy and germination in wheat grain (*Triticum aestivum* L.). Planta 238:121–138
Ji T, Baik BK (2016) Storage conditions affecting increase in falling number of soft red winter wheat grain. Cereal Chem 93:263–267
Ji T, Penning B, Baik BK (2018) Pre-harvest sprouting resistance of soft winter wheat varieties and associated grain characteristics. J Cereal Sci 83:110–115

Jiang W, Lee J, Jin Y-M et al (2011) Identification of QTLs for seed germination capability after various storage periods using two RIL populations in rice. Mol Cells 31:385–392

Kaneko M, Itoh H, Ueguchi-Tanaka M et al (2002) The α-amylase induction in endosperm during rice seed germination is caused by gibberellin synthesized in epithelium. Plant Physiol 128:264–1270

Kashiwakura Y, Kobayashi D, Jikumaru Y et al (2016) Highly sprouting-tolerant wheat grain exhibits extreme dormancy and cold imbibition-resistant accumulation of abscisic acid. Plant Cell Physiol 57(4):715–732

Kato K, Maruyama-Funatsuki W, Yanaka M et al (2017) Improving preharvest sprouting resistance in durum wheat with bread wheat genes. Breed Sci 67:466–471

King RW (1993) Manipulation of grain dormancy in wheat. J Exp Bot 44:1059–1066

King RW, Richards RA (1984) Water uptake in relation to pre-harvest sprouting damage in wheat: ear characteristics. Aust J Agric Res 35:327–336

King RW, von Wettstein-Knowles P (2000) Epicuticular waxes and regulation of ear wetting and pre-harvest sprouting in barley and wheat. Euphytica 112:157–166

Kiszonas AM, Morris CF (2018) Evaluation of commercial α-amylase enzyme-linked immunosorbent assay (ELISA) test kits for wheat. Cereal Chem 95:206–210

Kulwal P, Ishikawa G, Benscher D et al (2012) Association mapping for pre-harvest sprouting resistance in white winter wheat. Theor Appl Genet 125:793–805

Kumar A, Kumar J, Singh R et al (2009) QTL analysis for grain colour and pre-harvest sprouting in bread wheat. Plant Sci 177:114–122

Laethauwer SD, Riek JD, Stals I et al (2013) α-amylase gene expression during kernel development in relation to pre-harvest sprouting in wheat and triticale. Acta Physiol Plant 35:2927–2938

Lee G-A, Jeon Y-A, Lee H-S et al (2016) Variation in pre-harvest sprouting resistance, seed germination and changes in abscisic acid levels during grain development in diverse rice genetic resources. Plant Genet Resour:1–10

Lee G-A, Jeon Y-A, Lee H-S et al (2017) New genetic loci associated with preharvest sprouting and its evaluation based on the model equation in rice. Front Plant Sci 8:1–9

Li CD, Tarr A, Lance RCM et al (2003) A major QTL controlling seed dormancy and pre-harvest sprouting/grain α-amylase in two-rowed barley (*Hordeum vulgare* L.). Aust J Agric Res 54:1303–1313

Li C, Ni P, Francki M et al (2004) Genes controlling seed dormancy and pre-harvest sprouting in a rice-wheat-barley comparison. Funct Integr Genomics 4:84–93

Li Y-C, Ren J-P, Chob M-J et al (2009) The level of expression of thioredoxin is linked to fundamental properties and applications of wheat seeds. Mol Plant 2(3):430–441

Li W, Xu L, Bai X et al (2011) Quantitative trait loci for seed dormancy in rice. Euphytica 178:427–435

Lin M, Zhang D, Liu S et al (2016) Genome-wide association analysis on pre-harvest sprouting resistance and grain color in U.S. winter wheat. BMC Genomics 17:794

Liu P-P, Montgomery TA, Fahlgren N et al (2007) Repression of AUXIN RESPONSE FACTOR10 by microRNA160 is critical for seed germination and post-germination stages. Plant J 52:133–146

Liu S, Sehgal SK, Li J et al (2013) Cloning and characterization of a critical regulator for preharvest sprouting in wheat. Genet 195:263–273

Lombard V, Ramulu HG, Drula E, Coutinho PM, Henrissat B (2014) The carbohydrate-active enzymes database (CAZy) in 2013. Nucleic Acids Res 42:D490–D495

Longxing T, Xi W, Huijuan T et al (2007) Physiological analysis on pre-harvest sprouting in recombinant inbred rice lines. Front Agric China 1(1):24–29

Lu C, Han M-H, Guevara-Garcia A et al (2002) Mitogen-activated protein kinase signaling in post germination arrest of development by abscisic acid. Proc Natl Acad Sci USA 99(24):15812–15817

Lunn GD, Kettlewell PS, Major BJ et al (2002) Variation in dormancy duration of the U.K. wheat cultivar Hornet due to environmental conditions during grain development. Euphytica 126:89–97

MacGregor AW, Gordon AG, Meredith WOS et al (1972) Site of α-amylase in developing barley kernels. J Inst Brew 78:174–178

Maity A, Pramanik P (2013) Climate change and seed quality: an alarming issue in crop husbandry. Curr Sci 105:1336–1338

Mares D, Mrva K (2008) Late-maturity α-amylase: low falling number in wheat in the absence of preharvest sprouting. J Cereal Sci 47:6–17

Martinez SA, Tuttle KM, Takebayashi et al (2016) The wheat ABA hypersensitive *ERA8* mutant is associated with increased preharvest sprouting tolerance and altered hormone accumulation. Euphytica 212:229–245

Marzougui S, Sugimoto K, Yamanouchi U et al (2012) Mapping and characterization of seed dormancy QTLs using chromosome segment substitution lines in rice. Theor Appl Genet 124:893–902

Masojć P, Milczarski P (2009) Relationship between QTLs for preharvest sprouting and alpha-amylase activity in rye grain. Mol Breed 23:75–84

Miura K, Lin S, Yano M et al (2002) Mapping quantitative trait loci controlling seed longevity in rice (*Oryza sativa* L.). Theor Appl Genet 104:981–986

Morad MM, Rubenthaler GL (1983) Germination of soft white wheat and its effect on flour fractions, breadbaking, and crumb firmness. Cereal Chem 60:413–417

Mori M, Uchino N, Chono M et al (2005) Mapping QTLs for grain dormancy on wheat chromosome 3A and the group 4 chromosomes, and their combined effect. Theor Appl Genet 110:1315–1323

Nakamura S (2018) Grain dormancy genes responsible for preventing pre-harvest sprouting in barley and wheat. Breed Sci 68:295–304

Nakamura S, Abe F, Kawahigashi H et al (2011) A wheat homolog of MOTHER OF FT and TFL1 acts in the regulation of germination. Plant Cell 23:3215–3229

Nakamura S, Pourkheirandish M, Morishige H et al (2016) *Mitogen-activated protein kinase kinase 3* regulates seed dormancy in barley. Curr Biol 26:775–781

Nielsen MT, McCrate AJ, Heyne EG, Paulsen GM (1984) Effect of weather variables during maturation on preharvest sprouting of hard white winter wheat. Crop Sci 24:779–782

Nielsen PK, Bønsager BC, Fukuda K, Svensson B (2004) Barley α-amylase/subtilisin inhibitor: structure, biophysics and protein engineering. Biochim Biophys Acta 1696:157–164

Nonogaki H (2014) Seed dormancy and germination-emerging mechanisms and new hypotheses. Front Plant Sci 5:1–14

Nonogaki M, Nonogaki H (2017) Prevention of preharvest sprouting through hormone engineering and germination recovery by chemical biology. Front Plant Sci 8(90):1–7

Olaerts H, Courtin CM (2018) Impact of preharvest sprouting on endogenous hydrolases and technological quality of wheat and bread: a review. Compr Rev Food Sci Food Saf 17:698–713

Paterson AH, Sorrells ME, Obendorf RL (1989) Methods of evaluation for preharvest sprouting resistance in wheat breeding programs. Can J Plant Sci 69:681–689

Patwa N, Basak J (2018) Plant microRNAs – Tiny yet significant regulators of disease. Res J Biotechnol 13(6):85–92

Patwa N, Nithin C, Bahadur RP et al (2019) Identification and characterization of differentially expressed Phaseolus vulgaris miRNAs and their targets during mungbean yellow mosaic India virus infection reveals new insight into *Phaseolus*-MYMIV interaction. Genomics. 111:1333–1342

Penfield S, MacGregor DR (2017) Effects of environmental variation during seed production on seed dormancy and germination. J Exp Bot 68(4):819–825

Ral JP, Whan A, Larroque O et al (2016) Engineering high α-amylase levels in wheat grain lowers falling number but improves baking properties. Plant Biotechnol J 14:364–376

Ramaih S, Guedira M, Paulsen GM (2003) Relationship of indole acetic acid and tryptophan to dormancy and preharvest sprouting of wheat. Funct Plant Biol 30:939–945
Ray DK, Mueller ND, West PC, Foley JA (2013) Yield trends are insufficient to double global crop production by 2050. PLoS One 8:e66428
Reddy LV, Metzger RJ, Ching TM (1985) Effect of temperature on seed dormancy of wheat. Crop Sci 25:455–259
Rodriguez V, Gonzalez-Martin J, Insausti P et al (2001) Predicting pre-harvest sprouting susceptibility in barley: a model based on temperature during grain filling. Agron J 93:1071–1079
Rodriguez MV, Mendiondo GM, Cantoro R et al (2012) Expression of seed dormancy in grain sorghum lines with contrasting pre-harvest sprouting behavior involves differential regulation of gibberellin metabolism genes. Plant Cell Physiol 53(1):64–80
Rodriguez MV, Barrero J, Corbineau F et al (2015) Dormancy in cereals (not too much, not so little): about the mechanisms behind this trait. Seed Sci Res 25:99–119
Shao M, Bai G, Rife TW et al (2018) QTL mapping of pre-harvest sprouting resistance in a white wheat cultivar Danby. Theor Appl Genet 131:1683–1697
Shorter SC, Munro CA, Hodgkinson J (2005) Predicting pre-harvest sprouting susceptibility in New Zealand wheat cultivars. Euphytica 143:309–312
Shu K, LiuX-d XQ et al (2016) Two faces of one seed: hormonal regulation of dormancy and germination. Mol Plant 9:34–45
Simsek S, Ohm JB, Lu H et al (2014) Effect of pre-harvest sprouting on physicochemical properties of starch in wheat. Foods 3:194–207
Singh M, Singh J (2012) Seed development-related expression of ARGONAUTE4_9 class of genes in barley: possible role in seed dormancy. Euphytica 188:123–129
Singh M, Singh S, Randhawa H et al (2013) Polymorphic homoeolog of key gene of RdDM pathway, ARGONAUTE4_9 class is associated with pre-harvest sprouting in wheat (*Triticum aestivum* L.). PLoS One 8(10):1–9
Skerritt JH, Heywood RH (2000) A five-minute field test for on-farm detection of pre-harvest sprouting in wheat. Crop Sci 40:742–756
Springthorpe V, Penfield S (2015) Flowering time and seed dormancy control use external coincidence to generate life history strategy. elife 4:1–17
Steinbach HS, Benech-Arnold RL, Sanchez RA (1997) Hormonal regulation of dormancy in developing sorghum seeds. Plant Physiol 113:149–154
Strand E (1989) Studies on seed dormancy in small grain species. II. Wheat. Nor J Agric Sci 3:101–115
Takeuchi Y, Lin S, Sasaki T et al (2003) Fine linkage mapping enables dissection of closely linked quantitative trait loci for seed dormancy and heading in rice. Theor Appl Genet 107:1174–1180
Thevenot C, Lauriere C, Mayer C et al (1992) α-amylase changes during development and germination of maize kernels. J Plant Physiol 140:61–65
Thomason WE, Hughes KR, Griffey CA et al (2009) Understanding pre-harvest sprouting of wheat. Virginia cooperative extension. Virginia polytechnic institute and state university
Torada A, Koike M, Ogawa T et al (2016) A causal gene for seed dormancy on wheat chromosome 4A encodes a MAP kinase kinase. Curr Biol 26:782–787
Trethowan RM (1995) Evaluation and selection of bread wheat (*Triticum aestivum* L.) for preharvest sprouting tolerance. Aust J Agric Res 46:463–474
Trethowan RM, Rajaram S, Ellison FW (1996) Pre-harvest sprouting tolerance of wheat in the field and under rain simulation. Aust J Agric Res 47:705–716
Tuttle KM, Martinez SA, Schramm EC, Takebayashi Y, Seo M, Steber CM (2015) Grain dormancy loss is associated with changes in ABA and GA sensitivity and hormone accumulation in bread wheat, *Triticum aestivum* (L.). Seed Sci Res 25:179–193
Ullrich SE, Clancy JA, del Blanco IA et al (2008) Genetic analysis of preharvest sprouting in a six-row barley cross. Mol Breed 21:249–259
Van der Maarel MJEC, van der Veen B, Uitdehaag JCM et al (2002) Properties and applications of starch-converting enzymes of the α-amylase family. J Biotechnol 94:137–155

Verity JCK, Hac L, Skerritt JH (1999) Development of a field enzyme-linked immunosorbent assay (ELISA) for detection of α-amylase in pre-harvest sprouted wheat. Cereal Chem 76:673–681

Wan J, Jiang L, Tang J et al (2006) Genetic dissection of the seed dormancy trait in cultivated rice (*Oryza sativa* L.). Plant Sci 170:786–792

Wang H, Fernandez MR, McCaig TN et al (2003) Kernel discolouration and downgrading in spring wheat varieties in western Canada. Can J Plant Pathol 25:350–361

Xi W, Yu H (2010) Mother of FT and TFL1 regulates seed germination and fertility relevant to the brassinosteroid signaling pathway. Plant Signal Behav 5:1315–1317

Yanagisawa A, Nishimura T, Amano Y et al (2005) Development of winter wheat with excellent resistance to pre-harvest sprouting and rain damage. Euphytica 143:313–318

Ye H, Feng J, Zhang L et al (2015) Map-based cloning of seed *dormancy1-2* identified a gibberellin synthesis gene regulating the development of endosperm-imposed dormancy in rice. Plant Physiol 169:2152–2165

Yu N, Laurenz R, Siler L, Ng PKW, Souza E, Lewis JM (2015) Evaluation of α-amylase activity and falling number around maturity for soft white and soft red wheat varieties in Michigan. Cereal Res Commun 43:672–681

Zhang B, Wang Q (2015) MicroRNA based biotechnology for plant improvement. Cell Physiol 230:1–15

Zhang Y, Miao X, Xia X (2014) Cloning of seed dormancy genes (*TaSdr*) associated with tolerance to pre-harvest sprouting in common wheat and development of a functional marker. Theor Appl Genet 127:855–866

Zhou SM, Yin J, Ren JP et al (2006) Study on molecular identification and pre-harvest sprouting characteristic of the transgenic anti-*trxs*-gene wheat line 00T89. Chin J Biotechnol 22(3):438–444

Chapter 3
Plant Nutrients for Crop Growth, Development and Stress Tolerance

Mukhtar Ahmed, Mirza Hasanuzzaman, Muhammad Ali Raza, Amjad Malik, and Shakeel Ahmad

Abstract Abiotic stress is a problem of grave concern for the growth and productivity of plants in modern times. Abiotic stresses, such as drought, salinity, extreme temperatures and extreme radiation, are responsible for huge crop losses globally. Plants face a combination of different abiotic stresses under field conditions that are lethal to plant growth and production. Exposure of plants to biotic and abiotic stress induces a disruption in plant metabolism implying physiological costs and thus leads to a reduction in fitness and, ultimately, in productivity. Abiotic stress is one of the most important features of and has a huge impact on growth, and consequently, it is responsible for severe losses in the field. The resulting growth reductions can reach more than 50% in most plant species. One of the physiological processes greatly affected by these stresses in plants is photosynthesis. The decline in photosynthetic capacity of plants due to these stresses is directly associated with reduction in yield. Application of nutrients to overcome nutrient stress positively affects plant growth, yield and quality. Transition/heavy metals such as zinc, manganese and copper are essential minerals for healthy plant growth. Micronutrients

M. Ahmed (✉)
Department of Agricultural Research for Northern Sweden, Swedish University of Agricultural Sciences, Umeå, Sweden

Department of Agronomy, Pir Mehr Ali Shah Arid Agriculture University, Rawalpindi, Pakistan
e-mail: ahmadmukhtar@uaar.edu.pk

M. Hasanuzzaman
Department of Agronomy, Faculty of Agriculture, Sher-e-Bangla Agricultural University, Dhaka, Bangladesh

M. A. Raza
College of Agronomy, Sichuan Agricultural University, Chengdu, China

A. Malik
Faculty of Education, University College of the North, Thompson, MB, Canada

Biology Department, Brandon University, Brandon, MB, Canada

S. Ahmad
Department of Agronomy, Bahauddin Zakariya University, Multan, Pakistan

R. Roychowdhury et al. (eds.), *Sustainable Agriculture in the Era of Climate Change*, https://doi.org/10.1007/978-3-030-45669-6_3

are essential for balanced nutrition in plants under abiotic stress conditions. The present chapter describes the role of macro- and micronutrients under stress conditions. In this chapter, we have discussed a variety of macro- and micronutrients which are beneficial for plant physiological development under stress conditions.

Keywords Abiotic and biotic stress · Growth · Productivity · Nutrients · Physiology

1 Introduction

Agricultural production in future will be under stress due to the multiple factors such as climate change and global population. Demands for the agricultural products in future will be on rise. Therefore, it is necessary to boost up the yield of the crop by managing available resources effectively (Hasanuzzaman et al. 2019a; Ahmed and Stockle 2016; Ahmed 2020). However, due to the climate change, plants are suffering from different abiotic stresses such as salinity, drought, temperature extremes, flooding, toxic metals, high-light intensity, UV radiation, herbicides and ozone. These abiotic stresses are major threats to agriculture and food security (Hasanuzzaman et al. 2012a). Different nutrients can help to trigger different kinds of defence-related reactions which can help the plant to establish stress tolerance. Thus, application of nutrients could be an option to coup different type of abiotic stresses such as salinity, drought, high temperature, chilling, toxic metals, flooding, high-light intensity, UV-B radiation and O_3 (Hasanuzzaman et al. 2018a). Some nutrients generate different kinds of free radicals which can possess antioxidant properties and also produce a signal to activate reactive oxygen species (ROS) scavenging mechanisms to control abiotic stresses in plants (Hasanuzzaman et al. 2018a).

Study of elements or nutrients necessary/essential for the plant growth and metabolism is called plant nutrition. In the absence of these essential nutrients, plants would not be able to complete their normal life cycle, and these nutrients could be the part of some essential plant constituents or metabolites. This could be in accordance with the Liebig's law of the minimum or Liebig's law or the law of the minimum which states that growth is controlled by the limiting factor. Similarly, a mineral element is considered essential if it fulfils the criteria of essentiality given by Arnon (*Criteria of essentiality of inorganic micronutrients for plants*). There are 17 elements (carbon (C), hydrogen (H), oxygen (O), nitrogen (N), phosphorus (P), potassium (K), calcium (Ca), magnesium (Mg), sulphur (S), iron (Fe), manganese (Mn), zinc (Zn), copper (Cu), boron (B), molybdenum (Mo), nickel (Ni) and chlorine (Cl)) known as mineral nutrients essential for the completion of a productive life cycle in plants. Plants take C and S from the air as CO_2 and SO_2, while oxygen and hydrogen are provided as water. Other nutrients are taken from the soil. Nutrients are further classified into macro- and micronutrients based upon their demands by the plants. Nutrients required higher than 1–150 g kg^{-1} (>1000 mg kg^{-1} dry weight) of plant dry matter are called macronutrients, and these include N, P, K, Ca, Mg and S. However, nutrients which are required at the concentration of 0.1–100 mg kg^{-1} (<100 mg kg^{-1} dry weight) of plant dry matter are called as micronutrients. These include Fe, Zn, Mn, Cu, B, Mo and Cl. Elements like Al, Si, Co, Na and Se are not essential according to the criteria but widely taken up by the plants to perform different metabolic functions. Thus, elements which help to enhance biomass and yield but not required for survival are called as beneficial or functional elements. Mineral nutrient availability to the plants helps to maintain different metabolic processes, cell osmotic relations, turgor-related process, energy transfer reactions and enzyme-catalysed reactions (Hasanuzzaman et al. 2018a). Similarly, judicious use of these nutrients helps to establish good source-sink relationship which leads to the higher crop yield. In this chapter, we reviewed about mineral nutrient availability, uptake mechanisms and roles in plant growth particularly under various types of abiotic stresses.

2 Nitrogen (N)

Nitrogen (N) belongs to group V of the periodic table existing in the form of diatomic inert gas (N_2). It constitutes 78% of the atmosphere and mainly occurs in all organisms as proteins (amino acids), nucleic acids (DNA and RNA) and energy carrier (ATP). The nitrogen cycle describes the dynamics of N from air to biosphere, organic compounds and back to the atmosphere. N is one of the major nutrients and acts as a catalyst to support biochemical reactions in the plant. It is the main constituent of biomolecule chlorophyll (Fig. 3.1) (Croft and Chen 2018). In the presence of light, water and CO_2, chlorophyll molecules generate energy carriers called ATP which is required for different metabolic reactions in the plant body. N is also present in plant tissue in the range of 1–6%. Plants cannot use atmospheric N directly; it needs to be converted into a different form before it can be taken by pants from its rhizosphere. N availability to plant is possible by industrial, atmospheric and biological N fixation. The biggest and expensive source of N availability is industrial through the formation of urea $(NH_2)_2CO$. Urea is an uncharged molecule is decomposed into ammonium (NH_4^+) by enzyme urease within a week or two depending upon soil temperature. However, if it is applied at the soil surface, it volatilized readily. The fate of N fertilizer has been elaborated in Fig. 3.2. The hydrolysis and volatilization of urea have been described by Cameron et al. (2013). The equations are:

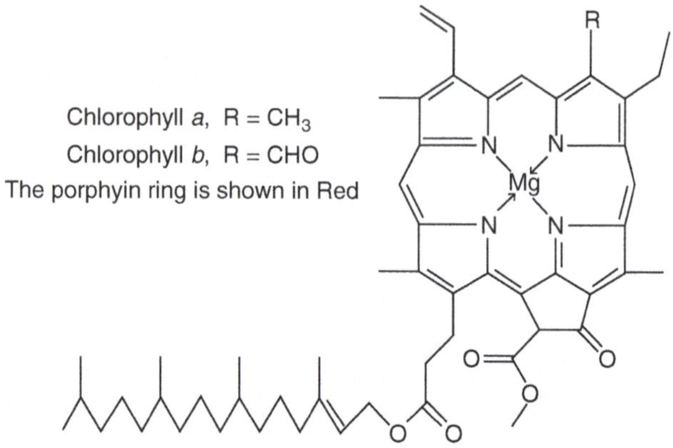

Fig. 3.1 Structure of chlorophyll showing importance of N (Modified from: Croft and Chen 2018)

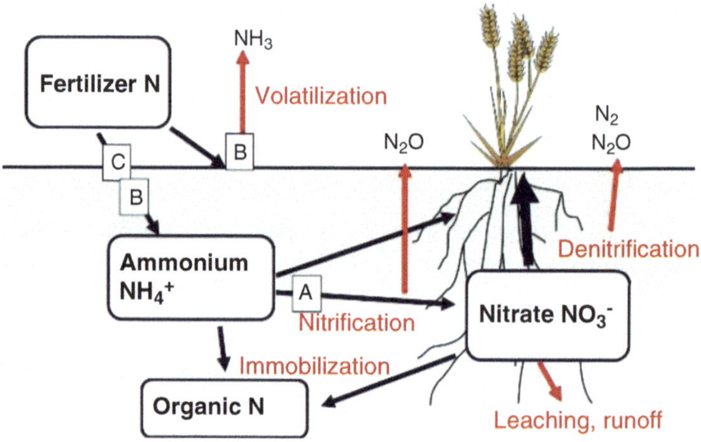

Fig. 3.2 Fate of N fertilizer

$$(NH_2)_2CO + 2H_2O \xrightarrow[\text{urease}]{\text{soil}} (NH_4)_2CO_3$$

$$(NH_4)_2CO_3 + 2H^+ \rightarrow 2NH_4^+ + CO_2\uparrow + H_2O$$

$$NH_4^+ + OH^- \leftrightarrow NH_3\uparrow + H_2O$$

Sommer et al. (2004) reported that about 50% of NH_3 is mainly volatilized from agriculture and its biggest source is urea fertilizer depending upon soil types and climatic conditions. Plants mainly uptake N as NH_4^+ and NO_3^- ions, but their availability depends upon soil conditions and root systems. Generally, NO_3^- uptake occurs at the root level through two transport systems which occur in coordination with each other from the soil solution to whole plant body (Daniel-Vedele et al. 1998; Tsay et al. 2007). Availability of N is linked with precipitation, temperature, wind, soil type and pH. According to Maathuis (2009), plants adapted to reducing soils and low pH take N as NH_4^+, while plants under higher pH and aerobic conditions prefer NO_3^-. N is taken up by three ways, i.e. mass flow, diffusion and root interception (Table 3.1). Mass flow is governed by potential gradient which regulates water movement into soil plant atmosphere. Quantity of ion transported through mass flow is determined by transpiration rate and soil solution concentration. Thus, mass flow efficiency is related to nutrient supply capacity of soil and plant demand which is further characterized by plant species and root activity. N is mainly taken up by mass flow. Mathematically, mass flow (MF; kg m^{-2} s^{-1})) could be represented as:

$$MF = WC$$

where $\underline{W}$ is water flow ($m^3 m^{-2}$ s^{-1}) and C is the average solute concentration (kg m^{-3}).

2.1 Mechanism of the Nitrogen Cycle

Plants make use of nitrogen present in the atmosphere through nitrogen fixation. This phenomenon is complex which is accomplished by multiple soil residing or symbiotic bacteria termed as nitrogen fixers. There are some bacteria which fix nitrogen having a symbiotic relationship with plants especially legume family. N metabolism can be divided into two phases: (i) aerobic metabolism which includes nitrate assimilation and nitrifications and (ii) anaerobic metabolism which consists

Table 3.1 Transport process in different nutrients

Nutrient	Mass flow	Diffusion	Root interception
Nitrogen (N)	✓		
Phosphorus (P)		✓	
Potassium (K)	✓	✓	
Calcium (Ca)	✓		✓
Magnesium (Mg)	✓		✓
Sulphur (S)	✓	✓	
Boron (B)	✓		
Copper (Cu)	✓		
Iron (Fe)	✓	✓	✓
Manganese (Mn)	✓		✓
Zinc (Zn)	✓	✓	✓
Molybdenum (Mo)	✓		

of nitrate respiration (dissimilation), nitrification and nitrogen fixation. Generally, N cycle can be divided into three distinct phases:

$$N_2 - \text{fixation} \rightarrow \text{nitrification} \rightarrow \text{denitrification}$$

2.1.1 N_2 Fixation

This process is somewhat complex contrary to as it seems apparently simple. The underlying mechanisms which would fix nitrogen into ammonium are having multiple forms. The first one is that N is directly converted into ammonium (NH_4^+). NH_4^+ is produced by the nitrogen fixers which would encode for the nitrogenase enzyme. This ammonium is either taken up into plant biomass or further respired by aerobic or anaerobic bacteria into other suitable nitrogen stable or unstable compounds.

$$N_2 \rightarrow NH_4^+ \rightarrow \text{Organic N}\left(\text{nitrogenase enzyme}\right)$$

$$N_2 \rightarrow NH_4^+\left(\text{N fixers}\right)$$

The second mechanism of ammonium formation is the reduction of NO_2^- into NH_4^+.

$$NO \rightarrow NH_4^+\left(\text{bacteria and fungi}\right)$$

2.1.2 Nitrification

Nitrification is the transformation of ammonium nitrogen to nitrites (NO_2^-) and nitrates (NO_3^-) through oxidation. Chemoautotrophic organisms play a big role in metabolism. *Nitrosomonas* and *Nitrobacter* are main sets of nitrifying bacteria which are involved in the removal of hydrogen ions (oxidation) from ammonium to form nitrite to obtain energy (E), and *Nitrobacter* oxidizes nitrite to nitrate for the same purpose. By three different mechanisms, this process is accomplished. Overall reaction is

$$\text{Organic N} \rightarrow NH_4^+ \rightarrow NO_2^- \rightarrow NO_3^-$$

Firstly, ammonium is converted into nitrite the unstable intermediate form.

$$NH_4^+ \rightarrow NO_2^-$$

The second mechanism is the transformation of this unstable transient form into nitrate the stable one.

$$NO_3^- \rightarrow NO_3^-$$

Thirdly the direct mechanism is involved in the all the way transformation of ammonium into nitrate excluding the nitrite conversion step.

$$NH_4^- \rightarrow NO_2^-$$

2.1.3 Denitrification

Returning back the nitrogen to the atmosphere in free form is accomplished by the phenomenon denitrification. The multiple forms of nitrogen in the cycle are NO_3^-, NO_2^-, NO, N_2O and NH_4^+. Microbes in absence of oxygen would respire the bound forms of nitrogen by the action of different denitrifying enzymes ultimately giving back the nitrogen to the atmosphere in N_2 form. Thus, a nitrogen cycle keeps on continuing the conversion of atmospheric nitrogen and returning it to air in a précised manner including the microbes as the noble responsible. Nitrate reduction (denitrification) and nitrification are the redox reactions that synthesize energy-releasing compounds. However, nitrogen reduction is a highly complicated process when compared with nitrification. The processes responsible for the nitrogen reduction are classified as assimilatory and de-assimilatory reduction processes.

$$NO_2^- \rightarrow NO \rightarrow N_2O \rightarrow N_2$$

2.1.4 Mineralization

Mineralization includes conversion of organic nitrogen from plant and animal residues to inorganic compounds such as ammonia (NH_3) by soil microorganisms also known as ammonification. The addition of proteins such as dried blood to the soil causes liberation of ammonium and the rest of the N retained in tissues of microbes. Similarly, microbial biomass increases with the inclusion of carbohydrates such as proteins mixed with cellulose. The addition of ammonium in the soil is affected by nitrification, uptake of ammonium, loss of nitrate by leaching, the rate of mineralization of organic nitrogen in the soil, volatilization of ammonia and uptake of ammonium by microbes for growth. However, the low concentration of ammonium does not show low mineralization but may indicate the high rate of volatilization and nitrification.

2.2 Fate of N in Soil

Soil organic nitrogen is transformed into ammonium NH_4^+ and nitrate NO_3^- both taken up by plants readily and an intermediate form nitrite (NO_2^-). Nitrite a negatively charged ion is very mobile and moves readily with percolating water. This is known as leaching of N in the soil. Ammonium is fixed in the soil and is released back to the mineral pool by soil weathering. It can sometimes volatilize into the atmosphere in the form of NH_3 gas. Thus, the nitrogen in organic or inorganic form applied to plants is converted to most suitable forms for uptake which are further leached down in water or eroded in soil by water erosion or upon weather conditions may runoff as well depending upon the soil type. Some N is lost to the environment through the gaseous form as well (Fig. 3.2).

2.3 Abiotic Stress and Mode of Action

Abiotic stress such as drought, heat and cold reduces N uptake efficiency and hinders the uptake and assimilation pathway. Less moisture in the soil will result in a lesser portion of N available or absorbed by the plant. This will ultimately reduce plant growth and development. But the long duration of stress will leave the lesser impact of N deficiency rather that stress will have a direct impact. Drought is one of the most devastating abiotic stresses which resulted in 13–94% decrease in yield (Farooq et al. 2009). An array of physiological and biochemical adaptations is shown by plants to have drought resistance which has three strategies, i.e. drought escape, avoidance and tolerance. N supply can regulate drought tolerance in plants (Ding et al. 2018).

2.4 Assimilatory Nitrate Reduction

Nitrate (NO_3^-) is used as a nitrogen source for cell growth in assimilatory nitrate reduction (ANR). By the action of the reductase enzyme, the nitrate is reduced to nitrite (NO_2^-), and the enzyme nitrite reductase further reduces it to ammonium (NH_4^+). This final product is then incorporated into the carbon skeletons either by glutamine synthetase or by glutamine synthase pathway.

2.5 Dissimilatory Nitrate Reduction

Dissimilatory NO_3^- reduction to NH_4^+ contributes to N conservation in the soil-plant system. It is also known as respiratory nitrate reduction. In this process, the oxygen molecule in the nitrate acts as a hydrogen acceptor in exothermic reactions.

The steps involved in nitrate reduction under this process are potentially the same as to those when molecular oxygen is used instead of nitrate (Lu et al. 2013).

3 Potassium (K)

K is regarded as a primary essential macronutrient, which is required in large amount for different functions and regulation of plant growth and development. Being major nutrient, it comes after nitrogen in terms of percentage requirement for crops. The description of potassium cycle is presented in Fig. 3.3. K can reach 10% of the plant dry weight. The concentration of K^+ ion in the cytoplasm is 100 mM (from 40 to 200 mM) and remains constant (Wang and Wu 2010).

Fig. 3.3 Potassium cycle

3.1 Sources and Availability of K in the Soil Environment

Potassium attains a position of seventh most abundant element in the earth's crust. In soil, it varies from 0.5 to 2.5% of the soil mass and exists as mineral form. Main sources of K in soil include feldspars (orthoclase) and micas which release K on weathering. However, if weathering has been prolonged, it results in the leaching of K (Fig. 3.3). There are three main available forms of potassium in soil, i.e. exchangeable K (1%) adsorbed or released from clay, non-exchangeable K (98.99%) and dissolved or solution form (K^+) (0.01%), and it is the only useable form for uptake by plants (Hillel 2008).

3.2 Sources and Availability of K in Plant

Potassium ion (K^+) is one of the most abundant cations present in plant tissues which contain 1–10% of dry matter (Epstein and Bloom 2005). It performs various physiological functions in plant, i.e. maintenance of turgor pressure, the release of an enzyme, the formation of proteins, helping in photosynthesis and balancing anion to maintain the pH of plant cells. It also plays a crucial role in nutrient transport.

3.3 Uptake of Potassium (K) by Plants

The mechanism of potassium uptake by plant depends on its availability and exchange capacity of cations. In higher plants, a large number of genes encode for K^+ transport. Plants uptake potassium in the form of K^+ ions by cation exchange capacity (CEC) (Fig. 3.4). Usually plant vacuole uptakes large amount of Na^+ ions which are later exchanged by K+ ions through CEC through concentration gradient (Wang and Wu 2010).

3.4 Biotic/Abiotic Stress and Role of K

Major stress factors affecting agricultural production are biotic and abiotic, both causing significant reduction in the quantity and quality of crop. It has been reported that biotic stress resulted in 28.2%, 37.4%, 31.2%, 40.3%, 26.3 and 28.8% losses in wheat, rice, maize, potatoes, soybeans and cotton, respectively (Oerke 2006; Ahmad and Hasanuzzaman 2020). Abiotic stress resulted in 65.8%, 82.1%, 69.3% and 54.1% for maize, wheat, soybeans and potatoes, respectively (Bray et al. 2000). Potassium can perform a significant role to minimize these losses (Fig. 3.5). K plays

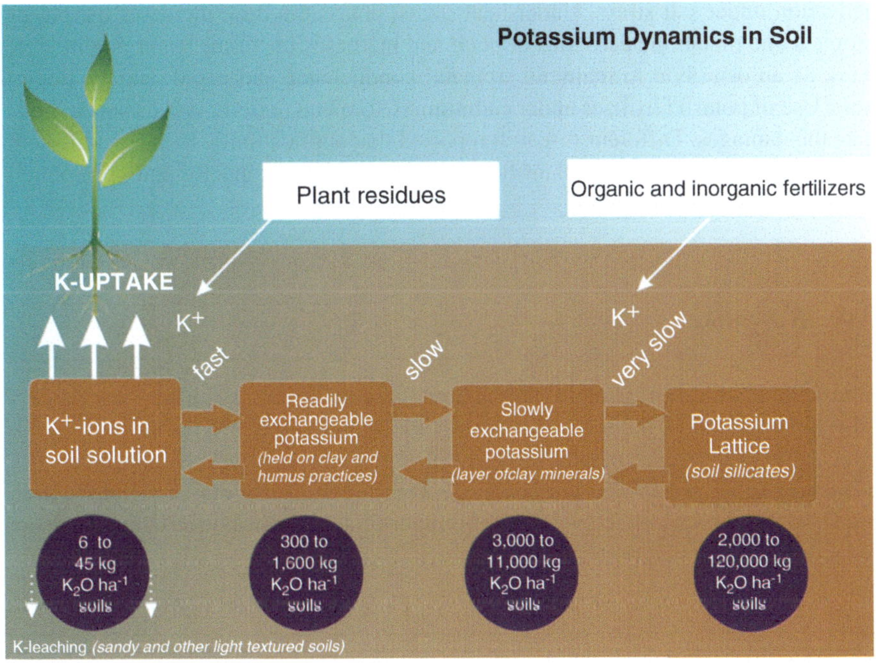

Fig. 3.4 Potassium dynamics in soil

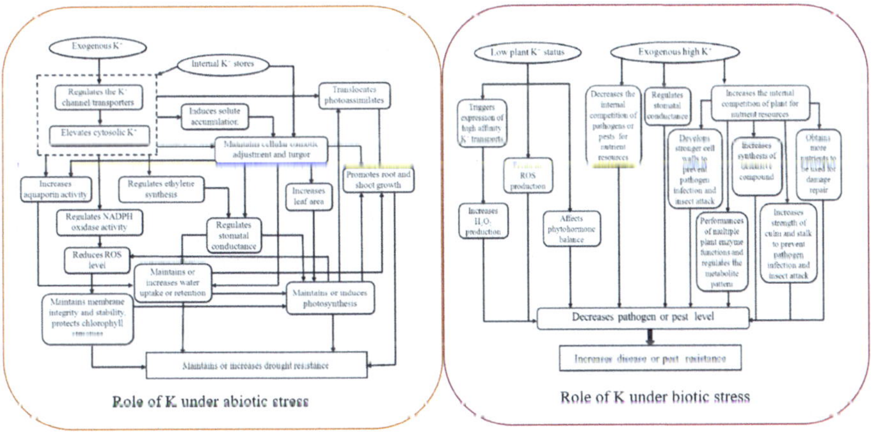

Fig. 3.5 Role of K under abiotic and biotic stress (Modified from: Wang et al. 2013)

a crucial role in maintaining ion homeostasis, enzyme activity and cellular integrity. Application of K can increase the plant dry matter and improve the root surface under drought condition. K input can increase the osmotic adjustment, retain turgidity, regulate membrane potential and help in protein synthesis and enzyme activation under salt stress. Under heat stress, potassium dose increases the physiological and metabolic activities of plant and increases the tissue water potential. K work as an osmolyte to maintain stomatal conductance and avoid damage due to heat. Use of potash fertilizer under cadmium (Cd) metal toxicity proves best to mitigate the damages. Different research reported that application K helps to cope with waterlogging by increasing plant height, photosynthetic capacity and chlorophyll contents (Ashraf et al. 2011).

3.5 K Uptake

Potassium is an essential plant macronutrient for all living organisms. It is usually taken up by plants in the form of K^+ ions. One main feature of K^+ is the high rate at which it is taken up by plants. Mobility of K ions throughout the entire plant is a consequence of membrane permeability. The direction of potassium ions within plants is often transported towards the younger tissue, redistribution occurring from older to younger plant parts.

4 Phosphorus (P)

P is the second most important plant nutrient that has significant role in plant growth development and yield. It is one of the main constituents of plant cell and its deficiency results in declined agricultural production. P uptake by plant is poor because of its low mobility in soil. P availability in fertile soil is even less because of its high absorption rate in the rhizosphere compared to replenishment in soil solution (Hernández et al. 2007). The detail of P cycle is presented in Fig. 3.6.

4.1 Phosphorus in Atmosphere

The movement of P is not being affected by the atmosphere as it is mostly available in solid form at room temperature and pressures. Only phosphine (PH_3) exists in gaseous form in local conditions.

4.2 Phosphorus in Soil

Phosphorus occurs in abundance in the form of the orthophosphate ion $(PO4)^{3-}$ in the soil. In earth's crust, the majority of P is found in rocks and minerals. P is released through the weathering of rocks and minerals in a soluble form which is available to plants, and it is transformed into organic compounds. A small part of P

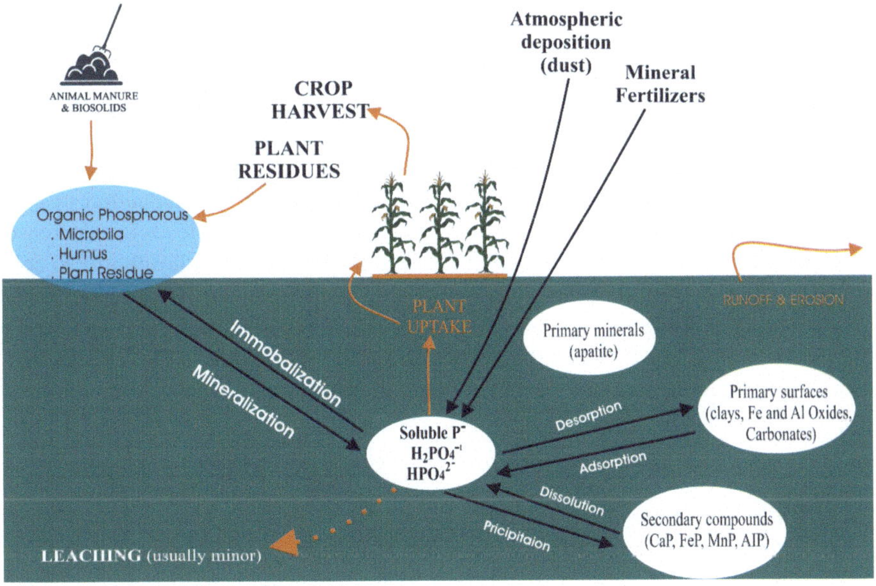

Fig. 3.6 Phosphorus cycle

may be deposited back to oceans through runoff. P in the soil becomes deficient with time (thousands of years) which leads to ecosystem retrogression (Peltzer et al. 2010).

4.3 *Phosphorous Availability*

P availability in soil solution is limited as it is very stable and insoluble in all its natural form. Plant uptakes phosphorous in the form of $H_2PO_4^{-2}$ and HPO_4^{-2}. Quantity uptaken by the plant is truly pH-dependent. The higher the pH, more HPO_4^{-2} will be uptaken by the plant. Phosphorous in the soil is mobile to some extent. Plants uptake P against concentration gradient because it is present in fewer amounts in soil solution. More energy is consumed so low temperature and excess of water inhibit root activity which in return inhibit phosphorous uptake. However, plants may take up some amount of organic phosphorous.

4.4 *Mode of Action of P Under Abiotic Stress*

The application of phosphatic fertilization increases the stress-tolerating ability of plants and eventually results in adjustments of physiological, morphological and biochemical processes that increase plant growth. Extension of root system is observed when phosphorus is applied under drought stress conditions. An increase in the yield of chickpea was observed when phosphorus was applied under salt stress conditions leading to improvement in the soil quality for better symbiosis efficiency. Saline soils (chloride) can be reclaimed through P fertilizer application, and improvement in crop yield can be achieved. Increase in temperature resulted in the increase of available P. Generally, at low temperature, the rate of phosphorus application is increased to ensure sufficient P uptake. Microbial reduction of P is also enhanced with an increase in temperature. An increase in pH leads to a decrease in the availability of phosphorus. High pH results in binding of phosphorus to calcium leaving less free phosphate ions available in solution. pH also affects the forms in which phosphorus exists in soil solution. In maize roots, phosphorus uptake was increased by a factor of 3 when the pH was lowered from 8.0 to 4.0. No change was observed in the concentration of HPO_4^{2-} (van de Wiel et al. 2016).

5 Sulphur (S)

S, an important nonmetallic element for a living organism, is helpful for different metabolic reactions. It can be called 'the fourth major nutrient'. S is required by plants same as P. S concentration in plant tissue is 0.12–0.35%. S is mainly taken by plants as $SO4^{-2}$. Different types of transporter help in the uptake of S. The sulphur cycle has been described in Fig. 3.7.

5.1 Sources of Sulphur in Soil

Organic matter (OM) is the main source of S. OM is mineralized to form SO_4^{-2} ion which is then used by roots. Availability of S from OM is a time-consuming process and it takes years. S is also present in different minerals which on weathering releases S.

5.2 Sources of Sulphur in Air

S in the air is the main cause of acid rains. Industrial combustion and automobile industry resulted in the formation of SO_2 gas which on further oxidation forms SO_3. Therefore, it is essential to use environmental protection measures to prevent its release to air as oxides of S (SOX). Different chemicals such as $Al_2(SO_4)_3$, $(NH_4)_2SO_4$, $NH_4S_2O_3$, $CaSO_4$, $MgSO_4 \cdot 7H_2O$, $FeSO_4$, K_2SO_4, $KMgSO_4$, S and H_2SO_4 could be used as a good source of S.

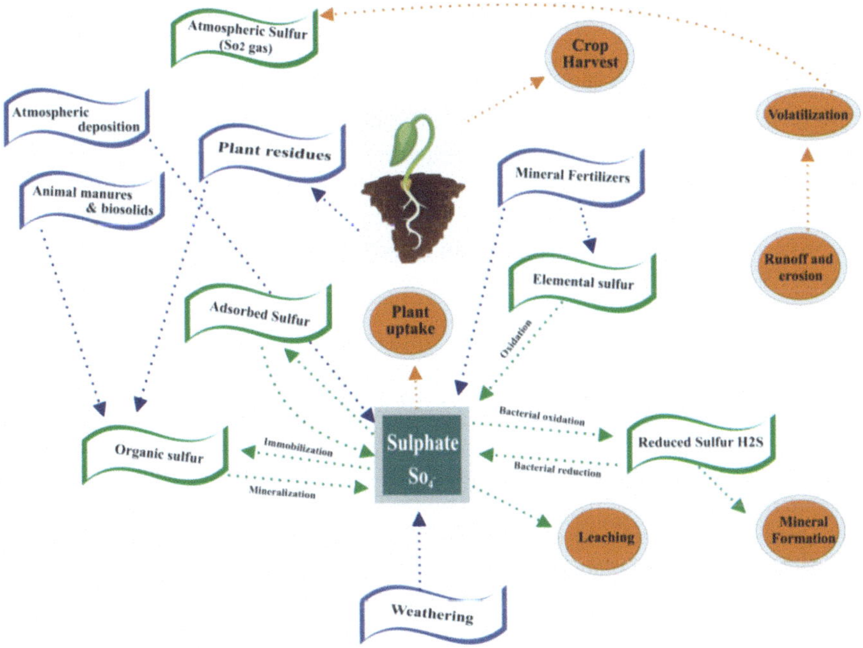

Fig. 3.7 Sulphur cycle

5.3 S Assimilation

S in the form of SO_4^{-2} combines with ATP to adenosine 5′ phosphosulphate which then changes to S^{-2} after series of reactions. S^{-2} is then readily incorporated into amino acids (cystine) (Daur et al. 2019).

6 Calcium (Ca)

Calcium is regarded as the secondary micronutrient for plants. It is essential for many plant functions, e.g. cell division and elongation, enzyme activity, proper development of cell wall, starch metabolism and nitrate uptake and metabolism. Ca acts as a messenger when plants are under stress (Rahman et al. 2016).

6.1 Nutrient Form and Presence in Soil

The third most important available nutrient in the soil and its concentration is about 3.6% in the earth's crust. The average concentration in soil is about 1.37%. It is present in the form of common minerals and rocks like feldspar, apatite, dolomite, limestone and gypsum. On weathering of these minerals, Ca comes into the soil. Calcium is a constituent of many other compounds like calcium sulphate, calcium gluconate, calcium chloride and calcium carbonate. It is available to plants in the form of Ca^{2+} (Ray and George 2010).

6.2 Nutrient Form and Presence in Plants

Ca^{2+} ions are present in leaves' veins and the thylakoid part of plant leaves. It is taken up by plants as Ca^{2+} ions through roots. The deficiency of calcium affects the photosynthetic rate and causes chlorosis of leaves. Calcium sensor transfers the biological responses of plants.

6.3 Mode of Action Under Abiotic Stress

Plants face biotic and abiotic stresses on a daily basis. To cope with these stresses, plants should have to sense the environmental changes to develop the specific response mechanism. Calcium plays an important role (as a secondary messenger) in signalling the environmental changes. Under stress, the concentration of cytosolic free calcium in plants increases which acts as a signalling process in plants. Calcium-dependent protein kinases (CDPKs) which are calcium-dependent are considered as a major primary sensor in plants (Schulz et al. 2013).

6.4 Availability to Plants

Plant roots absorbed the calcium, or calcium is available to plants in the form of Ca^{2+}. Roots absorb it from the soil solution, and then it is transported to the aerial parts. Root hairs play an important role in the absorption of the calcium due to large surface area. Calcium is a major cation, so its absorption also depends on the CEC (cation exchange capacity) of the roots. Between root CEC and Ca levels in shoot, a positive relationship has been reported (Ray and George 2010). Calcium cycle has been further elaborated in Fig. 3.8.

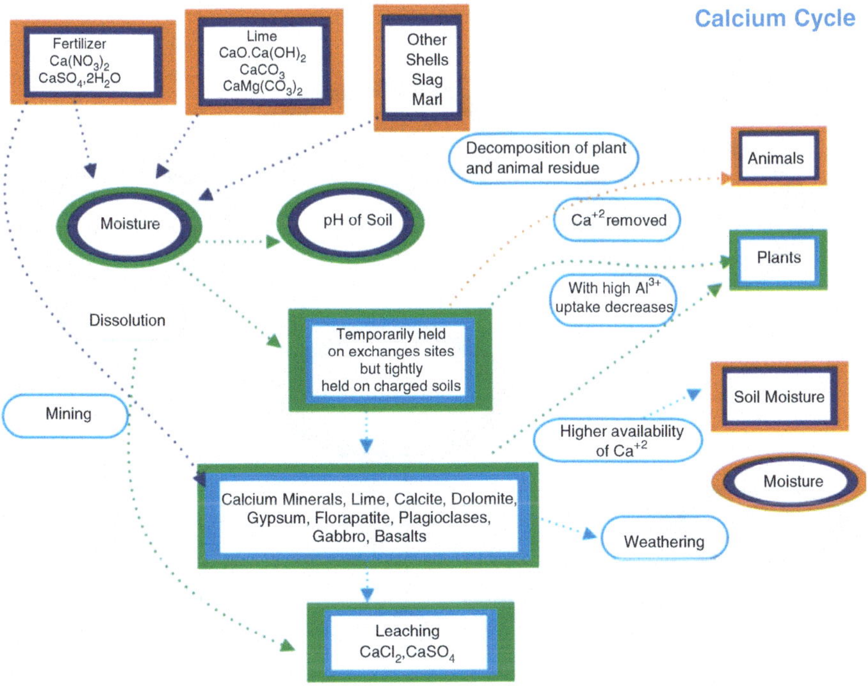

Fig. 3.8 Calcium cycle

7 Magnesium (Mg)

Magnesium is an essential secondary nutrient element for plants. It plays an important role in the growth and development of all plants. Its main role in plant is in the photosynthesis process. Magnesium is the building block of chlorophyll, and it imparts the leaf as green colour. Magnesium deficiency can cause chlorosis (yellowing of leaves), and prevailing condition can cause necrosis. The details of Mg are elaborated in Fig. 3.9.

7.1 Magnesium Uptake by Plants

In the soil solution, magnesium is present in ionic form when dissolved and is taken up by the plants as Mg^{2+}. The uptake of magnesium by plants is possible by two main processes: passive uptake, driven by transpiration stream, and diffusion wherein magnesium ions move from high gradient level to lower gradient level. Magnesium uptake by the plant depends upon the Mg^{2+} and the strength of soil to replenish the soil with magnesium.

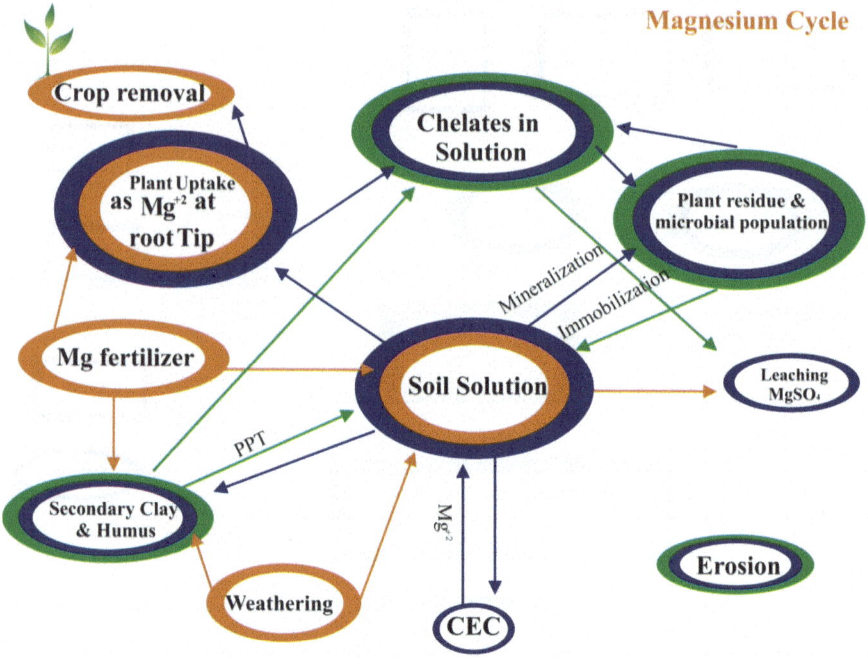

Fig. 3.9 Magnesium cycle

7.2 *Sources of Magnesium*

Magnesium is widely available on the earth's crust. It is found in a wide range of minerals. These minerals break or weather and produce magnesium that becomes available for plants to uptake. Dolomitic limestone is the rich source of magnesium, but it is very expensive to be applied to overcome magnesium deficiency. Dolomite limestone contains 8–10% Mg content. Broadcasting or incorporating this magnesium source into the soil is a very effective way to fulfil the magnesium deficiency inexpensively.

7.3 *Mechanism of Magnesium Uptake*

Magnesium is very mobile in soil and reaches to the plant root quickly by mass flow or by diffusion method. Plant absorbs magnesium in the form of magnesium ions (Mg^{2+}). The concentration of other nutrients such as calcium and potassium cations also influences the uptake of magnesium by plants. It is essential that during fertilization appropriate ratios of calcium/magnesium, magnesium/potassium and calcium+magnesium/potassium should be applied. This ratio and balanced concentration lead to high yield and excellent quality of the food crop. Magnesium is also very mobile within plants, and its deficiency symptoms appear on older leaves and then on younger leaves.

8 Boron (B)

Boron an essential micronutrient is available in the earth's crust in low concentration (Roychowdhury et al. 2019). B is mostly available as water-soluble borate minerals which were further used to produce borax ($Na_2B_4O_7$ $10H_2O$) and kernite [$Na_2B_4O_6(OH)_2.3H_2O$]. Boric acid (H_3BO_3) and sodium tetraborate ($Na_2B_4O_7.xH_2O$) are other main products of B. The largest deposit of B minerals is found in Turkey. B's main functions include strengthening cell wall, cell division, hormone production and fruit and seed development. Balanced nutrition of B is essential as its functions interrelate with N, P, K and Ca. Therefore, its deficiencies result in anatomical, physiological and biochemical changes in the plant. Some examples include a hollow heart in peanuts and rosetting (stunting) in legumes such as alfalfa and clovers. However, a higher concentration of B is toxic to plants. Healthy plants' growth and development are dependent upon B nutrition. It has been used in small quantities. The biggest source of B is boric acid mainly used as liquid fertilizer. B requirement is higher during reproductive growth as it helps in pollination and development of seed. B availability in the soil is not enough to meet the demands of plants. It is available as organic and inorganic (calcium, magnesium and sodium borates) form. Microorganism helps to convert inorganic form to organic form, but on the death of microbes, it oxidizes to inorganic boron again. Availability of B in the soil is dependent upon pH, leaching, organic matter, water, Ca in soil and N stress. B has the highest vertical movement in the soil profile due to variability in pH (Oliveira Neto et al. 2009; Bell and Rerkasem 2012).

8.1 B Uptake by Plants

Boric acid or borate ($H_2BO_3^-$) is the main form of B which plant takes from the soil. B is the only element taken up by the plant as an uncharged molecule. B uptake is a passive process and not facilitated by transport proteins. However, this concept was modified by Miwa and Fujiwara (2010). They concluded that suberized endodermal cells are unable to take solutes directly from root apoplast but NIP5;1 increases the permeability of boric acid to the cell membrane and influx of B into root accelerated. BOR1 exports B from out of cell to xylem against the concentration gradients. Therefore, BOR1 and NIP5;1 act in coordination with each other to supply B under limited conditions, while under higher levels of B, downregulation of NIP5;1 and BOR1 occurs (Fig. 3.10).

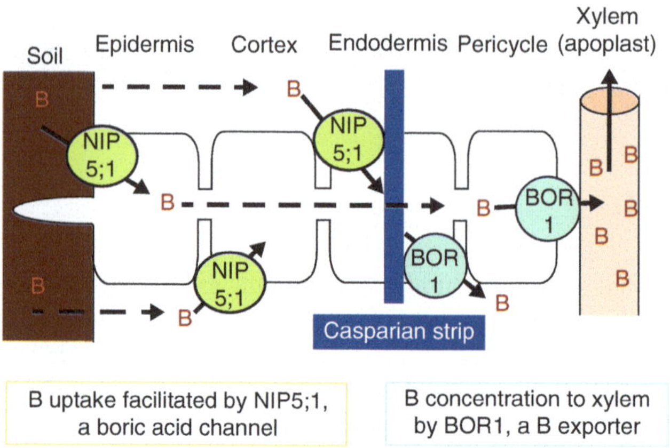

Fig. 3.10 Boron transport model (Modified from: Miwa and Fujiwara 2010)

8.2 *Mode of Action Under Abiotic Stress*

B helps to build resistance in plants against stress by maintaining integrity of cell membrane and cell wall. Deficiency of B resulted in poor vegetative and reproductive growth. B is also helpful for N fixation. B availability and salinity stress have a strong connection, and it has been observed that under saline soils, B uptake was less. Similarly, interactive effect of B and salinity is negative on photosynthesis (Alpaslan and Gunes 2001; Mohamed et al. 2015; Sairam et al. 2005). Boron dynamics have been described in Fig. 3.11.

9 Chlorine

Chlorine member of group VIIA halogens (salt former) is mainly present in diatomic gaseous form (Cl_2). It forms monovalent anion called chloride (Cl^-) which is a beneficial nutrient for plants (Raven 2016). Cl^- as a beneficial nutrient was proved by Franco-Navarro et al. (2015) who suggested 1–5 mol m^{-3} of Cl^- ions to *Nicotiana tabacum* to fulfil its micronutrient demands. In atmosphere, it is present as chlorine gas which is added to the soil through precipitation, fertilizer inputs or via chlorinated water application to the soil bed. Chlorine in the form of chloride ion is highly soluble in soil water and found predominantly in dissolved form. Its mobility in soil solution is high. In areas with greater rainfall and well-drained soils, it may be leached through soil profile, making it less available for the plant uptake. In heavy clays, where it is repelled by other anions (nitrate, sulphate, boron and molybdenum), it is barely adsorbed to soil particles and thus more dissolved in water (Fig. 3.12). All those anions interact with each other antagonistically, and excess of one can decrease another (Heinen et al. 2017).

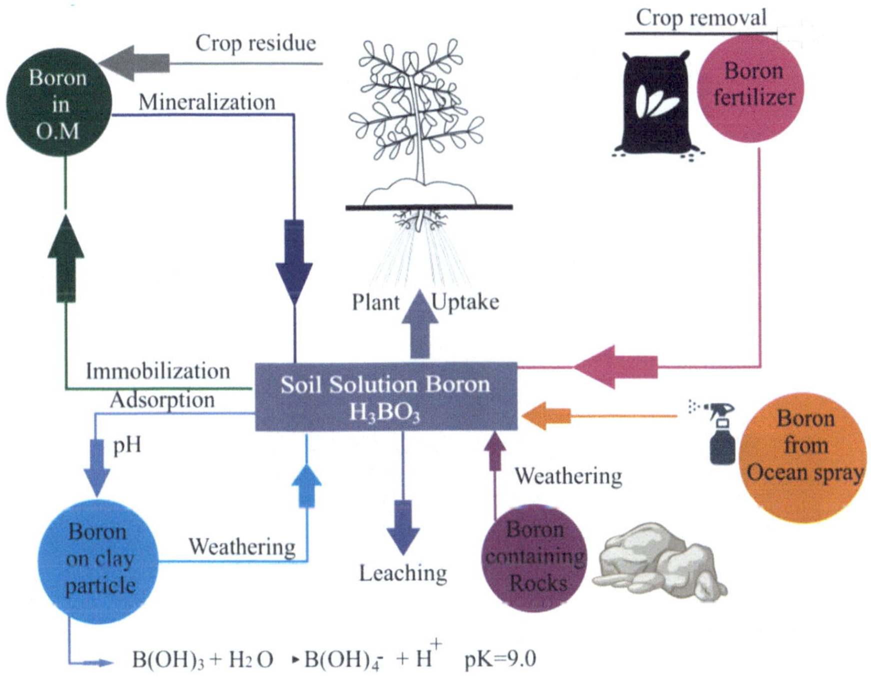

Fig. 3.11 Boron cycle

Soil water movement towards the plant roots took chloride anion closer to the root surface to be taken up by the plant. From root surface till stele in vascular bundle, it is transported via symplastic mode of translocation. The goal of chloride ion is to reach to the xylem of shoot, which is facilitated by H^+ co-transporters proteins present in stele (Wegner 2013). Further distribution takes place through shoot phloem, depositing it in the cytosol as a storage nutrient in vacuoles or incorporation in the oxygen-evolving complex of the photosystem II. The requirement for chlorine varies from species to species, genotypes and its amount in the environment (Raven 2016).

Chlorine plays binary roles. It can save plants from stress as well as can induce stress when taken up by plants in amounts higher than required. As a micronutrient, it has a role in osmosis and maintenance of turgor in plant tissues; it can be deposited in vascular structures, in guard cells or in growing tissues, in few cases and up to 150 mM in the vacuole. Here, Cl^- serves as an activator of osmosis that drives the water movement, resulting in increased turgor which facilitates compound movement, effects source-sink division and initiates turgor-based movements. It performs a similar function as malate during stomatal closure and adjustment of opening in guard cells. It maintains a counterbalance in relation to potassium ions. It, along with potassium and malate, reduces water potential of guard cells and causes the water to move in and make the cells turgid. That function is controlled by malate

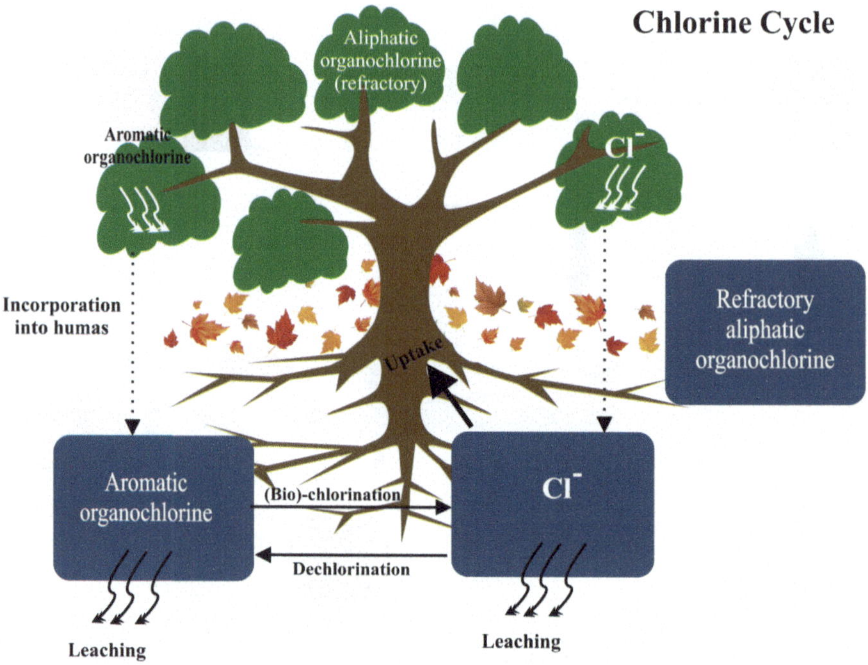

Fig. 3.12 Chlorine cycle

mainly. But when there is malate, it appears that chloride ions are also there. Chloride can help to improve water and nitrogen use efficiency as reported by Raven (2016). Chlorine is thought to be involved in building a positive membrane potential by activating H^+ ATPase, which favours the intake of anions into vacuoles and serves as osmosis activators and then causes inward movement of water to create turgidity. By this method, Cl^- affects the expanding growth of roots and shoots (Franco-Navarro et al. 2015).

Chlorine is said to have a structural role in photosynthetic activity of plants. It reduces disintegration of peptides of photosystem and maintains them by oxidation of water, which tends to stimulate photosynthesis by tenfold (Hasegawa et al. 2002). Chloride toxicity/salinity induces stress in the plants in terms of their disturbed physiological functions and disruptions of many biosynthetic processes, mainly inhibition of photosynthetic capacity, stunted root and shoot growth, affected tuber development in tuber crops, necrotic lesions on leaves and margins and anion-anion antagonistic effect. Increase in the concentrations of chloride in surrounding of plants, an antagonism between nitrate and Cl^-, is triggered, which inhibits nitrate transporters and catalyses root xylem which has larger amounts of nitrates. Chloride toxicity increases the nitrate concentration in roots but is decreased in shoots, reducing growth and yield. High concentrations of chloride disrupt the optimum pH for enzyme functioning resulting in inhibition of protein synthesis by catalysis of ribosomal enzymes (Abbaspour et al. 2013; Mohammadkhani and Abbaspour 2018). Chlorine stress affects postharvest quality by elevating productions of ethylene which induces catalysis of calyx and development of necrotic lesions (Besada et al. 2016).

10 Cobalt (Co)

Cobalt is a naturally occurring micronutrient having an atomic number of 27, molecular mass of 58.93, melting point of 1493 °C and density of 8.9 g/cm^3. Co is similar in properties to Ni + 2 as it enters the membrane through the same carriers. Co concentration in plants ranges from 0.1 to 10 mg kg^{-1} with some exceptions. Co helps to increase growth in legumes, retardation of leaf senescence and seed drought resistance (Kaur et al. 2016). It is the main constituent of vitamin B12. Co plays major role in N fixation as it is part of cobalamin which is required for activities of nitrogen-fixing enzymes of cyanobacteria and rhizobia bacteria. Gad (2006) reported the importance of Co on pea crop and concluded that the dose of 8 mg resulted in increased growth, plant nutrients, nodule size and number and seed pod yield and quality. Collegial effect of 10 mg Co was observed on sweet potato root growth and yield quality (starch, sugars, L-ascorbic acid and contents of N^+, P^{+3}, K^+, Mn^{2+}, $Zn^{2+}$and Cu^{2+}) compared to the control and other supplemented concentrations (Gad and Kandil 2008). Increased nitrogenase activity, growth, yield and quality were observed due to a supplemental dose of 8 mg Co to peanut (Gad 2012). Priming results with combination of ascorbic acid (2 mg L^{-1}) + $CoCl_2$ (150 mg L^{-1}) resulted in the significant increase in N protein contents compared to control (Gheeth et al. 2013). However, higher concentration of Co is related to catalase and peroxidase activities which might increase catabolism rather than anabolism (Zhu et al. 2000). Co helps to keep fruit fresh (Bulantseva et al. 2001) and delays ageing in flowers (marigold, chrysanthemum and rose) (Mitra 2015). Co in higher concentration suppresses Cd^{+2} uptake while it acts synergistically with Cr + 2 and Zn + 2 (Palit et al. 1994).

10.1 Presence of Cobalt in the Atmosphere, Soil and Sources

Cobalt is a native metal as it not found in free state on the earth because of chlorine in the ocean and oxygen in the atmosphere. In the upper layer of the earth's crust, oxygen and chlorine are abundant enough to prevent the native metal cobalt formation. A little number of cobalt compounds can be shown in soil, rocks, plants and animals (Fig. 3.13). Cobalt occurs as a component of nickel and copper minerals in compound form, while it occurs as a major metallic component with sulphur and arsenic in safflorite, glaucodot, sulphidic cobaltite and skutterudite minerals. The average content of cobalt is about 40 ppm in the atmosphere and about 1–40 ppm in the soil. Cobalt does not react with boron, carbon, arsenic, phosphorus and sulphur. The different sources of cobalt are cobalt nitrate, cobalt chloride and cobalt sulphate (Barałkiewicz and Siepak 1999).

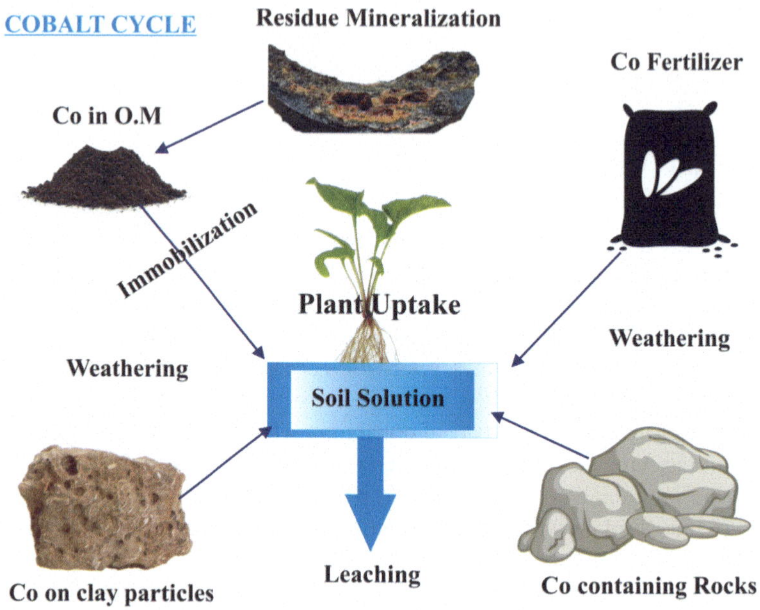

Fig. 3.13 Cobalt cycle

10.2 *Uptake and Accumulation of Cobalt by Plants*

Co^{2+} uptake takes place through root transporter, i.e. iron transporter 1 (IRT1). Transport occurs through active transport and passive diffusion (Jarosick et al. 1988). Uptake of cobalt is affected by several factors such as temperature, carbon dioxide concentration, pH and the light and dark cycles. Jayakumar and Jaleel (2009) tested the Co accumulation capacity of soybean. Co was given to soybean by soil drenching methods in pots. The result of this study showed more concentration of cobalt in plants due to 100–200 mg kg^{-1} compared to low concentration of cobalt, i.e. 50 mg kg^{-1}application in the soil. Similarly, uptake and accumulation of cobalt in plants depend on many factors which include absolute content of cobalt in a soil, level of soil fertility, acidic and alkaline conditions, reductive oxidation conditions and content of organic matter.

10.3 *Mode of Action Under Abiotic Stress*

Cobalt is rarely found in soil, but it plays a significant role in support of rhizobium and other soil bacteria. Under abiotic stresses, cobalt plays a supportive effect in plants by forming more nodulation and gives potential growth and yield.

10.4 Optimum Dose for Plant

Chickpea response to cobalt sulphate ($CoSO_4.7H_2O$) and cobalt chloride ($CoCl_2.6H_2O$) at 0, 100, 200, 400, 800 and 1600 g $ha^{-1}$was studied under field conditions. Crop was fertilized with Co at seedling stage, and result showed that both sources of Co are injurious at higher level of 400–1600 g ha-1. Cobalt sulphate showed a negative effect on all growth, yield and quality parameters, while cobalt chloride at 100–200 g ha^{-1} resulted in positive impacts. They further recommended 100 kg ha^{-1} good for chickpea crop. Awomi et al. (2012) recommended 0.5 ppm as an optimum dose of cobalt.

11 Copper (Cu)

Copper is an active redox transition metal and helpful for different physiological processes in plant. It exists as Cu^{2+} and Cu^{+}. It is the main structural component of regulatory proteins which participates in respiration and photosynthesis. Cu is a cofactor in many enzymes and plays a significant role in signalling mechanism. Therefore, Cu is an essential micronutrient for plant normal growth and development. Cu deficiency resulted in impaired growth. However, Cu higher concentration is highly toxic. Copper functions in combination with large number of enzymes playing a role in respiration and photosynthesis. Excessive Cu resulted in impaired growth of crop parts, nutritional deficiency, necrosis and chlorosis (Yruela 2005). Soil contains Cu in the range of 2–100 ppm. It is present as soluble (hydroxy and carbonate) and insoluble (oxide and sulphide) forms, but the availability of soluble form is dependent upon soil pH, clay and organic matter. Cu contents in plant lie in the range of 20–30 mg kg^{-1} dry weight of plant. Critical deficiency level is 3–5 mg kg^{-1} dry weight, and in young grain plant, it has been reported to be 1.5 mg kg^{-1} dry weight. Cu uptake in plants is similar to Fe uptake as described by Ryan et al. (2013). It is taken as Cu^{+2} from the soil either by active or passive transport. Cu is mainly accumulated in cell wall, cytoplasm, vacuole and chloroplast. Further detail about Cu is available in Fig. 3.14.

12 Iron (Fe)

Plants need nutrients to grow and survive, and Fe is one of the important nutrients crucial for healthy plant life but is required in small quantity for plant growth and development. Fe is an integral part of chlorophyll without which plants will not be able to produce food. Major functions of plant life including chlorophyll biosynthesis to energy transfer are controlled by Fe. Fe acts the same in plants as what it does in an animal system, helping to carry nutrients to all plant body parts. Fe is a transition metal with variable oxidation states (Fe^{+2} and Fe^{+3}). It has an atomic number of 26.

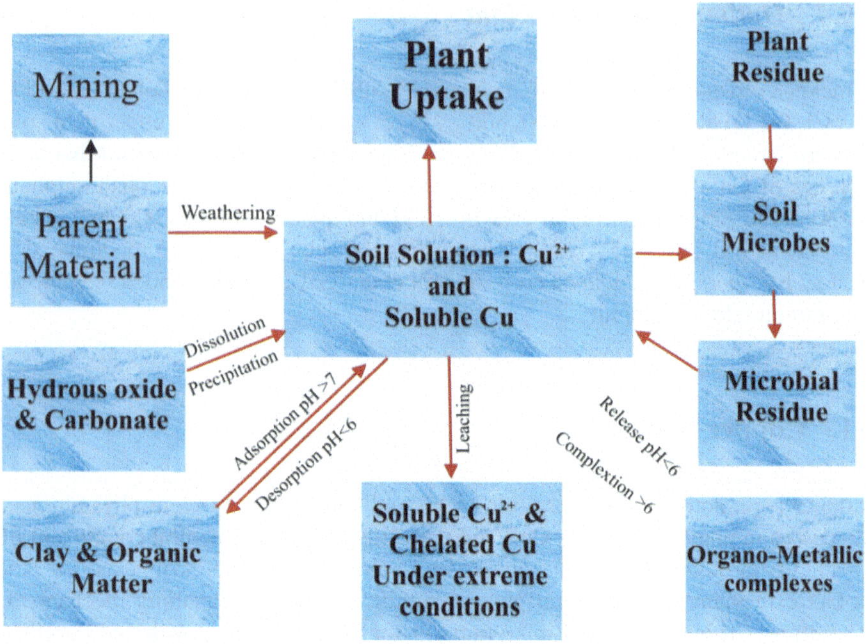

Fig. 3.14 Copper cycle

12.1 Presence in Soil/Atmosphere

Hematite (Fe_2O_3) is the biggest source of Fe, and it gives red colour to soil. Concentrations of ferric (Fe^{+3}) and ferrous (Fe^{+2}) iron are in the order of 10^{-15} molar, and it is linked with pH. One unit increase in pH resulted in the decrease in the activity of Fe^{+3} by 1000 times due to the formation of $Fe(OH)_3$. Fe is also present in decomposed plant material; therefore, compost and dead leaves could be a good source of Fe. George (2011) reported that in soil where organic matter is present Fe is available in reduced state (Fe^{+2}) (Fig. 3.15).

12.2 Uptake by Plant

Plant uptake Fe by its movement from bulk soil pore soil solution toward the root due to transpiration. It can also infiltrate into the root by diffusion. Fe has a significant role in biochemical and physiological process. It is an integral part of enzymes (cytochromes) and electron transport chain (ETC). Thus, it is required for a wide spectrum of biological functions as reported by Rout and Sahoo (2015). Plant acquires Fe by two strategies (strategy I (SI) and strategy II (SII)) as suggested by Marschner and Römheld (1994). SI has three components (enhanced net excretion

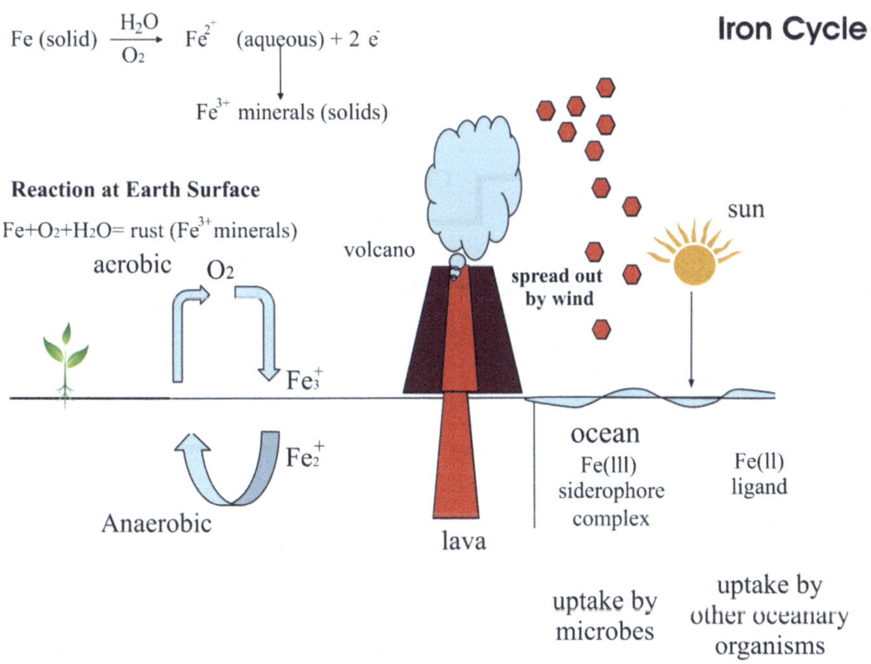

Fig. 3.15 Iron cycle

of protons, enhanced release of reductants/chelators and plasma membrane-bound inducible reductase (turbo)). This strategy prevails in monocot and dicot but with exception to Poaceae species plants (Roychowdhury et al. 2018). SII prevails in grass works by increased biosynthesis and secretion of phytosiderophores which form chelates with ferric (Fe^{+3}). The detail of these two strategies is presented in Figs. 3.16 and 3.17.

12.3 Role in Abiotic Stress Tolerance

Iron plays a significant role to have abiotic stress tolerance in plants by producing antioxidative enzymes. Enhanced drought tolerance was reported in plant due to Fe nutrition (Sultana et al. 2001; Pourgholam et al. 2013). Takahashi et al. (2001) stated that grasses can uptake Fe chelated by soluble siderophores. Under stress, different families of genes (Nramp, YSL) help in the translocation of Fe and defend plants against stress (salinity, drought and heavy metal). Fe activates antioxidants which act as a scavenger of reactive oxygen species. Excess of Fe is also not good as it creates disturbance in the homeostasis of plant cell which results in the decline in the photosynthesis, respiration and accumulation of ions. Iron plaque biofilm matrices can capture large amount of Fe at soil-root interface. Nikolic and Römheld (1999)

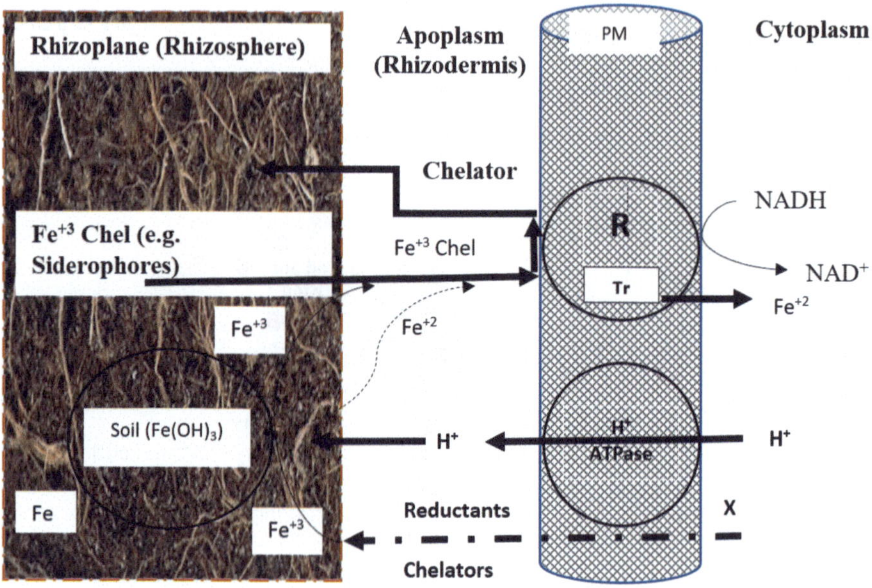

Fig. 3.16 Model for root response to Fe deficiency in non-graminous monocots and dicots (strategy I: R, inducible reductase; Tr, transporter for Fe^{+2}; St, stimulated proton pumping ATPase; X, increased production of reductants/chelators) (Modified from: Marschner and Römheld 1994)

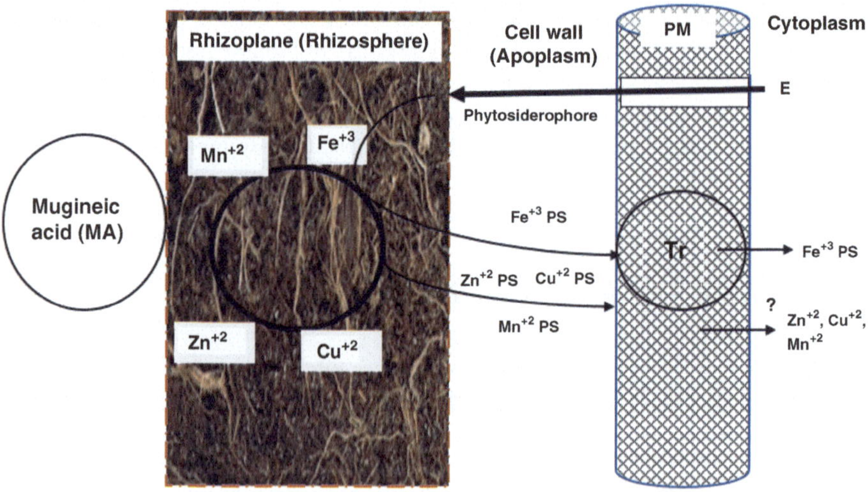

Fig. 3.17 Root response model for Fe deficiency in Poaceae family (strategy II: E, enhanced synthesis/secretion of phytosiderophores; Tr, translocator for Fe^{+3} phytosiderophores in the plasma membrane (PM)) (Modified from: Marschner and Römheld 1994)

studied the response of sunflower and faba bean on preculture varying solution of Fe. Leaf disc was analysed after 2–3 weeks of preculture, and results showed that Fe^{+3} reduction is prerequisite of Fe uptake into leaf cells. Rotaru (2011) and Elanz et al. (2011) studied the application of Fe nutrition under drought and concluded that Fe helps to build drought tolerance in plants. Similarly, Fe also helps to increase yield of wheat plants (Mahmood et al. 1990).

12.4 Iron Toxicity

Iron toxicity prevails in flooded soils due to higher concentrations of Fe^{+2} in the soil solution. Toxicity effects will be more on root, but it will also be seen in tissues if uptake is excessive.

12.5 Mode of Implementation

Chelated forms of Fe are recommended to apply as compared to inorganic forms. Since inorganic salts are less soluble, mostly precipitation of Fe occurs from these salts. This issue prevails in such soils where pH is high and bicarbonates are in larger quantity. Foliar application of Fe is considered to be good as it results in uniform application. Molecular genetic approaches, for example, by increasing phytosiderophore production, can help to increase uptake of Fe by plants.

13 Molybdenum (Mo)

13.1 Presence in Atmosphere and Soil and Source

Molybdenum (Mo) is the micronutrient that is required for biological processes. According to Rashid et al. (2002), molybdenum released in atmosphere takes place by process of weathering, industrial release and agricultural uses of Mo. Its concentration is lower in rural areas, whereas in urban areas, it is present in greater amount due to the combustion of fossil fuel which itself is a source of Mo. Mo is present in earth in the concentration of 2 mg kg^{-1}, whereas in soil, it is in the range of 0.2–5 mg kg^{-1}. It can be obtained from molybdenite and wulfenite. It can be attained from copper and tungsten mining. In plants, Mo can be obtained from legumes such as peas, lentils and beans as they are the richest sources. Good sources are nuts and grain products also.

Rashid and Rafique (2017) reported that Pakistan soil is mainly calcareous and has severe B deficiency (35–56%). Deficiency is more severe in rainfed conditions compared to irrigated. Similarly, the requirement of B varies from plant to plant; for example in cotton leaves, B demand is 53 mg kg^{-1}, while in young sorghum plant, it is 17 mg kg^{-1}. B fertilization increases crop yield by 14% (cotton and wheat) and 15–25% (rice). Furthermore, B fertilizer usage in Pakistan is negligible (92 Mg B per year) compared with potential demand (2245 Mg B per year). The main reasons for this low usage are availability of quality B fertilizer and ignorance about B use benefits.

13.2 Uptake by the Plants

Plants uptake Mo in the form of an anion molybdate (MoO_4^{-2}). It is immobile in plants, whereas in soil, it is somewhat mobile. It helps in nitrogen fixation. Plants need Mo for the conversion of inorganic form of phosphorous to the organic ones. It is also used in biological nitrogen fixation in crops especially legumes. Availability of Mo depends upon the pH. If pH decreases, Mo becomes unavailable to the plants. Acidic soils are deficient in Mo.

13.3 Mode of Action

Molybdenum plays a physiological role in plants. Molybdenum combines with active sites to form molybdenum complex. It is also involved in metabolism of nitrogen, purine, sulphur and phytohormones. It is also a growth promoter. Exogenous Mo application can cause abiotic stress tolerance in food crops. Application of Mo can be used to overcome the effect of drought, cold and salt stress.

14 Nickel

Nickel (Ni) belongs to transition metal and ductile in nature. Ni^{+2} is the accessible form of nickel that is used by plants. The use of nickel has been traced as far back as 3500 BCE. Nickel having the positive 2 valency ion which is instantly oxidized and become unavailable is why total Ni concentration is not a useful measure for Ni bioavailability. Nickel has five valencies, i.e. 0, +1, +2, +3 and + 4, and among these, only +2 form is considered to be accessible for plants. It is mainly found combined with sulphur or oxygen as sulphides or oxides. It was first isolated and classified as a chemical element by a Swedish mineralogist and chemist. It is found in nearly all types of soil and is also emitted through volcanic eruptions. It is the 22nd most abundant element on this earth (Hussain et al. 2013, 2017). Ni is an essential element for many bacteria as it helps to synthesize several enzymes (urease, glyoxalase-I, hydrogenases, some superoxide dismutase, carbon monoxide dehydrogenase and methyl-coenzyme M reductase) (de Macedo et al. 2016).

14.1 Plant Uptake

Ni is required by plants in small quantities, especially in legumes, as it has a role in N metabolism. However, its availability is a function of pH. de Macedo et al. (2016) conducted experiment on base-cation saturation ratios (BCS ratios) (50 and 70%), Ni amendments (0, 0.1, 0.5, 1.0 and 10.0 mg dm^{-3} of soil) on Ni uptake, N accumulation in leaves and grains and organic acid changes in soybean grown under greenhouse conditions. Results showed that soybean physiological performance was good in BCS70. Ni fertilization resulted in better urease activity, biological nitrogen fixation, higher organic acids, N accumulation and Ni concentration. A positive correlation was observed between node colour and Ni supplementation which resulted in higher N accumulation in leaves. Generally, plants use two transport systems known as low affinity and high affinity to uptake Ni. As far as the low-affinity transport system is concerned, plant can soak up Ni^{+2} ions at low concentration of approx. 4.4 ppb which is around 0.6 ounces nickel per million gallons of water. With the help of high-affinity transport system, plants take up to 1–8 ppm of Ni^{+2} ions. Nickel is instantly re-translocated within the plant body in the structure of complex with organic acids. It is not only absorbed by roots for entry into the plants; leaves are also the pathways through which nickel in the form of radioisotope can enter.

14.2 Mode of Action Under Abiotic Stress

Nickel helps in tolerance against abiotic stress like heavy metals and salt stress. Nickel is known to induce the antioxidant system in plants. Nickel is an important constituent of various enzymes where it manages with either the S-ligands or O-ligands; however, the only enzyme that possesses the nickel as an integral component is urease as stoichiometric quantities are concerned.

15 Selenium (Se)

Selenium (Se) is a naturally occurring trace element and occurs fundamentally in all soils. Se amount in most soils varies from 0.01 to 2.0 mg/kg, with a mean of 0.4 mg/kg, although its higher amount (>10 mg/kg) can occur in seleniferous areas. With parent material and environmental conditions, soil selenium concentration and bioavailability vary, and the dissemination of Se in soil is usually heterogeneous. Selenium is a nonmetal, its symbol is Se, and its atomic weight is 34. It has

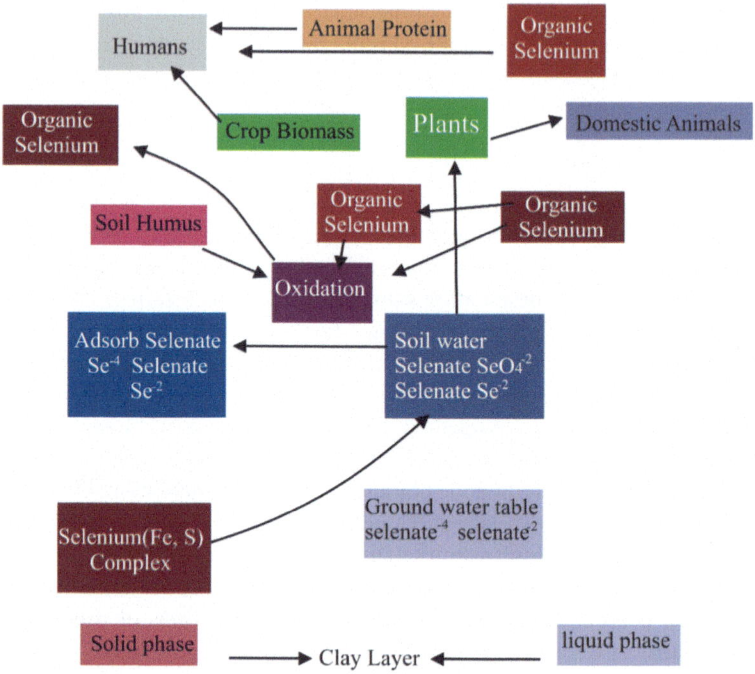

Fig. 3.18 The selenium cycle

properties similar to sulphur, tellurium and arsenic, and it was discovered in the year 1817. The word selenium is originated from the Greek word 'Selene'; its meaning is the moon (Hasanuzzaman et al. 2010; Wan et al. 2018). The detailed description of Se fate is shown in Fig. 3.18.

15.1 Nutrient Form of Se

The main natural sources of selenium are volcanoes and metal selenide, which are formed as a result of volcanic activity. Selenium occurs naturally as a microelement; it is vital for life and basically takes place in all soils and biomaterials. The availability of Se is dependent upon pH and soil supply. Se exists in soil as Se^{-2}, Se^0, SeO_3^{2-} SeO_4^{-2} and organic Se.

15.2 Se Uptake by the Plant

Se uptake by the plant is mainly in selenate form in alkaline well-oxidized soil, while under well-drained soils with acidic pH, it exists as selenite. However, under compacted soil, selenide becomes the dominant form. Se uptake is similar to sulphate transporters (SULTR). Plant species and nutritional demand can differentiate between S and Se. Plants can be divided into three categories depending upon Se concentrations in their tissues (Hasanuzzaman et al. 2011). These include non-accumulators (5 mg kg $^{-1}$ Se), accumulators (1000 mg Se kg $^{-1}$ Se) and hyperaccumulators (1000–15,000 mg kg^{-1} Se) (Ahmad et al. 2016).

15.3 Functions Under Abiotic Stress

Lower concentration of selenium is found to be useful for a variety of abiotic stresses in plants which include cold/drought stress produced due to different concentrations of metal and desiccation, salt, etc. (Chu et al. 2010; Hasanuzzaman et al. 2011, 2012b, c, 2014a, b; Shekari et al. 2017; Naim et al. 2017; Hawrylak-Nowak et al. 2018a, b). Selenium is important for response of immune system. When stress conditions are present, plant produces reactive oxygen species (ROS) by the help of selenium and that disturbs the membranes of cells and proteins, etc. Selenium increases the antioxidant compounds in plant which provide protection against the damage of vital organs. Selenium increases the functioning of antioxidant enzymes under stress conditions (Yao et al. 2010a; Hasanuzzaman and Fujita 2011a, b). Selenium behaves as a focal point of few selenoproteins, which include glutathione peroxidase, thioredoxin reductase and iodothyronine deiodinases; hence, it is critical in finding free radicals to protect from stress due to oxidation, and it strengthens the resistant system. Interrupted photosynthesis due to chloroplasts occurs when plants are exposed to environmental stress.

15.3.1 Drought

The maximum selenium application is promising for wheat seedling growth during drought spell. Physiological responses and growth of seedlings vary dependent on the Se amount. Additional application of selenium did not have an impact on the chlorophyll content, while the combination of drought and selenium significantly increased activities of accumulation of chlorophyll content, but the enhanced level has not responded to the lipid peroxidation of the membrane in plants (Hasanuzzaman and Fujita 2011a, b).

15.3.2 Chilling

Agricultural productivity has been decreased significantly due to often exposure to cold stress in growing periods of crops. Selenium enhances the antioxidant capacity and resistance in plants; however, the role of selenium has not been fully understood in plants. The role of selenium in wheat plant seedlings under chilling is reported as defensive. Treatments of selenium can decrease the rate of O_2 production under cold stress in the wheat plants. Moreover, Se treatments could enhance anthocyanins, flavonoids and phenolic compounds of plants liable to cold stress that have the capacity to quest free radicals of seedlings as compared to control treatment (Chu et al. 2010).

15.3.3 UV Radiation

Sufficient levels of selenium could increase the antioxidant capacity and decrease lipid peroxidation of the membrane in aboveground portions of wheat seedlings exposed to higher UV radiations. Moreover, Se treatments significantly decrease the rate of O_2 production of wheat seedlings developed under increased UV radiation. Se treatments could decrease oxidative stress and active O_2 content in aboveground and belowground parts of wheat seedlings exposed to stressful situations. The damage of increased UV on wheat seedlings lessens to some level through selenium application.

15.3.4 Optimum Dose for Plants and Mode of Application

Antioxidative capacity of a plant and its requirement as micronutrient depends on Se level. Selenium optimum range found in soil is 0.01 to 2 mg kg^{-1}. Selenium can be applied in the form of fertilizer in the soil; it can also be applied as foliar sprays directly on the leaves. Foliar applications provide best result. Foliar application (e.g. selocide) of selenium to soil is better as compared to application of selenium fertilizers because it creates difficulty in its transport from root to shoot.

16 Silicon (Si)

After oxygen, the soil contains silicon (Si) as the second most rich (roughly 31%) constituent. However, it is not presently considered to be a necessary factor in agricultural production, regardless of its abundance. Si exists as silicic acid, silica and silicates in soils due to its strong affinity for oxygen. Monosilicic acid, $Si(OH)_4$, is the main form of Si available in soil solution, and it can be absorbed by root system readily. Desilication of primary silicate minerals (lithogenic Si) and weathering are the vital source of available Si in the soil while releasing dissolved Si in the form of silicic acid. Also, Si is pollution-free and non-corrosive, and it is a useful and high-quality fertilizer for sustainable agriculture. Dynamics of Si have been further elaborated in Fig. 3.19.

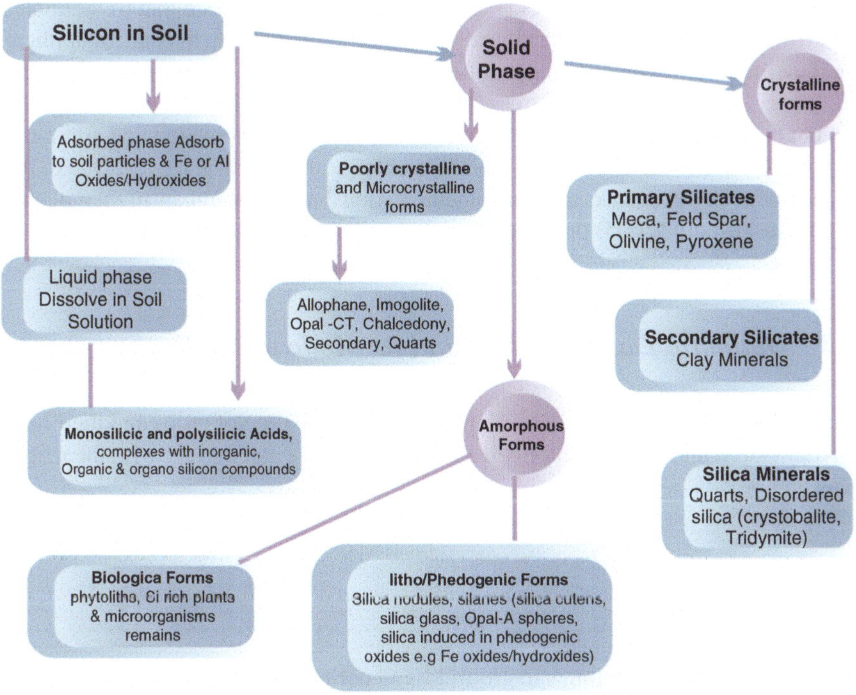

Fig. 3.19 The silicon dynamics

16.1 Silicon Uptake by the Plant

Si uptake mechanisms have been proposed by different modes for plants of varying Si accumulation levels; these modes are passive, active and rejective uptake (Ma et al. 2001).

16.2 Mode of Action Under Abiotic Stress

Abiotic stress has unfavourable effect on crop growth and productivity. Si showed the capacity to alleviate diverse abiotic stresses such as drought, salinity, nutritional imbalance and heavy metal toxicity (Ma et al. 2001; Hasanuzzaman et al. 2014b, 2017, 2018b, c, 2019b; Ahmed et al. 2011a, b, 2012, 2013, 2014a, b, 2016). It has been reported that Si fertilization can also enhance plant tolerance to drought and salinity stress by adjusting the concentration of solutes, i.e. proline, carbohydrates, glycine betaine, total free amino acids, total soluble sugars, polyols and antioxidant compounds like total phenolics which reduces the osmotic distress due to ion toxicity created by NaCl stress (Na+ and Cl-). It has been reported that Si facilitates root growth through increasing cell wall extensibility in the growth zone. Si application may induce stress resistance by affecting phytohormone homeostasis (Etesami and Jeong, 2018). Under salt stress, Si application reduced ethylene in sorghum (Yin et al. 2016). Daszkowska-Golec and Szarejko (2013) uncovered that adding Si marginally increased ABA concentration in water-stressed wheat leaves. Due to the enhancement of resistance to heavy metal toxicity and drought in higher plants, Epstein and Bloom (2005) proposed Si as "quasi-essential" for growth.

16.3 Optimum Dose for Plant and Its Mode of Application

There are diverse application doses depending upon the type of crops and their habits. Application methods commonly used are foliar spray and mixing with the irrigation water.

17 Silver (Ag)

Silver metallic element belongs to group 11 and period 5 of the periodic table. It has high thermal and electrical conductivity and available in +1 oxidation state. Silver is used as silver nanoparticles (AgNPs) with size ranging from 1 to 100 nm in plant growth assessment. The effect of silver nanoparticles is varying for different crop species. AgNPs have bioremediation and biomedicine properties. It is considered to be a great toxic concern for both plants and ecosystem. Silver nanoparticles (AgNPs) can act as biotic stress factors and pose environment contamination hazard which is to be taken seriously. AgNPs have antimicrobial properties, but they have shown toxic effects to fish (Asharani et al. 2008), algae (Miao et al. 2009; Navarro et al. 2008), animal (van der Zande et al. 2011) and human cell. The impact of AgNPs on vascular plants were studied by earlier researchers, and they reported detrimental effects on plant growth (Gubbins et al. 2011; Jiang et al. 2012; Kumari et al. 2009; Stampoulis et al. 2009) (Fig. 3.20). Decreased biomass (41%) and transpiration rate (57%) were observed in *Cucurbita pepo* (zucchini) due to the application of 100 nm AgNPs at 100 and 500 mg L^{-1} (Stampoulis et al. 2009). Cytotoxic and genotoxic impacts on *Allium cepa* root tip cells were reported by Kumari et al. (2009). Similarly, inhibition of the growth of *Lemna minor* (Gubbins et al. 2011) and decreased plant biomass, plant tissue nitrate-nitrogen content and chlorophyll a/b and chlorophyll fluorescence (Fv/Fm) of aquatic macrophyte (*Spirodela polyrhiza,* greater duckweed) (Jiang et al. 2012) were documented due to the application of AgNPs. Annual ryegrass root growth was significantly affected by the use of AgNPs

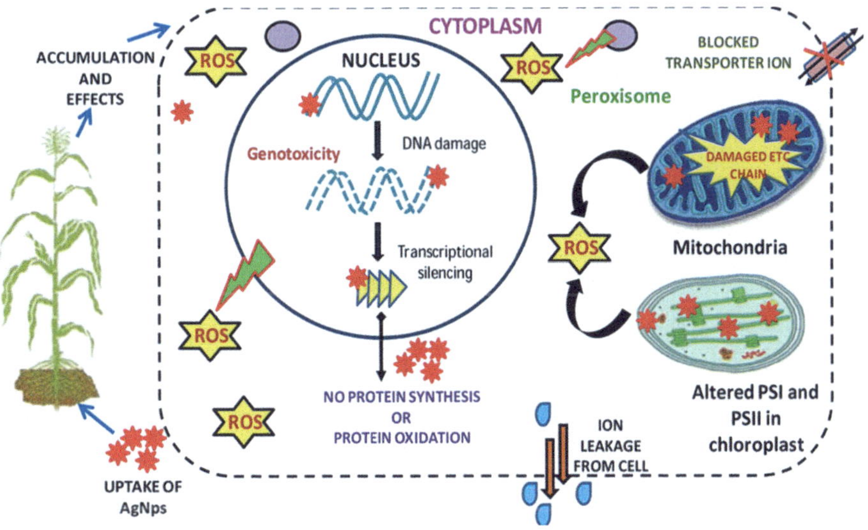

Fig. 3.20 Phytotoxicity of AgNps on cell organelles of plant cell (Modified from: Tripathi et al. 2017)

as reported by Yin et al. (2011). Phytotoxicity of AgNps on plant cell organelles has been shown in Fig. 3.17. Silver nanoparticles are used for their novel properties such as high density, good stability, etc. Nanosilver, when applied to plants, has been found effective in root disease incidence inhibition. Silver in the form of microparticles may also have positive impact on overall plant growth. Furthermore, silver cycle has been shown in Fig. 3.21 to show the impacts of AgNPs on plant.

18 Sodium (Na)

Sodium is a mineral that is not much needed for plants. It aids in metabolism and synthesis of chlorophyll. It helps in the opening and closing of stomate; therefore, it could be considered as partial replacement of potassium. The available form of sodium in soil is NaCl. It is taken up by plants in the form of Na^+ ion. The exchangeable form plus that already present in the soil solution account for nearly all of that available to the crop in course of growing season. A substantial amount of particulate sodium exists in the atmosphere because emission sources are widely spread on the earth's surface, and that can be the seawater and terrestrial substances. Plants being able to grow in high-sodium concentrations are concerned with the capability of plant to transport, extrude and compartmentalize and the mobility of sodium ions.

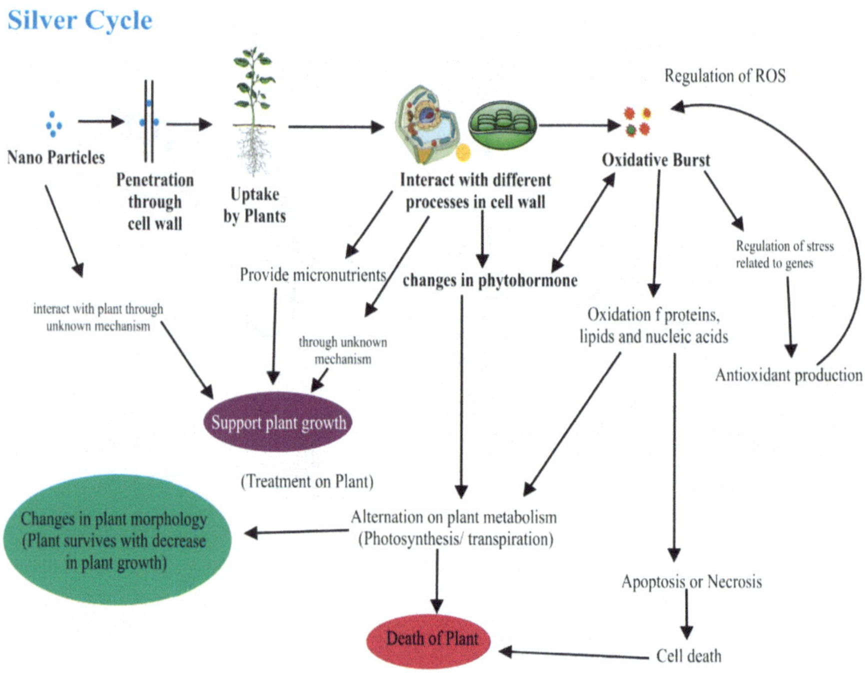

Fig. 3.21 Silver cycle showing role of nanoparticles

The entry of Na^{+1} ions into the plants is through the efflux and influx at roots. Sodium is very soft silvery-white metal and the sixth abundant element on earth comprising 2.8% of the earth's crust. It occurs in atmosphere in compounds as common salt, sodium hydroxide, sodium borate, sodium nitrate and sodium sulphate. Sodium salts are present in the soil in the form of carbonates, bicarbonates, chloride and sulphate. The proportion of each salt in different parts of the soil is extremely variable. Different soils show diverse results, and some salts like sulphates of sodium and carbonates are absorbed by soil to a greater amount than sodium chloride. Figure 3.22 shows the movement of sodium ions into the plant; further this figure illustrates the radial transport and longitudinal movement of Na^{+1} ions from the soil through xylem, roots and recirculation throughout the plants (Apse and Blumwald, 2007). Essentiality and critical leaf concentration of sodium and all other nutrients have been elaborated in Table 3.2.

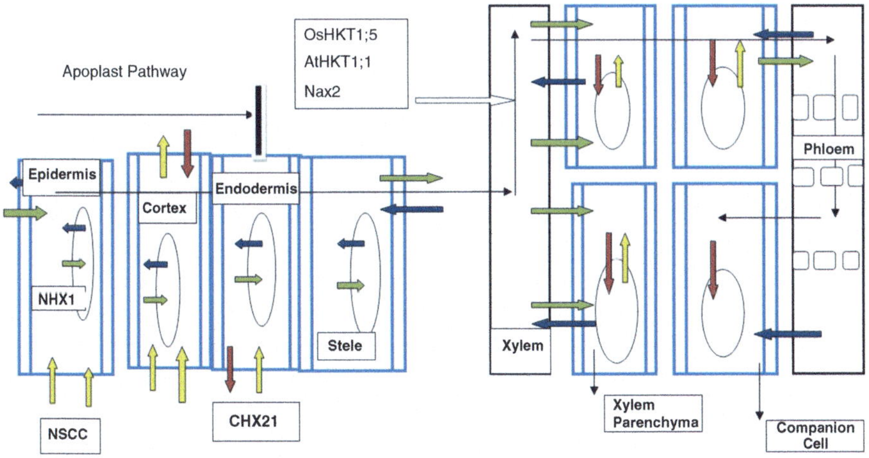

Fig. 3.22 Sodium ion (Na^{+1}) in the plant cell (Modified from: Apse and Blumwald 2007)

Table 3.2 Essentiality and critical leaf concentration of nutrients

		Critical leaf concentrations (mg g^{-1} dry matter)	
Nutrients	Plant essentiality	Sufficiency	Toxicity
N	Yes	15–40	–
K	Yes	5–40	>50
P	Yes	2–5	>10
Ca	Yes	0.5–10	>100
Mg	Yes	1.5–3.5	>15
S	Yes	1.0–5.0	–
Cl	Yes	0.1–6.0	4.0–7.0
B	Yes	5–100 × 10^{-3}	0.1–1.0
Fe	Yes	50–150 × 10^{-3}	>0.5
Mn	Yes	10–20 × 10^{-3}	0.2–5.3
Cu	Yes	1–5 × 10^{-3}	15–30 × 10^{-3}
Zn	Yes	15–30 × 10^{-3}	100–300 × 10^{-3}
Ni	Yes	0.1 × 10^{-3}	20–30 × 10^{-3}
Mo	Yes	0.1–1.0 × 10^{-3}	1
Na	Yes	–	2.5
Se	Yes	–	10–100 × 10^{-3}
Co	Yes	–	10–20 × 10^{-3}
Si	Beneficial	–	–
Al	Beneficial	–	40–200 × 10^{-3}

19 Zinc (Zn)

Zinc is an important element for the healthy growth of plants, animals and humans. Zn is taken up by plants as Zn^{2+}. Zinc insufficiency hinders plant growth, development, yield and quality. In excess of 3 billion individuals worldwide are experiencing Fe and Zn insufficiencies, and this condition is especially across the board in zones where populace is intensely subject to an unvaried eating regimen of grain-based products, in which Fe and Zn are put away solely in the husk and are thusly lost amid processing and cleaning (Cakmak and Kutman 2018). Zn helps to maintain different physiological processes in plants. An inadequacy of zinc can cause a decrease in net photosynthesis by 50–70% relying upon the plant species and the seriousness of lack. This diminished effectiveness of photosynthesis could be expected due to lessening in the action of the compound carbonic anhydrase. Zinc is a constituent of carbonic anhydrase; however, the carbonic anhydrase in dicotyledons is a bigger particle and contains more zinc than carbonic anhydrase in monocotyledons (e.g. the grains). Compounds associated with the arrangement of sucrose, for example, aldolase, are antagonistically influenced by zinc inadequacy. A decrease in the level of sucrose in sugar beet and maize is because of lower action of sucrose synthetase action. Zn also helps to maintain membrane integrity of plants, auxin metabolism and reproduction. The overall benefits of macro- and micronutrients have been shown in Fig. 3.23. Similarly, a detailed review about the importance of nutrients on abiotic stress tolerance has been presented in Table 3.3.

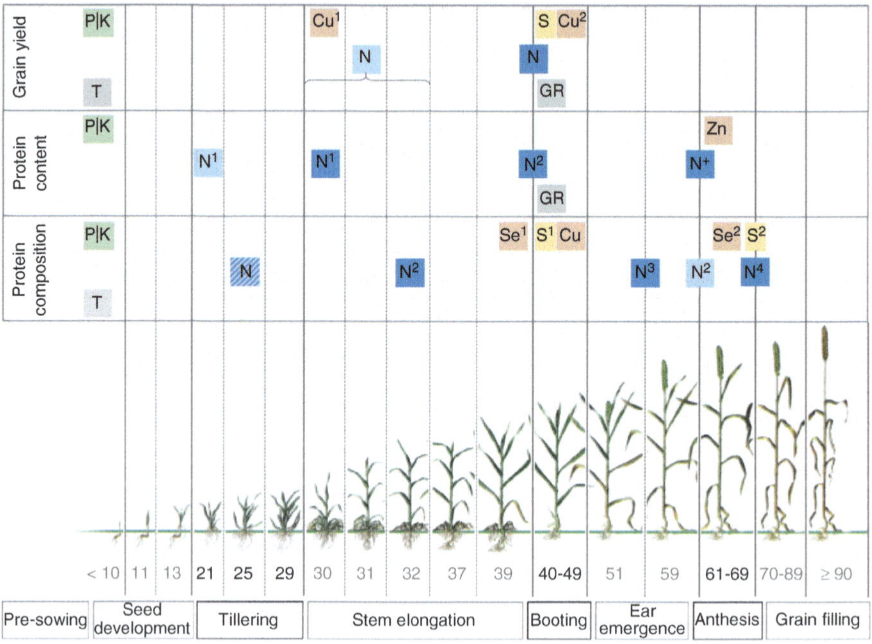

Fig. 3.23 Macro- and micronutrient benefits on yield and quality of wheat crop (Modified from: Hellemans et al. 2018)

Table 3.3 Control of abiotic stress by the application of nutrients

Type of stress	Nutrients	Method	Conclusion/result	References
Drought	K	K was sprayed with carboxymethyl Cellulose and a surfactant at 3 different growth stages (i.e. tillering, flower initiation and grain filling) in wheat	All stages showed increase in tolerance at drought conditions by increasing the osmotic potential, water potential and turgor potential of wheat plants	Raza et al. (2013)
	K	Three levels of K were applied in *Hibiscus*	K-applied plants showed higher water contents in leaves and sustained transpiration, photosynthesis and stomatal conductance	Egilla et al. (2005)
	K	Foliar application of K in tobacco	Transpiration and photosynthesis activities increased resulted in reduced water use efficiency	Bahrami and Hajiboland (2017)
	K	External K2CO3 was applied in hydroponic experiment	Shoot K^+, chlorophyll content and water potential were enhanced	Wei et al. (2013)
	NO	Different stressed conditions produced NO which increased the antioxidant activity and created tolerance in plants	NO played an essential role in tolerance of different abiotic stresses, i.e. drought, heavy metal salt, UV-B and temperature (high and low)	Siddiqui et al. (2011)
Salinity and drought	NaCl	Osmopriming of sugarcane	Salinity and drought stress were reduced	Patade et al. (2009)
Drought (mild and severe)	Se	25 µM Se was used. Seedlings of rapeseed were treated with Se and then exposed to mild and severe stress	Result showed Se increased the tolerance by increasing the DNA methylation and antioxidant activity	Hasanuzzaman and Fujita (2011a, b)
	Ni	Application of Ni under different stressed conditions	Under stressed conditions, Ni activated antioxidant activities	Fabiano et al. (2015)
Drought stress	Si	Drought stress was maintained by 75% and 50% of relative water content to wheat	Si improved the water contents and potential of stressed plants	Gong et al. (2005)

(continued)

Table 3.3 (continued)

Type of stress	Nutrients	Method	Conclusion/result	References
Salinity stress	Jasmonic acid	Foliar application of 2 mM JA on the seedlings of wheat	Antioxidant enzyme activities and antioxidative compounds were increased	Qiu et al. (2014)
Salinity	Salicylic acid (SA)	0.5 mM SA was applied to the two cultivars of mungbeans under salt stress conditions	Result showed that SA decreased the detrimental effect of salt stress on photosynthesis activity	Nazar et al. (2011a, b)
Salt stress, iron, Cd and As toxicity	Si	Application of Si under different stressed conditions	Result showed the positive responses in the alleviation of stressed conditions	Guntzer et al. (2012)
Salinity	Zn finger protein	Zn finger protein gene in potato was analysed under salt stress	Zn finger protein gene enhanced salt tolerance	Tian et al. (2010)
Heavy metal stress	Se	Addition of 50 µM Se to the culture where Cd toxicity was created	Cd-induced stress was controlled in red seaweed *Gracilaria dura* by regulating DNA methylation and antioxidants	Kumar et al. (2016)
Heavy metal (Cd)	P	Different ratios of phosphate were applied in *Trema micrantha* under Cd toxicity	Phosphate favoured plant growth by reducing the effect of Cd toxicity	Soares et al. (2007)
	Ca	*Arabidopsis* and *Trifolium repens* seedlings were exposed to Cd toxicity then treated with Ca	Ca reduced 1/3 of Cd toxicity in the seedlings of *Arabidopsis* as compared to Cd-treated seedlings alone and decreased Cd accumulation in *Trifolium repens*	Suzuki, (2005) and Wang and Song (2009)
	K	Cd stress was given to K-deficient rice seedlings	K deficiency in rice seedlings protected them from Cd stress	Liu et al. (2013)
	NO	Cd stress was given to different crops	NO application resulted in reduced Cd accumulation	Gill et al. (2013)

(continued)

Table 3.3 (continued)

Type of stress	Nutrients	Method	Conclusion/result	References
Ultraviolet-B (UV-B) stress	Se	Wheat roots and seedlings were exposed to ultraviolet-B (UV-B) stress. Lower (0.5 mg per kg) and higher (3 mg per kg) dose of Se were applied	It was concluded that optimum quantity of Se increased the antioxidant enzymes and hence increased the tolerance to radiation stress	Yao et al. (2010a, b)
Radiation	Si	Application of silicon under radiation-stressed plants	It was concluded that Si has positive effect on plants under stressed conditions	Tripathi et al. (2014)
High temperature	Se	Foliar application of sodium selenate (75 mg per L) was done 63 DAS in sorghum which was exposed to high temperature	It was concluded that application of Se has lessen the reactive oxygen species and enhanced the antioxidant enzyme activity	Djanaguiraman et al. (2010)
Cold	Se	0.5, 1, 2, 3 mg Se per kg was applied to wheat seedlings under cold stress	It was demonstrated that different levels of Se increase the antioxidant capacity of seedlings	Chu et al. (2010)
Defense system	S	Application of S at different concentrations under abiotic stressed conditions	Sulphur helps to improve the defence mechanism of plants under stressed conditions	Rausch and Wachter (2005) and Nazar et al. (2011a, b)

20 Conclusion

Macro- and microelements are not essential for most plants, but, when supplied at low dosages, they help to improve their growth, development and yield quality by stimulating different molecular, biochemical and physiological mechanisms triggering adaptive responses to challenging environments. When plants are exposed to environmental cues such as drought, heavy metal toxicity, low temperatures, saline soils, pest insects or pathogens, beneficial elements may induce tolerance, resistance or defence responses that allow plants to achieve acclimation to such stressors. Enhancement of nutrient uptake, synthesis of antioxidants and osmoprotectants, stimulation of secondary metabolism and signalling cascades and reduction of senescence are among the responses boosted by plant nutrients when applied at low dosages. Micronutrients applied in trace amounts act as supplements that recharge and reprogram the antioxidant system and lead to the osmotic balance.

References

Abbaspour N, Kaiser B, Tyerman S (2013) Chloride transport and compartmentation within main and lateral roots of two grapevine rootstocks differing in salt tolerance. Trees 27(5):1317–1325

Ahmed, M, Asif M, Goyal A (2012) Silicon the non-essential beneficial plant nutrient to enhanced drought tolerance in wheat. Chapter 2 Published by InTech Janeza Trdine 9, 51000 Rijeka, Croatia

Ahmed M, Hassen F, Qadeer U, Aslam MA (2011a) Silicon application and drought tolerance mechanism of sorghum. Afr J Agric Res 6(3):594–607

Ahmed M, Fayyaz ul H, Khurshid Y (2011b) Does silicon and irrigation have impact on drought tolerance mechanism of sorghum? Agric Water Manag 98(12):1808–1812. https://doi.org/10.1016/j.agwat.2011.07.003

Ahmed M, Kamran A, Asif M, Qadeer U, Ahmed ZI, Goyal A (2013) Silicon priming: a potential source to impart abiotic stress tolerance in wheat: A review. Australian Journal of Crop Science 7(4):4

Ahmed M, Asif M, Hassan F-U (2014a) Augmenting drought tolerance in sorghum by silicon nutrition. Acta Physiol Plant 36(2):473–483. https://doi.org/10.1007/s11738-013-1427-2

Ahmed M, Fayyaz ul H, Asif M (2014b) Amelioration of Drought in Sorghum (Sorghum bicolor L.) by Silicon. Communications in Soil Science and Plant Analysis 45 (4):470–486. https://doi.org/10.1080/00103624.2013.863907

Ahmed M, Stockle CO (2016) Quantification of climate variability, adaptation and mitigation for agricultural sustainability: Springer Nature Singapore Pvt. Ltd., pp. 437. https://doi.org/10.1007/978-3-319-32059-5. https://www.springer.com/gp/book/9783319320571

Ahmad S, Hasanuzzaman M (2020) Cotton Production and Uses. Springer Nature Singapore Pvt. Ltd., pp. 641. https://doi.org/10.1007/978-981-15-1472-2.) (https://link.springer.com/book/10.1007/978-981-15-1472-2)

Ahmed M (2020) Systems Modeling: Springer Nature Singapore Pte Ltd. https://doi.org/10.1007/978-981-15-4728-7_6

Ahmad R, Waraich EA, Nawaz F, Ashraf MY, Khalid M (2016) Selenium (Se) improves drought tolerance in crop plants – a myth or fact? J Sci Food Agric 96(2):372–380

Alpaslan M, Gunes A (2001) Interactive effects of boron and salinity stress on the growth, membrane permeability and mineral composition of tomato and cucumber plants. Plant Soil 236(1):123–128

Apse MP, Blumwald E (2007) Na^+ transport in plants. FEBS Lett 581(12):2247–2254

Asharani PV, Wu YL, Gong ZY, Valiyaveettil S (2008) Toxicity of silver nanoparticles in zebrafish models. Nanotechnology 19:25

Ashraf M, Akram NA, Al-Qurainy F, Foolad MR (2011) Chapter five - drought tolerance: roles of organic osmolytes, growth regulators, mineral nutrients. In: Donald LS (ed) Advances in agronomy, vol 111. Academic, pp 249–296

Awomi TA, Singh AK, Kumar M, Bordoloi LJ (2012) Effect of phosphorus, molybdenum and cobalt nutrition on yield and quality of mungbean (*Vigna radiata* L.) in acidic soil of Northeast India. Indian J Hill Farm 25(2):22–26

Bahrami-Rad S, Hajiboland R (2017) Effect of potassium application in drought-stressed tobacco (*Nicotiana rustica* L.) plants: comparison of root with foliar application. Ann Agric Sci 62(2):121–130

Barałkiewicz D, Siepak J (1999) Chromium, nickel and cobalt in environmental samples and existing legal norms. Polish J Environ Stud 8(4):201–208

Bell RW, Rerkasem B (eds) (2012) Boron in soils and plants: proceedings of the international symposium on boron in soils and plants held at Chiang Mai, Thailand, vol. 76, 7–11 September, 1997. Springer, New York

Besada C, Gil R, Bonet L, Quiñones A, Intrigliolo D, Salvador A (2016) Chloride stress triggers maturation and negatively affects the postharvest quality of persimmon fruit. Involvement of calyx ethylene production. Plant Physiol Biochem 100:105–112

Bray EA, Bailey-Serres J, Weretilnyk E (2000) Responses to abiotic stresses. In: Buchanan B, Gruissem W, Jones R (eds) Biochemistry and molecular biology of plants. American Society of Plant Physiologists, Rockville, pp 1158–1203

Bulantseva EA, Glinka EM, Protsenko MA, Sal'kova EG (2001) A protein inhibitor of poly galacturonase in apple fruits treated with amino ethoxy vinyl glycine and cobalt chloride. Prikl Biokhim Mikrobiol 37:100–104

Cakmak I, Kutman UB (2018) Agronomic biofortification of cereals with zinc: a review. Eur J Soil Sci 69(1):172–180

Cameron KC, Di HJ, Moir JL (2013) Nitrogen losses from the soil/plant system: a review. Ann Appl Biol 162:145–173

Chu J, Yao X, Zhang Z (2010) Responses of wheat seedlings to exogenous selenium supply under cold stress. Biol Trace Elem Res 136(3):355–363

Croft H, Chen JM (2018) 3.09 – Leaf Pigment Content. In Comprehensive Remote Sensing, ed. S. Liang, 117–142. Oxford: Elsevier

Daniel-Vedele F, Filleur S, Caboche M (1998) Nitrate transport: a key step in nitrate assimilation. Curr Opin Plant Biol 1:235–239

Daszkowska-Golec A, Szarejko I (2013) Open or close the gate - stomata action under the control of phytohormones in drought stress conditions. Front Plant Sci 4:138

Daur I, Alghabari F, Alzamanan S, Rizwan S, Ahmad M, Waqas M, Shafqat W (2019) Heat stress and plant development: role of sulphur metabolites and management strategies AU - Ihsan, Muhammad Zahid. Acta Agric Scand Sect B Soil Plant Sci:1–11

de Macedo FG, Bresolin JD, Santos EF, Furlan F, Lopes da Silva WT, Polacco JC, Lavres J (2016) Nickel availability in soil as influenced by liming and its role in soybean nitrogen metabolism. Front Plant Sci 7:1358

Ding L, Lu Z, Gao L, Guo S, Shen Q (2018) Is nitrogen a key determinant of water transport and photosynthesis in higher plants upon drought stress? Front Plant Sci 9:1143

Djanaguiraman M, Prasad PV, Seppanen M (2010) Selenium protects sorghum leaves from oxidative damage under high temperature stress by enhancing antioxidant defense system. Plant Physiol Biochem 48(12):999–1007

Egilla JN, Davies FT, Boutton TW (2005) Drought stress influences leaf water content, photosynthesis, and water-use efficiency of Hibiscus rosa-sinensis at three potassium concentrations. Photosynthetica 43(1):135–140

Elanz E, Mohsen R, Ahmed B (2011) Influence of salt stress on cations concentration quality and quantity of sunflower cultivars. J Food Agric Environ 2:469–476

Epstein I, Bloom EJ (2005) Mineral nutrition of plants: principles and perspectives, 2nd edn. Sinauer, Sunderland

Etesami H, Jeong BR (2018) Silicon (Si): review and future prospects on the action mechanisms in alleviating biotic and abiotic stresses in plants. Ecotoxicol Environ Saf 147:881–896

Fabiano C, Tezotto T, Favarin JL, Polacco JC, Mazzafera P (2015) Essentiality of nickel in plants: a role in plant stresses. Front Plant Sci 6:754

Farooq M, Wahid A, Kobayashi N, Fujita D, Basra S (2009) Plant drought stress: effects, mechanisms and management sustainable agriculture. Springer, Berlin, pp 153–188

Franco-Navarro JD, Rosales MA, Cubero-Font P, Colmenero-Flores JM, Brumós J, Talón M (2015) Chloride regulates leaf cell size and water relations in tobacco plants. J Exp Bot 67(3):873–891

Gad N (2006) Increasing the efficiency of nitrogen fertilization through cobalt application to pea plant. Res J Agric Biol Sci 2:433–442

Gad N (2012) Role and importance of cobalt nutrition on groundnut (*Arachis hypogaea*) production. World Appl Sci J 20:359–367

Gad N, Kandil H (2008) Response of sweet potato (*Ipomoea batatas* L.) plants to different levels of cobalt. Aust J Basic Appl Sci 2:945–955

George H (2011) Iron (Fe) nutrition of plants. http://edis.ifas.ufl.edu

Gheeth RHM, Moustafa YMM, Abdel-Hakeem WM (2013) Enhancing growth and increasing yield of peas (*Pissum sativum* L.) by application of ascorbic acid and cobalt chloride. J Nov Appl Sci 2:106–115

Gill SS, Hasanuzzaman M, Nahar K, Macovei A, Tuteja N (2013) Importance of nitric oxide in cadmium stress tolerance in crop plants. Plant Physiol Biochem 63:254–261

Gong H, Zhu X, Chen K, Wang S, Zhang C (2005) Silicon alleviates oxidative damage of wheat plants in pots under drought. Plant Sci 169(2):313–321

Gubbins EJ, Batty LC, Lead JR (2011) Phytotoxicity of silver nanoparticles to *Lemna minor* L. Environ Pollut 159:1551–1559

Guntzer F, Keller C, Meunier JD (2012) Benefits of plant silicon for crops: a review. Agron Sustain Dev 32:201–213

Hasanuzzaman M, Fujita M (2011a) Selenium pretreatment upregulates the antioxidant defense and methylglyoxal detoxification system and confers enhanced tolerance to drought stress in rapeseed seedlings. Biol Trace Elem Res 143:1758–1776

Hasanuzzaman M, Fujita M (2011b) Selenium pretreatment upregulates the antioxidant defense and methylglyoxal detoxification system and confers enhanced tolerance to drought stress in rapeseed seedlings. Biol Trace Elem Res 143:1758–1776

Hasanuzzaman M, Hossain MA, Fujita M (2010) Selenium in higher plants: physiological role, antioxidant metabolism and abiotic stress tolerance. J Plant Sci 5(4):354–375

Hasanuzzaman M, Hossain MA, Fujita M (2011) Selenium-induced up-regulation of the antioxidant defense and methylglyoxal detoxification system reduces salinity-induced damage in rapeseed seedlings. Biol Trace Elem Res 143(3):1704–1721

Hasanuzzaman M, Hossain MA, Fujita M (2012a) Exogenous selenium pretreatment protects rapeseed seedlings from cadmium-induced oxidative stress by upregulating the antioxidant defense and methylglyoxal detoxification systems. Biol Trace Elem Res 149(2):248–261

Hasanuzzaman M, Hossain MA, Teixeira da Silva JA, Fujita M (2012b) Plant response and tolerance to abiotic oxidative stress: antioxidant defense is a key factor. In: Venkateswarlu B, Shanker AK, Shanker C, Maheswari M (eds) Crop stress and its management: perspectives and strategies. Springer, New York, pp 261–315

Hasanuzzaman M, Nahar K, Alam MM, Fujita M (2014a) Modulation of antioxidant machinery and the methylglyoxal detoxification system in selenium supplemented *Brassica napus* seedlings confers tolerance to high temperature stress. Biol Trace Elem Res 161:297–307

Hasanuzzaman M, Nahar K, Fujita M (2014b) Silicon and selenium: two vital trace elements in conferring abiotic stress tolerance to plants. In: Parvaiz A, Rasool SL (eds) Emerging technologies and management of crop stress tolerance vol. 1–biological techniques. Academic/Elsevier, New York, pp 375–420

Hasanuzzaman M, Nahar K, Anee TI, Fujita M (2017) Exogenous silicon attenuates cadmium-induced oxidative stress in *Brassica napus* L. by modulating AsA-GSH pathway and glyoxalase system. Front Plant Sci 8:–1061

Hasanuzzaman M, Fujita M, Nahar K, Hawrylak-Nowak B (2018a) Plant nutrients and abiotic stress tolerance. Springer, Singapore, p 590

Hasanuzzaman M, Nahar K, Anee TI, Khan MIR, Fujita M (2018b) Silicon-mediated regulation of antioxidant defense and glyoxalase systems confers drought stress tolerance in *Brassica napus* L. S Afr J Bot 115:50–57

Hasanuzzaman M, Nahar K, Rohman MM, Anee TI, Huang Y, Fujita M (2018c) Exogenous silicon protects *Brassica napus* plants from salinity-induced oxidative stress through the modulation of AsA-GSH pathway, thiol-dependent antioxidant enzymes and glyoxalase systems. Gesunde Pflanzen 70:185–194

Hasanuzzaman M, Shabala S, Fujita M (2019a) Halophytes and climate change: adaptive mechanisms and potential uses. CABI, Wallingford, p xi

Hasanuzzaman M, Alam MM, Nahar K, Mohsin SM, Bhuyan MHMB, Parvin K, Hawrylak-Nowak B, Fujita M (2019b) Silicon-induced antioxidant defense and methylglyoxal detoxification works coordinately in alleviating nickel toxicity in *Oryza sativa* L. Ecotoxicology 28:261–276

Hasegawa K, Kimura Y, Ono TA (2002) Chloride cofactor in the photosynthetic oxygen-evolving complex studied by fourier transform infrared spectroscopy. Biochemistry 41(46):13839–13850

Hawrylak-Nowak B, Dreslerb S, Rubinowskaa K, Matraszek-Gawrona R, Wocha W, Hasanuzzaman M (2018a) Selenium biofortification enhances the growth and alters the physiological response of lamb's lettuce grown under high temperature stress. Plant Physiol Biochem 127:446–456

Hawrylak-Nowak B, Hasanuzzaman M, Matraszek-Gawron R (2018b) Mechanisms of selenium-induced enhancement of abiotic stress tolerance in plants. In: Hasanuzzaman M, Fujita M, Oku H, Nahar K, Hawrylak-Nowak B (eds) Plant nutrients and abiotic stress tolerance. Springer, Singapore

Heinen M, Dimkpa CO, Bindraban PS (2017) Effects of nutrient antagonism and synergism on yield and fertilizer use efficiency AU – Rietra, René P. J. J. Commun Soil Sci Plant Anal 48(16):1895–1920

Hellemans T, Landschoot S, Dewitte K, Van Bockstaele F, Vermeir P, Eeckhout M, Haesaert G (2018) Impact of crop husbandry practices and environmental conditions on wheat composition and quality: a review. J Agricd Food Chem 66:2491–2509

Hernández G, Ramírez M, Valdés-López O, Tesfaye M, Graham MA, Czechowski T, Schlereth A, Wandrey M, Erban A, Cheung F, Wu HC, Lara M, Town CD, Kopka J, Udvardi MK, Vance CP (2007) Phosphorus stress in common bean: root transcript and metabolic responses. Plant Physiol 144:752–767

Hillel D (2008) Soil fertility and plant nutrition. In: Hillel D (ed) Soil in the environment. Academic, San Diego, pp 151–162

Hussain MB, Ali S, Azam A, Hina S, Farooq MA, Ali B et al (2013) Morphological, physiological and biochemical responses of plants to nickel stress: a review. Afr J Agric Res 8:1596–1602

Hussain J, Husain I, Arif M, Gupta N (2017) Studies on heavy metal contamination in Godavari river basin. Appl Water Sci 7(8):4539–4548

Jarosick J, Vara PZ, Koneeny J, Obdrzalek M (1988) Dynamics of cobalt 60 uptake by roots of pea plants. Sci Total Environ 71:225–229

Jayakumar K, Jaleel CA (2009) Uptake and accumulation of cobalt in plants: a study based on exogenous cobalt in soybean. Bot Res Int 2:310–314

Jiang H, Li M, Chang F, Li W, Yin L (2012) Physiological analysis of silver nanoparticles and $AgNO_3$ toxicity to *Spirodela polyrhiza*. Environ Toxicol Chem 31:1880–1996

Kaur N, Siddique KHM, Nayyar H (2016) Beneficial elements for agricultural crops and their functional relevance in defence against stresses AU - Kaur, Simranjeet. Arch Agron Soil Sci 62(7):905–920

Kumar S, Trivedi PK (2016) Heavy metal stress signaling in plants, in Plant Metal Interaction-Emerging Remediation Techniques, ed. P. Ahmad (Amsterdan: Elsevier), 585–603. https://doi.org/10.1016/B978-0-12-803158-2.00025-4

Kumari M, Mukherjee A, Chandrasekaran N (2009) Genotoxicity of silver nanoparticles in *Allium cepa*. Sci Total Environ 407:5243–5246

Liu CH, Chao YY, Kao CH (2013) Effect of potassium deficiency on antioxidant status and cadmium toxicity in rice seedlings. Bot Stud 54(1):2

Lu WW, Zhang HL, Shi WM (2013) Dissimilatory nitrate reduction to ammonium in an anaerobic agricultural soil as affected by glucose and free sulfide. Eur J Soil Biol 58:98–104

Ma JF, Miyake Y, Takahashi E (2001) Chapter 2 Silicon as a beneficial element for crop plants. Stud Plant Sci 8.17–39

Maathuis F (2009) Physiological functions of mineral nutrients. Curr Opin Plant Biol 12:250–258

Mahmood T, Qureshi RH, Aslam M, Qadir M, Ashraf M (1990) Effect of salinity on germination, growth and chemical composition of some members of Gramineae, in 3rd National Congress of Soil Science, March 20–22 (Lahore)

Marschner H, Römheld V (1994) Strategies of plants for acquisition of iron. Plant Soil 165:261–274

Miao AJ, Schwehr KA, Xu C, Zhang SJ, Luo ZP et al (2009) The algal toxicity of silver engineered nanoparticles and detoxification by exopolymeric substances. Environ Pollut 157:3034–3041

Mitra GN (2015) Cobalt (Co), Selenium (Se), Vanadium (V), Cadmium (Cd), Lead (Pb) and Titanium (Ti). In: Garrity G et al (eds) Regulation of nutrient uptake by plants. Springer, New Delhi, pp 189–195

Miwa K, Fujiwara T (2010) Boron transport in plants: co-ordinated regulation of transporters. Ann Bot 105(7):1103–1108

Mohamed AKSH, Qayyum MF, Shahzad AN, Gul M, Wakeel A (2015) Interactive effect of boron and salinity on growth, physiological and biochemical attributes of wheat (*Triticum aestivum*). Int J Agric Biol 18:238–244

Mohammadkhani N, Abbaspour N (2018) Absorption kinetics and efflux of chloride and sodium in the roots of four grape genotypes (Vitis L.) differing in salt tolerance. Iran J Sci Technol Trans A Sci 42(4):1779–1793

Naim MA, Matin MA, Anee TI, Hasanuzzaman M, Chowdhury IF, Razafindrabe BH, Hasanuzzaman M (2017) Exogenous selenium improves growth, water balance and chlorophyll content in indica and japonica rice exposed to salinity. Transylv Rev XXV(16):4047–4057

Navarro E, Piccapietra F, Wagner B, Marconi F, Kaegi R et al (2008) Toxicity of silver nanoparticles to *Chlamydomonas reinhardtii*. Environ Sci Technol 42:8959–8964

Nazar R, Iqbal N, Masood A, Syeed S, Khan NA (2011a) Understanding the significance of sulfur in improving salinity tolerance in plants. Environ Exp Bot 70(2–3):80–87

Nazar R, Iqbal N, Syeed S, Khan NA (2011b) Salicylic acid alleviates decreases in photosynthesis under salt stress by enhancing nitrogen and sulfur assimilation and antioxidant metabolism differentially in two mungbean cultivars. J Plant Physiol 168(8):807–815

Nikolic M, Römheld V (1999) Mechanism of Fe uptake by the leaf symplast: is Fe inactivation in leaf a cause of Fe deficiency chlorosis? Plant Soil 215(2):229–237

Oerke EC (2006) Crop losses to pests. J Agric Sci 144:31–43

Oliveira Neto W, Muniz AS, Silva MAG, Castro C, Borkert CM (2009) Boron extraction and vertical mobility in Paraná State Oxisol, Brazil. Rev Bras Cienc Solo 33:1259–1267

Palit S, Sharma A, Talukder G (1994) Effect of cobalt on plants. Bot Rev 60:149–181

Patade VY, Bhargava S, Suprasanna P (2009) Halopriming imparts tolerance to salt and PEG induced drought stress in sugarcane. Agric Ecosyst Environ 134(1–2):24–28

Peltzer DA, Wardle DA, Allison VJ, Baisden WT, Bardgett RD, Chadwick OA, Condron LM, Parfitt RL, Porder S, Richardson SJ, Turner BL, Vitousek PM, Walker J, Walker LR (2010) Understanding ecosystem retrogression. Ecol Monogr 80:509–529

Pourgholam M, Nemati N, Oveysi M (2013) Effect of zinc and iron under the influence of drought on yield and yield components of rapeseed (*Brassica napus*). Ann Biol Res 4:186–189

Qiu Z, Guo J, Zhu A, Zhang L, Zhang M (2014) Exogenous jasmonic acid can enhance tolerance of wheat seedlings to salt stress. Ecotoxicol Environ Saf 104:202–208

Rahman A, Nahar K, Hasanuzzaman M, Fujita M (2016) Calcium supplementation improves Na^+/K^+ ratio, antioxidant defense and glyoxalase systems in salt-stressed rice seedlings. Front Plant Sci 7. https://doi.org/10.3389/fpls.2016.00609

Rashid A, Rafique E (2017) Boron deficiency diagnosis and management in field crops in calcareous soils of Pakistan: a mini review. Bor Dergisi 2(3):142–152

Rashid A, Rafique E, Ryan J (2002) Establishment and management of boron deficiency in field crops in Pakistan. In: Goldbach H, Brown PH, Rerkasem B, Thellier M, Wimmer MA, Bell RW (eds) Boron in plant and animal nutrition. Springer, Boston, pp 339–348

Rausch T, Wachter A (2005) Sulfur metabolism: a versatile platform for launching defence operations. Trends Plant Sci 10(10):503–509

Raven JA (2016) Chloride: essential micronutrient and multifunctional beneficial ion. J Exp Bot 68(3):359–367

Ray JG, George KJ (2010) Calcium accumulation in grasses in relation to their root cation exchange capacity. J Agron 9:70–74

Raza MAS, Saleem MF, Shah GM, Jamil M, Khan IH (2013) Potassium applied under drought improves physiological and nutrient uptake performances of wheat (*Triticum aestivum* L.). J Soil Sci Plant Nutr 13(1):175–185

Rotaru V (2011) The effect of phosphorus and iron on plant growth and nutrient status of two soybean (*Glycine max* L.) cultivars under suboptimal water regime of soil. Lucrări Stiintifice Suplimentseria Agronomie 54:11–16

Rout GR, Sahoo S (2015) Role of iron in plant growth and metabolism. Rev Agric Sci 3:1–24

Roychowdhury R, Khan MH, Choudhury S (2018) Arsenic in rice: an overview on stress implications, tolerance and mitigation strategies. In: Hasanuzzaman M, Nahar K, Fujita M (eds) Plants under metal and metalloid stress. Springer, Singapore, pp 401–415

Roychowdhury R, Khan MH, Choudhury S (2019) Physiological and molecular responses for metalloid stress in rice - a comprehensive overview. In: Hasanuzzaman M, Fujita M, Nahar K, Biswas J (eds) Advances in rice research for abiotic stress tolerance. Woodhead Publishing/ Elsevier, pp 341–369

Ryan BM, Kirby JK, Degryse F, Harris H, McLaughlin MJ, Scheiderich K (2013) Copper speciation and isotopic fractionation in plants: uptake and translocation mechanisms. New Phytol 199:367–378

Sairam RK, Srivastava GC, Agarwal S, Meena RC (2005) Differences in antioxidant activity in response to salinity stress in tolerant and susceptible wheat genotypes. Biol Plant 49:85–91

Schulz P, Herde M, Romeis T (2013) Calcium-dependent protein kinases: hubs in plant stress signaling and development. Plant Physiol 163(2):523–530

Shekari L, Kamelmanesh MM, Mozafariyan M, Hasanuzzaman M, Sadeghi F (2017) Role of selenium in mitigation of cadmium toxicity in pepper grown in hydroponic condition. J Plant Nutr 40(6):761–772

Siddiqui MH, Al-Whaibi MH, Basalah MO (2011) Role of nitric oxide in tolerance of plants to abiotic stress. Protoplasma 248(3):447–455

Soares CRFS, Siqueira JO, Carvalho JGD, Guilherme LRG (2007) Phosphate nutrition and arbuscular mycorrhiza on amelioration of cadmium toxicity in Trema [*Trema micrantha* (L.) Blum.]. Revista árvore 31(5):783–792

Sommer SG, Schjoerring JK, Denmead OT (2004) Ammonia emission from mineral fertilizers and fertilized crops. Adv Agron 82:558–622

Stampoulis D, Sinha SK, White JC (2009) Assay-dependent phytotoxicity of nanoparticles to plants. Environ Sci Technol 43:9473–9479

Sultana N, Ikeda T, Kashem MA (2001) Effect of foliar spray of nutrient solutions on photosynthesis, dry matter accumulation and yield in seawater-stressed rice. Environ Exp Bot 46:129–140

Suzuki N (2005) Alleviation by calcium of cadmium-induced root growth inhibition in Arabidopsis seedlings. Plant Biotechnol 22(1):19–25

Takahashi M, Nakanishi H, Kawasaki S, Nishizawa NK, Mori S (2001) Enhanced tolerance of rice to low iron availability in alkaline soils using barley nicotianamine aminotransferase genes. Nat Biotechnol 19:466

Tian ZD, Zhang Y, Liu J, Xie CH (2010) Novel potato C_2H_2-type zinc finger protein gene, *StZFP1*, which responds to biotic and abiotic stress, plays a role in salt tolerance. Plant Biol 12(5):689–697

Tripathi DK, Singh VP, Gangwar S, Prasad SM, Maurya JN, Chauhan DK (2014) Role of silicon in enrichment of plant nutrients and protection from biotic and abiotic stresses. In: Improvement of crops in the era of climatic changes. Springer, New York, pp 39–56

Tripathi DK, Tripathi A, Shweta SS, Singh Y, Vishwakarma K, Yadav G, Sharma S, Singh VK, Mishra RK, Upadhyay RG, Dubey NK, Lee Y, Chauhan DK (2017) Uptake, accumulation and toxicity of silver nanoparticle in autotrophic plants, and heterotrophic microbes: a concentric review. Front Microbiol 8.7

Tsay YF, Chiu CC, Tsai CB, Ho CH, Hsu PK (2007) Nitrate transporters and peptide transporters. FEBS Lett 581:2290–2300

van de Wiel CCM, van der Linden CG, Scholten OE (2016) Improving phosphorus use efficiency in agriculture: opportunities for breeding. Euphytica 207(1):1–22

van der Zande M, Peters RJB, Peijnenburg AA, Bouwmeester H (2011) Biodistribution and toxicity of silver nanoparticles in rats after subchronic oral administration. Toxicol Lett 205:S289–S289

Wan J, Zhang M, Adhikari B (2018) Advances in selenium-enriched foods: from the farm to the fork. Trend Food Sci Technol 76:1–5
Wang CQ, Song H (2009) Calcium protects *Trifolium repens* L. seedlings against cadmium stress. Plant Cell Rep 28(9):1341–1349
Wang YI, Wu WH (2010) Plant sensing and signaling in response to K+-deficiency. Mol Plant 3(2):280–287
Wang M, Zheng Q, Shen Q, Guo S (2013) The critical role of potassium in plant stress response. Int J Mol Sci 14(4):7370–7390
Wegner LH (2013) Root pressure and beyond: energetically uphill water transport into xylem vessels? J Exp Bot 65(2):381–393
Wei J, Li C, Li Y, Jiang G, Cheng G, Zheng Y (2013) Effects of external potassium (K) supply on drought tolerances of two contrasting winter wheat cultivars. PLoS One 8(7):e69737
Yao X, Chu J, Ba C (2010a) Responses of wheat roots to exogenous selenium supply under enhanced ultraviolet-B. Biol Trace Elem Res 137(2):244–252
Yao X, Chu J, Ba C (2010b) Antioxidant responses of wheat seedlings to exogenous selenium supply under enhanced ultraviolet-B. Biol Trace Elem Res 136(1):96–105
Yin LY, Cheng YW, Espinasse B, Colman BP, Auffan M et al (2011) More than the ions: the effects of silver nanoparticles on *Lolium multiflorum*. Environ Sci Technol 45:2360–2367
Yin L, Wang S, Tanaka K, Fujihara S, Itai A, Den X et al (2016) Silicon-mediated changes in polyamines participate in silicon-induced salt tolerance in *Sorghum bicolor* L. Plant Cell Environ 39:245–258
Yruela I (2005) Copper in plants. Braz J Plant Physiol 17(1):145–156
Zhu H, Shipp E, Sanchez RJ, Liba A, Stine JE, Hart PJ, Gralla EB, Nersissian AM, Valentine JS (2000) Cobalt (2) binding to tomato copper chaperone for superoxide dismutase: implications for the metal ion transfer mechanism. Biochemistry 39:5413–5542

Chapter 4
Role of Micronutrients in Biochemical Responses of Crops Under Abiotic Stresses

Shyam Narain Pandey

Abstract Essential micronutrients regulate and change the functioning of plants. The plants show tolerance in diverse ways under abiotic stress conditions, like drought, mineral deficiency, elevated salts concentrations in soil, etc., greatly contributed by the cellular biochemical reactions under suitable availability of micronutrients. Micronutrients such as zinc, copper, manganese, iron, molybdenum, boron and chloride support physiology and biochemical constituents in plants to neutralize the abiotic stresses. Under abiotic stresses, the micronutrients in plants, being constituents of biomolecules, function as an activator of many enzymes, electron carriers, etc. Also, their functions in metabolism regulations, reproduction, and protection against abiotic and biotic stress conditions are very significant. During diverse stress conditions, a large number of reactive oxygen species (ROS) are generated, and they damage the cellular metabolism of plants. The sufficient micronutrient status protects the plants against ROS by activating the enzymes and biomolecules to become a constituent or activator and directly favour the growth and cellular metabolism. This chapter describes the better interactions of plants and micronutrients for maintaining growth and metabolism and providing tolerance against abiotic stresses through changing the biochemical constituents in plants.

Keywords Abiotic stresses · Cellular metabolism · Micronutrients · Oxidative stress

1 Introduction

Today, global food demand is increasing due to the continuously growing population. Despite the development of agricultural technology for high production of food grain, abiotic stresses cause havoc on crop production. The major abiotic stresses in the world are drought, salinity, temperature (including global climatic change), mineral deficiency, heavy metals toxicity, etc. (Hussain and Shi 2014;

S. N. Pandey (✉)
Department of Botany, University of Lucknow, Lucknow, Uttar Pradesh, India

R. Roychowdhury et al. (eds.), *Sustainable Agriculture in the Era of Climate Change*, https://doi.org/10.1007/978-3-030-45669-6_4

Gautam et al. 2017). All plants require essential elements for their structure and functions. In recent decades, environmental stress conditions increased, and in this situation, the role of nutrients, particularly micronutrients, is very significant in the survival of plants (Pandey 2018). The micronutrients such as zinc, iron, copper, manganese, molybdenum, boron, chlorine, etc. along with other macronutrients play a direct role in the growth and metabolism of plants (Sharma 2006). Plants cannot survive in the absence of these micronutrients (Arnon and Stout 1939) even in normal growing conditions. The major abiotic stress conditions such as drought, salt and nutrients deficiency/toxicity are increasing day by day due to various anthropogenic activities. A large number of enzymes and organic molecules in cellular metabolism supported the defence system of plants against various stresses (Pandey et al. 2018). A large number of enzymes and biomolecules involved in plant defence system require micronutrient elements as constituents or activators (Pandey 2014). In food chains, plant nutrition effects both on plants growth and survival of other living organisms (Karthika et al. 2018). Micronutrients play their role in various biotic and abiotic stresses, because they are electron carriers, regulate cellular metabolism and play a role in osmotic regulation. Micronutrients maintain the structural integrity of the plasma membrane during abiotic stress conditions (Brown et al. 2002). Micronutrient such as iron is a cofactor of choline monooxygenase, a catalyst that helps in the biosynthesis of an osmoprotectant glycine betaine, which maintains the integrity of plasma membrane and stabilize the structure of proteins during salt and drought stress conditions (Gorham 1995).

Micronutrients play their protective role in plants during stress conditions such as drought, salt and minerals deficiency/toxicity, heavy metals, etc. Under such stress conditions, production of reactive oxygen species (ROS) creates oxidative stress damage to cellular structure and functions (Pandey 2018). Micronutrients such as iron, manganese, copper and zinc are participants in detoxification of ROS. Also, they play a role in signal transduction (Apel and Hirt 2004). In the antioxidant system of enzymes, many micronutrients are a cofactor for protecting cellular structure and functions from the excessive generation of ROS (Pandey and Gautam 2009). Under abiotic stresses, during electron transport system (in chloroplast and mitochondria), excessive ROS are produced. An important enzyme superoxide dismutase (SOD), which has micronutrient cofactors such as Cu-Zn SOD, Mn-SOD and Fe-SOD, supports the defence system against the ROS (Sharma 2006; Pandey 2014). These micronutrients containing SOD show high activity in mitochondria (Mn-SOD), chloroplast (Fe-SOD and Cu-Zn SOD) and cytosol (the activity of Cu-Zn SOD). The activity of SOD converts superoxide ions (O_2^-) into hydrogen peroxide (H_2O_2). The high concentrations of H_2O_2 are inducing lipid peroxidation and damage the cellular metabolic systems (Pandey 2018). So H_2O_2 concentration is prevented by its conversion into water (H_2O). This conversion of H_2O_2 to H_2O is catalysed by Fe-containing enzymes catalase (CAT) and ascorbate peroxidase (APX). Ascorbate peroxidase scavenges H_2O_2 in the chloroplast in combination with the ascorbate-glutathione cycle (Asada 1997). The enzyme catalase breaking down the H_2O_2 is present in peroxisomes. In peroxisomes, the conversion of glycolate to glyoxylate also contributes to accumulation of H_2O_2, thus becoming a cofactor of the above antioxidative enzymes (catalase, APX and SOD), the micronutrients involved in the cellular defence system against oxidative stress (Van Wiren et al. 1996). In cellular defence system, antioxidative enzymes decrease their activity due to limited availability of metal cofactors by various stresses in soil, which weakens the antioxidant defence system (Pandey 2014). Micronutrients are involved in various biosynthetic pathways (Pandey et al. 2018). Biosynthesis of secondary metabolites is formed with catalytic actions of enzymes. These enzymes contain micronutrient as cofactors. Also, they play a role in lignifications of the cell wall, genes regulation, hormonal balance, ionic homeostasis and cell signalling (Stiles 2013). A large number of enzymes present in the various compartment of the cell contain various micronutrients for performing their normal functions (Tej and Zeiger 2002). Even normal growth plants require a sufficient level of micronutrients. But availability of nutrients is greatly affected by soil physical, chemical and biological properties which determine the dynamics of the equilibrium of micronutrients in soil and its labile pool, from which micronutrients are absorbed by plant roots. From the labile pool, the acquisition of micronutrients is affected by the physical factors of the environment such as light intensity, pH, temperature, moisture, etc. and soil cultural practices. In addition, the genetic composition and stress disposition of plants influence the acquisition of micronutrients (Brady and Weil 1999; Karthika et al. 2018).

Salt stress conditions (salinity and sodicity) are a major threat to crops responsible to yield less by 4–6.5% annually across the world. In salt-affected soils, high concentration of sodium and chloride ions impairs the availability of essential nutrients to plant roots for absorption (Chakraborty et al. 2016), and these results in a reduction of crop productivity (Pandey 2014). Similarly, the stress in soil due to excessive heavy metals activity affects plant growth and contaminates our delicate food chain (Naaz and Pandey 2010; Roychowdhury and Tah 2011; Basu et al. 2012; Roychowdhury et al. 2018, 2019). Soil salinity causes nutrient loss, poor soil characteristics and essential nutrients imbalance. High concentration of salts in saline soils affects osmotic potential. Mostly, micronutrients such as Zn, Fe, Cu and Mn are less available for plants under high soil pH (alkaline) range, low soil moisture, high temperature and low organic matter content conditions (Brady and Weil 1999; Pandey et al. 2018). The status of some micronutrients influences plants-water relations. The role of chlorine in osmoregulation involves cell osmotic relations and turgor-dependent extension growth of cells (Sharma 2006).

2 Zinc (Zn)

Zinc is an essential micronutrient for the structural and functional role of all living organisms (Pandey et al. 2002). Mostly, plants absorb zinc from the soil as Zn^{++} ion. The uptake of zinc from the soil and translocation to plant shoot portion are concentration-dependent, and some workers have reported carrier-mediated transport of Zn^{2+} (Norvell and Welch 1993). In adverse soil conditions such as calcareousness, low temperature, salinity and alkalinity, zinc is less available in the soil for plants root absorption (Gautam et al. 2017). The zinc-deficient soil conditions induce the plants for the secretion of phytosiderophores which mobilize zinc (Treeby et al. 1989) and help their uptake in plants. Plant genotypes are variable in the secretion of phytosiderophores and efficiency of zinc uptake (Van Wiren et al. 1996). Several identified transporters such as *ZIP* (Guerinot 2000) and CDF (cation diffusion facilitator) family (Williams et al. 2000) facilitated zinc transport in plants. Zinc transporter genes (ZIP 1, 2, 3 and 4) have been isolated from *Arabidopsis,* and these genes are expressed in plant roots due to zinc deficiency (Guerinot 2000). A large number of proteins contain zinc as their structural component for performing their catalytic functions of more than 250 enzymes. These metal chelates with polypeptides are synthesized during heavy metals stress conditions in plants (Gautam et al. 2017). Zinc provides stability of various proteins involved in regulatory functions, and such proteins are zinc fingers, RING finger domains and zinc clusters. Under excessive heavy metals concentrations in plant tissues, metallo-polypeptides help tolerance in plants against heavy metals in tissues (Pandey et al. 2002).

2.1 *Enzymes*

Zinc does not function during oxidation-reduction reactions due to its single oxidation state (Zn^{2+}). Some important enzymes playing a vital role in cellular metabolism contain zinc as cofactor or activator, and these are DNA-dependent RNA polymerases, carbonic anhydrases, superoxide dismutase, carboxypeptidase A, etc. Zinc is used as a substrate by the enzyme H^+ ATPase and functions as Zn-ATPase. Most enzymes contain zinc as constituents catalyse hydrolysis reactions (Sharma 2006). The carbonic anhydrase catalyses the reversible conversion of carbon dioxide to bicarbonate (HCO_3^-) and is therefore critical to photosynthesis in C_4 plants growing in drought and high-temperature conditions. In C_4 plants in carboxylation reaction, bicarbonate is used as substrate and enzyme is PEPCase. Carbonic anhydrase enzyme is localized in mesophyll chloroplast in C_3 plants, while cytosol of mesophyll cell in C_4 plants. The cytosol is also the site for phosphoenolpyruvate carboxylase (PEPCase) using bicarbonate as substrate (Hatch and Burnell 1990). The active sites of SOD contain copper and zinc. SOD (Cu-Zn SOD) protects plant cells against ROS. Zinc is also a constituent of some DNA-dependent RNA polymerases involved in transcription (Petranyl et al. 1978). A large number of enzymes required zinc as a cofactor in animals, bacteria and fungi. The availability of zinc is influenced under various stress conditions or zinc deficiency, and the activity of all zinc-related enzymes is adversely affected. The decrease in the activity of carbonic anhydrase in leaves is a strong indicator of zinc deficiency (Rengel 1995). Zinc-deficient plants also show decreased activity of SOD (Pandey et al. 2002). The low activity of catalase and ascorbate peroxidase in plants due to zinc deficiency has been reported (Yu et al. 1998; Pandey et al. 2002). The role of glutathione reductase (GR) is increased in response to environmental stresses and xenobiotics which provide tolerance to the plants against these stresses. The decrease in GR activity in response to zinc deficiency in bean plant has been reported by Cakmak and Marschner (1993). Under stress conditions, zinc protects plants by an inhibitory effect on plasma membrane-bound NADPH oxidase (catalyses the production of superoxide ions).

2.2 Physiological Role

Plants show zinc deficiency grown under soil salinity, sodic or calcareousness stresses performed poor photosynthetic functions (Karthika et al. 2018). Low zinc in cells changes chloroplast structure, electron transport and carbon dioxide fixation (Sharma et al. 1994). The production of ROS increases due to zinc deficiency in plant tissues, and ROS damage in the thylakoid membranes consequently decreases the photosynthetic active area in leaf (Cakmak and Engels 1999; Henriques 2001). Carbonic anhydrase catalyses the production of the substrate (HCO_3^-) in C_4 plants and maintains CO_2 concentration in C_3 plants at the site of carboxylation by the conversion of HCO_3^- to CO_2 under the role of RuBP carboxylase (Hatch and Burnell 1990). The activity of carbonic anhydrase is inhibited in plants under the influence of zinc deficiency (Badger and Price 1994). In zinc-deficient plants, decrease in catalase and peroxidase activity, photosynthetic rate and accumulation of essential oil in leaves have been studied (Pandey 2018; Srivastava et al. 1997). Marschner et al. (1996) observed increased accumulation of carbohydrates in leaves of zinc-deficient plants. The increased accumulation of starch in zinc-deficient condition correlates the role of zinc in sucrose biosynthesis (Shrotri et al. 1980) and starch metabolism. Zinc is also involved in auxin metabolism by the help of tryptophan-dependent auxin biosynthesis. Tryptophan is a significant precursor of auxin which is a growth-promoting plant hormone. Zinc deficiency reduces the protein content, while increases the non-protein nitrogen in plants (Bisht et al. 2002) indicates the role of zinc in protein synthesis. The metabolism of nucleic acids and ribosome and zinc as a constituent of a transcription factor may also affect zinc deficiency which leads to decrease in protein synthesis. Low zinc status in plants suppressed the various stages of reproductive development (Pandey and Gautam 2009) such as flowering, male and female gametogenesis, seed development, development of anther, etc. (Sharma 2006; Pandey 2014). Zinc is a critical nutrient element for microsporogenesis and pollen fertility. Pollen tube accumulated zinc at their part that participated in fertilization. Zinc finger protein (TFIIIA type) controls the development of the flower. The zinc finger proteins (anther-specific) play a role in microsporogenesis (Kobayashi et al. 1998). Zinc-deficient plants accumulate high proline content which is indicative of water stress (Sharma and Sharma 1987; Pandey 2018).

2.3 *Oxidative Stress*

Plants produce high reactive oxygen species when influenced with zinc deficiency. Also, the power of defence system decreases which neutralizes the toxic effects of ROS (Cakmak 2000; Pandey and Gautam 2009; Singh and Pandey 2011). The activity of copper-Zn SOD, ascorbate peroxidase and catalase antioxidant enzymes is reduced in plants due to zinc deficiency (Yu and Rengel 1999; Naaz and Pandey 2009; Gautam et al. 2017). High production of free oxygen radicals oxidizes lipid in membrane causing leakage of cell sap due to structural changes in the plasma membrane. The toxic effect of ROS also leads damage to the photosynthetic apparatus (Cakmak and Engels 1999; Henriques 2001).

3 Copper (Cu)

Copper exists in Cu^{+} (cuprous) and Cu^{++} (cupric) oxidation states, particularly in oxidation-reduction reactions. In cellular metabolism, many copper-containing proteins act as enzymes and electron carriers. It is an important constituent of cytochrome C oxidase. In the mitochondrial electron transport system, copper as constituents of cytochrome C oxidase is the terminal component. The copper protein (plastocyanin) is an electron carrier in the light reaction phase of photosynthesis. Copper is a component of many proteins detoxifying the effect of reactive oxygen species highly produced during abiotic stress conditions. Copper is taken by plants roots through active transport proteins across the plasma membrane (Fox and Guerinot 1998). Copper transporter proteins COPT1, COPT2, COPT3, COPT4 and COPT5 have been identified from *Arabidopsis* (Kampfenkel et al. 1995; Sancenon et al. 2003). These copper transporters have been detected in leaf, stem, flowers and roots, but their expression was different in different parts of the plant. The cupric ions long-distance transports are facilitated by Cu^{++}-amino acid complex formation, and the amino acids involved in the transport process are asparagine, histidine and glutamic acid (Loneragan 1981). Under copper deficiency, transport from leaves to grains is poor (Loneragan 1981), particularly at the time of seed-filling stage.

3.1 *Physiological Functions*

Copper causes reduction of one atom of O_2 to water and another atom of O_2 to hydroxylation of substrates such as phenolase complex (Sharma 2006). It is a cofactor of a large number of enzymes in plants such as Cu-Zn SOD and cytochrome C oxidase. The metal receptor protein (metallochaperones) delivers copper metal Cu-proteins (Lippard 1999). Some important copper-containing enzymes are ascorbate oxidizer, SOD, cytochrome C oxidase, diamine oxidase, etc. Catechol oxidase (copper proteins) or polyphenol oxidases are important for secondary metabolism in plants. They play a role in the biosynthesis of lignin and alkaloids. Also, in the formation of phytoalexins, they have antifungal activity. Therefore, phenolases involved in metabolism provide strength to the cell wall for the protection of pathogens (Pandey et al. 2018). Laccase or p-diphenols oxidase contains four copper ions (both Cu^{+} and Cu^{++}) acting as a catalyst in the oxidation of p-diphenols to p-quinones. Laccase is involved in the synthesis of plastoquinones, useful in PS-II (Ayala et al. 1992). Tyrosinase enzyme was identified as mushrooms involved in the conversion of tyrosine to dopa-quinone. The Cu-Zn SOD enzyme involved in the conversion of ROS to H_2O_2 and oxygen, therefore, provides protection of cells from ROS. The enzymes SOD and ascorbate peroxidase are attached to the chloroplastic thylakoids in the vicinity of the PS-I. Ascorbate peroxidase is involved in the reduction of H_2O_2 to H_2O and protects cellular lipid peroxidation (Pandey 2018). Copper is a constituent of single-electron carriers protein plastocyanin found in chloroplasts. Plastocyanin is an electron carrier linking two pigment systems (PS-I to PS-II). Copper element that binds to polypeptides forms copper-chelating biomolecules (Maksymiec 1997). Copper deficiency causes a reduction in photosynthesis (Ayala et al. 1992; Pandey 2018). In some blue-green algae, plastocyanin synthesis is with the availability of copper. Under salinity and alkali soil conditions, low availability of Cu inhibits the Cu-Zn SOD activity and plastocyanin-mediated transport of electrons within the chloroplasts (Shikanai et al. 2003). Low copper in plants causes ultrastructural changes in chloroplast thylakoids in grana and decrease in chlorophyll content (Casimiro et al. 1990).

3.2 *Stress Neutralization*

Copper involves in the disproportion of superoxide ions (O_2^-) produced due to abiotic stresses because it is constituent of the enzyme Cu-Zn SOD for the conversion of H_2O_2 and molecular oxygen. Therefore, low copper weakens the oxidative defence system in plants. Copper plays a protective role in plants against pathogens through their involvement in the synthesis of lignin (Graham 1983). Copper-deficient plants are more vulnerable to pathogenic infections. During stress conditions, copper induces the formation of ROS and activates copper diamine oxidases that trigger defence mechanisms in the plant (Sharma 2006).

4 Iron (Fe)

Iron exists in ferric (Fe^{+++}) and ferrous (Fe^{++}) oxidation states and participates in oxidation-reduction reactions. It participates in the electron transport system, and it is a cofactor of several enzymes. They play important roles in fatty acids metabolism and biosynthesis of terpenoids, signalling molecules and growth hormones.

The role of iron is known for induction to produce and detoxification of reactive oxygen species. The biochemical reactions related to iron are very complex. Iron also plays a very important role in signal transduction and biosynthesis of Jasmonic acid (Sharma 2006). Under Fe deficiency conditions, mostly plants uptake reduction-based iron (Bienfait et al. 1983). The ferric ion chelate reduced into ferrous ion chelate by the catalytic activity of an enzyme known as Fe (III) chelate reductase. This reductase enzyme is induced in the plasma membranes of root epidermal cells under Fe deficiency (Sharma 2006). Yi and Guerinot (1996) described that the *Arabidopsis* mutants ferric reductase defective 1, 2 and 3 suppress the activity of Fe^{+++} chelate in *Arabidopsis* as reported by Robinson et al. (1999) and named it ferric reductase oxidase 2 (FRO2). The various genes involved in uptake in roots, as well as their distribution in shoot portions, are characterized (Robinson et al. 1999; Waters et al. 2002). Extracellular pH influences the solubility of Fe^{++} in the root zone area, which is observed maximum at pH 5.0. High pH reduces the Fe^{+++} reduction (Manthey et al. 1996; Zheng et al. 2003). Plant hormone ethylene helps in the uptake of iron in roots (Zaid et al. 2003) by changing the root morphology of iron-deficient plants (Schmidt et al. 2000). Several Fe transports proteins in plasma membranes, and several transporter genes of different gene families have been identified and characterized such as iron-regulated transporter (IRTI) belonging to ZIP family identified in *Arabidopsis* (Eide et al. 1996). Nramp family transporters are also involved in the uptake of iron (Thomine et al. 2000). The strategy of the plant to uptake the iron with the reduction of Fe^{3+} to Fe^{2+} by the catalytic activity of Fe chelate reductase is established only under iron deficiency condition. The second strategy is established for iron uptake by the formation of Fe complex of phytosiderophores. The mugineic acid forms complex with Fe^{3+} that is prerequisite for its recognition by the specific Fe^{+++} transport system in roots (Ma and Nomoto 1996; Ma et al. 1993). Mugineic acid synthesizes from methionine chelate $Fe^{3}+$ with their carboxyl groups and amino groups. The graminaceous plant species secrete mugineic acid (phytosiderophores, when soil is deficient in iron), but the secretion of specific mugineic acids (non-proteinogenic amino acids) is different (Sharma 2006). The mugineic acids (MA) are secreted by rice, corn and wheat. Soybean genotypes are 2-DMA, 3-HMA (3-hydroxyl mutagenic acid) and 3-epihydroxy mugineic acid (epi-HMA). These MAs provide tolerance to plants against iron chlorosis. The transport of iron to the shoot portion takes place through xylem as ferric citrate organic complex (Cataldo et al. 1996).

4.1 Physiological Roles

A large number of iron (Fe)-containing enzymes and carrier proteins are present in electron transport systems. These enzymes are essential for the biosynthesis of secondary metabolites, plant hormones and some other biomolecules significant as signalling molecules and developmental activities. Some enzymes are di-iron enzymes in which two iron atoms forming an Oxbridge are bound to two histidine and four carboxylate residues such as mitochondrial alternative oxidase (Sharma 2006). In non-haem iron enzymes, Fe cofactor is present as Fe-sulphur cluster which enables iron-sulphur proteins to be as donors or acceptors of electrons. The iron clusters are Fe-S, 2Fe-2S (two Fe ions are coordinated to two inorganic sulphide ions) and 4Fe-4S (contains four Fe ions and four sulphide ions). The enzymes containing Fe as activator are catalase, cytochrome C oxidase and peroxidases, and they are iron bound to the apoprotein as iron porphyrin. There, Fe ion is complexed to the four nitrogen atoms of the four pyrrole groups. Some non-haem iron enzymes are superoxide dismutase (SOD), lipoxygenase, alternative oxidase and 2-oxoglutarate-dependent enzymes. Some important non-haem iron enzymes with Fe-S cluster cofactor are nitrite reductase aconitase, sulphite reductase, formate dehydrogenase, succinate dehydrogenase, NADH-Q oxidoreductase and FDX-thioredoxin reductase. The various groups of enzymes containing Fe involved in detoxification of reactive oxygen species are SOD, catalase, peroxidase, alternative oxidase, etc. A large number of Fe proteins are electron carriers, a significant part of mitochondrial and chloroplastic electron transport system including ferredoxins and cytochromes. Leghaemoglobin in leguminous plants has a high affinity for O_2 molecule. The phytoferritin serves as a store of cell's excess iron in plants (Sharma 2006). Haem protein cytochromes are localized in chloroplasts, in mitochondrial inner membrane and in the endoplasmic reticulum, and cytochromes are involved in reversible reactions of two iron oxidation states ($Fe^{+++} = Fe^{++}$). Ferredoxins (2Fe-2S)-type iron-sulphur clusters function in electron transport in between iron oxidation states, mainly localized in chloroplasts and root plastids. Chloroplastic ferredoxin is reduced by taking electrons from P700, and it is an electron donor for several reductive reactions such as glutamine to glutamate and reduction of nitrite and sulphite. The phytoferritins in plants constitute a major sink of iron accommodating about 4500 iron atoms (Smith 1984). Iron plays a significant role in photophosphorylation and, as a part of ferredoxin and ferredoxin-thioredoxin reductase, functions in the photoreduction of thioredoxin. In the case of some plants, the ferredoxin thioredoxin regulates the activity of Rubisco (Zhang and Portis 1999) and also functions in the Calvin cycle. The iron proteins NADH-cytochrome b_5 reductase, cytochrome b_5 and a desaturase are membrane-bound proteins involved in fatty acid metabolism (Sharma 2006).

4.2 *Free Radical Production*

The reduction of O_2 and H_2O_2 generates superoxide ions which are dissimulated by SOD and make two ROS causing lipid peroxidation damaging cellular components (Pandey 2018). Superoxide ions react to H_2O_2 to produce hydroxyl radicals

$$Fe^{2+} + H_2O_2 \rightarrow Fe^{3+} + OH^{\circ} + OH^{-}$$

These ROS damage proteins, lipid and DNA and induce mutations. Ferrous ions also react with oxygen molecules to generate $Fe^{2+}O$ (ferryl) and $Fe^{2+}O_2$ (per ferryl), which are more toxic compounds. The activity of lipoxygenase (iron enzyme) in the hydroxylation of linoleic acid catalyses the production of singlet oxygen species ($^1O_2$).

4.3 Protective Role

In plants, iron also participates in the defence system against salt stress, oxidative stress and various pathogens attack. Although iron ions promote the production of ROS, iron-containing enzymes are participating in detoxification of ROS (Roychowdhury et al. 2018, 2019). The iron-SOD detoxify the superoxide ion (O^{2-}) conversion into H_2O_2. Other enzymes catalase and ascorbate peroxidase (haem enzyme) detoxify the H_2O_2 by converting them into H_2O. In the iron-deficient plant, decrease in the activity of ascorbate peroxidase and consequently increase in H_2O_2 concentration in sunflower have been reported (Ranieri et al. 2001). Quinol reacting with oxygen generates ROS, and iron enzyme alternative oxidase reduces quinol by providing an alternate pathway in the mitochondrial electron transport system, thus preventing the production of ROS. Iron plays a role as an osmoprotectant through biosynthesis of glycine betaine against high temperature and high salinity (Gorham 1995). Also, iron is involved in the biosynthesis of lignin, which provides mechanical strength to cell wall against pathogens (Caruso et al. 2001).

5 Manganese (Mn)

Manganese (Mn) exists in various oxidation states such as Mn^{2+}, Mn^{3+}, Mn^{4+} and Mn^{5+}, out of which the most prominent is Mn^{2+}. It forms a rapid bond with oxygen-containing species. Mn behaves as an activator of enzymes, plays a catalytic role in cellular metabolism and also functions in CO_2 fixation in CAM and C_4 plants. It also plays a significant role in scavenging oxygen free radicals. Mn^{2+}ions are absorbed by plant roots through facilitated diffusion. The acidification outside the plasma membrane by H^+ efflux increases uptake of Mn (Yan et al. 1992). Several genes have been identified and characterized, and they are encoding Mn transporter proteins (Hall and Williams 2003). The P-type ATPase is involved in Mn transport. The transporter ECA1 in *Arabidopsis* functions as an endoplasmic reticulum-bound Ca^{2+}/Mn pump, which also provides tolerance to the plants to excess manganese (Wu et al. 2002). Mn anti-portal Sh MTP1 confers tolerance to manganese in plants through internal sequestration (Delhaize et al. 2003). Nramp family transporter participates in manganese transport, and Nramp 3 plays a role in long-distance transport of manganese (Thomine et al. 2003). The translocation of manganese to developing grains is slowly observed in wheat.

5.1 Physiological Role

Over 30 enzymes contain Mn as constituents. Some important enzymes with Mn are Mn-SOD (manganese superoxide dismutase), PEPCK (phosphoenolpyruvate carboxyl kinase), NAD + malic enzyme, IDH (isocitrate dehydrogenase), NADP + malate enzyme, PEPCase (phosphoenolpyruvate carboxylase), glutamine synthetase, enolases, etc. PEPCK catalyzes the decarboxylation of oxaloacetate to PEP in the chloroplast of bundle sheath in C_4 plant. The released carbon dioxide is the starting point of the Calvin cycle.

$$\text{Oxaloacetate} + \text{ATP} \rightarrow \text{Phosphoenolpyruvate} + \text{ADP} + \text{CO}_2$$

The significant role of Mn is the oxidation of water in PS II (Ono and Onone 1991). Many enzymes in C_4 plants containing Mn^{2+} as cofactor help in photosynthesis in C_4 plants. Manganese is also involved in the decarboxylation of C_4 acids to generate carbon dioxide in bundle sheath cells. Such produced CO_2 is again fixed in a C_3 compound by Rubisco (Sharma 2006). Manganese-activated enzymes catalyse the biosynthesis of secondary metabolites in plants. It is also involved in biosynthesis pathway systems of pigments as an activator of many enzymes. The manganese plays a significant role in the biosynthesis of aromatic amino acids (tyrosine, tryptophan and phenylalanine). These amino acids are precursors of flavonoids, indole and lignin (Burnell 1988). Manganese is the activator of arginase and plays a role in the synthesis of polyamines. Polyamines are important for plant growth and development. Deficiency of Mn decreases photosynthesis in plants because Mn is an important constituent of PS II (Yachandra et al. 1993), and leaf ultrastructure changes are induced by Mn deficiency (Polle et al. 1992). Plants show an increase in soluble carbohydrate and a decrease in fatty oil content under manganese deficiency (Campbell and Nable 1988). Increase in soluble nitrogenous compounds due to Mn deficiency has been reported in plants because Mn is an activator of arginase and allantoate amidohydrolase (Winkler et al. 1985). Also, Mn deficiency is related to the accumulation of arginine and uriedes and reduction in nucleic acids (Chatterjee et al. 1994).

5.2 Protective Role

Manganese-SOD (superoxide dismutase) localized in mitochondria is an important enzyme of the antioxidant system; thus, Mn is involved in the defence system against reactive oxygen species. It prevents the accumulation of superoxide ions by dismutation of them and converts it to hydrogen peroxide (Sharma 2006). Low Mn in plants shows oxidative stress. Overexpression of Mn-SOD in transgenic tobacco increases tolerance in plant against oxidative stress as reported by Slooten et al. (1995). The Mn-SOD also contributes tolerance to plants against drought stress.

Increased Mn-SOD expression induced under drought stress has been observed in wheat seedlings as reported by Wu et al. (1999). The concentrations of tissue manganese are related to disease in plants, and it supports the resistance in plants to pathogenic diseases (Graham 1983). Manganese contributes protection and strength in plants to fungal infection through its involvement in the biosynthesis of phenolics and lignin (Morab et al. 2003).

6 Molybdenum (Mo)

Molybdenum (Mo) is a second transition series metal that exists in several oxidation states, and the most stable is the Mo VI (hexavalent) form. Molybdenum has an affinity for sulphur-containing groups. It participates in oxidation-reduction reactions due to its variable and easy convertibility oxidation states. Molybdenum is a cofactor of several enzymes. Mo plays a role in dinitrogen fixation and assimilation of nitrate in plants, and it is also essential for free-living and symbiotic bacterial nitrogen fixation processes. Plants take Mo as MoO_4^{2-} (molybdate ion). Molybdenum follows some transporter path as transport of phosphate. A higher concentration of sulphate decreases the uptake of molybdenum (Marschner 1995). For redox reactions in plants and microbes, Mo is a cofactor of several enzymes (Sharma 2006). Xanthine oxidase and nitrate reductase are Mo-containing enzymes. These enzymes involve in catalysing the reactions for nitrogen and nitrogenous compounds. Aldehyde oxidase contains iron and MoCo as prosthetic groups catalyse the terminal reaction of the biosynthetic pathway of abscisic acid and auxins (Taylor 1991). Molybdenum-containing enzymes also contain additional cofactors such as Fe-S, iron, haem, etc. (Koshiba et al. 1996).

6.1 Physiological Roles

The molybdoenzyme nitrate reductase is very significant to catalyse the conversion of nitrate to nitrite, and the enzyme plays a role in the synthesis of organic nitrogenous compounds. The bacterial and eukaryotic nitrate reductase enzymes belong to two different families of molybdoenzymes (Sharma 2006). The enzyme nitrate reductase distribution differs in different plant species but is localized in the cytosol in all plant species. The molybdoprotein xanthine dehydrogenase has two similar and catalytically independent subunits. Each subunit contains a molybdoprotein, two Fe-S clusters and one flavin cofactor which catalyses oxidative degradation of xanthine to uric acid with the use of NAD^+ as an electron acceptor (Sharma 2006). Sulphite oxidase works in plants and human beings, and it is an essential liver enzyme. Sulphite oxidase is a molybdoprotein but lacks the redox-active centres (Mendel and Hansch 2002). Aldehyde oxidase contains FAD (flavin), iron and MoCo as prosthetic groups, it catalyses the oxidation of the abscisic aldehyde to abscisic acid (Seo and Koshiba 2002) and indole-3-acetic acetaldehyde to indole-3-acetic acid, and it regulates growth in plants. Dinitrogenase (Mo-Fe protein) and dinitrogenase reductase (Fe protein) are nitrogenase (prokaryotic enzyme) that catalyses the nitrogen fixation process. Several other non-molybdoenzymes such as catalase (Agarwala et al. 1986), cytochrome oxidase (Chatterjee et al. 1985) and succinate dehydrogenase (Agarwala et al. 1986) are influenced by Mo status in plants. Deficiency of Mo induces changes in several organic compounds (Sharma 2006).

7 Boron (B)

Boron (B) is metalloid that exists in three-valence status with high affinity to oxygen. Boron-diol complexes are formed between as-diol furanoid groups of sugars (apiose and fucose) and boric acid. Therefore, boron plays an important role in the structural and functional process for plant cell walls and membranes (Roychowdhury et al. 2019). Boron bridging the hydroxyl groups gives a large number of boron-containing compounds in plants (Dembitsky et al. 2002). The important role of boron is cross-linking of cell wall polysaccharides which has been established (O' Neill et al. 2004). A most significant form of boron in soils is boric acid (B $[OH]_3$) that exists mostly at pH 7.5 in the cytoplasm (Woods 1996). The uptake of boron from the soil is a passive process with formation of boron complexes which create concentration gradient. A investigation has also been reported for the uptake of boron is active or passive or both (Pfeffer et al. 1998). The transport of boron for long distances is facilitated by transpiration stream through the xylem, whereas retranslocation of boron to sites that do not lose water such as fruits, inflorescence, etc. involves phloem (Brown et al. 2002).

7.1 *Physiological Role*

Boron plays an important structural role in the cell wall as it binds with dial groups of polysaccharides. It is also highly accumulated in the cell wall as an integral com ponent of polysaccharide complexes (Moth et al. 1993). Boron influences the uptake of ions from roots. Its effect on membrane potential (Em) is demonstrated by Schon et al. (1990). Boron-deficient plants show high efflux of K^+, sucrose and amino acids (Cakmak et al. 1995) due to structural changes in the cell wall. Short-term boron deficiency increases the accumulation of actin and tubulin involved in alteration in the polymerization pattern of the cytoskeletal assemblies as investigated in maize root apices by Yu et al. (2002). Boron increases photosynthetic oxygen evolution and photosynthetic efficiency in leaves. The supply of boron in plants affects the metabolism of carbohydrate, affecting the activity of related enzymes. The pentose phosphate pathway (alternative to glycolysis) is influenced by boron concentration in the plant (Sharma 2006). The activity of glucose-6-phosphate dehydrogenase and 6-phosphogluconate dehydrogenase increases in response to boron deficiency in plants (Lee and Aronoff 1967). In boron-deficient plants, the activity of acid phosphatase increases (Agarwala et al. 1991).

8 Chlorine (Cl)

Chlorine is a halogen element (group VII) with a single oxidation state as Cl^- ions. These ions are mostly bound to organic molecules or exchange sites. Among chlorine-containing organic compounds, 4-chloroindole acetic acid shows high auxin activity (Flowers 1988). The absorption of the free anion of chloride is by plant roots and also by shoot portion of plants in gaseous form or as chloride ions (White and Broadley 2001). Chloride uptake through plasma membrane occurs either by active transport (by H^+ ATPases) or by facilitated diffusion. The active transport with ATP hydrolysis involves a Cl^-/nH^+ symporter (Sharma 2006).

8.1 Physiological Roles

In manganese-linked oxygen evolution complex (of photosystem II), chlorine is a structural constituent. Chlorine deficiency in plants decreases photosynthesis and evolution of O_2 by chloroplasts of higher plants (Kelley and Izawa 1978). Chlorine deficiency inhibits extension growth of cells and cell division, and consequently, leaf area for photosynthesis is reduced (Terry 1977). Stomatal functioning indirectly affects photosynthesis process because proton pumping across vacuolar membrane induces an influx of chloride simultaneous with K^+, followed by water movement in guard cells, increase turgor and the opening of stomata, and affects the exchange of carbon dioxide (Schnabl 1980). Chlorine does not function as a catalyst, directly. It stimulates asparagine synthetase, which is important to catalyse the glutamate-dependent synthesis of asparagine. Asparagine synthetase also catalyses the amination of aspartate to asparagine. The synthesis of asparagine resulted in the storage of nitrogen and its transportation from source to sink (Rognes 1980).

8.2 Osmoregulation

Chlorine functions as osmoregulator in plants and therefore affects plant-water relations. The osmoregulatory functions such as cell division in shoot and root apices, stomatal functioning and development of stigma require chlorine. It causes turgor-induced extension cell growth in the apical portion of root and shoot. The rapid growth of stigma induces extension growth of cells which increases cell turgidity followed by quick mobilization of Cl^- and K^+ ions from the neighbouring cells as reported by Heslop-Harrison and Roger (1986). Such osmoregulatory functions of chlorine are very significant in reproductive biology. A carbon dioxide-induced chlorine efflux from guard cells into the apoplastic fluid is linked to the activation of the anion of protein channels in the plasma membrane of guard cells (Hanstein and Felle 2002) observed in faba bean. The concentrations of chlorine for osmoregulation process are variable from plant to plant, and osmotic responses are also variable under abiotic stress conditions (Sharma 2006; Roychowdhury et al. 2013; Ganie et al. 2014; Roychowdhury 2014; Hasanuzzaman et al. 2015; Pandey 2018).

9 Conclusions

The role of micronutrients in plant's biochemistry is very significant, particularly during biotic and abiotic stress conditions, because they play a significant role in the regulation of enzymatic reactions and biosynthesis of organic molecules. During abiotic stress conditions, the strength of the defence system, new synthesis and induction of biomolecules, induction of signalling, etc. are facilitated by appropriate concentration of essential micronutrients in plant tissues. The critical normal range of micronutrients not only promotes the growth but also protects cellular metabolism against abiotic stresses such as drought, mineral deficiency, salinity, adverse temperature, heavy metals and chemical toxicity.

Acknowledgements I acknowledge late Prof. C. P. Sharma (former head of Department of Botany and Dean Faculty of Science, University of Lucknow, India) for various discussions during the research and study on the physiological role of micronutrients in plants. I also acknowledge Ms. Isha Verma and Mithlesh Kumar for providing help in the preparation of this chapter.

References

Agarwala SC, Nautiyal BD, Chaterjee C (1986) Manganese, copper and molybdenum nutrition of papaya. J Hortic Sci 61:397–405

Agarwala SC, Abidi A, Sharma CP (1991) Variable boron supply and sugarbeet metabolism. Proc Natl Acad Sci (India) 61:109–114

Arnon DI, Stout PR (1939) The essentiality of certain elements in minute quantity for plants with special reference to copper. Plant Physiol 14:371–375

Asada K (1997) The role of ascorbate peroxidase and monodehydroascorbate reductase in H_2O_2 scavenging in plants. In: Scandalios JG (ed) Oxidation states and the molecular biology of Antioxidative defenses. Cold Spring Harbour Laboratory Press, Cold Spring Harbour, pp 715–735

Apel K, Hirt H (2004) REACTIVE OXYGEN SPECIES: Metabolism, Oxidative Stress, and Signal Transduction. Annual Review of Plant Biology 55:373–399

Ayala MB, Lopez GJ, Lachica M, Sandmann G (1992) Changes in carotenoids and fatty acids in photosystem II of Cu-deficient pea plants. Physiol Plant 84:1–5

Bisht SS, Nautiyal BD, Sharma CP (2002) Zinc nutrition dependent changes in tomato (*Lycopersicon esculentum* Mill) metabolism. J Plant Biol 29:159–163

Badger M, Price DG (1994) The role of carbonic anhydrase in photosynthesis. Annu Rev Plant Physiol Plant Mol Biol 45:369–392

Basu A, Roychowdhury R, Bhattacharyya SS, Tah J (2012) Estimation of major heavy metals (Fe, Cu and Zn) in the fruit part of *Cucumis sativus* L. World J Sci Technol 2(7):01–03

Bienfait HF, Bino RJ, Van Der Blick AM, Duivenvoordan JF, Fontain IM (1983) Characterization of ferric reducing activity in roots of Fe-deficient *Phaseolus vulgaris*. Physiol Plant 59:196–202

Brady NC, Weil RR (1999) The nature and properties of soils, 12th edn. Prentice Hall Inc. International (UK), London

Brown PH, Bellalovi N, Wimmer MA, Bassil ES, Ruiz J, Hu H, Pfeffer H, Dannel F, Romheld V (2002) Boron in plant biology. Plant Biol 4:205–223

Burnell JN (1988) The biochemistry of manganese in plants. In: Graham RD, Hannam RJ, Uren NC (eds) Manganese in soils and plants. Klumer Academic, Dordrecht, pp 125–137

Chatterjee C, Nautiyal N, Agarwala SC (1985) Metabolic changes in mustard plants associated with molybdenum deficiency. New Phytol 100:511–518

Chatterjee C, Nautiyal N, Agarwala SC (1994) Influence of changes in manganese and magnesium supply on some aspects of wheat physiology. Soil Sci and Plant Nutr 40:191–197

Cakmak I (2000) Possible roles of zinc in protecting plant cells from damage by reactive oxygen species. New Phytol 146:185–205

Cakmak I, Engels C (1999) Role of mineral nutrients in photosynthesis and yield formation. In: Rengel Z (ed) Mineral nutrition of crops: fundamental mechanisms and implications. Haworth Press, New York, pp 141–168

Cakmak I, Marschner H (1993) Effect of zinc nutritional status on activities of Superoxide radical and hydrogen peroxide scavenging enzymes in bean leaves. Plant Soil 155:127–130

Chakraborty K, Bose J, Shabala L, Eyles A, Shabala S (2016) Evaluating relative contribution of osmotolerance and tissue tolerance mechanisms toward salinity stress tolerance in three species. Physiologia Plantarum 158:135–151

Campbell LC, Nable RO (1988) Physiological functions of manganese in plants. In: Graham RD, Hannam RJ, Uren NC (eds) Manganese in soils and plants. Klumer Academic Publishers, Dordrecht, pp 139–154

Casimiro A, Barroso J, Pais MS (1990) Effect of copper deficiency on photosynthetic electron transport in wheat plants. Physiol Plant 79:459–464

Cakmak I, Kurz H, Marschner H (1995) Short-term effects of boron, germanium and high light intesity on membrane permeability in boron-deficient leaves of sunflower. Physiol Plant 95:11–18

Caruso C, Chilosi G, Leonardi L, Bertini L, Magro P, Buonocore V, Caporale C, (2001) A basic peroxidase from wheat kernel with antifungal activity. Phytochemistry 58:743–750

Cataldo DA, McFadden KM, Guerinot ML (1996) A novel iron-regulated metal transporter from plants identified by functional expression in yeast. Proc Natt Acad Sci U S A 93:5624–5628

Delhaize E, Katnoka T, Hebb DM, White RG, Rayan RR (2003) Genes encoding proteins of the cation diffusion facilitator family that confer manganese tolerance. Plant Cell 15:1131–1142

Dembitsky VM, Smoum R, Al-Quntar AA, Ali HA, Pergament I, Srebnik M, (2002) Natural occurrence of boron-containing compounds in plants, algae and microorganisms. Plant Sc 163:931–942

Eide D, Broderius M, Feit J, Guerniot ML (1996) A novel iron regulated metal transporter from plants identified by functional expression in yeast. Proc Natl Acad Sci U S A 93:5624–5628

Flowers TJ (1988) Chloride as a nutrient and as an osmoticum. In: Tinker B, Läuchii A (eds) Advances in plant nutrition. Prager, New York, pp 55–78

Fox TC, Guerinot ML (1998) Molecular biology of cation transport in plants. Annu Rev Plant Physiol Plant Mol Biol 49:669–696

Ganie SA, Karmakar J, Roychowdhury R, Mondal TK, Dey N (2014) Assessment of genetic diversity in salt-tolerant rice and its wild relatives for ten SSR loci and one allele mining primer of *salT* gene located on 1st chromosome. Plant Syst Evol 300(7):1741–1747

Gautam S, Rathoure AK, Chhabra A, Pandey SN (2017) Effects of nickel and zinc on biochemical parameters in plants- a review. Octa J Environ Res 5:14–21

Gorham J (1995) Betains in higher plants-biosynthesis and role in stress metabolism. In: Wallsgrove RM (ed) Amino acids and their derivatives in higher plants. Cambridge University Press, Cambridge, pp 171–203

Guerinot ML (2000) The ZIP family of metal transporters. Biochimica et Biophysica Acta (BBA) - Biomembranes 1465:190–198

Graham RD (1983) Effect of nutrient stress on susceptibility of plants to disease with particular reference to trace elements. Adv Bot Res 10:221–276

Hanstein SM, Felle HH (2002) CO_2-triggered chloride release from guard cells intact faba bean leaves. Kinetics of the onset of stomatal closure. Plant Physiol 130:940–950

Hasanuzzaman M, Roychowdhury R, Karmakar J, Dey N, Nahar K, Fujita M (2015) Recent advances in biotechnology and genomic approaches for abiotic stress tolerance in crop plants. In: Devarajan T, Jeyabalan S (eds) Genomics and proteomics: concepts, technologies and applications. Apple Academic Press, Canada, pp 333–366

Hatch MD, Burnell JN (1990) Carbonic Anhydrase Activity in Leaves and Its Role in the First Step of C Photosynthesis. Plant Physiology 93:825–828

Hall JL, Williams LE (2003) Transition metal transporters in plants. Journal of Experimental Botany 54:2601–2613

Henriques FS (2001) Loss of blade photosynthetic area and of chloroplasts photochemical capacity account for reduced CO_2 assimilation rates in zinc-deficient sugarbeet leaves. J Plant Physiol 158:915–919

Heslop-Harrison JS, Roger BJ (1986) Chloride and potassium ions and turgidity in the grass stigma. J Plant Physiol 124:55–60

Hussain SS, Shi B (2014) Role of miRNAs in abiotic and biotic stresses in plants. In: Emerging technologies and management of crop stress tolerance., Biological techniques, vol 1. Elsevier, pp 181–207

Kampfenkel K, Kushnir S, Babiychuk E, Inze D, Van Montagu M (1995) Molecular characterization of a putative *Arabidopsis thaliana* copper transporter and its yeast homologue. J Biol Chem 270:28479–28486

Karthika KS, Rashmi I, Sreekumar PM (2018) Biological functions, uptake and transport of essential nutrients in relation to plant growth. In: Plant nutrients and abiotic stress tolerance. Springer, Singapore, pp 1–49. https://doi.org/10.1007/978-981-10-9044-8_1

Kelley PM, Izawa S (1978) The role of chloride ion in photosystem II 1. Effects of chloride on photosystem II electron transport and hydroxylamine inhibition. Biochem Biophys Acta 502:198–210

Koshiba T, Saito E, Ono N, Yamamoto N, Sato M (1996) Purification and Properties of Flavin- and Molybdenum-Containing Aldehyde Oxidase from Coleoptiles of Maize. Plant Physiology 110:781–789

Kobayashi A, Sakamoto A, Kubo K, Rybka Z, Kanno Y, Takatsuji H (1998) Seven zinc-finger transcription factors are expressed sequentially during the development of anther in petunia. Plant J 13:571–576

Lee S, Aronoff S (1967) Boron in Plants: A Biochemical Role. Science 158 (3802):798–799

Lippard SJ (1999) Free copper ions in the cell. Science 284:748–749

Loneragan JF (1981) Distribution and moment of copper in plants. In: Loneragan JF, Robson AD, Graham RD (eds) Copper in soils and plants. Academic, London, pp 165–188

Marschner H (1995) Mineral Nutrition of Higher Plants. Academic Press, London

Ma JF, Nomoto K (1996) Effective regulation of iron acquisition in graminaceous plants. The role of mugineic acid as phytosiderophores. Physiol Plant 97:609–617

Ma JF, Kusano G, Kimura S, Nomoto K (1993) Specific recognition of mugineic acid ferric complex by barley roots. Phytochemistry 34:599–603
Manthey JJ, Tissarat B, Crowley DE (1996) Root responses of sterile grown onion plants to iron deficiency. J Plant Nutr 19:145–161
Moth T, Ishigaki K, Ohno K, Azuma J (1993) Isolation and characterization of a boron-polysaccharide complex from radish roots. Plant Cell Physiol 34:639–642
Marschner H, Kirkby EA, Cakmak I (1996) Effect of mineral nutritional status on shoot-root partitioning of photoassimilates and cycling of mineral nutrients. Journal of Experimental Botany 47:1255–1263
Mendel RR, Hansch R (2002) Molybdoenzymes and molybdenum cofactor in plants. J Exp Bot 53:1689–1698
Morab H, Koti RV, Chetti MB, Patil PV, Nalini AS (2003) Role of nutrients in inducing rust resistance in soybean. Indian J Plant Physiol 8:85–88
Maksymiec W (1997) Effect of copper on cellular processes in higher plants. Photosynthetica 34:321–342
Naaz S, Pandey SN (2009) Growth and biochemical responses of tomato irrigated with industrial effluent. Res Environ Life Sci 2:91–94
Naaz S, Pandey SN (2010) Effect of industrial waste water on the uptake of certain heavy metals on growth and biochemical responses of lettuce (Lactuca sativa L.). J Environ Biol 31:273–276
Norvell WA, Welch RM (1993) Growth and nutrient uptake by barley (*Hordeum vulgare* L. cv Herta): studies using an N- (2 hydroxyethy) ethylene dinitrilotriacetic acid buffered nutrient solution technique I. Zinc ion requirements. Plant Physiol 101:619–625
Ono T, Onone Y (1991) A possible role of redox-active histidin in the photoligation of manganese into photosynthetic O_2-evolving enzyme. Biochemistry 30:6183–6188
O'Neill MA, Ishii T, Albersheim P, Darvill AG, (2004) Rhamnogalacturonan II: Structure and Function of a Borate Cross-Linked Cell Wall Pectic Polysaccharide. Annual Review of Plant Biology 55:109–139
Pandey SN (2014) Effect of soil sodicity on growth, biochemical constituents and zinc content in wheat plants. J Biol Chem Res 31(1):317–324
Pandey SN (2018) Biomolecular functions of micronutrients toward abiotic stress tolerance in plants. In: Hasanuzzaman M, Fujita M, Oku H, Nahar K, Hawrylak-Nowak B (eds) Plant nutrients and abiotic stress tolerance. Springer, Singapore, pp 153–170
Pandey SN, Gautam S (2009) Effect of zinc supply on its uptake, growth and biochemical constituents in lentil. Indian J Plant Physiol 14:67–70
Pandey N, Pathak GC, Singh AK, Sharma CP (2002) Enzymic changes in response to zinc nutrition. J Plant Physiol 151:1151–1153
Petranyl P, Jendrisak JJ, Burgess RR (1978) RNA polymerase II from wheat germ contains tightly bound zinc. Biochem Biophys Res Commun 74:1031–1038
Pandey SN, Abid M, Abid Ali Khan MM (2018) Diversity, function and stress responses of soil microorganisms. In: Egamberdieva D, Ahmad P (eds) Plant microbiome:stress responses. Microorganisms for sustainability, vol 5. Springer, Singapore, pp 1–17
Pfeffer H, Dannel F, Römheld V (1998) Compartmentation of boron in roots and its translocation to the shoot of sunflower as affected by short term changes boron supply. In: Boron in Soils and Plants. Bell RW, Rerkasem B (eds.). Kluwer Academic Publishers, Dordrecht: 203–207
Polle A, Chakrabarti K, Chakarbarti S, Seifert F, Schramel P, Rennenberg H (1992) Antioxidant and manganese deficiency in needle of Norway spruce (*Picea abies* L.) trees. Plant Physiol 99:1084–1089
Ranieri AM, Cartanga A, Baldan B, Soldatini GF (2001) Iron deficiency affects peroxidase isoforms in sunflower. J Exp Bot 52:25–35
Robinson NJ, Protor CM, Connolly EL, Guerinot ML (1999) A ferric chelate reductase for iron uptake from soils. Nature 397:694–697
Rognes SE (1980) Anion regulation of lupin asparagine synthetase. Chloride activation of glutamine- utilizing reaction. Phytochemistry 19:2287–2293

Roychowdhury R (2014) Crop improvement in the era of climate change. IK International Publishing House, New Delhi, p 496

Roychowdhury R, Tah J (2011) Differential response by different parts of *Solanum melongena* L. for heavy metal accumulation. Plant Sci Feed 1(6):80–83

Roychowdhury R, Karmakar J, Karmakar J, Adak MK, Dey N (2013) Physio-biochemical and microsatellite based profiling of lowland rice (*Oryza sativa* L.) landraces for osmotic stress tolerance. Am J Plant Sci 4(12):52

Roychowdhury R, Khan MH, Choudhury S (2018) Arsenic in rice: an overview on stress implications, tolerance and mitigation strategies. In: Hasanuzzaman M, Nahar K, Fujita M (eds) Plants under metal and metalloid stress. Springer, Singapore, pp 401–415

Roychowdhury R, Khan MH, Choudhury S (2019) Physiological and molecular responses for metalloid stress in rice – a comprehensive overview. In: Hasanuzzaman M, Fujita M, Nahar K, Biswas J (eds) Advances in rice research for abiotic stress tolerance. Woodhead Publishing/ Elsevier, USA, pp 341–369

Rengel Z (1995) Carbonic anhydrase activity in leaves of wheat genotypes differing in zinc deficiency. J Plant Physiol 147:251–256

Sancenon V, Puigs MH, Thiele DJ, Peoarrubia L (2003) Identification of a copper transporter family in *Arabidopsis thaliana*. Plant Mol Biol 51:577–587

Schmidt W, Tillel J, Schikora A (2000) Role of hormone in induction of iron deficiency responses in *Arabidopsis* roots. Plant Physiol 122:1109–1118

Schnabl H (1980) Anion metabolism as correlated with volume changes in guard cell protoplasts. Z Naturforsch 35c:621–626

Shrotri CK, Tewari MN, Rathore VS (1980) Effect of zinc nutrition on sucrose biosynthesis in maize. Phytochemistry 19:139–140

Slooten L, Capiau K, Van Camp W, Van Montagu M, Sybesma C, Inze D (1995) Factors Affecting the Enhancement of Oxidative Stress Tolerance in Transgenic Tobacco Overexpressing Manganese Superoxide Dismutase in the Chloroplasts. Plant Physiol 107:737–750

Singh K, Pandey SN (2011) Effect of nickel-stresses in uptake, pigments and antioxidative responses of water lettuce, *Pistia stratioes* L. J Environ Biol 32:391–394

Schon MK, Novacky A, Blevins DG (1990) Boron induces hyperpolerization of sunflower root cell membranes and increases membrane permeability to K^+. Plant Physiol 93:566–571

Seo M, Koshiba T (2002) The complex regulation of ABA biosynthesis in plants. Trends Plant Sci 7:41–48

Sharma PN, Kumar N, Bisht SS (1994) Effect of zinc deficiency on chlorophyll contents, photosynthesis and water relations of cauliflower plants. Photosynthetica 30:353–359

Sharma CP (2006) Plant Micronutrients. Science Publishers, Enfield

Sharma CP, Sharma PN (1987) Mineral nutrient deficiencies affect plant water relations. J Plant Nutr 10:1637–1643

Shikanai T, Muller-Moul P, Munekage Y, Niyogi KK, Pilon M (2003) PAA1, P type ATPase of *Arabidopsis* functions in copper transport in chloroplasts. Plant Cell 15:1333–1346

Smith BN (1984) Iron in higher plants: storage and metabolic rate. J Plant Nutr 7:759–766

Stiles W (2013) Trace elements in plants. Cambridge University Press, New York

Srivastava NK, Mishra A, Sharma S (1997) Effect of zinc deficiency on netphotosynthetic rate, ^{14}C partitioning and oil accumulation in leaves of pepperment. Photosynthetica 33:71–79

Thomine S, Wang R, Ward JM, Crawford NM, Schroeder JI (2000) Cadmium and iron transport by members of a plant metal transporter family in Arabidopsis with homology to Nramp genes. Proceedings of the National Academy of Sciences 97 :4991–4996

Thomine S, Lelievre F, Debarbieux E, Schroeder JI, Barbier-Brygoo H (2003) At NRAMP3, a multispecific vacuolar metal transporter involved in plant responses to iron deficiency. Plant J 34:685–695

Tej L, Zeiger E (2002) Plant physiology. Sinauer Associates, 3rd ed. Publisher, pp 103–124. ISBN: 0-878930831-1

Terry N (1977) Photosynthesis, growth and role of chloride. Plant Physiol 60:69–75

Treeby M, Marschner H, Römheld V (1989) Mobilization of iron and other micronutrient cations from a calcareous soil by plant-borne, microbial, and synthetic metal chelators. Plant and Soil 114:217–226

Taylor IB, (1991) genetics of ABA synthesis. In: Abscisic Acid, Physiology and Biochemistry (Davis WJ, Jones HG, eds.). Bios Publishers, Oxford 23–37

Van Wiren W, Capiau K, Van Montgu M, Inze D, Slooten L (1996) Enhancement of oxidative stress tolerance in transgenic tobacco plants overproducing Fe- superoxide dismutase in chloroplasts. Plant Physiol 112:1703–1714

Waters BM, Blevins DG, Eide DJ (2002) Characterization of FRO1, a Pea Ferric-Chelate Reductase Involved in Root Iron Acquisition. Plant Physiology 129:85–94

Woods WG (1996) Review of possible boron speciation relating to its essentiality. The Journal of Trace Elem Exp Med 9:153–163

White PJ, Broadley MR (2001) Chloride in soils and its uptake and movement within in the plants. A review. Ann Bot 88:967–988

Winkler RG, Palacco JC, Blevims DG, Randall DD (1985) Enzymatic degradation of allantoate in developing soybeans. Plant Physiol 79:878–893

Williams LE, Pittman JK, Hall JL (2000) Emerging mechanisms for heavy metal transport in plants. Biochimica et Biophysica Acta (BBA) - Biomembranes 1465:104–126

Wu G, Wilen RW, Robertson AJ, Gusta LV (1999) Isolation, Chromosomal Localization, and Differential Expression of Mitochondrial Manganese Superoxide Dismutase and Chloroplastic Copper/Zinc Superoxide Dismutase Genes in Wheat. Plant Physiol 120:513–520

Wu Z, Liang F, Hong B, Young JC, Sussman MR, Harper JF, Sze H (2002) An Endoplasmic Reticulum-Bound Ca /Mn Pump, ECA1, Supports Plant Growth and Confers Tolerance to Mn Stress. Plant Physiol 130:128–137

Yachandra VK, De Rose VJ, Latimer MJ, Mukerji I, Sauer K, Klein MP (1993) Where plants make oxygen: a structural model for the photosynthetic oxygen-evolving manganese complex. Science 260:675–679

Yan F, Schubert S, Mengel K (1992) Effect of low root medium pH on the net proton release, root respiration and root growth of corn (*Zea mays* L.) and broad beans. Plant Physiol 99:415–421

Yi Y, Guerinot ML (1996) Genetic evidence that induction of root Fe (III) chelate reductase activity is necessary for iron uptake under iron deficiency. Plant J 10:835–844

Yu Q, Rengel Z (1999) Micronutrient deficiency influences plant growth and activities of superoxide dismutase in narrow-leafed lupins. Ann Bot 83:175–182

Yu G, Osborne LD, Rengel Z (1998) Micronutrient deficiency changes activities of superoxide dismutase and ascorbate peroxidase in tobacco plants. J Plant Nutr 21:1427–1437

Yu Q, Hlavacka A, Matoh T, Volkmann D, Menzel D, Goldbach HE, Baluška F (2002) Short-Term Boron Deprivation Inhibits Endocytosis of Cell Wall Pectins in Meristematic Cells of Maize and Wheat Root Apices. Plant Physiol 130:415–421

Zaid H, El Morbat R, Diem HG, Arahou M (2003) Does ethylene mediate cluster root formation under iron deficiency. Ann Bot 92:673–677

Zhang N, Portis AR (1999) Mechanism of light regulation of Rubisco: A specific role for the larger Rubisco activase isoform involving reductive activation by thioredoxin-f. Proceedings of the National Academy of Sciences 96:9438–9443

Zheng SJ, Tang C, Avakawa Y, Masaoka Y (2003) The responses of red clover (*Trifolium pratense* L.) to iron deficiency: a root Fe (III) chelate reductase. Plant Sci 164:679–687

Chapter 5
Phytomonitoring and Mitigation of Air Pollution by Plants

Nitesh Joshi, Ambika Joshi, and Bharati Bist

Abstract The urban environment is degrading globally at a more rapid pace than seen in the last few decades. Measures are taken in order to save the environment from rapid diffusing atmospheric pollution; however, the approach seems to be too slow either due to policies by the respective governments or lack of conviction. Air pollution monitoring is the initial step toward controlling the decay of bio-sustainable air. There are various methods to monitor air quality and its components using instrumental and chemical methods. These methods prove to be expensive and do not reflect the impact on living beings. Plants are stationary; hence, they participate and indicate the changes occurring in an environment. Several studies are done globally emphasizing the role of locally available vegetation as phytomonitor. In order to do so, various morphological, visual, and biochemical parameters are employed. The concept is based on the fact that different plant species respond differently to ambient air which can be used to quantify pollution. Different plants species also react in a varied way to different air pollutants. The pattern of air pollution also differs within and between the countries. That plants act as sinks of pollution is a well-known fact. Several researchers have enormously explained the biochemical pathways of air pollutants within the plants. The current work explores the practical case studies of phytomonitoring and the function of plants in mitigating air pollution. Plants from different locations around the industrial area were studied for their morphological and biochemical changes due to air pollution. Studies carried out to know the dust-capturing efficiencies of plants are discussed. The role of plants in mitigating airborne metals, either on the surface or as accumulators, is enumerated. For development of greenbelt, several tree species are also suggested here based on their resistance to air pollutants.

Keywords Air pollution · Phytomonitoring · Greenbelts · Biomonitors · Dust load · Dust chamber studies · Air Pollution Tolerance Index

---

N. Joshi (✉) · B. Bist
Department of Botany, Rizvi College of Arts, Science and Commerce, Mumbai, Maharashtra, India

A. Joshi
Department of Botany, Jai Hind College, Mumbai, Maharashtra, India

R. Roychowdhury et al. (eds.), *Sustainable Agriculture in the Era of Climate Change*, https://doi.org/10.1007/978-3-030-45669-6_5

Abbreviations

APTI	Air Pollution Tolerance Index
CPCB	Central Pollution Control Board
MPCB	Maharashtra Pollution Control Board
MOEF	Ministry of Environment and Forestry
SPM	Suspended Particulate Matter
DRI	Dust Retention Index

1 Air Pollution: An Introduction

Presence of any air pollutant in the environment in concentrations, which have a tendency to be detrimental to man, plants, property, and surroundings, contributes to air pollution (CPCB 2011).

1.1 Sources

Level of air pollution is soaring in developing countries than in already urbanized and industrialized ones. The main culprits considered for air pollution are industrialization, fast economic development, an elevated level of energy utilization, and a gradual increase in vehicular traffic due to transportation. Also, air pollution concentrations have seasonal, for example, higher concentrations during winter and diurnal pattern due to changes in temperature stratification during day- and nighttime (MOEF 1985).

1.1.1 Particulate Sources

Air pollutants are derived from natural and man-made sources. Natural sources comprise forest flames, volcanic emissions, sandstorm, pollen scattering, etc. Anthropogenic sources on the other hand consist of substances emitted during the burning of fossil fuels in factories and thermal power plants; suspended particulate matter; harmful gases like SO_2, NO_2, and CO; dust fall; etc.

India along with different developing countries is experiencing dreadful air quality due to modernization, lack of knowledge, increase in vehicles and use of poor performance fuels, poorly maintained transportation, and futile environmental laws (Joshi and Chauhan 2008).

1.1.2 Particulate Pollution

Particulates are present in the environment in various forms. These may range from submicron aerosols to evident dirt particles. The relative impact that particulates have on vegetation depends on their individual size, their cumulative quantity, and their chemistry. There is currently significant concern over small particulates smaller than the 10 μm diameter (PM_{10}) and 2.5 μm diameter ($PM_{2.5}$) (Quarg 1996).

1.1.3 Particulate Size

While small particulates are of most concerns for human health impacts, much larger particulates are important in affecting vegetation. Particulate can impact on plants in a number of ways, and those impacts mediated via smothering of leaves or altering soil chemistry are not dependent on particle size. When particulates block leaf stomata or enter cells, the size of stomata is important. These vary between species, but the diameter generally lies between 8 and 10 μm. Hence, relatively large particulates can play important physiological roles in plant responses. Particle size is also important in determining the distance that particles might travel before deposition. A study on unpaved roads by Everett (1980) showed that particles of more than 50 μm diameter in size were deposited within 8 m edge of the road, and particles of 20 μm diameter were found at 30 m from the road.

1.1.4 Levels of Deposition

The levels of deposition on roads depend upon the number of vehicles. Unpaved roads may be covered with limestone gravel and so generate alkaline particulates. However, in urban areas, many of the particulates are derived from motor vehicle exhausts which may contain a variety of heavy metal contaminants and be potentially toxic (Santelmann and Gorham 1988). Deposition of particulate may occur through a number of processes like sedimentation, diffusion, turbulence (especially around vegetation surfaces), and removal by rainfall or occult deposition (fog or mist).

1.1.5 Global Status

Escalating energy demands linked with economic expansion and industrialization in Asia, Africa, and Latin America have caused spectacular boost in air pollutant production. In developing countries, problems arise with the nearness of industrial area and thermal power plants to housing neighborhood (Singh 1995).

Air pollution kills millions of people every year with more deaths recorded in developing countries and in Asia (UNDP 1998). Very less information is available about pollutant concentrations in many human settlements where there may be noteworthy indirect effect of air pollution on human health through abridged crop production, food quality, and revenue.

Air pollution levels are rising globally. According to the World Health Organization (WHO) report (2005), nine out of ten people breathe air contaminated with pollutants. WHO estimates that around 7 million people die annually from contact to fine particulates present in the air that make a way deep into the respiratory and cardiovascular system causing life-threatening diseases.

1.2 Effect of Air Pollution on Plants

Air pollution impacts are both local and regional and are directly lethal to plants and animals (Mukherjee and Agrawal 2018). Emissions from industrial sources and automobiles have bad effect on urban air quality which ultimately affects the health of not only humans and animals but plants too. Urban air pollution has happened to be severe ecological trouble to vegetation and crops (Joshi and Chauhan 2008). Even though certain flora can survive the much higher level of pollution, too much pollution has an effect on them also. According to Talukdar et al. (2018), increase in pollution level is a chief threat to plants which can eliminate entire vegetation of an area. Air pollution not only destroys the local vegetation of an area but also affects the tolerant plant species which can endure air pollution up to certain level.

In India, around 60–70% and 20–30% of pollution in the metropolitan environment is caused by vehicles and industries, respectively (Agrawal 1985). Pollutants absorbed by leaves cause a reduction in chlorophyll contents resulting in reduced plants productivity. Hence, roadside plants become the main receiver of different trace element and gaseous discharge from automobiles (Swami and Abhishek 2018).

Trees experience the maximum contact and are subjective to pollutant concentration due to their permanent habit (Raina and Sharma 2003). The principal ways through which pollution affects vegetation include direct deposition onto leaf surfaces, blocking leaf stomata and/or uptake into leaf tissues, deposition onto soil, and an indirect effect via changes in soil chemistry.

1.3 Effect of Air Pollution on Animals and Human

Studies done all over the world have established that air pollution causes an increase in never-ending respiratory ailment and, in some cases, initiates lung and chest disease, tuberculosis, pneumonia, and cardiovascular diseases (Pope and Dockery 2006; McCormack et al. 2011). USEPA (the United States Environmental Protection Agency) has identified six major air pollutants, their sources, and their effects on human well-being (Table 5.1).

Table 5.1 Air pollutants with their sources and effects on humans

Pollutants	Sources	Effects
Nitrogen dioxide (NOx)	Combustion processes (heating, power generation, and vehicles)	Bronchitis in asthmatic children, reduced lung function
Particulate matter (PM2.5, PM10)	Vehicles, industrial sources, domestic fuel burning, road dust resuspension	Cardiovascular and respiratory diseases, lung cancer, ALRI (acute lower respiratory infections)
Carbon monoxide (CO)	Incomplete fuel combustion (as in motor vehicles)	Reduces the oxygen-carrying capacity of the blood; causes headaches, nausea, and dizziness; can lead to death at high levels
Sulfur dioxide (SO_2)	Burning of sulfur-containing fuels for heating, power, and vehicles	Affects respiratory system and lung function; coughing, mucus, secretion, asthma, and chronic bronchitis; causes acid rain
Lead (Pb)	Petrol and industry (such as smelting and paintworks)	Affects brain development in children; at very high doses leads to poisoning; may lead to brain and organ damage
Ozone (O_3)	Formed by the reaction of NOX and VOCs in sunlight	Breathing problems, asthma, reduced lung function

Modified source: MPCB (2014a, b, c)

According to Rao (1981), air pollution effects (gaseous pollutant, acid deposition, and particulates) on ecosystem levels may be summarized as (i) pollutant assimilation and buildup in vegetation, soil surface, and groundwater; (ii) damage to flora and fauna due to pollutant accumulation like leaf necrosis in plants and dental fluorosis in animals; (iii) change in species number, species density, and species diversity along with a change in survival; (iv) loss of constancy and decrease in the reproductive capability of species; (v) disintegration of links of biotic components; (vi) disturbance in biogeochemical cycle; and (viii) expansion of barren areas in the landscape.

1.4 *The Health of Trees in an Urban and Industrial Area*

Urban trees grow in a stressful environment with low nutrient content, moisture-deficient soil, and intensive air and water pollution along with disease and pest infection. Ozone and different air pollutants injure plants and can lead to acid rain. Acid rain can damage vegetation and change the chemical and substantial composition of the soil (Mukherjee 2015).

Solar radiations heat up soil surfaces, thereby escalating air temperature in regions having scattered tree canopy or less plantation; however, a tree shade can decrease the air temperature and also perk up air quality as release and amalgamation of many pollutants are temperature reliant (Joshi and Joshi 2013).

2 Biomonitors

Plants are essential to preserve ecological equilibrium; however, they too get affected by air pollution. They have been balancing nature from the day life emerged on the earth by amending the concentrations of CO_2 in the atmosphere, and now they are fated to initiate a more modern and perilous role of reducing air pollution through adsorption, assimilation, accumulation, detoxification, and metabolization (Rao 1979).

Bioindicator can be a flora or fauna which discloses the occurrence of a substance in its neighborhood by showing warning signs which can be differentiated from natural or anthropogenic stresses (Steubbing et al. 1989).

2.1 Plants as Indicators of Air Pollution

Falla et al. (2000) and Garrec and Haluwyn (2002) have modified the concept of biomonitoring and have described different terms as given below:

(i) **Bioindicators** (reactive or response indicators) are individual plants showing visible symptoms such as necrosis, chlorosis, physiological disturbances such as wilting of flowers and fruits, abridged number and thickness of flowers, epinasty of leaf lamina and petioles, and decline in growth.

(ii) **Biosensors or biomarkers** react to the presence of air pollutants with hidden effects. Those effects are non-visible changes at the cellular, molecular, biochemical, and physiological levels. Detection of effects needs to be carried out by using microscopic and physiological methods, as well as biochemical analysis.

(iii) **Bio-accumulators** (accumulative indicators) are floras that are in common less susceptible to air pollution, but they collect airborne dust particles and gases onto and into their tissues. Gaseous pollutants mostly enter the leaves; particulates are in general accumulated on the leaf surface.

(iv) The **ecological indicator** concept is mainly devoted to the plant population or community level and deals with modification in species composition, appearance and disappearance of species, and changes in density. Not only is the integration of climate and pollution effects over a long period important but also competition between species.

Sensitive plants are the ones which show apparent symptoms of effects of pollutants even in the lowest concentrations, whereas accumulator plants are those which readily collect particular air pollutants that can be analyzed in the plant matter frequently by physicochemical methods. This quantifies pollution load as that plant then acts as a receiver or absorber of pollutant, without showing any damage, and gets actively involved as a specific capturer of pollutants (Posthumus 1983).

Keller (1982) recognized two aspects of bioindicators in plants as (i) use of *visual symptoms of injury* in plants sensitive to a pollutant and (ii) use of *chemical tissue analysis* of accumulator plants.

2.2 Visual Symptoms

According to Posthumus (1982), some visible symptoms of plants in response to pollutants include (i) necrosis of the margins and tips of leaves in gladioli in the presence of fluoride accumulation, (ii) undersurface bronzing and leaf necrosis of *Urtica urens* L. (small nettle) in the presence of peroxyacetyl nitrate (PAN), (iii) necrosis indicating SO_2 pollution in *Medicago sativa* L. (alfalfa) showing intercostal chlorosis, (iv) speckle necrosis due to O_3 on the upper surface of the leaf of *Nicotiana tabacum* L. (tobacco), and (v) intercostal necrosis in the presence of NO_2 in *Spinacia oleracea* L. (spinach).

2.3 Biochemical Parameters

Many plants don't show visible injury symptoms, but this does not mean they are not affected by various pollutants. The physiological parameters thus play an important role as they indicate the invisible injury to plants. Biosynthesis in plants is adversely affected by pollution, so changes in peroxidase activity (Keller 1974), in catalase activity and protein content (Nandi et al. 1980), in ascorbic acid content (Rao 1981), or by the use of a grouping of parameters like total chlorophyll content, relative water content, ascorbic acid content, and leaf extract ph collectively called as Air Pollution Tolerance Index or APTI also serve as a better indicator of pollution (Singh and Rao 1983).

Verma and Singh (2006) noticed that obvious changes in photosynthetic pigments and protein content in leaf lamina were due to auto exhaust pollution. Correspondingly, as a consequence of increased dust deposition, a reduction in chlorophyll content was also recorded by Prajapati and Tripathi (2008).

2.4 Morphological Parameters for Air Pollution

Some plants can collect air pollutants that have extremely low concentrations and are difficult to determine precisely with physical and chemical techniques to a point that is easier to analyze (Temmerman et al. 2001). A number of such parameters used for phytomonitoring of air quality are mentioned below:

2.4.1 Plant Development

Manning and Fedder (1980) quoted the altered growth rates; changes in rates of maturation; reduction in flower, fruit, and seed formation; alterations in reproductive processes; and ultimately depression in productivity and yield as effects of pollution. Bist et al. (2016) also demonstrated the use of *Tithonia diversifolia* as a phytomonitor on a relative scale in an industrial region, and Joshi (1990) worked on the successful use of the same plant in Mumbai.

2.4.2 Macro Characters

The leaf is an organ which is sensitive to pollution, and any damage to it affects the vitality of the entire plant. Visible effects mainly are characterized as epinasty on some plant species like curling of strawberry leaves; the changed angles between stem and petiole by tomato plants have been reported by many workers. Ozone biomonitoring with tobacco Bel-W3 (*Nicotiana tabacum* L.) plant has been done for many years in different countries, for example, France (Vergé et al. 2002), the Netherlands (Posthumus 1982), Germany (Kerpen and Faensen-Thiebes 1985; Reiner et al. 1985), Estonia (Koppel and Sild 1995), Greece (Saitanis and Karandinos 2001), Spain (Peñuelas et al. 1999), Italy (Nali et al. 2001), and Brazil (Klumpp et al. 1996).

2.4.3 Micro Characters

Stomatal frequencies, the appearance of stomatal and epidermal cells, stomatal pore size and cuticular damage, and trichome density and length have been used as indicators of pollution in the past (Sharma and Butler 1973; Wagoner 1975; Garg and Varshney 1980; Rao 1985; Desai and Kapoor 2013). Biomonitoring of ozone injury using *Abies religiosa* trees has been done in Mexico by studying microscopic symptoms on the needles (Alvarez et al. 1998).

3 Phytomonitoring

One of the reasonably economical ways of screening pollution is by using plants that are susceptible to pollutants. Vegetation is significant to maintain ecological balance, but it also get affected by air pollution (Steubbing et al. 1989). Plant organs, when exposed to pollutants in the atmosphere, reflect changes in their health as they are the primary acceptors of air pollution (Joshi and Swami 2009). When exposed to subacute doses of air pollution, plants tend to indicate the same by a decrease in their growth (Banerjee and Chaphekar 1978; Rao 1981). The response of the vegetation to environmental stresses differs from species to species depending upon genetic makeup and the phenological phases of the plants, concentration of pollutants, and existing ecological conditions (Farooq and Beg 1980).

Plants which grow along the road of metropolitan cities function as a sieve for atmospheric pollutants, which extensively reduce their toxic effects and mitigate their blow on the surroundings. Leaf surface characteristics and the anatomical modifications occurring in leaves of many plant species may potentially be used as biological indicator for the occurrence of air pollution (Kushwaha 2018).

3.1 Concept

Various levels of association of the plant can be used for phytomonitoring, varying from a sole plant (leaf or even plant cell) to the plant organization and the ecosystem. Plants can also be used for examining pollution loads in different vicinity and time (Posthumus 1983). Phytomonitoring can be carried out in two ways (Steubbing 1982): observation and analysis of local flora (*passive biomonitoring*) and exposing indicator species under standard conditions (*active biomonitoring*). Passive monitoring includes the use of plants from their natural habitat where the reaction assessed is generally based on the leaf injury or accumulation of the substance deposited on the selected species. This technique is most frequently used in source identification or monitoring networks.

3.2 Lichens as Indicators of Pollution

The epiphytic lichen *Lecanora conizaeoides* was used in an industrial area in Denmark to monitor heavy metals and SO_2 (Pilegaard 1978) because of its high tolerance to air pollution. It is often the only species surviving in an area with high SO_2 concentrations. *Xanthoria parietina* was used in northeastern Italy to monitor airborne heavy metals (Nimis et al. 2000).

3.3 Plants as Indicators of Different Air Pollutants

Plants can efficiently be used as inexpensive and naturally existing monitoring systems (Steubbing 1982; Steubbing et al. 1989; Hawksworth 2001; Nash and Egan 1988). For example, in *Pinus* young needles, chlorosis indicates SO_2 pollution, necrosis indicates hydrogen fluoride pollution, bleaching indicates NO_2 pollution, while chlorotic mottle signifies Cl_2 pollution in the environment. Similarly, browning in moss leaves indicates fluoride accumulation. The plant species to be used in pollution monitoring should have certain features such as (1) easy identification of species in the field and easy handling for injury analysis; (2) species should have a broad range of distribution to be used in different regions; (3) sensitive species should be used to examine diverse group of pollutants in the area; and (4) species ought to show precise injury symptoms in response to particular types and concentrations of pollutants.

Sensitive species like mosses, lichens, ferns, algae, and aquatic plants are generally more useful in pollution screening as their array of pollutant specificity is generally much superior to that of higher vascular plants (Chaphekar 1972; Posthumus 1982; Steubbing 1982; Rao 1985; Joshi 1990).

3.3.1 Bryophytes

Thomas (1986) quoted that bryophytes are useful bio-accumulators of inorganic and particulate organic pollutants. Bryophytes are also important in biomonitoring of heavy metal deposits (Steinnes 1995); their growth and efficiency for element uptake have been studied by Sucharova and Suchara (1998).

3.3.2 Mosses

Feather mosses, *Pleurozium schreberi* and *Hylocomium splendens,* have been recognized as collectors of heavy metals in Scandinavia (Rühling and Tyler 1973) and Poland (Grodzinska 1978), and the epiphytic moss *Hypnum cupressiforme* in the UK (Goodman and Roberts 1971; Lee and Tallis 1973) and in Germany by Thomas (1983). *Pinnatella alopecuroides* and *Bryum* spp. has been identified as useful biomonitors of elemental pollution in and around Mumbai region (Chakrabortty and Paratkar 2006).

3.3.3 Higher Plants

Nearly all higher plants capture air pollutants with their aboveground plant parts. The leaves are the most important parts to absorb or adsorb pollutants, but in the case of trees, the bark can also be used for monitoring organic and inorganic pollutants (Schulz et al. 2000). *Acacia arabica* (Babul), *Citrus* sp., *Diospyros* sp., *Ficus benghalensis* (Banyan), *Ficus religiosa* (Peepal), *Lilium* sps. (Lily), *Polyalthia longifolia* (Ashok), *Tamarindus indica* (Imli), *Thuja occidentalis* (Cedar), *Ziziphus jujuba* (jujube), etc. are some of the plant species suggested by MOEF (2005) for their possible potential in pollution control.

3.4 Plant Parameters Used for Monitoring

3.4.1 Leaf Injury

The common misconception is that there is no injury to vegetation unless visible symptoms have developed. However, air pollutants have shown to reduce the plant development and yield before any visible symptoms appeared. Chaphekar (1990) stated that injury in case of plants initially starts at biochemical level (interference with photosynthesis, respiration, lipids and protein biosynthesis, etc.), sequentially progressing to the ultrastructure level (disorganization of cellular membranes) and then to the cellular level (cell wall, mesophyll, and nuclear breakdown). Thus, visible injury symptoms like chlorosis and necrosis of foliar tissues develop as a result of many subtle injuries getting magnified. *Nicotiana tabacum* L. cv. Bel-W3, a tobacco variety, has been used by many researchers globally since 1962 as an ozone-responsive bioindicator plant due to its supersensitive nature to exhibit characteristic symptom like grayish necrotic spots on laminar surfaces based on ambient levels of ozone (Heggestad 1991; Cheng and Sun 2013). Stomatal index, length and breadth of stomata, and trichome density have been shown to reduce in areas of pollution (Ahmed and Yunus 1985).

3.4.2 Morphological Indicators

The consequences of air pollution on plant growth and development with respect to height, leaf and fruit size, number of leaves, fruits, and dry matter accumulation values also provide a basis for monitoring pollutants. Chaphekar et al. (1980) located areas of comparatively poor air quality around fertilizer factory premises based on biomass values of *Commelina benghalensis*.

3.4.3 Biochemical Parameters

Biochemical injuries result when pollutant concentration exceeds the capacity of tissues to detoxify it through their normal metabolism, and reduction observed in plant growth and yield because of air pollution may be more serious than generally suspected (Treshow 1984). The protein content serves as a useful parameter for O_3 and SO_2 pollution (Agrawal 1985). The response of plants in terms of protein contents was found to be an indicator of auto exhaust fumes (Banerjee et al. 1983). Constant exposure to SO_2 decreases the ascorbic acid content in plants much before the emergence of visible damage (Keller and Schwager 1977). Chaudhary and Rao (1977) related pollution tolerance of plants with their ascorbic acid and concluded that the higher the ascorbic acid level, the greater was their tolerance. A decline in chlorophyll content has often been recommended as a marker of air pollution. In sensitive lichens, chronic exposure of too low concentrations (0.01 ppm) of SO_2 resulted in the loss of chlorophyll (Gilbert 1968).

3.5 Standardization of Plants

Plants used for active biomonitoring should be grown in pots filled with soil, offering most favorable conditions for plant growth, and supplied with a watering facility to avoid differences from indirect effects. Deficiency of essential elements should not occur. The necessity to use genetically standardized material and consistent culturing practices is also important. Such systems have been used by Chaphekar (1978), Posthumus (1982), Arndt (1982), Joshi (1990), etc. Standardization of techniques is also necessary to compare outcomes at different sites, months, or years.

4 Plants for Mitigation of Air Pollution and Greenbelt Development

Vegetation functions as a proficient gas exchange scheme. Their anatomical construction allows quick dispersal of water-soluble gases. This distinctiveness let the plant to breathe and photosynthesize and to eliminate pollutant from the air.

Hosker and Lindberg (1982) reported that air pollution levels are abridged when windblown particulates (PM2.5 and PM10) adhere to the foliage of plants. Likewise, gaseous air pollutants chiefly carbon dioxide can be impounded through stomata on leaf lamina (Currie and Bass 2008). Wei et al. (2017) suggested that plant leaves retain PM and act as biofilters in the environment.

Akbari (2002) considered that day temperature decrease due to trees would reduce the reliance on air-conditioning and lessen emissions of NO_x from coal-fired power plants causing an expected 10% decrease in smog originator or a drop of 350 tons of NO_x per day.

Vegetation acts as natural "air conditioners," at least with regard to the microclimate of the city (Rowntree 1986). Urban trees can improve environmental variables by preventing solar radiation from warming the surrounding buildings and exteriors, cooling the air by evapotranspiration, and dropping wind velocity (Akbari et al. 2001).

Nowak and Heisler (2010) suggested that vegetation influences air pollution by dropping air temperatures and air pollution and reducing ultraviolet radiation and carbon dioxide.

Importance of plantations as suggested by Moef (2005) is that trees are good SO_2 absorbers and parks with vegetation have lesser SO_2 level than urban roads. Street side shelterbelts can decrease airborne load generated by traffic on the backside of the plantation. Trees around the industrial area help to purify industrial emissions.

Total dust retention can be attained by a 30 m belt of plants where only a single line of trees can cause about 25% decline in airborne particles. Vegetation also helps in controlling pollution when other methods have been insufficient.

4.1 Stress Faced by Urban Trees

Urban trees are at a great disadvantage when compared to the trees growing in the wild. They have limited exposure to sunlight, have subnormal water supply, and have to adjust with the ambient atmosphere which generally varies. The limited quantum of sunlight is sometimes accentuated with smog. The small volume of soil available lacks in humus and is so compact that the air spaces between the soil particles scarcely allow either water or air to reach the root systems.

The ambient environment in which the urban tree thrives often includes toxins like SO_2 from industries and NO_2 from automobiles and contaminants like hydrogen fluorides, particulates, and other heavy metals. Tall buildings, narrow roads, limited water supply, vehicular traffic, and industries make it difficult for a tree to stay alive, but in spite of all these stresses, a tree in the urban area manages to survive (Joshi 1990).

4.2 Tolerant Plant Species

Individual plants vary with their susceptibility to air pollutants. Trees with good pollution tolerance are tolerant species, and those with low tolerance turn out to be sensitive species. Hence, tolerant species serve as pollution "sink" by providing a number of ecological benefits and should be planted in polluted zones (Chaphekar et al. 1980). Cultivation of tolerant and sensitive species in polluted habitats leads to improvement of the polluted environment by these plants acting as scavengers and pollutant indicators (Tiwari and Tiwari 2006). Thus, an assessment of plants in terms to their tolerance level to air pollution becomes important. The purpose of pollution reduction is best accomplished by pollution-tolerant flora; hence, for an industrial area where tree plantation is a site-specific activity, a thorough understanding of tolerance level of vegetation to pollution is necessary.

4.3 Forest as Sinks of Air Pollution

As per the norms of Moef (2005), three key norms for the choice of plants for controlling air pollution are (i) vegetation with thick foliage and large leaf lamina because leaves take up pollutants; (ii) evergreen trees are most efficient; and (iii) the plant species selected must be tolerant to pollutants, even during the initial stages of their development.

According to Moef (2005), some of the suitable plants having potential for pollution control include *Acacia arabica* (Babul), *Citrus species*, *Diospyros species*, *Ficus benghalensis* (Banyan), *Ficus religiosa* (Peepal), *Lilium* sp. (Lily), *Polyalthia longifolia* (Ashok), *Tamarindus indica* (Imli), *Thuja occidentalis* (Cedar), *Prosopis juliflora* (Mesquite), and *Ziziphus jujuba* (jujube). Smith (1987) in his book *Air Pollution and Forests* has given elaborative discussions on the role of trees in air pollution abatement as well as the impact on vegetation.

4.4 Dust Chamber Studies

The increased human activity in the field of development in India has created huge amounts of air pollutants (Abbasi 1998). Forty percent of the total air pollution problems in India are contributed by dust (Das et al. 1981; Lone et al. 2005). Chemical dust is more dangerous, and their phytotoxicity increases under humid conditions (Chaphekar 1972).

Plants intercept tons of dust on busy highways and absorb noise near the noisy factory area (Warren 1973). But the quantity of dust is by no means steady in the environment. It keeps varying with respect to difference in location, time of the year, weather, etc. Wind velocity, air temperature, inversions, cloud cover, and woodland openings all influence particulate movements (Fritschen et al. 1970; Edmonds and Driver 1974). The trees are the best sinks for particulate pollutants like radioactive trace elements, pollen, spore, salt particles, dust, etc. (William 1990). The greenbelts and shelterbelts are the best solutions to the problems of pollution in urban and industrial areas (Agrawal and Tiwari 1997).

In order to control dust pollution, nature of trees, type of canopy, branching pattern, the surface of the bark, phyllotaxy, venation, trichomes, etc. have a significant role (Mancharkar 2001). The canopies of alfalfa have been investigated for the uptake of ordinary pollutants like SO_2 and NO_2 (Hill 1971).

Various factors such as leaf shape, phyllotaxy, leaf texture, presence of trichomes, length of petioles, wind velocity, climatic conditions, height and canopy, etc. affect the dust interception/accumulation capacity of different plants (Younis et al. 2013). The influence of leaf distinctiveness on dust retention has been studied by numerous researchers (Somashekar et al. 1999; Garg et al. 2000).

External features of leaf such as cuticular waxes, epidermal cells, trichomes, stomata, and cuticular lignin give coarseness to leaves (Pal et al. 2002). Rough-leaf surfaces not only cause higher particulate deposition but also reduce the potential for these particulates to be washed away with rainfall. Also, leaves with longer life span have a greater time period to accumulate pollutants. Morphological and anatomical leaf features get changed by an intense dust fall (Somashekar et al. 1999; Gostin 2009; Sukumaran 2012).

In order to establish the finest plant for accumulating and capturing dust, examining of dust on plants under controlled conditions was carried out. A unique dust fumigation chamber was made based on the lines developed by Mancharkar (2001) and subsequently Joshi (2014) in India to study the plants under constrained conditions. Plants were rated according to their Dust Retention Indices (DRI), and the following plants were considered to be good dust capturers.

Mancharkar and Chaphekar (2005) have concluded that some plant species can behave as very good dust capturers as they showed higher values of DRI ranging from 70 to 90. Following plant species showed higher DRI values: *Abutilon indicium*, *Calotropis gigantea*, *Ficus benghalensis*, *Ficus glomerata*, *Gmelina arborea*, *Mangifera indica*, *Terminalia bellerica*, and *Thespesia populnea*.

The wind shows continuous variation in its speed and direction. The average wind speed can vary throughout the year from 3 km/h to 9 km/h (Tyagi et al. 2011). Also, the wind speed varies from location to location and ranged between 28 m/s and 44 m/s in Mumbai city (Kumar et al. 2012). Such chambers have been used in the past as a major tool in analyzing the efficiency of plants in monitoring and measuring pollutants under controlled conditions (Darley 1966; Hill 1971; Mancharkar 2001).

Dust Retention Index (DRI): For effective comparison between different species for their efficiency to retain dust that falls on the laminar surface, an index has been recommended which is called as Dust Retention Index – DRI (Mancharkar 2001). The index is calculated as:

$$\mathrm{DFI} / \mathrm{DRI} = \left(A / B\right) \times 100$$

where DFI = Dust Fall Index, DRI = Dust Retention Index, A = dust captured/retained by a leaf, and B = dust captured/retained by the same area of the greased plate.

In order to study the dust retaining capacities of different plant species, a fan was connected to the dust chamber and allowed to run for 3 min at the speed of 4 km/hr. After 3 min, the plant and greased plates were removed. The greased plate was reweighed, and three leaves were detached from the plant sapling. DRI of the leaves was calculated. This value is the dust-holding capacity of plants under controlled windy conditions, henceforth called as DRI or Dust Retention Index.

The plants were examined to understand dust under restricted conditions. The quantity of dust was estimated to establish dust-capturing and dust-holding capacities of the plants under study. The DFI is same as DRI (Mancharkar 2001) under still conditions in the chamber, while the DRI is dust measured under simulated conditions. The ability to capture dust on leaf surface was calculated in terms of DFI (Dust Fall Index), while the dust retained by the plant under windy conditions was termed as DRI (Dust Retention Index).

DFI and DRI values are percentage values of dust captured and dust retained on the leaf surfaces of different plant species. The foliage was then screened for dust fall within the dust chamber and was examined for their DFI and DRI. The values

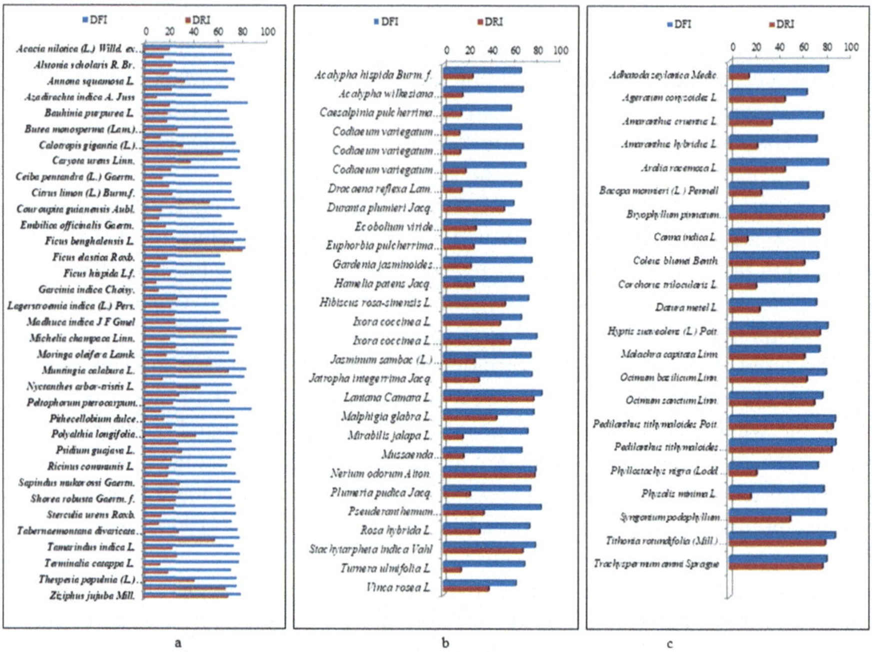

Fig. 5.1 DFI and DRI values for (**a**) trees, (**b**) shrubs, and (**c**) herbs

are represented in Fig. 5.1. For each plant species, Dust Fall Index and Dust Retention Index were calculated (Joshi 2014).

The current study revealed that trees, shrub, herbs, and climbers all engage in a significant task in phytomonitoring and phytoremediation. The best tree species to retain dust on its foliar surface was *Ficus benjamina* L. var. *nuda* (Miq.) M. F. Barrett. Similarly, the best shrub species to retain dust on its foliar surface was *Nerium odorum* Aiton, whereas the best dust retainer herb and climber were *Pedilanthus tithymaloides* Poit. and *Bougainvillea spectabilis* Wild., respectively (Fig. 5.1). *Ficus benjamina* L. had been discovered as an active indicator of air pollution and found to be sensitive at residential, industrial, and heavy traffic areas as it showed APTI<10 (Thambavani and Sabitha 2011). Also, *Bougainvillea* species had been reported as one of the tolerant species to air pollution (Radhapriya et al. 2012).

The work done by Joshi (2014) is a means for selecting the most appropriate plants for greenbelt development in Mumbai city on a morphological basis. *Bougainvillaea spectabilis* Willd., *Ficus benjamina* L. var. *nuda* (Miq.) M. F. Barrett, *Hyptis suaveolens* (L.) Poit., *Ixora coccinea* L., *Lantana camara* L., *Nerium odorum* Aiton., *Pedilanthus tithymaloides* Poit., and *Tithonia rotundifolia* (Mill.) S.F. Blake which require low maintenance but are best performers can be used as phytomonitors and phytoremediators on road dividers on highways and main roads. Similarly for beautiful avenues, *Trema orientalis* (L.) Blume, *Ficus benghalensis* L., *Mangifera indica* L., and *Ziziphus jujuba* Mill. can be cultivated on either side of highways and main roads. Plants like *Bryophyllum pinnatum* (Lam.) Oken, *Duranta plumieri* Jacq., *Stachytarpheta indica* Vahl, and *Trachyspermum ammi* Sprague can be cultivated in residential areas because they are suitable for controlling dust pollution but require regular attention watering.

Beckett et al. (1998) have emphasized that vegetation plays an essential role in enhancing air quality in urban areas. Trees either act as particle trap or create locally altered climatic condition (due to the action of transpiration) which reduces particulate concentrations. The city of Mumbai in India faces a severe problem of dust pollution as has been reported by various agencies in the past. The city has and is experiencing, just like other cities in the country, a surge in construction activity. Air is monitored in the city at a few selected locations by government agencies. There is also an awareness within the general citizens regarding the importance of plants in the city, and in their effort to make the city green, tree plantation drives along the roads and beautification of road dividers and traffic islands are carried out. These plantations include just not the trees but also ornamental shrubs and seasonal flowers. Thus, often questions are asked as to the type of plants to be selected for reducing pollution, and the air quality in areas not monitored is often written or spoken about. The work addresses the problem of dust pollution in the city and the possible role the existing vegetation can play in an urban ecosystem.

4.5 Air Pollution Tolerance Index (APTI)

The ambiance of an urban area is polluted by several pollutants, and plants growing there are open to a mixture of pollutants. So it is feasible to measure, on the whole, the effect of a great number of pollutants by evaluating changes in the plants. There is the likelihood of the synergistic action of pollutants by using vegetation as a marker for atmospheric pollution (Lakshmi et al. 2009). The effectiveness of plants in captivating pollutants is such that it can generate pockets of fresh air. Plants maturing in a contaminated environment often respond and show considerable changes in their structure, functioning, and biochemistry. This reaction of plants toward air can be evaluated by Air Pollution Tolerance Index (Gilbert 1968). Similar studies on APTI of plants has been done in the past by many researchers who suggested many tolerant plant species (Singh and Rao 1983; Tripathi et al. 2009; Patel and Kousar 2011; Govindaraju et al. 2012; Hallale and More 2013). Thus, the main aim of performing APTI is to suggest plant species for greenbelt expansion in order to avoid or reduce the air pollution troubles, in traffic noise reduction, and in mitigating pollution along roadsides and in industrial areas. The response of vegetation to pollutants at a physiological and biochemical level can be understood by evaluating the factors that decide sensitivity and tolerance (Suvarna et al. 2008). Singh and Rao (1983) made a brave attempt in proposing a technique of determining Air Pollution Tolerance Index (APTI) by synthesizing the values of four different biochemical parameters such as ascorbic acid, total chlorophyll content, leaf extract pH, and relative water content to establish the resistance and susceptibility of plants to air pollution. Various plants show significant difference in their receptiveness to air pollution. The plants with elevated and reduced APTI value can function as tolerant and sensitive species accordingly (Lohith et al. 2018). Plants with higher APTI value are more proficient to fight against air pollution and can be used to lessen pollution, while those with low index value show less tolerance and can be used to indicate levels of air pollution (Joshi and Bora 2011). In our work carried out in Tarapur industrial area by the Bist et al. (2017), fully grown leaf samples were collected from 30 tree species during two dry seasons, i.e., summer and winter (Fig. 5.2). The collected leaves were brought to the laboratory with care and were cleaned with distilled water to get rid of dust particles, and fresh weight was taken right away. Fresh leaf samples were then analyzed for ascorbic acid (Sadasivam and Manickam 2009), leaf extract pH (Singh and Rao 1983), total chlorophyll (Arnon 1949), and relative water content (Weatherly 1950).

In an industrialized surrounding, an array of pollutants is seen originating from various sources. So using only one biochemical parameter to estimate the vulnerability of plants to pollutants does not turn out to be a feasible method. But a combination of an assortment of biochemical parameters can give a more trustworthy outcome. Air Pollution Tolerance Index (APTI) study is thus important in understanding the tolerant and susceptible character of plants to pollutants in the surroundings. High APTI values indicate tolerant nature of the plant species, and less APTI value indicates the sensitivity of the plant species (Joshi et al. 2016a, b).

From the plant species collected for the experimental study, the range of plants tolerant to air pollution can be stated as *Putranjiva roxburghii*, *Mangifera indica*, *Ficus racemosa*, *Ficus hispida*, *Morinda citrifolia*, *Pongamia pinnata*, *Ficus benghalensis*, *Polyalthia longifolia*, *Cassia fistula*, and *Acacia auriculiformis*. Tolerant

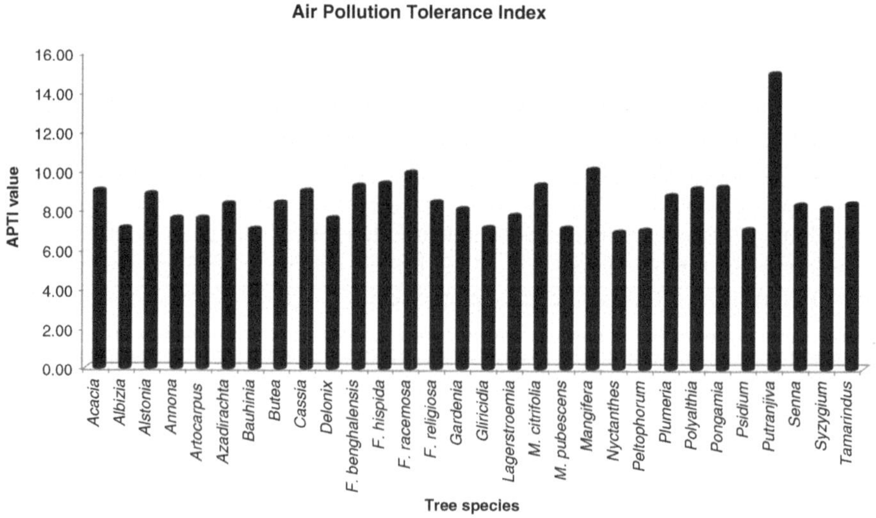

Fig. 5.2 Average APTI of 30 plant species from the industrial area of Tarapur

plant species can be used in greenbelt development as they tend to serve as barriers and act as a sink for air pollutants. These can thus be planted in and around industrial vicinity and traffic islands to control the level of air pollution. Similarly, the order of sensitive species can be given as *Nyctanthes arbor-tristis*, *Bauhinia purpurea*, *Peltophorum pterocarpum*, *Psidium guajava*, *Morinda pubescens*, *Albizia saman*, *Gliricidia sepium*, *Annona squamosa*, *Artocarpus heterophyllus*, and *Delonix regia*. Sensitive plant species on the other hand act as bioindicators of air pollution and thus can be planted in order to check the environmental health from time to time.

Urbanization and industrialization have led to a massive loss in the crop yield and to the financial system of many regions. Hence, APTI evaluation of different crops provides information regarding their tolerance capacity to air pollutants, and such crops may be suggested to the farmers in industrialized, urban, and semi-urban areas (Bakiyaraj and Ayyappan 2014).

Soaring pollution levels can lead to deforestation in the long run, and thus this kind of study proves useful in understanding the plant's susceptibility and resistance to pollution loads. Also, the trees with high tolerance to air pollutants can be used for planting in greenbelt areas so as to mitigate pollution from surrounding to some degree.

4.6 Guidelines for Greenbelt

Greenbelts are the lungs of the city as they function as a sink for a number of the detrimental gases released by automobiles and industries working in the metropolitan. Greenbelt development plan for a particular zone largely depends upon (i) nature and amount of pollution load, (ii) sinking capacity of the ecosystem, (iii) climatic factors, and (iv) soil and water quality (Horaginamani et al. 2012).

4.7 Objectives of Greenbelt Development

The objectives of greenbelt development range from the micro level air pollution abatement to improvement of the socioeconomic position of the area (Gupta et al. 2008). The main objective of greenbelt development is reduction in air and noise pollution. It also serves as a measure to decrease the soil erosion and to improve aesthetic value of the area. It enhances the socioeconomic status of the region by generating employment and also helps in environmental protection. The thickness of the greenbelt and the species used for planting varies from industry to industry depending on nature and concentration of air pollutants. Greenbelt also serves five purposes which include to ensure the unrestricted sprawl of large built-up areas, to prevent neighboring towns merging into one another, to assist in safeguarding the countryside from encroachment, to preserve the setting and special character of historic towns, and to assist in urban regeneration by encouraging the recycling of ruined and another urban land. Shannigrahi et al. (2004) and Gupta et al. (2008) recommended that an ideal tree for use in greenbelt should have quick development rate for canopy growth, thick and strong canopy with sturdy branches to withstand storm conditions, and large foliar area and dense foliage for better pollutant retention; perennial and evergreen trees are preferred for longevity of the greenbelt; indigenous tree species should be selected; and resistance to disease, pest, and specific air pollutants, ability to maintain hydrological and environmental balance of the area along with tolerance to urban soil conditions, and Air Pollution Tolerance Index (APTI) should be evaluated in order to identify the tolerance nature of tree species. Different plant species show different APTI scores; for an instance, a study revealed that the maximum APTI is observed in *Azadirachta indica* and minimum is *Psidium spicigera* (Horaginamani et al. 2012). Another study at Varanasi revealed that *Ficus benghalensis* L. and *F. religiosa* are outstanding performers in urban forest; *Polyalthia longifolia*, *F. glomerata* (Roxb.), *A. indicus,* and *Mangifera indica* are excellent performers in the marble industrial areas of Potwar region; and in the similar fashion, *Cassia fistula* L., *D. roxburghii*, *T. arjuna*, *P. guajava* L., *M. hortensis,* and *D. sissoo* perform well in urban forest with respect to their API (Pandey et al. 2015). When suggesting plants for greenbelt development in industrial complex, points to be considered (Tiwari and Tiwari 2006) are that both types of trees should be planted: one with high APTI value to serve as sink and other with low APTI value to serve as bioindicators. Plants with economic value should be selected. Plants releasing less pollen in the atmosphere should be chosen. Canopy structure should be dense and leaf area should be large. Aesthetic plants should be selected. The innermost belt should lie close to the industry with plants having high APTI value. The second belt should have a few trees with high APTI and majority with moderate APTI values. The third belt should have trees with moderate APTI values but with high aesthetic significance. A few sensitive species should be planted in all three vegetation belts to act as indicators. The distance of 1.0 m from plant to plant and row to row should be maintained. The choice of tree species for greenbelt development is based on the local extent of the pollution load along with soil quality, precipitation, temperature, and human interactions. In any greenbelt development, monoculture is not sensible due to its climatic factor and other ecological limitations.

Pollution-sensitive species like *Bauhinia purpurea, Delonix regia, Gliricidia sepium, Nyctanthes arbor-tristis, Peltophorum pterocarpum,* and *Psidium guajava* must be planted around the human settlements in order to examine pollutant levels as these plants act as bioindicators of pollution. Also, these tree species have thick canopies and stunning flowers which enhance their aesthetic value (Bist et al. 2017). Evergreen pollution-tolerant trees with dense foliage, beautiful flowers, leaves with larger surface area, and wide canopies like *Ficus religiosa, Manilkara zapota, Pongamia pinnata,* and *Putranjiva roxburghii* along with fairly tolerant tree species like *Alstonia scholaris, Azadirachta indica, Butea monosperma, Ficus religiosa, Gardenia jasminoides, Lagerstroemia speciosa, Plumeria obtusa, Senna siamea, Syzygium cumini,* and *Tamarindus indica* must be planted along the pollution sources so that they can soak up more pollutants and help in purifying the surroundings. Roadside plants serve as efficient bioindicators by capturing a considerable amount of harmful particles from the atmosphere, thereby cleaning the air quality (Rai and Panda 2015). Sources such as industrialized areas, factories, or highways become the root cause of air pollution in urban regions. Foliage barrier can help lessen the pollution buildup in cities by acting as sinks. Greenbelts can be used as a visual screen and partial noise blockade next to infirmary, educational organizations, playgrounds, parks, and housing societies close to main road networks.

4.8 Trees in an Industrial Area and Its Role

Trees function as natural "air conditioners" in urban area (Rowntree 1986). Urban plants can enhance ecological variables by preventing solar energy from warming up the surrounding buildings and exteriors, cooling the air by evapotranspiration, and reducing wind velocity (Akbari et al. 2001). Nowak and Heisler (2010) suggested that vegetation and parks influence air pollution by dropping air temperatures and UV radiation, maintaining carbon dioxide level, and ultimately reducing air pollution.

4.9 Monitoring Dust in a City

Mumbai city is heavily inhabited and polluted. The pollutants vary from NOx, CO, and SPM (Joshi and Chauhan 2008). This metropolitan shows a high range of SPM levels from190μg/m^3 to 500 μg/m^3 (Municipal Corporation of Greater Mumbai 2013). The hazardous emissions from industries and vehicles are mainly responsible for increasing air pollution in the city, whereas the rapid construction activities and ever-increasing traffic conditions make it more vulnerable to dust pollution.

Intense vehicular traffic leads to movement of suspended particulate matter in the surrounding air, which settles on ruderal plants. It is well-known fact that vegetation can be used efficiently for screening dust (Yunus et al. 1985; Joshi 1990). In the past for scrutinizing air pollution in the city, urban vegetation has been used successfully by many researchers (Chaphekar et al. 1980). It is a frequent practice in urbanized countries to use plants to clean out dust, soot, and particulates from the environment. Vegetation acts as biofilter and takes in huge quantities of particulates from the surroundings (Central Pollution Control Board 2007).

Urban air quality can be enhanced by planting vegetation along the streets (Beckett et al. 2000; Freer-Smith et al. 2005; Raupach et al. 2001). The foliage acts as steady absorbers for suspended particulate pollutants (Samal and Santra 2002). The dust accumulations differ with difference in time of year (Prajapati and Tripathi 2008).

In plants, the dust-retaining ability depends on its leaf shape, leaf arrangement, and leaf epidermal appendages such as trichomes, cuticle, and lengths of petioles, tree cover, and the existing climatic conditions with wind direction and velocity (Prajapati and Tripathi 2008). Pollutants commonly found in cities include dust from the roads which can be potentially injurious to ruderal vegetation, animals, birds, and the neighboring mankind (Bhattacharya et al. 2011).

An enormous amount of work is done using trees to study air pollution levels in urban areas by many researchers. Shetye and Chaphekar (1980) provided information about various locations on roads with high dust loads using *Erythrina indica*, *Mangifera indica*, *Thespesia populnea*, and *Polyalthia longifolia*. Giridhar (1984) investigated the ascorbic acid contents of some common trees of Mumbai and tried to trace the relationship between their ascorbic acid levels and pollution levels in the city. The growth performance of some common trees near a fertilizer factory emitting various pollutants was studied by Chaphekar et al. (1980). An increase in the intensity of dust causes the reduction in growth, i.e., to shoot length and chlorophyll content. Thus, it can be stated that shoot length and chlorophyll content are more reliable parameters for air quality indication (Nitesh and Bharati 2019). The experiments carried out by Hareesh et al. (2018) on *Phaseolus mungo* L. (black gram) revealed that a raise in different dust pollutants causes a decline in total chlorophyll and transpiration rate in vegetation which additionally decreases their yield.

According to their research, Chaudhary and Rathore (2018) stated that elevated dust fall had harmful effect on leaf functioning, biomass, and micromorphological attributes of tree species. Also, dust fall is usually maximum at the industrialized area as compared to traffic and housing area.

Faqih (2014) explored the potentials of urban plant species, commonly grown along roadsides in the city, as phytomonitors of dust. As a preliminary part of the research, a survey was conducted to study the plant diversity in the city. Four plant species *Bougainvillea spectabilis* Willd., *Ficus benjamina L. var. nuda* (Miq.) M. F. Barrett, *Nerium odorum* Aiton. and *Pedilanthus tithymaloides* Poit. were found to have relatively high dust-capturing capacities and were chosen for investigations on dust distribution at some traffic signals in the city. Considerable work has been done in the city by the authors, under a Major UGC Research Project entitled "Studies on monitoring Suspended Particulate Matter using urban plants and understanding their Green Belt Potentials." In plants like *Nerium odorum*, *Pedilanthus tithymaloides*, *Bougainvillaea spectabilis*, and *Ficus benjamina*, the elemental composition of dust was analyzed and electron diffusion spectrum was studied for urban dust (Joshi 2014).

Vehicular traffic was monitored using a Sony digital camera (Model No. DSC – W150). Traffic count was calculated on hourly basis by examining the film recording taken during field inspection. Vehicular count has been expressed in terms of total number of vehicles/minute (Kadiyali 1996). The amount of captured foliar dust correlated well with the vehicular count along different roads in the city of Mumbai. In a study carried out by Joshi (2014) on dust-capturing capacities of four plant species, viz., *Nerium odorum, Pedilanthus tithymaloides, Bougainvillaea spectabilis, and Ficus benjamina,* throughout the dry seasons at 67 different sites

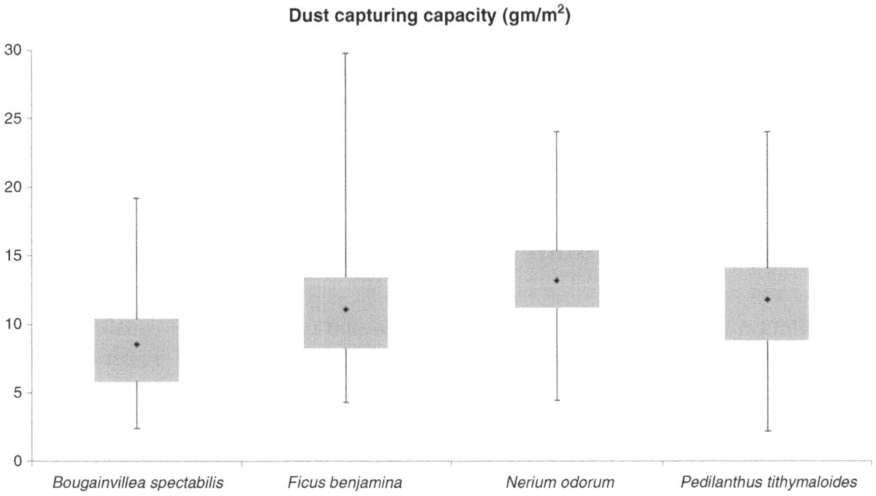

Fig. 5.3 Box plot expressing seasonal dust-capturing capacities of four different plants in g/m^2

spread over Mumbai city, the dust fall values showed the following trend: *Nerium odorum*>*Pedilanthus tithymaloides*>*Ficus benjamina*> *Bougainvillea spectabilis* (Fig. 5.3).

From the work, it was established that *Nerium odorum* Aiton. is the best plant species for phytomonitoring and phytoremediation. Also, its easy availability and maintenance make it suitable for growing all over the city on the roadsides and road dividers (Joshi 2014; Faqih 2015).

4.10 Basic Issues in Defining Guidelines

The four key steps in defining air quality guidelines are (i) identifying the appropriate sensitive species or ecosystem on which to base the guideline, (ii) identifying the response parameter of concern and deciding what size of the change in this parameter can be judged as adverse, (iii) identifying the method of characterizing pollutant exposure, and (iv) using appropriate experimental data, field data, or models to determine the threshold pollutant exposure for the identified species and response parameter. Complexities begin from three key reasons: (1) theoretical problems in defining "adverse effect," both logically and in terms of community judgment; (2) the bad effects of air pollution might be apparent in conditions where they are present in elevated concentrations and have remarkable effects, and also after reaching a "threshold" concentration, the effects steadily become more fine and are not easy to spot in field observations; and (3) constant exposures of the vegetation to air pollution cause effects that may be the consequence of long-term buildup of pollutants or long-term chemical changes in soil and flora. There are no clear experimental techniques of testing the consequences of these increasing effects directly. The unconventional approach adopted to identify dangerous loads of acidity to avoid long-term harm to ecosystems is to characterize a vital chemical concentration in the appropriate mediums likes soil or freshwater and then to build up statistical models to measure the level of deposition being exceeded.

5 Conclusion and Future Perspective

Air pollution is no more restricted to countries and continents. Air quality shows differences at microclimatic levels, irrespective of the class of cities or areas. The air in its purest form has become a rare commodity accepted in forests and mountains. Plants being stationary, thus, are the first one to respond any changes in the environment, be it air, water, or soil. Plants due to its diversity in morphological, biochemical, and growth responses express the nature of the environment and thus can be used as indicators. Plants known to be sensitive to pollutants can be used easily as monitors of air pollution. In cities, it becomes difficult to place instruments to measure pollutants at street levels or block levels due to the vulnerable nature of the region itself. Moreover, it also becomes economically an unviable proposition to do so. Phytomonitoring as a tool thus becomes the most promising tool in urban climatic conditions with minimum inputs and its aesthetic value also. Studies carried out in the city of Mumbai has proved the same. The very nature of some trees makes them ideal solutions for reducing particulate matter in tropical cities which has become a chronic problem. Development of greenbelts and sustainable urban development in cities has become an urgent need. It is in this area where green architecture plays an important role along with plant ecologist and urban ecologist.

References

Abbasi SA (1998) Environmental pollution and its control. Cogent Int Pondicherry 9:445

Agrawal M (1985) Plant factors as indicator of SO_2 and O_3 pollutants. Proceedings of international symposium on biological monitoring of the state environment (bio-indicator). Indian National Science Academy, New Delhi, pp 225–231

Agrawal S, Tiwari S (1997) Susceptibility level of few plants on the basis of Air Pollution Tolerance Index. Indian Forester 123:319–322

Ahmed KJ, Yunus M (1985) Leaf surface characteristics as indicators of air pollution. Symp. Biomonitoring State Environ. 254–257. In: Biological monitoring of the state of environment: bioindicators for I.C.S.U. by R.W. Press, 1986

Akbari H (2002) Heat island reduction: an overview—effects of trees and implementation issues. Presentation by Lawrence Berkeley Laboratory at the University of Pennsylvania, LAPR 760, November 13th, 2002

Akbari H et al (2001) Cool surface and shade trees to reduce energy use and improve air quality in urban areas. Sol Energy 70(3):259–310

Alvarez D et al (1998) Macroscopic and microscopic symptoms in Abies religiosa exposed to ozone in a forest near Mexico City. Environ Pollut 103(2):251–259
Arndt UL (1982) Comparability and standardization of biological processes. In: Steubing L, Jager HJ (eds) Monitoring of air pollutants. Dr. Junk Publishers, The Hague, pp 29–30
Arnon DI (1949) Copper enzyme in isolated chloroplasts. Polyphenol oxidase in Beta vulgaris. Plant Physiol 24:1–15
Bakiyaraj R, Ayyappan D (2014) Air pollution tolerance index of some terrestrial plants around an industrial area. Int J Mod Res Rev 2(1):1–7
Banerjee A, Sarjar RK, Mukherji S (1983) Reduction in soluble protein and chlorophyll contents in a few plants as indicators of automobile exhaust pollution. Int J Environ Stud 20:239–243
Banerjee AK, Chaphekar SB (1978) Observations of foliar injury to plants by SO_2. Geobios 5:8–11
Beckett KP et al (1998) Urban woodlands: their role in reducing the effects of particulate pollution. Environ Pollut 99:347–360
Beckett KP, Freer-Smith PH, Taylor G (2000) Particulate pollution capture by urban trees: Effect of species and windspeed. Global Change Biol 6:995–1003
Bhattacharya T et al (2011) Heavy metal concentration in street and leaf deposited dust in Anand city, India. R J Chem Sci 1(5):61–66
Bist B, Joshi N, Joshi A (2016) Dust retaining potential of Tithonia diversifolia (Hemsl.) A. Gray, International J of Life Sciences 4(2):235–240
Bist B, Joshi N, Mule P, Joshi A (2017) Choice of selection of tree species for green belt development in Tarapur Industrial area. IJBPAS 6(9):1619–1628
Central Pollution Control Board (CPCB) (2007) "Phytoremediation of particulate matter from ambient environment through dust capturing plant species", a report under Central Pollution Control Board. Ministry of Environment and Forests, Delhi. http://www.cpcb.nic.in/upload/NewItems/NewItem_87_NewItem_87_phyto_package.pdf
Central Pollution Control Board (CPCB) (2011) "Current air quality issues in India", a paper presented by Central Pollution Control Board. Ministry of Environment and Forests, Delhi
Chakrabortty S, Paratkar G (2006) Biomonitoring of trace element air pollution using mosses. Aerosol Air Qual Res 6(3):247–258
Chaphekar SB (1972) Effects of atmospheric pollutants on plants in Bombay. J Biol Sci 15–126
Chaphekar SB (1978) Biological indicators: The concept and new additions. Int J Ecol Environ Sci 4:45–52
Chaphekar SB (1990) Effects of air pollution on plants in bombay metropolitan region. Final Report to the Min. of Environ, and Forests, New Delhi.
Chaphekar SB et al (1980) Plants for air monitoring in industrial area. In: Furtado JI (ed) Tropical ecology and development. I.S.T.E, Kuala Lampur, pp 669–675
Chaudhary CS, Rao DN (1977) Study of some factors in plant controlling their susceptibility to SO_2 pollution. Proc India Natl Sci Acad Part B 46:211–236
Chaudhary, Rathore (2018) Phytomonitoring of dust load and its effect on foliar micro morphological characteristics of urban trees. Plantica 2(3):170–179
Cheng J, Sun E (2013) Factors affecting ozone sensitivity of tobacco Bel-W3 seedlings. Bot Stud 54:21
Currie BA, Bass B (2008) Estimates of air pollution mitigation with green plants and green roofs using UFORE model. Urban Ecosyst 11:409–422
Darley EF (1966) Studies on the effect of cement-kiln dust on vegetation. J Air Pollut Control Assoc 16:145–150
Das TM et al (1981) Trees as dust filters. Sci Today 19:19–21
Desai Y, Kapoor M (2013) Effect of building construction-dust on foliar micromorphology and biochemistry of Peltophorum pterocarpum (DC) Baker. Bionano Frontier 6(1):53–56
Edmond RL, Driver CH (1974) Dispersion and deposition of spore fumes and fluorescent particles. Phytopathology 64:1313–1321
Everett KR (1980) Distribution and properties of road dust along the northern portion of the haul road. In Environmental Engineering and Ecological Baseline Investigations along the Yukon

River-Purdhoe Bay Haul Road, ed. J. Brown & R. Berg. US Army Cold Regions Research and Engineering Laboratory, CRREL Report 80–19, pp. 101–28
Falla J et al (2000) Biological air quality monitoring: a review. Environ Monit Assess 64(3):627–644
Farooq M, Beg M (1980) Effect of aqueous sulphur dioxide on the membrane permeability of common Indian tree leaves. New Botanist 7:213–217
Faqih AG (2015) Studies on foliar dust deposition on urban plants in Mumbai. Thesis submitted to the University of Mumbai
Freer-Smith PH et al (2005) Deposition velocities to Sorbus aria, Acer campestre, Populus deltoides X trichocarpa 'Beaupre', Pinus nigra and X Cupressocyparis leylandii for coarse, fine and ultrafine particles in the urban environment. Environ Pollut 133:157–167
Fritschen J et al (1970) Dispersion of air traces into and within a forested area (3). Report No. OSD 1366, College of Forest Resources, Washington Univ., Seattle, WA:53
Garg KK, Varshney CK (1980) Effect of air pollution on the leaf epidermis at submicroscopic level. Exp Dermatol 368:1364–1366
Garg SS et al (2000) Ind J Environ Prot 20:326–328
Garrec JP, Van Haluwyn C (2002) Biomonitoring of the quality of the air. Editions Tec & Doc, Paris, p 117
Goodman GT, Roberts TM (1971) Plants and soils as indicators of metals in the air. Nature 231, 287–292.
Govindaraju M, Ganeshkumar RS, Muthukumaran VR, Visvanathan P (2012) Identification and evaluation of air-pollution-tolerant plants around lignitebased thermal power station for greenbelt development. Environ Sci Pollut Res 19:1210–1223
Gostin IN (2009) Air pollution effects on the leaf structure of some Fabaceae species. Not Bot Hort Agrobot Cluj 37:57–63
Gilbert OL (1968) Bryophytes as indicators of air pollution in the Tyne Valley. New Phytol 67:15
Giridhar BA (1984) Study of interactions between industrial air pollutants and plants. PhD thesis, Univ. Bombay
Grodzinska K (1978) Mosses as bioindicators of heavy metal pollution in Polish National Parks. Water Air Soil Pollut 9(1):83–97
Gupta RB et al (2008) Overview on attenuation of industrial air pollution by Green belt development. J Ind Pollut Control 24(1):1–8
Hallale BV, More PG (2013) Biomonitoring of air pollution around urban and industrial sites of Washim city (Maharashtra). Asian J Contemp Sci 2(1):13–15
Hareesh B et al (2018) Assessment of different dust pollutants effect on total chlorophyll content, transpiration rate and yield of blackgram (*Phaseolus mungo* L.). Int J Curr Microbiol App Sci 7(04):2890–2896. https://doi.org/10.20546/ijcmas.2018.704.329
Hawksworth DL (2001) The magnitude of fungal diversity: the 1.5 million species estimate revisited. Mycol Res 105:1422–1432
Heggestad HE (1991) Origin of Bel-W3, Bel-C and Bel-B tobacco varieties and their use as indicators of O_3. Environ Pollut 74:264–291
Hill AC (1971) Vegetation: a sink for atmospheric pollutants. J Air Pollut Cont Assoc 21:341–346
Horaginamani SM et al (2012) Air pollution tolerance of selected plant species considered for urban green belt development in Trichy. World J Environ Biosci 1(1):51–54
Hosker RP, Lindberg SE (1982) Review: atmospheric deposition and plant assimilation of gases and particles. Atmos Environ 16(5):889–910
Joshi NC (1990) Experiments in phytomonitoring of urban atmosphere. Dissertation, University of Mumbai
Joshi N (2014) Studies on monitoring suspended particulate matter using urban plants and understanding their green belt potentials. Final report to the University Grants Commission. Ref: No. F. No. 40-320/2011 (SR) Date: 1 Jul 2011
Joshi N, Bora M (2011) Impact of air quality on physiological attributes of certain plants. Report and Opinion, 3(2):42–47

Joshi PC, Chauhan A (2008) Performance of locally grown rice plants (Oryza sativa L.) exposed to air pollutants in a rapidly growing industrial area of district Haridwar, Uttarakhand, India. Life Sci J 5(3):41–45

Joshi N, Joshi A (2013) Dust monitoring potentials of ruderal vegetation of Mumbai. J Ind Pollut Contr 29(2):269–274

Joshi PC, Swami A (2009) Air pollution induced changes in the photosynthetic pigments of selected plant species. J Environ Biol 30(2):295–298

Joshi N et al (2016a) Determining Air Quality Index of an industrial area using plants. J Ecol Photon 111:446–456

Joshi N et al (2016b) Air pollution tolerance index of some trees species from the industrial area of Tarapur. Int J Life Sci Scienti Res 2(2):173–182

Kadiyali RL (1996) Traffic engineering and transport planning, vol 5. Khanna Publication, Delhi, p 499

Keller T (1974) The use of peroxidase activity for monitoring and mapping air pollution areas. Eur J For Pathol 4:11–19

Keller T (1982) Physiological bioindicators of an effect of air pollution on plants. In: Steubing L, Jager J (eds) Monitoring of air pollutants by plants. Methods and problems. Dr. Junk Publishers, The Hague

Keller T, Schwager H (1977) Air pollution and ascorbic acid. Eur J For Pathol 7:338–350

Kerpen J, Faensen-Thiebes A (1985) Bioindicator tobacco Bel–W3. With simultaneous exposure to ozone and climate chamber measurements. Dust Air Pollut Prev 45(3):127–130

Klumpp A et al (1996) Bio-indication of air pollution in the tropics – the active monitoring programme near Cubatao. Brazil Gefahrstoffe – Reinhaltung Luft 56:27–31

Koppel A, Sild E (1995) Bioindication of ozone in Estonia by using tobacco variety Bel W3. Water Air Soil Pollut 85:1515–1519

Kumar KS et al (2012) Assessment of design wind speeds for metro cities of India. The Seventh International Colloquium on Bluff Body Aerodynamics and Applications (BBAA7) Shanghai, China, September 2–6

Kushwaha U (2018) Dust pollution effects on the leaves anatomy of Catharanthus roseus and Nerium Oleander growing along the road side of Rewa city (M.P.). Int J Eng Sci 7(9):01–07

Lakshmi PS, Sravanti KL, Srinivas N. (2009) Air pollution tolerance index of various plant species growing in industrial areas. J Environ Sci 2:203–206

Lee JA, Tallis JH (1973) Regional and historical aspects of lead pollution in Britain. Nature 245(5422):216–218

Lohith K et al (2018) Assessment of air pollution impact on micromorphological and biochemical properties of Pentas lanceolata Forssk. and Cassia siamea Lam. Trop Plant Res 5(2):141–151. https://doi.org/10.22271/tpr.2018.v5.i2.019

Lone PM et al (2005) Study of dust pollution caused by traffic in Aligarh city. Indian J Environ Health 47(4):33–36

Maharashtra Pollution Control Board (MPCB) (2014a) Air quality status of Maharashtra. (Compilation of air quality data recorded by MPCB 2013–2014.) http://mpcb.gov.in/ereports/pdf/Air%20Quality%20Report_MPCB_2013-14.pdf

Maharashtra Pollution Control Board (MPCB) (2014b) Air quality status of Maharashtra. (Compilation of air quality data recorded by MPCB 2012–2013.) http://mpcb.gov.in/ereports/pdf/Air%20Quality%20Report_MPCB_2012-13_TERI.pdf

Maharashtra Pollution Control Board (MPCB) (2014c) Air quality status of Maharashtra. (Compilation of air quality data recorded by MPCB 2011–2012.) http://mpcb.gov.in/ereports/pdf/AirQualityReport_MPCB_2011-12.pdf

Mancharkar AR (2001) Assessment of characters of plants to optimize green belt efficiency. Dissertation, University of Mumbai

Mancharkar AR, Chaphekar SB (2005) Morphological basis of green belt efficiency. A.W. College, Otur

Manning WJ, Feder WA (1980) Biomonitoring air pollutants with plants. Applied Science Publishers, London, p 142

McCormack MC et al (2011) Indoor particulate matter increases asthma morbidity in children with non-atopic and atopic asthma. Ann Allergy Asthma Immunol 106:308–315

Ministry of Environment and Forests (MOEF) (1985) Air pollution and plants: a state of the art report. In: Subrahmanyam GV, Rao DN, Varshney CK, Biswas DK (eds). Department of Environment, Government of India, New Delhi, p 193

Ministry of Environment and Forests (MOEF) (2005) Manual on norms and standards for environment clearance of large construction projects. Government of India, New Delhi. http://envfor.nic.in/divisions/iass/Construction_Manual.pdf

Mukherjee A (2015) Importance of urban forestry with reference to Kolkata. IOSR J Humanit Soc Sci 2(8):89–94

Mukherjee A, Agrawal M (2018) Ecotoxicol Environ Saf 152:42–54

Municipal Corporation of Greater Mumbai (MCGM) (2013) Report on Environment Status Air Quality Monitoring and Research Laboratory, Santacruz Pers. comm

Nali C et al (2001) Monitoring and biomonitoring of surface ozone in Florence, Italy. Environ Monit Assess 69(2):159–174

Nandi PK et al (1980) Effect of ozone, sulphur dioxide and their mixture on germination of Phaseolus aureus seeds. Indian J Air Pollut Contr 3(2):50–55

Nash TH, Egan RS (1988) The biodiversity of lichens and bryophytes. In: Lichen, bryophytes and air quality. Thomas Nash III & Vilkmar Wirth (eds). Bibl. Carmer in der Gebr. Borntra. Verlag. Berlin, Stuttgart. Lichenol 30: 11–22

Nimis PL et al (2000) Biomonitoring of trace elements with lichens in Veneto (NE Italy). Sci Total Environ 255(1):97–111

Nitesh, Bharati B (2019) Impacts of ambient air quality of an industrial region on a member of asteraceae and its potentials as a phytomonitor. Environ Risk Assess Rem 3(1):8–14

Nowak DJ, Heisler GM (2010) Air quality effects of urban trees and parks. www.nrpa.org

Patel AM, Kousar H (2011) Assessment of relative water content, leaf extract pH, ascorbic acid content and total chlorophyll of some plant species growing in Shimoga. Plant Archives 11(2):935–939

Pal A, Kulshreshtha K, Ahmad KJ, Behl HM (2002) Do leaf surface characters play a role in plant surface resistance to auto exhaust pollution? Flora 197:47–55

Pandey AK, Pandey M, Mishra A, Tiwary SM, Tripathi BD (2015) Air pollution tolerance index and anticipated performance index of some plant species for development of urban forest. Urban forestry and Urban greening (14):866–871

Peñuelas J et al (1999) Dependence of ozone biomonitoring on meteorological conditions of different sites in Catalonia (N.E. Spain). Environ Monit Assess 56(2):221–224

Pilegaard K (1978) Airborne metals and SO_2 monitored by epiphytic lichens in an industrial area. Environ Pollut 17:81–92

Pope CA, Dockery DW (2006) Health effects of fine particulate air pollution: lines that connect. J Air Waste Manage Assoc 56:709–742

Posthumus AC (1982) Biological indicators of air pollution. In: Unsworth MH, Ormrod DP (eds) Effects of gaseous air pollution in agriculture and horticulture. Butterworth Scientific, London, pp 27–42

Posthumus AC (1983) Higher plants as indicator and accumulator of gaseous air pollution. Environ Monit Assess 3:263–274

Prajapati SK, Tripathi BD (2008) Seasonal variation of leaf dust accumulation and pigment content in plant species exposed to urban particulates pollution. J Environ Qual 37:865–870

Quarg (1996) Airborne particulate matter in the United Kingdom. Third report of the quality of urban air review group. Department of the Environment, London

Radhapriya P, Navaneetha GA, Malini P, Ramachandran A (2012) Assessment of air pollution tolerance levels of selected plants around cement industry, Coimbatore, India. J Environ Biol 33:635–641

Rai PK, Panda LS (2015) Roadside plants as bio indicators of air pollution in an industrial region, Rourkela, India. Int J Adv Res Technol 4(1):14–36

Raina AK, Sharma A (2003) Effect of vehicular pollution on the leaf micro-morphology, anatomy and chlorophyll contents of Syzygium cumini L. Indian J Environ Prot 23:897–902

Rao DN (1979) Plants as a pollution monitoring device. Fertilizer News 24:25–28

Rao DN (1981) Phytomonitoring of air pollutants. Proc. WHO workshop biol. indicators indices of environmental pollution. Osmania University, Hyderabad, Cent. Bd. Prev. Cont. Water Poll., 1–8

Rao DN (1985) Plant and air pollutant mixture. In: Air pollution and plants: a state-of-the-art report. Department of Environment Ministry of Environment and Forests, Government of India, New Delhi, p 67

Raupach MR et al (2001) The entrapment of particles by windbreaks. Atmos Environ 35:3373–3383

Reiner C et al (1985) The use of Nicotiana tabacum L. Bel-W3. Staub – Reinhaltung der Luft 45(2):59–61

Rowntree AR (1986) Ecology of the urban forest-introduction to part II. Urban Ecol 9:229–243

Rühling A, Tyler G (1973) Heavy metal deposition in Scandinavia. Water Air Soil Pollut 2(4):445–455

Sadasivam S, Manickam A (2009) Biochemical methods. New Age International Pvt. Ltd. Publishers, New Delhi, p 284

Saitanis CJ, Karandinos MG (2001) Instrumental recording and biomonitoring of ambient ozone in the Greek countryside. Chemosphere 44(4):813–821

Samal AK, Santra SC (2002) Ind J Environ Health 44:71–76

Santelmann MV, Gorham E (1988) The influence of airborne road dust on the chemistry of Sphagnum mosses. J Ecol 76:1219–1231

Schulz H, Schulz U, Huhn G, Schüürmann G (2000) Biomonitoring of airborne inorganic and organic pollutants by means of pine tree barks. I Deposition types and impact levels. J Appl Bot 74(5–6):248–253

Shannigrahi AS et al (2004) Anticipated air pollution tolerance of some plant species considered for green belt development in and around an industrial/urban area in India: an overview. Int J Environ Stud 61(2):125–137

Sharma GK, Butler J (1973) Leaf cuticular variation in Trifolium repens L. as indicator of environmental pollution. Environ Pollut 5(4):287–293

Shetye RP, Chaphekar SB (1980) Some estimations on dust fall in the city of Bombay, using plants. Vol. 4: pp. 61–70. In: Progress in Ecology. V. P. Agarwal and V.K. Sharma (Eds.). Today and Tomorrow's Printers and publishers, New Delhi

Singh ON (1995) Local air pollution in India. Pure Appl Chem 67:1462–1465

Singh SK, Rao DN (1983) Evaluation of plants for their tolerance to air pollution. In: Proceedings symposium on air pollution control, vol 1. Indian Association for Air Pollution Control, New Delhi, pp 218–224

Smith WH (1987) Air pollution and forests. Interaction between air contaminants and forest ecoscystems. Springer-Verlag New York Inc.

Somashekar RK et al (1999) Pollut ResPollut Res 18:445–451

Steinnes E (1995) A critical evaluation of the use of naturally growing moss to monitor the deposition of atmospheric metals. Sci Total Environ 160-161:243–249

Steubbing L (1982) Problems of bioindication and the necessity of standardization. In: Steubing L, Jager J (eds) Monitoring of air pollutants by plants. Dr. Junk Publishers, The Hague, pp 19–24

Steubbing L, Fangmier A, Both R (1989) Effects of SO_2, NO_2 and O_3 on population development and morphological and physiological parameters of nature herb layer species in a beech forest. Environ Pollut 58:281–302

Sucharova J, Suchara I (1998) Atmospheric deposition levels of chosen elements in the Czech Republic determined in the framework of the International Bryomonitoring Program 1995. Sci Total Environ 223:37–52

Sukumaran D (2012) Effect of particulate pollution on various tissue systems of tropical plants. Central Pollution Control Board (CPCB), Zonal Office, Kolkata

Suvarna LP et al (2008) Air pollution tolerance index of various plant species growing in industrial areas. The Ecoscan 2(2):203–206

Swami, Abhishek (2018) Impact of automobile induced air pollution on road side vegetation: A Review. ESSENCE Int J Env Rehab Conserv IX(1):101–116

Talukdar P et al (2018) Effect of air pollution on plant growth and pollen viability. World Sci News 109(2018):131–142

Temmerman LD et al (2001) Biomonitoring with plants – considerations for the future. EuroBionet – Urban Air Pollution, Bioindication and Environmental Awareness, pp 337–373

Thambavani SD, Sabitha MA (2011) Variation in air pollution tolerance indexand anticipated performance index of plants near a sugar factory: implications for landscape-plant species selection for industrial areas. Journal of research in Biology 7:494–502

Thomas W (1983) Using plants in spatial trace substance analysis for emission monitoring. Staub–Reinhalt Luft 43(4):142–148

Thomas W (1986) Representativity of mosses as biomonitor organisms for the accumulation of environmental chemicals in plants and soils. Ecotoxicol Environ Saf 11:339–346

Tiwari S, Tiwari M (2006) Air pollution tolerance indices of few plants growing near Raigarh (India). J Environ Res Develop 1(2):129–135

Treshow M (1984) Effects of air pollutants on plants. In: Englund, S. and Calvert, S. (eds), Hand-Book on Air Pollution Control Technology, pp. 7–24. John Wiley and Sons, Chichester and New York

Tripathi A, Mahima, Tiwari PB, Singh D (2009) Assessment of air pollution tolerance index of some trees in Moradabad city, India. J Environ Biol 30(4):545–550

Tyagi A et al (2011) Climate of Jaipur, meteorological centre, Jaipur India meteorological department, Ministry of Earth Sciences, Government of India

UNDP (1998) Human development report 1998. United Nations Development Program. Oxford University Press, New York

Vergé X et al (2002) Bioindicator reliability: the example of Bel W3 tobacco (Nicotiana tabacum L.). Environ Pollut 118:337–349

Verma A, Singh SN (2006) Biochemical and ultrastructural changes in plant foliage exposed to auto-pollution. Environ Monit Assess 120(1):585–602

Wagoner KF (1975) Leaf cuticular and morphological variations in Plantago lanceolata as indicators of environmental pollution. J Tenn Acad Sci 50(3):79–83

Warren, JL (1973) Green space for air pollution control. School of Forest Resources, Tech. Rep. No. 50, North Carolina State Univ., Raleigh, NC:118

Weatherly PE (1950) Studies in the water relations of the cotton plant. The field measurement of water deficits in leaves. New Phytol 49:81–87

Wei X et al (2017) Phytoremediation of air pollutants: exploiting the potential of plant leaves and leaf-associated microbes. Front Plant Sci 8:1318

William HS (1990) Interaction between air contaminants and forest ecology. In: Air pollution and forest, 2nd edn. Springer, New York. Resour. Tech. Rep.50. Raleigh NC from Novak (1988)

World health organization (WHO) (2005) Air quality guidelines, global update. WHO Regional Publications, Germany, 484

Younis U et al (2013) Variations in leaf dust accumulation, foliage and pigment attributes in fruiting plant species exposed to particulate pollution from Multan. Int J Agri Sci Res 3(3):1–12

Yunus M et al (1985) Dust falling on some common plants near Lucknow city. Env Pollut (Series B) 9:71–80

Chapter 6
Drought Stress and Its Mitigation and Management Strategies in Crop Plants

Avinash Chandra Rai and Krishna Kumar Rai

Abstract Drought stress is becoming a daunting challenge for the twenty-first-century agriculture worldwide, provoking significant threat to food security as well as to growth and productivity of the major crops as global warming progresses. So developing cultivars with enhanced tolerance to drought stress appears to be the only feasible option for strengthening food security and crop productivity under climate extremes. In recent years, substantial effort has been made to unravel signalling processes underlying plant drought stress response and has confirmed the involvement of complex transcription regulatory networks. Coextensively, unravelling genetic composition of crop plants and identification of signalling cascades along with primitive molecular mechanisms through the deployment of "OMICS" techniques can expedite the piercing of drought tolerance mechanisms. Parallel to this, use of integrated studies can remarkably assist in elucidating the complex mechanism underlying drought stress tolerance, thus easing the gap between the existence and future knowledge about drought stress tolerance in plants. Regulation of drought stress tolerance in plants via certain stress-responsive genes and transcription factors (TFs) is now well established that galvanizes plants to endure unfavorable conditions. Growing evidence has also propounded that the connection between small noncoding RNAs and epigenetic regulation is imperative in the induction of drought-induced transcriptional regulation and stress memory. In this study, we intend to provide profound knowledge about the putative roles of TFs, small RNAs, and epigenetic modifications in the regulation of drought stress response and to contrive management strategies to improve plant defense against drought.

Keywords Drought · Small RNAs · Epigenetics · Transcription factors · Stress memory

---

A. C. Rai
Institute of Plant Sciences, Agricultural Research Organization (ARO) – The Volcani Center, Rishon LeZion, Israel

K. K. Rai (✉)
Department of Botany, Institute of Sciences, Banaras Hindu University, Varanasi, Uttar Pradesh, India

R. Roychowdhury et al. (eds.), *Sustainable Agriculture in the Era of Climate Change*, https://doi.org/10.1007/978-3-030-45669-6_6

1 Introduction

The ever-growing global population is currently being tormented by the global food security pervaded by the magnification of processes leading to aggravated global warming (Chakraborty et al. 2014; Lesk et al. 2016). As an aftermath of global warming, drought has become the eminent abiotic threat to crop growth, productivity, and food security. Furthermore, global warming has also elicited uncertainty in the rainfall patterns and also debilitated precipitation, thus triggering a recurrent outbreak of drought episodes around the globe (Lobell et al. 2011). Consequential drought episodes radically affect crop growth and yield by arresting their growth, intruding their physiology, and impeding their reproduction (Yordanov et al. 2000). Recently, Fahad et al. (2017) have examined the results from several studies published from 1980 to 2017 and reported up to 45% reduction in total yield in wheat, maize, and tomato as a consequence of drought exposure. The continuous increase in the greenhouse gasses is one of the major factors evoking global warming, and in the past couple of centuries, earth has witnessed 30–150% rise in CO_2 and methane (Friedlingstein et al. 2010). The drought stress has severe repercussions on plant growth and productivity compared to any other abiotic factor.

Plants experience drought episodes when the root has a limited supply of water or there is excessive loss of water through transpiration (Anjum et al. 2011). The austerity of deterioration caused by drought stress depends upon the factors like rainfall duration and pattern, soil moisture-holding capacity, and loss of water through evapotranspiration, thus making it quite difficult to predict the actual cause (Challinor et al. 2014). Drought stress adversely affect photosynthesis, nutrient assimilation, and ion/osmotic homeostasis which ultimately affects their growth, reproduction, and productivity (Praba et al. 2009), and severity of which generally varies from one species to other, developmental stage, and other physical factors like decreased photosynthetic radiation absorption, water use efficiency, and reduced harvest index. Plants respond to these environmental changes, per se drought stress, either by adapting their physiological processes or by causing acclimation in their growth pattern in order to survive under harsh environmental conditions (Yang et al. 2009).

Over the past few years, several genes including transcription factors (TFs) and small RNAs (sRNA) have been identified which are exclusively induced in plants upon exposure to drought stress, and their putative function in the adaptation and acclimation to drought stress has been validated and confirmed with wide range of tools such as Differential Display PCR (DDPCR), Serial Analysis of Gene Expression (SAGE), and DNA microarray (Fahad et al. 2017). Furthermore, significant progress has been made by upper hand technologies like modern genetics and "OMICS" approaches (Fig. 6.1) such as transcriptomics, proteomics, metabolomics, and epigenomics in the identification of drought stress-responsive genes involved in maintaining ion/osmotic homeostasis, proline/glycine betaine biosynthesis, and detoxification (Joshi et al. 2016). In addition, TFs like DREB, AP2, ERF, RAV, WRKY, MYB, MYC, NAC, and bZIP have been found to have an astounding

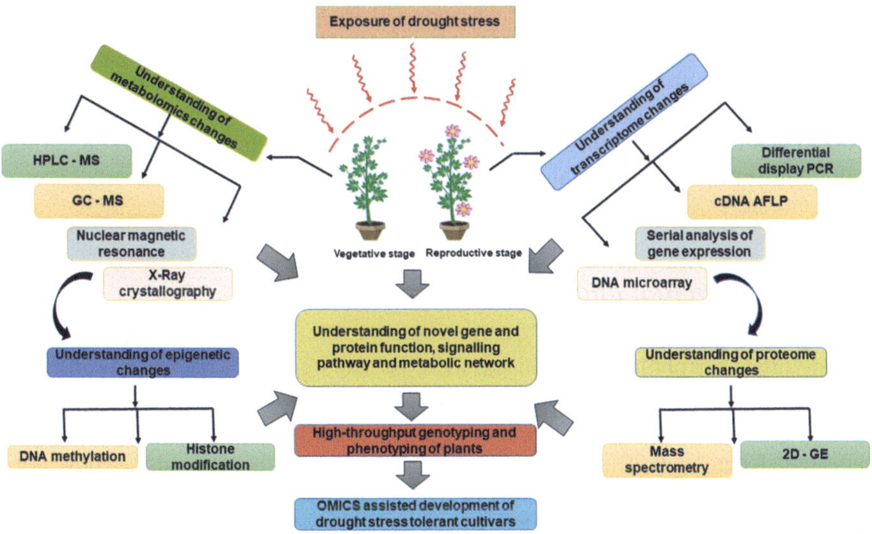

Fig. 6.1 Schematic representation of integrated "OMICS" approaches for the development of drought-tolerant/drought-resistant plants

role in regulating several downstream stress-responsive genes and strengthening plant defense against several abiotic stresses (Nuruzzaman et al. 2013). Several studies have reported that the involvement of small RNA, DNA, and histone methylation also plays a key role in the regulation of gene expression and plant growth under drought stress condition (Chinnusamy and Zhu 2009).

In this chapter, we have reviewed the recent progress made in the past few years to enhance our understanding of plant response to drought stress prevailing from physiological to gene expression level in plants. We have also summarized the functions of key TFs, the role of epigenetic mechanisms, small RNA, and their interactive networks that significantly contributed toward plant drought stress tolerance. Lastly, we have also discussed the fundamentals response of plants to drought stress along with management strategies that can be exploited to minimize the adverse effect of drought stress.

2 Morphological Responses of Plants Under Drought Stress

2.1 Growth

One of the early reverberations of drought on the plants is impaired germination and unproductive seed establishment. Several lines of studies have confirmed the resistive impact of drought stress on germination and seedling growth (Kaya et al. 2006). The devaluation in germination, seedling growth, vegetative and reproductive growth, and total biomass have been reported in many economically important crops such as rice, pea, and tomato as a consequence of drought exposure (Manickavelu et al. 2006). Growth in plants is usually accompanied by the processes like cell division and elongation which are significantly affected by the drought that ultimately reduces water flow from xylem to other cells, thus impairing cellular differentiation and plant growth (Hussain et al. 2008).

2.2 *Reproduction and Yield*

Among other morphological changes, there is also significant decrease in the size and in the number of flower and leaves, plant height, leaf, and shoot dry weights due to decrease in the turgor pressure and supply of nutrient assimilates under drought exposure (Khan et al. 2015). Length and severity of drought stress also expose significant yield losses in major field crops by causing shortening of anthesis time and reducing the grain-filling period (Estrada-Campuzano and Miralles 2008). The reduction in yield, which is basically a complex integration of several physiological processes, might be due to decrease in photosynthesis due to poor development of flag leaf, unequal assimilate portioning, and decrease in the pool of important biosynthetic enzymes such as starch synthase, sucrose synthase, α-amylase, and other starch-branching enzymes (Huang et al. 2012).

3 Physiological Responses of Plants Under Drought Stress

3.1 *Water*

Drought stress strongly influences plant-water relations, affects leaf canopy temperature, alters transpiration rate, and disturbs stomatal conductance (Farooq et al. 2009). Significant reduction in leaf relative water content and transpiration ratio has been observed in many plants exposed to drought stress that have eventually led to increase in their leaf canopy temperature (Monclus et al. 2006). Water use efficiency, i.e., the ratio of the dry matter produced and the amount of water consumed, is also an important physiological process which is adversely affected in many of the drought-exposed plants where drought-tolerant plants maintain higher water use efficiency compared to susceptible cultivars (Abbate et al. 2004). The elevation in the water use efficiency in drought-tolerant cultivars is mainly due to increase accumulation of biomass and low evapotranspiration due to closing of stomata.

3.2 Nutrient Assimilations

Nutrient assimilations are also indubitably disturbed under severe drought stress as uptake of several important nutrients, viz., nitrogen, silicon, calcium, and magnesium, by the roots via diffusion is hampered under drought stress which may eventually lead to stunted growth (Barber 1995). Stomatal closure is one of the premiere and predominant response of almost all plants exposed to drought stress so as to minimize the water loss by controlling transpiration rate. Decrease in turgor pressure and water potential together with low humidity are major factors that trigger stomatal closure, limiting CO_2 intake, reducing assimilation, and generating drought-induced oxidative damage (Anjum et al. 2011). Drought-induced stomatal closure in turn significantly devaluates the photosynthetic process which ultimately affects export rate of sucrose from source to sink. Severe drought stress also affects the ability of plants to efficiently utilize assimilates by altering the activities of source and sink, thus imposing severe threat to plant growth and productivity (Craven et al. 2016).

3.3 Photosynthesis

Drought stress in plants also has severe ramification on thylakoid membranes and photosynthetic pigments. Several studies have reported reduction in chlorophyll content occurring due to either the reduced chlorophyll synthesis or decreased accumulation as a consequence of drought exposure (Zhao et al. 2017a, b, c). The low biosynthesis of chlorophyll pigments under drought stress may be due to inactivation of important chlorophyll of various biosynthetic enzymes, for example, 5-aminolevulinate dehydratase which is an important enzyme of pyrrole biosynthesis pathway and ribulose 1,5 bisphosphate carboxylase oxygenase (RuBisCO) which is an important enzyme of Calvin cycle (Zhao et al. 2017a, b, c). For instance, 35–60% decrease of chlorophyll synthesis and chlorophyll a/b ratio has been observed in various plants such as *Brassica*, cucumber, and tomato due to the inhibition of 5-aminolevulinate dehydratase enzyme as direct consequence of drought stress (Liu et al. 2016). Decreased ATP synthesis due to impaired phosphorylation and low regeneration of nicotinamide adenine dinucleotide phosphate (NADP) could also be one of the major factors for reduced photosynthesis under drought condition.

4 Biochemical Responses of Plants Under Drought Stress

4.1 *Oxidative Damage*

Exposure of plants to drought stress instigates an array of biochemical changes which include rigidity of the cellular membranes, fluidity of the plasma membranes, lipid peroxidation, osmolytes production, reactive oxygen species (ROSs) generation, and activation of several enzymes involved in oxidative defense system (Roychowdhury et al. 2018, 2019; Qi et al. 2018). Among all, generation of ROSs provokes substantial damage to cellular operation by causing lipid peroxidation and damage to proteins as it was observed in the case of pea plants where drought-induced ROSs generation had catastrophic effect on protein/lipid membrane where the damage caused was four times higher compared to normal condition (Aswani et al. 2018). ROSs such as hydrogen peroxide (H_2O_2), superoxide radical ($O_2^{\bullet -}$), hydroxyl radical ($OH^{\bullet}$), and singlet oxygen ($^1O_2$) are mainly generated either by enzymatic or nonenzymatic processes in the chloroplast during photosynthesis and by partial reduction or oxidation of atmospheric oxygen with the components of the mitochondrial electron transport system (Mittler 2017). However, recent studies have shown that ROSs are ought to have dual function in plant biology where at basal level, they are involved in many of the important signalling processes and at higher level, they act as toxic by-products of aerobic metabolism (Mittler 2017).

4.2 *Enzymatic Antioxidants*

Plants have evolved several components that are involved in the modulation of ROS homeostasis and to cope up with oxidative stress (Rai et al. 2018a). These include several enzymatic antioxidants such as catalase (CAT), superoxide dismutase (SOD), ascorbate peroxidase (APX), glutathione reductase (GR), monodehydroascorbate reductase (MDHAR), dehydroascorbate reductase (DHAR), and nonenzymatic antioxidants like carotenoids, ascorbate (AsA), and glutathione (GSH) and accumulation of osmo-protectants such as proline and glycine betaine (Hasanuzzaman et al. 2015; Rai et al. 2018b; Roychowdhury et al. 2019). These enzymatic and nonenzymatic antioxidants protect plants either by directly scavenging the ROS or by increasing the pool of other antioxidants (Rai et al. 2018d). Therefore, increased level of enzymatic and nonenzymatic antioxidants by plants is of absolute requirement to counteract stress-induced oxidative damage of ROSs (Rai et al. 2018c). In addition, exogenous application of several phytohormones such as salicylic acid (SA), nitric oxide (NO), abscisic acid (ABA), and jasmonic acid (JA) has been variously used as a defense priming under several abiotic stresses (Rai et al. 2018a). These molecules when applied exogenously by seed soaking, foliar application, and irrigation stimulate several enzymatic and nonenzymatic antioxidants under stress conditions that eventually help plants to adapt to changing climatic conditions by regulating growth, development, and nutrient assimilation (Rai et al. 2018b).

5 Molecular Responses of Plants Under Drought Stress

Plethora of researches have exemplified molecular responses associated with drought-induced transcriptional signalling pathways in plants. In past few years, several stress-responsive genes and TFs have been identified which have the potential to mitigate the adverse effect of drought-induced oxidative stress (Kudo et al. 2017). The TFs proteins act by interacting specifically with the cis-elements present in the promoter region of the genes and, upon binding, stimulate the expression of stress-inducible genes of various signalling pathways (Anumalla et al. 2016; Joshi et al. 2016). These TFs are categorized into different families based on their conserved motifs that code their DNA binding domain (DBD), viz., APETALA 2 (AP2)/ethylene-responsive element binding factor (ERF); dehydration-responsive element binding protein (DREB); no apical meristem/*Arabidopsis* transcription activation factor, cup-shaped cotyledon (NAC); related to abscisic acid insensitive (ABI3)/VIVIPAROUS 1 (VP1) (RAV); WRKY; auxin response factor (ARF); and SQUAMOSA-promoter binding protein (SBP). The DBDs of the AP2/ERF, DREB, NAC, SBP, and WRKY are named as per the names of their respective TFs family, whereas DBDs of ABI3/VP1 and ARF family of TFs are named as B3 family (Kidokoro et al. 2015).

5.1 *AP2/ERF Family*

The AP2/ERF TFs family is one of the largest plant-specific TFs families which has been categorized into four subfamilies, viz., AP2, ERF, DREB, and RAV, which are active against diverse range of abiotic stresses per se heat, salt, cold, heavy metal, and drought as well as in flower and root development (Dossa et al. 2016), whereas members of RAV subfamily stimulate leaf senescence. The AP2/ERF family of TFs are characterized by the presence of AP2/ERF DNA binding domain which specifically binds to GCC box DNA element potentially involved in the regulation of ethylene transcription (Jisha et al. 2015). The TFs belonging to AP2/ERF family have been found to diverse function in several plant biological processes, i.e., from cell differentiation to growth and development and regulation of plant hormones under different abiotic stresses. Genome-wide identification studies have revealed the existence of over 1000 AP2/ERF family of TFs in Arabidopsis, rice, poplar, Chinese cabbage, and foxtail millet (Zhang et al. 2018). Among all the four subfamilies of AP2/ERF, ERF and DREB possessing single AP2/ERF domain are extensively studied which are exclusively induced under cold and drought stress, functioning in ABA-independent manner. Apropos, AP2 subfamily, possesses two repeated domains, and RAV subfamily members possess C-terminal B3 domain and N-terminal AP2/ERF domain (Kazan 2015). Both ERF and DREB subfamilies of TFs recognize slightly different DNA sequences, viz., ERF recognize GCC box sequence, 5′-AGCCGCC-3′, whereas DREB recognize dehydration-responsive element (DRE), 5′-GCCGAC-3′ (Rakocevic et al. 2018). Plant-overexpressing DREB/AP2/ERF TFs (Table 6.1) have shown upregulation of several defense-responsive genes, thus imparting increased tolerance toward drought stress in tomato, pea, and Arabidopsis (Joshi et al. 2016). Overexpression of these TFs has been reported to increase osmolyte such as proline and glycine betaine accumulation under drought stress. The member of RAV subfamily of TFs was named after the presence of third basic region comprising of 110 amino acids in ABI3/VP1 proteins. These TFs are further classified by the presence of B3 domain which specifically binds to 5′-TGTCTC-3′ DNA sequence and the structure of which has been determined by NMR and X-ray crystallography (Zhao et al. 2017a, b, c). The member of RAV subfamily of TFs possesses barrel-like seven stranded β-sheet structure and two short α-helices at N-terminus which is tightly packed in barrel-like structure. The structures of B3 subfamily of TFs exhibit strong similarity with the structure of non-catalytic DBD of restriction endonuclease *Eco*RII and *Bfi*I of *Escherichia coli* and *Bacillus firmus* (Labbo et al. 2018).

5.2 AREB/ABF Family

The abscisic acid-responsive element binding protein (AREB) and AREB binding factor (ABF) are one of the preeminent TFs that mediate the transcription of various stress-responsive genes during ABA signalling (Yoshida et al. 2015). The expression of ABRE TFs is efficiently controlled by the presence of conserved ABRE element, viz., PyACGTGG/TC in the promoter region. Several findings have revealed the critical function of AREB TFs along with ABF in the regulation of ABA-responsive gene expression in response to various abiotic stresses (Zandkarimi et al. 2015). The structure of AREB/ABFs is composed of four conserved sites for SnRK2 phosphorylation which is exclusively involved in the modulation of ABA synthesis in vascular tissues and their efficient transportation to target cellular destinations per se guard cells (Kuromori et al. 2015). For instance, five genes have been identified to code 9-cis-epoxy carotenoid dioxygenase, which is an important enzyme of ABA synthesis, and expression under drought stress in Arabidopsis after binding to AG-box recognition sequence presents 2250 bp upstream of transcriptional start site (Todaka et al. 2015). ABA is predominantly perceived by the presence of ABA-bound PYL/PYR/RCARs receptor involved in the activation of SnRK2, and once activated, SnRK2 triggers the phosphorylation of several downstream genes/proteins such as late embryogenesis proteins (LEA) that bind specifically to ABRE *cis*-element (Ma et al. 2017). However, in the absence of ABA, PYL/PYR/RCARs receptor inhibits ABA signalling by dephosphorylating SnRK2. The AREB/ABF have been widely reported to regulate ABA signalling during drought

Table 6.1 List of major transcription factors (TFs) family expressed under drought stress response in plants

Transcription factors (TFs) family	TFs ID	Plants/crops	Cellular and molecular function	References
AP2/ERF	*ERF3*	Sugarcane (*Saccharum officinarum* L.)	Ectopic expression enhances drought and salt tolerance	Ithape et al. (2017)
	DREB1A/ DREB2A	Arabidopsis (*Arabidopsis thaliana*)	Improved root architecture and enhanced drought tolerance	Lisei-de-Sa et al. (2017)
	DREB1F/ DREB2A/ EREBP1	Rice (*Oryza sativa*)	Enhanced ROS scavenging induced drought tolerance	Xiong et al. (2017)
	ERF3/ DREB2A/ DREB1/ CBF	Soybean (*Glycine max*)	Improved root surface area and distribution under drought stress	Zhao et al. (2017a, b, c)
	DREB1	Buckwheat (*Fagopyrum esculentum*)	Improved dehydration stress tolerance through ABA-dependent pathway	Fang et al. (2015)
	DREB2A	Mung bean (*Vigna radiata*)	Enhanced expression of DREB2A-regulated stress-inducible genes and improved drought tolerance	Chen et al. (2016)
	DREB2	Jesweit (*Erianthus arundinaceus*)	Pyramiding of EaDREB2 with DNA helicase enhances drought stress tolerance	Augustine et al. (2015)
	DREB2/ DREB3	Wheat (*Triticum aestivum*)	Overexpression causes strengthening of the antioxidant defense system in response to drought stress	Wani et al. (2018)
	ERF053	Daisy (*Chrysanthemum*)	Stimulated lateral shoot and root binding under drought stress	Nie et al. (2018)
	DREB	Moso bamboo (*Phyllostachys edulis*)	Overexpression causes enhanced expression of root leading to drought tolerance	Wu et al. (2015)
	ERF5	Tomato (*Solanum lycopersicum*)	Regulates drought tolerance by inducing autophagy activity	Jafarov and Gasimov (2017)

(continued)

Table 6.1 (continued)

Transcription factors (TFs) family	TFs ID	Plants/crops	Cellular and molecular function	References
AREB/ABF	*AREB1/ AREB2/ ABF3/ ABF4*	Arabidopsis (*Arabidopsis thaliana*)	Induce drought tolerance by trifurcating feed forward pathway	Sakuraba et al. (2015)
	AREB1	Black cottonwood (*Populus trichocarpa*)	Improve drought tolerance by inducing histone acetylation	Li et al. (2018)
NAC	*NAC1/ NAC5/ NAC022*	Rice (*Oryza sativa*)	Stimulated drought tolerance via ABA-mediated pathway	Hong et al. (2016)
	NAC2/ NAC69	Wheat (*Triticum aestivum*)	Improved drought tolerance by regulating cellular contexts	Gahlaut et al. (2016)
	NAC20	Soybean (*Glycine max*)	Enhanced drought tolerance by activating DREB/CBF-COR pathway	Agrawal et al. (2018)
	JUB1	Tomato (*Solanum lycopersicum*)	Improve drought tolerance by stimulating several downstream genes	Thirumalaikumar et al. (2018)
bZIP	*bZIP44/ bZIP62*	Soybean (*Glycine max*)	Induction of drought tolerance by modulating ROS synthesis	Kalia et al. (2017)
	bZIP23	Rice (*Oryza sativa*)	Improved drought tolerance by stimulating other stress-responsive genes	De Schutter et al. (2017)
	bZIP17	Maize (*Zea mays*)	Induced drought tolerance by promoting cell differentiation	Zhan et al. (2018)
	bZIP28	Arabidopsis (*Arabidopsis thaliana*)	Activates brassinosteroid signalling and promotes acclimation to drought stress	Kataoka et al. (2017)

stress promoting vegetative and reproductive growth in plants. Enhanced expression of AERB1 protein has been shown to improve drought tolerance in rice, soybean, and tomato (Yoshida et al. 2015). Arabidopsis plants expressing wheat AREB TFs exhibited improved sensitivity to ABA and drought adaptation (Wang et al. 2016a).

5.3 NAC TFs Family

The NAC family of TFs comprised of three proteins, viz., no apical meristem (NAM), Arabidopsis transcription activator factor (ATAF), and cup-shaped cotyledon (CUC), first identified in Arabidopsis (Hong et al. 2016). Structurally, the NAC TFs are composed of highly conserved DNA binding N-terminal domain and a highly variable C-terminal transcriptional activator region linked with nuclear localization sequence (NLS) (Rahman et al. 2016). The NAC TFs have been widely reported to regulate downstream signalling of various drought-responsive genes upon interacting with NAC core binding motif 5′-CACG-3′ in the promoter sequence. Several NAC TFs have been reported to confer drought tolerance in rice via regulating the expression of several downstream stress-responsive genes in ABA-independent manner (Fang et al. 2015). Genome-wide identification studies have revealed the presence of numerous NAC TFs among different plant species of which 152 are identified from maize and soybean, 164 in rice, and 128 in Arabidopsis (Mao et al. 2015). The function of NAC TFs in the regulation of abiotic stress tolerance in plants has been validated by distinct transcriptomic studies, a tissue- and developmental stage-specific gene expression analysis, thereby confirming their role in the regulation of complex signalling network under abiotic stresses (Yu et al. 2016). Several studies have highlighted the importance of NAC1 gene overexpression of which resulted in decrease evapotranspiration by facilitating the stomatal closure in the guard cells of Indian wild rice (Takasaki et al. 2015). Furthermore, enhanced expression of *Ta*NAC2 and *Ta*NAC29 ameliorated the adverse effect of drought and salt and cold stress tolerance in Arabidopsis by overexpressing several stress-responsive genes and increased accumulation of osmolytes (Huang et al. 2016a, b). Interestingly, improved tolerance to drought stress has also been observed in barley overexpressing HvSNAC1 gene which also provided resistance against leaf spot causing *Ramularia collo-cygni* (McGrann et al. 2015). In transgenic plants, DNA microarray analysis has revealed that about 65 downstream genes like Zn-finger, ATPase, NAC, and meristem protein were differentially expressed under various abiotic stress conditions. Therefore, tailoring NAC TFs with appropriate promoter will provide an efficient way to manipulate drought stress response in plants (Huang et al. 2016a, b).

5.4 WRKY TFs Family

WRKY TFs are also an important class of transcription regulator which regulate the expression of several downstream stress-responsive genes under biotic and abiotic stress conditions (He et al. 2016). WRKY class of TFs has been known to proportionate 60 amino acid long DBD containing WRKYGQK consensus sequence after which the WRKY TFs family has been named. These TFs also contain conserved Cys and His residues along with Zn binding site. WRKY TFs are classified into three groups, viz., group I possessing two copies of WRKY domains and groups II and III possessing single WRKY domain (Wang et al. 2018a, b). The members of all the three groups of WRKY TFs specifically recognize W-box DNA element having core sequence 5-'TTGAC(C/T)-3' (Chu et al. 2015). C-terminal end of group 1 WRKY TFs is involved in the interaction with W-box element; however, in groups II and III, both C- and N-terminal interact with W-box element (Wang et al. 2018a, b). For instance, Arabidopsis AtWRKY1 and AtWRKY4 belong to group I of WRKYC family whose crystal structures have been determined by using NMR and X-ray crystallography and have predicted the existence of C_2-H_2 motif in β-sheet which appears to bind with major groove of the DNA (Xia et al. 2018). In the WRKYGQK domain, Trp (W) is the key amino acid residue at the core of the structure, and rest of the RKYGQK are solely involved in DNA binding (Fei et al. 2019). The presence of Gly residue enables deep penetration into the major groove of the DNA by causing deviation in the strand. The group I WRKY TFs containing two WRKY domains have also been identified in non-photosynthetic primitive eukaryotes such as protists and slime molds as well as in green algae (Zhang et al. 2017). The presence of two WRKY domains might be due to duplication of single WRKY domain during evolution process. A body of literature has indicated that the involvement of C-terminal WRKY domain has specific DNA binding activity; however, recently, studies have reported that N-terminal WRKY domain is also specifically and symbolically involved in DNA binding activity and has been concealed by several selection pressure occurring during course of evolution (Abid et al. 2017).

5.5 bZIP TFs Family

The basic leucine zipper (*bZIP*) is composed of highly conserved bZIP domain containing extremely basic nuclear localization signal (NLS) at N-terminal end and a C-terminal leucine rich motif for dimerization (Zong et al. 2016). The members of bZIP TFs family are widely known to regulate plant growth and development under various abiotic stress conditions per se drought stress, and hitherto, 75 Arabidopsis, 89 rice, 92 sorghum, 125 maize, and 131 soybean have been isolated and characterized (Huang et al. 2016a, b). Studies have reported and confirmed that the bZIP TFs exert their function under the constitutive expression of zinc finger motif and are able to regulate the signalling of various downstream stress-responsive genes in ABA-dependent manner after specifically interacting with *cis*-element of ABRE TFs (Wang et al. 2017). Recently, a novel gene was identified in *T. halophila* that specifically belongs to bZIP TFs family member containing a functionally conserved Cys-2/His-2 region analogous to the DNA binding domain of Arabidopsis ZFP family. Several of bZIP TFs have reported to be induced in plants upon their encounter with various abiotic factors per se drought and salinity (Wang et al. 2017). For instance, researchers have reported the overexpression of OsbZIP16 in transgenic rice plants exhibiting strong tolerance to drought both at seedling and tillering stage (Liu et al. 2014). Additionally, several of the other downstream stress-responsive genes were also induced up to a greater extent in transgenic rice plants compared to corresponding wild-type plants. A contemporary report on Arabidopsis published that enhanced expression of TabZIP60 ameliorated the adverse effect of drought, salt, and freezing along with improved plants sensitivity to ABA (Zhang et al. 2015). Similarly, DNA microarray analysis has also revealed ectopic expression of various downstream stress-responsive genes in conjunction with several of the stress-related TFs, heat shock proteins, late embryogenesis abundant proteins, and dehydrins (Sornaraj et al. 2016). Concomitantly, these findings confirmed the predominant role of bZIP TFs in ABA signalling which can be exploited for materializing superior genotypes with enhanced drought tolerance.

5.6 SBP TFs Family

SQUAMOSA binding protein (SBP) is the class of TFs which is exclusively involved in the stimulation of flower development by specifically binding to the promoter region of floral meristem (Zhang and Li 2013). The DBD of SBP TFs constitute of 80 amino acid residues characterized by the presence of 10 Cys or His residues in their basic region. Several studies have that the functional domain of SBP TFs specifically interacts and binds to a consensus 5′-(C/C)/GTAC(A/G)-3′ (Hajyzadeh et al. 2015). The NMR crystal structure of SBP TFs has been reported in Arabidopsis which has confirmed the presence of eight Cys/His residues coordinately bonded with two Zn ions in such a way that four N-terminal and four C-terminal bind to one Zn ion each (Yamasaki et al. 2013). NMR titration model for SBP TFs has also demonstrated that the conserved basic amino acid residues are located within the DNA-protein binding site (Yamasaki 2016). Despite of the disparity in the structure of SBP TFs, the N-terminal and C-terminal Zn-coordinating units exhibit extensive similarities with the classical zinc figure protein (C_2H_2 ZnF) (Rai et al. 2014). In this context, analysis of Arabidopsis SPL12 gene had confirmed the N-terminal Zn binding domain is essential for the formation of tertiary structure of the protein compared to C-terminal protein (Duque et al. 2013). Similarly, mutational analysis has also confirmed that the C-terminal domain is also predominantly involved in the DNA binding mediated by the basic C-terminal loop (Li et al. 2016). However, determination of crystal structure of SBP/DNA complex by NMR or X-ray crystallography is still resolved. In Chlamydomonas, binding of Cu to the SBP domain causes hindrance in its DNA binding activity (Pilon 2017). The presumed model for Chlamydomonas CRR1 has confirmed the presence of Cu-sensor mechanism that causes flexibility of the C-terminal Zn binding site and owing to the difference in their coordination geometry, when bonded Cu ion causes a conformational change in the C-terminal basic region and thus interfering with the DNA binding ability (Pilon 2017). SBP family of TFs is also strongly related with their ability to counteract adverse effect of several abiotic stress in plants, so their further classification will indeed strengthen plant innate immunity.

6 Drought Stress Mitigation Strategies

Implementation of suitable agricultural practices in combination with advanced genetic improvement programs is generally considered imperative for overcoming adverse effect of abiotic stress (Cai et al. 2015). Good agricultural practices along with the exploitation of modern genetic tools could help breeders to develop high-yielding/stress-tolerant varieties as until now conventional breeding has yielded high-yielding varieties which were able to perform best under non-stressed condition (Karmakar et al. 2012; Roychowdhury et al. 2013; Hasanuzzaman et al. 2015; Yuan et al. 2015). In the recent years, attempt has been made to breed stress-tolerant varieties by using conventional as well as modern breeding strategies which are discussed below.

6.1 Conventional Breeding Strategies

Conventional breeding technique is one of the oldest techniques for development/identification of high-yielding/stress-tolerant crops with the ability to transfer agronomically important traits into other crops (Fita et al. 2015), and significant progress has been made by various research institutes around the globe. For instance, CIMMYT initiated a breeding program for the development of maize hybrids tolerant to drought stress and other common diseases, and the developed hybrids performed well under drought condition with good economic yield (Vivek et al. 2016). Similarly, crop research institute in Ghana has developed 16 inbred lines of maize with improved tolerance against drought stress (Bawa et al. 2015). Similar research programs have also been initiated by IRRI, Philippines, and ICRISAT for development of drought-tolerant cultivars in major cereals and legumes crops (Bharadwaj et al. 2017).

6.2 Modern Breeding Strategies

Tolerance to abiotic stresses in plants is a complex process which is controlled by several genes and loci present on the chromosomes called Quantitative Trait Loci (QTLs) (Desmae et al. 2017). In this context, QTL mapping followed by marker trait association and marker-assisted selection approaches can be used for exploiting genetic variation present for stress tolerance among cultivars (Roychowdhury 2014; Roychowdhury et al. 2014; Fletcher et al. 2016) that can facilitate the identification and location of stress-tolerant genes along the chromosomes and their function. However, selection of appropriate QTLs associated with drought tolerance is a bit difficult owing to high degree of interaction between environment and stress-responsive genes, thus making isogenization necessary for its classification immediately after its identification (Fan et al. 2015). Several attempts have been made to map QTL for drought tolerance in many crops; for example, in cotton, 30 QTLs have been identified under drought stress, and later 11 of the QTLs were directly linked with yield-enhancing attributes, and 17 of them were characterized and linked with fiber quality (Saleem et al. 2015). Once the QTLs have been identified, the next most important part is their proper utilization in the breeding program for the development of stress-tolerant cultivars (Baytar et al. 2018). Parallel to this, several QTLs responsible for the induction of drought stress tolerance in pearl millet have been identified by the breeders of ICRISAT and developed introgression lines by marker-assisted backcrossing approach (MABC) with improved drought tolerance and yield (Bharadwaj et al. 2017). These studies clearly portrayed that QTL mapping followed by marker-assisted breeding can expedite development of crops with enhance tolerance to drought stress.

7 Transcriptomics Approaches for the Identification of Novel Genes

Functional genomic tools have greatly accelerated the discovery of novel genes which are imperative for enhancing growth and productivity of plants under drought stress condition (Budak et al. 2015). Transcriptome sequencing is one of the best alternative method for targeted gene identification which can provide significant insight into its cellular and molecular function. Several efforts are being focused on the gene expression analysis, development of cDNA libraries, expressed sequence tags (ESTs) mining, and their in silico characterization in most of the economically important crops for enhanced tolerance to drought tolerance (Wang et al. 2018a, b). Significant work has been done in legume transcriptomics where extensive number of ESTs has been generated in soybean, Medicago, Lotus, and other crops as well as via Sanger sequencing approach (Wang et al. 2015). In the recent past year, several new low-cost sequencing platforms have become available with the potential to generate huge transcript datasets from stress-responsive plant tissues; for example, FLX/454 sequencing in conjunction with Sanger sequencing has generated 103,215 Transcript Assembly Contigs (TACs) (Mir et al. 2017). Similarly, Illumina sequencing technologies have generated 53,409 contigs and performed the assembly of several Mb of unique transcriptome sequence (Mir et al. 2017). Furthermore, FLX/454, Sanger sequencing and Illumina sequencing have also significantly contributed toward the identification of several stress-responsive genes, functional markers, and transcription factors that can be exploited in the crop breeding programs for enhanced stress tolerance (Table 6.2).

Table 6.2 Summary of differentially regulated genes as an epigenetic modification under drought stress

Crops	Regulatory genes	Signalling cascades	Major function	Epigenetic marks	Methylation status	References
Arabidopsis (*Arabidopsis thaliana*)	*ERD2, AT2G20920, AT2G34060*	Hormone signalling	Formation of transgenerational stress memory	Epi-type DMRs	DNA methylation	Ganguly et al. (2017)
Tomato (*Solanum lycopersicum*)	*SlAGO4A*	RNA-directed DNA methylation	Regulation of methylation process-associated genes	*DRM1*	DNA methylation	Huang et al. (2016a, b)
Rice (*Oryza sativa*)	*BTF3*	Antioxidant defense pathway	Modulation of developmental and stress-related genes	Hypo-/hypermethylation	DNA methylation	Harihar and Nataraja (2016)
Wheat (*Triticum aestivum*)	*GAPC1*	Glycolytic pathway	Regulation of drought stress-responsive genes	Hypomethylation and hypermethylation	DNA demethylation	Fei et al. (2017)
Faba bean (*Vicia faba*)	*VfMSAP 1–6*	Abscisic acid signalling pathway	Increases the expression of lipoxygenase (LOX), calcium-dependent protein kinase (CDPK), glycosyl hydrolase (GH), and phosphoenolpyruvate carboxylase (PEPC)	DNA methylation polymorphic fragments	Methylation and demethylation	Abida et al. (2017)
Tomato (*Lycopersicum esculentum*)	*Non-transposon Asr1*	Tissue-dependent sugar and hormone sensitivities	Drought conditions brought about higher CG methylation levels in the first exon	*Bar gene*	Asymmetric CNN methylation	Gonzalez et al. (2011)
Rice (*Oryza sativa*)	*H3K4me3*	Terpenoid biosynthesis-related pathways	Histone modification and chromatin modelling	*DH3M*	Tri-methylation	Zheng et al. (2014)

8 Proteomics Approaches for Identification of Novel Proteins

Proteomics mainly involve the analysis of cellular proteomes for identification of novel proteins present that are being synthesized in specific cells, tissue, and organs under different stress conditions (Wang et al. 2016a, b). With the advancement in the quantitative measurements and increased protein coverage, the proteomics analysis has become particularly important for analyzing protein composition, posttranslational modifications, and their modulation for better understanding of stress response mechanism at different developmental stage of plant system (Xie et al. 2016). The emerging areas of plant proteomics involve protein profiling, protein-protein interaction analysis, and protein mapping (Zandalinas et al. 2018). Translational and posttranslational proteomics have now been widely used in crop stress research for enhancing agricultural production based on the information obtained by comparative proteomics study (Kosova et al. 2016). Information from different species (ortho-proteomics) and genotypes (comparative proteomics) will help in the standardization and coordination of different proteomics approaches which ultimately will facilitate advance breeding programs. Furthermore, advancement in the mass spectrometry (MS) has greatly facilitated high-throughput classification and quantification of different proteins (Zandalinas et al. 2018). Parallelly, advances in two-dimensional gel electrophoresis (2D-GE) have allowed separation of proteins with higher efficiency and reproducibility. Proteomics studies have provided significant insight into plant stress responses particularly in drought stress in chickpea, common bean, and pea (Xie et al. 2016), and different protein reference maps have been generated for performing comparative proteomics study in other non-sequenced crops to facilitate breeding program for enhanced stress tolerance and better economic yield under different stress conditions.

9 Metabolomics Approaches for Identification of Novel Metabolites

Identification of various low-molecular-weight stress-responsive metabolites has been successfully achieved by metabolomics analysis at particular developmental stages in plants exposed to abiotic stresses (Chmielewska et al. 2016). Ultra-throughput analysis of metabolites is more worthwhile as compared to conventional genetic approaches for metabolic profiling as the metabolomics approach provides a deep understanding of functional metabolomics networks at particular developmental stages (Moradi et al. 2017). The mass spectrometry (MS) and nuclear magnetic resonance (NMR) are two main backbone of the metabolomics profiling that has led to the identification and characterization of hundreds of metabolites in plants (Catola et al. 2016). Besides these analytical techniques, application of gas chromatography MS (GC-MS) and liquid chromatography MS (LC-MS) has advanced the present knowledge and has expedite the rapid profiling of large number of primary metabolites (Kosova et al. 2016). Several of the metabolomics analysis have been done in model systems such as Arabidopsis and Medicago where a number of conserved and novel metabolites have been identified specifically under drought stress response (Pinasseau et al. 2017). In another instance, metabolomics profiling in Lotus has led to the identification of flavonoids under salt stress response (Michaletti et al. 2018). Metabolite profiling could also facilitate in the identification of new pathways and mechanisms for secondary metabolite synthesis pathway for high-throughput screening of metabolites to assist the metabolic engineering process in plants exposed to different abiotic stress conditions.

10 Epigenomics Approaches of Stress Tolerance

Expression of DNA sequence is largely influenced by chromatin modifications and is often associated with epigenetic marks such as DNA methylation, histone variants, and posttranslational modifications (Yadav et al. 2018). Several lines of studies have reported that changes in environmental and developmental stage can result in the epigenetic modifications within the DNA sequences which in turn can result in the development of multiple epigenomes (Kalavacharla et al. 2017). Comprehensive understanding of stress-induced epigenetic processes in response to plant stress tolerance can help in the establishment of "stress memory" and its heritability within plants (Fleta-Soriano and Munne-Bosch 2016). Development of stress memory for shorter durations either by acclimation or adaptation processes is well known in plants. The stress memory can be retained for shorter period if the stress memory is based on changes caused by stress-inducible RNAs, proteins, and metabolites. However, the memory can be retained for longer duration if it involves transcriptional reprogramming in morphology and phenology of the plants (Kumar 2018). Epigenetic modifications such as DNA methylation and histone modifications have emerged as an alternative for development and retention of stress memory for longer duration (Fortes and Gallusci 2017). For instance, enhanced expression of drought stress-responsive genes in Arabidopsis was associated with increased methylation of histone H3 and lysine 4 (H3K4) and increased acetylation of histone H3 and lysine 9 (H3K9). Study has also reported that transient chromatin changes help in the establishment of adaptation responses, thus leading to the development of transgenerational stress memory (Fortes and Gallusci 2017). However, till date, true nature of epigenetic modification and their mitotic and meiotic heritability is unclear. Therefore, comprehensive understanding of DNA methylation/demethylation dynamics, posttranslational histone modifications, and small RNAs could help in the establishment of stress memory which can be further utilized for enhancing yield and management of crop plants under adverse environmental conditions.

11 Conclusion and Future Perspectives

The global food demand is continuously increasing which could cause inadequacy in the demand-supply of the ever-growing global population. The only feasible option to bridge this gap is to make an effort toward breeding of high-yielding cultivars with improved resistance/tolerance to the various abiotic stresses especially in the context of drought stress. With the advancement in the "OMICS" techniques and cost-effective sequencing technologies, it has now become easier to pinpoint probable genes coding for distinct plant phenotypes, thus facilitating their selection on the basis of genotyping information. Furthermore, the information generated based on OMICS techniques can be integrated into the breeding program and thus allowing the breeders to move from time-consuming conventional breeding to accelerated, advanced, and knowledge-based breeding. In addition, the knowledge generated by genomics, transcriptomics, proteomics, metabolomics, and epigenomics has significantly impacted plant science community that has accelerated their understanding of crop response to different abiotic stresses along with improved agronomic traits to develop an effective strategy which will further assist in crop improvement programs in a less time-consuming cost-effective manner.

References

Abbate PE, Dardanelli JL, Cantarero MG, Maturano M, Melchiori RJM, Suero EE (2004) Climatic and water availability effects on water-use efficiency in wheat. Crop Sci 44(2):474–483

Abid G, Muhovski Y, Mingeot D, Saidi MN, Aouida M, Aroua I, Jebara M (2017) Identification and characterization of two faba bean (*Vicia faba* L.) WRKY transcription factors and their expression analysis during salt and drought stress. J Agric Sci 155(5):791–803

Abida G, Mingeotb D, Muhovskib Y, Mergeaic G, Aouidaa M, Abdelkarima S, Arouaa I, Ayedd ME, Hamdie MM, Sassif K, Jebaraa M (2017) Analysis of DNA methylation patterns associated with drought stress response in faba bean (*Vicia faba* L.) using methylation-sensitive amplification polymorphism (MSAP). Environ Exp Bot 142:34–44

Agarwal P, Agarwal PK, Gohil D (2018) Transcription factor-based genetic engineering for salinity tolerance in crops. In: Kumar V, Wani SH, Suprasanna P, Tran L-SP (eds) Salinity responses and tolerance in plants. Springer, Cham, pp 185–211

Anjum SA, Xie XY, Wang LC, Saleem MF, Man C, Lei W (2011) Morphological, physiological and biochemical responses of plants to drought stress. Afric J Agri Res 6:2026–2032

Anumalla M, Roychowdhury R, Geda CK, Bharathkumar S, Goutam KD, Mohandev TSS (2016) Mechanism of stress signal transduction and involvement of stress inducible transcription factors and genes in response to abiotic stresses in plant. Int J Recent Sci Res 7(8).12754–12771

Aswani V, Rajsheel P, Bapatla RB, Sunil B, Raghavendra AS (2018) Oxidative stress induced in chloroplasts or mitochondria promotes proline accumulation in leaves of pea (*Pisum sativum*): another example of chloroplast-mitochondria interactions. Protoplasma:1–9

Augustine SM, Narayan JA, Syamaladevi DP, Appunu C, Chakravarthi M, Ravichandran V, Subramonian N (2015) Overexpression of EaDREB2 and pyramiding of EaDREB2 with the pea DNA helicase gene (PDH45) enhance drought and salinity tolerance in sugarcane (Saccharum spp. hybrid). Plant Cell Rep 34:247–263

Barber SA (1995) Soil nutrient bioavailability: a mechanistic approach. Wiley, New York

Bawa A, Isaac AK, Abdulai MS (2015) SSR markers as tools for screening genotypes of maize (*Zea mays* L.) for tolerance to drought and *Striga hermonthica* (Del.) Benth in the Northern Guinea Savanna Zone of Ghana. Res Plant Biol 5:17–30

Baytar AA, Peynircioğlu C, Sezener V, Basal H, Frary A, Frary A, Doğanlar S (2018) Genome-wide association mapping of yield components and drought tolerance-related traits in cotton. Mol Breed 38:74

Bharadwaj C, Patil BS, Madan P, Manu B, Tapan K, Neeraj K, Nazish N (2017) Breeding chickpea for water limited environments: selection indices and strategies

Budak H, Hussain B, Khan Z, Ozturk NZ, Ullah N (2015) From genetics to functional genomics: improvement in drought signaling and tolerance in wheat. Front Plant Sci 6:1012

Cai X, Zeng R, Kang WH, Song J, Valocchi AJ (2015) Strategic planning for drought mitigation under climate change. J Wat Resour Plann Manag 14:04015004

Catola S, Marino G, Emiliani G, Huseynova T, Musayev M, Akparov Z, Maserti BE (2016) Physiological and metabolomic analysis of *Punica granatum* (L.) under drought stress. Planta 243:441–449

Chakraborty S, Pattanayak A, Mandal S, Das M, Roychowdhury R (2014) An overview of climate change: causes, trends and implications. In: Roychowdhury R (ed) Crop improvement in the era of climate change. IK Int Publ Hous, New Delhi, pp 1–29

Challinor AJ, Watson J, Lobell DB, Howden SM, Smith DR, Chhetri N (2014) A meta-analysis of crop yield under climate change and adaptation. Nat Clim Chan 4:287

Chen H, Liu L, Wang L, Wang S, Cheng X (2016) VrDREB2A, a DREB-binding transcription factor from Vigna radiata, increased drought and high-salt tolerance in transgenic *Arabidopsis* thaliana. J Plant Res 129:263–273

Chinnusamy V, Zhu JK (2009) Epigenetic regulation of stress responses in plants. Curr Opin Plant Biol 12:133–139

Chmielewska K, Rodziewicz P, Swarcewicz B, Sawikowska A, Krajewski P, Marczak Ł, Krystkowiak K (2016) Analysis of drought-induced proteomic and metabolomic changes in barley (*Hordeum vulgare* L.) leaves and roots unravels some aspects of biochemical mechanisms involved in drought tolerance. Front Plant Sci 7:1108

Chu X, Wang C, Chen X, Lu W, Li H, Wang X, Guo X (2015) The cotton WRKY gene GhWRKY41 positively regulates salt and drought stress tolerance in transgenic *Nicotiana benthamiana*. PLoS One 10:e0143022

Craven D, Isbell F, Manning P, Connolly J, Bruelheide H, Ebeling A, Beierkuhnlein C (2016) Plant diversity effects on grassland productivity are robust to both nutrient enrichment and drought. Philos Trans R Soc B 37:20150277

De Schutter K, Tsaneva M, Kulkarni SR, Rougé P, Vandepoele K, Van Damme EJ (2017) Evolutionary relationships and expression analysis of EUL domain proteins in rice (*Oryza sativa*). Rice 10:26

Desmae H, Ntare B, Motagi BN, Falalou H, Upadhyaya HD, Oteng-Frimpong R, Varshney RK (2017) Advances in groundnut breeding for drought prone West and Central Africa

Dossa K, Wei X, Li D, Fonceka D, Zhang Y, Wang L, Zhang X (2016) Insight into the AP2/ERF transcription factor superfamily in sesame and expression profiling of DREB subfamily under drought stress. BMC Plant Biol 16:171

Duque AS, de Almeida AM, da Silva AB, da Silva JM, Farinha AP, Santos D, de Sousa Araújo S (2013) Abiotic stress responses in plants: unraveling the complexity of genes and networks to survive. In: Duque et al (eds) Abiotic stress-plant responses and applications in agriculture. InTech, Rijeka, pp 49–101

Estrada-Campuzano G, Miralles DJ, Slafer GA (2008) Genotypic variability and response to water stress of pre-and post-anthesis phases in triticale. Europ J Agron 28:171–177

Fahad S, Bajwa AA, Nazir U, Anjum SA, Farooq A, Zohaib A, Ihsan MZ (2017) Crop production under drought and heat stress: plant responses and management options. Front Plant Sci 8:1147

Fan Y, Shabala S, Ma Y, Xu R, Zhou M (2015) Using QTL mapping to investigate the relationships between abiotic stress tolerance (drought and salinity) and agronomic and physiological traits. BMC Genomics 16:43

Fang Z, Zhang X, Gao J, Wang P, Xu X, Liu Z, Feng B (2015) A buckwheat (*Fagopyrum esculentum*) DRE-binding transcription factor gene, FeDREB1, enhances freezing and drought tolerance of transgenic *Arabidopsis*. Plant Mol Biol Rep 33:1510–1525

Farooq M, Wahid A, Kobayashi N, Fujita D, Basra SMA (2009) Plant drought stress: effects, mechanisms and management. In: Sustainable agriculture. Springer, Dordrecht, pp 153–188

Fei Y, Xue Y, Du P, Yang S, Deng X (2017) Expression analysis and promoter methylation under osmotic and salinity stress of TaGAPC1 in wheat (*Triticum aestivum* L). Protoplasma 254:987–996

Fei X, Hou L, Shi J, Yang T, Liu Y, Wei A (2019) Patterns of drought response of 38 WRKY transcription factors of *Zanthoxylum bungeanum* maxim. Int J Mol Sci 20:68

Fita A, Rodríguez-Burruezo A, Boscaiu M, Prohens J, Vicente O (2015) Breeding and domesticating crops adapted to drought and salinity: a new paradigm for increasing food production. Front Plant Sci 6:978

Fleta-Soriano E, Munné-Bosch S (2016) Stress memory and the inevitable effects of drought: a physiological perspective. Front Plant Sci 7:143

Fletcher RS, Herrmann D, Mullen JL, Li Q, Schrider DR, Price N, McKay JK (2016) Identification of polymorphisms associated with drought adaptation QTL in *Brassica napus* by resequencing. G3 Genes Genom Genet 8:2757–2771

Fortes AM, Gallusci P (2017) Plant stress responses and phenotypic plasticity in the epigenomics era: perspectives on the grapevine scenario, a model for perennial crop plants. Front Plant Sci 8:82

Friedlingstein P, Houghton RA, Marland G, Hackler J, Boden TA, Conway TJ, Le Quere C (2010) Update on CO_2 emissions. Nat Geosci 3:811

Gahlaut V, Jaiswal V, Kumar A, Gupta PK (2016) Transcription factors involved in drought tolerance and their possible role in developing drought tolerant cultivars with emphasis on wheat (*Triticum aestivum* L.). Theoret Appl Genet 129:2019–2042

Ganguly D, Crisp PA, Eichten SR, Pogson BJ (2017) The Arabidopsis DNA methylome is stable under transgenerational drought stress. Plant Physiol 175:1893–1912

Gonzalez RM, Ricardi MM, Iusem ND (2011) A typical epigenetic mark in a typical location: cytosine methylation at asymmetric (CNN) sites within the body of a non-repetitive tomato gene. BMC Plant Biol 11(1):94

Hajyzadeh M, Turktas M, Khawar KM, Unver T (2015) miR408 overexpression causes increased drought tolerance in chickpea. Gene 555:186–193

Harihar S, Nataraja KN (2016) Prediction of DNA methylation marks and related gene expression pattern in contrasting rice genotypes under drought stress. Mysore J Agric Sci 50:316–319

Hasanuzzaman M, Roychowdhury R, Karmakar J, Dey N, Nahar K, Fujita M (2015) Recent advances in biotechnology and genomic approaches for abiotic stress tolerance in crop plants. In: Devarajan T, Jeyabalan S (eds) Genomics and proteomics: concepts, technologies and applications. Apple Academic Press, Oakville, pp 333–366

He GH, Xu JY, Wang YX, Liu JM, Li PS, Chen M, Xu ZS (2016) Drought-responsive WRKY transcription factor genes TaWRKY1 and TaWRKY33 from wheat confer drought and/or heat resistance in Arabidopsis. BMC Plant Biol 16:116

Hong Y, Zhang H, Huang L, Li D, Song F (2016) Overexpression of a stress-responsive NAC transcription factor gene ONAC022 improves drought and salt tolerance in rice. Front Plant Sci 7:4

Huang B, Rachmilevitch S, Xu J (2012) Root carbon and protein metabolism associated with heat tolerance. J Exp Bot 63:3455–3465

Huang C, Zhou J, Jie Y, Xing H, Zhong Y, She W, Ma Y (2016a) A ramie (*Boehmeria nivea*) bZIP transcription factor BnbZIP3 positively regulates drought, salinity and heavy metal tolerance. Mol Breed 36:1–15

Huang W, Xian Z, Hu G, Li Z (2016b) SlAGO4A, a core factor of RNA-directed DNA methylation (RdDM) pathway, plays an important role under salt and drought stress in tomato. Mol Breed 36:28

Hussain M, Malik MA, Farooq M, Ashraf MY, Cheema MA (2008) Improving drought tolerance by exogenous application of glycine betaine and salicylic acid in sunflower. J Agron Crop Sci 194:193–199

Ithape DM, Maharana M, Tripathy SK (2017) Scope of genetic transformation in sugarcane – a review. Genom Appl Biol 8:1–7

Jafarov HR, Gasimov KG (2017) Expression pattern of SLWRKY33 and SLERF5 in tomato plants under elevated salt concentration and water deficit. Фактори експериментальної еволюції організмів 20:266–270

Jisha V, Dampanaboina L, Vadassery J, Mithöfer A, Kappara S, Ramanan R (2015) Overexpression of an AP2/ERF type transcription factor OsEREBP1 confers biotic and abiotic stress tolerance in rice. PLoS One 10:e0127831

Joshi R, Wani SH, Singh B, Bohra A, Dar ZA, Lone AA, Singla-Pareek SL (2016) Transcription factors and plants response to drought stress: current understanding and future directions. Front Plant Sci 7:1029

Kalavacharla VK, Subramani M, Ayyappan V, Dworkin MC, Hayford RK (2017) Plant epigenomics. In: Tollefsbol TO (ed) Handbook of epigenetics, 2nd edn. Academic Press, Elsevier, pp 245–258

Kalia R, Sareen S, Nagpal A, Katnoria J, Bhardwaj R (2017) ROS-induced transcription factors during oxidative stress in plants: a tabulated review. In: Khan MIR, Khan NA (eds) Reactive oxygen species and antioxidant systems in plants: role and regulation under abiotic stress. Springer, Singapore, pp 129–158

Karmakar J, Roychowdhury R, Kar RK, Deb D, Dey N (2012) Profiling of selected indigenous rice (*Oryza sativa* L.) landraces of Rarh Bengal in relation to osmotic stress tolerance. Physiol Mol Biol Plants 18(2):125–132

Kataoka R, Takahashi M, Suzuki N (2017) Coordination between bZIP28 and HSFA2 in the regulation of heat response signals in Arabidopsis. Plant Signal Behav 12:e1376159

Kaya MD, Okçu G, Atak M, Cıkılı Y, Kolsarıcı Ö (2006) Seed treatments to overcome salt and drought stress during germination in sunflower (*Helianthus annuus* L.). Europ J Aagron 24:291–295

Kazan K (2015) Diverse roles of jasmonates and ethylene in abiotic stress tolerance. Trends Plant Sci 20:219–229

Khan MB, Hussain M, Raza A, Farooq S, Jabran K (2015) Seed priming with $CaCl_2$ and ridge planting for improved drought resistance in maize. Turkish J Agricul Forestry 39:193–203

Kidokoro S, Watanabe K, Ohori T, Moriwaki T, Maruyama K, Mizoi J, Yamaguchi-Shinozaki K (2015) Soybean DREB 1/CBF-type transcription factors function in heat and drought as well as cold stress-responsive gene expression. Plant J 81:505–518

Kosová K, Urban MO, Vítámvás P, Prášil IT (2016) Drought stress response in common wheat, durum wheat and barley: transcriptomics, proteomics, metabolomics, physiology and breeding for an enhanced drought tolerance. In: Hossain MA, Wani SH, Bhattachajee S, Burritt DJ, L-SP T (eds) Drought stress tolerance plants, vol 2. Springer, Cham, pp 277–314

Kudo M, Kidokoro S, Yoshida T, Mizoi J, Todaka D, Fernie AR, Yamaguchi-Shinozaki K (2017) Double overexpression of DREB and PIF transcription factors improves drought stress tolerance and cell elongation in transgenic plants. Plant Biotechnol J 15:458–471

Kumar S (2018) Epigenomics of plant responses to environmental stress. Epigenomes 2:6

Kuromori T, Mizoi J, Umezawa T, Yamaguchi-Shinozaki K, Shinozaki K (2015) Stress signaling networks: drought stress. Mol Biol:1–23

Labbo AM, Mehmood M, Akhtar MN, Khan MJ, Tariq A, Sadiq I (2018) Genome-wide identification of AP2/ERF transcription factors in mung bean (*Vigna radiata*) and expression profiling of the *VrDREB* subfamily under drought stress. Crop Pasture Sci 69:1009–1019

Lesk C, Rowhani P, Ramankutty N (2016) Influence of extreme weather disasters on global crop production. Nature 529:84

Li D, Huang X, Liu Z, Li S, Okada T, Yukawa Y, Wu J (2016) Effect of AtR8 lncRNA partial deletion on Arabidopsis seed germination. Mol Soil Biol 7:1–7

Li S, Lin YCJ, Wang P, Zhang B, Li M, Chen S, Dai X (2018) Histone acetylation cooperating with AREB1 transcription factor regulates drought response and tolerance in *Populus trichocarpa*. Plant Cell tpc-00437

Lisei-de-Sá ME, Arraes FB, Brito GG, Beneventi MA, Lourenço-Tessutti IT, Basso AM, Mizoi J (2017) AtDREB2A-CA influences root architecture and increases drought tolerance in transgenic cotton. Agricult Sci 8:1195

Liu C, Mao B, Ou S, Wang W, Liu L, Wu Y, Wang X (2014) OsbZIP71, a bZIP transcription factor, confers salinity and drought tolerance in rice. Plant Mol Biol 84:19–36

Liu D, Kong DDFXK, Ali B, Xu L, Zhou WJ (2016) Influence of exogenous 5-aminolevulinic acid on chlorophyll synthesis and related gene expression in oilseed rape de-etiolated cotyledons under water-deficit stress. Photosynthetica 54:468–474

Lobell DB, Schlenker W, Costa-Roberts J (2011) Climate trends and global crop production since 1980. Science 333(6042):1204531

Ma QJ, Sun MH, Lu J, Liu YJ, You CX, Hao YJ (2017) An apple CIPK protein kinase targets a novel residue of AREB transcription factor for ABA-dependent phosphorylation. Plant Cell Environ 40:2207–2219

Manickavelu A, Nadarajan N, Ganesh SK, Gnanamalar RP, Babu RC (2006) Drought tolerance in rice: morphological and molecular genetic consideration. Plant Growth Regul 50:121–138

Mao H, Wang H, Liu S, Li Z, Yang X, Yan J, Qin F (2015) A transposable element in a NAC gene is associated with drought tolerance in maize seedlings. Nat Commun 6:8326

McGrann GR, Steed A, Burt C, Goddard R, Lachaux C, Bansal A, Brown JK (2015) Contribution of the drought tolerance-related stress-responsive NAC 1 transcription factor to resistance of barley to *Ramularia* leaf spot. Mol Plant Pathol 16:201–209

Michaletti A, Naghavi MR, Toorchi M, Zolla L, Rinalducci S (2018) Metabolomics and proteomics reveal drought-stress responses of leaf tissues from spring-wheat. Sci Rep 8:5710

Mir RR, Rather IA, Bhat MA, Parray GA, Varshney RK (2017) Molecular mapping of genes and QTLs in pigeonpea. In: Varshney RK, Saxena RK, Jackson SA (eds) The Pigeonpea Genom. Springer, Cham, pp 55–64

Mittler R (2017) ROS are good. Trends Plant Sci 22:11–19

Monclus R, Dreyer E, Villar M, Delmotte FM, Delay D, Petit JM, Brignolas F (2006) Impact of drought on productivity and water use efficiency in 29 genotypes of *Populus deltoides* × *Populus nigra*. New Phytol 169:765–777

Moradi P, Ford-Lloyd B, Pritchard J (2017) Metabolomic approach reveals the biochemical mechanisms underlying drought stress tolerance in thyme. Anal Biochem 527:49–62

Nie J, Wen C, Xi L, Lv S, Zhao Q, Kou Y, Zhou X (2018) The AP2/ERF transcription factor CmERF053 of chrysanthemum positively regulates shoot branching, lateral root and drought tolerance. Plant Cell Rep:1–12

Nuruzzaman M, Sharoni AM, Kikuchi S (2013) Roles of NAC transcription factors in the regulation of biotic and abiotic stress responses in plants. Front Microbiol 4:248

Pilon M (2017) The copper microRNAs. New Phytol 213:1030–1035

Pinasseau L, Vallverdú-Queralt A, Verbaere A, Roques M, Meudec E, Le Cunff L, Terrier N (2017) Cultivar diversity of grape skin polyphenol composition and changes in response to drought investigated by LC-MS based metabolomics. Front Plant Sci 8:1826

Praba ML, Cairns JE, Babu RC, Lafitte HR (2009) Identification of physiological traits underlying cultivar differences in drought tolerance in rice and wheat. J Agron Crop Sci 195(1):30–46

Qi J, Song CP, Wang B, Zhou J, Kangasjärvi J, Zhu JK, Gong Z (2018) Reactive oxygen species signaling and stomatal movement in plant responses to drought stress and pathogen attack. J Integr Plant Biol 60(9):805–826

Rahman H, Ramanathan V, Nallathambi J, Duraialagaraja S, Muthurajan R (2016) Over-expression of a NAC 67 transcription factor from finger millet (*Eleusine coracana* L.) confers tolerance against salinity and drought stress in rice. BMC Biotechnol 16:35

Rai AC, Singh I, Singh M, Shah K (2014) Expression of *ZAT12* transcripts in transgenic tomato under various abiotic stresses and modeling of ZAT12 protein *in silico*. Biometals 27:1231–1247

Rai KK, Rai N, Rai SP (2018a) Salicylic acid and nitric oxide alleviate high temperature induced oxidative damage in *Lablab purpureus* L plants by regulating bio-physical processes and DNA methylation. Plant Physiol Biochem 128:72–88

Rai KK, Rai N, Rai SP (2018b) Response of *Lablab purpureus* L. to high temperature stress and role of exogenous protectants in mitigating high temperature induced oxidative damages. Mol Biol Rep 45:1375–1395

Rai KK, Rai N, Rai SP (2018c) Recent advancement in modern genomic tools for adaptation of *Lablab purpureus* L to biotic and abiotic stresses: present mechanisms and future adaptations. Acta Physiol Plant 40:164

Rai KK, Rai N, Rai SP (2018d) Investigating the impact of high temperature on growth and yield of *Lablab purpureus* L. inbred lines using integrated phenotypical, physiological, biochemical and molecular approaches. Indian J Plant Physiol 23:209–226

Rakocevic M, Müller M, Matsunaga FT, Neumaier N, Farias JRB, Nepomuceno AL, Fuganti-Pagliarini R (2018) Daily heliotropic movements assist gas exchange and productive responses in DREB 1A soybean plants under drought stress in the greenhouse. Plant J 96:801–814

Roychowdhury R (2014) Crop improvement in the era of climate change. IK International Publishing House, New Delhi, p 496

Roychowdhury R, Karmakar J, Karmakar J, Adak MK, Dey N (2013) Physio-biochemical and microsatellite based profiling of lowland rice (*Oryza sativa* L.) landraces for osmotic stress tolerance. Am J Plant Sci 4(12):52

Roychowdhury R, Taoutaou A, Hakeem KR, Gawwad MR, Tah J (2014) Molecular marker-assisted technologies for crop improvement. In: Roychowdhury R (ed) Crop improvement in the era of climate change. IK International Publishing House, New Delhi, pp 241–258

Roychowdhury R, Khan MH, Choudhury S (2018) Arsenic in rice: an overview on stress implications, tolerance and mitigation strategies. In: Hasanuzzaman M, Nahar K, Fujita M (eds) Plants under metal and metalloid stress. Springer, Singapore, pp 401–415

Roychowdhury R, Khan MH, Choudhury S (2019) Physiological and molecular responses for metalloid stress in rice – a comprehensive overview. In: Hasanuzzaman M, Fujita M, Nahar K, Biswas J (eds) Advances in rice research for abiotic stress tolerance. Woodhead Publishing, Elsevier, pp 341–369

Sakuraba Y, Kim YS, Han SH, Lee BD, Paek NC (2015) The Arabidopsis transcription factor NAC016 promotes drought stress responses by repressing AREB1 transcription through a trifurcate feed-forward regulatory loop involving NAP. Plant Cell 27(6):tpc-15

Saleem MA, Malik TA, Shakeel A, Ashraf M (2015) QTL mapping for some important drought tolerant traits in upland cotton. J Animal Plant Sci 25:502–509

Sornaraj P, Luang S, Lopato S, Hrmova M (2016) Basic leucine zipper (bZIP) transcription factors involved in abiotic stresses: a molecular model of a wheat bZIP factor and implications of its structure in function. BBA-General Subjects 1860:46–56

Takasaki H, Maruyama K, Takahashi F, Fujita M, Yoshida T, Nakashima K, Shinozaki K (2015) SNAC-As, stress-responsive NAC transcription factors, mediate ABA-inducible leaf senescence. Plant J 84:1114–1123

Thirumalaikumar VP, Devkar V, Mehterov N, Ali S, Ozgur R, Turkan I, Balazadeh S (2018) NAC transcription factor JUNGBRUNNEN 1 enhances drought tolerance in tomato. Plant Biotechnol J 16:354–366

Todaka D, Shinozaki K, Yamaguchi-Shinozaki K (2015) Recent advances in the dissection of drought-stress regulatory networks and strategies for development of drought-tolerant transgenic rice plants. Front Plant Sci 6:84

Vivek BS, Krishna GK, Vengadessan V, Babu R, Zaidi PH, Kha LQ, Singh IS (2016) Use of genomic estimated breeding values results in rapid genetic gains for drought tolerance in maize. Plant Genom 10:1–8
Wang D, Yang C, Dong L, Zhu J, Wang J, Zhang S (2015) Comparative transcriptome analyses of drought-resistant and-susceptible *Brassica napus* L. and development of EST-SSR markers by RNA-Seq. J Plant Biol 58:259–269
Wang F, Guo Z, Li H, Wang M, Onac E, Zhou J, Zhou Y (2016a) Phytochrome A and B function antagonistically to regulate cold tolerance via abscisic acid-dependent jasmonate signaling. Plant Physiol 170:459–471
Wang X, Cai X, Xu C, Wang Q, Dai S (2016b) Drought-responsive mechanisms in plant leaves revealed by proteomics. Int J Mol Sci 17:1706
Wang C, Lu G, Hao Y, Guo H, Guo Y, Zhao J, Cheng H (2017) ABP9, a maize bZIP transcription factor, enhances tolerance to salt and drought in transgenic cotton. Planta 246:453–469
Wang B, Zhao M, Yao L, Babu V, Wu T, Nguyen HT (2018a) Identification of drought-inducible regulatory factors in *Lablab Purpureus* by a comparative genomic approach. Crop Pasture Sci 69:632–641
Wang CT, Ru JN, Liu YW, Yang JF, Li M, Xu ZS, Fu JD (2018b) The maize WRKY transcription factor ZmWRKY40 confers drought resistance in transgenic *Arabidopsis*. Int J Mol Sci 19:2580
Wani SH, Tripathi P, Zaid A, Challa GS, Kumar A, Kumar V, Bhatt M (2018) Transcriptional regulation of osmotic stress tolerance in wheat (*Triticum aestivum* L.). Plant Mol Biol 97:469–487
Wu H, Lv H, Li L, Liu J, Mu S, Li X, Gao J (2015) Genome-wide analysis of the AP2/ERF transcription factors family and the expression patterns of DREB genes in Moso Bamboo (*Phyllostachys edulis*). PLoS One 10:e0126657
Xia Z, Cai Z, Ma Q, Cheng Y, Liu J, Nian H (2018) GmWRKY16 enhances drought and salt tolerance through an ABA-mediated pathway in *Arabidopsis thaliana*. Front Plant Sci 9:1979
Xie H, Yang DH, Yao H, Bai GE, Zhang YH, Xiao BG (2016) iTRAQ-based quantitative proteomic analysis reveals proteomic changes in leaves of cultivated tobacco (*Nicotiana tabacum*) in response to drought stress. Biochem Biophys Res Commun 469:768–775
Xiong H, Yu J, Li J, Wang X, Liu P, Zhang H, Fu B (2017) Natural variation of OsLG3 controls drought stress tolerance in rice by inducing ROS scavenging. BioRxiv 228403
Yadav CB, Pandey G, Muthamilarasan M, Prasad M (2018) Epigenetics and epigenomics of plants. Adv Biochem Eng Biotechnol 164:237–261
Yamasaki K (2016) Structures, functions and evolutionary histories of DNA-binding domains of plant-specific transcription factors. In: Gonzalez DH (ed) Plant transcription factors. Academic Press, San Diego, pp 57–72
Yamasaki K, Kigawa T, Seki M, Shinozaki K, Yokoyama S (2013) DNA-binding domains of plant-specific transcription factors: structure, function and evolution. Trends Plant Sci 18:267–276
Yang F, Xu X, Xiao X, Li C (2009) Responses to drought stress in two poplar species originating from different altitudes. Biologia Plantar 53:511
Yordanov I, Velikova V, Tsonev T (2000) Plant responses to drought, acclimation and stress tolerance. Photosynthetica 38:171–186
Yoshida T, Fujita Y, Maruyama K, Mogami J, Todaka D, Shinozaki K, Yamaguchi-Shinozaki K (2015) Four *Arabidopsis* AREB/ABF transcription factors function predominantly in gene expression downstream of SnRK2 kinases in abscisic acid signalling in response to osmotic stress. Plant Cell Environ 38:35–49
Yu X, Liu Y, Wang S, Tao Y, Wang Z, Mijiti A, Ma H (2016) A chickpea stress-responsive NAC transcription factor, CarNAC5, confers enhanced tolerance to drought stress in transgenic *Arabidopsis*. Plant Growth Regul 79:187–197
Yuan XC, Wang Q, Wang K, Wang B, Jin JL, Wei YM (2015) China's regional vulnerability to drought and its mitigation strategies under climate change: data envelopment analysis and analytic hierarchy process integrated approach. Mitigat Adapt Strateg Global Chan 20:341–359
Zandalinas SI, Mittler R, Balfagón D, Arbona V, Gómez-Cadenas A (2018) Plant adaptations to the combination of drought and high temperatures. Physiol Plant 162:2–12

Zandkarimi H, Ebadi A, Salami SA, Alizade H, Baisakh N (2015) Analyzing the expression profile of AREB/ABF and DREB/CBF genes under drought and salinity stresses in grape (*Vitis vinifera* L.). PLoS One 10:e0134288

Zhan J, Li G, Ryu CH, Ma C, Zhang S, Lloyd A, Yadegari R (2018) Opaque-2 regulates a complex gene network associated with cell differentiation and storage functions of maize endosperm. Plant Cell 30:2425–2446

Zhang H, Li L (2013) SQUAMOSA promoter binding protein-like7 regulated microRNA408 is required for vegetative development in *Arabidopsis*. Plant J 74:98–109

Zhang L, Zhang L, Xia C, Zhao G, Liu J, Jia J, Kong X (2015) A novel wheat bZIP transcription factor, TabZIP60, confers multiple abiotic stress tolerances in transgenic Arabidopsis. Physiol Plant 153:538–554

Zhang T, Tan D, Zhang L, Zhang X, Han Z (2017) Phylogenetic analysis and drought-responsive expression profiles of the WRKY transcription factor family in maize. Agri Gene 3:99–108

Zhang B, Su L, Hu B, Li L (2018) Expression of AhDREB1, an AP2/ERF transcription factor gene from Peanut, is affected by histone acetylation and increases abscisic acid sensitivity and tolerance to osmotic stress in *Arabidopsis*. Int J Mol Sci 19:1441

Zhao G, Xu H, Zhang P, Su X, Zhao H (2017a) Effects of 2, 4-epibrassinolide on photosynthesis and Rubisco activase gene expression in *Triticum aestivum* L. seedlings under a combination of drought and heat stress. Plant Growth Regul 81:377–384

Zhao SP, Xu ZS, Zheng WJ, Zhao W, Wang YX, Yu TF, Chai SC (2017b) Genome-wide analysis of the RAV family in soybean and functional identification of GmRAV-03 involvement in salt and drought stresses and exogenous ABA treatment. Front Plant Sci 8:905

Zhao T, Aleem M, Sharmin RA (2017c) Adaptation to water stress in soybean: morphology to genetics. In: Andjelkovic V (ed) Plant, abiotic stress responses climate change. IntechOpen, pp 34–68

Zheng X, Chen L, Lou Q, Xia H, Li M, Luo L (2014) Changes in DNA methylation pattern at two seedling stages in water saving and drought-resistant rice variety after drought stress domestication. Ric Sci 21:262–270

Zong W, Tang N, Yang J, Peng L, Ma S, Xu Y, Xiong L (2016) Feedback regulation of ABA signaling and biosynthesis by a bZIP transcription factor targets drought resistance related genes. Plant Physiol 171:2810–2825

Chapter 7
Drought Stress: An Impact of Climate Change, Its Consequences and Amelioration Through Silicon (Si)

Sudeshna Das, Pratibha Rawat, Deepti Shankhdhar, and Shailesh Chandra Shankhdhar

Abstract The eighteenth century simultaneously with the onset of industrial revolution and introduction of steam engines witnessed a meteoric upsurge in urbanization coupled with harmful anthropogenic activities which in turn subsequently steered climatic changes prompting calamitous floods, drought, global warming and altered CO_2 concentration in air. Drought is one of the most imperative abiotic stresses wherein water availability recedes the optimum water requirement to an extent causing significant reduction in yield potential. The augmentation of harmful reactive oxygen species (ROS), such as hydrogen peroxide (H_2O_2) and the superoxide and hydroxyl radicals, leads to an alteration in various morphological, physiological, biochemical and molecular pathways consequently triggering the accumulation of numerous compatible osmolytes such as glycine-betaine, soluble sugars and proline and regulated expression of other genes. Drought, most importantly, influences the process of photosynthesis through the depletion of photosynthetic pigments, distortion of the structure and morphology of mesophyll cells altering the photosynthesis and gas-exchange processes as well as diminishing the reduction (electron transfer) efficiency of PSII. Silicon, in conditions of drought stress, has emerged to be a magical bullet compensating the yield loss to a great extent. Furthermore, it contributes to the amelioration of photosynthetic efficiency through the optimization of thylakoid membrane protein components. It is perceived to enhance seed germination in lentil and form a Si-cuticle double-layer diminishing transpiration rate and water flow rate in xylem vessels improving WUE up to 35% in case of maize under water stress. Silicon thus can be ascertained as a promising component to exterminate the hostile consequences of drought stress.

Keywords Drought · ROS · Photosynthesis · Silicon · WUE

S. Das · P. Rawat · D. Shankhdhar · S. C. Shankhdhar (✉)
Department of Plant Physiology, College of Basic Sciences & Humanities,
G. B. Pant University of Agriculture & Technology, Pantnagar, Uttarakhand, India

R. Roychowdhury et al. (eds.), *Sustainable Agriculture in the Era of Climate Change*, https://doi.org/10.1007/978-3-030-45669-6_7

Abbreviations

GDP	Gross domestic product
PSI	Photosystem I
PSII	Photosystem II
ROS	Reactive oxygen species
WUE	Water use efficiency

1 Climate Change: An Outline

Change in nature is inevitable, so is in the milieu of climate. However, the change which was en route to materialize in its own momentum accelerated with the genesis of the industrial revolution and introduction of steam engines in the second half of the eighteenth century, witnessing an expeditious upsurge in urbanization coupled with harmful anthropogenic activities. These inadvertent actions sequentially prompted enhanced concentration of greenhouse gases such as CO_2, N_2O and CH_4 and provoking global warming and heightening the global temperature (Chakraborty et al. 2014). The outcomes of global warming encompass an altered rainfall pattern triggering calamitous floods/drought, melting of glaciers, increased sea level (Khan et al. 2009) and varied CO_2 concentration in air. The soil-plant-air continuum is an abode of several microorganisms mandatory for the systematic functioning of the nutrient cycle in the biosphere as well as for the preservation and continuation of the natural and man-made ecosystems. Climate change is a threat to those essential organisms which are unable to adapt themselves to the rapidly changing environment and are thus either endangered or on a verge of extinction ensuing disruption of ecosystem functioning and reduction in ecosystem services. Numerous experiments have reflected that soil warming influences nutrient availability and enhances soil N mineralization, nitrate leaching and organic matter decomposition.

These configurations of climate change have an incredible potential to challenge and influence the standard food production procedure, diminishing the global food produced essential to feed the escalating world population (Roychowdhury 2014). The weather throughout the course of the crop growing season influences the crop growth and development, water utilization efficiency and yield under optimum surroundings. Surprisingly, the slightest fluctuation from the conventional weather necessary for the crop growing period is known to critically diminish the efficiency of the applied inputs aiming for better yield and crop health (Mall et al. 2006).

Food security, a fundamental need that can never be discounted by any society, therefore, has been put to the foreground by the scientific community (Mall et al. 2006). The natural agricultural systems are hugely complex which intensifies the complication of speculating the upcoming trajectory of agriculture in the changing world (Khan et al. 2009). For a country like India where agriculture and its allied sectors contribute to almost 18% of its GDP, research concerning the influence of climate change on agriculture and the appropriate measures to be undertaken seems to be a soaring priority; if not, consequences can be expected to be severe and widespread. One significantly paramount concern of climate warming outcomes is water inadequacy, drought intensification and expansion, along with the sluggish deterioration of agricultural system, anticipated influencing five billion people by 2050 (UNESCO 2018; Watts 2018).

2 Drought: The Scarcity of Water

Of all the detrimental outcomes of climate change, drought is one major destructive situation known to hamper plant growth and its development. Drought is usually a meteorological term depicting a stretch of time span devoid of rain resulting in water-deficit conditions in the soil. Drought can be referred to as an environmental event wherein water availability recedes the optimum water requirement to an extent causing significant reduction in yield potential (Ceccarelli et al. 2007; Blum 2011). Global and regional drought is adjudged to be one of the several predominant reasons for undernourishment, malnutrition and hunger, originating from reduced agricultural production (Kogan et al. 2019). The challenge, therefore, shifts towards having higher production from the restricted amount of water supply and practicing the slogan of the CGIAR Challenge Program on Water and Food, i.e. to yield 'more crop per drop' (Passioura 2006).

3 Impacts of Drought Stress

The harmful effects of drought stress are noticeable in plants ranging from the morphological to the molecular level and are perceptible at every phenological stage of crop growth at any period of time when water-deficit conditions arise (Farooq et al. 2009). The effects of drought stress on plants are elaborated and discussed (Table 7.1).

3.1 Morphological Impacts of Drought Stress

The most primary and noticeable outcomes of drought stress are impaired seed germination followed by inadequate crop establishment (Harris et al. 2002). Reduced number of stomata, thickened leaf cell walls, cutinized leaf surfaces and shrinkage of leaves are some of the frequently monitored reliable drought stress morphological parameters (Hussain et al. 2008). Grain size and grain number in crops are found to be drastically reduced in conditions of pre-anthesis drought (Dickin and Wright 2008). Leaf senescence in numerous crops is conjoined with

Table 7.1 General drought-induced modifications in plants

Characters	Drought-induced modifications	References
Morphology	Impaired seed germination and poor crop establishment	Harris et al. (2002)
	Reduced stomata number, thick cell walls and cutinized leaf surfaces	Hussain et al. (2008)
	Diminished leaf area index (LAI) and biomass	Upadhyaya and Panda (2019)
	Reduced grain size and number	Dickin and Wright (2008)
Plant-water relations	Lowered relative water content (RWC), leaf water potential and rate of transpiration	Farooq et al. (2009)
	Reduced WUE	Costa et al. (1997)
Nutrient uptake	Increased N content	Garg (2003)
	Decreased P content	Garg (2003)
	Unaltered K content	Garg (2003)
	Obstructed Zn, Cu, Mn uptake	Upadhyaya et al. (2012, 2013)
Physiology and biochemical	Depleted turgor pressure	Taiz and Zeiger (2006)
	Enhanced cytoplasmic osmolality	Marsal et al. (2002)
	Barred cell elongation and expansion and mitosis	Hussain et al. (2008)
	Retarded enzyme activity	Bota et al. (2004), Zhou et al. (2007)
	Disrupted structure of the photosynthetic apparatus	Kaiser et al. (2015), Wang et al. (2019)

drought stress-promoted modifications such as altered length, plant height, leaf area and biomass (Upadhyaya et al. 2012). Various yield components are detected to deviate from their usual pattern in case of drought stress. Parameters, namely, primary/secondary branches per panicle, panicle length, seed setting rate as well as grain weight per panicle, are observed to substantially reduce in case of drought in *O. sativa* L. Minute modifications induced due to drought conditions during the panicle development period severely hamper crop yield as development of spikelet and panicle at the reproductive stage is a salient feature governing grain yield potential (Wei et al. 2017).

The leaf area index (LAI) of the plant is observed to reduce which consequently tends to cut down the fresh as well as dry biomass production antagonistically affecting the crop plants (Upadhyaya and Panda 2019). Shoots of drought-induced *O. sativa* L. seedlings were shorter in height, its leaves dehydrated and chlorosis affected in contrast to the control which was taller and hydrated (Wang et al. 2019). Root length and shoot length in *Brassica napus* L. are detected to upsurge and diminish, respectively, with increasing degrees of drought stress. Root-shoot length ratio and dry weight and fresh weight of roots as well as shoots were sternly arrested under drought stress (Karmakar et al. 2012; Roychowdhury et al. 2013; Khan et al. 2019).

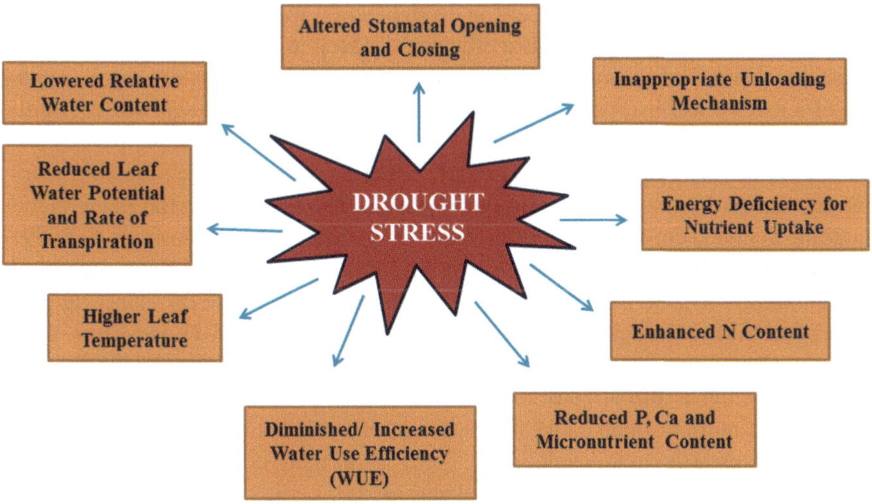

Fig. 7.1 Modifications in plant-water relations and nutrient uptake under drought conditions

3.2 Impact on Water and Nutrient Relations

Plant-water relations are another pivotal aspect influenced during drought conditions; its end-results delineated in Fig. 7.1. Characteristics, namely, relative water content (RWC), leaf water potential, rate of transpiration, stomatal resistance, canopy temperature and leaf temperature, are a few which modulate plant-water relations. Although reduced availability of water obstructs the efficient parameters of the plant-water relations, it is precisely the stomatal opening and closing that is most acutely altered (Farooq et al. 2009).

Experiments have pinpointed the fact that leaf development stages of crops such as wheat have higher relative water content which in due course of time was found to decrease with the advancement of leaf maturation and dry matter accumulation. A drop in relative water content, leaf water potential and rate of transpiration with a simultaneous increase in leaf temperature was observed when plants of rice and wheat were introduced to drought stress (Siddique et al. 2001). Additionally, it is the fluctuations of leaf temperature that can be considered as a vital factor regulating the leaf water status of the plant experiencing drought stress (Farooq et al. 2009).

Exposures of crop stand to drought stress in case of crops such as *Pinus ponderosa* and *Artemisia tridentate* (DeLucia and Heckathorn 1989), lucerne (*Medicago sativa*) (Lazaridou et al. 2003), wheat (*Triticum aestivum*) (Abbate et al. 2004) and clover (*Trifolium alexandrinum*) (Lazaridou and Koutroubas 2004) have led to the advocation of the observation that water-use efficiency under circumstances of drought increased when compared to the well-watered conditions. This could be attributed to the fact that water-deficit conditions are known to reduce the leaf area, hinder stomatal conductance and encourage stomatal closure so as to cut down the transpiration rate. On the contrary, crops such as potato (*Solanum tuberosum*) have huge adverse effect under early-season drought stress tremendously hampering the growth and development, water-use efficiency as well as the biomass accumulation (Costa et al. 1997). Likewise, examining the case of *Hibiscus rosa-sinensis*, water-use efficiency was observed to decline in plants undergoing drought stress (Egilla et al. 2005).

One of the many outcomes that drought usually results in is the impeding nutrient acquisition by the plant root, their improper transportation to the plant shoots entailing declined concentration of nutrients in the plant tissues (Farooq et al. 2009). Inappropriate unloading mechanism and curtailed transpirational flow, as well as nutrient uptake, can be quoted as some of the factors responsible for scanty nutrient absorption during drought stress (Garg 2003; McWilliams 2003). However, apart from the altered transpirational flow and unloading mechanism, it is the deficiency of energy for the assimilation of ions, such as SO_4^{2-}, NO_3^- /NH_4^+ and PO_4^{3-}, which hampers the nutrient uptake and its translocation to the shoots. These ions can be utilized for growth and developments only post their conversion into energy-dependent processes (Grossman and Takahashi 2001).

The intimate relation between water and nutrient requirement pinpoints towards the likelihood that fertilizer application may increase the crop efficiency in utilizing the available water demonstrating a crucial interaction amid nutrient uptake and soil moisture deficits. Studies have unveiled the aspect that crop yields inhibited by drought stress can be markedly enhanced by boosting the plant nutrient efficiency (Garg 2003). Habitually, water-deficit conditions incite an increment in nitrogen content and a reduction in the phosphorous content, whereas no definitive influence on the potassium content is witnessed (Garg 2003). Correspondingly, P and PO_4^{3-} contents in the plant tissues declined drastically under water-deficit conditions conceivably owing to diminished PO_4^{3-} mobility as a consequence of scarce moisture availability (Peuke and Rennenberg 2004). Drought stress has also been perceived to obstruct the uptake of nutrients such as Ca, Zn, Cu, Mn, Mg as well as B (Upadhyaya et al. 2012, 2013).

The inability of the plant to uptake nutrients such as Zn, vital for several physiological processes in plants, leads to stunted growth, altered membrane integrity and severely reduced crop production. Investigations have highlighted that nitrogen and potassium uptake was cut down in cotton plants encountering drought stress (McWilliams 2003). Chen et al. (2011) with their experiments and trials in rice testified an enhancement in K and Fe contents in rice. Alterations in the process of absorption and translocation of essential elements by plants under drought-inflicted conditions, in turn, lead to oxidative stress in plants (Fahad et al. 2015; Roychowdhury et al. 2018, 2019).

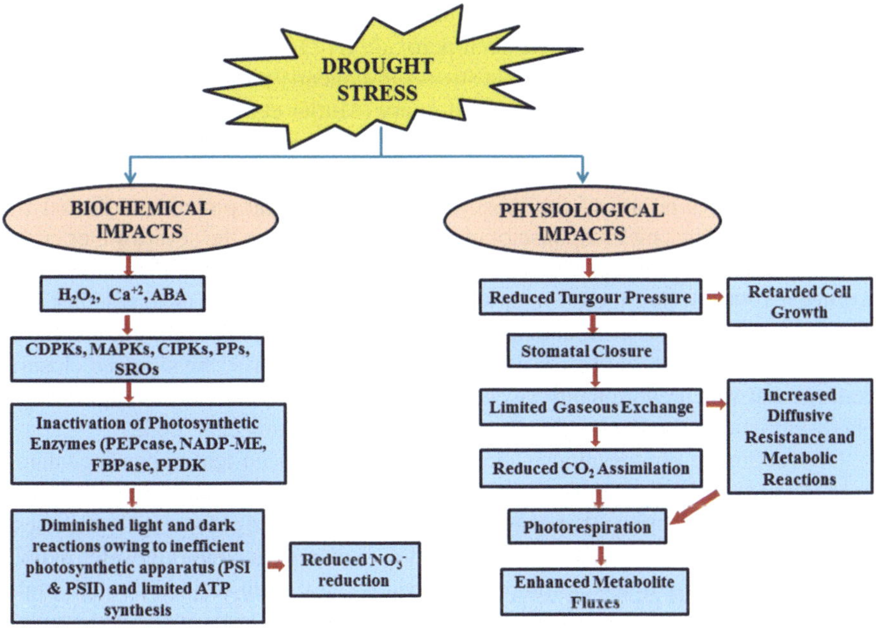

Fig. 7.2 Physiological and biochemical impacts on plants encountering drought stress. ABA, abscisic acid; H_2O_2, hydrogen peroxide; CDPKs, calcium-dependent protein kinases; PPs, protein phosphatases; SROs, similar to RCD-ONE; MAPKs, mitogen-activated protein kinases; CIPKs, CBL-interacting protein kinases; PPDKs, pyruvate phosphate dikinases; PEPCase, phosphoenolpyruvate carboxylase; FBPase, fructose 1,6-bisphosphatase; NADP-ME, NADP malic enzyme. (Modified from Zargar et al. 2017)

3.3 Physiological and Biochemical Impacts of Drought Stress

Drought is known to influence varied physiological processes in plants, thus constraining growth, development and productivity of crops (Fig. 7.2). The most imperative physiological process affected by drought stress is cell growth caused by depletion in turgor pressure (Taiz and Zeiger 2006). Subjected to acute water deficiency, cell elongation in higher plants is restrained through the intervention of water flow from the xylem to the adjacent elongating cells (Nonami 1998). Stunted plant height and decreased leaf area along with retarded crop growth is the outcome of disabled cell elongation and expansion and mitosis (Hussain et al. 2008).

Drought stress enhances the osmolality in the cytosolic compartment (Marsal et al. 2002). Deviations in the concentrations of several metabolites disturbing the carbohydrate and amino acid metabolism occur as a result of decreased water availability to plants (Lee et al. 2008, 2009; Hasanuzzaman et al. 2015). Drought stress is known to incite the production of reactive oxygen species (ROS) and proline (Rejeb et al. 2014) controlled by the hormonal signaling pathway (Herrera-Vásquez et al. 2015) along with the increment of osmolytes such as the free amino acid glycine-betaine and soluble sugars (Fariduddin et al. 2009). The elevated production of toxic ROS, including hydrogen peroxide (H_2O_2) as well as the superoxide and hydroxyl radicals, during drought stress subsequently prompts the generation of oxidative stress leading to cell death and injury (Mittler et al. 2004). The ability of a plant to strike a balance amid the production and eradication of ROS determines its survival (Golshan et al. 2011).

Photosynthesis is one of the most salient physiological processes affected by drought stress. The chief response of essentially all plants to the severe water crisis is the stomatal closure to check the transpirational water loss (Mansfield and Atkinson 1990). It has, therefore, always been a debatable topic as to whether it is the stomatal closure or the metabolic impairment that mainly delimits photosynthesis during drought stress (Tezara et al. 1999). It was, however, the stomatal closure concept that was predominantly accepted to be the chief factor for the diminished photosynthetic rate under mild to moderate drought (Yokota et al. 2002).

Obstruction in photosynthesis emerges through reduced leaf expansion, damaged photosynthetic machinery, premature leaf senescence and accompanying declined food production (Wahid and Rasul 2005). Hindered chlorophyll syntheses, as well as enhanced chlorophyll decomposition in plants encountering water-deficit conditions, resulting in lower chlorophyll content, have been verified through trials (Wu et al. 2008). Subsequent studies, experiments and field trials conducted eventually unfolded the fact that in addition to stomatal closure, impaired enzyme activity, as well as disrupted structure of photosynthetic apparatus and proteins under water-deficit conditions, added up to the disturbing process of photosynthesis in plants. Inspections have also revealed the fact that under drought stress, the mesophyll cell structure and morphology were highly distorted (Hu et al. 2018).

Rubisco, a key enzyme in photosynthesis, under severe drought conditions has a reduction in its activity (Bota et al. 2004). Levels of this enzyme in the leaves managed by the rate of its synthesis and degradation witness a speedy depletion in the production of its small subunit transcripts, indicating its overall diminished synthesis (Vu et al. 1999). The activity of Rubisco is influenced in vivo either by reaction with Mg^{2+} and CO_2 in order to carbamylate a lysine residue located in its catalytic site or through binding inhibitors within the catalytic site. Binding of an inhibitor within the catalytic site either obstructs the activity of the enzyme or the lysine residue carbamylation, essential for its activity (Taiz and Zeiger 2006).

Diminution in Rubisco activity in tobacco (*Nicotiana tabacum*) under drought stress, through experiments conducted, could be attributed to the existence of tight-binding inhibitors rather than alterations in the enzyme's activation by CO_2 and Mg^{2+} (Parry et al. 2002). A rapid drop in photosynthesis during drought conditions is followed by the declined maximum velocity of ribulose-1,5-bisphosphate carboxylation by Rubisco, speed of ribulose-1,5-bisphosphate regeneration and Rubisco and stromal fructose bisphosphatase activities (Zhou et al. 2007). Rubisco carboxylation efficiency in case of severe drought conditions was observed to diminish immensely, and it was perceived to perform more as an oxygenase rather than a carboxylase enzyme. At times of water stress, activities of the enzymes such as phosphoenolpyruvate carboxylase, fructose-1,6-bisphosphatase, nicotinamide adenine dinucleotide phosphatemalic enzyme, pyruvate orthophosphate dikinase and Rubisco reduced linearly with diminishing leaf water potential (Farooq et al. 2009).

Concentrating on the structure of photosynthetic machinery, it is perceived that the thylakoid membrane is a highly significant fraction of plant photosynthetic apparatus, delicate to the changes in the external environment. It, therefore, swiftly responds to hostility by altering structure and functions to attain an optimum photosynthetic performance (Kaiser et al. 2015). Trials have pinpointed the fact that quantities of the principal thylakoid membrane protein complexes diminished under drought stress induced through PEG (Wang et al. 2019).

Psa O is known to bring to equilibrium the excitation energy between PSI and PSII (Jensen et al. 2004). An element of the core complex of PSII, PSII 47 kDa protein, binds chlorophyll and facilitates the catalysation of the primary light-induced photochemical processes of PSII. Assessments have detected the fact that quantities of Psa O protein and PSII 47 kDa protein declined under PEG-induced drought stress highlighting that the photosystem of rice seedlings was damaged under drought stress impeding the transmission of light and energy. These studies unravelled the fact that function of PSII would be directly or indirectly affected by drought stress (Wang et al. 2019).

Nitrogen metabolism is another physiological process modified by drought stress. Modulation of plant metabolism during drought stress is through reduced activity of nitrate reductase (Xu et al. 2015a, b). Decreased nitrate reductase activity concomitantly influences photosynthesis, it being a crucial component in attenuating the stomatal limitation to photosynthesis arising under water-limiting conditions. Additionally, it maintains enhanced Rubisco activity as well as heightens the nitrate and ammonia assimilation (Zhong et al. 2017). Glutamate dehydrogenase gene when overexpressed checks the accumulation of toxic ammonia as well as upsurges the production of proline, glutamate and soluble sugar conferring drought tolerance to rice (Zhou et al. 2015). The NH_4^+ and NO_3^- ratio in drought-stressed plants was detected to enhance which could be accredited to the relatively lower energy requirement of NH_4^+ when compared to NO_3^-. Decreased NO_3^- uptake rather than NH_4^+ uptake and N assimilation could be another probable explanation for the elevated NH_4^+-NO_3^- ratio in drought-induced plants (Meng et al. 2016). Moreover, curtailed NO_3^- uptake under water-stressed conditions can be well explained on the basis of reduced expression of NRT genes and NO_3^- concentration in diverse tissues. The alterations in NO_3^- assimilation and photosynthesis prompted by water stress may additionally advocate that nitrate reduction in leaves was restricted by declined photosynthesis, which is known to be an important source for supplying reducing power for the nitrate reduction process generated during photosynthetic electron transport (Chen et al. 2018).

4 Amelioration Through Silicon (Si)

Promoted from the category of 'beneficial' element to 'essential' element, silicon is the youngest member of the essential element family. Despite Si being the second most abundant element in the Earth's crust, its content in soil varying from <1–45% dry weight (Sommer et al. 2006), it has largely been overlooked by the plant physiologists for decades, and thus its potential in terms of crop growth, protection and production remains highly concealed. Despite its profusion in nature in the form of silicate minerals, secondary aluminosilicates and various forms of silicon dioxide, it can only be absorbed by the plant in monomeric form, i.e. silicic acid H_4SiO_4. Plants take up Si (as silicic acid) in amounts that are equivalent to the number of absorbed macronutrients such as N, K and Ca, and the Si is translocated to the shoots where it is stored primarily in the external walls of the epidermal cells as amorphous silica in particles known as phytoliths (Shakoor et al. 2014; Roychowdhury et al. 2019). The most striking feature of silicon is that it is the only known nutrient that is proved not to be detrimental when accumulated in excess (Epstein and Bloom 2005; Roychowdhury et al. 2019).

A consequence of Si on crop growth, yield and quality has been well documented (Liang et al. 2015). Field application of silicate fertilizers on a long-term basis in soil has shown to have a drastic impact on yield enhancement of various crops as depicted in Table 7.2.

The highest percent of yield augmentation is observed in the case of peach followed by cabbage and wheat. The least response of silicate fertilizers was calculated in cases of ginseng, sugar beet and banana. Silicon has been found to be effective in alleviating drought stress as depicted in Fig. 7.3 and enhancing drought tolerance in a wide range of crops which include rice (Agarie et al. 1998b), maize (Gao et al. 2006), sunflower (Gunes et al. 2008), cucumber (Hattori et al. 2008), pepper (Lobato et al. 2009), soybean (Hamayun et al. 2010; Lee et al. 2010), sorghum (Sonobe et al. 2010; Ahmed et al. 2011) and wheat (Gong and Chen 2012). Higher germination percentage which is as high as 2.5-fold, increased fresh and dry weight of the seedlings, enhanced levels of osmolytes, triggered activities of hydrolytic enzymes and declining quantities of ROS and lipid peroxidation in seedlings were affirmed when

Table 7.2 Effects of slow-released potassium silicate on 20 crops yield in 26 provinces of China (Liu et al. 2011)

Crops tested	Yield increase (%)	Crops tested	Yield increase (%)
Wheat	13.8	Chili pepper	8.4
Potato	12.3	Peach	18.1
Peanut	6.7	Banana	4.8
Radish	11.2	Citrus	12.3
Soybean	5.1	Longan	10.7
Green bean	6.0	Tea	11.0
Sugar beet	4.7	Ginseng	3.2
Tomato	8.8	Papaya	9.7
Cabbage	15.2	Grapevine	6.5

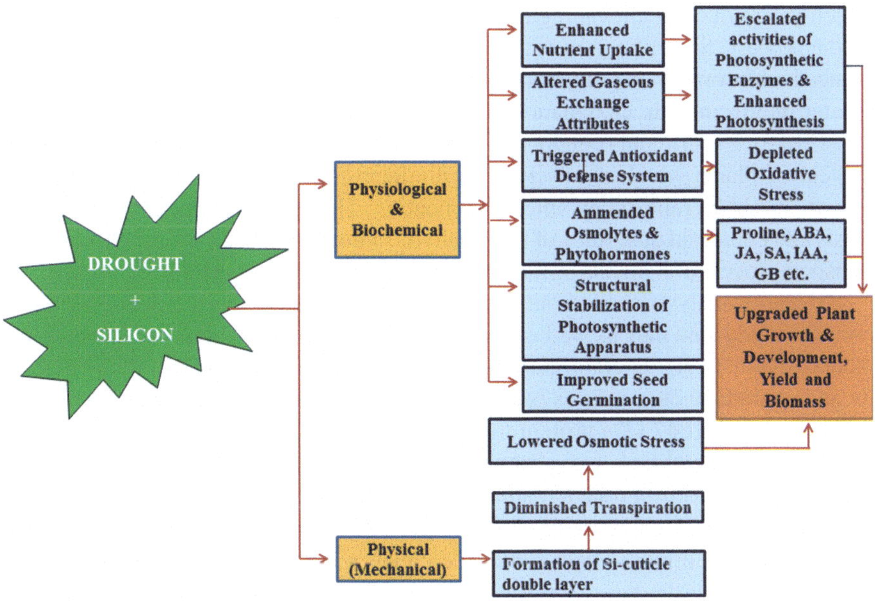

Fig. 7.3 Beneficial outcomes of Si application on drought-inflicted crops. (Modified from Rizwan et al. 2015)

extensive experiments on lentil under water-deficit conditions with silicon application were conducted (Biju et al. 2017).

Si nutrition for drought-affected crops in case of maize is known to impede the excessive leaf transpiration as well as diminish the water loss by subsiding the water flow rate in xylem improving WUE up to 35%. This could be attributed to the formation of a Si-cuticle double-layer diminishing transpiration rate and the increment of the appetency to water in xylem vessels owing to the deposition of Si on the cell wall that rate of water flow in xylem decreases (Gao et al. 2006). Under water-deficit conditions, Si fertilization contributes to maintaining the leaf water content at the higher side by diminishing the leaf water loss and augmenting the root water uptake (Yin et al. 2014). Similar results in wheat were observed by Maghsoudi et al. (2016).

Drought is sometimes coupled with high temperatures (Epstein 1994; Hasanuzzaman et al. 2013). A study led by Agarie et al. (1998) validated the contribution of Si in reducing the electrolyte leakage occurring as a result of high temperature. Electrolyte leakage owing to increased temperature was found to be lower in the leaves of the plants treated with Si as compared to those that were not treated with Si. The investigators accredited this outcome to the association of Si to the thermal stability of lipids in cell membranes, even though the definite mechanism of the outcome was not well understood and, thus, required further elaborate analysis.

Since it has been confirmed that photosynthesis is one of the most integral plant processes altered by drought, it is, therefore, of utmost importance to verify the effectiveness of Si in alleviating the deleterious implications that the stress implies on the process. Si fertilization in case of drought-affected plants has helped achieve reduced chlorophyll degradation, enhance stomatal conductance and maintain greater photosynthetic enzymatic activities and net photosynthetic rate (Gong and Chen 2012). Experiments conducted by Yin et al. (2014) concluded that chlorophyll content of plants which is observed to diminish under drought, hampering plant growth, could be relieved by silicon application. The structural integrity of chloroplasts and enhanced quantities of chlorophyll is known to encourage the capture and transport process of light during drought stress in various crops (Zhu and Gong 2014).

Extensive analysis carried out by Wang et al. (2019) is in accordance with the previous study, and they as well established the fact that Si addition could assist the plant to maintain higher chlorophyll content and greater thylakoid membrane stability and upgrade photosynthesis performance under drought stress. The upregulation of Psa O and PSII 47 kDa protein through Si application led to the stabilization of the thylakoid membrane components. The analysts also stated that Si treatment could enhance electron transfer, strengthen the activity of reaction centre PSII and boost the transmission capacity of electron transfer chain of rice seedlings during water-deficit conditions, which could help enhance the adaptive capability of rice to drought stress. Furthermore, the study revealed that quantities of the core thylakoid membrane protein complexes known to decline under PEG-induced drought stress could be revived and enhanced in varying degrees through supplementation of Si, especially extents of PSI and LHCI stating that Si has a larger impact on PSI than PSII in ameliorating drought stress in rice seedlings.

Quantity of glyceraldehyde-3-phosphate dehydrogenase in drought-stressed Si-treated rice seedlings was observed declining, which is known to have a critical role in plant glycolysis, gluconeogenesis, Calvin cycle and resistance of oxidative stress caused by ROS (Bedhomme et al. 2012). Thus its reduction signals that Si application would cut down the oxidative stress and the production of ROS. Furthermore, Si is deduced to control the oxidative damage in crop plants such as wheat under drought stress via adjustment of the plant's antioxidant defence system (Tayyab et al. 2018).

5 Conclusions

Drought can, thus, be perceived as major abiotic stress and a sequel to climate change taking a toll on the environment as well as agriculture and food production. Drought out-turns several harmful effects such as hampered seed germination, poor crop stand, and decreased number of stomata, grain number and grain size. Thickened cell walls, cutinized leaf surfaces and shrunken leaves are some of the other morphological impacts of drought. Drought modulates plant-water relations observed through descending relative water content, leaf water potential and rate of transpiration and obstructs nutrient uptake in plant curtailing plant growth and development. Drought, at the physiological level, is deduced to be a major constraint to cell elongation and expansion, mitosis, enzyme activities, production of photosynthetic pigments and nitrogen metabolism which, in turn, steers the damage to photosynthetic machinery and thylakoid membrane complex and proteins.

Silicon, in conditions of drought stress, has emerged to be a magical bullet compensating the yield loss to a great extent. Furthermore, it contributes to the amelioration of reduced photosynthetic efficiency through the optimization of thylakoid membrane protein components. It is perceived to enhance seed germination in lentil and form a Si-cuticle double-layer diminishing transpiration rate and water flow rate in xylem vessels improving WUE up to 35% in case of maize under water stress. Silicon thus can be ascertained as a promising component to exterminate the hostile consequences of drought stress. Future physiological and molecular researches pertaining to the effectiveness of Si will for bye aid in exploiting its prospective in combating drought stress.

References

Abbate PE, Dardanellib JL, Cantareroc MG, Maturanoc M, Melchiorid RJM, Sueroa EE (2004) Climatic and water availability effects on water-use efficiency in wheat. Crop Sci 44:474–483

Agarie S, Hanaoka N, Ueno O, Miyazaki A, Kubota F, Agata W, Kaufman PB (1998) Effects of silicon on tolerance to water deficit and heat stress in rice plants (*Oryza sativa* L.), monitored by electrolyte leakage. Plant Prod Sci 1:96–103

Agarie S, Uchida H, Agata W, Kubota F, Kaufman PB (1998b) Effects of silicon on transpiration and leaf conductance in rice plants (*Oryza sativa* L.). Plant Prod Sci 1:89–95

Ahmed M, Qadeer U, Aslam MA (2011) Silicon application and drought tolerance mechanism of sorghum. Afr J Agric Res 6:594–607

Bedhomme M, Adamo M, Marchand C, Couturier J, Rouhier N, Lemaire SD, Zaffagnini M, Trost P (2012) Glutathionylation of cytosolic glyceraldehyde-3-phosphate dehydrogenase from the model plant *Arabidopsis thaliana* is reversed by both glutaredoxins and thioredoxins *in vitro*. Biochem J 3:337–347

Biju S, Fuentes S, Gupta D (2017) Silicon improves seed germination and alleviates drought stress in lentil crops by regulating osmolytes, hydrolytic enzymes and antioxidant defense system. Plant Physiol Biochem 119:250–264

Blum A (2011) Plant breeding for water-limited environments. Springer, New York, pp 53–55

Bota J, Flexas J, Medrano H (2004) Is photosynthesis limited by decreased Rubisco activity and RuBP content under progressive water stress? New Phytol 162:671–681

Ceccarelli S, Grando S, Baum M (2007) Participatory plant breeding in water-limited environments. Environ Exp Agric 43:411–435

Chakraborty S, Pattanayak A, Mandal S, Das M, Roychowdhury R (2014) An overview of climate change: causes, trends and implications. In: Roychowdhury R (ed) Crop improvement in the era of climate change. IK International Publishing House, New Delhi, pp 1–29

Chen W, Yao X, Cai K, Chen J (2011) Silicon alleviates drought stress of rice plants by improving plant water status, photosynthesis and mineral nutrient absorption. Biol Trace Elem Res 142:67–76

Chen H, Zhang Q, Lu Z, Xu F (2018) Accumulation of ammonium and reactive oxygen mediated drought-induced rice growth inhibition by disturbed nitrogen metabolism and photosynthesis. Plant Soil 431:107–117

Costa LD, Vedove GD, Gianquinto G, Giovanardi R, Peressotti A (1997) Yield, water use efficiency and nitrogen uptake in potato: influence of drought stress. Potato Res 40:19–34

DeLucia EH, Heckathorn SA (1989) The effect of soil drought on water-use efficiency in a contrasting Great Basin desert and Sierran montane species. Plant Cell Environ 12:935–940

Dickin E, Wright D (2008) The effects of winter waterlogging and summer drought on the growth and yield of winter wheat (*Triticum aestivum* L.). Eur J Agron 28:234–244

Egilla JN, Davies FT Jr, Boutton TW (2005) Drought stress influences leaf water content, photosynthesis, and water-use efficiency of *Hibiscus rosa-sinensis* at three potassium concentrations. Photosynthetica 43:135–140

Epstein E (1994) The anomaly of silicon in plant biology. Proc Natl Acad Sci U S A 91:11–17

Epstein E, Bloom AJ (2005) Mineral nutrition of plants: principles and perspectives, 2nd edn. Sinauer Associates. Inc, Sunderland

Fahad S, Hussain S, Matloob A, Khan FA, Khaliq A, Saud S, Hassan S, Shan D, Khan F, Ullah N, Faiq M, Khan MR, Tareen AK, Khan A, Ullah A, Ullah N, Huang J (2015) Phytohormones and plant responses to salinity stress: a review. Plant Growth Regul 75:391–404

Fariduddin Q, Yusuf M, Hayat S, Ahmad A (2009) Effect of 28-homobrassinolide on antioxidant capacity and photosynthesis in *Brassica juncea* plants exposed to different levels of copper. Environ Exp Bot 66:418–424

Farooq M, Wahid A, Kobayashi N, Fujita D, Basra SMA (2009) Plant drought stress: effects, mechanisms and management. In: Sustainable agriculture. Springer, Dordrecht, pp 153–188

Gao X, Zou C, Wang L, Zhang F (2006) Silicon decreases transpiration rate and conductance from stomata of maize plants. J Plant Nutr 29:1637–1647

Garg BK (2003) Nutrient uptake and management under drought: nutrient-moisture interaction. Curr Agric 27:1–8

Golshan M, Habibi D, Beladi SM, Maleki MJ (2011) Copper and lead tolerance strategies in mustard (*Sinapis arvensis*) Egyptian clover (*Trifolium alexandrinum*) and hairy vetch (*Vicia villosa*): role of some antioxidant enzymes. Am Eurasian J Agric Environ Sci 11:122–128

Gong H, Chen K (2012) The regulatory role of silicon on water relations, photosynthetic gas exchange, and carboxylation activities of wheat leaves in field drought conditions. Acta Physiol Plant 34:1589–1594

Grossman A, Takahashi H (2001) Macronutrient utilization by photosynthetic eukaryotes and the fabric of interactions. Annu Rev Plant Phys 52:163–210

Gunes A, Pilbeam DJ, Inal A, Coban S (2008) Influence of silicon on sunflower cultivars under drought stress, I: growth, antioxidant mechanisms, and lipid peroxidation. Commun Soil Sci Plant Ana 39:1885–1903

Hamayun M, Sohn EY, Khan SA, Shinwari ZK, Khan AL, Lee IJ (2010) Silicon alleviates the adverse effects of salinity and drought stress on growth and endogenous plant growth hormones of soybean (*Glycine max* L.). Pak J Bot 42:1713–1722

Harris D, Tripathi RS, Joshi A (2002) On-farm seed priming to improve crop establishment and yield in dry direct-seeded rice. Direct seeding: research strategies and opportunities. International Research Institute, Manila, pp 231–240

Hasanuzzaman M, Nahar K, Alam M, Roychowdhury R, Fujita M (2013) Physiological, biochemical, and molecular mechanisms of heat stress tolerance in plants. Int J Mol Sci 14(5):9643–9684

Hasanuzzaman M, Roychowdhury R, Karmakar J, Dey N, Nahar K, Fujita M (2015) Recent advances in biotechnology and genomic approaches for abiotic stress tolerance in crop plants. In: Devarajan T, Jeyabalan S (eds) Genomics and proteomics: concepts, technologies and applications. Apple Academic Press, Burlington, pp 333–366

Hattori T, Sonobe K, Inanaga S, An P, Morita S (2008) Effects of silicon on photosynthesis of young cucumber seedlings under osmotic stress. J Plant Nutr 31:1046–1058

Herrera-Vásquez A, Salinas P, Holuigue L (2015) Salicylic acid and reactive oxygen species interplay in the transcriptional control of defense genes expression. Front Plant Sci 6:171. https://doi.org/10.3389/fpls.2015.00171

Hu W, Tian SB, Di Q, Duan SH, Dai K (2018) Effects of exogenous calcium on mesophyll cell ultrastructure, gas exchange, and photosystem II in tobacco (*Nicotiana tabacum* Linn.) under drought stress. Photosynthetica 56:1204–1211

Hussain M, Malik MA, Farooq M, Ashraf MY, Cheema MA (2008) Improving drought tolerance by exogenous application of glycinebetaine and salicylic acid in sunflower. J Agron Crop Sci 194:193–199

Jensen PE, Haldrup A, Zhang SP, Scheller HV (2004) The PSI-O subunit of plant photosystem I is involved in balancing the excitation pressure between the two photosystems. J Biol Chem 23:24212–24217

Kaiser E, Morales A, Harbinson J, Kromdijk J, Heuvelink E, Marcells LF (2015) Dynamic photosynthesis in different environmental conditions. J Exp Bot 9:2415–2426

Karmakar J, Roychowdhury R, Kar RK, Deb D, Dey N (2012) Profiling of selected indigenous rice (*Oryza sativa* L.) landraces of Rarh Bengal in relation to osmotic stress tolerance. Physiol Mol Biol Plants 18(2):125–132

Khan SA, Kumar S, Hussain MZ, Kalra N (2009) Climate change, climate variability and Indian agriculture: impacts vulnerability and adaptation strategies. In: Climate change and crops. Springer, Berlin/Heidelberg, pp 19–38

Khan MN, Zhang J, Luo T, Liu J, Ni F, Rizwan M, Fahad S, Hu L (2019) Morpho-physiological and biochemical responses of tolerant and sensitive rapeseed cultivars to drought stress during early seedling growth stage. Acta Physiol Plant 41:25

Kogan F, Guo W, Yang W (2019) Drought and food security prediction from NOAA new generation of operational satellites. Geomat Nat Haz Risk 10:651–666

Lazaridou M, Koutroubas SD (2004) Drought effect on water use efficiency of berseem clover at various growth stages. In New directions for a diverse planet: Proceedings of the 4th international crop science congress Brisbane, Australia, 26 Sept–1 Oct 2004

Lazaridou M, Kirilov A, Noitsakis B, Todorov N, Katerov I (2003) The effect of water deficit on yield and water use efficiency of lucerne. In: Optimal forage systems for animal production and the environment, Proceedings of the 12th symposium of the European Grassland Federation. Pleven, Bulgaria, pp 344–347

Lee BR, Jin YL, Jung WJ, Avice JC, Morvan-Bertrand A, Qurry A, Kim TH (2008) Water-deficit accumulates sugars by starch degradation-not by de novo synthesis-in white clover leaves (*Trifolium repens*). Physiol Plant 134:403–411

Lee BR, Jin YL, Avice JC, Cliquet JB, Ourry A, Kim TH (2009) Increased proline loading to phloem and its effects on nitrogen uptake and assimilation in water-stressed white clover (*Trifolium repens*). New Phytol 182:654–663

Lee SK, Sohn EY, Hamayun M, Yoon JY, Lee IJ (2010) Effect of silicon on growth and salinity stress of soybean plant grown under hydroponic system. Agrofor Syst 80:333–340

Liang Y, Nikolic M, Bélanger R, Gong H, Song A (2015) Effect of silicon on crop growth, yield and quality. In: Silicon in agriculture. Springer, Dordrecht, pp 209–223

Liu JM, Han C, Sheng XB, Liu SK, Qi X (2011) Potassium containing silicate fertilizer: its manufacturing technology and agronomic effects. In: Oral presentation at 5th international conference on Si Agriculature, Beijing, pp 13–18

Lobato AKS, Luz LM, Costa RCL, Santos Filho BG, Meirelles ACS, Oliveira Neto CF, Laughinghouse HD, Neto MAM, Alves GAR, Lopes MJS, Neves HKB (2009) Silicon exercises influence on nitrogen compounds in pepper subjected to water deficit. Res J Biol Sci 4:1048–1055

Maghsoudi K, Emam Y, Pessarakli M (2016) Effect of silicon on photosynthetic gas exchange, photosynthetic pigments, cell membrane stability and relative water content of different wheat cultivars under drought stress conditions. J Plant Nutr 39:1001–1015

Mall RK, Singh R, Gupta A, Srinivasan G, Rathore LS (2006) Impact of climate change on Indian agriculture: a review. Clim Chang 78:445–478

Mansfield TJ, Atkinson CJ (1990) Stomatal behaviour in water stressed plants. In: Alscher RG, Cumming JR (eds) Stress responses in plants: adaptation and acclimation mechanisms. Wiley-Liss, New York, pp 241–264

Marsal J, Gelly M, Mata M, Arbonés A, Rufat J, Girona J (2002) Phenology and drought affects the relationship between daily trunk shrinkage and midday stem water potential of peach trees. J Hortic Sci Biotechnol 77:411–417

McWilliams D (2003) Drought strategies for cotton, cooperative extension service circular 582, College of Agriculture and Home Economics, New Mexico State University, USA
Meng S, Zhang C, Su L, Li Y, Zhao Z (2016) Nitrogen uptake and metabolism of Populus simonii in response to PEG-induced drought stress. Environ Exp Bot 123:78–87
Mittler R, Vanderauwera S, Gollery M, Breusegem FV (2004) Abiotic stress series. Reactive oxygen gene network of plants. Trends Plant Sci 9:490–498
Nonami H (1998) Plant water relations and control of cell elongation at low water potentials. J Plant Res 111:373–382
Parry MAJ, Andralojc PJ, Khan S, Lea PJ, Keys AJ (2002) Rubisco activity: effects of drought stress. Ann Bot 89:833–839
Passioura J (2006) The drought environment: physical, biological and agricultural perspectives. J Exp Bot 58:113–117
Peuke AD, Rennenberg H (2004) Carbon, nitrogen, phosphorus, and sulphur concentration and partitioning in beech ecotypes (*Fagus sylvatica* L.): phosphorus most affected by drought. Trees 18:639–648
Rejeb KB, Abdelly C, Savouré A (2014) How reactive oxygen species and proline face stress together. Plant Physiol Biochem 80:278–284
Rizwan M, Ali S, Ibrahim M, Farid M, Adrees M, Bharwana SA, Rehman MZ, Qayyum MF, Abbas F (2015) Mechanisms of silicon-mediated alleviation of drought and salt stress in plants: a review. Environ Sci Pollut Res 22:15416–15431
Roychowdhury R (2014) Crop improvement in the era of climate change. IK International Publishing House, New Delhi, p 496
Roychowdhury R, Karmakar J, Karmakar J, Adak MK, Dey N (2013) Physio-biochemical and microsatellite based profiling of lowland rice (*Oryza sativa* L.) landraces for osmotic stress tolerance. Am J Plant Sci 4(12):–52
Roychowdhury R, Khan MH, Choudhury S (2018) Arsenic in rice: an overview on stress implications, tolerance and mitigation strategies. In: Hasanuzzaman M, Nahar K, Fujita M (eds) Plants under metal and metalloid stress. Springer, Singapore, pp 401–415
Roychowdhury R, Khan MH, Choudhury S (2019) Physiological and molecular responses for metalloid stress in rice – a comprehensive overview. In: Hasanuzzaman M, Fujita M, Nahar K, Biswas J (eds) Advances in rice research for abiotic stress tolerance. Woodhead Publishing, Elsevier, Duxford, pp 341–369
Shakoor SB, Bhat MA, Mir SH (2014) Phytoliths in plants: a review. Res Rev J Bot Sci 3:10–24
Siddique MRB, Hamid A, Islam MS (2001) Drought stress effects on water relations of wheat. Bot Bull Acad Sinica 41:35–39
Sommer M, Kaczorek D, Kuzyakov Y, Breuer J (2006) Silicon pools and fluxes in soils and landscapes—a review. J Plant Nutr Soil Sci 169:310–329
Sonobe K, Hattori T, An P, Tsuji W, Eneji AE, Kobayashi S, Kawamura Y, Tanaka K, Inanaga S (2010) Effect of silicon application on sorghum root responses to water stress. J Plant Nutr 34:71–82
Taiz L, Zeiger E (2006) Plant physiology, 4th edn. Sunderland, Sinauer Associates Inc. Publishers
Tayyab M, Islam W, Zhang H (2018) Promising role of silicon to enhance drought resistance in wheat. Commun Soil Sci Plant Ana 49:2932–2941
Tezara W, Mitchell VJ, Driscoll SD, Lawlor DW (1999) Water stress inhibits plant photosynthesis by decreasing coupling factor and ATP. Nature 401:914–917
UNESCO (2018) Climate Change and Water Security. https://en.unesco.org/themes/addressing-climate-change/climate-change-and-water-security
Upadhyaya H, Panda SK (2019) Drought stress responses and its management in rice. In: Advances in rice research for abiotic stress tolerance. Woodhead Publishing, pp 177–200
Upadhyaya H, Dutta BK, Sahoo L, Panda SK (2012) Comparative effect of Ca, K, Mn and B on post-drought stress recovery in tea [*Camellia sinensis* (L.) O Kuntze]. Am J. Plant Sci 3:443
Upadhyaya H, Dutta BK, Panda SK (2013) Zinc modulates drought induced biochemical damages in tea [*Camellia sinensis* (L.) O Kuntze]. J. Agric. Food Chem 61:6660–6670

Vu JCV, Gesch RW, Allen LH, Boote KJ, Bowes G (1999) CO2 enrichment delays a rapid, drought-induced decrease in Rubisco small subunit transcript abundance. J Plant Physiol 155:139–142

Wahid A, Rasul E (2005) Photosynthesis in leaf, stem, flower and fruit. In: Pessarakli M (ed) Handbook of photosynthesis, 2nd edn. CRC Press, Florida, pp 479–497

Wang Y, Zhang B, Jiang D, Chen G (2019) Silicon improves photosynthetic performance by optimizing thylakoid membrane protein components in rice under drought stress. Environ Exp Bot 158:117–124

Watts J (2018) Water shortages could affect 5bn people by 2050, UN report warns. The Guardian, March 19. https://www.theguardian.com/environment/2018/mar/19/water-shortages-could-affect-5bn-people-by-2050-un-report-warns

Wei H, Chen C, Ma X, Zhang Y, Han J, Mei H, Yu S (2017) Comparative analysis of expression profiles of panicle development among tolerant and sensitive rice in response to drought stress. Front Plant Sci 8:437

Wu F, Bao W, Li F, Wu N (2008) Effects of drought stress and N supply on the growth, biomass partitioning and water-use efficiency of *Sophora davidii* seedlings. Environ Exp Bot 63:248–255

Xu K, Chen S, Li T, Ma X, Liang X, Ding X, Liu H, Luo L (2015a) OsGRAS23, a rice GRAS transcription factor gene, is involved in drought stress response through regulating expression of stress-responsive genes. BMC Plant Bio 1:141

Xu Z, Jiang Y, Zhou G (2015b) Response and adaptation of photosynthesis, respiration, and antioxidant systems to elevated CO_2 with environmental stress in plants. Front Plant Sci 6:701

Yin LN, Wang SW, Liu P, Wang WH, Cao D, Deng XP, Zhang SQ (2014) Silicon-mediated changes in polyamine and 1-aminocyclopropane-1-carboxylic acid are involved in silicon-induced drought resistance in *Sorghum bicolor*, L. Plant Physiol Biochem 80:268–277

Yokota A, Kawasaki S, Iwano M, Nakamura C, Miyake C, Akashi K (2002) Citrulline and DRIP-1 protein (ArgE homologue) in drought tolerance of wild watermelon. Ann Bot 89:825–832

Zargar SM, Gupta N, Nazir M, Mahajan R, Malik FA, Sofi NR, Shikari AB, Salgotra RK (2017) Impact of drought on photosynthesis: molecular perspective. Plant Gene 11:154–159

Zhong C, Cao X, Hu J, Zhu L, Zhang J, Huang J, Jin Q (2017) Nitrogen metabolism in adaptation of photosynthesis to water stress in rice grown under different nitrogen levels. Front Plant Sci 8

Zhou Y, Lam HM, Zhang J (2007) Inhibition of photosynthesis and energy dissipation induced by water and high light stresses in rice. J Exp Bot 58:1207–1217

Zhou Y, Zhang C, Lin J, Yang Y, Peng Y, Tang D, Zhao X, Zhu Y, Liu X (2015) Over-expression of a glutamate dehydrogenase gene, MgGDH, from Magnaporthe grisea confers tolerance to dehydration stress in transgenic rice. Planta 241:727–740

Zhu Y, Gong H (2014) Beneficial effects of silicon on salt and drought tolerance in plants. Agron Sustain Dev 2:455–472

Chapter 8
Ion Transporter Genes from Wild Relatives of Cereals Hold the Key for the Development of Salinity Tolerance

Buddhadev Sarkar and Swarnendu Roy

Abstract Crop productivity is often threatened by salinity in arid and semi-arid regions where irrigation is indispensable for agriculture. Excess salt in soil and water causes ion toxicity that can hamper the growth and yield of crop plants affecting various metabolic pathways. Almost all the cereal crops are sensitive to increasing salinity levels, and therefore there is a requirement for the development of salt-tolerant varieties. Though conventional breeding and interspecific hybridization offer some sort of solution to overcome this problem, genetic engineering for the development of novel salt-tolerant variety is more desirable. Salt tolerance is mediated through ion exclusion, compartmentalization of Na^+ within vacuole or restricting Na^+ transport especially to the aerial parts, in which ion transporters play a crucial role in excluding Na^+ from the cell and thus reduce ion toxicity. Most recently, the identification and functional characterization of transporter genes have received greater attention in the development of salt-tolerant crops for sustainable food production. Many wild grasses like *Sporobolus airoides*, *Spartina alterniflora*, *Distichlis palmeri*, *Pennisetum glaucum*, *Porteresia coarctata*, *Puccinellia tenuiflora*, etc. growing mostly in coastal areas or in saline environment show higher salinity tolerance as compared to glycophytes. These wild grasses can adapt to increasing salinity because of the presence of some unique strategies like the exclusion of excess salts by salt glands and maintenance of K^+/Na^+ ratio in leaves mainly due to the presence of highly efficient ion transporters. Targeting the genes responsible for ion transport in these grasses can thus be a useful aid for improving the salinity tolerance of cereal crops. This chapter covers a wide range of salt-related ion transporter genes from the wild relatives of cereal crops to assess the scope of these genes in the production of elite varieties of cereal crops that can survive in highly saline environment.

Keywords Ion exclusion · Salt-tolerant · Salinity · Ion transporters · Cereal crops

B. Sarkar · S. Roy (✉)
Plant Biochemistry Laboratory, Department of Botany, University of North Bengal, Darjeeling, West Bengal, India
e-mail: swarnendubotany@nbu.ac.in

R. Roychowdhury et al. (eds.), *Sustainable Agriculture in the Era of Climate Change*, https://doi.org/10.1007/978-3-030-45669-6_8

1 Introduction

Salinity is one of the severe abiotic stresses that limits plant growth and causes a significant loss to crop yield. It is estimated that more than 800 million hectares of land is globally affected by salt stress, in which about 20% of irrigated land is included that accounts for one-third of world's food production (Rozema and Flowers 2008; Shrivastava and Kumar 2015). Increased soil salinization can rapidly transform a fertile land into barren land due to the deposition of salt in soil and can thus directly impact on the vegetation of that area. Plant growth is retarded by soil salinity, which induces osmotic stress firstly and is followed by ion toxicity. When plants are exposed to high salinity, cell metabolism is immediately affected because of the ion-specific toxicity (Blumwald et al. 2000). Under NaCl stress, plants generally face hyperionic stress due to the accumulation of toxic Na^+ and Cl^- in cells that impart ionic imbalance. The increase in Na^+ concentration within the cell can, in turn, inhibit K^+ uptake and can retard the biological activity of many enzymes resulting in less productivity and growth of the plant (James et al. 2011). Salt stress in plants is also characterized by the overproduction of ROS such as superoxide radicals (O_2^-), hydrogen peroxide (H_2O_2) and hydroxyl radicals (OH^-), which imparts a negative effect on various cellular metabolisms and exerts oxidative damage of different cellular structures like proteins, lipid, DNA, etc. (Bartels and Sunkar 2005; Roychowdhury et al. 2018, 2019).

For the maintenance of normal physiology and metabolism under salt stress, plants strategize to increase their tolerance level through the accumulation of osmolytes, overexpression of antioxidative enzymes and sequestration and removal of toxic ions (Munns and Tester 2008; Roy and Chakraborty 2014). Among these, removal of toxic ion is the most determining factor as excess accumulation of Na^+ imposes serious imbalances in plant metabolism (Flowers and Lauchli 1983; Tsunekawa et al. 2009). In salt-tolerant plants, the accumulation of Na^+ in the cell is regulated by the rate of net uptake of Na^+ by root, the net translocation from root to shoot and lastly the transportation of Na^+ from shoot to root through the phloem (Tester and Davenport 2003). The presence of specialized glandular structures in the form of salt glands and bladders in several plants also accounts for removal of excess amount of Na^+ from the cells, thereby aiding in maintaining an optimal K^+/Na^+ ratio (Agarie et al. 2007).

On the basis of salinity tolerance, plants could be broadly classified into glycophytes and halophytes. Unfortunately, all the important cereal crops are glycophytic in nature, and their growth is retarded even at low concentrations of salt (50 mM NaCl). Among the cereals, rye is the most tolerant, followed by barley, maize and rice (Hoang et al. 2016). However, in the order Poales, approximately 8% of the species that include near and distant relatives of cereal crops are halophytic or salt-tolerant in nature and therefore offer a promising system for harbouring salt-tolerant genes for cereal crop improvement (Flowers and Colmer 2008). It has been realized that the ability of these grasses to maintain ion homeostasis in stressed environments is one of the most important mechanisms of salt tolerance. Mostly, in the chloridoid and panicoid grasses, ion homeostasis is mainly achieved by efflux of excess Na^+ through bicellular salt glands (Marcum 1999). Several ion transporter genes like SOS1 are involved in the transcription of plasma membrane specific Na^+/H^+ ion transporters for the secretion of Na^+ across the plasma membrane (Qiu et al. 2002). Also, the loading of excess Na^+ into the vacuoles of senescing and maturing leaves helps to maintain the turgor pressure of the cell maintaining a high K^+ in the cytosol (Storey 1995). Presence of Na^+/H^+ ion transporters like NHX in the tonoplast plays an important role in the uploading of toxic ions from the cytosol to vacuoles (Jeschke 1984). Many ion transporters from salt-tolerant grasses like *Agropyron elongatum*, *Aeluropus littoralis*, *Pennisetum glaucum*, *Puccinellia tenuiflora*, *Spartina alterniflora*, etc. have been deciphered in the last few decades, and their functionality in conferring salinity tolerance has been well explained (Shabala and Mackay 2011).

The understanding of salt-tolerance attributes of ion transporter genes from wild grasses has encouraged the evaluation of these genes in conferring salinity tolerance in cereal crops (Bhatnagar-Mathur et al. 2008). In the past decade, plasma membrane and vacuolar Na^+/H^+ antiporters have been the centre of attention in transgenic studies for the alleviation of Na^+-induced toxicity in plants (Zhang and Blumwald 2001; Vera-Estrella et al. 2005; Liu et al. 2008). This review, therefore, attempts to provide an in-depth status of the ion transporter genes from wild grasses and their potentiality to confer salinity tolerance in crop plants and also to assess the present status of salt-tolerant transgenic cereal crops.

2 Ion Transporters and Their Mechanism of Action in Salinity Tolerance

Several ion transporters, pumps and channel proteins located on the plasma membrane or vacuolar membrane are involved in ionic homeostasis and thus confer salinity tolerance in plants. These ion transporters include high-affinity potassium transporters (HKTs), salt overly sensitive (SOS) and non-selective cation channels (NSCCs) present on the plasma membrane. These are the carrier-type transporter that is involved in Na^+ transportation at the cellular level. Other transporters such as Na^+/H^+ antiporters (NHXs), cation transporter (CAX) and H^+ pumps are present on the vacuolar membrane that facilitates compartmentalization of Na^+ toxic ions into the vacuoles (Hasanuzzaman et al. 2015; Volkov 2015; Shabala et al. 2015; Anumalla et al. 2016).

A number of plants like rice, wheat and *Arabidopsis* show the presence of HKTs in their plasma membrane that acts either as a Na^+/K^+ symporter or Na^+ uniporter (Waters et al. 2013). HKT protein family can be divided into two groups: the first group is HKT1 that transports only Na^+, and the other group is HKT2 that is involved in the transportation of both Na^+ and K^+ (Corratge-Faillie et al. 2010). In *Arabidopsis*, HKT1 was shown to be involved in recirculation of Na^+ from shoot to root by loading the excess ion into the phloem and by releasing it into the roots (Berthomieu et al. 2003). It was also reported that HKT1 was involved in Na^+ unloading from xylem vessels to xylem parenchyma cells that contribute to salt tolerance in *Arabidopsis* (Sunarpi et al. 2005).

To date, SOS1, a Na^+/H^+ antiporter, is the best-known ion transporter, which is involved in Na^+ efflux from the cytosol to the apoplast of the cortical cell. It was reported by Shi et al. (2000) that SOS1 found in the root apex of *Arabidopsis* is the only transporter involved in the Na^+ exclusion from the cytosol to the apoplast, functioning as Na^+/H^+ antiporter. In transgenic plants with SOS1 gene, salt tolerance was found to be enhanced by the overexpression of SOS1 (Yang et al. 2009; Yue et al. 2012; Hasanuzzaman et al. 2015; Anumalla et al. 2016), whereas the loss in function of SOS1 showed hypersensitivity to salt even in the salt-tolerant wild relative of *Arabidopsis* – i.e. *Thellungiella salsuginea* (Oh et al. 2009). Similarly, the SOS1 gene isolated from *Salicornia brachiata*, when overexpressed in tobacco plant, showed better performance under saline environment due to extracellular Na^+ effluxing (Yadav et al. 2012). It was reported from *Arabidopsis* that the activity of SOS1 was regulated by two other proteins, viz. SOS2 and SOS3 (Shi et al. 2000). SOS3 is a calcium-binding protein which in turn induces SOS2, a serine/threonine protein kinase to activate phosphorylation in SOS1, thereby increasing Na^+/H^+ exchange activity of the SOS1 transporter (Halfter et al. 2000; Liu et al. 2000; Luan et al. 2009).

Non-selective cation channels (NSCCs) play a major role in the regulation of Na^+ transportation when the salt concentration is high in the soil. It has been demonstrated in several studies that NSCCs catalyse primary Na^+ fluxes under salinity stress (Kronzucker and Britto 2011). According to Tester and Davenport (2003), there are two classes of NSCCs, viz. cyclic nucleotide-gated channels (CNGCs) and glutamate-activated channels (GLRs). Inhibition of cation channels is associated with the amelioration of Na^+ toxicity in plants, and external Ca^{2+} plays an important role in this. Ca^{2+} and intracellular and extracellular Na^+ can all have regulatory roles by blocking the K^+-selective outwardly rectifying channel (KOR) and thus preventing excessive Na^+ influx and cellular K^+ loss (Demidchik and Maathuis 2007). The rice CNGC1 was reported to decline Na^+/K^+ ratio by decreasing Na^+ influx and was observed to be more downregulated in salt-tolerant rice variety than the salt-sensitive variety under salinity stress (Senadheera et al. 2009). NSCC members of glutamate-activated channels from *Arabidopsis thaliana*, *AtGLR3;7*, *AtGLR1;1* and *AtGLR 1;4* when expressed in *Xenopus* oocytes, were shown to exhibit Na^+ permeability, and it was also revealed that these channel proteins could function in the transportation of Na^+, K^+ and Ca^{+2} (Roy et al. 2008; Tapken and Hollmann 2008).

Halophytes and salt-tolerant plants can tolerate salinity by lowering net Na^+ influx and resist salt stress by the accumulation of excess Na^+ inside its vacuole and efflux of excess Na^+ outside the cell (Hanana et al. 2007; Munns and Tester 2008). In saline conditions, vacuolar Na^+ sequestration reduces the toxicity of cytoplasm and lowers the osmotic potential to maintain turgor pressure. Various transporters such as Na^+/H^+ antiporters (NHXs), cation antiporters (CAXs) and auto-inhibited Ca^{+2} ATPase (ACA) are present in the vacuolar membrane for the storage and translocation of Na^+ from the cytosol during salt stress. There are also some proton pumps, viz. V-H^+-PPase and the vacuolar H^+-ATPase in tonoplast that also aids in Na^+ compartmentalization. Na^+ loading in vacuoles is controlled by NHX proteins which are operated by electrochemical gradient of proton generated across the vacuolar membrane. NHX can function as both Na^+/H^+ and K^+/H^+ exchangers under salt conditions (Apse et al. 2003; Leidi et al. 2010). Most of the *NHX* genes are expressed under salt stress and help in maintaining K^+ homeostasis in the cytosol for the growth and development of the plant in saline environments (Pardo et al. 2006; Xu et al. 2013). Maathuis and Amtmann (1999) reported that the Na^+ sequestration helped the plants under salt stress in different aspects such as maintenance of ionic homeostasis in the cytosol, enhancing the efflux of Na^+ and controlling the osmotic potential of tonoplasm. NHX1, a $Na^+,K^+/H^+$ exchanger, is the best-known transporter for vacuolar Na^+ sequestration, and most of the NHX transporter such as AtNHX1, AtNHX2, AtNHX3, AtNHX4 and OsNHX1 are located on the tonoplast, which was reported from *Arabidopsis* and rice, and the overexpression of these genes has been shown to improve salinity tolerance (Apse et al. 1999; Chen et al. 2007). The transgenic rice variety carrying the *NHX* gene exhibited overexpression of OsNHX1 under saline conditions which differentiated it from the wild-type cultivar by its ability to confer higher growth rate and salt tolerance (Fukuda et al. 2004).

Cation transporters (CAXs) are located on the vacuole and belong to a multigene family of ion transporters (Kamiya et al. 2005; Shigaki et al. 2006; Martinoia et al. 2007). During Ca^{2+} stress, the upregulation of *Arabidopsis CAX1* and *CAX3* was reported, and most importantly the expression of *CAX3* was upregulated and enhanced during Na^+ stress as observed in the studies with yeasts (Hirschi 1999; Shigaki and Hirschi 2000). CAX1 and CAX3 showed differential regulation of H^+ pumps that can affect H^+ coupled ion transport and also helped in the regulation of cellular pH. On the basis of amino acid sequences, CAX family is divided into two groups: Type IA includes *CAX1*, *CAX3* and *CAX4*, whereas Type IB includes *CAX2*, *CAX5* and *CAX6* (Manohar et al. 2011).

Apart from the ion transporters, there are proton pumps that can play an important part in conferring salinity stress tolerance. There are three types of proton pumps such as P-type ATPase, vacuolar H^+-pyrophosphatase (V-H^+-PPase) and vacuolar H^+-ATPase that produces electrochemical gradients throughout the plasma membrane. The P-type ATPase pumps out the cytoplasmic H^+ into extracellular space while a pH gradient is maintained between cytoplasm and vacuole by the acidification of lumen through the V-H^+-PPase and the vacuolar H^+-ATPase (Sze et al. 1999).

3 Cereal Crops and Their Responses to Salinity

Plants attempt a large number of physiological and biochemical strategies to improve their salt tolerance. According to Roy and Chakraborty (2014), salt-tolerance strategies include ion exclusion, osmotic tolerance and tissue tolerance.

When exposed to saline soils, plants first experience the osmotic stress which affects plant growth. After that Na^+ accumulation causes ion toxicity which upon reaching threshold level negatively impacts ion homeostasis. Plants generally tolerate salinity through osmotic balance, ionic homeostasis or tissue tolerance (Hasanuzzaman et al. 2015; Anumalla et al. 2016). On the basis of salt sensitivity, plants can be divided into two groups: glycophytes, the salt-susceptible plants, and halophytes which tolerate more salinity and can survive in high salt concentration (Tuteja et al. 2011). Most of the cereal crops are glycophytes and are highly affected by soil salinity. Various genotypes of crop plants, however, show varying degrees of growth responses to salinity (Eynard et al. 2005). According to Munns and Tester (2008), plant growth can be affected by salinity in two phases: first is the osmotic phase where the growth of young leaves is inhibited, and second is the ionic phase which speeds up the senescence of leaves. Salinity inhibits shoot elongation by inducing stomatal closure which in turn increases the temperature of leaves (Rajendran et al. 2009; Sirault et al. 2009). The basic mechanism of the ion transporters in cereal crops under salinity stress is illustrated in Fig. 8.1.

In rice, salinity shows a negative effect on yield components and slows down the heading (Grattan et al. 2002). Salinity can interfere with the germination of the seed, seedling growth, shoot growth, dry and fresh weight of shoot, size of the leaf, numbers of tillers per plant, number of spikelet and productivity (Lauchli and Grattan 2007; Moradi and Ismail 2007; Roychowdhury et al. 2013; Gupta and Huang 2014). At the flowering stage, the pollen viability can be reduced by salt stress, thus affecting the grain yield (Singh et al. 2004). Chinnusamy et al. (2005) reported that salt tolerance is controlled by multiple genes and known as a complex quantitative trait. Rice is a salt-susceptible cereal plant, mainly at the young stage (Lutts et al. 1995), and also at the mature stage, the productivity is restricted by salinity (Todaka et al. 2012). Maize is also a salt-sensitive crop and exhibits a strong inhibition in shoot growth in the first phase of salt stress (Pitann et al. 2009; Wakeel et al. 2011). According to Omoto et al. (2012), assimilation of carbon in maize is very sensitive during salinity stress. The decrease in photosynthetic pigment and stomatal conductance and impaired activities of photosynthetic enzymes are some key factors which hamper the carbon fixation capacity of maize in the saline environment (Kaya et al. 2010; Qu et al. 2012). In maize, during the reproductive phase, salinity causes a decrease in grain weight and numbers, resulting in an overall reduction in grain yield (Kaya et al. 2013). Under high salt concentration, wheat plant shows a wide range of physiological responses such as decreased water potential in leaf, loss in turgor pressure, stomatal closure, an increase of oxidative stress and production of toxic metabolites (Kingsbury and Epstein 1984; Taiz et al. 2015). Salt stress also declined the content of chlorophyll and carotenoid pigments (Ali et al. 2008) and enhances the formation of ROS (Wahid et al. 2007). It was also reported by Kong et al. (2001) and Sairam et al. (2002) that the proline, soluble sugars, glycine-betaine and abscisic acid content increased during salt stress. Barley is the most tolerant among the cereal crops under salinity stress, but its growth and yield is also affected when it is exposed to a higher level of salt concentrations. Comparing the different genotypes of barley under salt stress indicated that salt

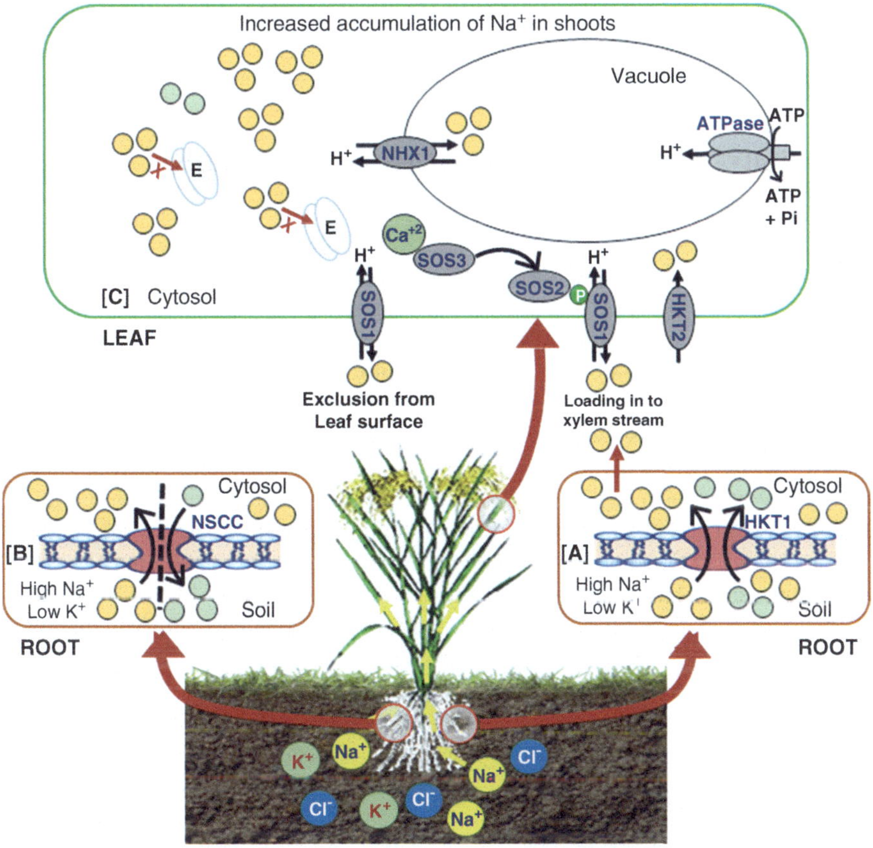

Fig. 8.1 Illustration of salinity stress responses in a typical cereal plant. The role of few ion transporters which are either upregulated or downregulated under salinity stress is shown. Owing to the susceptibility of cereals to salinity, most of the ion transporters depicted are expected to be downregulated. **[A]** NSCCs (nonspecific cation channels) are expected to increase the uptake of Na^+ ions through roots. **[B]** Also HKT1 (high-affinity K^+ transporters) due to high concentration of Na^+ in soil functions in Na^+ uptake rather than K^+ uptake. **[C]** In the leaves, SOS pathway involving SOS3, SOS2 and SOS1 (plasma membrane transporters) plays an important role in Na^+ ion exclusion. But in cereal crops, due to the absence of salt glands, the exclusion of Na^+ is minimal, whereas the SOS1 functions in recirculating the excess Na^+ ion back into the conducting tissues. As a whole Na^+ ion concentration in plant tissues increases at a high rate inhibiting the normal metabolism. Also the vacuolar proton antiporters (NHX1) which are dependent on ATPases to function are also downregulated under salt stress; therefore the loading of excess Na^+ ions into the vacuoles is also not possible

tolerance in barley is determined by the abscisic acid pathway (Kamboj et al. 2015). Bakht et al. (2007) observed that the reduction in shoot length of barley and oats was maximum at higher concentration of NaCl (150 mM). Higher level of salt stress increased ion concentration in rooting medium which resulted in the decrease of water potential which in turn was responsible for the shortening of shoot length (Munns et al. 2002).

4 Salinity-Induced Transporters from Wild Relatives of Cereals

The native grasses growing in the coastal area face environmental conditions much different from the plants growing inland. Though there are many plants which can tolerate salt stress up to a threshold level, salt-tolerant grasses specifically are different in their distinct morphology and leaf structure allowing them to exclude excess salt from the interior. Vetiver grass (*Vetiveria zizanioides* L.) which is a salt-tolerant perennial grass found mostly in tropical and subtropical areas is different from its close relatives in having strong and extensive root system. The root system specifically provides a superior advantage to the plant in adapting and overcoming a wide range of stresses like flood, drought, high temperatures, acidity and salinity (Truong et al. 2002; Zhou and Yu 2009; Ghotbizadeh and Sepaskhah 2015). The wild vetiver grass has a high salinity threshold of about 100 mM NaCl proving its ability to tolerate high salt stress which is achieved by the plant's ability to maintain an optimum K^+/Na^+ ratio in its leaves (Liu et al. 2016). It has been observed in *Porteresia coarctata* - a wild rice variety, which is a highly salt-resistant species growing in the coastal and sub-coastal regions of India, ameliorates Na^+ toxicity by the unicellular salt hairs present on the adaxial surface of leaves (Flowers et al. 1990; Ganie et al. 2014). There is a C4 perennial halophytic grass *Aeluropus littoralis* that grows in salty areas without any toxicity symptoms owing to their ability to adjust the osmolarity of cytoplasm to match the osmolarity of sea water (Touchette 2007; Zouari et al. 2007). Also, most of the other grasses like *Sporobolus*, *Zoysia* and *Cynodon* secrete Na^+ through salt gland (Marcum 1999). Greater retention of K^+ in shoots is observed in *Puccinellia* and *Thinopyrum* by maintaining a negative membrane potential in the root cell (Teakle et al. 2013). A list of salt-tolerant grasses though not exhaustive is provided in Table 8.1, depending upon their ability to grow in high salt concentrations and presence of salt-tolerant attributes specially the salt glands.

Among the multimodal strategies of salinity tolerance, management of ionic balance inside the cell seems to be of prime importance. The ion homeostasis is achieved due to the presence of several transporters in the plasma membrane and vacuolar membrane in the shoots and roots of the plants. Transcriptomic analysis in the halophytic turfgrass *Sporobolus virginicus* has revealed that several ion transporters like high-affinity potassium transporter (HKT), ABC transporters (for Cl^- transport) and H^+-ATPase were upregulated in both the shoots and roots of the plant under salinity stress (Yamamoto et al. 2015). This indicated efficient machinery of the ion transporters in loading the excess ions into the vacuoles for maintaining an optimum concentration of ions in the cytosol. Class II HKTs which are known to

Table 8.1 List of wild grasses that could be a source of ion transporter genes for cereal crop improvement

Name of grasses species	Family	Subfamily	Attributes
Aegilops triuncialis L.	Poaceae	Pooideae	Presence of salt gland, mainly distributed in coastal or sub-coastal areas
Aeluropus spp. (*A. lagopoides* (L.) Thwaites; *A. littoralis* (Gouan) Parl.)	Poaceae	Panicoideae	Salt secreted through salt gland; increase of osmolyte like proline, soluble sugars and amino acid; habitat includes coastal, desert
Agrostis spp. (*A. pallens* Trin.; *A. stolonifera* L.)	Poaceae	Pooideae	Increase in concentration of proline, asparagines, glutamine, serine and glycine-betaine in response to salinity; found in seashore, open meadows, grasslands, coastal strand, woodland or forest
Buchloe dactyloides (Nutt.) Engelm	Poaceae	Chloridoideae	Salt gland present; a greater sodium (Na^+) concentration was found in the root tissue than in the shoot tissue; habitat coastal
Brachiaria mutica (Forrsk.) Stapf	Poaceae	Panicoideae	Mainly distributed in coastal areas
Cenchrus pennisetiformis Steud.	Poaceae	Panicoideae	Maintain a low Na^+/Ca^+ ratio; habitat desert and grassland
Chloris spp.(*C. virgata* Sw.; *C. gayana* Kunth)	Poaceae	Chloridoideae	Salt gland present to secret Na^+ and K^+; found in coastal, desert and grassland area
Cynodon dactylon (L.) Pers.	Poaceae	Chloridoideae	Presence of salt gland, mostly glycophytic with occasional distribution in sub-coastal areas, also found in desert, grassland and forest
Dactyloctenium aegyptium (L.) Wild	Poaceae	Chloridoideae	Presence of salt gland; found in coastal, desert and natural grassland area
Desmostachya bipinnata (L.) Stapf	Poaceae	Panicoideae	Distributed in coastal, deserts, wetlands and grassland
Digitaria ciliaris (Retz.) Koeler	Poaceae	Panicoideae	Distributed in coastal, wetland, grassland, also abundant in roadside and wasteland
Distichlis spp. (*D. palmeri* (Vasey) Fassett ex I.M.Johnst.; *D. spicata* (L.) Greene)	Poaceae	Chloridoideae	Excreting salt via salt glands; found along seashore, inland and desert
Eragrostis tef (Zucc.) Trotter	Poaceae	Chloridoideae	Found in inland swamps
Halopyrum mucronatum (L.) Stapf	Poaceae	Panicoideae	Found in coastal regions
Holcus lanatus L.	Poaceae	Pooideae	Found in inland and sea cliff
Imperata cylindrica	Poaceae	Panicoideae	Presence of salt gland, invasive species with wide distribution from inlands to coastal areas

(continued)

Table 8.1 (continued)

Name of grasses species	Family	Subfamily	Attributes
Leptochloa fusca (L.) Kunth	Poaceae	Chloridoideae	Salt gland is present; accumulate greater concentration of Na^+ and Cl^- in shoot rather than root; habitat coastal
Lolium perenne L.	Poaceae	Pooideae	Found in freshwater wetlands and coastal beaches
Panicum spp. (*P. repens* L.; *P. turgidum* Forssk.; *P. virgatum* L.)	Poaceae	Panicoideae	Maintain low concentration of both Na^+ and Cl^- in shoot and show Na^+ sequestration in root; *P. repens* found in wetlands and coastal areas; *P. turgidum* found in desert; *P. virgatum* found in plateau
Paspalum spp. (*P. distichum* L.; *P. vaginatum* L.)	Poaceae	Panicoideae	Accumulation of osmolytes; abundant in seashore
Pennisetum spp. (*P. divisum* (Forssk.ex J.F.Gmel.) Henrard; *P. glaucum* (L.) R.Br.)	Poaceae	Panicoideae	Found in coastal or large river biome
Porteresia coarctata (Roxb.) Tateoka	Poaceae	Oryzoideae	Presence of salt gland and found in coastal area
Puccinellia spp. (*P. distans* (Jacq.) Parl.; *P. tenuiflora* (Griseb.) Scribn. & Merr.)	Poaceae	Pooideae	Maintain low Na^+ in shoot; habitat grassland or river valley
Spartina alterniflora Loisel.	Poaceae	Chloridoideae	Tolerate salt stress by osmotic adjustment; dominant in saline marsh and coastal regions
Sporobolus spp. (*S. airoides* (Torr.) Torr.; *S. heterolepis* (Gray.) A. Gray)	Poaceae	Chloridoideae	Salt gland is present; found near the coast, foothills and desert mountains
Uniola paniculata L.	Poaceae	Chloridoideae	Found in coastal area
Zizania aquatica L.	Poaceae	Oryzoideae	Distribute from inland to coastal areas
Zoysia spp. (*Z. matrella* (L.) Merr.; *Z. tenuifolia* Thiele)	Poaceae	Chloridoideae	Salt glands clustered into groups of two or three; found in coastal dunes, riverbanks and grasslands

cotransport Na^+-K^+ in plants have been elucidated from *Sporobolus virginicus* which are known to maintain ion homeostasis in low K^+ saline environments (Tada et al. 2014). The two transporters named as SvHKT2;1 and SvHKT2;2 were shown to be upregulated by high NaCl concentrations and low K^+ ion, respectively, thereby regulating ion homeostasis by both outward and inward movements of K^+ and Na^+ ions. Similarly, transcriptomic analysis of the smooth cord grass – *Spartina alterniflora* has revealed the importance of several transporter proteins like K^+ transporters, H^+ transporters and Na^+/H^+ which played a crucial role in the ion homeostasis and adaptation of this grass in the saline environment (Bedre et al. 2016).

Wang et al. (2005) investigated the K^+ selectivity over Na^+ ion in *Puccinellia tenuiflora* in comparison to wheat and found that the accumulation of K^+ ion in the former was up to 24% higher than the latter under different NaCl treatments. This revealed the presence of high-affinity K^+ transporters in *Puccinellia tenuiflora* that was induced under NaCl stress, thereby increasing the uptake and accumulation of K^+ ion in the shoots. The gene *PtAKT1* was found to encode for these channel proteins that maintain a highly selective absorption capacity for K^+ ion under high salt stress, which was also confirmed by the presence of a high number of *PtAKT1* transcripts in roots of plants subjected to a NaCl concentration greater than 150 mM (Wang et al. 2015). A vacuolar antiporter LfNHX1 from *Leptochloa fusca* – a halophytic forage grass was shown to be upregulated in response to different levels of salinity which played an important role in the salt tolerance of the grass (Panahi et al. 2013). Real-time PCR analysis of the *LfNHX1* gene from the plant treated with 200 and 500 mM NaCl revealed the enhancement in expression of this gene by 4–8 times in comparison to the control plants. In *Aeluropus littoralis* upregulation of ion transporter genes like *SOS1* (encoding plasma membrane antiporters), *NHX1* (encoding vacuolar transporters) and *VHA* (encoding proton pumps) located in the shoots and roots accounted for increased tolerance at high NaCl concentrations (>200 mM) by regulating the cytosolic Na^+ ion threshold (Moshaei et al. 2014). Similarly in *Aeluropus lagopoides*, differential expression of ion transporter genes, viz. *SOS1*, tonoplast *NHX1*, *HKT* and *V-ATPases*, in shoots and roots of the plants subjected to high NaCl accounted for ion homeostasis which was pivotal in conferring salinity tolerance (Sanadhya et al. 2015a, b). The expression level study of the plasma membrane (CdSOS1) and vacuolar (CdNHX) transporters from *Cynodon dactylon* and its counterparts in rice revealed the role of these transporters in salt tolerance of the former (Roy and Chakraborty 2018). *CdSOS1* expression was found to increase up to 2.5 and 5 fold in plants exposed to 100 and 200 mM NaCl concentrations, respectively. On the other hand, *CdNHX* expression increased up to 1.5 and 2 fold. In comparison, *SOS1* and *NHX* expressions in rice plants were not upregulated under NaCl treatments. The list of ion transporters that have been characterized by different wild grasses has been summarized in Table 8.2.

5 Transgenic Approaches with Wild Ion Transporter Genes

During salinity stress plant growth is positively induced by the ion transporter proteins, and thus the identification of these proteins and their related gene is essential (Maathuis 2007). The function of salt-tolerant genes and proteins can be monitored by evaluating the salt uptake at the root level, distribution pattern and the cellular influx of Na^+ and K^+ ions, long-distance transport and compartmentation at cellular as well as tissue level (Flowers and Colmer 2008). There are some useful approaches to improve salt tolerance of plants for introduction and propagation under saline conditions such as genotype screening, the introduction of salt tolerance gene and conventional breeding methods. As conventional breeding takes a long time and can

Table 8.2 List of ion transporters from wild relatives and their mechanism of action

Ion transporters	Generic type	Source species	Mechanism of action	Reference
High-affinity potassium transporters	HKT1	*Puccinellia tenuiflora*, *Phragmites australis*, *Aegilops cylindrica*	Transports only Na^+, recirculates Na^+ from shoot to root	Zhang et al. (2017); Takahashi et al. (2007), Arabbeigi et al. (2018)
	HKT2	*Aeluropus lagopoides*	Transports both Na^+ and K^+	Sanadhya et al. (2015a, b)
Salt overly sensitive (SOS)	SOS1	*Puccinellia tenuiflora*, *Cynodon dactylon*, *Leptochloa fusca*, *Aegilops cylindrica*	Na^+/H^+ antiporter; efflux Na^+ from the cytosol to the apoplast, expressed in root parenchyma and involve in Na^+ loading into the xylem sap during moderate salt stress	Zhang et al. (2017); Roy and Chakraborty (2018); Taherinia et al. (2015); Arabbeigi et al. (2018)
Vacuolar antiporters (NHXs)	NHX1	*Puccinellia tenuiflora*, *Cynodon dactylon*, *Aeluropus lagopoides*, *Leptochloa fusca*, *Panicum virgatum*, *Lolium perenne*, *Aegilops cylindrica*	Involve in Na^+ sequestration in vacuole and maintain K^+ homeostasis	Zhang et al. (2017), Roy and Chakraborty (2018), Sanadhya et al. 2015a, b, Adabnejad et al. (2015), Huang et al. (2018), Tang et al. (2013), Arabbeigi et al. (2018)
H^+ pumps	V-H^+-PPase	*Lolium temulentum*	H^+ transport, increase acidity in vacuole	Baldwin and Dombrowski (2006)
	Vacuolar H^+-ATPase	*Spartina alterniflora*, *Aeluropus lagopoides*, *Lolium temulentum*	Involved in H^+ transport to vacuole and increase acidity	Baisakh et al. (2012); Sanadhya et al. (2015a, b), Baldwin and Dombrowski (2006)
	Plasma membrane H^+-ATPase	*Aeluropus littoralis*, *Spartina patens*, *Zostera marina*, *Lolium temulentum*	It provides the driving force for Na^+ ion exclusion	Olfatmiri et al. (2014), Wu and Seliskar (1998), Muramatsu et al. (2002), Baldwin and Dombrowski (2006)

reduce plant vigour, the transgenic approach seems to be the most effective and appropriate technique for the development of salt-tolerant cultivars. Many workers have focused on the ion transporter genes as it regulates the Na^+ inclusion and compartmentalization, plays an important role in the maintenance of cellular toxicity and helps in improving salinity tolerance. Therefore, screening the ion transporter genes in the germplasm of the cereal crops and the near and distant relatives of these plants seems to be of great importance in designing salt-tolerant cultivars of cereal crops.

Overexpression of genes encoding vacuolar NHXs in plants treated with salt has shown that these ion transporters could improve the salt tolerance in plants including rice, wheat, tobacco, cotton and tomato up to a certain extent (Apse et al. 1999; Zhang and Blumwald 2001; Zhang et al. 2001; Agarwal et al. 2013). A number of Na^+/H^+ antiporters have been isolated and characterized by wild salt-tolerant grasses which have been used in transgenic experiments to exhibit their potentiality in increasing the salinity tolerance levels of the transgenic plants. It was observed that the *AlTMP2* gene isolated from halophytic C4 grass *Aeluropus littoralis*, when expressed in transgenic tobacco, improved tolerance against salinity, osmotic, H_2O_2, heat and freezing stresses during seed germination and seedling stage by improving ion homeostasis and membrane integrity and by regulating some genes related to salt stress (Ben Romdhane et al. 2018). Another gene *AlNHX1* from *A. littoralis* exhibited improved growth in severe saline condition (150 mM NaCl) in transgenic plants which accumulated a higher level of K^+ and a lower level of Na^+ in the leaves than wild-type plants and thus helped in enhancing salt tolerance (Liu et al. 2014). Qiao et al. (2007) isolated a homolog of Na^+/H^+ antiporters, *AeNHX1* from the root of *Agropyron elongatum* which promoted salt tolerance in *Arabidopsis* and *Festuca* by improving osmotic adjustment and photosynthesis in the transgenic plants under salt stress. The *PvNHX1* gene expression in switchgrass (*Panicum virgatum* L.), when compared to the wild type or transgenic control, showed a better growth-related phenotype and also maintained K^+ homeostasis, resulting in more K^+ accumulation (Huang et al. 2017). Baisakh et al. (2012) isolated and introduced the *SaVHAc1* gene from the halophyte grass *Spartina alterniflora* in rice where it is observed that the *SaVHAc1*-expressing plants increased K^+/Na^+ ratio to protect the cytoplasm from toxic Na^+ and maintained higher chlorophyll retention. Zhang et al. (2017) analysed the expression pattern of *PtSOS1*, *PtHKT1;5* and *PtNHX1* from *Puccinellia tenuiflora* under 25 and 150 mM NaCl and observed that the expression of *PtSOS1* in roots significantly increased by 6 fold, while *PtHKT1;5* increased by 1.2-fold under 25 mM NaCl; in contrast the *PtSOS1* increased by 1.4-fold, whereas the PtHKT1;5 increased by 2.2-fold in roots under 150 mM NaCl; on the other hand, compartmentalization of Na^+ into vacuole was mediated by *PtNHX1*. The details of the transgenic experiments are provided in Table 8.3.

6 Success Story So Far and the Way Ahead

Many transgenic cereal crops have been produced in the past two decades with enhanced salt tolerance, but the success of these plants in agricultural fields is often less talked about. Though a large amount of government and public funding has been allocated for the development of salt-tolerant designer crops, desired success till today has not been achieved. Among the cereal crops, only transgenic maize carrying the *Bt* gene for insect resistance has been known to be commercialized in

Table 8.3 List of transgenic experiments validating the wild genes in improvement of salinity stress tolerance in cereals and other plants

Source grass	Transgene	Characteristic feature	Target organism	Characters induced for salinity tolerance	Reference
Aeluropus lagopoides	*AlHKT2;1*	Low-affinity K^+ transporter	Yeast	Increased the accumulation of K^+ ion	Sanadhya et al. (2015a, b)
Aeluropus littoralis	*AlNHX1*	Na^+/H^+ antiporter	*Glycine max*	The transgenic line exhibited improved growth by lowering the accumulation of Na^+ and increasing the concentration of K^+ in leaves under saline condition (150 mM NaCl)	Liu et al. (2014)
Agropyron elongatum	*AeNHX1*	Root-specific vacuolar Na^+/H^+ antiporter	*Arabidopsis*, *Festuca*	Overexpression of *AeNHX1* promoted salt tolerance improving osmotic adjustment and photosynthesis under salt stress	Qiao et al. (2007)
Leptochloa fusca	*LfNHX1*	Na^+/H^+ antiporter	*Nicotiana tabacum*	Overexpression of *LfNHX1* gene conferred salinity tolerance by increasing the germination rate and root growth	Rauf et al. 2014
Panicum virgatum	*PvNHX1*	$Na^+(K^+)/H^+$ antiporter	*Panicum virgatum*	Transgenic switchgrass overexpressing *PvNHX1* showed better plant height and leaf development compared to the wild-type plants	Huang et al. (2017)
Pennisetum glaucum	*PgNHX1*	Na^+/H^+ antiporter	*Oryza sativa*	Overexpression of *PgNHX1* in transgenic rice plants developed a more extensive root system and promoted setting of flowers and seeds in the presence of 150 mM NaCl	Verma et al. (2007)
Phragmites australis	*PhaHAK1*	High-affinity K^+ transporter	Yeast	Enhanced uptake of K^+ content	Takahashi et al. (2007)
	PhaNHA1	Plasma membrane Na^+/H^+ antiporter, i.e. SOS1	Yeast	Expression of this gene resulted in a lower cellular Na^+ content	Takahashi et al. (2009)

(continued)

Table 8.3 (continued)

Source grass	Transgene	Characteristic feature	Target organism	Characters induced for salinity tolerance	Reference
Puccinellia tenuiflora	*PutHKT2*	High-affinity K^+ transporter	Yeast, *Arabidopsis thaliana*	In yeast, *PutHKT2* increased the uptake of K^+ in low K^+ concentration medium or NaCl. In *Arabidopsis*, overexpression of this gene helped in maintaining ion homeostasis	Ardie et al. (2009)
	PutAKT1	K^+ channel protein	*Arabidopsis thaliana*	Enhanced K^+ under NaCl treatment and lowered the accumulation of Na^+ in both shoot and root	Ardie et al. (2010)
Spartina alterniflora	*SaVHAc1*	Vacuolar H^+-ATPase subunit c1	*Oryza sativa*	Establishment of ion homeostasis by increasing K^+/Na^+ ratio and other cations to protect the plants from toxic Na^+. Higher chlorophyll retention than the wild-type plants	Baisakh et al. (2012)

many countries like the USA, Brazil, China, etc. (James 2016). This presents a gloomy picture at present when none of the cereal crops with salt-tolerant genes are in agricultural fields. This is mainly due to the failure in performance of these transgenics in field trials when compared to the laboratory results.

Molecular breeding for imparting salt tolerance in the crop germplasm could also prove to be a significantly viable option. One of the major achievements in imparting salt tolerance has been through the introgression of *saltol* quantitative trait locus (QTL), which is present in the chromosome 1 of rice and is derived from the near-isogenic lines of the salt-tolerant cultivar of rice 'Pokkali' (Linh et al. 2012). The genetic analysis of the *saltol* region was characterized to be containing shoot K^+ content 1 (*SKC1*), which was eventually cloned to be identified as the gene *OsHKT1;5* which encodes for a Na^+ transporter regulating K^+ homeostasis (Ren et al. 2005). Recently the marker-assisted transfer of this *saltol* QTL into the aromatic rice Pusa Basmati 1 (PB1) has conferred enhanced tolerance to the cultivar at seedling stage (Singh et al. 2018). In 2013, Grain Research and Development Corporation (GRDC), Government of Australia, initiated a project for the development of salinity-tolerant wheat and barley in the light of huge losses incurred by the Australian farmers due to soil salinity (Roy et al. 2013). This initiative was started to locate the genes responsible for shoot Na^+ exclusion and Cl^- accumulation for enhancing tissue tolerance to salinity and to provide molecular markers to the breeders for developing salt-tolerant cultivars. Similar target-oriented initiatives could lead to the development of salt-tolerant cereal crops in the near future. One of the major objectives of marker-assisted breeding has been the identification of QTLs related to ion accumulation and exclusion. In this context, at least 14 QTLs from rice and 30 QTLs from barley have been reported to be involved in the regulation of ion movement (Ashraf and Foolad 2012). Some of these QTLs are *Nax1* for Na^+ exclusion, *SKC1/OsKKT8* for K^+/Na^+ homeostasis, *Qkr1* for root K^+ uptake and so on.

Therefore, it could be clearly seen that the urge for the search of novel genes and QTLs from near and distant relatives of crop plants has increased quite significantly. The future of cereal crop improvement thus undoubtedly lies in developing a good gene pool of ion transporter genes that can significantly enhance the tolerance ability of the cereal crops when introgressed into the crop genome.

7 Conclusions

Though the current status of salt-tolerant cereal crops presents a shoddy picture, however, the quest for the development of truly salt-tolerant cereal crops is highly desirable keeping in mind the challenges to enhance the crop productivity and to feed the ever-increasing world population. The complexity of salinity responses in the plants has made the picture even worse, and it has also been realized that the transfer of a single gene is not sufficient enough to make the crop plants salt-tolerant in the natural environment. However, there has been a continuous attempt to develop salt-tolerant varieties of cereal crops, and many of the successes in the in vitro conditions have the potentiality to be transformed at the larger scale. The development of modern biological tools has made the scientific community more equipped than ever, and it is high time to use the knowledge gathered during the past few decades to design the cereal crops to sustain their growth and productivity in saline environments. In this context, the ion transporter genes from the wild grasses which are near and distant relatives of cereal crops are likely to play a significant role in the near future for the development of salt-tolerant cereal crops.

References

Adabnejad H, Kavousi HR, Hamidi H, Tavassolian I (2015) Assessment of the vacuolar Na^+/H^+ antiporter (NHX1) transcriptional changes in *Leptochloa fusca* L. in response to salt and cadmium stresses. Mol Biol Res Commun 4(3):133–142

Agarie S, Shimoda T, Shimizu Y, Baumann K, Sunagawa H, Kondo A, Ueno O, Nakahara T, Nose A, Cushman JC (2007) Salt tolerance, salt accumulation, and ionic homeostasis in an epidermal bladder-cell-less mutant of the common ice plant *Mesembryanthemum crystallinum*. J Exp Bot 58:1957–1967

Agarwal PK, Shukla PS, Gupta K, Jha B (2013) Bioengineering for salinity tolerance in plants: state of art. Mol Biotechnol 54:102–123

Ali Q, Athar HR, Ashraf M (2008) Modulation of growth, photosynthetic capacity and water relations in salt stressed wheat plants by exogenously applied 24-epibrassinolide.Plant. Growth Regul 56(2):107–116

Anumalla M, Roychowdhury R, Geda CK, Bharathkumar S, Goutam KD, Mohandev TSS (2016) Mechanism of stress signal transduction and involvement of stress inducible transcription factors and genes in response to abiotic stresses in plant. Int J Recent Sci Res 7(8):12754–12771

Apse MP, Aharon GS, Snedden A, Blumwald E (1999) Salt tolerance conferred by overexpression of a vacuolar Na^+/H^+ antiport in *Arabidopsis*. Science 285:1256–1258

Apse MP, Sottonsanto JB, Blumwald E (2003) Vacuolar cation/H^+ exchange, ion homeostasis and leaf development are altered in a T-DNA insertional mutant of AtNHX1, the *Arabidopsis* vacuolar Na^+/H^+ antiporter. Plant J 36:229–239

Arabbeigi M, Arzani A, Majidi MM, Sayed-Tabatabaei BE, Saha P (2018) Expression pattern of salt tolerance-related genes in *Aegilops cylindrica*. Physiol Mol Biol Plants 24(1):61–73

Ardie SW, Xie L, Takahashi R, Liu S, Takano T (2009) Cloning of a high-affinity K^+ transporter gene *PutHKT2;1* from *Puccinellia tenuiflora* and its functional comparison with *OsHKT2;1* from rice in yeast and *Arabidopsis*. J Exp Bot 60(12):3491–3502

Ardie SW, Liu S, Takano T (2010) Expression of the AKT1-type K^+ channel gene from *Puccinellia tenuiflora*, *PutAKT1*, enhances salt tolerance in *Arabidopsis*. Plant Cell Rep 29(8):865–874

Ashraf M, Foolad MR (2012) Crop breeding for salt tolerance in the era of molecular markers and marker-assisted selection. Plant Breed 132:1–20

Baisakh N, Rao MVR, Rajasekaran K, Subudhi P, Janda J, Galbraith D, Vanier C, Pereira A (2012) Enhanced salt stress tolerance of rice plants expressing a vacuolar H^+-ATPase subunit C1 (*SaVHAC1*) gene from the halophyte grass *Spartina alterniflora* Loisel. Plant Biotechnol J 10(4):453–464

Bakht J, Shafi M, Khan AA, Hussain S, Tariq M, Akbar H, Jan AU, Rehamn MU (2007) Growth performance of oat and barley at early seedling stage under saline environment. Sarhad J Agric 23:565–569

Baldwin JC, Dombrowski JE (2006) Evaluation of *Lolium temulentum* as a model grass species for the study of salinity stress by PCR-based subtractive suppression hybridization analysis. Plant Sci 171:459–469

Bartels D, Sunkar R (2005) Drought and salt tolerance in plants. Crit Rev Plant Sci 24:23–58

Bedre R, Mangu VR, Srivastava S, Sanchez LE, Baisakh N (2016) Transcriptome analysis of smooth cordgrass (*Spartina alterniflora* Loisel), a monocot halophyte, reveals candidate genes involved in its adaptation to salinity. BMC Genomics 17:657

Ben Romdhane M, Riahi L, Jardak R, Ghorbel A, Zoghlami N (2018) Fingerprinting and genetic purity assessment of F1 barley hybrids and their salt-tolerant parental lines using nSSR molecular markers. 3. Biotech 8(1):57–65

Berthomieu P, Conejero G, Nublat A, Brackenbury WJ, Lambert C, Savio C, Uozumi N, Oiki S, Yamada K, Cellier F, Gosti F, Simonneau T, Essah PA, Tester M, Very AA, Sentenac H, Casse F (2003) Functional analysis of AtHKT1 in *Arabidopsis* shows that Na^+ recirculation by the phloem is crucial for salt tolerance. EMBO J 22(9):2004–2014

Bhatnagar-Mathur P, Vadez V, Sharma KK (2008) Transgenic approaches for abiotic stress tolerance in plants: retrospect and prospects. Plant Cell Rep 27:411–424

Blumwald E, Aharon GS, Apse MP (2000) Sodium transport in plant cells. Biochim Biophys Acta 1465(1–2):140–151

Chen H, An R, Tang JH, Cui XH, Hao FS, Chen J, Wang XC (2007) Over-expression of a vacuolar Na^+/H^+ antiporter gene improves salt tolerance in an upland rice. Mol Breed 19:215–225

Chinnusamy V, Jagendorf A, Zhu JK (2005) Understanding and improving salt tolerance in plants. Crop Sci 45(2):437–448

Corratge-Faillie C, Jabnoune M, Zimmermann S, Very A-A, Fizames C, Sentenac H (2010) Potassium and sodium transport in non-animal cells: the Trk/Ktr/HKT transporter family. Cell Mol Life Sci 67(15):2511–2532

Demidchik V, Maathuis FJM (2007) Physiological roles of nonselective cation channels in plants: from salt stress to signalling and development. New Phytol 175:387–404

Eynard A, Lal R, Wiebe K (2005) Crop response in salt-affected soils. J Sustain Agric 27(1):5–50

Flowers TJ, Colmer TD (2008) Salinity tolerance in halophytes. New Phytol 179(4):945–963

Flowers TJ, Lauchli A (1983) Sodium versus potassium: substitution and compartmentation. In: Lauchli A, Bieleski RL (eds) Inorganic plant nutrition. Springer, Berlin, pp 651–681

Flowers TJ, Flowers SA, Hajibagheri MA, Yeo AR (1990) Salt tolerance in the halophytic wild rice, *Porteresia coarctata* Tateoka. New Phytol 114(4):675–684

Fukuda A, Nakaumura A, Tagiri A, Tanaka H, Miyao A, Hirochika H, Tanaka Y (2004) Function, intracellular localization and the importance in salt tolerance of a vacuolar Na^+/H^+ antiporter from rice. Plant Cell Physiol 45:146–159

Ganie SA, Karmakar J, Roychowdhury R, Mondal TK, Dey N (2014) Assessment of genetic diversity in salt-tolerant rice and its wild relatives for ten SSR loci and one allele mining primer of *salT* gene located on 1st chromosome. Plant Syst Evol 300(7):1741–1747

Ghotbizadeh M, Sepaskhah AR (2015) Effect of irrigation interval and water salinity on growth of vetiver (*Vetiveria zizanioides*). Int J Plant Prod 9:17–38

Grattan SR, Zeng L, Shannon MC, Roberts S (2002) Rice is more sensitive to salinity than previously thought. Calif Agric 56(6):189–195

Gupta B, Huang B (2014) Mechanism of salinity tolerance in plants: physiological, biochemical, and molecular characterization. Int J Genom 2014(1):701596

Halfter U, Ishitani M, Zhu JK (2000) The Arabidopsis SOS2 protein kinase physically interacts with and is activated by the calcium-binding protein SOS3. Proc Natl Acad Sci 97:3735–3740

Hanana M, Cagnac O, Yamaguchi T, Hamdi S, Ghorbel A, Blumwald E (2007) A grape berry (*Vitis vinifera* L.) cation/proton antiporter is associated with berry ripening. Plant Cell Physiol 48:804–811

Hasanuzzaman M, Roychowdhury R, Karmakar J, Dey N, Nahar K, Fujita M (2015) Recent advances in biotechnology and genomic approaches for abiotic stress tolerance in crop plants. In: Devarajan T, Jeyabalan S (eds) Genomics and proteomics: concepts, technologies and applications. Apple Academic Press, Burlington, pp 333–366

Hirschi KD (1999) Expression of *Arabidopsis CAX1* in tobacco: altered calcium homeostasis and increased stress sensitivity. Plant Cell 11(11):2113–2122

Hoang TML, Tran TN, Nguyen TKT, Williams B, Wurm P, Bellairs S, Mundree S (2016) Improvement of salinity stress tolerance in rice: challenges and opportunities. Agronomy 6:54. (23 pages)

Huang Y, Guan C, Liu Y, Chen B, Yuan S, Cui X, Zhang Y, Yang F (2017) Enhanced growth performance and salinity tolerance in transgenic switchgrass via overexpressing vacuolar Na^+ $(K^+)/H^+$ antiporter gene (*PvNHX1*). Front Plant Sci 8:458

Huang Y, Cui X, Cen H, Wang K, Zhang Y (2018) Transcriptomic analysis reveals vacuolar Na^+ $(K^+)/H^+$ antiporter gene contributing to growth, development, and defense in switchgrass (*Panicum virgatum* L.). BMC Plant Biol 18(1):57

James C (2016) Global status of commercialized Biotech/GM crops: ISAAA Brief No 52. http://www.isaaa.org

James RA, Blake C, Byrt CS, Munns R (2011) Major genes for Na^+ exclusion, Nax1 and Nax2 (wheat HKT1;4 and HKT1;5), decrease Na^+ accumulation in bread wheat leaves under saline and waterlogged conditions. J Exp Bot 62:2939–2947

Jeschke WD (1984) K^+-Na^+ exchange at cellular membranes, intracellular compartmentation of cations, and salt tolerance. In: Staples RC, Toenissen RH (eds) Salinity tolerance in plants. Wiley, New York, pp 37–66

Kamboj A, Ziemann M, Bhave M (2015) Identification of salt-tolerant barley varieties by a consolidated physiological and molecular approach. Acta Physiol Plant 37:1716

Kamiya T, Akahori T, Maeshima M (2005) Expression profile of the genes for rice cation/H^+ exchanger family and functional analysis in yeast. Plant Cell Physiol 46:1735–1740

Kaya C, Tuna AL, Okant AM (2010) Effect of foliar applied kinetin and indole acetic acid on maize plants grown under saline conditions. Turk J Agric For 34:529–538
Kaya C, Ashraf M, Dikilitas M, Tuna AL (2013) Alleviation of salt stress induced adverse effects on maize plants by exogenous application of indoleacetic acid (IAA) and inorganic nutrients—a field trial. Aust J Crop Sci 7:249–254
Kingsbury RW, Epstein E (1984) Selection for salt-resistant spring wheat. Crop Sci 24(2):310–315
Kong Y, Zhou G, Wang Y (2001) Physiological characteristics and alternative respiratory pathway under salt stress in two wheat cultivars differing in salt tolerance. Russ J Plant Physiol 48(5):595–600
Kronzucker HJ, Britto DT (2011) Sodium transport in plants: a critical review. New Phytol 189:54–81
Lauchli A, Grattan S (2007) Plant growth and development under salinity stress. In: Jenks MA, Hasegawa PM, Jain SM (eds) Advances in molecular breeding toward drought and salt tolerant crops. Springer, Dordrecht, pp 1–32
Leidi EO, Barragan V, Rubio L, El-Hamdaoui A, Ruiz MT, Cubero B, Fernández JA, Bressan RA, Hasegawa PM, Quintero FJ, Pardo JM (2010) The AtNHX1 exchanger mediates potassium compartmentation in vacuoles of transgenic tomato. Plant J 61(3):495–506
Linh LH, Linh TH, Xuan TD, Ham LH, Ismail AM, Khanh TD (2012) Molecular breeding to improve salt tolerance of rice (*Oryza sativa* L.) in the red river delta of Vietnam. Int J Plant Genom 2012:949038
Liu J, Ishitani M, Halfter U, Kim CS, Zhu JK (2000) The *Arabidopsis thaliana SOS2* gene encodes a protein kinase that is required for salt tolerance. Proc Natl Acad Sci 97:3730–3734
Liu H, Wang Q, Yu M, Zhang Y, Wu Y, Zhang H (2008) Transgenic salt tolerant sugar beet (*Beta vulgaris* L.) constitutively expressing an *Arabidopsis thaliana* vacuolar Na^+/H^+ antiporter gene, *AtNHX3*, accumulates more soluble sugar but less salt in storage roots. Plant Cell Environ 31(9):1325–1334
Liu J, Zhang S, Dong L, Chu J (2014) Incorporation of Na^+/H^+ antiporter gene from *Aeluropus littoralis* confer salt tolerance in soybean (*Glycine max* L.). Indian J Biochem Biophys 51(1):58–65
Liu WG, Liu JX, Yao ML, Ma QF (2016) Salt tolerance of a wild ecotype of vetiver grass (*Vetiveria zizanioides* L.) in southern China. Bot Stud 57(1):–27
Luan S, Lan W, Lee SC (2009) Potassium nutrition, sodium toxicity, and calcium signaling: connections through the CBL-CIPK network. Curr Opin Plant Biol 12(3):339–346
Lutts S, Kinet JM, Bouharmont J (1995) Changes in plant response to NaCl during development of rice (*Oryza sativa* L.) varieties differing in salinity resistance. J Exp Bot 46(12):1843–1852
Maathuis FJM (2007) Monovalent cation transporters; establishing a link between bioinformatics and physiology. Plant Soil 301(1–2):1–15
Maathuis FJM, Amtmann A (1999) K^+ nutrition and Na^+ toxicity: the basis of cellular K^+/Na^+ ratios. Ann Bot 84(2):123–133
Manohar M, Shigaki T, Mei H, Park S, Marshall J, Aguilar J, Hirschi KD (2011) Characterization of *Arabidopsis* Ca^{2+}/H^+ exchanger CAX3. Biochemistry 50(28):6189–6195
Marcum KB (1999) Salinity tolerance mechanism of grasses in the sub-family Chloridoideae. Crop Sci 39(4):1153–1160
Martinoia E, Maeshima M, Neuhaus HE (2007) Vacuolar transporters and their essential role in plant metabolism. J Exp Bot 58:83–102
Moradi F, Ismail AM (2007) Responses of photosynthesis, chlorophyll fluorescence and ROS scavenging systems to salt stress during seedling and reproductive stages in rice. Ann Bot 99(6):1161–1173
Moshaei MR, Nematzadeh GA, Askari H, Nejad ASM, Pakdin A (2014) Quantitative gene expression analysis of some sodium ion transporters under salinity stress in *Aeluropus littoralis*. Saudi J Biol Sci 21(5):394–399
Munns R, Tester M (2008) Mechanisms of salinity tolerance. Annu Rev Plant Biol 59:651–681

Munns R, Husain S, Rivelli AR, James RA, Condon AG, Lindsay MP, Lagudah ES, Schachtman DP, Hare RA (2002) Avenues for increasing salt tolerance of crops, and the role of physiologically based selection traits. Plant Soil 247(1):93–105

Muramatsu Y, Harada A, Ohwaki Y, Kasahara Y, Takagi S, Fukuhara T (2002) Salt-tolerant ATPase activity in the plasma membrane of the marine angiosperm *Zostera marina* L. Plant Cell Physiol 43(10):1137–1145

Oh DH, Leidi E, Zhang Q, Hwang SM, Li Y, Quintero FJ, Jiang X, D'Urzo MP, Lee SY, Zhao Y, Bahk JD, Bressan RA, Yun DJ, Pardo JM, Bohnert HJ (2009) Loss of halophytism by interference with SOS1 expression. Plant Physiol 151(1):210–222

Olfatmiri H, Alemzadeh A, Zakipour Z (2014) Up-regulation of plasma membrane H^+-ATPase under salt stress may enable *Aeluropus littoralis* to cope with stress. Mol Biol Res Commun 3(1):67–75

Omoto E, Taniguchi M, Miyake H (2012) Adaptation responses in C4 photosynthesis of maize under salinity. J Plant Physiol 169:469–477

Panahi B, Ahmadi FS, Mehrjerdi MZ, Moshtaghi N (2013) Molecular cloning and the expression of the Na^+/H^+ antiporter in the monocot halophyte *Leptochloa fusca* (L.) Kunth. NJAS – Wag J Life Sci 64–65(5):87–93

Pardo JM, Cubero B, Leidi EO, Quintero FJ (2006) Alkali cation exchangers: roles in cellular homeostasis and stress tolerance. J Exp Bot 57:1181–1199

Pitann B, Zorb C, Muhling KH (2009) Comparative proteome analysis of maize (*Zea mays* L.) expansins under salinity. J Plant Nutr Soil Sci 172:75–77

Qiao WH, Zhao XY, Li W, Luo Y, Zhang XS (2007) Overexpression of AeNHX1, a root-specific vacuolar Na^+/H^+ antiporter from *Agropyron elongatum*, confers salt tolerance to *Arabidopsis* and Festuca plants. Plant Cell Rep 26:1663–1672

Qiu QS, Guo Y, Dietrich MA, Schumaker KS, Zhu JK (2002) Regulation of SOS1, a plasma membrane Na^+/H^+ exchanger in *Arabidopsis thaliana*, by SOS2 and SOS3. Proc Natl Acad Sci 99:8436–8441

Qu C, Liu C, Gong X, Li C, Hong M, Wang L, Hong F (2012) Impairment of maize seedling photosynthesis caused by a combination of potassium deficiency and salt stress. Environ Exp Bot 75:134–141

Rajendran K, Tester M, Roy SJ (2009) Quantifying the three main components of salinity tolerance in cereals. Plant Cell Environ 32(3):237–249

Rauf M, Shahzad K, Ali R, Ahmad M, Habib I, Mansoor S, Berkowitz GA, Saeed NA (2014) Cloning and characterization of Na^+/H^+ antiporter (LfNHX1) gene from a halophyte grass *Leptochloa fusca* for drought and salt tolerance. Mol Biol Rep 41(3):1669–1682

Ren ZH, Gao JP, Li LG, Cai XL, Huang W, Chao DY, Zhu MZ, Wang ZY, Luan S, Lin HX (2005) A rice quantitative trait locus for salt tolerance encodes a sodium transporter. Nat Genet 37(10):1141–1146

Roy S, Chakraborty U (2014) Salt tolerance mechanisms in Salt Tolerant Grasses (STGs) and their prospects in cereal crop improvement. Bot Stud 55(1):31

Roy S, Chakraborty U (2018) Role of sodium ion transporters and osmotic adjustments in stress alleviation of *Cynodon dactylon* under NaCl treatment: a parallel investigation with rice. Protoplasma 255(1):175–191

Roy SJ, Gilliham M, Berger B, Essah PA, Cheffings C, Miller AJ, Davenport RJ, Liu LH, Skynner MJ, Davies JM, Richardson P, Leigh RA, Tester M (2008) Investigating glutamate receptor-like gene co-expression in *Arabidopsis thaliana*. Plant Cell Environ 31:861–871

Roy SJ, Huang W, Wang XJ, Evrard A, Schmockel SM, Zafar ZU, Tester M (2013) A novel protein kinase involved in Na^+ exclusion revealed from positional cloning. Plant Cell Environ 36(3):553–568

Roychowdhury R, Karmakar J, Karmakar J, Adak MK, Dey N (2013) Physio-biochemical and microsatellite based profiling of lowland rice (*Oryza sativa* L.) landraces for osmotic stress tolerance. Am J Plant Sci 4(12):52

Roychowdhury R, Khan MH, Choudhury S (2018) Arsenic in rice: an overview on stress implications, tolerance and mitigation strategies. In: Hasanuzzaman M, Nahar K, Fujita M (eds) Plants under metal and metalloid stress. Springer, Singapore, pp 401–415

Roychowdhury R, Khan MH, Choudhury S (2019) Physiological and molecular responses for metalloid stress in rice – a comprehensive overview. In: Hasanuzzaman M, Fujita M, Nahar K, Biswas J (eds) Advances in rice research for abiotic stress tolerance. Woodhead Publishing, Elsevier, Duxford, pp 341–369

Rozema J, Flowers T (2008) Crops for a salinized world. Science 322(5907):1478–1480

Sairam RK, Rao KV, Srivastava GC (2002) Differential response of wheat genotypes to long term salinity stress in relation to oxidative stress, antioxidant activity and osmolyte concentration. Plant Sci 163(5):1037–1046

Sanadhya P, Agarwal P, Agarwal PK (2015a) Ion homeostasis in a salt-secreting halophytic grass. AoB Plants 7:plv055

Sanadhya P, Agarwal P, Khedia J, Agarwal PK (2015b) A low-affinity K^+ transporter *AlHKT2;1* from recretohalophyte *Aeluropus lagopoides* confers salt tolerance in yeast. Mol Biotechnol 57(6):489–498

Senadheera P, Singh RK, Maathuis FJ (2009) Differentially expressed membrane transporters in rice roots may contribute to cultivar dependent salt tolerance. J Exp Bot 60:2553–2563

Shabala SN, Mackay AS (2011) Ion transport in halophytes. Adv Bot Res 57:151–187

Shabala S, Bose J, Fuglsang AT, Pottosin I (2015) On a quest for stress tolerance genes: membrane transporters in sensing and adapting to hostile soils. J Exp Bot 67(4):1015–1031

Shi H, Ishitani M, Wu SJ, Kim CS, Zhu JK (2000) The *Arabidopsis thaliana* salt tolerance gene *SOS1* encodes a putative Na^+/H^+ antiporter. Proc Natl Acad Sci 97(12):6896–6901

Shigaki T, Hirschi KD (2000) Characterization of CAX-like genes in plants: implications for functional diversity. Gene 257:291–298

Shigaki T, Rees I, Nakhleh L, Hirschi KD (2006) Identification of three distinct phylogenetic groups of CAX cation/proton antiporters. J Mol Evol 63:815–825

Shrivastava P, Kumar R (2015) Soil salinity: a serious environmental issue and plant growth promoting bacteria as one of the tools for its alleviation. Saudi J Biol Sci 22:123–131

Singh RK, Mishra B, Singh KN (2004) Salt tolerance rice varieties and their role in reclamation programme in Uttar Pradesh. Indian Farm 2004:6–10

Singh V, Singh AP, Bhadoria J, Giri J, Singh J, Vineet TV, Sharma PC (2018) Differential expression of salt-responsive genes to salinity stress in salt-tolerant and salt-sensitive rice (*Oryza sativa* L.) at seedling stage. Protoplasma 255(6):1667–1681

Sirault XRR, James RA, Furbank RT (2009) A new screening method for osmotic component of salinity tolerance in cereals using infrared thermography. Funct Plant Biol 36(10–11):970–977

Storey R (1995) Salt tolerance, ion relations and the effect of root medium on the response of citrus to salinity. Aust J Plant Physiol 22(1):101–114

Sunarpi HT, Motoda J, Kubo M, Yang H, Yoda K, Horie R, Chan WY, Leung HY, Hattori K, Konomi M, Osumi M, Yamagami M, Schroeder JI, Uozumi N (2005) Enhanced salt tolerance mediated by AtHKT1 transporter-induced Na^+ unloading from xylem vessels to xylem parenchyma cells. Plant J 44(6):928–938

Sze H, Li X, Palmgren MG (1999) Energization of plant cell membranes by H^+-pumping ATPases: regulation and biosynthesis. Plant Cell 11:677–689

Tada Y, Komatsubara S, Kurusu T (2014) Growth and physiological adaptation of whole plants and cultured cells from a halophyte turf grass under salt stress . AoB plants 6: plu041

Taherinia B, Kavousi HR, Dehghan S (2015) Isolation and characterization of plasma membrane Na^+/H^+ antiporter (SOS1) gene during salinity stress in kallar grass (*Leptochloa fusca*). Eurasia J Biosci 9:12–20

Takahashi R, Liu S, Takano T (2007) Cloning and functional comparison of a high-affinity K^+ transporter gene *PhaHKT1* of salt-tolerant and salt-sensitive reed plants. J Exp Bot 58(15–16):4387–4395

Takahashi R, Liu S, Takano T (2009) Isolation and characterization of plasma membrane Na(+)/H(+) antiporter genes from salt-sensitive and salt-tolerant reed plants. J Plant Physiol 166(3):301–309

Tang J, Yu X, Luo N, Xiao F, Camberato JJ, Jiang Y (2013) Natural variation of salinity response, population structure and candidate genes associated with salinity tolerance in perennial ryegrass accessions. Plant Cell Environ 36(11):2021–2033

Tapken D, Hollmann M (2008) *Arabidopsis thaliana* glutamate receptor ion channel function demonstrated by ion pore transplantation. J Mol Biol 383:36–48

Taiz L. Zeiger E, Moller IM, Murphy A (2015) Plant Physiology and Development. 6th Edition, Sinauer Associates, Sunderland, CT

Teakle NL, Bazihizina N, Shabala SN, Colmer TD, Barrett-Lennard EG, Rodrigo-Moreno A, Lauchli AE (2013) Differential tolerance to combined salinity and O_2 deficiency in the halophytic grasses *Puccinellia ciliata* and *Thinopyrum ponticum*: the importance of K^+ retention in roots. Environ Exp Bot 87:69–78

Tester M, Davenport RJ (2003) Na^+ tolerance and Na^+ transport in higher plants. Ann Bot 91:503–527

Todaka D, Nakashima K, Shinozaki K, Yamaguchi-Shinozaki K (2012) Towards understanding transcriptional regulatory networks in abiotic stress responses and tolerance in rice. Rice 5(1):6

Touchette BW (2007) Seagrass-salinity interactions: physiological mechanisms used by submersed marine angiosperms for a life at sea. J Exp Mar Biol Ecol 350(1–2):194–215

Truong P, Gordon I, Armstrong F, Shepherdson J (2002) Vetiver grass for saline land rehabilitation under tropical and Mediterranean climate. In: Eighth national conference productive use of saline lands, Perth, Australia (conference paper)

Tsunekawa K, Shijuku T, Hayashimoto M, Kojima Y, Onai K, Morishita M, Ishiura M, Kuroda T, Nakamura T, Kobayashi H, Sato M, Toyooka K, Matsuoka K, Omata T, Uozumi N (2009) Identification and characterization of the Na^+/H^+ antiporter Nhas3 from the thylakoid membrane of *Synechocystis* sp. PCC 6803. J Biol Chem 284(24):16513–16521

Tuteja N, Gill SS, Tuteja R (2011) Plant responses to abiotic stresses: shedding light on salt, drought, cold and heavy metal stress. In: Tuteja N (ed) Omics and plant abiotic stress tolerance. Bentham Science Publishers Ltd, Sharjah, pp 39–64

Vera-Estrella R, Barkla BJ, Garcia-Ramirez L, Pantoja O (2005) Salt stress in *Thellungiella halophila* activates Na^+ transport mechanisms required for salinity tolerance. Plant Physiol 139(3):1507–1517

Verma D, Singla-Pareek SL, Rajagopal D, Reddy MK, Sopory SK (2007) Functional validation of a novel isoform of Na^+/H^+ antiporter from *Pennisetum glaucum* for enhancing salinity tolerance in rice. J Biosci 32(3):621–628

Volkov V (2015) Salinity tolerance in plants. Quantitative approach to ion transport starting from halophytes and stepping to genetic and protein engineering for manipulating ion fluxes. Front Plant Sci 6:873

Wahid A, Perveen M, Gelani S, Basra SMA (2007) Pretreatment of seed with H_2O_2 improves salt tolerance of wheat seedlings by alleviation of oxidative damage and expression of stress proteins. J Plant Physiol 164(3):283–294

Wakeel A, Sumer A, Hanstein S, Yan F, Schubert S (2011) In vitro effect of Na^+/K^+ ratios on the hydrolytic and pumping activity of the plasma membrane H^+-ATPase from maize (*Zea mays* L.) and sugar beet (*Beta vulgaris* L.) shoot. Plant Physiol Biochem 49:341–345

Wang SM, Zhao GQ, Gao YS, Tang ZC, Zhang CL (2005) *Puccinellia tenuiflora* exhibits stronger selectivity for K^+ over Na^+ than wheat. J Plant Nutr 27(10):1841–1857

Wang P, Guo Q, Wang Q, Zhou XR, Wang SM (2015) PtAKT1 maintains selective absorption capacity for K^+ over Na^+ in halophyte *Puccinellia tenuiflora* under salt stress. Acta Physiol Plant 37:1–10

Waters S, Gilliham M, Hrmova M (2013) Plant high-affinity potassium (HKT) transporters involved in salinity tolerance: structural insights to probe differences in ion selectivity. Int J Mol Sci 14:7660–7680

Wu J, Seliskar DM (1998) Salinity adaptation of plasma membrane H^+-ATPase in the salt marsh plant *Spartina patens*: ATP hydrolysis and enzyme kinetics. J Exp Bot 49(323):1005–1013

Xu Y, Zhou Y, Hong S, Xia Z, Cui D, Guo J, Xu H, Jiang X (2013) Functional characterization of a wheat NHX antiporter gene *TaNHX2* that encodes a K^+/H^+ exchanger. PLoS One 8:e78098

Yadav NS, Shukla PS, Jha A, Agarwal PK, Jha B (2012) The *SbSOS1* gene from the extreme halophyte *Salicornia brachiata* enhances Na^+ loading in xylem and confers salt tolerance in transgenic tobacco. BMC Plant Biol 12:188

Yamamoto N, Takano T, Tanaka K, Ishige T, Terashima S, Endo C, Kurusu T, Yajima S, Yano K, Tada Y (2015) Comprehensive analysis of transcriptome response to salinity stress in the halophytic turf grass *Sporobolus virginicus*. Front Plant Sci 6:241

Yang Q, Chen ZZ, Zhou XF, Yin HB, Li X, Xin XF, Hong XH, Zhu JK, Gong Z (2009) Overexpression of SOS (Salt Overly Sensitive) genes increases salt tolerance in transgenic *Arabidopsis*. Mol Plant 2(1):22–31

Yue Y, Zhang M, Zhang J, Duan L, Li Z (2012) *SOS1* gene overexpression increased salt tolerance in transgenic tobacco by maintaining a higher K^+/Na^+ ratio. J Plant Physiol 169(3):255–261

Zhang HX, Blumwald E (2001) Transgenic salt-tolerant tomato plants accumulate salt in foliage but not in fruit. Nat Biotechnol 19:765–768

Zhang HH, Hodson JN, Williams JP, Blumwald E (2001) Engineering salt-tolerant *Brassica* plants: characterization of yield and seed oil quality in transgenic plants with increased vacuolar sodium accumulation. Proc Natl Acad Sci 98:12832–12836

Zhang WD, Wang P, Bao Z, Ma Q, Duan LJ, Bao AK, Zhang JL, Wang SM (2017) SOS1, HKT1;5, and NHX1 synergistically modulate Na^+ homeostasis in the halophytic grass *Puccinellia tenuiflora*. Front Plant Sci 8:576

Zhou Q, Yu BJ (2009) Accumulation of inorganic and organic osmolytes and their role in osmotic adjustment in NaCl-stressed vetiver grass seedlings. Russ J Plant Physiol 56:678–685

Zouari N, Saad RB, Legavre T, Azaza J, Sabau X, Jaoua M, Masmoudi K, Hassairi A (2007) Identification and sequencing of ESTs from the halophyte grass *Aeluropus littoralis*. Gene 404:61–69

Chapter 9
Role of Heat Shock Proteins (HSPs) and Heat Stress Tolerance in Crop Plants

Zeba Khan and Durre Shahwar

Abstract Environmental stresses can alter the physiological mechanisms of plants by dysfunctioning proteins which is a major consequence of less crop productivity. Heat stress (high temperature (HT)) is one of the major abiotic stresses that causes damage to the plant tissues. Such stress limits the crop production by altering developmental growth, metabolism and productivity. Better crop yield and sustainable food quality in HT become the potent agricultural goal in this ongoing climate change era. In the molecular level, the plant's survival strategy in HT condition is to change the expression of gene during transcription/translation process resulting in the generation of heat shock proteins (HSPs). Sometimes it alters the level of heat shock factors (HSFs) that control the expression of heat shock and non-heat shock gene. These HSPs are ubiquitous and are produced in all organisms from prokaryotes to eukaryotes under stress condition and are highly conserved and constitutively function as molecular chaperons, which multipurpose function. Apart from the heat stress, HSPs are known to be expressed during other stressful conditions such as during tissue damage, inflammation, hypoxia, UV light, cold stress, water stress, etc. HSPs have been classified in plants on the basis of approximate molecular weight and their intracellular localization. Plants growing under HT stress can perceive HT stimulus and transmit the signal causing effect in terms of physio-biochemical and gene regulatory responses. Under HT, oxidative stress is generated as a result of the production of reactive oxygen species (ROS) which alters the synthesis of macromolecules and nucleic acids. Thus, HSPs play an important role in maintaining cell membrane integrity, ROS scavenging and production of antioxidants, osmolytes. In this chapter, we are discussing physiological and biochemical responses to heat stress and the adopted molecular approaches for the mitigation strategy of the plant under HT.

Keywords Chaperons · Heat shock proteins · ROS · Stress tolerance · Crop productivity

---

Z. Khan (✉) · D. Shahwar
Department of Botany, Aligarh Muslim University, Aligarh, Uttar Pradesh, India

R. Roychowdhury et al. (eds.), *Sustainable Agriculture in the Era of Climate Change*, https://doi.org/10.1007/978-3-030-45669-6_9

Abbreviations

HS	Heat stress
HSF	Heat shock factor
HSPs	Heat shock transcription factors
HSPs	Heat shock proteins
HT	High temperature
QTL	Quantitative trace loci
ROS	Reactive oxygen species
TFs	Transcription factors

1 Introduction

One of the stresses exerted by the environment on the plant is heat stress which produces certain heat shock proteins (HSPs). In today's scenario, increase in the mean temperature due to global warming is raising serious concerns in the agriculture around the world causing an estimated annual loss of crop yield up to billions of dollars (Mittler et al. 2012). Heat shocks cause severe damage to the crop by altering protein synthesis, inactivating major enzymes and membrane damage. High temperature limits the plant growth and induces oxidative stress, culminating into yield loss and nutrient quality.

Heat stress causes improper folding of polypeptide chains of non-native proteins leading to unfavourable protein aggregations resulting in dysfunctioning of proteins (Moriwaki et al. 1999). Plants have developed molecular chaperons for stabilizing protein by assembling multi-protein complexes, folding or unfolding of proteins, control of cell cycle and signalling, transport of proteins into respective compartments at the subcellular level to protect it against stress or apoptosis (Lindquist and Craig 1988).

HSPs, heat shock transcription factor (HSF) and chaperons are necessary for balance and restoration of protein structure and sustain the condition of the plant under high-temperature stress (Boston et al. 1996). Lethal doses of high temperature induce heat shock response which protects the organism from cellular damage and enables the renewal of normal cellular and physiological activities allowing a higher level of thermotolerance in plants (Schöffl et al. 1998). Heat shock response is also associated with many environmental stresses. High temperature or heat stress can cause excessive generation of ROS and oxidative damage (Hasanuzzaman et al. 2012a, b, c, 2013a; Asthir 2015). The plant enables to tolerate heat stress under high-temperature condition by making some changes in the plant body such as plants accumulate different metabolites like antioxidants, osmoprotectants and HSPs by altering their metabolism in various ways (Bokszczanin and Fragkostefanakis 2013; Hasanuzzaman et al. 2013c). Heat stress also causes alterations at molecular level where genes are involved in the expression of factor such as osmoprotectants (glycine betaine, GB; proline, Pro; trehalose, Tre; etc.), phytohormones (gibberellic acids, GA; abscisic acid, ABA; brassinosteroids, BR; jasmonic acids, JA; salicylic acid, SA; etc.), detoxifying enzymes, signalling molecules (e.g. nitric oxide, NO), transporters, regulatory proteins, free radical scavengers, polyamines (spermidine, Spd; putrescine, Put; and spermine, Spm) and protectants in trace elements (silicon, Si; selenium, Se; etc.). These factors involved in signalling cascades and transcriptional control process have been found effective to counter effect of HT stress (Wang et al. 2004; Krasensky and Jonak 2012; Semenov and Halford 2009; Hasanuzzaman et al. 2010, 2012b, c, 2013a, b; Waraich et al. 2012; Barnabás et al. 2008). HSPs or molecular chaperons are an important class of protein which plays an important role in stress signal transduction for protection and repair protein damage and maintains plant homeostasis (Boston et al. 1996).

2 Classification of Heat Shock Proteins (HSPs)

HSPs are a family of proteins that are produced by cells under the stressful condition in almost all organisms. HSPs are ubiquitous, obtained in both prokaryotic and eukaryotic cells. It plays an essential role in cell functions because of its high level of conservation. They have been extensively studied by the researchers on the basis of cellular localization, regulation and functions (Hightower 1991; Benjamin and McMillan 1998). Italian scientist (Ritossa 1962) described the gene expression in chromosome of *Drosophila melanogaster* following exposure to heat shock. It was the beginning of discovery of heat shock proteins, but nowadays heat shock proteins are considered to be induced by different kinds of stresses such as low temperature, UV light, wound healing or other abiotic stresses (Vierling 1991; Boston et al. 1996). It has been shown that during the cell growth and development, HSPs are responsible for folding of protein, assembly, translocation and degradation (Lindquist and Craig 1988; Wang et al. 2004). Whitley et al. (1999) reported that some HSPs accomplish chaperon functions under stress conditions within the cell, due to which they are also known to be HSPs because of their aggregation when denatured by high temperature. There are five major families of HSPs in plants and animals conservatively recognized as molecular chaperons.

The molecular size of HSP families ranges from 10 to more than 100 kDa, located in various cellular compartments. The HSPs were grouped into five families (Kregel 2002) on the basis of their molecular weight such as HSP100, HSP90, HSP70, HSP60 and small HSP (sHSP) (Wang et al. 2004; Kotak et al. 2007; Gupta et al. 2010; Al-Whaibi 2011) (Fig. 9.1, Table 9.1). Some of these HSPs are mainly found in cytoplasm and are expressed in response to biotic and abiotic stresses (Vierling 1991; Boston et al. 1996). Moreover, these HSPs are also found in several cytoplasmic organelles such as endoplasmic reticulum, mitochondria, chloroplasts and the nucleus which plays a dynamic role in protein homeostasis (Vierling 1991; Boston et al. 1996).

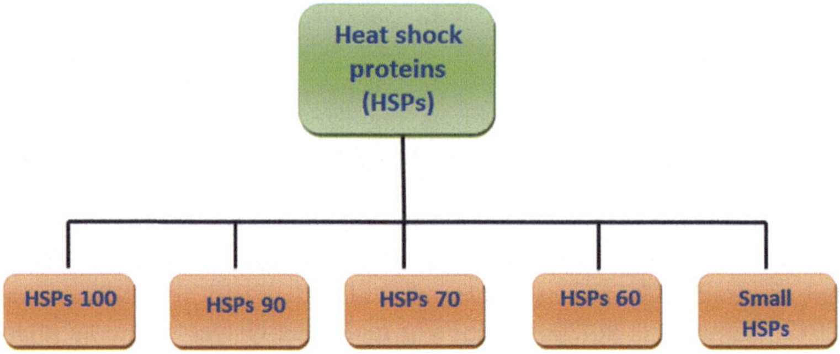

Fig. 9.1 Classification of heat shock proteins (HSPs)

Table 9.1 Sites and functions of HSP families (Kregel 2002)

HSP family/ classification	Cellular location/site	Function/role
HSP 20 (sHSPs)	Cytosol, endoplasmic reticulum, mitochondria, chloroplast	Preventing aggregation, co-chaperons
HSP 27 (sHSPs)	Cytosol, nucleus	Microfilament, antiapoptotic, stabilization
HSP 60	Mitochondria	Prevents aggregation of denatured proteins, refolds proteins, proapoptotic
HSP 70 family		Antiapoptotic
HSP 72 (HSP 70)	Nucleus, cytosol	Cytoprotection, protein folding
HSP 73 (Hsc 70)	Nucleus, cytosol	Molecular chaperons
HSP 75 (mHSP 70)	Mitochondria	Molecular chaperons
HSP 78 (GRP78)	Endoplasmic reticulum (ER)	Molecular chaperons, cytoprotection
HSP 90	Cytosol, nucleus, endoplasmic reticulum (ER)	Translocation of protein, regulation of steroid hormone receptors
HSP100/104	Cytosol	Folding of protein

2.1 HSP100

Heat shock protein with the molecular weight of 100 to 104 is categorized into HSP 100 family present in both prokaryotes and eukaryotes and located in the cytoplasm with HSP 40, HSP 70 and HSP 90. The bacterial HSP 100 proteins are known to be considered as Clp proteins which are made up of two-subunit protease system (Squires and Squires 1992). The larger subunit ClpA functions as chaperon, whereas smaller subunit ClpP is protease. HSP100 are categorized into two major classes of protein: class 1 protein containing two ATP-binding sites and class 2 proteins containing only one ATP-binding site (Schirmer et al. 1996; Miernyk 1999). Under heat stress condition, HSP 100 proteins reactivate the aggregated proteins by resolubilization of nonfunctional protein and also help to break irreversible polypeptides (Bosl et al. 2006; Kim et al. 2007; Gupta et al. 2010). A member of this family is essential for high-temperature tolerance but not for germination and growth in plants (Queitsch et al. 2000; Hong and Vierling 2001; Lin et al. 2014). The yeast homolog HSP 104 is necessary for induced thermotolerance in yeast. HSP 100 of yeast have deletion which can be complemented with the wild type of plant HSP 100 homologs from several crops like *Arabidopsis*, wheat, soybean, tobacco, etc. so it can be concluded that roles of plant HSP 100 proteins and yeast are conserved in thermotolerance (Lee et al. 1994). Wells et al. (1998) and Ling et al. (2000) suggested that plant HSP 101 not only plays an important role in HT but also has been also involved as RNA-binding protein for mediating translational enhancement of TMV (tobacco mosaic virus) RNA and ferredoxin mRNA (Ling et al. 2000). According to Lee et al. (2006), HSP 100 family is also involved in housekeeping functions, necessary for development of chloroplast. It seems that this class also participates in facilitating the normal situation after severe stress in organism (Gurley 2000).

2.2 HSP90

HSP90 is found both in prokaryotic and eukaryotic cells, and it is most abundant in the cytoplasm and endoplasmic reticulum and induced under heat stress. It possesses ATP-binding domain at N-terminal, substrate-interacting middle domain and dimerization domain at C-terminal (Jackson 2013). HSP90 assists in proper folding of proteins and also plays an important role in protein trafficking, protein degradation, signal transduction and cell cycle control (Efeoglu 2009). Being important for proper functioning of many signalling pathways, HSP90 plays an important role in regulation of cellular response to environmental and other external stimuli, on exposure to stress by redirecting cellular metabolism in such a way so as to endure tolerance (Wang et al. 2004). HSP90 genes have been isolated from different plants such as *Arabidopsis* tomato (Koning et al. 1992), maize (Marrs et al. 1993) and *Brassica napus* (Krishna et al. 1995) and *thaliana* (Krishna and Gloor 2001). HSP 90 is subdivided into HSP 90α and HSP90 β. HSP 90 cohered with protein kinases, intermediate filaments, steroid receptors, actin microfilaments and microtubules in a specific manner. This class of HSPs plays another key role to regulate the glucocorticoid receptor (GR) activity (Pratt et al. 2004). Matsumiya et al. (2009) and Gupta et al. (2010) recorded that HSP90 connects with different intracellular proteins such as tubulin, calmodulin, kinases, actin and receptor proteins under physiological conditions. It has been reported that HSP 90 is key for normal growth and development in *Arabidopsis* and *Nicotiana benthamiana* (Liu et al. 2004; Sangster and Queitsch 2005; Sangster et al. 2007). Cytoplasmic HSP90 is also considered accountable for resistance against pathogen by reacting with resistance protein (R) which is signal receptor of the pathogen. Liu et al. (2004) studied reaction between HSP90 and resistance protein in *A. thaliana* and tobacco species, viz. *Nicotiana tabacum* and *N. benthamiana*. HSP90-silenced *N. benthamiana* plants exhibited growth and phenotypic alterations such as chlorotic leaves, change in flowering time and morphology.

2.3 HSP70

HSP 70 family is highly conserved classes of HSPs and present in the cytosol. Apart from being induced by high temperature and other stresses, constitutively expressed HSP70 has also important functions under no stress (Efeoğlu 2009). It has a GRP78 family which helps protein folding and refolding and enables to block protein degradation in ER. HSP 70 aids in transmembrane transport of newly synthesized protein by stabilizing them in partially folded state and removes denatured proteins and also prohibits their aggregation (Su and Li 2008). HSP 70 contains two domains of which one is of a ~45 kDa amino terminal ATP-binding domain and another one is a ~25 kDa carboxyl terminal substrate-binding domain. HSP70 functions both co- and post-translationally in higher eukaryotes. Rouch et al. (2004) reported that HSP70 and small HSPs work as molecular chaperon exhibiting coherent role in protecting plant cell from high-temperature stress.

According to Zhang et al. (2008), HSP70 and small HSP17.6 play an important role in the development of cross-adaptation to temperature stress induced by heat acclimation (HA) or cold acclimation (CA) pretreatment in grape plants. HSP70 binds to protein precursor for transfer via membranes of cell organelles such as chloroplast (Jackson-Constan et al. 2001; Soll 2002). HSP70 B is the type of HSP70 protein situated in the matrix of chloroplasts and helps in reconstruction and photoprotection of photosystem II during photoinhibition (Schroda et al. 1999). Su and Li (2008) studied that presence of HSP70 is essential in the stroma of chloroplast for the differentiation of germinating seeds during heat tolerance. HSP70 also plays an important role during biotic stress responses. Heat shock cognate 70 (HSC70) chaperon present in cytoplasm and nucleus is highly homologous to HSP70 and regulates immune responses along with SGT1 in case of *Arabidopsis thaliana.* HSP70 also plays an important function in microbial pathogenesis. Boevink and Oparka (2005) and Hafren et al. (2010) reported that HSP 70 controls the reproduction and movement of virus which induces infection. Chen et al. (2008) studied that *N. benthamiana* is infected by *Potato virus* X, *Tobacco mosaic virus* and *Watermelon mosaic virus* enhanced by cytoplasmic HSP70.

2.4 HSP60

HSP60 is located in mitochondria with mitochondrial protein import and macromolecular assembly. HSP 60 is also referred to as P60 lymphocyte protein, mitochondrial matrix protein P1 and HSPD1. In some literature it is also called as chaperonins as they are important in helping plastid protein such as rubisco (Wang et al. 2004) and assist in attaining functional configuration of imported proteins to the chloroplast

(Jackson-Constan et al. 2001). HSP60 chaperonins assist the proper folding of imported proteins, also protect from aggregation and misfolding and promote correct refolding and assembly of unfolded polypeptides in mitochondrial matrix produced under stress. Wang et al. (2004) reported that HSP 60 is generally facilitating plastid proteins such as Rubisco. HSP60 has been involved in stress response as well. The upregulation of HSP60 during stress allows regulation of cellular processes occurring in the cell. Many researchers have shown that HSP60 and other HSPs are necessary for survival of cells under toxic or stressful conditions (Rossi et al. 2002).

2.5 *Small HSPs (sHSPs)*

sHSPs are class of heat shock proteins which are ubiquitously produced both in prokaryotic and eukaryotic cells in response to heat stress, having size ranging from 15 to 42 kDa. In *A. thaliana*, 13 sHSPs are classified into 6 classes on the basis of intracellular localization and sequence relatedness, an additional 6 open reading frame encoding proteins which are distantly related to plastidial class of sHsps (Scharf et al. 2001). The C-terminal region of these proteins has a common alpha-crystallin domain containing 80–100 amino acid residues (Seo et al. 2006). It also has an NH2-terminal end which is quite divergent and belongs to different classes. Gupta et al. (2010) reported that small HSPs also work as chaperons like other HSPs, preventing undesirable interaction of protein, and facilitate refolding of denatured proteins. sHSPs also involve in protecting by hampering the thermal aggregation of protein (van Montfort et al. 2001). Unlike other HSPs, the activity of small HSPs is independent of ATP (Miernyk 1999). According to Kotak et al. (2007) and Liberek et al. (2008), HSP20 is the type of sHSPs which sustained denatured proteins in a folding state and enables ATP-dependent disaggregation through the HSP70/90 chaperon system. sHSPs are also known for viral infection similar to HSP70/HSP40 (Verchot 2012). Several researchers reported that it also plays an important role in plant disease resistance. For example, HSPs 20 is the member of small HSPs, which associate with I-2 and provide resistance to *Fusarium oxysporum* (Simons et al. 1998) by the accumulation of I-2 (Verchot 2012). Another HSP20 from *N. tabacum* (NtsHSP) is known to be involved in resistance against biotic stress, for example, disease resistance in plants (Maimbo et al. 2007). Mogk et al. (2003) reported that sHSP 18.1 from *Pisum sativum* and sHSP 16.6 from *Synechocystis*, isolated under lab condition, bind to unfolded proteins and further enable refolding of proteins by HSP70/HSP100 complexes. sHSPs play an important role in thermotolerance as well as other biotic and abiotic stresses which occur as a result of genetic changes produced in the absence of any environmental stress at specific developmental stages such as germination, pollen development embryogenesis and fruit maturation in the expression of sHSPs in transgenic plants (Harndahl et al. 1999; Sun et al. 2001, 2002).

3 Role of HSPs in Encountering High-Temperature Stress

High-temperature stress induces expression of defence genes which normally do not express under normal condition (Morimoto 1993). Gupta et al. (2010) reviewed that gene expression and synthesis of proteins inside the cells are triggered by stress. However, a factor that causes stress firstly blocks cell metabolism including DNA replication, transcription, mRNA export and translation, until cells start to recover (Biamonti and Caceres 2009).

Expression/transcription of HSP genes is regulated by regulatory proteins known as heat stress transcription factors (HSFs) which are present in an inactive form inside the cell cytoplasm (Nover and Baniwal 2006). HSFs contribute in all phases of heat stress response including maintenance and recovery. According to Baniwal et al. (2004) and Hu et al. (2009), HSFs are also known as transcriptional activators for heat shock response. There are 15 known HSFs in *Arabidopsis thaliana* (Nover et al. 2001), whereas in tomato >21 HSFs are known to play an important role in heat stress response (Scharf et al. 1998). Out of 21 only 2 of these HSFs such as HSF A2 and HSF B1 are heat inducible, but their expressions are controlled by HsfA1which is known to be a master regulator of the HSR (Mishra et al. 2002).

HsfA2 is a dominant HSF during high-temperature stress because it is recognized as 'work horse' of the stress response. When plant feels high-temperature stress, the inactive HSFs located in the cytosol are detached from HSP and activated and polymerized into triplets/trimers. After this phenomenon, these trimers' HSF gets phosphorylated in the cytoplasm and transported into the nucleus where they stick to the heat shock element located in the promoter region on the HSP gene. After transcription and translation of mRNA, level of HSPs in the cytoplasm increases (Usman et al. 2014) (Fig. 9.2). Role of HSPs under heat stress in different crop plants has been enlisted in Table 9.2.

4 HSPs/Chaperons Induced Heat Stress Response Mechanism in Plants

Heat stress induces protein denaturation which is a constant direct or indirect consequence of not only heat stress but any physical stress, as stresses are defined as factors impeding normal cellular functions carried out by proteins. Among the HSP/chaperon 'client proteins', many of them are primary metabolism enzymes and signal transduction proteins which play essential roles in the folding of proteins and in turn proper functioning of cells under stress. Plants respond to unfavourable changes through developmental, physiological and biochemical ways, and these responses require the expression of stress-responsive genes, which are regulated by a network of transcription factors (TFs), called as heat stress transcription factors (HSFs), that form an active trimeric complex, move to the site of the target genes and rapidly initiate transcription, recruited under stress conditions such as cold, osmotic,

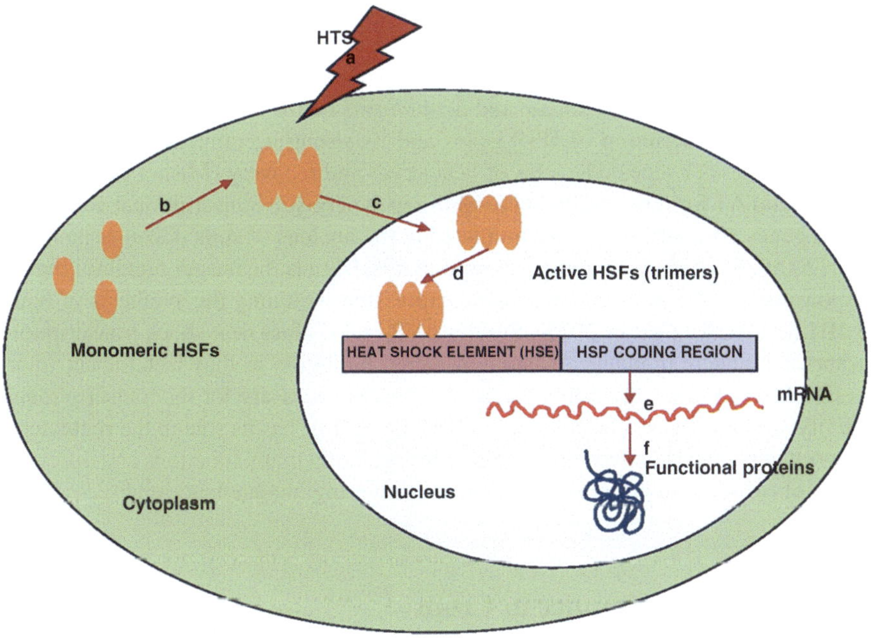

Fig. 9.2 Illustrative diagram exhibiting the molecular mechanism of heat shock proteins (HSPs)

Table 9.2 Role of HSPs in different crops

Types of HSP	Stress types	Organisms	References
HSP70	Heat stress	Rice, wheat, foxtail, millet, *Chrysanthemum*, barley	Hu et al. (2009), Rollins et al. (2013), Singh et al. (2016), Zhang et al. (2017)
HSP90	Heat stress	Wheat, alfalfa, soybean	Li et al. (2013), Xu et al. (2013)
HSP100 and HSP101	Heat stress	Rice, maize, *Brassica*	Young et al. (2004)
HSP60	Heat stress	Grasses	Xu et al. (2011)
HSP17.8 and HSP17.2	Heat stress	*Rosa chinensis*, *Camellia sinensis*	Jiang et al. (2009), Wang et al. (2017)
HSP17.6	Heat stress	*Brassica napus*	Young et al. (2004)

drought, salt, UV, high light, oxidative stress and pathogen infection (Swindell et al. 2007). These HSFs are present in the cytoplasm as single and free because there is no binding activity with DNA in the absence of stress, however when stress starts, the factors aggregate in triplet and accumulate in the nucleus. Inducers of molecular chaperons, or heat shock transcription factors are grouped into three classes A, B and C, according to the structural differences in their triplets aggregation, as plant HsfA such as HsfA1 and HsfA2 in *Lycopersicon esculentum*, plant Hsf B such as Hsf B1 in *L. esculentum* and plant Hsf C (Tripp et al. 2009). $HSFA_2$ has been found to be the dominant HSF in tomato and *Arabidopsis* on the basis of its high activator potential for transcription of HSP genes and its strong accumulation under long-term heat stress or repeated cycles of heat stress and recovery (Mishra et al. 2002). HSFA2 and A1 form heterodimers resulting in synergistic transcriptional activation of HS genes after HSFA2 are accumulated in the nucleus of cells (Chan-Schaminet et al. 2009). Studies in the tomato showed that HsfA1a is the master regulator that is responsible for the induced stress gene expression including the synthesis of both HsfB1 and HsfA2 (Asthir 2015). Studies on B and C class heat shock transcription factors are meagre in plants and are considered inhibitory as they lack the activator motif AHA (aromatic, hydrophobic, acidic) that is necessary for the transcriptional activity of class A HSFs (Jacob et al. 2017). Each HSF has its role in the regulatory network in plants. However, all cooperate in regulating many functions and different stages of response to periodical heat stress (triggering, maintenance and recovery).

5 Heat Stress Tolerance in Crops

In the natural environment, plants are usually exposed to the combined effect of temperature and light. But excess exposure of plants to a temperature above the range of optimal can cause disturbance to the overall life cycle of the plant.

5.1 Effect of Heat Stress on Germination and Plant Growth

Heat stress acts as a prime factor in altering the rate of plant development. Each species has a specific tolerance temperature range from the boundaries of observable growth. Reduced germination percentage, plant emergence, growth inhibition, abnormal seedlings, poor seedling vigour and reduced radical and stunted seedlings are major impacts caused by heat stress recognized in a variety of cultivated plant species (Kumar et al. 2011). High temperatures may also show visual stress symptoms like discolouration of fruits and leaves, scorching of the twigs and leaves senescence (Vollenweider and Gunthardt-Goerg 2005). It has been studied that the increase in temperature to optimum thresholds stimulates biochemical mechanisms, affecting the rate of development and declining the lengths of growing seasons which could have consequently an adverse impact on yield (Cleland et al. 2007). Studies in maize showed elevated temperature beyond 37 °C impaired protein synthesis in embryo leading inhibition in germination and complete halt of coleoptile growth at 45 °C (Riley 1981; Akman 2009). According to Sung et al. (2001), HSP70s may be needed in two important aspects of protein metabolism during germination. First, as a result of seed desiccation during HT, proteins that are unfolded or misfolded could be susceptible to aggregation during seed imbibition (i.e. rehydration of proteins), and presence of these HSP70 chaperons in every compartment of the cells soon after rehydration is necessary to minimize the toxic effects of protein aggregation. Second, the initiation of active synthesis and translocation of proteins must be protected for the proper functioning of metabolic activities during germination.

5.2 *Effect of Heat Stress on Yield*

Stress especially HS is the principal cause of reduced production of crops globally and in turn responsible for lowering the average yield of major crops by more than 50% (Mahajan and Tuteja 2005) by constraining grain filling. Heat stress can occur very abruptly and even short episodes of exposure to high temperature can cause a considerable decline in yield. The reasons behind the decreased yield under high temperature are impaired meiosis, pollen germination and pollen tube growth, reduced number of pollen grains, proembryo and unfertilized embryo (Cao et al. 2008). The reproductive phase is more sensitive to high temperature, as it interferes with anther dehiscence and gametophyte development, which resulted in a final yield reduction (Nadeem et al. 2018). High heat exposure during pod and seed set stages results in a substantial decrease in crop yield by a reduction in seed weight and effecting the economic production. Elevated temperature speeds up the rate of seed filling by reducing the duration of this stage ultimately affecting the yield potential (Kaushal et al. 2016). Many crop plants show early maturation, which is intimately linked with smaller yield losses under high temperatures, which may be attributed to the engagement of an escape mechanism (Adams et al. 2001). Heat stress negatively affects the nutritional quality of cereals and oilseeds as it substantially reduces the oil, starch and protein contents (Maestri et al. 2002). Studies on *Sorghum* show that high-temperature exposure (>36–38 °C) during flowering time for 10–15 days resulted in reduced pollen germination, failure of fertilization and flower abortion (Prasad et al. 2008; Singh et al. 2015). Similarly, it has been found that mean temperature beyond 34 °C enhances the rate of senescence in wheat resulting in a significant reduction in yield (Lobell and Gourdji 2012). In tomato (*Lycopersicum esculentum* Mill.), HS influences meiosis, fertilization and growth of fertilized embryo consequently resulting in a noticeable reduction in the yield (Camejo et al. 2005).

5.3 *Effect of Heat Stress on Plant Physiology*

The increase in temperature up to a certain level shows a positive response towards plant growth, photosynthesis, respiration and enzyme activity; after that these parameters tend to decline due to change in the cellular state, lipid composition, membrane fluidity and organelle properties of the plant (Hasanuzzaman et al. 2013d) (Fig. 9.3). Plants may experience different types of stress at different devel-

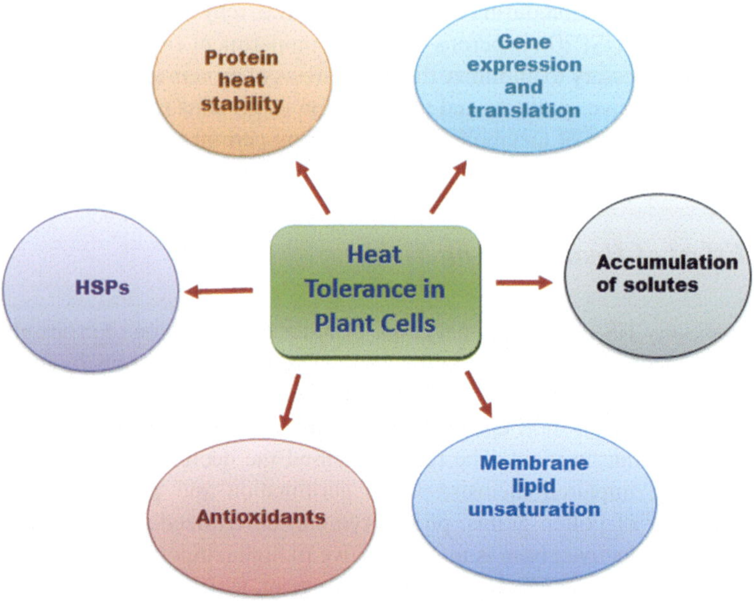

Fig. 9.3 Affecting factors for the cellular level of heat tolerance in crop plants

opmental stages, and their mechanisms of response to stress may vary in different tissues (Queitsch et al. 2000). High-temperature stress tends to reduce the relative water content of the crop and induces oxidative stress-mediated damage to the cell structures and associated proteins causing physiological drought.

5.3.1 Photosynthesis

High temperature, whether transitory or constant, affects plant growth and productivity by reducing photosynthesis. Photosynthesis is one of the most crucial heat-sensitive physiological processes in plants. Efficient photosynthesis involves photosynthetic pigments and photosystems, the electron transport system, CO_2 fixation pathways and glycolic metabolism. Damage to any of these components may reduce photosynthetic capacity (Ashraf and Harris 2013). Transitory or constant high-temperature stress reduces photosynthesis and limits plant growth and productivity (Wang et al. 2001); however moderate heat stress causes reversible-irreversible damage to the photosynthetic apparatus, resulting in greater inhibition of plant growth (Yan et al. 2011). Photosynthesis incorporates numerous components, including CO_2 reduction pathways, photosynthetic photosystems and the electron transport system, as its one of the most intricate physiological processes in plants (Ashraf and Harris 2013). High temperature alters leaf stomatal conductance and intercellular CO_2 concentration, resulting in the closure of stomata which might be another reason for impaired photosynthesis that affects the intercellular CO_2 (Greer and Weedon 2012). Ribulose-1,5-bisphosphate (RuBP) carboxylase/oxygenase (Rubisco) catalyses the first step in net photosynthetic CO_2 assimilation and photorespiratory carbon oxidation. The enzyme Rubisco is inefficient as a catalyst for the carboxylation of RuBP and is subject to competitive inhibition by O_2, inactivation by loss of carbamylation and dead-end inhibition by RuBP which make Rubisco rate limiting for photosynthesis. The decline in Rubisco activity results in a loss in photosynthesis. It is also suggested that high temperatures inactivate Rubisco activase (Salvucci and Crafts-Brandner 2004), which is essential for the activity of Rubisco. The repress of carboxylation processes by HT is likely a cause for the Rubisco decline that causes photosynthesis to plummet at higher temperatures. Under such conditions, inhibition of the repair of photodamaged PSII as a response to limitations in fixation of CO2 might accelerate photoinhibition.

Organelle-localized low-molecular-weight HSPs have been identified to be produced at high temperatures, and their association with membranes upon heat stress was reported (Adamska and Kloppstech 1991), suggesting that HSPs play a key role in protecting photosynthetic electron transport, under high-light conditions. Dual role of LMW HSP21 in tomato has been described as protecting PSII from oxidative damage (Neta-Sharir et al. 2005). In vivo experiments have demonstrated that small HSPs could associate with thylakoids and protect O_2 evolution and OEC proteins of PSII against heat stress. The protein possesses the unique property to associate with thylakoid membranes during heat stress and support the stability of thylakoid membranes.

Heat stress induces changes in the structure of Chl-protein complexes, and the integrity of thylakoid membranes is also compromised due to destacking of thylakoid membrane (Gounaris et al. 1983), and thus inactivation of enzyme activity constitutes the early effects of thermal stress. These structural transformations are also reflected by the ion leakage from heat-stressed leaves (Inaba and Grandall 1988; Wahid and Shabbir 2005) and changes in energy distribution between the two photosystems (state transition mechanism) due to high-temperature stress in vivo (Mohanty et al. 2002). The heat-induced changes in the structure of Chl-protein complexes and the inactivation of enzyme activity constitute the early effects of thermal stress. Among the various components of the photosynthetic machinery, the oxygen-evolving complex of PSII is most sensitive to heat. This might be due to the loss of two out of four Mn ions from oxygen-evolving complex (Enami et al. 1994, 1998). The loss of photosynthetic activity is partly caused by the inactivation of the acceptor side of PSII and reduction of the rate of electron transport in the chloroplasts. The detrimental effects of heat on chlorophyll and the photosynthetic apparatus are also associated with the production of injurious reactive oxygen species (ROSs) (Camejo et al. 2006; Guo et al. 2007). High temperature modifies the activities of carbon metabolism enzymes, starch accumulation and sucrose synthesis, by down-regulating specific genes in carbohydrate metabolism (Ruan et al. 2010).

5.3.2 Water Loss

High temperature is generally associated with reduced water availability to the plants, and this water shortage results in the inhibition of the photosynthesis process, and the reserves become critical because respiration continues to demand substrates; as a result, the cellular activity is hampered and nutrient supply to the generative organs are diminished. Generally, plants try to stabilize their tissue water content irrespective of temperature when moisture is available in ample quantity; however, rise in temperature proves fatal under limited supply of water (Machado and Paulsen 2001). Wahid and Close (2007) observed a rapid reduction in leaf tissue water contents in sugarcane exposed to high temperature despite the fact that the ample quantity of water was available in the soil. This water loss is more during the day as compared to night because vital physiological activities like transpiration and photosynthesis also occur. Tsukaguchi et al. (2003) reported that under heat stress, water loss was more common during the daytime because of increased transpiration than night time, causing stress in snap bean (*Phaseolus vulgaris* L.). Sita et al. (2017) reported that heat stress affects plant-water relations due to the rapid loss of water from the soil, which affects soil temperature and transpiration. Heat stress also reduces the growth of the roots which limits the supply of water and nutrients to the above-ground parts of the plant (Wahid et al. 2007; Huang et al. 2012).

5.3.3 Oxidative Stress

A well-known consequence of elevated temperatures in plants is oxidative stress caused by a heat-induced imbalance of photosynthesis and respiration (Fitter and Hay 1987), which mediates damage to cell structures, including lipids, membranes, proteins and DNA. High-temperature stress causes multi-step injuries to the plasma membrane and produces activated oxygen species, where membrane lipids and pigments peroxidation compromise membrane permeability and function, resulting in cellular damage. ROS cause damage to a wide range of cellular components such as the photosynthetic apparatus and various other components, hindering thus metabolic activities and affecting plant growth and yield by limiting metabolic flux activities (Sairam and Tyagi 2004; Xu et al. 2006). Subsequently, ROS production contributes to the transduction of the heat signal and expression of heat shock genes (Königshofer et al. 2008).

To counteract the injurious effects of over-produced ROS under heat stress, plants have evolved complex antioxidative detoxification system which includes antioxidant enzymes such as superoxidase (SOD), catalase (CAT), guaiacol peroxidase (GPOX), glutathione reductase (GR), ascorbate peroxidase (APX), peroxidase (POX) and non-enzymatic antioxidants (Harsha et al. 2016); secondary metabolites such as phenolics including flavonoids, anthocyanins and steroids also play a significant role in combating abiotic stress responses associated with heat tolerance (Wahid 2007). These antioxidant enzymes either directly scavenge the ROS or protect plants indirectly by managing non-enzymatic defence (Anjum et al. 2011).

Rivero et al. (2001) observed accumulation of soluble phenolics in tomato followed by increased phenylalanine ammonia-lyase activity and decreased peroxidase and polyphenol oxidase activity, in response to heat stress and possible acclimation. It has also been observed that osmolyte production under high temperature is associated with protein stability and, in turn, stabilizes the structure of the membrane bilayer (Sung et al. 2003; Mirzaei et al. 2012). In response to ROS, increased content of malondialdehyde has been reported which is a pure indicator of drought-induced oxidative damage (Moller et al. 2007). For example, in pea, the lipid and protein peroxidation was increased four times under stress as compared to normal conditions (Moran et al. 1994).

Thermal stress results in the generation of ROS and production of heat shock proteins. HSPs can function in protection as molecular chaperons to prevent but no reverse protein denaturation and aggregation, as membrane stabilizers and, possibly, as site-specific antioxidants. Under thermal stress the protein structure gets distorted; these HSPs fold the protein into shapes essential for their normal functioning. Different HSPs have been identified to be produced under heat stress. For example, HSP68, localized in mitochondrion, has been found to express under heat stress in cells of potato, maize, tomato, soybean and barley (Neumann et al.. 1993). According to Wahid et al. (2007), another HSP 101 localized in maize nucleus belongs to the Campylobacter invasion antigen B (CiaB) protein sub-family, whose members promote the renaturation of protein aggregates, and is essential for the induction of thermotolerance. Therefore, to counteract the negative effect of ROS, plants generate higher levels of the antioxidants (Sharma and Dubey 2005).

5.4 *Effect of Heat Stress on the Molecular Response in Plants*

High temperatures cause transcriptional repression of genes involved in cell growth, such as histones and DNA polymerases, and deregulation of DNA methylation and transposon activation (Sakata and Higashitani 2008; Pecinka et al. 2010; Smith and Workman 2012). At the molecular level, heat stress causes alterations in the expression of genes involved in direct protection HT (Aprile et al. 2009; Shinozaki and Yamaguchi-Shinozaki 2007). These include genes responsible for the expression of osmoprotectants, detoxifying enzymes, transporters and regulatory proteins (Semenov and Helford 2009; Moreno and Ollerana 2011). Protein thermostability is believed to be provided by chaperons, a specific class of proteins which assist in proper post-translational folding and in maintaining proteins in a functional state; heat stress results in misfolding of newly synthesized proteins and the denaturation of existing proteins. (Ellis 1990). Increased production of HSPs occurs when plants experience either abrupt or gradual increases in temperature resulting in heat stress (Nover et al. 2001).

Heat stress also leads to the transient activation of repetitive elements or silenced gene clusters close to the centromeric regions as well as the transient loss of epigenetic gene silencing (Lang-Mladek et al. 2010; Pecinka et al. 2010). Such gene silencing mechanisms are thought to be involved in transcriptional repression by hetero-chromatinization of repetitive DNA regions in plants (Khraiwesh et al. 2012). Heat stress is known to swiftly alter the pattern of gene expression, inducing the HSP complement and inhibiting expression of many genes expressed under normal temperature conditions (Yost and Lindquist 1988). Heat stress inhibits splicing, and it was hypothesized that HSP-encoding mRNAs can be processed properly due to the absence of introns in the corresponding genes. However, subsequently, it was shown that some HSP-encoding genes have introns and, under heat stress conditions, their mRNAs are correctly spliced nevertheless (Osteryoung et al. 1993; Visioli et al. 1997). It is also supposed that other proteins different from HSPs can contribute to heat tolerance including ubiquitin (Sun and Callis 1997), cytosolic Cu/Zn-SOD (Herouart and Inze 1994) and Mn-POD (Brown et al. 1993), whose expressions are stimulated upon heat stress. Wahid and Close (2007) identified three low-molecular-weight dehydrin proteins in sugarcane leaves with increased expression in response to heat stress; the functions of these proteins are apparently related to protein degradation pathway, minimizing the adverse effects of dehydration and oxidative stress at high temperature (Schoffl et al. 1999).

Modern breeding approaches involving a wide range of markers associated with the QTLs of high-temperature tolerance have been identified; however, their actual role in marker-assisted selection is very limited. A set of 4 QTLs in *Arabidopsis* and 11 QTLs in maize were identified which play an important role in thermotolerance (Frova and Sari-Gorla 1994). Mason et al. (2010) identified QTLs in wheat related to different traits involved in heat tolerance such as grain filling duration and leaf senescence.

6 Conclusions and Future Prospects

High temperature has become major abiotic stress across the world because it affects greatly plant growth, reproduction and productivity. Understanding the molecular mechanisms of plant response to heat stress is a prerequisite since plants show a wide range of responses to heat stresses. Increase in mean temperature of the earth culminates into global warming and release of greenhouse gases in the environment which affects the plant response depicted by a variety of alterations in the growth, physiology and morphology. Therefore plant response and adaptation to the increasing temperature need to be better understood for the important crop. Detailed research defining the mechanisms of thermotolerance and understanding the nature of heat shock signalling proteins and gene expression will be valuable for developing tolerant varieties. Temperature varies in different regions all around the world, and plants show adaptation to HT according to their developmental growth and their ability to adapt to variable climatic condition. Plants accumulate antioxidant, osmoprotectant, metabolites, HSPs and different metabolic pathways to hamper the heat stress.

Future research should be focused on developing HT-resistant varieties and high-temperature-tolerant transgenic crops. At the field level, heat stress effect can be decreased by managing cultural practices such as timing and methods of sowing, irrigation and selection of cultivars and species. Exogenous application of protectants such as osmoprotectants, phytohormones and mineral elements has shown beneficial effects on plant growing under HT, due to growth-promoting effect of these compounds, and also to develop such type of plants which synthesize these compounds may prove a promising technique in producing thermotolerance crop plant.

References

Adamska I, Kloppstech K (1991) Evidence for the localization of the nuclear-coded 22-kDa heat shock protein in a subfraction of thylakoid membranes. Eur J Biochem 198:375–381

Adams S, Cockshull K, Cave C (2001) Effect of temperature on the growth and development of tomato fruits. Ann Bot 88:869–877 10.1006/anbo.2001.1524

Akman Z (2009) Comparison of high temperature tolerance in maize, rice and sorghum seeds, by plant growth regulators. J Anim Vet Adv 8:358–361

Al-Whaibi MH (2011) Plant heat-shock proteins: a mini review. J King Saud Univ 23:139–150

Anjum SA, Wang LC, Farooq M, Hussain M, Xue LL, Zou CM (2011) Brassinolide application improves the drought tolerance in maize through modulation of enzymatic antioxidants and leaf gas exchange. J Agron Crop Sci 197:177–118

Aprile AM, Mastrangelo AM, De Leonardis G, Galiba E, Roncaglia F, Ferrari L, De Bellis L, Turchi G, Giuliano, L Cattivelli (2009) Transcriptional profiling in response to terminal drought stress reveals differential responses along the wheat genome, BMC Genomics 10 (279), 279

Ashraf M, Harris PJC (2013) Photosynthesis under stressful environments: an overview. Photosynthetica 51:163–190

Asthir B (2015) Protective mechanisms of heat tolerance in crop plants. J Plant Interact 10:202–210

Baniwal SK, Bharti K, Chan KY, Fauth M, Ganguli A, Kotak S, Mishra SK, Nover L, Port M, Scharf K, Tripp L, Weber C, Zielinski D, Von Koskull-Doring P (2004) Heat stress response in plants: a complex game with chaperones and more than 20 heat stress transcription factors. J Biosci 29:471–487

Barnabás B, Jäger K, Fehér A (2008) The effect of drought and heat stress on reproductive processes in cereals. Plant Cell Environ 31:11–38

Benjamin IJ, McMillan DR (1998) Stress (heat shock) proteins. Circ Res 83:117–132

Biamonti G, Caceres JF (2009) Cellular stress and RNA splicing. Trends Biochem Sci 34:146–153

Boevink P, Oparka KJ (2005) Virus-host interactions during movement processes. Plant Physiol 138:1815–1821

Bokszczanin KL, Fragkostefanakis S (2013) Perspectives on deciphering mechanisms underlying plant heat stress response and thermotolerance. Front Plant Sci 4:315–335

Bosl B, Grimminger V, Walter S (2006) The molecular chaperone HSP104--a molecular machine for protein disaggregation. J Struct Biol 156:139–148

Boston RS, Viitanen PV, Vierling E (1996) Molecular chaperones and protein folding in plants. Plant Mol Biol 32:191–222

Brown JA, Li D, Ic M (1993) Heat shock induction of manganese peroxidase gene transcription in *Phanerochaete chrysosporium*. Appl Environ Microbiol 59:4295–4299

Camejo D, Rodriguez P, Morales MA, Dell'amico JM, Torrecillas A, Alarcon JJ (2005) High temperature effects on photosynthetic activity of two tomato cultivars with different heat susceptibility. J Plant Physiol 162:281–289

Camejo D, Jiménez A, Alarcón JJ, Torres W, Gómez JM, Sevilla F (2006) Changes in photosynthetic parameters and antioxidant activities following heat-shock treatment in tomato plants. Funct Plant Biol 33:177–187

Cao YY, Duan H, Yang LN, Wang ZQ, Zhou SC, Yang JC (2008) Acta Agron Sin 34:2134–2142

Chan-Schaminet KY, Baniwal SK, Bublak D, Nover L, Scharf KD (2009) Specific interaction between tomato HsfA1 and HsfA2 creates hetero-oligomeric superactivator complexes for synergistic activation of heat stress gene expression. J Biol Chem 284:20848–20857

Chen Z, Zhou T, Wu X, Hong Y, Fan Z, Li H (2008) Influence of cytoplasmic heat shock protein 70 on viral infection of Nicotiana benthamiana. Mol Plant Pathol 9:809–817

Cleland EE, Chuine I, Menzel A, Mooney HA, Schwartz MD (2007) Shifting plant phenology in response to global change. Trends Ecol Evol 22:357–365

Efeoglu B (2009) Heat shock proteins and heat shock response in plants. Gazi Uni J of Sci 22:67–75

Ellis RJ (1990) Molecular chaperones: the plant connection. Science 250:954–959

Enami I, Kitamura M, Tomo T, Isokawa Y, Ohta H, Katoh S (1994) Is the primary cause of thermal inactivation of oxygen evolution in spinach PS II membranes release of the extrinsic 33 kDa protein or of Mn? Biochim Biophys Acta 1186:52–58

Enami I, Kamo M, Ohta H, Takahashi S, Miura T, Kusayanagi M, Tanabe S, Kamei A, Motoki A, Hirano M, Tomo T, Satoh K (1998) Intramolecular cross-linking of the extrinsic 33-kDa protein leads to loss of oxygen evolution but not its ability of binding to photosystem II and stabilization of the manganese cluster. J Biol Chem 273:4629–4634

Frova C, Sari-Gorla M (1994) Quantitative trait loci (QTLs) for pollen thermotolerance detected in maize. Mol Gen Genet MGG 245(4):424–430

Fitter AH, Hay RKM (1987) Environmental Physiology of Plants. Academic Press, London

Greer DH, Weedon MM (2012) Modeling photosynthetic responses to temperature of grapevine (*Vitis vinifera* cv. Semillon) leaves on vines grown in a hot climate. Plant Cell Environ 35(6):1050–1064

Gounaris K, Whitford D, Barber J (1983) The effect of thylakoid lipids on an oxygen-evolving Photosystem II preparation. FEBS Lett 163(2):230–234

Guo TR, Zhang GP, Zhang YH (2007) Physiological changes in barley plants under combined toxicity of aluminum, copper and cadmium. Colloids Surf B Biointerfaces 57:182–188

Gupta SC, Sharma A, Mishra M, Mishra RK, Chowdhuri DK (2010) Heat shock proteins in toxicology: how close and how far? Life Sci 86:377–384

Gurley WB (2000) HSP101: a key component for the acquisition of thermotolerance in plants. Plant Cell 12:457–460

Hafren A, Hofius D, Ronnholm G, Sonnewald U, Makinen K (2010) HSP70 and its cochaperone CPIP promote potyvirus infection in Nicotiana benthamiana by regulating viral coat protein functions. Plant Cell 22:523–535

Harndahl U, Hall RB, Osteryoung KW, Vierling E, Bornman JF, Sundby C (1999) The chloroplast small heat shock protein undergoes oxidation-dependent conformational changes and may protect plants from oxidative stress. Cell Stress Chaperones 4:129–138

Harsha A, Sharma YK, Joshi U, Rampuria S, Singh G, Kumar S, Sharma R (2016) Effect of short-term heat stress on total sugars, proline and some antioxidant enzymes in moth bean (*Vigna aconitifolia*). Ann Agric Sci 61:57–64

Hasanuzzaman M, Hossain MA, Fujita M (2010) Selenium in higher plants: Physiological role, antioxidant metabolism and abiotic stress tolerance. J Plant Sci 5:354–375

Hasanuzzaman M, Hossain MA, da Silva JAT, Fujita M (2012a) Plant responses and tolerance to abiotic oxidative stress: antioxidant defenses is a key factor. In: Bandi V, Shanker AK, Shanker C, Mandapaka M (eds) Crop stress and its management: perspectives and strategies. Springer, Berlin, pp 261–316

Hasanuzzaman M, Hossain MA, Fujita M (2012b) Exogenous selenium pretreatment protects rapeseed seedlings from cadmium-induced oxidative stress by upregulating the antioxidant defense and methylglyoxal detoxification systems. Biol Trace Elem Res 149:248–261

Hasanuzzaman M, Nahar K, Alam MM, Fujita M (2012c) Exogenous nitric oxide alleviates high temperature induced oxidative stress in wheat (*Triticum aestivum*) seedlings by modulating the antioxidant defense and glyoxalase system. Aust J Crop Sci 6:1314–1323

Hasanuzzaman M, Nahar K, Fujita M (2013a) Plant response to salt stress and role of exogenous protectants to mitigate salt-induced damages. In: Ahmad P, Azooz MM, Prasad MNV (eds) Ecophysiology and responses of plants under salt stress. Springer, New York, pp 25–87

Hasanuzzaman M, Gill SS, Fujita M (2013b) Physiological role of nitric oxide in plants grown under adverse environmental conditions. In: Tuteja N, Gill SS (eds) Plant acclimation to environmental stress. Springer, New York, pp 269–322

Hasanuzzaman M, Nahar K, Alam M, Roychowdhury R, Fujita M (2013c) Physiological, biochemical, and molecular mechanisms of heat stress tolerance in plants. Int J Mol Sci 14:9643–9684

Hasanuzzaman M, Nahar K, Fujita M (2013d) Extreme temperatures, oxidative stress and antioxidant defense in plants. In: Vahdati K, Leslie C (eds) Abiotic stress—plant responses and applications in agriculture. In Tech, Rijeka, pp 169–205

Herouart DVMM, Inze D (1994) Developmental and environmental regulation of the *Nicotiana plumbaginifolia* cytosolic Cu/Zn-superoxide dismutase promoter in transgenic tobacco. Plant Physiol 104:873–880

Hightower LE (1991) Heat shock, stress proteins, chaperones, and proteotoxicity. Cell 66:191–197

Hong SW, Vierling E (2001) HSP101 is necessary for heat tolerance but dispensable for development and germination in the absence of stress. Plant J 27:25–35

Hu W, Hu G, Han B (2009) Genome-wide survey and expression profiling of heat shock proteins and heat shock factors revealed overlapped and stress specific response under abiotic stresses in rice. Plant Sci 176:583–590

Huang B, Rachmilevitch S, Xu J (2012) Root carbon and protein metabolism associated with heat tolerance. J Exp Bot 63:3455–3465

Inaba M, Grandall P (1988) Electrolyte leakage as an indicator of high-temperature injury to harvested mature green tomatoes. J Am Soc Hortic Sci 113:96–99

Jackson SE (2013) HSP90: structure and function. Top Curr Chem 328:155–240

Jackson-Constan D, Akita M, Keegstra K (2001) Molecular chaperones involved in chloroplast protein import. Biochim Biophys Acta 1541:102–113

Jacob P, Heribert H, Bendahmane A (2017) The heat-shock protein/chaperone network and multiple stress resistance. Plant Biotechnol J 15:405–414

Jiang C, Xu J, Zhang H, Zhang X, Shi J, Li M, Ming F (2009) A cytosolic class I small heat shock protein, RcHSP17.8, of *Rosa chinensis* confers resistance to a variety of stresses to *Escherichia coli*, yeast and *Arabidopsis thaliana*. Plant Cell Environ 32:1046–1059

Kaushal N, Bhandari K, Siddique KHM, Nayyar H (2016) Food crops face rising temperatures: an overview of responses, adaptive mechanisms, and approaches to improve heat tolerance. Cogent Food Agric 2:1134380

Khraiwesh B, Zhu JK, Zhu J (2012) Role of miRNAs and siRNAs in biotic and abiotic stress responses of plants. Biochim Biophys Acta 1819:137–148

Kim HJ, Hwang NR, Lee KJ (2007) Heat shock responses for understanding diseases of protein denaturation. Mol Cells 23:123–131

Könighofer H, Tromballa HW, Löppert HG (2008) Early events in signaling high-temperature stress in tobacco BY2 cells involve alterations in membrane fluidity and enhanced hydrogen peroxide production. Plant Cell Environ 31:1771–1780

Koning AJ, Rose R, Comai L (1992) Developmental expression of tomato heat-shock cognate protein 80. Plant Physio 100:801–811

Kotak S, Larkindale J, Lee U, von Koskull-Doring P, Vierling E, Scharf KD (2007) Complexity of the heat stress response in plants. Curr Opin Plant Biol 10:310–316

Kumar B, Verma SK, Singh HP (2011) Effect of temperature on seed germination parameters in Kalmegh (Andrographis paniculata Wall. ex Nees.). Ind Crop Prod 34(1):1241–1244

Krasensky J, Jonak C (2012) Drought, salt, and temperature stress-induced metabolic rearrangements and regulatory networks. J Exp Bot 63:1593–1608

Kregel KC (2002) Heat shock proteins: modifying factors in physiological stress responses and acquired thermotolerance. J Appl Physiol 92:2177–2186
Krishna P, Gloor G (2001) The HSP90 family of proteins in Arabidopsis thaliana. Cell Stress Chape 6:238–246
Krishna P, Sacco M, Cherutti JF, Hill S (1995) Cold-induced accumulation of HSP90 transcripts in *Brassica napus*. Plant Physio 107:915–923
Lang-Mladek C, Popova O, Kiok K, Berlinger M, Rakic B, Aufsatz W et al (2010) Transgenerational inheritance and resetting of stress-induced loss of epigenetic gene silencing in Arabidopsis. Mol Plant 3:594–602
Lee YJ, Nagao RT, Key JL (1994) A soybean 101-kDa heat stress protein complements yeast HSP104 deletion mutant in acquiring thermotolerance. Plant Cell 6:1889–1897
Lee U, Rioflorido I, Hong SW, Larkindale J, Waters ER, Vierling E (2006) The Arabidopsis ClpB/ HSP100 family of proteins: chaperones for stress and chloroplast development. Plant J 49:115–127
Li W, Wei Z, Qiao Z, Wu Z, Cheng L, Wang Y (2013) Proteomics analysis of alfalfa response to heat stress. PLoS One 8:e82725
Liberek K, Lewandowska A, Zietkiewicz S (2008) Chaperones in control of protein disaggregation. EMBO J 27:328–335
Lin MY, Chai KH, Ko SS, Kuang LY, Lur HS, Charng YY (2014) A positive feedback loop between HEAT SHOCK PROTEIN101 and HEAT STRESS-ASSOCIATED 32-KD PROTEIN modulates long-term acquired thermotolerance illustrating diverse heat stress responses in rice varieties. Plant Physiol 164:2045–2053
Lindquist S, Craig E (1988) The heat-shock proteins. Annu Rev Genet 22:631–677
Ling J, Wells DR, Tanguay RL, Dickey LF, Thompson WF, Gallie DR (2000) Heat shock protein HSP101 binds to the Fed-1 internal light regulatory element and mediates its high translational activity. Plant Cell 12:1213–1227
Liu Y, Burch-Smith T, Schiff M, Feng S, Dinesh-Kumar SP (2004) Molecular chaperone HSP90 associates with resistance protein N and its signaling proteins SGT1 and Rar1 to modulate an innate immune response in plants. J Biol Chem 279:2101–2108
Lobell DB, Gourdji SM (2012) The influence of climate change on global crop productivity. Plant Physiol 160:1686–1697
Machado S, Paulsen GM (2001) Combined effects of drought and high temperature on water relations of wheat and sorghum. Plant Soil 233:179–187
Maestri E, Klueva N, Perrotta C, Gulli M, Nguyen HT, Marmiroli N (2002) Molecular genetics of heat tolerance and heat shock proteins in cereals. Plant Mol Biol 48:667–681
Mahajan S, Tuteja N (2005) Cold, salinity and drought stresses: an overview. Arch Bioche Biophy 444:139–158
Mason RE, Mondal S, Beecher FW, Pacheco A, Jampala B, Ibrahim AMH, Hays DB (2010) QTL associated with heat susceptibility index in wheat (Triticum aestivum L.) under short-term reproductive stage heat stress. Euphytica 174(3):423–436
Maimbo M, Ohnishi K, Hikichi Y, Yoshioka H, Kiba A (2007) Induction of a small heat shock protein and its functional roles in Nicotiana plants in the defense response against *Ralstonia solanacearum*. Plant Physiol 145:1588–1599
Marrs KA, Casey ES, Capitant SA, Bouchard RA, Dietrich PS, Mettler IJ, Sinibaldi RM (1993) Characterization of two maize HSP90 heat shock protein genes: expression during heat shock, embryogenesis and pollen development. Dev Genet 14:27–41
Matsumiya T, Imaizumi T, Yoshida H, Satoh K, Topham MK, Stafforini DM (2009) The levels of retinoic acid-inducible gene I are regulated by heat shock protein 90-alpha. J Immunol 182:2717–2725
Miernyk JA (1999) Protein folding in the plant cell. Plant Physiol 121:695–703
Mirzaei M, Pascovici D, Atwell BJ, Haynes PA (2012) Differential regulation of aquaporins, small GTPases and V-ATPases proteins in rice leaves subjected to drought stress and recovery. Proteomics 12:864–877

Mishra SK, Tripp J, Winkelhaus S, Tschiersch B, Theres K, Nover et al (2002) In the complex family of heat stress transcription factors, HsfA1 has a unique role as master regulator of thermotolerance in tomato. Genes Dev 16:1555–1567

Mittler R, Finka A, Goloubinoff P (2012) How do plants feel the heat?. Trends Biochem Sci 37(3):118–125

Moriwaki M, Yamakawa T, Washino T, Kodama T, Igarashi Y (1999) Delayed recovery of β-glucuronidase activity driven by an Arabidopsis heat shock promoter in heat-stressed transgenic Nicotiana plumbaginifolia. Plant Cell Rep 19(1):96–100

Mogk A, Schlieker C, Friedrich KL, Schonfeld HJ, Vierling E, Bukau B (2003) Refolding of substrates bound to small HSPs relies on a disaggregation reaction mediated most efficiently by ClpB/DnaK. J Biol Chem 278:31033–31042

Mohanty P, Vani B, Prakash S (2002) Elevated temperature treatment induced alteration in thylakoid membrane organization and energy distribution between the two photosystems in *Pisum sativum*. Z Naturforsch 57:836–842

Moller IM, Jensen PE, Hansson A (2007) Oxidative modifications to cellular components in plants. Ann Rev Plant Biol 58:459–481

Moran JF, Becana M, Iturbe-Ormaetxe I, Frechilla S, Klucas RV, Aparicio-Trejo P (1994) Drought induces oxidative stress in pea plants. Planta 194:346–352

Moreno AA, Orellana A (2011) The physiological role of the unfolded protein response in plants. Biol Res 44(1):75–80

Morimoto RI (1993) Cells in stress: the transcriptional activation of heat shock genes. Science 259:1409–1410

Nadeem M, Li J, Wang M, Shah L, Lu S et al (2018) Unraveling field crops sensitivity to heat stress: mechanisms, approaches, and future prospects. Agronomy 8:128

Neta-Sharir I, Isaacson T, Lurie S, Weiss D (2005) Dual role for tomato heat shock protein 21: protecting photosystem ii from oxidative stress and promoting color changes during fruit maturation. Plant Cell 17:1829–1838

Nover L, Baniwal SK (2006) Multiplicity of heat stress transcription factors controlling the complex heat stress response of plants. In: Proceedings of international symposium on environmental factors, cellular stress and evolution, p 15

Nover L, Bharti K, Doring P, Mishra SK, Ganguli A, Scharf K (2001) Arabidopsis and the heat stress transcription factor world: how many heat stress transcription factors do we need? Cell Stress Chape 6:177

Osteryoung KW, Sundberg H, Vierling E (1993) Poly(A) tail length of a heat shock protein RNA is increased by severe heat stress, but intron splicing is unaffected. Mol Gen Genet MGG 239(3):323–333

Pecinka A, Dinh HQ, Baubec T, Rosa M, Lettner N, Scheid OM (2010) Epigenetic regulation of repetitive elements is attenuated by prolonged heat stress in Arabidopsis. Plant Cell 22:3118–3129

Prasad PVV, Pisipati SR, Mutava RN, Tunistra MR (2008) Sensitivity of grain sorghum to high temperatures stress during reproductive development. Crop Sci 48:1911–1917

Pratt WB, Galignana MD, Harrell JM, Deranco DB (2004) Role of hsp90 and the hsp90-binding immunophilins in signalling protein movement Cell Signal 16(2004).857–872

Queitsch C, Hong SW, Vierling E, Lindquest S (2000) Heat shock protein 101 plays a crucial role in thermotolerance in *Arabidopsis*. Plant Cell 12.479–492

Riley GJP (1981) Effects of high temperature on protein synthesis during germination of maize (*Zea mays* L.). Planta 151.75–80

Ritossa F (1962) A new puffing pattern induced by temperature shock and DNP in drosophila. Experientia 18:571–573

Rivero RM, Ruiz JM, Garcia PC, Lopez-Lefebre LR, Sanchez E, Romero L (2001) Resistance to cold and heat stress: accumulation of phenolic compounds in tomato and watermelon plants. Plant Sci 160:315–321

Rollins JA, Habte E, Templer SE, Colby T, Schmidt J, Von Korff M (2013) Leaf proteome alterations in the context of physiological and morphological responses to drought and heat stress in barley (*Hordeum vulgare* L.). J Exp Bot 64:3201–3212
Rossi MR, Somji S, Garrett SH, Sens MA, Nath J, Sens DA (2002) Expression of HSP 27, HSP 60, hsc 70, and HSP 70 stress response genes in cultured human urothelial cells (UROtsa) exposed to lethal and sublethal concentrations of sodium arsenite. Environ Health Perspect 110:1225–1232
Rouch JM, Bingham SE, Sommerfeld MR (2004) Protein expression during heat stress in thermo-intolerance and thermotolerance diatoms. J Exp Mar Biol Ecol 306:231–243
Ruan YL, Jin Y, Yang YJ, Li GJ, Boyer JS (2010) Sugar input, metabolism, and signaling mediated by invertase: roles in development, yield potential, and response to drought and heat. Mol Plant 3:942–955
Sairam R, Tyagi A (2004) Physiology and molecular biology of salinity stress tolerance in plants. Curr Sci 86:407–421
Sakata T, Higashitani A (2008) Male sterility accompanied with abnormal anther development in plants–genes and environmental stresses with special reference to high temperature injury. Int J Plant Dev Biol 2:42–51
Salvucci ME, Crafts-Brandner SJ (2004) Relationship between the heat tolerance of photosynthesis and the thermal stability of Rubisco activase in plants from contrasting thermal environments. Plant Physiol 134:1460–1470
Sangster TA, Queitsch C (2005) The HSP90 chaperone complex, an emerging force in plant development and phenotypic plasticity. Curr Opin Plant Biol 8:86–92
Sangster TA, Bahrami A, Wilczek A, Watanabe E, Schellenberg K, McLellan C, Kelley A, Kong SW, Queitsch C, Lindquist S (2007) Phenotypic diversity and altered environmental plasticity in Arabidopsis thaliana with reduced HSP90 levels. PLoS One 2:e648
Scharf K, Höhfeld I, Nover L (1998) Heat stress response and heat stress transcription factors. J Biosci 23:313–329
Scharf KD, Siddique M, Vierling E (2001) The expanding family of Arabidopsis thaliana small heat stress proteins and a new family of proteins containing alpha-crystalline domains (Acd proteins). Cell Stress Chape 6:225–237
Schirmer EC, Glover JR, Singer MA, Lindquist S (1996) HSP100/Clp proteins: a common mechanism explains diverse functions. Trends Biochem Sci 21:289–295
Schöffl F, Prändl R, Reindl A (1998) Regulation of the heat-shock response. Plant physio 117:1135–1141
Schoffl F, Prandl R, Reindl A (1999) Molecular responses to heat stress. In: Shinozaki K, Yamaguchi-Shinozaki K (eds) Molecular responses to cold, drought, heat and salt stress in higher plants. R.G. Landes Co, Austin, pp 81–98
Schroda M, Vallon V, Wollman F, Beck CF (1999) A chloroplast-targeted heat shock protein 70 (HSP70) contributes to the photoprotection and repair of photosystem II during and after photoinhibition. Plant Cell 11:11165–11178
Semenov MA, Halford NG (2009) Identifying target traits and molecular mechanisms for wheat breeding under a changing climate. J Exp Bot 60:2791–2804
Seo JS, Lee YM, Park HG, Lee JS (2006) The intertidal copepod Tigriopus japonicus small heat shock protein 20 gene (HSP20) enhances thermotolerance of transformed Escherichia coli. Biochem Bioph Res Co 340:901–908
Sharma P, Dubey RS (2005) Drought induces oxidative stress and enhances the activities of antioxidant enzyme in growing rice seedling. Plant Growth Regul 46:209–221
Shinozaki K, Yamaguchi-Shinozaki K (2007) Gene networks involved in drought stress response and tolerance. J Exp Bot 58:221–227
Simons G, Groenendijk J, Wijbrandi J, Reijans M, Groenen J, Diergaarde P, Van der Lee T, Bleeker M, Onstenk J, de Both M, Haring M, Mes J, Cornelissen B, Zabeau M, Vos P (1998) Dissection

of the fusarium I2 gene cluster in tomato reveals six homologs and one active gene copy. Plant Cell 10:1055–1068

Smith KT, Workman JL (2012) Chromatin Proteins: Key Responders to Stress. PLoS Biol 10(7):e1001371

Singh V, Nguyen TC, van Oosterom EJ, Chapman SC, Jordan DR, Hammer GL (2015) Sorghum genotypes differ in high temperature responses for seed set. Field Crop Res 171:32–40

Singh RK, Jaishankar J, Muthamilarasan M, Shweta S, Dangi A, Prasad M (2016) Genome-wide analysis of heat shock proteins in C4 model, foxtail millet identifies potential candidates for crop improvement under abiotic stress. Sci Rep 6:32641

Sita K, Sehgal A, HanumanthaRao B, Nair RM, Vara Prasad PV, Kumar S, Gaur PM, Farroq M, Siddique KHM, Varshney RK et al (2017) Food legumes and rising temperatures: effects, adaptive functional mechanisms specific to reproductive growth stage and strategies to improve heat tolerance. Front Plant Sci 8:1–30

Soll J (2002) Protein import into chloroplasts. Curr Opin Plant Biol 5:529–535

Squires C, Squires CL (1992) The Clp proteins: proteolysis regulators or molecular chaperones? J Bacteriol 174:1081–1085

Su PH, Li HM (2008) Arabidopsis stromal 70-kD heat shock proteins are essential for plant development and important for thermotolerance of germinating seeds. Plant Physiol 146:1231–1241

Sun CW, Callis J (1997) Independent modulation of Arabidopsis thaliana polyubiquitin mRNAs in different organs of and in response to environmental changes. Plant J 11:1017–1027

Sun W, Bernard C, van de Cotte B, van Montagu M, Verbruggen N (2001) At-HSP17.6A, encoding a small heat-shock protein in Arabidopsis, can enhance osmotolerance upon overexpression. Plant J 27:407–415

Sun W, Motangu MV, Verbruggen N (2002) Small heat shock proteins and stress tolerance in plants. Biochim Biophys Acta 1577:1–9

Sung DY, Vierling E, Guy CL (2001) Comprehensive expression profile analysis of the Arabidopsis HSP70 gene family. Plant Physiol 126:789–800

Sung DY, Kaplan F, Lee KJ, Guy CL (2003) Acquired tolerance to temperature extremes. Trends Plant Sci 8:179–187

Swindell WR, Huebner M, Weber AP (2007) Transcriptional profiling of Arabidopsis heat shock proteins and transcription factors reveals extensive overlap between heat and non-heat stress response pathways. BMC Genomics 8:125

Tripp J, Mishra SK, Scharf K-D (2009) Functional dissection of the cytosolic chaperone network in tomato mesophyll protoplasts. Plant Cell Environ 32:123–133

Tsukaguchi T, Kawamitsu Y, Takeda H, Suzuki K, Egawa Y (2003) Water status of flower buds and leaves as affected by high temperature in heat-tolerant and heat-sensitive cultivars of snap bean (*Phaseolus vulgaris* L.). Plant Prod Sci 6:24–27

Usman MG, Rafii MY, Ismail MR, Malek MA, Latif MA, Oladosu Y (2014) Heat shock proteins: functions and response against heat stress in plants. Int J Sci Technol Res 3:204–218

Van Montfort RL, Basha E, Friedrich KL, Slingsby C, Vierling E (2001) Crystal structure and assembly of a eukaryotic small heat shock protein. Nat Struct Biol 8:1025–1030

Verchot J (2012) Cellular chaperones and folding enzymes are vital contributors to membrane bound replication and movement complexes during plant RNA virus infection. Front Plant Sci 3:275

Visioli G, Maestri E, Marmiroli N (1997) Differential display-mediated isolation of a genomic sequence for a putative mitochondrial LMW HSP specifically expressed in condition of induced thermotolerance in *Arabidopsis thaliana* (L.) Heynh. Plant Mol Biol 34:517–527

Vierling E (1991) The roles of heat shock proteins in plants. Annu Rev Plant Physiol & Plant Mol Biol 42:579–620

Vollenweider P, Gunthardt-Goerg MS (2005) Diagnosis of abiotic and biotic stress factors using the visible symptoms in foliage. Environ Pollut 137:455–465

Wahid A, Shabbir A (2005) Induction of heat stress tolerance in barley seedlings by pre-sowing seed treatment with glycinebetaine. Plant Growth Regul 46:133–141
Wahid A (2007) Physiological implications of metabolite biosynthesis for net assimilation and heat-stress tolerance of sugarcane (Saccharum officinarum) sprouts. J Plant Res 120(2):219–228
Wahid A, Close J (2007) Expression of dehydrins under heat stress and their relationship with water relations of sugarcane leaves. Biol Plant 51:104–109
Wahid A, Gelani S, Ashraf M, Foolad M (2007) Heat tolerance in plants: An overview. Environ Exp Bot 61(3):199–223
Wang WX, Vinocur B, Shoseyov O, Altman A (2001) Biotechnology of plant osmotic stress tolerance: physiological and molecular considerations. Acta Hort 560:285–292
Wang W, Vinocur B, Shoseyov O, Altman A (2004) Role of plant heat-shock proteins and molecular chaperones in the abiotic stress response. Trends Plant Sci 9:244–252
Wang M, Zou Z, Li Q, Sun K, Chen X, Li X (2017) The CsHSP17.2 molecular chaperone is essential for thermotolerance in Camellia Sinensis. Sci Rep 7:1237
Waraich EA, Ahmad R, Halim A, Aziz T (2012) Alleviation of temperature stress by nutrient management in crop plants: a review. J Soil Sci Plant Nutr 12:221–244
Wells DR, Tanguay RL, Le H, Gallie DR (1998) HSP101 functions as a specific translational regulatory protein whose activity is regulated by nutrient status. Genes Dev 12:3236–3251
Whitley D, Goldberg SP, Jordan WD (1999) Heat shock proteins: a review of the molecular chaperones. J Vasc Sur 29:748–751
Xu S, Li J, Zhang X, Wei H, Cui L (2006) Effects of heat acclimation pretreatment on changes of membrane lipid peroxidation, antioxidant metabolites and ultrastructure of chloroplasts in two cool season turfgrass species under heat stress. Environ Exp Bot 56:274–285
Xu Y, Zhan C, Huang B (2011) Heat shock proteins in association with heat tolerance in grasses. Int J Prot 529648
Xu J, Xue C, Xue D, Zhao J, Gai J, Guo N, Xing H (2013) Overexpression of GmHSP90s, a heat shock protein 90 (HSP90) gene family cloning from soybean, decrease damage of abiotic stresses in Arabidopsis thaliana. PLoS One 8:e69810
Yan K, Chen P, Shao H, Zhang L, Xu G (2011) Effects of short-term high temperature on photosynthesis and photosystem II performance in Sorghum. J Agro Crop Sci 197:400–408
Yost H, Lindquist S (1988) Translation of unspliced transcripts after heat shock. Science 242(4885):1544–1548
Young LW, Wilen RW, Bonham-Smith PC (2004) High temperature stress of Brassica napus during flowering reduces micro- and megagametophyte fertility, induces fruit abortion, and disrupts seed production. J Exp Bot 55:485–495
Zhang JH, Wang LJ, Pan QH, Wang YZ, Zhan JC, Huang WD (2008) Accumulation and subcellular localization of heat shock proteins in young grape leaves during cross-adaptation to temperature stresses. Sci Horticul 117:231–240
Zhang Y, Pan J, Huang X, Guo D, Lou H, Hou Z, Su M, Liang R, Xie C, Mingshan You M, Li B (2017) Differential effects of a post-anthesis heat stress on wheat (*Triticum aestivum* L.) grain proteome determined by iTRAQ. 7:Sci Rep, 3468

Chapter 10
Assessment of Irradiation Stress in Crop Plants with Modern Technical Advances

Indraneel Saha, Debabrata Dolui, Arijit Ghosh, Bipul Sarkar, Arnab Kumar De, and Malay Kumar Adak

Abstract It is their necessity for sustaining the crop productivity especially under the adverse condition of environmental impacts to meet the ever-increasing human population. The unfavourable environmental conditions especially with climate change force the selection of crop genotypes with a sustainable practice for better and higher productivities. Light is the crucial environmental inputs that set a bottleneck for the growth of the plants through low and/or excess irradiation. Photosynthesis is a multifaceted physiological phenomenon that solely depends on the quality and quantity of irradiance under any ecological niche that concerns the carbon dioxide (CO_2) fixation and regulation of photorespiration. With different hierarchies of light requirement in photosynthesis, the induced photooxidation is the key to regulate the constitution of the photosynthetic organelle and its proper functions. Photochemistry with its successful utilization of irradiance is often limited by the loss of energy with higher wavelength as fluorescence. It is the realization of the plant scientists to consider the fluorescence as a valve to reduce the photooxidation and its consequences to the loss of carbon through photorespiration. Identification of specific genes imparts the structural and functional integrities of both the photosystems, their modalities in regulation and variations in C_3 and C_4 species at the insights for photosynthetic improvements at the molecular level. Moreover, the promising direction for CO_2 photoassimilation leading to the mechanism of carbon concentration also opens up the possibilities of ribulose-1,5-bisphosphate carboxylase/oxygenase (RuBisCO) in proper functioning, particularly under high irradiation stress. Undoubtedly, it is next to impossible to transfer the C_4 trait into C_3 into precise manner; however, improvisation of CO_2 concentration on RuBisCO is fixed. Therefore, the modern state of the art in increasing photosynthetic efficiency is based on proper utilization of irradiation by more into photochemistry, reducing the photooxidation, prone to RuBisCO more into carboxylation and partitioning of photoassimilates into cellular constituents which are the spaces for crop improvement. Thus, the review is intended to improve the techniques and approaches for photo

I. Saha · D. Dolui · A. Ghosh · B. Sarkar · A. K. De · M. K. Adak (✉)
Plant Physiology and Plant Molecular Biology Research Unit, Department of Botany, University of Kalyani, Kalyani, India

R. Roychowdhury et al. (eds.), *Sustainable Agriculture in the Era of Climate Change*, https://doi.org/10.1007/978-3-030-45669-6_10

synthetic activities with possible and feasible directions by which sustenance crop growth and productivity may be assured under irradiation stress.

Keywords Photosynthesis · Photoprotection · Irradiation stress · Crop improvement

1 Introduction

The basic life process for plant growth and development is the involvement of light energy harness and its consecutive allocation of dry matter. Since the second of the twentieth century's plant productivity vis-à-vis, the crop yield has been tremendous in the quest for better genotypes and agricultural practices. Notwithstanding the limitation within genotypic potential, researchers are still in question in any moderation on the manipulation of regulatory activities concerning the crop yield. Sustainable crop research with improved water and nitrogen use efficiency, herbicides tolerance and sequestering of carbon would be of those choices that contribute towards better productivity (Singh et al. 2016). Moreover, the advancement of understanding the genetic or genotypic constituents, more often necessary in the regulation and thereby its manipulation of genes, has also been another approach (Sticklen 2006). The ongoing changes of climate and their impacts on plant sustainability pose another question, and that has been shouted with development of climate resilience crops (Henry 2014). Thereby, the whole operative measure is likely to be based on solar energy utilizing and its proper turnover to dry weather acquisition.

Biomass production is the principal as well as limiting factor releasing adequate growth and development of any plant species where photosynthetic carbon assimilation is the driver for the whole process. Energy-rich compounds and the building block of cellular constituents are the two major attributes that the photosynthesis becomes essential for green plants. A significant plethora of citation has revealed the vast area of genetic variation in photosynthesis throughout the diversity of crop species (Goron and Raizada 2015). Both single leaves and the canopy photosynthesis would be granted for selection or screening indices of genotype with better photosynthesis following adequate productivity. The other direction of photosynthetic research aims to moderate the photorespiratory depletion of CO_2 which is about 1/3rd of the total fixed carbon in photosynthesis (Hagemann et al. 2013). Photorespiration undoubtedly a wasteful process, however, with few significant contributions is still in a process of quest where plant biologists minimize the inevitable loss of CO_2 (Sharma et al. 2017). On behalf of gene regulation, the molecular insight pertaining to the regulatory elements and their binding proteins is another cascade to research on photorespiratory depletion of CO_2. It is the miRNA (microRNA) and their identification as well as transgenesis that command any of regulatory paths with photorespiratory control in crop species. Therefore, plant scientists have been still in untiring effort to control the photosynthesis towards a contribution to sustainable agricultural development. For the crop photosynthesis, researchers are also paying their serious attention to different edaphic factors, environmental fluctuations and their interwork complications to effect on photosynthesis. As a whole, from the consideration from basic and applied research, photosynthesis would be still the most respectful domain to work out.

Within this stipulated chapter, the various approaches in favour of productivity improvement through photosynthesis would be described. This would be accompanied by the knowledge gaps with technical details, limitations of genetic material and environmental impositions on photosynthetic activities and possible limitations.

2 Enhancement of Photosynthetic Efficiency Through Modulation of Energy Harnessing with Light Reaction

Initially, the light absorption with photosynthetically active radiation (PAR) and its imposition through electron transport chain (ETC) for enhancement of photosynthetic rate are the major concerns of photosynthetic research. With the visible spectrum, a major range of the solar spectrum (900–1200 $\mu Em^{-2} s^{-1}$) is required for green plant photosynthesis according to the spectral requirement of photosynthetic pigment. However, this may vary with the course of evolution of photosynthetic plants which exist in varied ecological niches to maximize the light absorption but not to extend the absorption spectrum (Slattery et al. 2017). Green plant photosynthesis evolved oxygen (O_2) is characterized by the employment of two photosystems within a similar range of photon absorption efficiency. Higher plants have evolved chlorophyll-based (chlorophyll-a) absorption of visible light and others the far-red absorption. This is noncontiguous to bacterial photosynthesis where a single major photosynthetic pigment is represented with a wider range of light absorption. Therefore, improving any of the photosystems in higher plants must have the scope for enhancement of photosynthetic abilities. The photosynthetic organelle or apparatus are also tuned with the absorption of maximum light, still with PAR in green plants. Therefore, high light irradiation may proportionate to increase the absorption of light. With the categories of plant species for photosynthesis, C_3 ones are maximally saturated below 1/4th of the total light irradiation on leaves (Schuler et al. 2016). Excess light beyond the saturation is of no use, rather destructive in nature, and undergoes dissipation. In fields, a fraction of irradiation apparent on canopy by upper foliages can absorb more but not in use. In the bottom canopy, lower leaves remain non-saturated. So, a saturation of light intensities comes only from roofing leaves of the canopy (Guenni et al. 2018). Thus, it creates a thirst in development of plant canopies with minimum light-harvesting complexes in those parts of the canopies, and thereby a fair chance exists incompatible light transduction throughout the canopy (Anten and Bastiaans 2016). On the physiological level, the leaf area, leaf area index, leaf area duration, phyllotaxy and light transmission through leaves are the focused attributes for engineering of the plants in improvement of canopy photosynthesis with reference to cereal crops particularly; with rice there had been evident major factor for green evolution with sort stature and the plant height rather than erect and slender and that actually meant for improved canopy photosynthesis (Saha et al. 2018a, b). This type of canopy agriculture has realized adequate light transmission through foliages resulting in the high saturation of almost leaves vis-à-vis adequate CO_2 fixation rate. In more studies, rice genotypes with dwarf mutants could release a significant amount of grain yield (Li et al. 2017). The mutant was called *Os*dwarf4–1 having few quantitative trait loci or QTLs that phenotypically identified distinct leaf morphology and their distribution. The NARROW LEAF 1 (NAL1) QTL has been in analysis in indigenous tall *India* which is not much even inadequate yield potential. The gene product of such QTL has many other pleiotropic effects on plant vegetative characters and contributing yield attributes like spikelet numbers, effective panicles, number of panicles and also translocation efficiencies (Taguchi-Shiobara et al. 2015). The concept of light energy utilization through the canopy in optimized mode has created the smart foliage/canopy throughout the rice genotypes along with its subspecies: *indica*, *japonica* and *javanica* (Castillo et al. 2016). Still, this hypothesis demands the focus of letting the smart canopy into agronomic practices even with transgenic module approaches, however, quite difficult but possible. It was Cardona who has contributed the best in the adaptation of the light phase of photosynthesis to grace into the smart canopy. At the metabolic level, the light use efficiency must meet with the downstream source-sink paraphernalia. This simply interprets the diversion of reduced carbon in the light reaction must be allocated and partitioned into other metabolites including cellular constituents of shrink tissues compatible to light energy utilization. Such an imbalance or disproportionate light energy through photosynthetic apparatus often links the loss of energy by heat or non-photochemical quenching. The latter may protect the plants from oxidative stress for sustenance even with a compromise with decreased light use efficiency. It is the sink sizes or capacities and activities that can minimize the situations along with the intermediates of reduced carbon residues in the Calvin cycle (Wang et al. 2018). Still, the increased rate of photoassimilates and their translocations also regulate the photosynthetic carbon reduction by the accumulation of excess sugar in leaves.

The light-independent phase of photosynthesis could be modulated for CO_2 fixation ability. Ambient and intercellular CO_2 could be the most vital ones to turn over the efficiency of photosynthesis following plant productivity. It is the inherent property of plant genotypes that demarks C_3 and C_4 types from the CO_2 concentration mechanism (Tripathi et al. 2015). Basically, the tissue differentiation coordinating the CO_2 concentration mechanism without any loss or at least to minimize the loss of CO_2 due to photorespiration is expected. From the kinetic properties and concentration mechanisms, C_3 and C_4 are different light/CO_2 compensation mechanisms, requirements, yields, etc. to characterize C_3 and C_4 types. According to all these categories are typically responsive to act under high/low light and CO_2 concentration. The key vis-à-vis rate-limiting enzymes of these pathways, RuBisCO is the driver to differentiate plant types either to divert C_3 or C_4. It is the nondiscriminating ability for CO_2 and O_2 by RuBisCO to set apart the photorespiratory process either present or absent in C_3 and C_4, respectively (Busch and Sage 2017). Therefore, any mechanism underlined to increase the discriminatory perception for RuBisCO and thereby its concomitant effect on photosynthesis must be a choice to alter the carbon fixation (Hanson 2016). Still, the RuBisCO's origin in anoxic/hypoxic environment and its evolution for nondiscrimination for CO_2 and O_2 leading to photorespiration has well been accepted. The evolution of RuBisCO from earlier geological time scale with anoxygenic/anaerobic condition is still perpetuated in some plant species, however, under species ecological condition. Thus, ancient RuBisCO must not be a target for suppression of oxidation reaction (Munne-Bosch and Pinto-Marijuan 2016). In the present era, an abundance of O_2 (209,445 ppm) is still in a controlled condition. C_3 plants have to sacrifice almost 1/4th RuBP (ribulose-1,5-bisphosphate) moiety to be oxygenated. The product of phosphoglycolate, a C_2 residue, becomes inhibitory to RuBP carboxylase, and thereby it undergoes hydrolysis to successive compounds through a photorespiratory cycle in mitochondria and peroxisomes. In bonus, a significant amount of reducing $NADP+H^+$ and energy potential are depleted in photorespiration along with the release of CO_2 and NH_3 (ammonia). The alignment situation is that through ongoing elevated CO_2 and temperature of the environment, C_3 plants become prone to photorespiration that may be accountably more than 1/4th of the CO_2 fixation (South et al. 2018). This also creates another hazard that may contribute to the increase of 29 Gt (Gigatonnes) of atmospheric carbon on a yearly basis (Hagemann and Bauwe 2016). The RuBisCO is also associated to pose another bottleneck to its delay timid reaction rate. The enzymatic turnover of RuBisCO for a response to 1–10 per second is in contrast to other photosynthetic enzymatic proteins with average values of 50–100 per second (Nowicka et al. 2018). Therefore, the RuBisCO activity in green plants needs to be sustained, and thus its synthesis as compared to overall leaf proteins may exceed greater than 50% (Vitlin and Feiz 2018). This is another cause that C_3 plants are high nitrogen responsive to allocate a significant amount in favour of RuBisCO synthesis. So, the various mechanisms of carbon fixation are likely to sustain the CO_2 assimilation as well as to regulate or command on photorespiratory CO_2 loss. The advent of recombinant DNA (deoxyribonucleic acid) technology a fair attempt has been taken to increase the CO_2 fixation either to enrich the RuBisCO active site or RuBisCO residing tissue with elevated CO_2 concentration, and thereby a situation is lowering the photorespiratory oxygenation reaction (Kebeish et al. 2015). Still, this is possible through single-cell/tissue photosynthesis and equally impossible to introduce the so-called Kranz anatomy. Another constraint may exist to select the tissue-specific promoter to let functioning the concern transgene(s). The selection of typical vector, mostly the binary ones, has been more choice to introduce such a specific sequence for increasing CO_2 concentration mechanism protein. Still, another setback arises for the discrimination of different isogenies for the same reaction. This is typically exampled by a selection of phosphoenolpyruvate carboxylase, NADP-ME (nicotinamide adenine dinucleotide phosphate-malic enzyme) and NADP-malate dehydrogenase to reintroduce those specific enzymes into concerned tissue to function by their overexpression. In addition to moderation of light saturation for major proteins like RuBisCO, it is also important to consider the photoinhibition and its protection. It is true that absorbed solar energy may exceed if the capacity of Calvin cycle reactions is limited under excess light and becomes detrimental. With the increase of photosynthetic photon flux transmission (PPFT), the gross photosynthesis may saturate both the quantum efficiency for CO_2 uptake and release of O_2 which is proportional within the linear phases of saturation kinetics. Chlorophyll fluorescence is the important and key player to minimize the photoinhibition by excess irradiation over the capacity of downstream reaction in CO_2 reduction (Guidi et al. 2019). The quantum yield of oxygenic photosynthesis for photosystem II or PSII is the precise ratio of variable/maximum chlorophyll fluorescence as representative by F_v/F_m. Photosynthetic rates are limited by a number of edaphic factors, and that improvised with sink strength directly under irradiance in excess over the demand of CO_2 assimilation. This essentially needs to cover the photosynthetic system from photooxidation vis-à-vis photoinhibition. So, photoprotection and its genetical engineering also necessitate the scope for increasing photosynthetic rate. Knowing fully the well different vegetative characters of canopy-like leaf area, leaf area index, leaf area duration, phyllotaxy, leaf angle, chloroplast tracking and even the antenna pigment of light-harvesting complex (LHC), the plant architecture should be such that can avoid in few ways the excess energy by thermal dispassion, conduction and convection (Gilmore, 1999). Few other reactions within chloroplast behave as a photochemical sink where excess photon would be utilized just as photoassimilates sink. Another facet for photooxidation is the development of reactive oxygen species (ROS) and thereby its induction in oxidative stress defence or anti-defence.

3 Physiological Principles of Photoprotection and Photoinhibition

The physiological basis for photoprotection vis-à-vis photosynthesis from excess irradiation undoubtedly the photoprotection is a real requirement to sustain the photosynthetic rate with its normal rate. Chlorophyll molecule after being encircled into a singlet state can transmit the light energy to PSII and PSI (Hall et al. 2016). A charge separation along with the electron transport reduces $NADP^+$ to NADPH + H^+, and ATP (adenosine triphosphate) synthesis with the gradient of pH over the membrane is accomplished. This is the linear electron flow and more flexible under normal light condition. Still, the excess light could be wasted away by quenching either through heat emission or fluorescence or as long wavelength fluorescence light as by default photoprotective process. The possibilities exist in the manipulation of other photoprotective pathways that may support stress tolerance following sustains of photosynthetic productivity. Still, a lot of things are yet to be understood, the actual gene regulation or photoprotection either by de novo synthetic pathway or an induced phenomenon. The PSII-induced overexcitation is initially managed by inhibition by F_v/F_m ratio. The maximum (F_m) and variable (F_v) fluorescence thus may be correlated with a few functions with PSII like oxygen evolution (Ouzounidou et al. 1997). It is due to light gradient the photoinhibition LHC is restricted on above tier palisade parenchyma tissues. This is measured by F_v/F_m ratio of photoinhibition leaves under dark and irradiated tissue. On irradiated leaf face, F_v/F_m ratio comes down in comparison to negligible to dark adopted leaf face (Neill and Gould 2003). The photochemical quenching (qP) and light energy quenching due to δpH is measured by qE or energy-dependent quenching. Altogether qP and qE are the actual measures for photochemical efficiency. The photoinhibition is understood with the significantly decreased value of those parameters and thereby also reflects the decrease of quantum efficiency of oxygen evolution. This is also variable for sun and shade plants where F_v/F_m value is approximately below 0.8 for some plants. In contrast, plants under low light adaptation had the same value but 4% less than the earlier one. Technically any value below 0.70 is considered to be effective for photoinhibition (Malnoë 2018).

Under varying conditions of environmental fluctuation, growth and photosynthetic rates are limited mostly by the water stress, nutrient deficiency and high temperature which are the factors to enforce the low sink straight of the light energy, and thereby light could be higher/excess than that required for CO_2 reduction (Saha et al. 2018a, b). These may begin the photosynthetic responses to avoid high stress of photoinhibition. Therefore, photoprotection essentially involves the mechanism in the regulation of light absorption as well as light desiccation of light energy. From the physiological point of view, plants could adopt the changes in leaf area, leaf angle, etc., and for chloroplast, the adjustment in LHC and antenna size could be involved. Light energy in excess and its desiccations also involved some reactions to act as a photochemical sink. Different antioxidation cascades are evoked to remove the ROS. Photoinhibition is characterized with a light-induced decline in maximum photosynthesis (A_{max}) as well as a decrease in quantum yield in terms of both the CO_2 (ΦCO_2) and O_2 (ΦO_2) (Maxwell et al. 1998). Photoinhibition would be moderate or severe. The first one is the transient in occurrence but to face in $\Phi CO_2/\Phi O_2$ and F_m/F_v. Under severe photoinhibition, the reduction of those is not transient but sustained and thereby A_{max} is declined. This is more realized from the light response curve where the slope represents the quantum yield which is maximum for CO_2 assimilation/O_2 evolution. Under severe photoinhibition, there is a significant downregulation of the A_{max}, and that essentially requires the photoprotective responses (Idris et al. 2018). In some cases of abiotic stresses, the sustained photoinhibition is attributed to photoprotective responses which are more characterized in some evergreen leaves.

4 Photoacclimation Through the Properties of Photochemical Quenching: A Way to Genetical Engineering

The ratio of saturation of the photosynthetic component to light intensity may otherwise conceptualize with photo-acclimatization. The enzymes for photosynthesis electron carriers and light-harvesting complexes (LHCs) have their maximum capacity to be saturated under available light within PAR (Walker et al. 2018).

This is an economy of the photosynthetic process where high light intensity along with electron transport following CO_2 reduction coupled with degradation of few proteins in LHCs. Few crop species are found to be photo-acclimatized and also identified a few factors contributing to improved photosynthesis (Orr et al. 2016). The antenna pigments and their sizes are examples of those to adjust adequate light absorption but not to be over-oxidized. Plants have to compromise to acclimate light intensities by some alteration in protein pigments complex under such a condition. This is more evident from their various wild plant types against the mutant (Vialet-Chabrand et al. 2017). Under excess light chlorophyll may not be able for photochemistry and thereby the chlorophyll attains triplet state in higher amount. Under this condition, triplet chlorophyll can directly induce the generation of singlet oxygen ($^1O_2$) by direct transfer of electron. Thereby sequential configurations of ROS are developed. Photochemistry can also induce the return of chlorophyll to the ground state with release of higher wavelength light, and this is called chlorophyll fluorescence. Thermal dissipation is another way by which de-oxidation of singlet chlorophyll comes down to the ground state. Both chlorophyll fluorescence and heat dissipation are a measure of photoprotection and called non-photochemical quenching (NPQ) (Ballottari et al. 2016). The NPQ is attributed by a number of factors like qE which is a measure of δpH formation under the thylakoid membrane. Another component qZ (zeaxanthin-dependent quenching) is responsible for carrying out the xanthophyll cycle (Sacharz et al. 2017). Chlorophylls in LHCII, P_{700}, are always able to form triplet state. This triplet state is efficiency related to the induced formation of xanthophyll residues: zeaxanthin (Z) and antheraxanthin (A). The level of concentration of A and Z against F/M gives a non-linear relationship. Thus, a decrease in F_v/F_m value can be read as photoprotective downregulation of PSII. This is actually done through the xanthophyll cycle, where increased energy dissipation in the antenna pigments is the major attribution. The higher plants have some few options to regulate photoinhibition. Initially, qE is required for carrying out the xanthophyll cycle (Johnson et al. 2008). It is a reversible inter-conversion of two components of xanthophyll: zeaxanthin and violaxanthin. These two residues are required to energize the thylakoid membrane under the light. Under light saturation within the thylakoid membrane, protein concentration and the H^+ concentration increase. This also induces the violaxanthin de-epoxidase to convert violaxanthin to zeaxanthin and thereby also increases the de-epoxidation condition of the xanthophyll cycle (Müller-Moulé et al. 2002). Zeaxanthin epoxidase converts zeaxanthin to violaxanthin. How and why the zeaxanthin could affect the qE and whether it is directly quenching $^1O_2$ in exited chlorophyll or even whether it is on the allosteric mode of regulation is still uncertain. Another mode of regulation of photoinhibition through NPQ has been done by involvement of PSII binding proteins (PsbS) (Li et al. 2002). Mutants lacking PsbS are inactive to operate qE and thereby photoinhibition is altered. PsbS is an integral LHC protein and it acts as a sensor of pH in chloroplast lumen. This is also required for immediate induction and its relaxation under extreme light. NPQ can also modulate the qE; however, it is a much slower rate exclusively when PsbS is not present attached to LHC. This was isolated from *Chlamydomonas* sp. and absent for higher plants (Gerotto and Morosinotto 2013).

Any case of operation of qE could improve the photoprotection, and thus crop protection in the stressed condition is revelled. Therefore, PsbS with its overexpression could be a possible strategy to increase the photoprotection (Horton 2012).

So, it comes to the notion that contributing soluble proteins for carbon reduction, enzymes in electron transport paths and antenna pigments with light-harvesting complexes are essential contributions in efficiency in photochemistry following its use in reduced carbon residues. Plants shifted to excess light may increase the CO_2 assimilation along with an overburden of light harvest complex proteins oxidation (Ruhil et al. 2015). This ensures an increase in photosynthetic carbon fixation over the A_{max} as well as prone to photooxidative stress. Crop plants are there which have to compromise the photoacclimation along with an increased yield. Therefore, contributing attributes should be addressed more regarding the balance of these two. The ratio of antenna pigments to light-harvesting complex sizes would be those to adjust the compromisation of plants for productivity under the light in excess. Therefore, photoacclimation may be granted the plants' own engineering capacity to manipulate any machinery to downregulate photooxidation but to sustain the growth and development. On dynamic photoinhibition plants often sacrifice in growth against light intensities by possible changes in photosynthetic apparatus. The naturally available wild forms of any genotypes when considered with photoacclimation essentially refer to the dynamic photoinhibition which is reversible in nature and that facilitates the plants in determining growth and yield through sustained photosynthetic rate.

5 Engineering of PSII Proteins for Its Repair

PSII appears to be the most sensitive zone for photooxidation. Its molecular structures are a concern to receive the over-illumination and thereby oxidative damages. The integral membrane proteins of thylakoid like PsbS had the initial defence to repair the overoxidation by several antioxidants. These membrane-bound antioxidants continue the xanthophyll cycle with alternative changes in zeaxanthin and violaxanthin. Downstream to these reactions, the major LHC proteins and PSII-adhered proteins like CD24, CD26, CD29, etc. are other PSII inactivation pathways against higher illumination (Schuster et al. 1986). During rapid transport of electrons from PSII to PSI, it is the quinone gate that causes the accumulation of protons into lumen site. Under continuous electron flow through two quinone gates (PQA and PQB), the stomatal site becomes more alkaline as compared to lumen site. Therefore, these variations of δpH would be maintained for normal cases where PSII is optimized with illumination within the PAR.

The heterodimeric proteins of D1 and D2 in the PSII photochemistry centre happen to be the most targeted residues under intense illumination (Fig. 10.1) (Chen et al. 2016).

This is an irreversible inhibition or inactivation which is inevitable for photooxidation. Therefore, a repair cycle is involved that initiates a partial excision of the

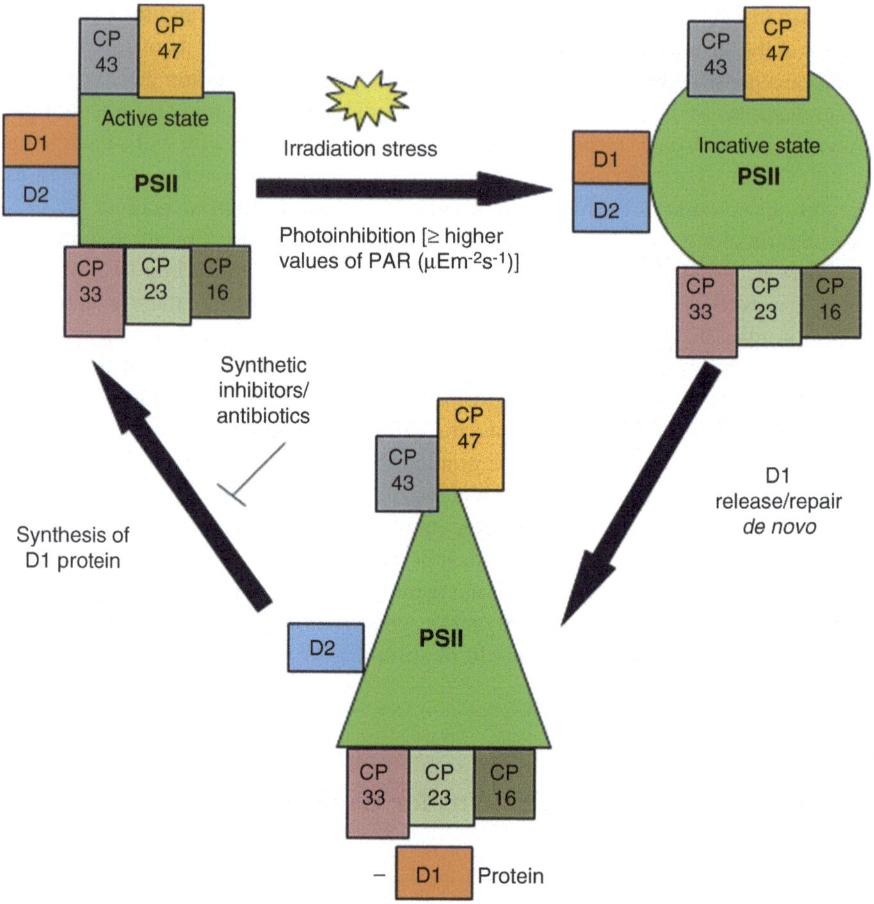

Fig. 10.1 Major paths for photoinhibition through inactivation of protein

PSII reaction centre to remove the damaged D1 protein. This is undergone a de novo synthesis for retrieval and reassembly of D1 protein. Therefore, photooxidation of D1 protein could be monitored a fall of quantum yield of PSII and its sustenance. This undoubtedly indicates photoinhibition. The reassembly of D1 protein would be equivalent to non-photochemical quenching. The overall phenomenon for photoinhibition and its recovery is also dependent on stressful conditions where the plants were adopted. Under stressful conditions, this mode of NPQ is also contributory like resistance to temperature variations including high or chilling sensitivity (Park et al. 1996). Repair of D1 protein is thus a target of biotechnological platforms where few abiotic stress tolerances are developed through either mutations or gene manipulations to minimize the NPQ. This holds also the sink capacity for electrons from photochemical activity. PSII must have the potentiality to decrease the CO_2 reduction and thereby the photooxidation is minimized. Rapid-growing short-duration varieties and timid-growing few perennial species would be the possible sources of photorespiration minimization where low photosynthetic capacities always achived with higher capacity of non-photochemical quenching under intense illumination, photorespiration, water-water cycle and alternative mode of electron transport. The cyclic electron cycle encompasses the flow of electron with PSI that yields ATP generation but never involved in the production of NADPH + H^+, rather electrons from NADP(H)/reduced ferredoxin to plastoquinone that increases the pH of the lumen. Cyclic electron transport can support the ATP or NADPH production for the Calvin-Benson cycle (Alric et al. 2010). It also provides an alteration of lumen pH as well as a higher value of qE that is the photoprotective role. Under the condition of transition or inhibition of linear electron transport, this is favoured to adopt the alternative cyclic electron cycle. It has also got relevance in the regulation of photosynthesis. It was found in the *Arabidopsis* mutant with low NPQ traits which is more susceptible to photoinhibition (Sarvikas et al. 2006). But plants can be survived under low light intensities. A protein called PGR5 (proton gradient regulation 5) in thylakoid has directly correlated for its expression to the cyclic electron flow when the plants are transferred onto high irradiance (Long et al. 2008). Though its actual function is unknown, still, the mutant of pgr5 has the same phenotypic natures than the wild. The correlation between changing physiological environment and shifting of the cyclic flow of metabolites in photosynthetic carbon reduction cycle remains unproven.

6 Water Splitting Complex and Its Engineering in Subdued Photooxidation

Another source for photooxidation is after the name of Mehler who proposed a water-water cycle in the exercise of PSI and II, however, in a concomitant manner. When light causes the reduction of O_2 into the water at PSI through electron generation, waters are split in PSII. In sequential reactions, a number of antioxidative proteins and supporting proteins for antioxidation are sequentially involved. Superoxide dismutase is the prime one which is followed by ascorbate-mediated peroxidase; ascorbate-metabolizing proteins like monodehydroascorbate reductase and dehydroascorbate reductase; and non-thiol redox residue maintaining proteins like glutathione reductase as well as NADPH + H^+-producing proteins and ferredoxin-$NADP^+$ reductase. These all proteins constitute the antioxidation system in the elimination of ROS, and thereby peroxidation of chloroplast membrane is reduced (Table 10.1) (Bandyopadhyay et al. 1999). In addition, PSII excision energy and electron load in PSI are also dissipated. It is the water stress coupled with the rapid transfer from dark to illumination; the water-water cycle is more active. Under this condition, the electron transport system is prone to more oxidized and generate ROS. The PSI under this condition remains in an oxidized form supporting cyclic electron cycle (Foyer et al. 1994). This also favours the activation of qE induced by stomatal alkaline pH when ATP-to-NADPH ratio is high that will activate the

Table 10.1 Distributions of ROS in plant tissue, type of reactions, cellular effects and plants defence under irradiation in excess

Cell organelle	Type of reaction(s)	Cellular effects	Defence
Cell wall/ apoplastic spaces	Wall-bound peroxidase, amine oxidase	Hydrolysis of cellular microfibrils, peroxidation of membrane lipids	Peroxidase, guaiacol peroxidase, ascorbate peroxidase
Plasma membrane/ tonoplast	NAD(P) oxidase, xanthine oxidase, oxidoreductase, ubiquinone oxidoreductase	Over-reduction of membrane-bound ETS, reduction of unsaturated fatty acids, over-reduction of quinone pool	Overexpression of alternative oxidase: $2H^+ + ½ O_2 = H_2O$
Intermembrane space/matrix	Cyt-b/c_1 complex (complex III), NADH-H complex (complex I)	Spillage of electrons from complexes II and III, reduction of reduced ubiquinone (UQH_2)	Alternative oxidase-mediated H_2O production, intermembrane space-bound NADP (H) oxidase
Chloroplast	PSII: the linkage between QA and QB PSI: Fd-NAD(P) oxidase, iron sulphur cluster (2Fe-2S/4 Fe-4S)	Chlorophyll proteins degradation, overexpression of chlorophylas, oxidation of LHC-I	Membrane-bound antioxidants: tocopherol, ascorbate; overexpression of PsbS, xanthophyll cycle: zeaxanthin⇌violaxanthin
Mitochondria	Complex I: $NADH + H^+ \rightarrow NAD^+$ $FMNH_2 \rightarrow$ FMN (flavin mononucleotide)/flavin oxidoreductase Complex II: succinate dehydrogenase, FAD (flavin adenine dinucleotide) and FeS Complex III: cytochrome c oxidoreductase Complex IV: cytochrome c oxidase	Over-reduction of complex II, development of superoxide $O_2 \rightarrow O_2^{-}$ $O_2^{\cdot -} \rightarrow H_2O_2$	Superoxide dismutase (SOD): Cu-SOD, Zn-SOD, Mn/Fe-SOD Antioxidation with Halliwell-Asada pathway: $H_2O_2 \rightarrow H_2O$ Ascorbate/monodehydroascorbate GSSG→GSH Glutathione reductase $NADPH \rightarrow NADP^+$
Peroxisome/ glyoxysome	Xanthine oxidase, Cu-oxidase, flavin oxidase, dismutation of oxygen into superoxide	Xanthine oxidase: hypoxanthine + $O_2 \rightarrow O_2^{\cdot -} + NO + O_2 \rightarrow$ peroxynitrile	Apoplastic SOD, glycolic oxidase

subsequent steps of CO_2 reduction with equivalent nature and function of cyclic electron flow to protect the plants from photooxidation. Thereby, ample scopes are there to select the plant species with regard to photooxidation under oxidative stress out of water deficit and others.

7 Conclusions

Light is the most important edaphic factor to consider; its precise role in photosynthetic regulation has been addressed in this communication. The complexity of light, its specific wavelengths and quantum and temporary and spatial demarcation for photosynthesis are based on photochemistry and quenching to check the over-reduction of ETC following ROS generation. The issue of photoinhibition vis-à-vis the photoprotection clearly indicates as a natural variation among the crop species. The suboptimal environment which has always been adhered to crop sustenance has a serious limitation for gaining of reduced carbon. Therefore, plant genotypes would be selected with their optimum carbon gain following dry matter and apportioned into yield should be considered. The physiological screening and genetical engineering independently or in combination for a crop species and its repeated trial in real field condition would be the measure for crop improvement.

References

Alric J, Lavergne J, Rappaport F (2010) Redox and ATP control of photosynthetic cyclic electron flow in *Chlamydomonas reinhardtii* (I) aerobic conditions. Biochim Biophys Acta 1797(1):44–51

Anten NP, Bastiaans L (2016) The use of canopy models to analyze light competition among plants. In: Hikosaka K, Niinemets Ü, Anten NPR (eds) Canopy photosynthesis: from basics to applications. Springer, Dordrecht, pp 379–398

Ballottari M, Truong TB, De Re E, Erickson E, Stella GR, Fleming GR, Bassi R, Niyogi KK (2016) Identification of pH-sensing sites in the light harvesting complex stress related 3 protein essential for triggering non-photochemical quenching in Chlamydomonas reinhardtii. J Biol Chem 291(14):7334–7346

Bandyopadhyay U, Das D, Banerjee RK (1999) Reactive oxygen species: oxidative damage and pathogenesis. Curr Sci 77:658–666

Busch FA, Sage RF (2017) The sensitivity of photosynthesis to O_2 and CO_2 concentration identifies strong Rubisco control above the thermal optimum. New Phytol 213(3):1036–1051

Castillo CC, Tanaka K, Sato YI, Ishikawa R, Bellina B, Higham C, Chang N, Mohanty R, Kajale M, Fuller DQ (2016) Archaeogenetic study of prehistoric rice remains from Thailand and India: evidence of early japonica in South and Southeast Asia. Archaeol Anthropol Sci 8(3):523–543

Chen YE, Liu WJ, Su YQ, Cui JM, Zhang ZW, Yuan M, Zhang HY, Yuan S (2016) Different response of photosystem II to short and long-term drought stress in *Arabidopsis thaliana*. Physiol Plant 158(2):225–235

Foyer CH, Lelandais M, Kunert KJ (1994) Photooxidative stress in plants. Physiol Plant 92(4):696–717

Gerotto C, Morosinotto T (2013) Evolution of photoprotection mechanisms upon land colonization: evidence of PSBS-dependent NPQ in late Streptophyte algae. Physiol Plant 149(4):583–598
Gilmore AM (1999) How higher plants respond to excess light: energy dissipation in photosystem II. In: Singhal GS, Renger G, Sopory SK, Irrgang KD, Govindjee (eds) Concepts in photobiology. Springer, Dordrecht, pp 513–548
Goron TL, Raizada MN (2015) Genetic diversity and genomic resources available for the small millet crops to accelerate a new green revolution. Front Plant Sci 6:157
Guenni O, Romero E, Guédez Y, Bravo de Guenni L, Pittermann J (2018) Influence of low light intensity on growth and biomass allocation, leaf photosynthesis and canopy radiation interception and use in two forage species of Centrosema (DC.) Benth. Grass For Sci 73(4):967–978
Guidi L, Landi M, Lo Piccolo E (2019) Chlorophyll fluorescence, photoinhibition and abiotic stress: does it make any difference the fact to be a C3 or C4 species? Front Plant Sci 10:174
Hagemann M, Bauwe H (2016) Photorespiration and the potential to improve photosynthesis. Curr Opin Chem Biol 35:109–116
Hagemann M, Fernie AR, Espie GS, Kern R, Eisenhut M, Reumann S, Bauwe H, Weber AP (2013) Evolution of the biochemistry of the photorespiratory C2 cycle. Plant Biol 15(4):639–647
Hall J, Renger T, Müh F, Picorel R, Krausz E (2016) The lowest-energy chlorophyll of photosystem II is adjacent to the peripheral antenna: emitting states of CP47 assigned via circularly polarized luminescence. Biochim Biophys Acta 1857(9):1580–1593
Hanson DT (2016) Breaking the rules of RuBisCO catalysis. J Exp Bot 67(11):3180
Henry RJ (2014) Genomics strategies for germplasm characterization and the development of climate resilient crops. Front Plant Sci 5:68
Horton P (2012) Optimization of light harvesting and photoprotection: molecular mechanisms and physiological consequences. Philos Trans R Soc Lond B Biol Sci 367(1608):3455–3465
Idris A, Linatoc AC, Aliyu AM, Muhammad SM, Bakar MF (2018) Effect of light on the photosynthesis, pigment content and stomatal density of Sun and shade leaves of Vernonia Amygdalina. Int J Eng Technol 7(4.30):209–212
Johnson MP, Davison PA, Ruban AV, Horton P (2008) The xanthophyll cycle pool size controls the kinetics of non-photochemical quenching in Arabidopsis thaliana. FEBS Lett 582(2):262–266
Kebeish R, Kreuzaler F, Metzlaff M, Niessen M, Peterhaensel C, Van Rie J (2015) Method for increasing photosynthetic carbon fixation in rice. United States patent application US 14/576,874. WO2010012796A1
Li XP, Müller-Moulé P, Gilmore AM, Niyogi KK (2002) PsbS-dependent enhancement of feedback de-excitation protects photosystem II from photoinhibition. Proc Natl Acad Sci 99(23):15222–15227
Li G, Jain R, Chern M, Pham NT, Martin JA, Wei T, Schackwitz WS, Lipzen AM, Duong PQ, Jones KC, Jiang L (2017) The sequences of 1504 mutants in the model rice variety kitaake facilitate rapid functional genomic studies. Plant Cell 29(6):1218–1231
Long TA, Okegawa Y, Shikanai T, Schmidt GW, Covert SF (2008) Conserved role of proton gradient regulation 5 in the regulation of PSI cyclic electron transport. Planta 228(6):907
Malnoë A (2018) Photoinhibition or photoprotection of photosynthesis? Update on the (newly termed) sustained quenching component qH. Environ Exp Bot 154:123–133
Maxwell K, Badger MR, Osmond CB (1998) A comparison of CO_2 and O_2 exchange patterns and the relationship with chlorophyll fluorescence during photosynthesis in C3 and CAM plants. Funct Plant Biol 25(1):45–52
Müller-Moulé P, Conklin PL, Niyogi KK (2002) Ascorbate deficiency can limit violaxanthin de-epoxidase activity *in vivo*. Plant Physiol 128(3):970–977
Munne-Bosch S, Pinto-Marijuan M (2016) Free radicals, oxidative stress and antioxidants. Encycl Appl Plant Sci 2:16–19
Neill SO, Gould KS (2003) Anthocyanins in leaves: light attenuators or antioxidants? Funct Plant Biol 30(8):865–873
Nowicka B, Ciura J, Szymańska R, Kruk J (2018) Improving photosynthesis, plant productivity and abiotic stress tolerance–current trends and future perspectives. J Plant Physiol 231:415–433

Orr DJ, Alcântara A, Kapralov MV, Andralojc PJ, Carmo-Silva E, Parry MA (2016) Surveying RuBisCO diversity and temperature response to improve crop photosynthetic efficiency. Plant Physiol 72(2):707–717

Ouzounidou G, Moustakas M, Strasser RJ (1997) Sites of action of copper in the photosynthetic apparatus of maize leaves: kinetic analysis of chlorophyll fluorescence, oxygen evolution, absorption changes and thermal dissipation as monitored by photoacoustic signals. Funct Plant Biol 24(1):81–90

Park YI, Chow WS, Anderson JM (1996) Chloroplast movement in the shade plant Tradescantia albiflora helps protect photosystem II against light stress. Plant Physiol 111(3):867–875

Ruhil K, Ahmad A, Iqbal M, Tripathy BC (2015) Photosynthesis and growth responses of mustard (*Brassica juncea* L. cv. Pusa Bold) plants to free air carbon dioxide enrichment (FACE). Protoplasma 252(4):935–946

Sacharz J, Giovagnetti V, Ungerer P, Mastroianni G, Ruban AV (2017) The xanthophyll cycle affects reversible interactions between PsbS and light-harvesting complex II to control non-photochemical quenching. Nat Plants 3(2):16225

Saha I, De AK, Ghosh A, Sarkar B, Dey N, Adak MK (2018a) Preliminary variations in physiological modules when sub1A QTL is under soil-moisture deficit stress. Am J Plant Sci 9(04):732

Saha I, De AK, Sarkar B, Ghosh A, Dey N, Adak MK (2018b) Cellular response of oxidative stress when sub1A QTL of rice receives water deficit stress. Plant Sci Today 5(3):84–94

Sarvikas P, Hakala M, Pätsikkä E, Tyystjärvi T, Tyystjärvi E (2006) Action spectrum of photoinhibition in leaves of wild type and npq1-2 and npq4-1 mutants of *Arabidopsis thaliana*. Plant Cell Physiol 47(3):391–400

Schuler ML, Mantegazza O, Weber AP (2016) Engineering C4 photosynthesis into C3 chassis in the synthetic biology age. Plant J 87(1):51–65

Schuster G, Dewit M, Staehelin LA, Ohad I (1986) Transient inactivation of the thylakoid photosystem II light-harvesting protein kinase system and concomitant changes in intramembrane particle size during photoinhibition of Chlamydomonas reinhardtii. J Cell Biol 103(1):71–80

Sharma M, Gupta SK, Deeba F, Pandey V (2017) Effects of reactive oxygen species on crop productivity: an overview. In: Singh VP, Singh S, Tripathi DK, Prasad SM, Chauhan DK (eds) Reactive oxygen species in plants: boon or bane – revisiting the role of ROS, vol 11. Wiley, Hoboken, p 117

Singh S, Tripathi DK, Dubey NK, Chauhan DK (2016) Global explicit profiling of water deficit-induced diminutions in agricultural crop sustainability: key emerging trends and challenges. In: Ahmad P (ed) Water stress and crop plants: a sustainable approach. Wiley, Chichester

Slattery RA, VanLoocke A, Bernacchi CJ, Zhu XG, Ort DR (2017) Photosynthesis, light use efficiency, and yield of reduced-chlorophyll soybean mutants in field conditions. Front Plant Sci 8:549

South PF, Cavanagh AP, Lopez-Calcagno PE, Raines CA, Ort DR (2018) Optimizing photorespiration for improved crop productivity. J Integr Plant Biol 60(12):1217–1230

Sticklen M (2006) Plant genetic engineering to improve biomass characteristics for biofuels. Curr Opin Biotechnol 17(3):315–319

Taguchi-Shiobara F, Ota T, Ebana K, Ookawa T, Yamasaki M, Tanabata T, Yamanouchi U, Wu J, Ono N, Nonoue Y, Nagata K (2015) Natural variation in the flag LEAF morphology of rice due to a mutation of the NARROW LEAF 1 gene in *Oryza sativa* L. Genetics 201(2):795–808

Tripathi A, Chauhan DK, Singh GS, Kumar N (2015) Effect of elevated CO_2 and temperature stress on cereal crops. In: Azooz MM, Ahmad P (eds) Plant-environment interaction: responses and approaches to mitigate stress, vol 2. Wiley, Hoboken, p 184

Vialet-Chabrand S, Matthews JS, Simkin AJ, Raines CA, Lawson T (2017) Importance of fluctuations in light on plant photosynthetic acclimation. Plant Physiol 173(4):2163–2179

Vitlin Gruber A, Feiz L (2018) RuBisCO assembly in the chloroplast. Front Mol Biosci 5:24

Walker BJ, Drewry DT, Slattery RA, VanLoocke A, Cho YB, Ort DR (2018) Chlorophyll can be reduced in crop canopies with little penalty to photosynthesis. Plant Physiol 176(2):1215–1232

Wang Z, Wu L, Sun K, Chen T, Jiang Z, Cheng T, Goddard WA III (2018) Surface ligand promotion of carbon dioxide reduction through stabilizing chemisorbed reactive intermediates. J Phys Chem Lett 9(11):3057–3061

Chapter 11
Antioxidants: Responses and Importance in Plant Defense System

Narendra Kumar, Hukum Singh, and Satish Kant Sharma

Abstract Antioxidants are chemical substances that prevent cell damage during different stress. Antioxidants are found in all the form of foods, including fruits and vegetables. Although oxidation reactions are crucial, this might cause damage to life. Antioxidants (glutathione, vitamin E, C, etc.) and various other enzymes (superoxide dismutase, catalase, peroxidase, etc.) are maintained by a range of plant and animal systems. Free radicals produced by the oxidative reactions cause damage and finally cell death by multiple reactions. Dietary foods, as well as traditional herbal materials, are the main sources of antioxidants in daily life, which are important for the prevention of different diseases like coronary heart disease cancer, altitude sickness, etc. Recently, pharmaceutical and food industries are producing antioxidants synthetically for increasing the shelf life of the products. It is important to focus on the natural sources of antioxidants for the replacement of the synthetic one which leads to minimizing the damage of living cells.

Keywords Antioxidants · SOD · Peroxidases · Vitamins · Stresses

Abbreviations

BHA	Butylated hydroxyanisole
BHT	Butylated hydroxytoluene
CAT	Catalase
EDTA	Ethylenediaminetetraacetic acid
GPx	Glutathione peroxidase
GR	Glutathione reductase
GSSG	Oxidized glutathione
H_2O_2	Hydrogen peroxide
HO•	Hydroxyl radical
NO	Nitric oxide

N. Kumar · H. Singh (✉) · S. K. Sharma
Forest Ecology and Climate Change Division, Forest Research Institute, Dehradun, Uttarakhand, India

R. Roychowdhury et al. (eds.), *Sustainable Agriculture in the Era of Climate Change*, https://doi.org/10.1007/978-3-030-45669-6_11

O_2^-	Superoxide anion
PG	Propyl gallate
POX	Peroxidases
ROS	Reactive oxygen species
SOD	Superoxide dismutase
TBHQ	Tertiary butylhydroxyquinone

1 Introduction

Plants and animals on earth have evolved various mechanisms for facing the unfavorable environmental circumstances since life is evolved. One of the challenges of living organisms was to survive in the early oxidizing environment after the evolution of oxygen. In such conditions, some molecules act as a savior of the organisms from a hostile oxidizing environment, and these molecules are later known as an antioxidant. These molecules are naturally occurring substances which are defined as "substance with low concentration compared to the oxidizable substrate inhibits the oxidation of that substrate."

The physiological role of antioxidants is to protect plant and animal body from damage caused by harmful molecules called free radicals (Halliwell and Gutteridge 1995). Different organisms of the present era also face a high amount of oxidants in different stages of their life due to various ecological, physiological, and metabolic processes. In such conditions, antioxidants detoxify the oxidants and help organisms to attain their normal life. Oxidation means wherein loses of an electron by an atom within a molecular entity or increases their oxidation state. Besides, it can also be defined as gaining of oxygen and losing of hydrogen by the organic substrate (Silverstein 2010; Yadav et al. 2019). Oxidants (oxidizing agents) are molecules that accept electrons in a reaction or stimulate other substances to lose electrons. Common oxidizing agents are oxygen, ozone, potassium permanganate, copper compounds, ferric chloride, free radicals, hydrogen peroxide, halogens, nitric acid, peroxy acids, and acetone. Oxidizing agents cause damage in all components of the cell that are lined by a lipid membrane, DNA, RNA, cellular proteins, and lipids. Reactive oxygen species are also categorized as reactive oxygen intermediates (ROIs) (viz., superoxide anion (O_2^-), hydrogen peroxide (H_2O_2), and hydroxyl radical (HO^*)) involved in reducing the form of atmospheric oxygen (O_2). These are highly active species of chemicals (Boris et al. 2017; Sharma et al. 2018; Singh et al. 2018). In the organisms, these ROS are produced through various endogenous sources including oxidative phosphorylation which is occurring in mitochondria and endoplasmic reticulum. Apart from that, these are also generated by the interaction of different compounds such as pollutants, radiation, and xenobiotic. These are considered as an unavoidable by-product of natural metabolism and generated during respiration and photosynthesis. However, their concentration increased during various stress conditions. Elevated concentration of the ROS in the cell causes oxidative stress, and this condition is directly linked with damage of nucleic acids (DNA and RNA), proteins, and lipids. Usually, there is an integral mechanism in the cell for the scavenging of these compounds. During various stress conditions, integral defense mechanisms get hampered and ROS concentration increased. ROS are not always associated with deleterious effects, but in small quantity, they can also act as a signaling molecule for various cellular mechanisms. Besides, for normal functioning of the cell, the balance of different ROS levels is mandatory, and it is achieved by various scavenging processes (Mittler 2002; Del Rio et al. 2006; Singh et al. 2017a, b).

Antioxidants help to defend against the damage caused by reactive oxygen species and limit to maintain optimum health. They can avoid oxidation, which directly causes cell damage and leads to aging. Nature has gifted enough mechanism related to defense to each cell component against the undesirable effects of oxidants. For instance, there is a buffering system in each cell toward some compounds, viz., glutathione peroxidase (GPx), superoxide dismutase (SOD), catalase (CAT), glutathione reductase (GR), thiols, and thioredoxin. Besides, vitamin E (α-tocopherol) is also an important nutrient which plays a key role in chain-breaking and stops the increase of free radicals in cell membranes (Li et al. 2018; Roychowdhury et al. 2018; Roychowdhury et al. 2019). Besides, ascorbic acid, carotenoids, flavonoids, related polyphenols, α-lipoic acid, etc. are also involved in the protection mechanism in the cell (Shebis et al. 2013, Kumar et al. 2015; Singh et al. 2017a, b).

2 Classification of Antioxidants

These are broadly classified into two categories based on their origin which is synthetic and natural. Naturally occurring antioxidants are those which are synthesized by organisms themselves or extracted from another natural source, whereas synthetic antioxidants are synthesized outside living systems (Fig. 11.1).

2.1 *Synthetic Antioxidants*

Synthetic antioxidants are intentionally added in small quantity in food items during processing to inhibit oxidation of lipids. Synthetic antioxidants should be stable, nontoxic, and not allow any unwanted characters to the food. Phenol compounds, i.e., butylated hydroxyanisole (BHA), butylated hydroxytoluene (BHT), tertiary butylhydroxyquinone (TBHQ), ethoxyquin, and propyl gallate (PG), are most commonly used synthetic antioxidants. The specific amounts of phenolic compounds are used for antioxidant because these become prooxidant due to their higher reactivity. Apart from these homologs and derivatives of naturally occurring compounds such as tocopherols (vitamin E homologs), erythorbic acid (D-ascorbic acid) and ascorbyl palmitate (a derivative of ascorbic acid) are some synthetic antioxidants. Regulations of these compounds in various food items vary from country to country (Reische et al. 2008; Singh et al. 2017a, b).

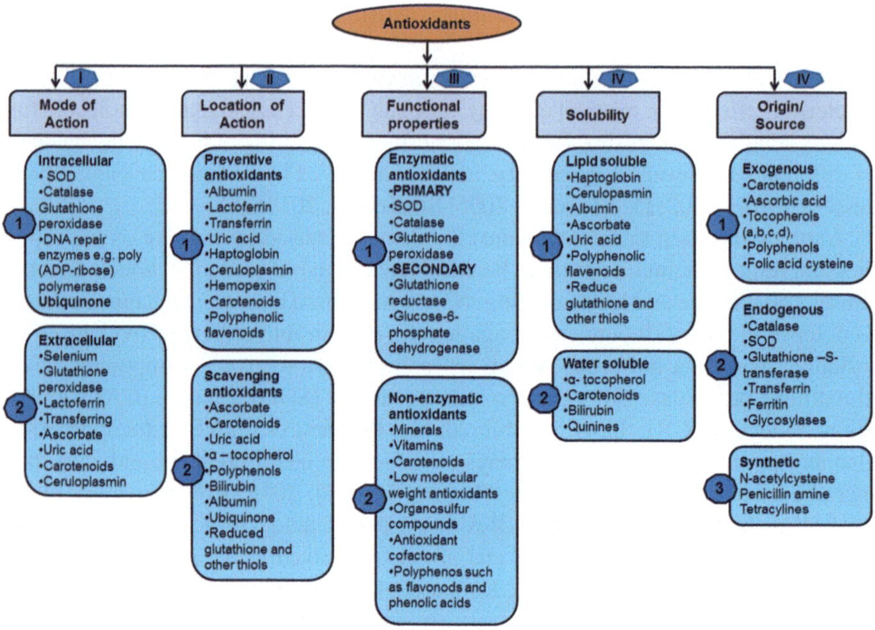

Fig. 11.1 Overview of the classification of antioxidants

2.2 *Natural Antioxidants*

These are produced in the living systems from other natural sources and for their activity; they depend on their mode of action and physical and chemical properties. Tocopherols and ascorbic acids are the most important and commercially exploited natural antioxidant. Other sources of natural antioxidants are carotenoids, flavonoids, phospholipids, protein hydrolysates, the product of Maillard reaction, sterols, and amino acids. Natural antioxidants are further classified, i.e., enzymatic and non-enzymatic antioxidants (Hurrell 2003; Singh et al. 2014).

2.2.1 Enzymatic Antioxidants

These are the enzyme which plays an important role in scavenging free radicals. In a biological system, these enzymes eliminate cellular free radicals and also keep the concentration of ROS at low levels. These constitute the main mechanism for the scavenging of ROS. These enzymatic antioxidants are classified again into two categories, i.e., primary enzymatic antioxidants and secondary enzymatic antioxidants.

Primary Antioxidant

Primary antioxidants are also considered as chain-breaking antioxidants. These are the compounds which have a potential to neutralize lipid free radicals. Neutralization is occurring either by low reactivity compounds or preventing their radical state. Generally, these reactive and stable compounds can't contribute to the propagation process of lipid that is why chain termination reaction occurs. These antioxidants are efficient at induction period, and in the presence of primary antioxidant, oxidation process begins later. These include gallic acid, phenol, TPC, flavonoids, etc. (Reische et al. 2008; Singh and Verma 2013a, b). Enzymes fall under these categories are mainly CAT, SOD, and GPx. Apart from enzymes, phenol compounds, i.e., butylated hydroxyanisole (BHA), butylated hydroxytoluene (BHT), tertiary butylhydroxyquinone (TBHQ), ethoxyquin, and propyl gallate (PG), also act as primary antioxidants. Some natural compounds such as tocopherols and carotenoids also function as primary antioxidants (Singh et al. 2010a, b; Bunaciu et al. 2012; Kumar et al. 2016).

Superoxide dismutase enzyme is produced in both eukaryotic and prokaryotic cells which eliminates the superoxide radical (O_2^-) and leads to maintaining body cells from the free radical damage. Superoxide anions convert hydrogen peroxide by the enzymatic action of SOD and compete for the nitric oxide (NO) by inactivating NO to form peroxynitrite (Bunaciu et al. 2012; Kumar et al. 2016, 2019). Another enzyme, i.e., CAT, is also found in prokaryotic and eukaryotic cells which converts H_2O_2 into their simpler compound, i.e., water and oxygen. In the food industry, catalase is used commercially with glucose peroxidase for preservation purposes by removing the oxygen. Besides, glutathione peroxidase (GPx), a selenium-dependent enzyme catalyzes the overall reaction of H_2O_2 with the help of reduced glutathione (GSH), which ultimately produces oxidized glutathione (GSSG). Also, it again recycled back in their reduced form with the help of the enzymes nicotinamide adenine dinucleotide phosphatase (NADPase) and glutathione reductase (GR) (Young and Woodside 2001; Meena et al. 2011; Singh et al. 2018).

Secondary Antioxidant

Secondary antioxidants are those which slow the rate of oxidation through various mechanisms. These antioxidants act as chelators for the metal ions, and they are not able to switch free radicals into stable molecules and have the ability to provide H^+ to primary antioxidants. These antioxidants also play an important role as oxygen scavengers. Also, these are involved in the deactivation of singlet oxygen and decompose hydrogen peroxide. Secondary antioxidant includes two enzymes such as glutathione reductase (GR) and glucose-6-phosphate dehydrogenase (G6PDH). Among them, G6PDH play a key role in the generation of NADPH, while GR enzyme is required for the recycling of GSH (Bunaciu et al. 2012; Sharma et al. 2018).

Despite enzymes, various other molecules also act as secondary antioxidants. Some of these are discussed here. Trace metals (Fe, Cu, Mg, Cr, Zn, Al, etc.) have two or more valence states and are known as redox-active transition metals.

Table 11.1 List of nonenzymatic antioxidant with their function

Nonenzymatic antioxidants	Types	Function
Minerals	Iron, selenium, zinc, copper, and manganese	As cofactors
		Required for the proper functioning of the enzymes
Vitamins	Vitamin C, vitamin A, and vitamin E	**Vitamin A** plays an important role in maintenance of epithelial cells and skin and is useful in night vision
		Vitamin C prevents DNA damage caused by free radicals
		Vitamin E – membrane protected from lipid peroxidation
Carotenoid	Lycopene, β-carotene, zeaxanthin, and lutein	Singlet oxygen quenching
Polyphenols	Flavonoid	Regulates the metabolism

These act as a catalyst for various free radical reactions. Some compounds reduce peroxidation activity of these metals by chelating them and are known as chelators. Chelators either reduce redox potential of metals or sterically hinder the formation of metal hydrogen peroxide complex, e.g., malic acid, citric acid, oxalic acid, tartaric acid, EDTA, and succinic acid. On the other hand, some other compounds such as ascorbic acid, erythorbic acid, and sulfites, which are also known as oxygen scavengers, donate hydrogen atom, react with oxygen, and remove it from the reaction mixture (Reische et al. 2008; Kumar et al. 2016).

2.2.2 Nonenzymatic Antioxidants

Nature has endowed cells with various other molecules for defense against ROS. These molecules are vitamins, carotenoids, polyphenols, etc. These are not synthesized in the human body, but their requirement is necessary for the proper metabolism. Some of the known nonenzymatic antioxidants and their functions are listed in Table 11.1 (Vaisi-Raygani et al. 2007).

3 Sources of Antioxidants

Several natural antioxidants and their dietary sources are summarized in Table 11.2.

Table 11.2 List of natural antioxidants with their dietary source

Antioxidants	Dietary sources
Flavonoids with antioxidant effect:	Fruits: apples, blackberries, blueberries, citrus fruits, grapes, pears, pomegranate, raspberries, strawberries
Anthocyanidins, aurones, chalcones, flavanones (naringenin), flavanols (procyanidin), flavan-3-ol (epicatechin, catechin), flavones (apigenin, luteolin), flavonols (isorhamnetin, kaempferol, myricetin, quercetin, quercetin glycosides, rutin), isoflavonoids (anisole, coumestrol, daidzein, genistein)	Vegetables: beetroot, brinjal, broccoli, celery, endives, leek, lettuce, onion (white and red), pepper, spinach, tomatoes
	Legumes: horse gram, green gram, lupin peas, soya beans, white and black beans
	Spices: cardamom, cinnamon, cloves, coriander, cumin
	Beverages: cocoa, tea, wine (red and white wines, sherry)
Carotenoids with antioxidant effects: astaxanthin, bixin, canthaxanthin, capsorubin, α-carotene, β-carotene, γ-carotene, crocin, β-cryptoxanthin, lutein, lycopene, zeaxanthin	Fruits: apples, apricot, banana, blackberries, blueberries, cherries, grapefruits, grapes, jackfruit, kiwi fruit, lemon, mango, melon, orange, papaya, peach, pears, pineapple, plum, strawberries, watermelon
Vitamins: Vitamin C, vitamin E (α tocopherols, tocotrienols), nicotinamide	Amla (Indian gooseberry), lemon, oranges
	Oil: groundnut oil, olive oil, palm oil, cashew nuts, germinated pulses, resins
Other compounds: curcumin, caffeine, chlorophyllin, sesaminol	Coffee, cocoa, colas, green vegetables, tea, turmeric, zinger

4 Significance of Antioxidants in Disease

Antioxidants play a vital role in various diseases, which cause abiotic stress in plants. Biologically, these stresses can be controlled by the many bio-inoculants which can enhance the ROS inside the plant body during stress (Kumar et al. 2017). Antioxidants play a key role with free radicals, and thereafter chain reaction is terminated before the damage of essential molecules. They process the defense mechanism as follows (Hajhashemi et al. 2010).

1. Elimination of reactive oxygen species is catalyzed by the enzymatic action which catalyzes, e.g., SOD and CAT.
2. Eliminated antioxidants provide oxygen to free radicals, such as vitamin E and vitamin C.
3. Some of the proteins can minimize prooxidants availability, and these proteins are metallothionein, haptoglobins, transferrin, and hemopexin.
4. Some other proteins are also involved and protected by other related mechanisms, e.g., heat shock protein and cold shock protein (Hasanuzzaman et al. 2013).

5 Role of Abiotic Stress on Antioxidant Production

Abiotic stress conditions (viz., temperature, drought, low temperature, salinity, freezing, varying light intensity, level of nutrient anaerobic stresses, and ozone (O_3)) adversely affect plant growth, development, and finally overall production worldwide. These changes might be due to the variation in the climate as well as the water shortage, respectively (Hu and Xiong 2014). Also, the current changing climatic scenario indicates the ambient temperature gradually increases, and in the future, rise in temperature with high frequency and amplitude directly or indirectly affects the crop yield (Chakraborty et al. 2014; IPCC 2014). Food security is a major task for increasing world population; scientists are facing a big problem to tackle these issues. To cope up with these problems, necessities are required in the form of "second green revolution" to improve the crop productivity and stability under harsh condition by using various genomic approaches (Eckardt et al. 2009; Kumar et al. 2017, 2019).

Plants change their physiological and metabolic responses (viz., stress-responsive gene, protein as well as generations of ROS) toward adverse condition to ensure the tolerance to various stresses. There are some reactive oxygen species produced by the excitation of molecular oxygen such as H_2O_2, $OH*$, O_2*^-, and $^1O_2$. There are several reports in this regard. In the case of wheat, the interaction of reactive oxygen species was reported negative under salt and water stress (Yousfi et al. 2010), while in case of rice, oxidative stress tolerance increases the cellular level of enzyme substrates which are small, water-soluble antioxidant molecule and functional as in millimolar concentration (Singh et al. 2010a, b; Kumar et al. 2011; Bunaciu et al. 2012).

Oxidative stress leads to enhanced SOD activity in all aerobic organisms against the defense. It is related to the metalloenzymes and catalyzes the O_2^{*-} to H_2O_2 and O_2. There are three distinct types of SOD found in different subcellular compartments, i.e., Mn-SOD (prokaryotic organisms and mitochondria of eukaryotic cells), Fe-SOD (prokaryotic organism's chloroplast stroma), and structurally unrelated Cu/Zn-SOD (cytosol, chloroplasts, peroxisome, mitochondria, and Gram-negative bacteria) (Matamoros et al. 2003). Cu/Zn-SOD dimer present in eukaryotes is cyanide sensitive, while Mn-SOD dimer and Fe-SOD tetramer are not cyanide sensitive. SOD is recorded in various organs of oats (*Avena sativa* L.), corn (*Zea mays* L.), and peas (*Pisum sativum* L.). It was also reported that under salt stress condition, SOD activity was higher in different plants, viz., *C. arietinum* (Kukreja et al. 2005), mulberry (Harinasut et al. 2003), and *Lycopersicon esculentum* (Gapinska et al. 2008). In another study, elevated temperature in sorghum increases oxidant production while decreases the antioxidant activities (Djanaguiraman et al. 2010).

Among antioxidant enzymes, catalase (tetrameric heme-containing enzyme) was the first enzyme to be isolated and described as multiple isozymes that catalyze the H_2O_2 molecule into water and oxygen. It is highly efficient for H_2O_2 and shows interaction with organic peroxides. H_2O_2-degrading enzymes are present in the several types of plants, while in cell, H_2O_2 production occurs in the peroxisomes;

however, catalase (CATs) has a lower affinity for H_2O_2 than APX and does not need any reducing equivalent for the reaction. Catalase has a property to scavenge H_2O_2 which are generated in the peroxisomes during the process of β-oxidation and photorespiratory oxidation. Similarly, other enzyme systems such as xanthine oxidase (XOD) are attached to SOD. H_2O_2 which generated in the peroxisomes are implicated in numerous stress conditions. H_2O_2 are generated rapidly through the catabolic processes when a cell is in a stressed condition and required more energy. Thereby, the enzyme catalase degraded H_2O_2 efficiently (Del Carlo et al. 2004). CAT activity is declined by the types, intensity and duration, and type of stresses. In case of Cd stress, increased catalase activity was recorded in *B. juncea* (Mobin and Khan 2007), *O. sativa* (Hsu and Kao 2004), and *T. aestivum* (Khan et al. 2007) and declined in *A. thaliana* (Cho and Seo 2005), *Glycine max* (Balestrasse et al. 2001), and *Capsicum annuum* (Leon et al. 2002).

Peroxidases as an antioxidant are implicated in several physiological and biochemical processes. These enzymes are found as multiple isozymes in plant tissues and are found in the plasmalemma and tonoplast of the cell. These enzymes are found in the cell wall as ionically bound forms as well as soluble forms. In the presence of H_2O_2, various compounds are oxidized by the peroxisomes. It plays a key role in the defense system of the plant by the conversion of H_2O_2 to water and oxygen (Del Rio et al. 2006; Kumar et al. 2008, 2016). Many physiological functions have been reported related to the peroxidases, such as oxidation of different toxic reductants, auxin catabolism, removal of H_2O_2, biosynthesis, insect attack, defense against the pathogen, etc. Also, peroxidases are mainly focused on the physiological injuries of the plant which are caused by different stress especially in temperature stress, and the activity was enhanced (Kumar et al. 2016, 2018).

6 Uses of Antioxidants in Technology

6.1 Food Preservatives

Antioxidants play an important role as food preservatives which protect food from deterioration. There are two factors for the deterioration of food, i.e., sunlight and toxic form of oxygen. For preservation, either food can be kept in dark or wax can be used such as in case of cucumbers and apple. However, oxygen is very important for storing plant materials in anaerobic condition and for plant respiration (Paul and Pandey 2014). Besides that, it can be also used in fresh vegetables and fruit packaging, and the level of oxygen should be 8% for the preservation so that the fruits and vegetables remain fresh. There are different classes of antioxidant which are categorized as preservatives, and they are considered as a natural antioxidant (ascorbic acid and tocopherols) and synthetic antioxidant (PG, TBHQ, BHT, etc.) (Ioannou and Ghoul 2013).

6.2 Industrial Uses

Antioxidants are also included in industrial purposes as lubricants for prevention of oxidation process and fuel stabilizers. Besides that, it can be used in gasoline's purposes for the prevention of polymerization that is helpful for the formation of engine-fouling residues (Finley et al. 2011).

7 Functions and Benefits of Antioxidants

7.1 Protection from ROS

Reactive oxygen species are the unavoidable and toxic species which can spoil the photosynthetic machinery and considered as a by-product of aerobic metabolism. Despite their roles as a signaling molecule during plant stress (Jones 2006), light-driven ROS production is still depicted as harmful. Therefore, an effective antioxidant system is essential to cope with the adverse effects of ROS. Enhancement of antioxidant level in plants shows greater resistance toward oxidative damage. The important reactive oxygen scavenger proteins in plants are given below:

7.2 Protection from Environmental Stress

Environmental stress includes drought, UV-B radiation, high salinity, metal toxicity, herbicides, air pollutants, temperature, light, topography, and restricted oxygen concentration. During such harsh stress conditions, quenching activity related to antioxidants and ROS production is disturbed, and these disturbances lead to oxidative damage. It was reported that under salt stress condition, the activities of antioxidant enzymes such as SOD, CAT, APX, POD, GS, and GR increase. Also, overexpression of genes is responsible for the increased activity and amount of antioxidants under different stress (Akram et al. 2017). Under field conditions, crops often exert a variety of abiotic stress which ultimately affects the crop productivity, and therefore, critical evaluation of the antioxidant system and their mechanisms is needed for sustainable agriculture practices.

7.3 Signaling in Plant Cells

Reactive oxygen species play a key role as a signaling molecule and regulating numerous cellular processes. Besides that, they are also involved in plant-microbe interaction. However, a toxic form of reactive oxygen species limits the various cellular processes of plants which harm plant growth and development. A higher level of ascorbate oxidase and peroxidase in tomato plants was reported to activate the genes of hormone signaling. They were also involved in seed germination and fruit ripening. Moreover, increased SOD activity has been correlated with decreased barley-pathogen interactions (Nath et al. 2017). Besides that, jasmonic acid, an important signaling molecule, regulates reoxygenation process in plants by the means of antioxidant enzymes through transcriptional activation. Recently, the significance of hydrogen peroxide signaling in the photosynthetic electron transport chain has been studied (Yuan et al. 2017).

7.4 Senescence and Programmed Cell Death

Senescence is the response of enhancement level of ROS due to less capacity to accommodate in plants. Among antioxidants, ascorbate directly affects the plant metabolism if low level of ascorbate in the cell enhances the ROS production and ultimately disturbs photosynthetic apparatus. Once apparatus is disturbed, it leads to a decline in the photosynthetic activity in those plants which have ascorbic acid-deficient tissues and results in plant senescence (Sade et al. 2018). Antioxidant and ROS are also important in programmed cell death signaling (Dauphinee et al. 2017).

8 Conclusion

All organisms face a high amount of oxidants in their different life stages due to various biotic and abiotic conditions. In these conditions, oxidants are detoxified through antioxidants, and organisms attain normal life. ROS which are highly active species of chemicals mainly help in detoxification. During various stress, ROS are generated in organisms endogenously or through interactions with exogenous sources. Antioxidants are most important for defense against damage and help to prevent oxidation. These chemicals are synthesizing naturally as well as synthetically. Naturally, they are synthesized in the living systems, and their activity is based on physiological, physical, and chemical properties. Synthetically, they are added in small quantity in various foods during processing to inhibit oxidation of lipids, and they are stable and nontoxic. Enzymatic antioxidants are primarily chain-breaking, whereas secondarily, they slow oxidation rate and nonenzymatic antioxidants (viz., vitamins, carotenoids, polyphenols, etc.) directly act on ROS. Nowadays, these molecules are used in various industries as a preservative in food industries. For future perspectives, the antioxidants can be tested as anticancer in various cell lines, apoptosis-inducing activity, etc.

References

Akram NA, Shafiq F, Ashraf M (2017) Ascorbic acid-a potential oxidant scavenger and its role in plant development and abiotic stress tolerance. Front Plant Sci 8:613. https://doi.org/10.3389/fpls.2017.00613

Balestrasse K, Bgardey L, Gallego SM, Tomaro ML (2001) Response of antioxidant defence system in soybean nodules and roots subjected to cadmium stress. Aust J Plant Physiol 28:497–504

Boris NI, Maria MBM, Marina AK (2017) Formation mechanisms of superoxide radical and hydrogen peroxide in chloroplasts, and factors determining the signalling by hydrogen peroxide. Funct Plant Biol 45:102–110

Bunaciu AA, Aboul-Enein HY, Fleschin S (2012) FTIR spectrophotometric methods used for antioxidant activity assay in medicinal plants. Appl Spect Rev 47:245–255

Chakraborty S, Pattanayak A, Mandal S, Das M, Roychowdhury R (2014) An overview of climate change: causes, trends and implications. In: Roychowdhury R (ed) Crop improvement in the era of climate change. IK International Publishing House, New Delhi, pp 1–29

Cho U, Seo N (2005) Oxidative stress in Arabidopsis thaliana exposed to cadmium is due to hydrogen peroxide accumulation. Plant Sci 168:113–120

Dauphinee AN, Fletcher JI, Denbigh GL (2017) Remodelling of lace plant leaves: antioxidants and ROS are key regulators of programmed cell death. Planta 246(1):133–147

Del Carlo M, Sacchetti G, Di Mattia C, Compagnone D, Mastrocola D, Liberatore L, Cichelli A (2004) Contribution of the phenolic fraction to the antioxidant activity and oxidative stability of olive oil. J Agric Food Chem 52(13):4072–4079

Del Rio LA, Sandalio LM, Corpas FJ, Palma JM, Barroso JB (2006) Reactive oxygen species and reactive nitrogen species in peroxisomes. Production, scavenging, and role in cell signaling. Plant Physiol 141:330–335

Djanaguiraman M, Prasad PVV, Seppanen M (2010) Selenium protects sorghum leaves from oxidative damage under high temperature stress by enhancing antioxidant defense system. Plant Physiol Biol 48:999–1007

Eckardt NA, Cominelli E, Galbiati M, Tonelli C (2009) The future of science: food and water for life (meeting report). Plant Cell 21:368–372

Finley JW, Kong AN, Hintze KJ, Jeffery EH, Ji LL, Lei XG (2011) Antioxidants in foods: state of the science important to the food industry. J Agric Food Chem 59(13):6837–6846

Gapinska M, Sklodowska M, Gabara B (2008) Effect of short- and long-term salinity on the activities of antioxidative enzymes and lipid peroxidation in tomato roots. Acta Physiol Plant 30:11–18

Hajhashemi V, Vaseghi G, Pourfarzam M, Abdollah A (2010) Are antioxidants helpful for disease prevention? Res Pharm Sci 5:1–8

Halliwell B, Gutteridge JC (1995) The definition and measurement of antioxidants in biological systems. Free Rad Biol Med 18:125–126

Harinasut P, Poonsopa D, Roengmongkol K, Charoensataporn R (2003) Salinity effects on antioxidant enzymes in mulberry cultivar. Sci Asia 29:109–113

Hasanuzzaman M, Nahar K, Alam M, Roychowdhury R, Fujita M (2013) Physiological, biochemical, and molecular mechanisms of heat stress tolerance in plants. Int J Mol Sci 14(5):9643–9684

Hsu YT, Kao CH (2004) Cadmium toxicity is reduced by nitric oxide in rice leaves. Plant Growth Regul 42:227–238

Hu H, Xiong L (2014) Genetic engineering and breeding of drought resistance crops. Annu Rev Plant Boil 65:715–741

Hurrell RF (2003) Influence of vegetable protein sources on trace element and mineral bioavailability. J Nutr 133:2973S–2977S

Ioannou I, Ghoul M (2013) Prevention of enzymatic browning in fruit and vegetables. Eur Sci J9:310–341

IPCC (2014) Synthesis report. In: Pachauri RK, Meyer LA (eds) Contribution of Working Groups I, II and III to the fifth assessment report of the Intergovernmental Panel on Climate Change. IPCC, Geneva, p 151

Jones DP (2006) Redefining oxidative stress. Antioxid Redox Signal 8:1865–1879

Khan NA, Singh SS, Nazar R (2007) Activities of antioxidative enzymes, sulphur assimilation, photosynthetic activity and growth of wheat (*Triticum aestivum*) cultivars differing in yield potential under cadmium stress. J Agron Crop Sci 193:435–444

Kukreja S, Nandval AS, Kumar N, Sharma SK, Sharma SK, Unvi V, Sharma PK (2005) Plant water status, H_2O_2 scavenging enzymes, ethylene evolution and membrane integrity of *Cicer arietinum* roots as affected by salinity. Biol Plant 49:305–308

Kumar P, Kamle M, Singhand J, Rao DP (2008) Isolation and characterization of peroxidase from the leaves of *Ricinus communis*. Int J Biotechnol Biochemist 4:283–292

Kumar S, Kaur R, Kaur N, Bhandhari K, Kaushal N, Gupta K, Bains TS, Nayyar H (2011) Heat-stress induced inhibition in growth and chlorosis in mungbean (*Phaseolus aureus*) is partly mitigated by ascorbic acid application and is related to reduction in oxidative stress. Acta Physiol Plant 33:2091. https://doi.org/10.1007/s11738-011-0748-2

Kumar N, Kumar N, Shukla A, Shankhdhar SC, Shankhdhar D (2015) Impact of terminal heat stress on pollen viability and yield attributes of rice (*Oryza sativa* L.). Cereal Res Commun 43:616–626

Kumar N, Shankhdhar SC, Shankhdhar D (2016) Impact of elevated temperature on antioxidant activity and membrane stability in different genotypes of rice (*Oryza sativa* L.). Ind J Plant Physiol 21:37–43

Kumar N, Suyal DC, Sharma IP, Verma A, Singh H (2017) Elucidating stress proteins in rice (*Oryza sativa* L.) genotype under elevated temperature: a proteomic approach to understand heat stress response. 3 Biotech 7:205

Kumar A, Tewari S, Singh I, Pandey R, Kumar D, Anand R (2018) Effect of nutrient sources on growth, yield and quality of turmeric under Harad (*Terminalia chebula*) based agroforestry system. Ind J Agrofor 20(2):1–6

Kumar N, Jeena N, Singh H (2019) Elevated temperature modulates rice pollen structure: a study from foothill Himalayan Agro-ecosystem in India. 3 Biotech 9:175

Leon AM, Palma JM, Corpas FJ, Gomez M, Romero-Puertas MC, Chatterjee D, Mateos RM, del Rio LA, Sandalio LM (2002) Antioxidant enzymes in cultivars of pepper plants with different sensitivity to cadmium. Plant Physiol Biochem 40:813–820

Li Z, Wang F, Zhao Q, Liu J, Cheng F (2018) Involvement of NADPH oxidase isoforms in the production of O_2^- manipulated by ABA in the senescing leaves of early-senescence-leaf (*esl*) mutant rice (*Oryza sativa*). PLoS One 13:1–17

Matamoros MA, Clemente MR, Sato S, Asamizu E, Tabata S, Ramos J, Moran JF, Stiller J, Gresshoff PM, Becana M (2003) Molecular analysis of the pathway for the synthesis of thiol tripeptides in the model legume, *Lotus japonicas*. Mol Plant-Microbe Interact 16:1039–1046

Meena D, Singh H, Chaudhari SK (2011) Elucidating strontium response on growth dynamics and biochemical change in *Phaseolus mungo* L. Int J Agric Environ Biotech 4:107–113

Mittler R (2002) Oxidative stress, antioxidants and stress tolerance. Trends Plant Sci 7:405–410

Mobin M, Khan NA (2007) Photosynthetic activity, pigment composition and antioxidative response of two mustard (*Brassica juncea*) cultivars differing in photosynthetic capacity subjected to cadmium stress. J Plant Physiol 164:601–610

Nath M, Bhatt D, Prasad R, Tuteja N (2017) Reactive oxygen species (ROS) metabolism and signaling in plant-mycorrhizal association under biotic and abiotic stress conditions. In: Varma A, Prasad R, Tuteja N (eds) Mycorrhiza – eco-physiology, secondary metabolites, nanomaterials. Springer, Cham, pp 223–232. https://doi.org/10.1007/978-3-319-57849-1_12

Paul V, Pandey R (2014) Role of internal atmosphere on fruit ripening and storability – a review. J Food Sci Tech 51:1223–1250

Reische DW, Lillard DA, Eitenmiller RR (2008) Antioxidants. In: Akoh CC, Min DB (eds) Food lipids: chemistry, nutrition, and biotechnology. CRC Press, New York, pp 409–430

Roychowdhury R, Khan MH, Choudhury S (2018) Arsenic in rice: an overview on stress implications, tolerance and mitigation strategies. In: Hasanuzzaman M, Nahar K, Fujita M (eds) Plants under metal and metalloid stress. Springer, Singapore, pp 401–415

Roychowdhury R, Khan MH, Choudhury S (2019) Physiological and molecular responses for metalloid stress in rice – a comprehensive overview. In: Hasanuzzaman M, Fujita M, Nahar K, Biswas J (eds) Advances in rice research for abiotic stress tolerance. Woodhead Publishing, Elsevier, Cary, pp 341–369

Sade N, Rubio-Wilhelmi MDM, Umnajkitikorn K, Blumwald E (2018) Stress-induced senescence and plant tolerance to abiotic stress. J Exp Bot 69:845–853

Sharma R, Singh H, Kaushik M, Nautiyal R, Singh O (2018) Adaptive physiological response, carbon partitioning and biomass production of *Withania Somnifera* (L.) Dunal grown under elevated CO_2 regimes. 3 Biotech 8:267. https://doi.org/10.1007/s13205-018-1292-1

Shebis Y, Iluz D, Kinel-Tahan Y, Dubinsky Z, Yehoshua Y (2013) Natural antioxidants: function and sources. Food Nutr Sci 4:643–649

Silverstein TP (2010) Oxidation and reduction: too many definitions? J Chem Educ 88:279–281

Singh H, Verma A (2013a) Characterization and screening of high nitrogen efficient rice genotype to curtail environmental pollution. Int J Agric Environ Biotech 6:23–30

Singh H, Verma A (2013b) Physiological responses of rice cultivars to various nitrogen levels. Int J Agric Environ Biotech 6:383–388

Singh H, Verma A, Krishnamoorthy M, Shukla A (2010a) Consequence of diverse nitrogen levels on leaf pigments in five rice genotypes under field emergent circumstance. Int J Bio-resour Stress Manag 1:189–193

Singh H, Verma A, Shukla A (2010b) Comparative study of yield and yield components of hybrid and inbred genotypes of rice (*Oryza Sativa* L.). Int J Agric Environ Biotech 3:355–360

Singh H, Verma A, Ansari MW, Shukla A (2014) Physiological response of rice (*Oryza sativa* L.) genotypes to elevated nitrogen applied under field conditions. Plant Signal Behav 9:e29015. https://doi.org/10.4161/psb.29015

Singh H, Savita, Sharma R, Sinha S, Kumar M, Kumar P, Verma A, Sharma SK (2017a) Physiological functioning of *Lagerstroemia speciosa* L. under heavy roadside traffic: an approach to screen potential species for abatement of urban air pollution. 3 Biotech 7:1–10

Singh H, Verma A, Kumar M, Sharma R, Gupta R, Kaur M, Negi M, Sharma SK (2017b) Phytoremediation: a green technology to clean up the sites with low and moderate level of heavy metals. Aust Biochem 2:1–8

Singh H, Sharma R, Savita SMP, Kumar M, Verma A, Ansari MW, Sharma SK (2018) Adaptive physiological response of *Parthenium hysterophorus* to elevated atmospheric CO_2 concentration. Indian For 144:1–14

Vaisi-Raygani A, Rahimi Z, Zahraie M, Noroozian M, Pourmotabbed A (2007) Enzymatic and non-enzymatic antioxidant defense with Alzheimer disease. Acta Med Iran 45(4):271–276

Yadav SK, Singh H, Ginwal HS, Barthwal S (2019) Elevated CO_2 enhanced growth and physiological process of Populus deltoides Bartr. ex Marsh. Indian For 145:23–27

Young IS, Woodside J (2001) Antioxidants in health and disease. J Clin Pathol 54:176–186

Yousfi S, Serret MD, Voltas J, Araus JL (2010) Effect of salinity and water stress during the reproductive stage on growth, ion concentrations, D13C, and d15N of durum wheat and related amphiploids. J Exp Bot 61:3529–3542

Yuan LB, Dai YS, Xie LJ, Yu LJ, Zhou Y, Lai YX, Yang YC, Xu L, Chen QF, Xiao S (2017) Jasmonate regulates plant responses to post-submergence re-oxygenation through transcriptional activation of antioxidant synthesis. Plant Physiol 173:1864–18806

Chapter 12
Reactive Oxygen Species and Antioxidant Defence Systems in Plants: Role and Crosstalk Under Biotic Stress

Koushik Biswas, Sinchan Adhikari, Avijit Tarafdar, Roshan Kumar, Soumen Saha, and Parthadeb Ghosh

Abstract The plant defence system always acts under the influence of many biotic and abiotic factors, and it has a prime key role in enhancing crop yield potentiality by improving quality as well as disease resistance. Similar to abiotic stress, biotic stress plays a critical role in accelerating reactive oxygen species (ROS) production in the plant system. ROS are generated in response to stress-responsive stimuli and generally regarded as oxidative stress indicative marker. ROS were initially recognized as toxic by-products of stress metabolism causing oxidative stress damage. However, the importance of ROS as signal transduction molecules in the regulation of various biological processes involved in growth, development and different pathways during plant adaptation to various biotic stresses has also been deciphered. To restrict pathogen invasion, a delicate spatio-temporal balance between ROS-producing and ROS-scavenging pathways is essential for the utilization of ROS as signalling molecules and highly crucial for sustainable agriculture. Besides this, plants also developed another survival mechanism that deals with antioxidant systems having two different components, namely, enzymatic and non-enzymatic. These two antioxidant systems coordinately work together, and the generated functional proteins are overexpressed during pathogen attack to ensure the molecular

K. Biswas
Department of Biotechnology, Visva-Bharati University, Bolpur, West Bengal, India

S. Adhikari · P. Ghosh (✉)
Department of Botany, University of Kalyani, Kalyani, Nadia, West Bengal, India

A. Tarafdar
Department of Botany, University of Kalyani, Kalyani, Nadia, West Bengal, India

Legumes Pathology, ICRISAT, Hyderabad, India

R. Kumar
Department of Human Genetics & Molecular Medicine, Central University of Punjab, Bathinda, Punjab, India

S. Saha
Department of Sericulture, Raiganj University, Raiganj, West Bengal, India

R. Roychowdhury et al. (eds.), *Sustainable Agriculture in the Era of Climate Change*, https://doi.org/10.1007/978-3-030-45669-6_12

defence response in resistant and/or tolerant plant species. The role of ROS in correlation to antioxidant systems in plant cells is gaining attention due to the fact that ROS are having valuable contribution towards increasing the response in plant defence systems at the molecular level. In the present chapter, we have described some newer aspects of ROS signalling networks in plants under biotic stress along with the use of synthetic biology approaches in increasing a better defence system, especially in crop plants. The molecular network between ROS signalling and antioxidant systems during defence response is also addressed. Further, the chapter also describes the novel techniques being investigated in order to understand the mechanism of signalling networks and metabolic networks in plant systems under biotic stress condition.

Keywords Reactive oxygen species · Antioxidants · Biotic stress · Homeostasis · Pathogenesis · Defence

1 Introduction

As we know, owing to their sessile lifestyle, plants always remain under a constant threat of different biotic and/or abiotic stresses, being effected within their natural habitats by single or combined stress (Choi et al. 2013; Padaria et al. 2015, 2016). To overcome the effect of these stresses, plants exhibit certain inimitable and convergent physiological as well as molecular responses that crosstalk each other by activating several signalling pathways (Bostock 2005; Mauch-Mani and Mauch 2005). In favourable environmental conditions, biotic stress alone such as pathogen (including fungi, bacteria, virus and nematodes) infection and herbivore pest attack in crops could have a devastating impact on agriculture as it impedes the plant growth and reduces the total yield under a challenging condition in an open field (Kumar et al. 2016; Suzuki et al. 2014; James et al. 2015). Therefore, it is crucial to understand the effect of environmental stresses in particular crops and the utility of the respective defence strategies in plants to combat incompatible situations.

'Oxidative burst', the biphasic production of reactive oxygen intermediates (ROI), is among the temporal events that occurred following successful recognition of plant pathogen, while generation of intracellular reactive oxygen species (ROS) such as superoxide radicals (O_2^-), hydrogen peroxide (H_2O_2) and hydroxyl radicals, i.e. neutral hydroxyl ion ($OH^•$), is the primary key process of plant defence after a pathogen attack (Suzuki and Mittler 2006; Bailey-Serres and Mittler 2006). Apart from biotic stress, production of ROS is a regular physiological response in a plant in action to abiotic stress defence also. In the beginning, ROS were thought to be the executioner of cell death (Torres and Dangl 2005). However, recent researches acknowledge that ROS are having pivotal roles in plant defence as a source of common signalling molecules generated in response to different biotic and abiotic stresses, developmental process and programmed cell death (Fujita et al. 2006).

Regardless, ROS can be produced in different plant cell organelles like mitochondria, chloroplasts and peroxisomes, by-products of the different metabolic and physiological process such as photosynthesis and respiration (Shadel and Horvath 2015). The generation of O_2^- in the plant for disease resistance was first reported by Doke (1983). The superoxide radicals and/or H_2O_2 and their dismutation product were found to be detected repeatedly in resistant potato seedlings during an incompatible interaction with late blight pathogen *Phytophthora infestans* (Auh and Murphy 1995; Grant et al. 2000). However, in the case of herbivore attack in the plant, the origin of ROS in insect-infested tissue is not that clear yet. It is assumed that insect causes wounding in the plant while feeding which leads to the generation of ROS into the surrounding tissue of the damaged area. For instance, a shift in the oxidative status was noticed during *Helicoverpa* infestation in soybean (*Glycine max*) where an increase in superoxide (O_2^-) and hydroxyl ($OH^•$) radical is formed (Ebel and Mithöfer 1998). It is also assumed that enzyme such as glucose oxidase from the salivary gland of *Helicoverpa* generates H_2O_2, which might contribute to enhancing the ROS concentration at the attacking site (Mithofer et al. 2004).

ROS have long been appreciated for their detrimental effects and damage-promoting activities, but there was no clear perception of their roles as signalling molecules. Inside the plant cell, among all organelles, mitochondria are known to be the main site of ROS activity. In addition to metabolism, mitochondria are directly involved with the ion homeostasis, apoptosis and innate defence of the plant (Shadel and Horvath 2015). Oxidation of unsaturated fatty acids occurs in mitochondria during the process of oxidative phosphorylation and produce ROS when oxygen is utilized to generate ATP from organic fuel molecules. During the period of fatty acid oxidation, oxylipin, a lipid derivative compound, is produced in plant cells, which represents as new endogenous signals involved in multiple defence reaction in response to pathogen and herbivore attack (Mithofer et al. 2004).

Generally, ROS signalling molecules are generated by respiratory-burst nicotinamide adenine dinucleotide phosphate-reduced oxidase during environmental stresses to activate stress-responsive pathways and induce defence mechanism. The resulting ROS can damage macromolecules, including DNA, proteins and lipids. Several reports indicate that ROS-mediated signalling for cellular plant defence is controlled by an elusive balance between production and scavenging (Mithofer et al. 2004; Torres et al. 2006). Although the involvement of enzymes in localizing and scavenging of ROS production is fairly well defined in plants, the temporal localization of ROS production is likely to be extremely critical to deciphering during inter- and intracellular transduction of ROS signal. A wide range of enzymatic systems, namely, peroxidase (POD), catalase (CAT), ascorbate peroxidase (APX), glutathione reductase (GR), glutathione peroxidase (GPX), superoxide dismutase (SOD) and metabolites, namely, glutathione (GSH), ascorbate (ASC), carotenoids and tocopherol, give protection against the toxic effect of ROS (Mithofer et al. 2004; Kadioglu et al. 2012; Tarafdar et al. 2018; Roychowdhury et al. 2018; Roychowdhury et al. 2019). The transcriptome analysis of different plants revealed the induction of a group of genes that encode ROS-scavenging enzymes in response to environmental stresses (Mittler et al. 2004). The scavenging enzymes such as

APX, SOD and GPX were reported to be used in genetically engineered plants to make them tolerant to biotic stress (reviewed by Caverzan et al. 2016).

In this chapter, we have enlightened some newer aspect of ROS-mediated signalling pathways in plants under biotic stress with an emphasis on how they are involved in defence responses. In addition, we have discussed here on the use of different approaches related to synthetic biology for improving plant defence system. Further, we have also described in this chapter about the novel techniques being investigated in order to understand the dual signalling and metabolic networks in plant system under biotic stress condition.

2 ROS Localization, Homeostasis and Detection

In plants, ROS are generally normal by-products of oxygen-dependent metabolic pathways such as respiration, photosynthesis and photorespiration. Sequential reduction of molecular oxygen as a result of electron leakage from membrane-linked (chloroplastic, mitochondrial and plasma membrane) electron transport leads to the generation of highly toxic ROS (Alscher et al. 1997; Arora et al. 2002; Bhattacharjee 2005) at the basal level in non-stressed plants cells, whose production is intensified under the stimulus of abiotic and biotic cues.

The ROS capable of inducing oxidative damage includes singlet oxygen ($^1O_2$), superoxide (O_2^-), hydrogen peroxide (H_2O_2), hydroxyl radical ($OH^•$), alkoxy radical ($RO^•$), peroxy radical ($ROO^•$), organic hydroperoxide (ROOH), etc. Reduction of O_2 to H_2O promotes the generation of highly reactive ROS, namely, O_2^-, H_2O_2 and $OH^•$. These ROS are supposed to be generated from only 1–2% of the total O_2 consumed by plants (Bhattacharjee 2005). Superoxide is generated at the onset of cellular oxidative burst as a result of the one-electron reduction of molecular oxygen and remains reactive in the interior of the cellular membrane. The superoxide anion helps in the generation of secondary ROS either through enzymatic catalysis or by using metal depending on the cellular location (Valko et al. 2005). Hydrogen peroxide is formed via dismutation of superoxide. Hydrogen peroxide is a relatively stable oxidant which can readily cross the lipid bilayer of cell membranes and can diffuse some distances from its site of production (Dat et al. 2000). Among the ROS, hydroxyl radical ($OH^•$) is the most reactive and believed to be formed directly through Haber-Weiss reaction. Although ROS localization is specific to the type of stress to which the plant is subjected, chloroplast, mitochondria and peroxisomes serve as the primary sites for ROS generation (Gill and Tuteja 2010). There are secondary sites as well like the endoplasmic reticulum, cell membrane, cell wall and the apoplast. Cellular localization and basic properties of ROS in plant tissues are depicted in Table 12.1.

In parallel to the ROS production, plants employ defence strategies relying on the function of several enzymes and metabolites to quench ROS for protection of the cellular compartments where they are localized. ROS-scavenging pathways are found in different cellular compartments and work together to modulate the

Table 12.1 Important ROS in plant tissues and their basic properties

ROS	Source	Half-life	Migration capacity	Reacts with			References
				DNA	Protein	Lipid	
Superoxide ($O_2^{\cdot-}$)	Membranes, chloroplast (Mehler reaction) and mitochondria	1–4 μs	30 nm	No	Yes (Fe centre)	Hardly	Dat et al. (2000) and Bhattacharjee (2010)
Hydrogen peroxide (H_2O_2)	Membranes, chloroplast, mitochondria and peroxisome	1 ms	1 μm	No	Yes (cysteine)	Hardly	Varnova et al. (2002) and Imlay (2008)
Hydroxyl radical ($OH^{\cdot}$)	Chloroplast, membranes and mitochondria	1 μs	1 nm	Rapidly	Rapidly	Rapidly	Halliwell and Gutteridge (1999) and Møller et al. (2007)
Singlet oxygen ($^1O_2$)	Chloroplast, membranes and mitochondria	1 μs	30 nm	Yes (guanine)	Trp, His, Tyr, Met and Cys	PUFA	Foyer and Harbinson (1994) and Halliwell and Gutteridge (1999)

accumulation of ROS in plant cells. The enzymatic components of ROS-scavenging mechanisms include SOD, CAT, POD, APX, GPX, GR, monodehydroascorbate reductase (MDHAR), dehydroascorbate reductase (DHAR) and peroxiredoxin (PRX). SOD belongs to the group of metalloenzymes and acts as the first line of defence against oxidative stress. SOD catalyses the dismutation of $O_2^{\bullet -}$ to O_2 and H_2O_2. CAT, APX and GPX catalyse the disproportionation of H_2O_2 into H_2O and O_2. Unlike CAT, APX uses ascorbic acid (AsA) and/or a glutathione (GSH), regenerating AsA-GSH cycle catalysed by MDHAR, DHAR and GR. AsA-GSH cycle plays a central in combating oxidative stress. GPX and PRX also detoxify H_2O_2 through ascorbate-independent thiol-mediated pathways using GSH and thioredoxin (TRX) as nucleophile, respectively (Noctor et al. 2014). Besides a series of enzymes, certain redox metabolites serve as non-enzymatic antioxidants and scavenge ROS either in combination with the antioxidative enzymes or independently. Non-enzymatic components of the antioxidative defence system include AsA, GSH, tocopherol, carotenoids and phenolic compounds. Besides traditional antioxidant machinery, soluble sugars such as disaccharides and raffinose family oligosaccharides and fructans are strongly related to ROS homeostasis in plants (Keunen et al. 2013).

Several methods are available for the detection of ROS accumulation. However, the difficulty in detection of ROS is associated with their short lifetimes together with the capability of living cells to scavenge ROS. Superoxide anions can be detected with the histochemical stain nitro blue tetrazolium (NBT). This is a semi-quantitative microscopic assay wherein the presence of superoxide radical reduction of water-soluble NBT is observed in the form of water-insoluble formazan (Xia et al. 2009; L'Haridon et al. 2011; Großkinsky et al. 2012; Kawarazaki et al. 2013). Dihydroethidium (DHE), a cell-permeable blue fluorescent stain, can also be used for the detection of superoxide. The oxidation of DHE forms red fluorescent oxyethidium, which intercalates with nucleic acids (Lehotai et al. 2011). H_2O_2 generation in plant tissue can be visualized by DAB (3–3′ diaminobenzidine) staining (Asai et al. 2010; L'Haridon et al. 2011; Zhang et al. 2012; Simon et al. 2013). Titanium oxalate (Becana et al. 1986) or ABTS [(2, 2′-Azino-bis-(3-ethylbenzothiazoline-6-sulfonate)] (Messner and Boll 1994) also detects H_2O_2 spectrophotometrically. Małolepsza and Rozalska (2005) described the utility of 2-deoxyribose (DOR) for the estimation of hydroxyl radicals. Spectrophotometrical quantification of thiobarbituric acid-reactive degradation products using DOR has also been demonstrated (Von Tiedemann 1997). Dansyl-based fluorescence sensors such as DanePy or HO-1889-NH (Hideg et al. 2002) and Singlet Oxygen Sensors Green (SOGS) (Plancot et al. 2013) are efficiently used for the detection of singlet oxygen. Chemiluminescent methods to determine ROS in plant tissues include the use of luminals (Kunz et al. 2006; Dubreuil-Maurizi et al. 2010; L'Haridon et al. 2011). The utility of broad-spectrum fluorescent probes such as fluorescein diacetate (H2DCF-DA, CM-H2DCF-DA) and dihydrorhodamine 123 in the detection of ROS is exemplified in several studies (Guo et al. 2010; Benikhlef et al. 2013).

3 ROS Signalling Under Biotic Stress

The plant has developed a multichannel defence response against biotic stimulus by expressing different types of transcriptionally activated genes, which are further finely tuned by surrounding temperature, humidity and other microclimates. These further developed a complex signalling network for the recognition of biotic factors (both beneficial and harmful). Among these various types of network systems associated with ROS regeneration, NADPH oxidase is considered as a key enzyme to control the pathway. This enzyme is activated by Ca^{2+} ion, triggers O_2^- synthesis and which is further converted to H_2O_2 in plasma membrane under the influence of avirulent microbial entities or other abiotic factors (Fraire-Velázquez et al. 2011). Recent molecular studies on defence response identify that a number of novel defence-responsive or transcription factors including kinase play a key role in signalling cascades and ROS regeneration.

It has been postulated that the amount of intracellular Ca^{2+} ion and ROS component is elevated and acts as signal transducer under abiotic and biotic stimulus especially feeding of plant tissues by herbivores (Arimura and Maffei 2010). Thus, the Ca^{2+} ion influx is reached up to a level in the cytoplasm where it is involved in signal transduction through other parallel biochemical pathways by interacting with calcium Ca^{2+} sensor protein, namely, calcium-dependent protein kinase (CDPKs) and calcineurin B-like proteins (CBLs) (Kim et al. 2009).

If we are concerning about the most powerful and highly active calcium-based sensor molecules, CDPKs consisting of calmodulin-like- and protein kinase-conserved domain and mutagen-activated protein kinases (MAPKs) belonging to Ser/Thr protein kinase family are considered as prime member category.

It has been well documented that hypersensitive response (HR) is triggered in the plant as the first level of defence response when the influx of Ca^{2+} and H^+ ion is balanced at a higher rate with the efflux of K^+ and Cl^- ion. As a result, the signal is transduced to generate ROS ($O_2^-$and H_2O_2) and reactive nitrogen species (RNS) (NO) (Delledonne et al. 1998; Hancock et al. 2002). The generated ROS is involved in multiple activities as a part of HR, such as stimulation of cell wall lignification by cross linking mechanism and other microcidal activities. H_2O_2 and NO as the part of ROS and RNS, respectively, can induce the expression of some vital defence-responsive genes, namely, pathogenesis-related (PR) genes, phenylalanine ammonia lyase (PAL) and glutathione S-transferase (GST), when both of the species are simultaneously available in proper balance (Delledonne et al. 1998; Desikan et al. 1998).

Under oxidative stress situations, ROS absorb electrons from target molecules which lead to detrimental effects on various macromolecules constituting cells. The negative effects on unsaturated lipids, proteins and DNA also modify their structure and functions and, in some cases, inactivation of these molecules. These types of molecular trauma subsequently promote the cellular dysfunction and finally result in cell death (Taylor et al. 2002). Another important characteristic of ROS in signalling pathway is their capability to act as a secondary messenger. This functionality depends on preferential and strong interaction among cell sensors that regulate different biological process such as transcription, translation and post-translational modification. This complete sensor system eventually modulates not only the growth and development of the plant but also the programmed cell death in response to biotic and abiotic stimuli (Kovtun et al. 2000).

4 ROS Generation and Their Roles During Plant Pathogenesis

The production of ROS by the consumption of molecular oxygen is termed as oxidative burst, recognized as the earliest cellular responses during successful pathogen recognition and activation of plant defences. Avirulent pathogens, sensed by the disease resistance (R) gene products, often induce a biphasic ROS generation with a small, transient first phase, followed by a prolonged phase of much higher magnitude which associates with disease resistance. However, a three-phasic ROS generation has also been observed in some cases of fungal pathogen depending on the complicated development of ROS generation and influence of the host genotype (Shetty et al. 2003; Hückelhoven and Kogel 2003). In contrast, virulent pathogens that avoid host recognition induce only the first transient phase of this response (Bolwellet al. 2002). Elicitors of defence responses, often referred to as microbe or pathogen-associated molecular patterns (PAMPs), are also capable of triggering oxidative burst (Chisholm et al. 2006). ROS generation in plants by plasma membrane NADPH oxidases and cell wall peroxidases following pathogen infections is a well-recognized process (Allan and Fluhr 1997; Lamb and Dixon 1997; Bolwell et al. 2002; Torres et al. 2006; Galletti et al. 2008). Plant membrane-bound NADPH oxidase (NOX), also known as respiratory burst oxidase (RBO), has the capacity to transfer electrons from cytoplasmic NADPH to molecular oxygen in the apoplast site and generate superoxide anions, which subsequently dismutate to H_2O_2 and O_2^-. Besides ROS-scavenging activity, peroxidases in the cell wall have been demonstrated to act as NADH oxidases leading to the formation of O_2^- and H_2O_2. Peroxidase activity has been demonstrated to increase during incompatible reactions (Baker and Orlandi 1995). Contribution of cellular organelles in ROS generation during plant-pathogen interactions has also been documented (Kariola et al. 2005; Kuźniak and Skłodowska 2005; Amirsadeghi et al. 2007). Possible mechanisms of ROS generation and their roles in response to pathogenesis are depicted in Fig. 12.1.

ROS generated in response to pathogens and elicitors have been hypothesized to have implications in many defensive processes related to pathogen interactions with their hosts. Antimicrobial effects of ROS, particularly H_2O_2, have been proposed during the plant defence response (Custers et al. 2004; Walters 2003; Shetty et al. 2007). However, the amount of extracellular H_2O_2 produced, nature of the elicitor,

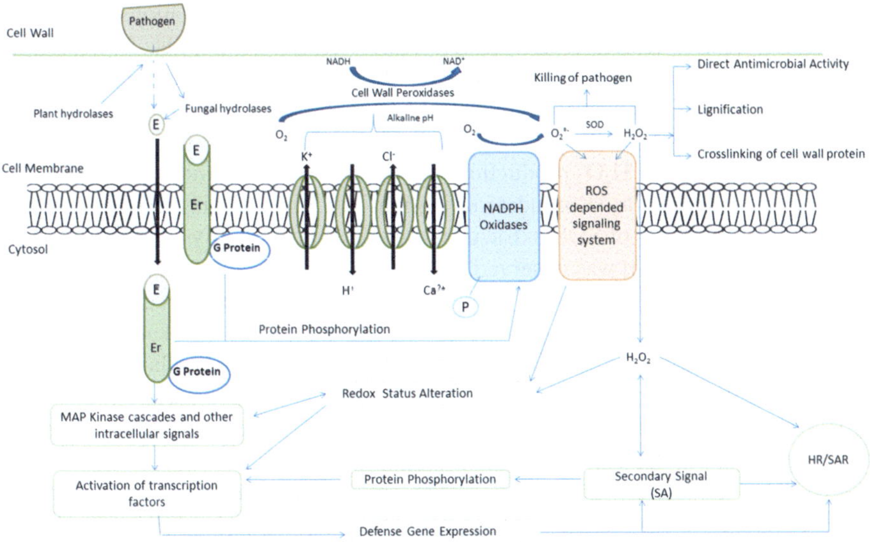

Fig. 12.1 Schematic representation of ROS production and functions in response to pathogens. (Modified from: Jabs (1999) and Wojtaszek (1997))

developmental stages of the plant cells, plant species and life cycle stages of the pathogen determine the actual sensitivity of the pathogen to the ROS generation in a plant-pathogen interaction (Legendre et al. 1993; Levine et al. 1994; Nürnberger et al. 1994; Małolepsza 2005; Shetty et al. 2007). The evidence from the studies evaluating the effect of H_2O_2 against bacteria and fungi also indicated a lack of direct antimicrobial effect of H_2O_2 in the plant defence, possibly due to the parallel production of antioxidants in plants which is necessary for manipulation of the host tissue preventing H_2O_2 from reaching concentrations that are lethal to the pathogens (Miguel et al. 2000; Shetty et al. 2008). The combined action of ROS and ROS detoxification enzymes has also been observed during compatible interactions involving virus (Allan et al. 2001; Clarke et al. 2002). Although ROS generation is usually associated with disease resistance responses, the involvement of ROS in successful pathogenesis have also been documented. Necrotrophic pathogens are believed to stimulate ROS generation for their advantage. ROS production in the infected tissue helps to induce cell death, which facilitates pathogenesis of fungal necrotrophs (Govrin and Levine 2000).

ROS produced as a result of oxidative burst during early development of plant-pathogen interaction is involved in the process of cell wall strengthening to pathogen attack. Cell wall strengthening process is characterized by lignification, papillae formation, cross-linking of hydroxyproline-/proline-rich proteins and other cell wall polymers. Lignification of the cell wall during pathogenesis is a well-evidenced process, and its association with H_2O_2 production has been shown in several systems (Olson and Varner 1993; Repka 2002). Papillae are the cell wall appositions formed by the accumulation of callose, proteins and phenolic compounds to act as an important barrier to pathogen penetration. Cytological investigations have indicated the close association of H_2O_2 production with effective papillae formation at the site of fungal infection (Hückelhoven et al. 1999). Callose synthesis was observed at the H_2O_2 generation site of suspension-cultured rice cells infected with *Acidovorax avenae* (Iwano et al. 2002). Induction of cell wall protein cross-linking process with enhanced H_2O_2 production was observed in pearl millet infected with *Sclerospora graminicola* (Kumudini and Shetty 2002). Similarly, H_2O_2-mediated oxidative cross-linking of SDS-extractable structural proteins [hydroxyproline-rich glycoproteins (HRGPs)] was observed in the cell walls of bean (*Phaseolus vulgaris* L.) and soybean (*Glycine max* L.) suspension cells treated with crude fungal elicitors (Bradley et al. 1992; Brisson et al. 1994). Together, these cell wall strengthening processes inhibit the transport of pathogenic toxin inside the plant cells, limit the nutrient availability to pathogens and thus limit the infection spread.

Hypersensitive response (HR), a well-known aspect of resistance, is characterized by rapid, localized host cell death at the sites of infection (Lam et al. 2001). In plants, HR is induced by the activation of R genes, demonstrated in different plant species against important microbial pathogens (van't Slot and Knogge 2002; Hammond-Kosack and Parker 2003). The HR is highly effective against biotrophic pathogens as the host cell death is associated with the shortage of nutrient supply (Thordal-Christensen et al. 1997; Mellersh et al. 2002; Greenberg and Yao 2004). However, HR is not generally effective against necrotrophic pathogens since cell death facilitates pathogenesis leading to enhanced colonization of fungal necrotrophs (Govrin and Levine 2000; Mayer et al. 2001; Greenberg and Yao 2004). Involvement of ROS, specifically H_2O_2, formed during the oxidative burst in the elicitation of the HR has been documented in numerous patho-systems (Levine et al. 1994; Baker and Orlandi 1995; Low and Merida 1996; Thordal-Christensen et al. 1997). Furthermore, diverse approaches and systems such as infiltration of antioxidants (Li et al. 2005) and ROS inhibitors (Levine et al. 1994; Sasabe et al. 2000; Li et al. 2005), treatment with pathogen elicitors (Levine et al. 1994; Sasabe et al. 2000; Greenberg and Yao, 2004; Montillet et al. 2005) and use of specific mutated lines of *Arabidopsis* and other plants (Jabs et al. 1996; Dat et al. 2003; Hoeberichts et al. 2003; Lorrain et al. 2003; Mateo et al. 2004; Montillet et al. 2005; Torres et al. 2005; Van Breusegem and Dat 2006) have been adopted to identify the correlation between accumulation of ROS and HR. Although different models have been proposed to explain how ROS may elicit and regulate HR (Delledonne et al. 2002; Torres et al. 2005; Van Breusegem and Dat 2006), the exact role and mechanism of ROS in elicitation of the HR remain unclear as exemplified in different studies (Sasabe et al. 2000; Delledonne et al. 2001; Shetty et al. 2003, 2007; Montillet et al. 2005). Elucidation of the simple correlation between ROS and HR is also complicated by the complex cooperation of ROS with plant hormones such as SA, JA, ET and ABA which also influence the activation of HR (Hoeberichts et al. 2003; Torres et al. 2005; Van Breusegem and Dat 2006).

ROS, especially H_2O_2, in conjunction with salicylic acid (SA), was proposed to facilitate the establishment of a systemic defence response known as SAR (systemic acquired resistance) (Durrant and Dong 2004). The establishment of SAR is associated with the expression of defence-related genes, including pathogenesis-related (PR) proteins (Durrant and Dong 2004). It appears that ROS mediates SAR by interacting with non-expressor of PR-1 (NPR1) and regulates the systemic responses by controlling NPR1 redox state (Mou et al. 2003; Kanzaki et al. 2003). SAR provides long-lasting, broad-spectrum resistance to infections caused by a broad range of pathogens/elicitors (Ryals et al. 1996). Initially, the signalling system is activated at the site of primary infection, which subsequently translocated to the distal uninfected tissues. The implication of SA in H_2O_2 generation has been established by the findings showing the inhibitory effect of SA on catalase activity (Chen et al. 1993). Thus, the action of SA in SAR is likely mediated by elevated amounts of H_2O_2. On the other hand, H_2O_2-mediated stimulation of SA synthesis was also observed, suggesting that H_2O_2 may act upstream of SA (Leon et al. 1995). The influence of SA and its functional analogues in the induction of SAR has been demonstrated by several studies (Ward et al. 1991; Delaney et al. 1994; Friedrich et al. 1995; Lawton et al. 1995; Friedrich et al. 1996). Taken together, this evidence strongly suggests that H_2O_2 and SA act in a complex coordinated manner of inactivation of SAR, although their involvement in signal transduction pathways as a signalling molecule remains to be determined.

5 Antioxidant Enzymes as Biotic Stress Response Against ROS Generation in Plants

The ROS defence mechanism comprises enzymatic components including the antioxidant machinery, SOD, CAT, MDHAR, DHAR, APX, GR and GPX as well as non-enzymatic components such as ascorbic acid, GSH, carotenoids, phenolics, flavonoids,α-tocopherol and amino acid cum osmolyte proline (Das and Roychoudhury 2014).

Pathogens (bacteria, fungi and viruses), herbivores, insects and rodents are the biotic stress for the plant (Caverzan et al. 2016). In the case of biotic stress, plant first recognizes the pathogen by the receptor present on it. After recognition, constitutive basal defence mechanisms start that leads to an activation of the signalling cascade. Subsequently, kinase cascades and specific ion channels are activated; phytohormones like abscisic acid (ABA), ethylene, jasmonic acid (JA) and salicylic acid (SA) accumulate; ROS starts generating; and a reprogramming of the genetic machinery leads to adequate defence reactions and an increase in plant tolerance in order to minimize the biological damage caused by the stress (Rejeb et al. 2014). ROS at higher concentration cause oxidative damage to membranes (lipid peroxidation), proteins and RNA and DNA molecules and can even activate programmed cell death (Caverzan et al. 2016).

The ascorbate-glutathione cycle (AsA-GSH) is the main ROS-scavenging pathway that takes place in mitochondria, chloroplasts, cytosol, apoplast and peroxisomes in plants. The AsA-GSH cycle involves successive oxidation and reduction

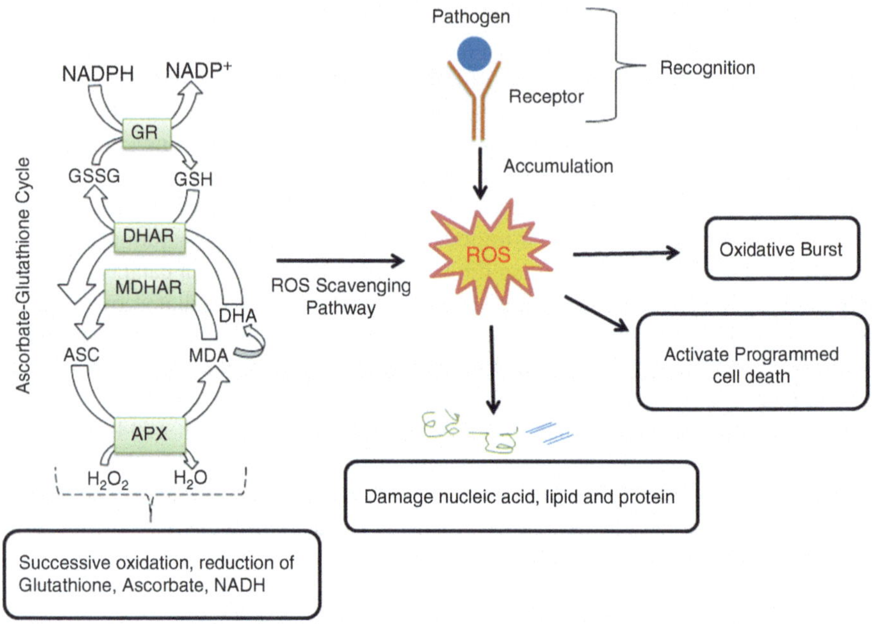

Fig. 12.2 ROS-induced oxidative damage and involvement of ROS-scavenging enzymes in response to biotic stress in plants

of glutathione, ascorbate and NADPH catalysed by GR, APX, DHAR and MDHAR (Caverzan et al. 2016). ROS-induced oxidative damage and involvement of ROS-scavenging enzymes in response to biotic stress in plants have been shown in Fig. 12.2.

6 Crosstalk Between ROS-Producing and ROS-Scavenging Pathways During Biotic Stress

As mentioned earlier, faster ROS generation plays a vital role in disease resistance responses. In order to protect the cell from injury or death caused by excess ROS, an alternative ROS-scavenging enzyme system comprising the players like CAT, SOD, GPX and APX is getting activated to regulate the steady-state level of ROS (Apel and Hirt 2004). In this context, it can be mentioned that a large number of genes encoding ROS-scavenging enzymes were identified and found to be induced when wide-scale transcriptome analyses was conducted in plants treated with various biotic and abiotic stress factors (Mittler et al. 2004). In recent years, several ROS-scavenging enzymes have been adopted to engineer biotic and abiotic stress-tolerant plants. For example, APX1 enzyme was found to be a central component of the ROS-scavenging gene products, and this was confirmed in *Arabidopsis* mutant lacking cytosolic *APX* (*cAPX*) which leads to the elevated level of H_2O_2 concentration, protein oxidation and growth retardation simultaneously. Interestingly, transcription analyses of a large set of pathogen-responsive genes, especially disease resistance, were highly induced in plants overexpressing cytosolic *APX1* gene subjected to several environmental stresses (Caverzanet al. 2016).

Ascorbate functions as a primary antioxidant by quenching O_2, O_2^- and $OH^{\bullet}$ where it acts as electron donor and participates in the regeneration of oxidized α-tocopherol and violaxanthin de-epoxidase. In most of the cases, it was also observed that ascorbate in conjugation with glutathione functions as co-substrate in the AsA-GSH cycle at the time of H_2O_2 detoxification. In this cycle, ascorbate is oxidized by APX to generate monodehydroascorbate (MDHA) and which is further reduced to form H_2O. From this point, MDHA is again oxidized itself to yield dehydroascorbate (DHA) (Asada 1992). Another important ROS-scavenging enzyme, SOD, is primarily considered as the first line of antioxidant defence. This enzyme functions as a catalysing agent for dismutation of O_2^- to H_2O_2 via cyclic reduction and oxidation of transition metal ions such as copper, zinc or manganese. Surprisingly, SOD comprises its multiple isoforms only in the case of plants rather than other animals (McCord and Fridovich 1969). In plants, another ubiquitous enzyme, namely, CAT, also removes the toxic H_2O_2 by converting it to nontoxic O_2 and water. It was also revealed that three genes encode for individual subunits to form at least six different homo- and heteromeric isoforms of this enzyme in *Arabidopsis* (Frugoli et al. 1996). Another important ubiquitous enzyme, peroxiredoxin (PRX), is under the category of non-haem-containing peroxidases. This gene family has a specific functionality on the structure of cysteine residues which results in the detoxification of H_2O_2. Four independent subgroups, namely, 1-Cys-PRX, 2-Cys-PRX, PRX-Q and type II PRX, were found to be the primary classes under PRX gene family. In recent years, glutathione peroxidase (GPX) was taken under the consideration of the fifth group of PRX displaying thioredoxin-dependent functional activity (Navrot et al. 2006). Concurrently, in order to suppress the activity of potentially highly toxic H_2O_2 derivative, i.e. $OH^{\bullet}$, all the antioxidative defence enzyme components are coordinately get activated and clean the continuously intoxicated cellular environment. The parallel evolvement of ROS regeneration with ROS-scavenging systems and their coordinated action during stress conditions thus make a redox homeostasis to maintain all the cellular machinery active in plants.

7 Coordination Between ROS and Other Regulatory Pathways Under the Combined Influence of Biotic and Abiotic Stresses

As plants grow in open field conditions, they are always exposed to multiple biotic and abiotic stresses concurrently. To overcome the effects of those combined stresses, plants have adopted a wide range of defence mechanisms. The defence mechanisms involved in each stress are not independent but synchronized. Infection and establishment of the pathogen into the plant tissue is the foremost step towards the development of any disease in plants, and it is well documented that the pathogen infection is completely dependent on the environmental factors (Sharma et al. 2015, 2017; Madgwick et al. 2011). So environmental stress may either have a direct or indirect effect on community interactions between plants and pathogens leading to either negative or positive effects on plant defence responses (Sinha et al. 2016; Sharma and Ghosh 2017). Recent studies revealed that several molecules such as hormones, kinases and transcription factors are potential key players involved in crosstalk between stress signalling pathways. Hormone signalling pathways are directly regulated by salicylic acid, jasmonic acid, abscisic acid and ethylene and indirectly by ROS signalling pathways (Fujita et al. 2006). The involvement of regulatory molecules in the ROS signalling pathway under combined stress (biotic and abiotic) is elucidated in Fig. 12.3.

Attempts have been made to identify convergence points among the stress signalling pathways in plants for better understanding of the physiological and molecular responses that interact each other to cope combined effect of the stresses (Padaria et al. 2015; Pandey et al. 2015; Tarafdar et al. 2018). In instance, several research on combined stress (in a combination of pathogen infection and environmental stress) scenarios revealed that soil moisture stress could affect the pathogen infection in plants either positively or negatively (Ghosh et al. 2017; Tarafdar et al. 2018). Pathogens growth and aetiology are often dependent on the water availability in a particular medium where they are growing. For example, an increase in

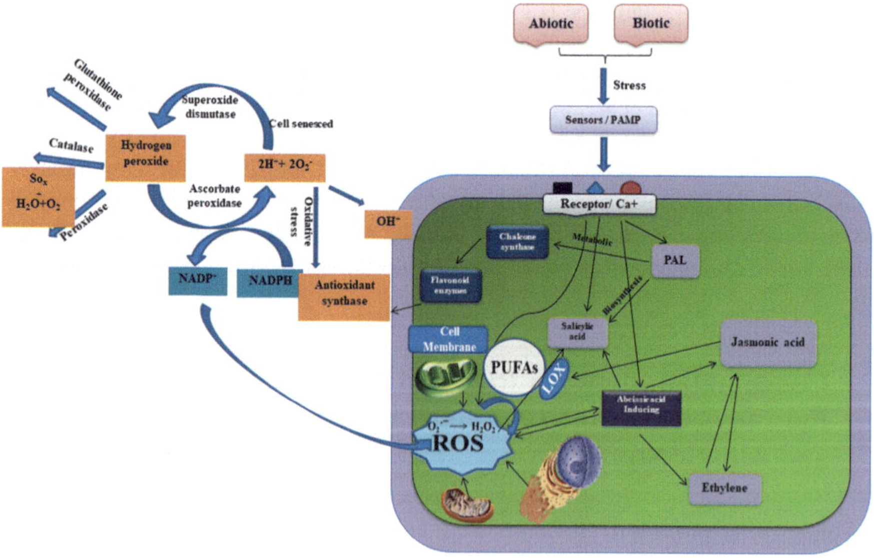

Fig. 12.3 Coordination between ROS and other regulatory pathways under the combined influence of biotic and abiotic stresses

susceptibility of plants to bacterial pathogens under low soil moisture stress has been reported (Mohr and Cahill 2003). In another case, it is found that high soil moisture stresses significantly increase the incidence of collar rot disease (*Sclerotium rolfsii*) in chickpea (Tarafdar et al. 2018) but, on the contrary, lower the dry root rot disease (*Rhizoctonia bataticola*) in similar condition. Although it has been reported that low soil moisture stress improves the defence response in plants against pathogens (Sinha et al. 2016), it is also required to ensure that whether the same conditions favour for the growth of other non-specific pathogen or not. A particular environmental condition may not favour the growth of a specific pathogen of a crop, but a similar condition can be favourable for other pathogens which can easily infect the same crop (Sharma and Ghosh 2017). Therefore, it is crucial to understand the effect of combined stress and respective defensive strategies adopted by the plants to overcome the same. In recent research, it is found that the expression of genes like lipoxygenase (*LOX*), *PO* and *SOD* with exception of *CAT* involved in ROS metabolism pathway was higher in chickpea plants challenged with a combined stress comprising of collar rot disease (biotic) and low soil moisture (abiotic) stress than in chickpea plants that were grown in only biotic stress condition (Tarafdar et al. 2018). Apart from this, it is also found in the same research that the expression of the *LOX* gene in resistant chickpea cultivar is higher in abiotic stress but not in combined stress. The results indicated that in low soil moisture condition, some plant improves their defence mechanism and gain more ability to withstand their biotic stresses (Tarafdar et al. 2018). This coordination may simply suggest that plants are having their own strategies to avoid biotic and abiotic stresses simultaneously by producing proteins that are involved in defence responses. Thus, ROS represent a significant point of convergence among the pathways that are involved in combined stresses. Nevertheless, till now, only a little information is available yet about ROS participation in the crosstalk between biotic and abiotic stresses, and an extensive investigation is needed in future to understand the set of defence mechanisms deployed in the plants. Dissection of data generated through transcriptome, proteome and metabolome analyses in plants would help to elucidate the ROS network components that participate in multiple stress signalling pathways.

8 Modulation of ROS and Involvement of Antioxidant Defence During Nitrogen-Fixing Symbiosis

The role of ROS and antioxidant systems discussed up to this point mainly addresses their involvements in pathogenesis, where initial defence response often manifested as HR. Plant pathogenesis and symbiosis are the variants on a common theme (Baron and Zambryski 1995). The symbiosis between legumes and rhizobia is initiated by the production of Nod factors (NFs) by the bacteria upon sensing the flavonoids present in the root exudates. NFs participate in the bacterial infection as well as trigger nodule organogenesis by dedifferentiation and division of root cortical cells. Within the developing nodules, the bacteria differentiate into bacterioid, able to reduce atmospheric N_2 into NH_4^{2+}. Unlike infection by a pathogen, a close compatible association is established with symbiotic bacteria. However, the existence of HR indicates that the bacteria are initially perceived as invaders by the plant as evidenced by the existence of HR but then successfully overcome the plant defence response due to the production of defence-related protein homologous to plant. Legume root nodules have a high capacity to produce ROS which mainly arises from elevated transcription rates, strong reducing conditions and leghaemoglobin autoxidation process (Becana et al. 2000). Increasing evidence support that successful symbiotic and nodulation efficiency is highly dependent on ROS accumulation and/or GSH synthesis.

ROS production during symbiosis is regulated in a spatio-temporal manner. In *P. vulgaris*, fast and transient ROS changes have been observed at the tip of root hairs within minutes after NFs treatments (Cárdenas et al. 2008). Inhibition of this ROS production prevents root hair curling and formation of infection threads (Peleg-Grossman et al. 2007). This strongly suggests that ROS production is associated with the development of proper symbiotic interaction. During the infection process, oxidation of NBT has been detected in infection thread and infected cells indicating the generation of O_2^- (Santos et al. 2001; Ramu et al. 2002). However, ROS production was not observed in *M. truncatula* when inoculated with *S. meliloti* nodDIABC mutant that is unable to produce NFs, suggesting the role of NFs in oxidative burst (Ramu et al. 2002).

In comparison to O_2 production in the very early stages of symbiotic interaction, H_2O_2 production appears to be inhibited by NFs (Shaw and Long 2003). This also explains the overall increase in H_2O_2 concentration during infection by *S. meliloti* $nodC^-$ mutant, which is defective in NFs biosynthesis (Bueno et al. 2001). The positive role of H_2O_2 in the development of nodule and regulation of infection thread rigidity has been observed; in the similar way, a *S. meliloti* strain, overexpressing a catalase gene, produced a delayed nodulation phenotype associated with enlargement of infection thread (Jamet et al. 2007). Moreover, involvement of H_2O_2 during nodule initiation has been also confirmed by the pharmacological experiment (D'Haeze et al. 2003). Accumulation of ROS (O_2^-, H_2O_2) has been also observed in the latter stages of nodulation process, highlighting the importance of prolonged production of these radicals during nodule development. Cerium chloride staining indicates the presence of H_2O_2 in the ultra-thin section of 6 weeks mature nodules as an electron dense precipitate (Santos et al. 2001; Rubio et al. 2004). Furthermore, H_2O_2 production has been also observed around senescing symbiosomes as a dense cerium precipitate, indicating the involvement of H_2O_2 in nodule senescence process (Alesandrini et al. 2003; Rubio et al. 2004).

GSH is one of the major antioxidant molecules implicated in antioxidant defence system through AsA-GSH cycle. The crucial role of GSH during plant-pathogen interaction has been demonstrated. Presence of hGSH instead of, or in addition to, GSH is one of the specific features of legumes (Frendo et al. 1999; Matamoros et al. 1999). The synthesis of hGSH is catalysed by homoglutathione synthetase (hGSHS) which adds β-alanine to the carboxy-terminal of γ-EC to form hGSH (Frendo et al. 2001).

GSH and hGSH are more abundant in nodules than the other organs of the plant (Frendo et al. 1999; Matamoros et al. 1999).

Treatment with BSO, an inhibitor of γ-ECS, and genetic transformation experiment with GSHS and hGSHS antisense constructs causing the deficiency in GSH and hGSH were carried out in *M. truncatula* inoculated with *S. meliloti,* and the results appeared to inhibit the formation of root nodules reinforcing the importance of GSH and hGSH in early development of nodules (Frendo et al. 2005). Both GSH and hGSH may have a role in the formation and subsequent development of the nodule meristem as they have been shown to be very important in the proper functioning of root tip meristem (Vernoux et al. 2000). The high abundance of both thiols in the meristematic and infected zones of nodules along with the evidence demonstrating the presence of thiol peptides in three to four fold higher concentration in effective than in ineffective nodules clearly indicates that part of GSH and hGSH may be synthesized by the microsymbiont (Frendo et al. 1999; Matamoros et al. 1999; Dalton et al. 1993). Moreover, thiol content and synthesis are found to decline during both natural and induced senescence in a large number of legumes such as soybean (Evans et al. 1999), pea (Groten et al. 2006) or common bean (Loscos et al. 2008). This strongly suggests that N_2-fixing ability of the nodule senescent zone is highly correlated with the diminution of GSH/hGSH pools. Taken together, the above-described data confirms the importance of ROS accumulation and GSH synthesis in establishing and maintaining symbiosis between rhizobia and legumes.

9 Recent Trends in Synthetic Biology to Improve Resistance Response

A recent development in synthetic biology to improve resistance response gives a wide opportunity to the research world to redesign and fabricate the biological components (genetic codes) to optimize them in such a way that they can provide a unique level of protection to the plants against biotic entities and thus make us think and rebuild sustainable agriculture. The effect of single or multiple biotic stresses in plants leads to a variable degree of stress response which can cope up by switching on the resistance mechanisms in some plant species. Under the influence of various types of biotic stresses such as pathogen attack or feeding of plant tissues by insects, plants respond by expressing a wide range of gene families related to defence mechanism. Among them, an oxidative burst of different ROS molecules activates the functioning of a series of antioxidant genes. As the plant grows, this efficient antioxidant system to fight against the toxic substances is deposited during stress and sometimes coordinately works with other defence machinery like pathogenesis-related (PR) genes, activation of resistance gene analogue (RGA) or modulating the synthesis of some phytohormones (jasmonic Acid, salicylic acid, etc.). The overexpression of such antioxidant genes for scavenging the deleterious ROS components in plants by implying latest gene transfer technologies may be capable of buffering the ROS concentration consistently to promote resistance against biotic agents in elite crop species. The recent development in synthetic biology opens a number of possible root of gene transfer, namely, physical and chemical means of gene transfer (particle bombardment method, PEG and liposome-mediated method), *Agrobacterium*- and virus-mediated and even latest CRISPR-CAS 9-based genome-editing approach, ensuring us the gene transfer with precise and cent percent accuracy. Most of the applications in transgenic plants development by incorporating antioxidant genes are typically focused on the enhancement of resistance against abiotic or environment stresses. Biotic stress resistance of transgenic plants governing antioxidant transgenes fished out from other plants or organisms sources has been reported in less frequency than abiotic stress resistance in the last three decades of scientific research development. Here, we have listed some presentable and significant genetic modification events which give us a brief outlook of behavioural pattern of antioxidant genes and their transfer to susceptible plant species to promote biotic stress-resilient transgenic crops (Table 12.2).

10 Conclusion

The present chapter suggests that plant responses to various biotic factors considering other natural conditions are fully orchestrated by a group of complex networking systems in combination with signalling molecules. A number of circumstantial experimental evidence come into a single point of the conclusion that various types of enzymes (mostly antioxidant and regulatory) and functional proteins play key roles in response to most of the biotic stress factors through ROS and other hormone signalling pathways. However, present representation on the crosstalk between different ROS-responsive enzymes and various regulatory pathways in response to biotic stresses is limited. The beginning section of the chapter outlines the appearance of numerous ROS molecules and their function as a marker of stress conditions followed by triggering stress signalling in various way to prevent further damages. This incident is also further associated with the regeneration of various ROS-scavenging molecules, mainly antioxidant enzymatic components. The specific functions, as well as combined activities of these antioxidant enzymes along with other non-antioxidant metabolites, lead to a final compression to cellular damages caused by ROS molecules regenerated due to pathogen attack and other biotic stimuli. Besides, of these damaging activities, ROS also triggers numerous degree of hypersensitive response as a preliminary level of defence response. It also acts as a signal transmitter cum sensor molecule in the establishment of the complex interaction between disease resistance (R) genes and pathogen avirulent (AVR) genes escalating disease response. The latter part of our review mainly highlights the crosstalk between ROS regeneration and scavenging alleys during natural biotic stress involvement in plants. The key mechanism of these complex associations fully controls the spatial and temporal responses, which by default play a signalling cascade

Table 12.2 List of some vital antioxidant genes studied and introduced to numerous crop species displaying resistance to various biotic stresses

Plant/organism sources	Genes encoding antioxidant enzymes	Transgenic plant and target disease(s)/pathogen(s)	References
Aspergillus niger	H_2O_2-generating glucose oxidase	Transgenic potato resistant to *Phytophthora infestans* (late blight), verticillium wilt and *Alternaria solani* (early blight)	Wu et al. (1997)
Tobacco (*Nicotiana tabacum*)	Anionic peroxidase	Transgenic tobacco showing their leaf resistant to *Helicoverpa zea* (corn earworm)	Dowd and Lagrimini (1997)
Pseudomonas pyrrocinia	Chloroperoxidase (CPO-P)	Transgenic tobacco resistant to *Colletotrichum destructivum*	Rajasekaran et al. (2000)
Tobacco	Anionic peroxidase	Transgenic tomato less susceptible to *Helicoverpa zea*	Behle et al. (2002)
Tomato (*Solanum lycopersicum*)	Cytosolic *Cu/Zn-SOD*	Transgenic sugar beets resistant to fungus *Cercospora beticola*	Tertivanidis et al. (2004)
Rice (*Oryza sativa*)	Ten different class III peroxidases (*POX*)	Transgenic transcripts upregulated showing resistance to rice blast	Sasaki et al. (2004)
Wheat (*Triticum aestivum*) and rice	Rice cationic peroxidase (*POC1*) in combination with wheat class IV chitinase + β 1,3 glucanase	Transgenic carrot resistant to grey mould rot, *Botrytis cinerea* and soft rot *Sclerotinia sclerotiorum*	Wally et al. (2009)
Spinach (*Spinacia oleracea) and* pea (*Pisum sativum*)	Cytosolic Cu/ Zn-superoxide dismutase (*cytsod*) and cytosolic ascorbate peroxidase (*cytapx*)	Transgenic tobacco against bacterial wildfire (*Pseudomonas syringae*) and crown gall (*Agrobacterium tumefaciens*)	Faize et al. (2012)
Sugarcane (*Saccharum officinarum*) variety (Yacheng 05–179, Liucheng 03–182)	APX enzyme	Overexpressed in smut resistant cv. Yacheng 05–179 and downregulated in susceptible cv. Liucheng 03–182 inoculated with *Sporisonium scitamineum*	Wang et al. (2015)
Sugarcane variety (ROC22)	L-ascorbate peroxidase 6 gene (*APX6*)	Transgenic *Nicotiana benthamiana* resistant to *Pseudomonas solanacearum* and *Fusarium solani*	Liu et al. (2018)

of the cell-to-cell intercommunication for serving the metabolism highly active under adverse situations. Notwithstanding the interaction between ROS-producing and ROS-scavenging network systems, the association of ROS with other regulatory pathways is also well defined and needs to be explored more to comprehend the key secrets of their delicate balance at the molecular level. More rigorous work effort is needed to clarify how the inbuilt mechanism of ROS production and functions is strictly associated with various biological incidences and environmental factors. In the near future, the current understanding on ROS formation, functions and interaction with other regulatory pathways under biotic stress will make a clear vision on the development of a transgenic approach, which enables the overexpression of multidimensional antioxidant gene cassettes in order to intensify the tolerance of plants against multiple stresses at single or different time points.

References

Alesandrini F, Mathis R, Van de Sype G, Hérouart D, Puppo A (2003) Possible roles for a cysteine protease and hydrogen peroxide in soybean nodule development and senescence. New Phytol 158:131–138

Allan AC, Fluhr R (1997) Two distinct sources of elicited reactive oxygen species in tobacco epidermal cells. Plant Cell 9:1559–1572

Allan AC, Lapidot M, Culver JN, Fluhr R (2001) An early tobacco mosaic virus-induced oxidative burst in tobacco indicates extracellular perception of the virus coat protein. Plant Physiol 126:97–108

Alscher RG, Donahue JL, Cramer CL (1997) Reactive oxygen species and antioxidants: relationships in green cells. Physiol Plant 100:224–233

Amirsadeghi S, Robson CA, Vanlerberghe GC (2007) The role of the mitochondrion in plant responses to biotic stress. Physiol Plant 129:253–266

Apel K, Hirt H (2004) Reactive oxygen species: metabolism, oxidative stress, and signal transduction. Annu Rev Plant Biol 55:373–399

Arimura GI, Maffei ME (2010) Calcium and secondary CPK signalling in plants in response to herbivore attack. Biochem Biophys Res Commun 400:455–460

Arora A, Sairam RK, Srivastava GC (2002) Oxidative stress and antioxidative system in plants. Curr Sci:1227–1238

Asada K (1992) Ascorbate peroxidase-a hydrogen peroxide-scavenging enzyme in plants. Physiol Plant 85:235–241

Asai S, Mase K, Yoshioka H (2010) Role of nitric oxide and reactive oxygen species in disease resistance to necrotrophic pathogens. Plant Signal Behav 5:872–874

Auh CK, Murphy TM (1995) Plasma membrane redox enzyme is involved in the synthesis of O_2^- and H_2O_2 by Phytophthora elicitor-stimulated rose cells. Plant Physiol 107:1241–1247

Bailey-Serres J, Mittler R (2006) The roles of reactive oxygen species in plant cells. Plant Physiol 141:311

Baker CJ, Orlandi EW (1995) Active oxygen in plant pathogenesis. Annu Rev Phytopathol 33:299–321

Baron C, Zambryski PC (1995) The plant response in pathogenesis, symbiosis, and wounding: variations on a common theme? Annu Rev Genet 29:107–129

Becana M, Aparicio-Tejo P, Irigoyen JJ, Sanchez-Diaz M (1986) Some enzymes of hydrogen peroxide metabolism in leaves and root nodules of *Medicago sativa*. Plant Physiol 82:1169–1171

Becana M, Dalton DA, Moran JF, Iturbe-Ormaetxe I, Matamoros MA, Rubio MC (2000) Reactive oxygen species and antioxidants in legume nodules. Physiol Plant 109:372–381

Behle RW, Dowd PF, Tamez-Guerra P, Lagrimini LM (2002) Effect of transgenic plants expressing high levels of a tobacco anionic peroxidase on the toxicity of *Anagrapha falcifera* Nucleopolyhedrovirus to *Helicoverpa zea* (Lepidoptera: Noctuidae). J Econ Entomol 95:81–88

Benikhlef L, L'Haridon F, Abou-Mansour E, Serrano M, Binda M, Costa A, Lehmann S, Métraux JP (2013) Perception of soft mechanical stress in Arabidopsis leaves activates disease resistance. BMC Plant Biol 13:133
Bhattacharjee S (2005) Reactive oxygen species and oxidative burst: roles in stress, senescence and signal transduction in plants. Curr Sci 10:1113–1121
Bhattacharjee S (2010) Sites of generation and physicochemical basis of formation of reactive oxygen species in plant cell. In: Gupta SD (ed) Reactive Oxygen Species and Antioxidants in Higher Plants. Science Publishers: New York, NY, USA, pp 1–30
Bolwell GP, Bindschedler LV, Blee KA, Butt VS, Davies DR, Gardner SL, Gerrish C, Minibayeva F (2002) The apoplastic oxidative burst in response to biotic stress in plants: a three-component system. J Exp Bot 53:1367–1376
Bostock RM (2005) Signal crosstalk and induced resistance: straddling the line between cost and benefit. Annu Rev Phytopathol 43:545–580
Bradley DJ, Kjellbom P, Lamb CJ (1992) Elicitor-and wound-induced oxidative cross-linking of a proline-rich plant cell wall protein: a novel, rapid defence response. Cell 70:21–30
Brisson LF, Tenhaken R, Lamb C (1994) Function of oxidative cross-linking of cell wall structural proteins in plant disease resistance. Plant Cell 6:1703–1712
Bueno P, Soto MJ, Rodríguez-Rosales MP, Sanjuan J, Olivares J, Donaire JP (2001) Time-course of lipoxygenase, antioxidant enzyme activities and H_2O_2 accumulation during the early stages of Rhizobium-legume symbiosis. New Phytol 152:91–96
Cárdenas L, Martínez A, Sánchez F, Quinto C (2008) Fast, transient and specific intracellular ROS changes in living root hair cells responding to Nod factors (NFs). Plant J 56:802–813
Caverzan A, Casassola A, Patussi Brammer S (2016) Reactive oxygen species and antioxidant enzymes involved in plant tolerance to stress. In: Shanker A (ed) Abiotic and biotic stress in plants. Recent advances and future perspectives. InTechOpen, London, pp 463–480
Chen Z, Silva H, Klessig DF (1993) Active oxygen species in the induction of plant systemic acquired resistance by salicylic acid. Science 262:1883–1886
Chisholm ST, Coaker G, Day B, Staskawicz BJ (2006) Host-microbe interactions: shaping the evolution of the plant immune response. Cell 124:803–814
Choi HK, Iandolino A, da Silva FG, Cook DR (2013) Water deficit modulates the response of *Vitis vinifera* to the Pierce's disease pathogen *Xylella fastidiosa*. Mol Plant-Microbe Interact 26:643–657
Clarke SF, Guy PL, Burritt DJ, Jameson PE (2002) Changes in the activities of antioxidant enzymes in response to virus infection and hormone treatment. Physiol Plant 114:157–164
Custers JH, Harrison SJ, Sela-Buurlage MB, Van Deventer E, Lageweg W, Howe PW, Van Der Meijs PJ, Ponstein AS, Simons BH, Melchers LS, Stuiver MH (2004) Isolation and characterisation of a class of carbohydrate oxidases from higher plants, with a role in active defence. Plant J 39:147–160
D'Haeze W, De Rycke R, Mathis R, Goormachtig S, Pagnotta S, Verplancke C, Capoen W, Holsters M (2003) Reactive oxygen species and ethylene play a positive role in lateral root base nodulation of a semiaquatic legume. Proc Natl Acad Sci U S A 100:11789–11794
Dalton DA, Langeberg L, Treneman NC (1993) Correlations between the ascorbate-glutathione pathway and effectiveness in legume root nodules. Physiol Plant 87:365–870
Das K, Roychoudhury A (2014) Reactive oxygen species (ROS) and response of antioxidants as ROS-scavengers during environmental stress in plants. Front Environ Sci 2:53
Dat J, Vandenabeele S, Vranová E, Van Montagu M, Inzé D, Van Breusegem F (2000) Dual action of the active oxygen species during plant stress responses. Cell Mol Life Sci 57:779–795
Dat JF, Pellinen R, Beeckman T, Van De Cotte B, Langebartels C, Kangasjärvi J, Inzé D, Van Breusegem F (2003) Changes in hydrogen peroxide homeostasis trigger an active cell death process in tobacco. Plant J 33:621–632
Delaney TP, Uknes S, Vernooij B, Friedrich L, Weymann K, Negrotto D, Gaffney T, Gut-Rella M, Kessmann H, Ward E, Ryals J (1994) A central role of salicylic acid in plant disease resistance. Science 266:1247–1250

Delledonne M, Xia Y, Dixon RA, Lamb C (1998) Nitric oxide functions as a signal in plant disease resistance. Nature 394:585–588

Delledonne M, Zeier J, Marocco A, Lamb C (2001) Signal interactions between nitric oxide and reactive oxygen intermediates in the plant hypersensitive disease resistance response. Proc Natl Acad Sci U S A 98:13454–13459

Delledonne M, Murgia I, Ederle D, Sbicego PF, Biondani A, Polverari A, Lamb C (2002) Reactive oxygen intermediates modulate nitric oxide signalling in the plant hypersensitive disease-resistance response. Plant Physiol Biochem 40:605–610

Desikan R, Reynolds A, Hancock TJ, Neill JS (1998) Harpin and hydrogen peroxide both initiate programmed cell death but have differential effects on defence gene expression in Arabidopsis suspension cultures. Biochem J 330:115–120

Doke N (1983) Involvement of superoxide anion generation in the hypersensitive response of potato tuber tissues to infection with an incompatible race of *Phytophthora infestans* and to the hyphal wall components. Physiol Plant Pathol 23:345–357

Dowd PF, Lagrimini LM (1997) Examination of different tobacco (*Nicotiana* spp.) types under- and overproducing tobacco anionic peroxidase for their leaf resistance to *Helicoverpa zea*. J Chem Ecol 23:2357–2370

Dubreuil-Maurizi C, Trouvelot S, Frettinger P, Pugin A, Wendehenne D, Poinssot B (2010) β-Aminobutyric acid primes an NADPH oxidase-dependent reactive oxygen species production during grapevine-triggered immunity. Mol Plant-Microbe Interact 23:1012–1021

Durrant WE, Dong X (2004) Systemic acquired resistance. Annu Rev Phytopathol 42:185–209

Ebel J, Mithöfer A (1998) Early events in the elicitation of plant defence. Planta 206:335–348

Evans PJ, Gallesi D, Mathieu C, Hernandez MJ, de Felipe M, Halliwell B, Puppo A (1999) Oxidative stress occurs during soybean nodule senescence. Planta 208:73–79

Faize M, Burgos L, Faize L, Petri C, Barba-Espin G, Diaz-Vivancos P, Clemente-Moreno MJ, Alburquerque N, Hernandez JA (2012) Modulation of tobacco bacterial disease resistance using cytosolic ascorbate peroxidase and Cu, Zn-superoxide dismutase. Plant Pathol 61:858–866

Foyer CH, Harbinson J (1994) Oxygen metabolism and the regulation of photosynthetic electron transport. In: Foyer CH, Mullineaux P (eds) Causes of Photooxidative Stress and Amelioration of Defense Systems in Plants. CRC Press Inc, Boca Raton, FL, pp 1-42

Fraire-Velázquez S, Rodríguez-Guerra R, Sánchez-Calderón L (2011) Abiotic and biotic stress response crosstalk in plants. In: Shanker A (ed) Abiotic stress response in plants-physiological. Biochemical and genetic perspectives. InTech, Rijeka, pp 3–54

Frendo P, Gallesi D, Turnbull R, Van de Sype G, Hérouart D, Puppo A (1999) Localisation of glutathione and homoglutathione in *Medicago truncatula* is correlated to a differential expression of genes involved in their synthesis. Plant J 17:215–219

Frendo P, Jiménez MJ, Mathieu C, Duret L, Gallesi D, Van de Sype G, Hérouart D, Puppo A (2001) A *Medicago truncatula* homoglutathione synthetase is derived from glutathione synthetase by gene duplication. Plant Physiol 126:1706–1715

Frendo P, Harrison J, Norman C, Jiménez MJ, Van de Sype G, Gilabert A, Puppo A (2005) Glutathione and homoglutathione play a critical role in the nodulation process of *Medicago truncatula*. Mol Plant-Microbe Interact 18:254–259

Friedrich L, Vernooij B, Gaffney T, Morse A, Ryals J (1995) Characterization of tobacco plants expressing a bacterial salicylate hydroxylase gene. Plant Mol Biol 29:959–968

Friedrich L, Lawton K, Ruess W, Masner P, Specker N, Rella MG, Meier B, Dincher S, Staub T, Uknes S, Métraux JP (1996) A benzothiadiazole derivative induces systemic acquired resistance in tobacco. Plant J 10:61–70

Frugoli JA, Zhong HH, Nuccio ML, McCourt P, McPeek MA, Thomas TL, McClung CR (1996) Catalase is encoded by a multigene family in *Arabidopsis thaliana* (L.) Heynh. Plant Physiol 112:327–336

Fujita M, Fujita Y, Noutoshi Y, Takahashi F, Narusaka Y, Yamaguchi-Shinozaki K, Shinozaki K (2006) Crosstalk between abiotic and biotic stress responses: a current view from the points of convergence in the stress signalling networks. Curr Opin Plant Biol 9:436–442

Galletti R, Denoux C, Gambetta S, Dewdney J, Ausubel FM, De Lorenzo G, Ferrari S (2008) The AtrbohD-mediated oxidative burst elicited by oligogalacturonides in Arabidopsis is dispensable for the activation of defence responses effective against *Botrytis cinerea*. Plant Physiol 148:1695–1706

Ghosh R, Tarafdar A, Sharma M (2017) Rapid and sensitive diagnoses of dry root rot pathogen of chickpea (*Rhizoctonia bataticola* (Taub.) Butler) using loop-mediated isothermal amplification assay. Sci Rep 7:42737

Gill SS, Tuteja N (2010) Reactive oxygen species and antioxidant machinery in abiotic stress tolerance in crop plants. Plant Physiol Biochem 48:909–930

Govrin EM, Levine A (2000) The hypersensitive response facilitates plant infection by the necrotrophic pathogen *Botrytis cinerea*. Curr Biol 10:751–757

Grant M, Brown I, Adams S, Knight M, Ainslie A, Mansfield J (2000) The *RPM1* plant disease resistance gene facilitates a rapid and sustained increase in cytosolic calcium that is necessary for the oxidative burst and hypersensitive cell death. Plant J 23:441–450

Greenberg JT, Yao N (2004) The role and regulation of programmed cell death in plant-pathogen interactions. Curr Biol 6:201–211

Großkinsky DK, Koffler BE, Roitsch T, Maier R, Zechmann B (2012) Compartment-specific antioxidative defence in Arabidopsis against virulent and avirulent *Pseudomonas syringae*. Phytopathology 102:662–673

Groten K, Dutilleul C, van Heerden PD, Vanacker H, Bernard S, Finkemeier I, Dietz KJ, Foyer CH (2006) Redox regulation of peroxiredoxin and proteinases by ascorbate and thiols during pea root nodule senescence. FEBS Lett 580:1269–1276

Guo M, Guo W, Chen Y, Dong S, Zhang X, Zhang H, Song W, Wang W, Wang Q, Lv R, Zhang Z (2010) The basic leucine zipper transcription factor Moatf1 mediates oxidative stress responses and is necessary for full virulence of the rice blast fungus *Magnaporthe oryzae*. Mol Plant-Microbe Interact 23:1053–1068

Hammond-Kosack KE, Parker JE (2003) Deciphering plant–pathogen communication: fresh perspectives for molecular resistance breeding. Curr Opin Biotechnol 14:177–193

Hancock JT, Desikan R, Clarke A, Hurst RD, Neill SJ (2002) Cell signalling following plant/pathogen interactions involves the generation of reactive oxygen and reactive nitrogen species. Plant Physiol Biochem 40:611–617

Halliwell B, Gutteridge JM (1999) Free Radicals in Biology and Medicine. Oxford, UK, Oxford University Press

Hideg É, Barta C, Kálai T, Vass I, Hideg K, Asada K (2002) Detection of singlet oxygen and superoxide with fluorescent sensors in leaves under stress by photoinhibition or UV radiation. Plant Cell Physiol 43:1154–1164

Hoeberichts FA, Ten Have A, Woltering EJ (2003) A tomato metacaspase gene is upregulated during programmed cell death in *Botrytis cinerea*-infected leaves. Planta 217:517–522

Hückelhoven R, Kogel KH (2003) Reactive oxygen intermediates in plant-microbe interactions: who is who in powdery mildew resistance? Planta 216:891–902

Hückelhoven R, Fodor J, Preis C, Kogel KH (1999) Hypersensitive cell death and papilla formation in barley attacked by the powdery mildew fungus are associated with hydrogen peroxide but not with salicylic acid accumulation. Plant Physiol 119:1251–1260

Iwano M, Che FS, Goto K, Tanaka N, Takayama S, Isogai A (2002) Electron microscopic analysis of the H_2O_2 accumulation preceding hypersensitive cell death induced by an incompatible strain of *Pseudomonas avenae* in cultured rice cells. Mol Plant Pathol 3:1–8

Imlay JA (2008) Cellular defenses against Superoxide and Hydrogen Peroxide. Annu Rev Biochem 77(1):755–776

Jabs T (1999) Reactive oxygen intermediates as mediators of programmed cell death in plants and animals. Biochem Pharmacol 57:231–245

Jabs T, Dietrich RA, Dangl JL (1996) Initiation of runaway cell death in an Arabidopsis mutant by extracellular superoxide. Science 273:1853–1856

James D, Tarafdar A, Biswas K, Sathyavathi TC, Padaria JC, Kumar PA (2015) Development and characterization of a high temperature stress responsive subtractive cDNA library in Pearl Millet *Pennisetum glaucum* (L.) R. Br. Indian J Exp Biol 53:543–550

Jamet A, Mandon K, Puppo A, Hérouart D (2007) H_2O_2 is required for optimal establishment of the *Medicago sativa/Sinorhizobium meliloti* symbiosis. J Bacteriol 189:8741–8745

Kadioglu A, Terzi R, Saruhan N, Saglam A (2012) Current advances in the investigation of leaf rolling caused by biotic and abiotic stress factors. Plant Sci 182:42–48

Kanzaki H, Saitoh H, Ito A, Fujisawa S, Kamoun S, Katou S, Yoshioka H, Terauchi R (2003) Cytosolic HSP90 and HSP70 are essential components of INF1-mediated hypersensitive response and non-host resistance to *Pseudomonas cichorii* in *Nicotiana benthamiana*. Mol Plant Pathol 4:383–391

Kariola T, Brader G, Li J, Palva ET (2005) Chlorophyllase 1, a damage control enzyme, affects the balance between defence pathways in plants. Plant Cell 17:282–294

Kawarazaki T, Kimura S, Iizuka A, Hanamata S, Nibori H, Michikawa M, Imai A, Abe M, Kaya H, Kuchitsu K (2013) A low temperature-inducible protein AtSRC2 enhances the ROS-producing activity of NADPH oxidase AtRbohF. Biochim Biophys Acta 1833:2775–2780

Keunen EL, Peshev D, Vangronsveld J, Van Den Ende WI, Cuypers AN (2013) Plant sugars are crucial players in the oxidative challenge during abiotic stress: extending the traditional concept. Plant Cell Environ 36:1242–1255

Kim MC, Chung WS, Yun DJ, Cho MJ (2009) Calcium and calmodulin-mediated regulation of gene expression in plants. Mol Plant 2:13–21

Kovtun Y, Chiu WL, Tena G, Sheen J (2000) Functional analysis of oxidative stress- activated mitogen-activated protein kinase cascade in plants. Proc Natl Acad Sci U S A 97:2940–2945

Kumar R, Pankaj VP, Tarafdar A, Biswas K, Kumar S (2016) Soil microbes and their interaction with plants. In: Mitra R, Barman A (eds) Plant pathogen interaction: recent trends. Sharma Publications & Distributors, New Delhi, pp 1–46

Kumudini BS, Shetty HS (2002) Association of lignification and callose deposition with host cultivar resistance and induced systemic resistance in pearl millet to *Sclerospora graminicola*. Aust Plant Pathol 31:157–164

Kunz C, Vandelle E, Rolland S, Poinssot B, Bruel C, Cimerman A, Zotti C, Moreau E, Vedel R, Pugin A, Boccara M (2006) Characterization of a new, non-pathogenic mutant of *Botrytis cinerea* with impaired plant colonization capacity. New Phytol 170:537–550

Kuźniak E, Skłodowska M (2005) Fungal pathogen-induced changes in the antioxidant systems of leaf peroxisomes from infected tomato plants. Planta 222:192–200

L'Haridon F, Besson-Bard A, Binda M, Serrano M, Abou-Mansour E, Balet F, Schoonbeek HJ, Hess S, Mir R, Léon J, Lamotte O (2011) A permeable cuticle is associated with the release of reactive oxygen species and induction of innate immunity. PLoS Pathog 7:e1002148

Lam E, Kato N, Lawton M (2001) Programmed cell death, mitochondria and the plant hypersensitive response. Nature 411:848–853

Lamb C, Dixon RA (1997) The oxidative burst in plant disease resistance. Annu Rev Plant Biol 48:251–275

Lawton K, Weymann K, Friedrich L, Vernooij B, Uknes S, Ryals J (1995) Systemic acquired resistance in Arabidopsis requires salicylic acid but not ethylene. Mol Plant-Microbe Interact 8:863–870

Legendre L, Rueter S, Heinstein PF, Low PS (1993) Characterization of the oligogalacturonide-induced oxidative burst in cultured soybean (*Glycine max*) cells. Plant Physiol 102:233–240

Lehotai N, Pető A, Bajkán S, Erdei L, Tari I, Kolbert Z (2011) *In vivo* and *in situ* visualization of early physiological events induced by heavy metals in pea root meristem. Acta Physiol Plant 33:2199–2207

Leon J, Lawton MA, Raskin I (1995) Hydrogen peroxide stimulates salicylic acid biosynthesis in tobacco. Plant Physiol 108:1673–1678

Levine A, Tenhaken R, Dixon R, Lamb C (1994) H_2O_2 from the oxidative burst orchestrates the plant hypersensitive disease resistance response. Cell 79:583–593

Li J, Zhang ZG, Ji R, Wang YC, Zheng XB (2005) Hydrogen peroxide regulates elicitor PB90-induced cell death and defence in non-heading Chinese cabbage. Physiol Mol Plant Pathol 67:220–230

Liu F, Huang N, Wang L, Ling H, Sun T, Ahmad W, Muhammad K, Guo J, Xu L, Gao S, Que Y, Su Y (2018) A novel L-ascorbate peroxidase 6 gene, *ScAPX6*, plays an important role in the regulation of response to biotic and abiotic stresses in sugarcane. Front Plant Sci 8:2262

Lorrain S, Vailleau F, Balagué C, Roby D (2003) Lesion mimic mutants: keys for deciphering cell death and defence pathways in plants? Trends Plant Sci 8:263–271

Loscos J, Matamoros MA, Becana M (2008) Ascorbate and homoglutathione metabolism in common bean nodules under stress conditions and during natural senescence. Plant Physiol 146:1282–1292

Low PS, Merida JR (1996) The oxidative burst in plant defence: function and signal transduction. Physiol Plant 96:533–542

Madgwick JW, West JS, White RP, Semenov MA, Townsend JA, Turner JA, Fitt BD (2011) Impacts of climate change on wheat anthesis and fusarium ear blight in the UK. Eur J Plant Pathol 130:117–131

Małolepsza U (2005) Spatial and temporal variation of reactive oxygen species and antioxidant enzymes in o-hydroxyethylorutin-treated tomato leaves inoculated with *Botrytis cinerea*. Plant Pathol 54:317–324

Małolepsza U, Rozalska S (2005) Nitric oxide and hydrogen peroxide in tomato resistance. Nitric oxide modulates hydrogen peroxide level in o-hydroxyethylorutin-induced resistance to *Botrytis cinerea* in tomato. Plant Physiol Biochem 43:623–635

Matamoros MA, Moran JF, Iturbe Ormaetxe I, Rubio MC, Becana M (1999) Glutathione and homoglutathione synthesis in legume root nodules. Plant Physiol 121:879–888

Mateo A, Mühlenbock P, Rustérucci C, Chang CC, Miszalski Z, Karpinska B, Parker JE, Mullineaux PM, Karpinski S (2004) LESION SIMULATING DISEASE 1 is required for acclimation to conditions that promote excess excitation energy. Plant Physiol 136:2818–2830

Mauch-Mani B, Mauch F (2005) The role of abscisic acid in plant-pathogen interactions. Curr Opin Plant Biol 8:409–414

Mayer AM, Staples RC, Gil-ad NL (2001) Mechanisms of survival of necrotrophic fungal plant pathogens in hosts expressing the hypersensitive response. Phytochemistry 58:33–41

McCord JM, Fridovich I (1969) Superoxide dismutase an enzymic function for erythrocuprein Hemocuprein. J Biol Chem 244:6049–6055

Mellersh DG, Foulds IV, Higgins VJ, Heath MC (2002) H_2O_2 plays different roles in determining penetration failure in three diverse plant-fungal interactions. The Plant J 29:257–268

Messner B, Boll M (1994) Cell suspension cultures of spruce (*Picea abies*): inactivation of extracellular enzymes by fungal elicitor-induced transient release of hydrogen peroxide (oxidative burst). Plant Cell 39:69–78

Miguel E, Poza-Carrión C, López-Solanilla E, Aguilar I, Llama Palacios A, García-Olmedo F, Rodríguez-Palenzuela P (2000) Evidence against a direct antimicrobial role of H_2O_2 in the infection of plants by *Erwinia chrysanthemi*. Mol Plant-Microbe Interact 13:421–429

Mithofer A, Schulze B, Boland W (2004) Biotic and heavy metal stress response in plants: evidence for common signals. FEBS Lett 566:1–5

Mittler R, Vanderauwera S, Gollery M, Van Breusegem F (2004) Reactive oxygen gene network of plants. Trends Plant Sci 9:490–498

Mohr PG, Cahill DM (2003) Abscisic acid influences the susceptibility of *Arabidopsis thaliana* to *Pseudomonas syringae* pv. tomato and *Peronospora parasitica*. Funct Plant Biol 30:461–469

Montillet JL, Chamnongpol S, Rustérucci C, Dat J, Van De Cotte B, Agnel JP, Battesti C, Inzé D, Van Breusegem F, Triantaphylides C (2005) Fatty acid hydroperoxides and H_2O_2 in the execution of hypersensitive cell death in tobacco leaves. Plant Physiol 138:1516–1526

Mou Z, Fan W, Dong X (2003) Inducers of plant systemic acquired resistance regulate NPR1 function through redox changes. Cell 113:935–944

Møller IM, Jensen PE, Hansson A (2007) Oxidative Modifications to Cellular Components in Plants. Annu Rev Plant Biol 58(1):459–481

Navrot N, Collin V, Gualberto J, Gelhaye E, Hirasawa M, Rey P, Knaff DB, Issakidis E, Jacquot JP, Rouhier N (2006) Plant glutathione peroxidases are functional peroxiredoxins distributed in several subcellular compartments and regulated during biotic and abiotic stresses. Plant Physiol 142:1364–1379

Noctor G, Mhamdi A, Foyer CH (2014) The roles of reactive oxygen metabolism in drought: not so cut and dried. Plant Physiol 164:1636–1648

Nürnberger T, Nennstiel D, Jabs T, Sacks WR, Hahlbrock K, Scheel D (1994) High affinity binding of a fungal oligopeptide elicitor to parsley plasma membranes triggers multiple defence responses. Cell 78:449–460

Olson PD, Varner JE (1993) Hydrogen peroxide and lignification. Plant J 4:887–892

Padaria JC, Thuy NT, Tarafdar A, Yadav R (2015) Development of a heat-responsive cDNA library from *Prosopis cineraria* and molecular characterisation of the *Pchsp17.9* gene. J Hortic Sci Biotechnol 90:318–234

Padaria JC, Yadav R, Tarafdar A, Lone SA, Kumar K, Sivalingam PN (2016) Molecular cloning and characterization of drought stress responsive abscisic acid-stress-ripening (*Asr1*) gene from wild jujube, *Ziziphus nummularia* (Burm. f.) Wight & Arn. Mol Biol Rep 43:849–859

Pandey P, Ramegowda V, Senthil-Kumar M (2015) Shared and unique responses of plants to multiple individual stresses and stress combinations: physiological and molecular mechanisms. Front Plant Sci 6:723

Peleg-Grossman S, Volpin H, Levine A (2007) Root hair curling and Rhizobium infection in *Medicago truncatula* are mediated by phosphatidylinositide-regulated endocytosis and reactive oxygen species. J Exp Bot 58:1637–1649

Plancot B, Santaella C, Jaber R, Kiefer-Meyer MC, Follet-Gueye ML, Leprince J, Gattin I, Souc C, Driouich A, Vicré-Gibouin M (2013) Deciphering the responses of root border-like cells of Arabidopsis and flax to pathogen-derived elicitors. Plant Physiol 163:1584–1597

Rajasekaran K, Cary JW, Jacks TJ, Stromberg KD, Cleveland TE (2000) Inhibition of fungal growth in plants and in vitro by transgenic tobacco expressing a bacterial nonheme chloroperoxidase gene. Plant Cell Rep 19:333–338

Ramu SK, Peng HM, Cook DR (2002) Nod factor induction of reactive oxygen species production is correlated with expression of the early nodulin gene *rip1* in *Medicago truncatula*. Mol Plant-Microbe Interact 15:522–528

Rejeb IB, Pastor V, Mauch-Mani B (2014) Plant responses to simultaneous biotic and abiotic stress: molecular mechanisms. Plan Theory 3:458–475

Repka V (2002) Hydrogen peroxide generated via the octadecanoid pathway is neither necessary nor sufficient for methyl jasmonate-induced hypersensitive cell death in woody plants. Biol Plantarum 45:105–115

Roychowdhury R, Khan MH, Choudhury S (2018) Arsenic in rice: an overview on stress implications, tolerance and mitigation strategies. In: Hasanuzzaman M, Nahar K, Fujita M (eds) Plants under metal and metalloid stress. Springer, Singapore, pp 401–415

Roychowdhury R, Khan MH, Choudhury S (2019) Physiological and molecular responses for metalloid stress in rice - a comprehensive overview. In: Hasanuzzaman M, Fujita M, Nahar K, Biswas J (eds) Advances in rice research for abiotic stress tolerance. Woodhead Publishing, Elsevier, pp 341–369

Rubio MC, James EK, Clemente MR, Bucciarelli B, Fedorova M, Vance CP, Becana M (2004) Localization of superoxide dismutases and hydrogen peroxide in legume root nodules. Mol Plant-Microbe Interact 17:1294–1305

Ryals JA, Neuenschwander UH, Willits MG, Molina A, Steiner HY, Hunt MD (1996) Systemic acquired resistance. Plant Cell 8:1809–1819

Santos R, Hérouart D, Sigaud S, Touati D, Puppo A (2001) Oxidative burst in alfalfa- *Sinorhizobium meliloti* symbiotic interaction. Mol Plant-Microbe Interact 14:86–89

Sasabe M, Takeuchi K, Kamoun S, Ichinose Y, Govers F, Toyoda K, Shiraishi T, Yamada T (2000) Independent pathways leading to apoptotic cell death, oxidative burst and defence gene expression in response to elicitin in tobacco cell suspension culture. Eur J Biochem 267:5005–5013

Sasaki K, Iwai T, Hiraga S, Kuroda K, Seo S, Mitsuhara I, Miyasaka A, Iwano M, Ito H, Matsui H, Ohashi Y (2004) Ten rice peroxidases redundantly respond to multiple stresses including infection with rice blast fungus. Plant Cell Physiol 45:1442–1452

Shadel GS, Horvath TL (2015) Mitochondrial ROS signalling in organismal homeostasis. Cell 163:560–569

Sharma M, Ghosh R (2017) Heat and soil moisture stress differentially impact chickpea plant infection with fungal pathogens. In: Senthil-Kumar M (ed) Plant tolerance to individual and concurrent stresses. Springer, New Delhi, pp 47–57

Sharma M, Ghosh R, Tarafdar A, Telangre R (2015) An efficient method for zoospore production, infection and real-time quantification of *Phytophthora cajani* causing Phytophthora blight disease in pigeonpea under elevated atmospheric CO_2. BMC Plant Biol 15:90

Sharma S, Kumar V, Tripathi RB (2017) Isolation of phosphate solubilizing microorganism (PSMs) from soil. J Microbiol Biotechnol Res 1:90–95

Shaw SL, Long SR (2003) Nod factor inhibition of reactive oxygen efflux in a host legume. Plant Physiol 132:2196–2204

Shetty NP, Kristensen BK, Newman MA, Møller K, Gregersen PL, Jørgensen HL (2003) Association of hydrogen peroxide with restriction of *Septoria tritici* in resistant wheat. Physiol Mol Plant Pathol 62:333–346

Shetty NP, Mehrabi R, Lütken H, Haldrup A, Kema GH, Collinge DB, Jørgensen HJ (2007) Role of hydrogen peroxide during the interaction between the hemibiotrophic fungal pathogen *Septoria tritici* and wheat. New Phytol 174:637–647

Shetty NP, Jørgensen HJ, Jensen JD, Collinge DB, Shetty HS (2008) Roles of reactive oxygen species in interactions between plants and pathogens. Eur J Plant Pathol 121:267–280

Simon UK, Polanschütz LM, Koffler BE, Zechmann B (2013) High resolution imaging of temporal and spatial changes of subcellular ascorbate, glutathione and H_2O_2 distribution during *Botrytis cinerea* infection in Arabidopsis. PLoS One 8:e65811

Sinha R, Gupta A, Senthil-Kumar M (2016) Understanding the impact of drought on foliar and xylem invading bacterial pathogen stress in chickpea. Front Plant Sci 7:902

Suzuki N, Mittler R (2006) Reactive oxygen species and temperature stresses: a delicate balance between signalling and destruction. Physiol Plant 126:45–51

Suzuki N, Rivero RM, Shulaev V, Blumwald E, Mittler R (2014) Abiotic and biotic stress combinations. New Phytol 1:32–43

Tarafdar A, Rani TS, Chandran US, Ghosh R, Chobe DR, Sharma M (2018) Exploring combined effect of abiotic (soil moisture) and biotic *(Sclerotium rolfsii* Sacc.) stress on collar rot development in chickpea. Front Plant Sci 9:1154

Taylor NL, Day DA, Millar AH (2002) Environmental stress causes oxidative damage to plant mitochondria leading to inhibition of glycine decarboxylase. J Biol Chem 277:42663–42668

Tertivanidis K, Goudoula C, Vasilikiotis C, Hassiotou E, Perl-Treves R, Tsaftaris A (2004) Superoxide dismutase transgenes in sugarbeets confer resistance to oxidative agents and the fungus *C. beticola*. Transgenic Res 13:225–233

Thordal-Christensen H, Zhang Z, Wei Y, Collinge DB (1997) Subcellular localization of H_2O_2 in plants.H_2O_2 accumulation in papillae and hypersensitive response during the barley-powdery mildew interaction. Plant J 11:1187–1194

Tiedemann AV (1997) Evidence for a primary role of active oxygen species in induction of host cell death during infection of bean leaves with *Botrytis cinerea*. Physiol Mol Plant Pathol 50:151–166

Torres MA, Dangl JL (2005) Functions of the respiratory burst oxidase in biotic interactions, abiotic stress and development. Curr Opin Plant Biol 8:397–403

Torres MA, Jones JD, Dangl JL (2005) Pathogen-induced, NADPH oxidase-derived reactive oxygen intermediates suppress spread of cell death in *Arabidopsis thaliana*. Nat Genet 37:1130–1134

Torres MA, Jones JD, Dangl JL (2006) Reactive oxygen species signalling in response to pathogens. Plant Physiol 141:373–378

Valko MM, Morris H, Cronin MT (2005) Metals, toxicity and oxidative stress. Curr Med Chem 12:1161–1208

Varnova E, Van Breusegem F, Dat J, Belles-Boix E, Inze D (2002) The role of reactive oxygen species in signal transduction. In: Scheel D, Wasternack C (eds) Plant Signal Transduction. Oxford University press, pp 41–73

Van Breusegem F, Dat JF (2006) Reactive oxygen species in plant cell death. Plant Physiol 141:384–390

Van't Slot KA, Knogge W (2002) A dual role for microbial pathogen-derived effector proteins in plant disease and resistance. Crit Rev Plant Sci 21:229–271

Vernoux T, Wilson RC, Seeley KA, Reichheld JP, Muroy S, Brown S, Maughan SC, Cobbett CS, Van Montagu M, Inzé D, May MJ (2000) The ROOT MERISTEMLESS1/CADMIUM SENSITIVE2 gene defines a glutathione-dependent pathway involved in initiation and maintenance of cell division during postembryonic root development. Plant Cell 12:97–109

Wally O, Jayaraj J, Punja Z (2009) Comparative resistance to foliar fungal pathogens in transgenic carrot plants expressing genes encoding for chitinase, b-1, 3-glucanase and peroxidise. Eur J Plant Pathol 123:331–342

Walters DR (2003) Polyamines and plant disease. Phytochemistry 64:97–107

Wang ZQ, Chen Y, Yang YT, Su YC, Chen SS, Wu QB et al (2015) Cloning and expression analysis of ascorbate peroxidase gene (*ScAPX*) in sugarcane (*Saccharum officinarum*). J Agric Biotechnol 23:170–180

Ward ER, Uknes SJ, Williams SC, Dincher SS, Wiederhold DL, Alexander DC, Ahl-Goy P, Métraux JP, Ryals JA (1991) Coordinate gene activity in response to agents that induce systemic acquired resistance. Plant Cell 3:1085–1094

Wojtaszek P (1997) Oxidative burst: an early plant response to pathogen infection. Biochem J 322:681–692

Wu GS, Shortt BJ, Lawrence EB, Leon J, Fitzsimmons KC, Levine EB, Raskin I, Shah DM (1997) Activation of host defence mechanisms by elevated production of H_2O_2 in transgenic plants. Plant Physiol 115:427–435

Xia XJ, Wang YJ, Zhou YH, Tao Y, Mao WH, Shi K, Asami T, Chen Z, Yu JQ (2009) Reactive oxygen species are involved in brassinosteroid-induced stress tolerance in cucumber. Plant Physiol 150:801–814

Zhang L, Oh Y, Li H, Baldwin IT, Galis I (2012) Alternative oxidase in resistance to biotic stresses: *Nicotiana attenuata* AOX contributes to resistance to a pathogen and a piercing-sucking insect but not *Manduca sexta* larvae. Plant Physiol 160:1453–1467

Chapter 13
Climate Change-Induced Heavy Metal (or Metalloid) Stress in Crop Plants and Possible Mitigation Strategies

Navin Kumar, Shilpi Paul, and Shekhar Mallick

Abstract Among the various changes which climate change would bring about, i.e., extreme weather, migration of population, shortage of food, and an outbreak of disease, the increase of heavy metal(loid)s in soils will also be another indirect effect as the soils would dry up. In such scenario, not only the productivity of crop plants would be compromised due to a changing environment, but also the food obtained will be having higher loads of heavy metal(loid)s. In addition, the plants would have to overcome the oxidative stress due to the excessive metal uptake. Abiotic stress due to toxic heavy metal(loid) like As, Cd, Pb, Hg, etc. remains an elusive and invisible challenge to the plants in general and crop plants in particular. Although plants has interconnected tolerance/s defense mechanisms, i.e., redox detoxification, glyoxalase, phytochelatin and metallothionein-mediated vacuolar sequestration, glutaredoxin-mediated pathways, under excess metal levels, the plant suffers from oxidative damage. The challenge becomes more acute under the extreme climatic conditions, demanding strategies to overcome such situations. Traditional breeding methods have a limited potential to improve plant genomes against environmental stresses, while the advent of recent biotechnological breakthroughs, i.e., genetically engineered varieties of different crops such as cotton, maize, rice, canola, and soybean and genome editing through CRISPER-Cas9, holds the answer. Several strategies can be adopted to achieve this, for example, engineering of tolerance pathways, like antioxidant enzymes, osmolyte accumulation, membrane-localized transporters, accumulation of essential elements, resistance against pests or pathogens and introgression of transporters and their overexpression, efficient nitrogen cycling, ascorbate-glutathione pathway, and regulation of stress-inducible phytohormones for climate resilient crops. In view of these above views, this chapter comprises (i) metal/metalloid and their source of induction in the environment, (ii) mode of uptake by the plants, (iii) affected crops,

N. Kumar · S. Mallick
Plant Ecology and Climate Change Science, National Botanical Research Institute, Lucknow, India

S. Paul (✉)
Science and Engineering Research Board, New-Delhi, India

R. Roychowdhury et al. (eds.), *Sustainable Agriculture in the Era of Climate Change*, https://doi.org/10.1007/978-3-030-45669-6_13

(iv) metabolic pathways, (v) current investigations, (vi) mitigation strategies, (vii) and overall conclusion. Although climate change does not directly influence plant's heavy metal(loid) uptake, in the environment, every aspect is interconnected. This chapter attempts to comprehensively account for all the strategies known for countering excess heavy metal(loid) uptake in plants.

Keywords Heavy metal(loid)s · Abiotic stress · Climate change · Glyoxalase · Metallothionein · Asada cycle · Phytohormones · Dehydrins

1 Introduction

Human population is projected to increase to 9.8 billion by 2050 from the present 7.7 billion as of February 2019 (un.org), and this would be accompanied by equivalent urbanization of forest and arable land for industrialization and dwelling. These events would directly or indirectly govern the environmental changes both at the local level, i.e., extreme drought and water logging, and also at the global level, i.e., climate change and biodiversity shifts (Chakraborty et al. 2014; Roychowdhury 2014). In addition, according to UN intergovernmental panel on climate change (IPCC), global temperature is projected to increase by 1.8–4.0 °C by the end of the twenty-first century with the higher warming projected for high latitudes. At a global scale, regions which fall between the equator and the 30° north and south latitudes, mid-latitudes regions, i.e., North and South America, northern Europe, and northern and central Asia, are projected to experience increased precipitation, while others, including southern Africa and Asia and the Mediterranean, are expected to experience substantial droughts (Noyes et al. 2009).

Although the heavy metal(loid) pollution and climate change are two different aspects which may not appear to be interconnected, not many studies are available (Roychowdhury and Tah 2011; Basu et al. 2012). However, few studies appear to indicate that with the advent of global climate rise attributed for climate change, the release of metals from a bound state to the available state would rise and subsequently increase in the uptake in plants. In the Dommel catchment in the Netherlands, increase in the levels of Cd and Zn in surface waters has been projected in the discharge runoff toward the end of the twenty-first century as a result of breakthrough of Cd and Zn in the soil system by 820% and 570%, respectively (Wijngaard et al. 2017). In another study on four vegetables in Northwest China, temperature played a major role in the level of Cd, Pb, Cu, and Zn accumulation (Yu et al. 2013). The freshwater contaminations with, for example, heavy metal(loid)s would also further exacerbate due to the disposal of chemical wastes and industrial effluents and over demand of freshwater and absence of insufficient treatment technologies (Singh and Mosley 2003).

The mobility of heavy metal(loid)s in the soil and surface water is greatly governed by soil biogeochemical processes, including sorption/desorption, complexation, dissolution/precipitation, and uptake/release by biota, control, and thus its residence time in soils and water. The mineral-concentrated water and high temperature consequently increase the bioavailability of metal/metalloids to plants in general and crops in particular. The soil-solution partitioning in the soil is controlled by pH, redox potential, clay and soil organic matter (SOM) content, and the concentration of complex organic or inorganic ligands and its competing cations. Future climate scenario is projected to increase the global surface temperature which will eventually enhance evapotranspiration in certain regions, while certain regions will encounter heavy precipitation events (Seneviratne et al. 2012). During floods, the concentrations of heavy metal(loid) increase due to desorption or resuspension, while during droughts, eutrophication increases heavy metal(loid) concentrations possibly due to a decrease in dilution. The climate change also alters the physiological as well as metabolic responses in the crops; for example, the high temperature enhances root exudation of organic compounds which accounts for carbon loss by liberating organic compounds from mineral associations (Keiluweit et al. 2015). Additionally, the change in temperature could reduce the capacity to retain heavy metal(loid)s in soil, which enhances its bioavailability (Fritioff et al. 2005). Raising climate temperature could also affect uptake balance of cations and anions from the rhizosphere (Marschner 2011; Hasanuzzaman et al. 2013). This process would likely decrease the pH of the pore water in the rhizosphere. In addition, the enhanced plant transpiration may enhance the metals/metalloids uptake (Qaderi et al. 2012; Yin et al. 2013). The metals such as Zn^{2+}, Cu^{2+}, Fe^{2+}, Mn^{2+}, Co^{2+}, and Ni^{2+} are essential micronutrients for plants development. The divalent metal cations are the prime centers of different enzymes, transcription factors, and the regulatory mechanisms, which are involved in the metal ion homeostasis in plants. However, above the optimum level, these essential metals and nonessential toxic metals, such as Hg^{2+}, Ag^{2+}, Cd^{2+}, and Pb^{2+}, are toxic to the crops (Williams et al. 2000). In crops, the uptake of metals is facilitated by the selective transporters and some housekeeping proteins, which maintain the cellular metal(loid) homeostasis (González-Guerrero et al. 2016). On the other hand, above the optimum level, the accumulation of metal(loid)s in crops elevates the level of reactive oxygen species (ROS). The ROS such as superoxide ($O2^-$), hydrogen peroxide (H2O2), and singlet oxygen ($^1O2$), directly disrupting the structure of the biomolecules, resulted into the inhibition of physiological as well as the biochemical process in the crops (Foyer and Shigeoka 2011). Owing to these toxic manifestations, crop growth and yields are reported to reduce. To counter this oxidative stress, plant adopts different stress ameliorative strategies. Antioxidants system (comprising of enzymatic and nonenzymatic antioxidants) directly counters the elevated level of ROS (Das and Roychoudhury 2014). Simultaneously, the biosynthesis of diverse cellular biomolecules also regulates the tolerance response in the crops. Compounds such as polyamines, mugineic acids, organic acids, phytochelatins (PC), dehydrins, and metallothioneins are involved in stress amelioration either by scavenging the ROS or sequestering the toxic cations itself. Certain cellular secondary metabolites, such as flavonoid and phenolic compounds, and amino acids, such as cysteine and proline, are also reported to take part in defense mechanism. Phytohormones such as salicylic acid, jasmonic acid, abscisic acid, auxin, and ethylene are elevated against metals/metalloids stress (Kumar et al. 2019; Ghasemzadeh and Ghasemzadeh 2011; Al-Whaibi 2011; Suzuki et al. 2006). On the other hand, during metal/metalloid stress, the reduced state of cell slightly changes to the oxidizing state where cellular level of reducing agents such as glutaredoxin (GRX), glutathione (GSH), and ascorbic acid elevates to maintain cellular reduced environment (Foyer and Noctor 2011). In this chapter, we briefly describe the different strategies adopted by the plants involved in the stress ameliorative response against metal/metalloid stress, which could possibly be explored for developing climate-resilient crops.

2 Antioxidants Against Metals/Metalloids Stress

The antioxidants system contains several enzymatic and nonenzymatic compounds, i.e., glutathione, ascorbic acid, α-tocopherol, carotenoids, and phenolic compounds, to counter oxidative stress induced by the elevated accumulation of metals/metalloids (Fig. 13.1). In the enzymatic antioxidant system, superoxide dismutase (SOD) is considered first line of defense, which converts $O2^-$ to less reactive and decomposable H2O2. Superoxide dismutase activity varies quite widely with metal(loid)s accumulation in different plants. In *Arabidopsis*, genes encoding the three classes of SOD (FeSOD, MnSOD, Cu/ZnSOD) respond to metal(loid)s differentially at the transcript level (Abercrombie et al. 2008). Transcripts for genes encoding a chloroplastic and a cytosolic Cu/ZnSOD induced more than twofold by metalloid exposure, while transcripts for a FeSOD were downregulated about fivefold (Abercrombie et al. 2008). Hydrogen peroxide produced in a plant cell is neutralized by enzyme

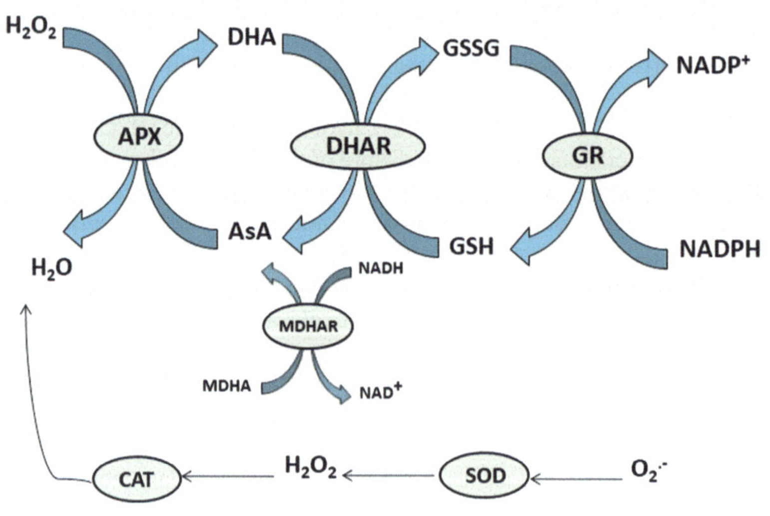

Fig. 13.1 The schematic diagram of enzymatic antioxidant defense mechanism

action such as catalase (CAT), ascorbate peroxidase (APX), glutathione peroxidase (GPX), etc. (Mittler 2002; Roychowdhury et al. 2018, 2019). In addition to enzymes, plants have a two-component system for regulating the balance of H2O2, broadly ROS, within cells. Nonenzymatic antioxidants include GSH, PC, ascorbate, carotenoids, and anthocyanin. Their accumulation enhances upon metal(loid) exposure (Schmöger et al. 2000; Song et al. 2010).

The second component of the two-component H_2O_2 neutralizing system consists of series of enzymes, viz., monodehydroascorbate reductase, dehydroascorbate reductase, and GSH reductase. These enzymes recycle the oxidized GSH and ascorbate for donating electron for H2O2 reduction. The enzymes involved in the recycling of oxidized GSH and ascorbate are also often induced upon exposure of plants to metalloid. Thus, the interdependent ascorbate-GSH cycle has an important role in maintaining ROS balance in plants (Foyer and Noctor 2011) even during metalloid exposure (Tripathi et al. 2012).

3 GABA Shunt Against Metals/Metalloids Stress

In plant, γ-aminobutyric acid (GABA) is a multifunctional, nonproteinogenic amino acid involved in signaling acting as metabolic intermediates. γ-aminobutyric acid was primarily discovered in the potato tubers (Steward 1949).

In plant cytoplasm, GABA is produced by evolutionarily conserved three-step pathway, viz., "GABA shunt" (Fig. 13.2). The GABA shunt is composed of a three-enzyme pathway, viz., cytosolic glutamate decarboxylase (GAD), mitochondrial GABA transaminase (GABA-T), and succinic semialdehyde dehydrogenase (SSADH) (Bouche and Fromm 2004). A TCA intermediate, i.e., succinate, synthesis is inhibited during oxidative stress. The two enzymes of TCA cycle, i.e., succinyl CoA ligase and α-ketoglutarate dehydrogenase, are oxidative stress sensitive, hence inhibit the synthesis of succinate. Thus, this step is bypassed through GABA shunt pathway (Bouche and Fromm 2004). The regulation of this conserved metabolic pathway seems to have particular significance in plants. Several experimental observations exhibited that the GABA shunt rapidly activates against the biotic and abiotic stress (Bown and Shelp 2016). GABA shunt does not only fulfill the intermediate of TCA cycle but is also associated with the various physiological responses: regulation of cytosolic pH, carbon flux into TCA cycle, nitrogen metabolism, deterrence toward insects, osmoregulation, and signaling (Ramesh et al. 2015). A study reported that the *ssadh* mutant of *Arabidopsis thaliana* was hypersensitive to the environmental stress Bouché et al. (2003). Kumar et al. (2017) also reported that the accumulation of arsenic enhances the level of GABA to counter stress through GABA shunt pathway.

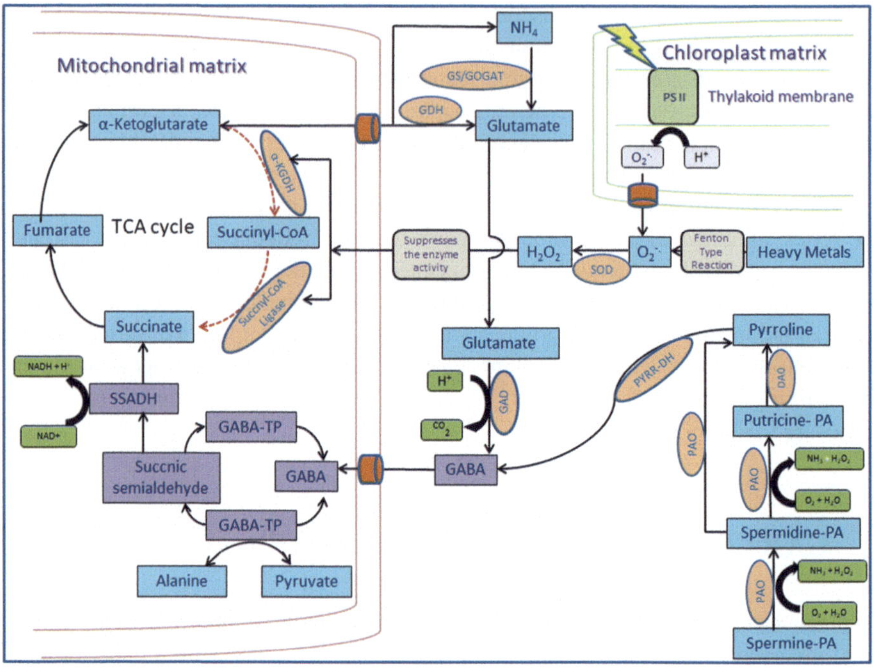

Fig. 13.2 The γ-aminobutyric acid (GABA) shunt pathway and its regulation in plants

4 Nitric Oxide Against Metal/Metalloid Stress

Nitric oxide (NO) is a gaseous free radical with a relatively short half-life (3–5 s) in a biological system, which is involved in the stress ameliorative activity. In plants, NO is synthesized through different pathways such as via nitrate reductase (NR), plasma membrane-bound nitrite: NO reductase (Ni-NOR), xanthine oxidoreductase (XOD), and arginine-dependent and polyamine-mediated synthesis (Stöhr et al. 2001; Rockel et al. 2002; Tun et al. 2006; Parankusam et al. 2017) (Fig. 13.3). The various abiotic stresses such as metalloids, drought, low and high temperatures, and UV and ozone exposure induce generation of ROS (Neill et al. 2002; Vranová et al. 2000). ROS causes several oxidative destructive processes both physiologically and morphologically; however, at the same time, it also activates various signaling pathways (Neill et al. 2002). Thus, maintenance of appropriate ROS levels is prerequisite for survival of plant. In fact, NO interacts with ROS in various ways and might serve antioxidant function during various stresses (Beligni and Lamattina 1999). Nitric oxide may directly act as an ROS scavenger being a free radical and can react with other free radicals such as ROS, or it may activate the antioxidant gene expression or may activate antioxidant enzymes. Modulation by NO against superoxide

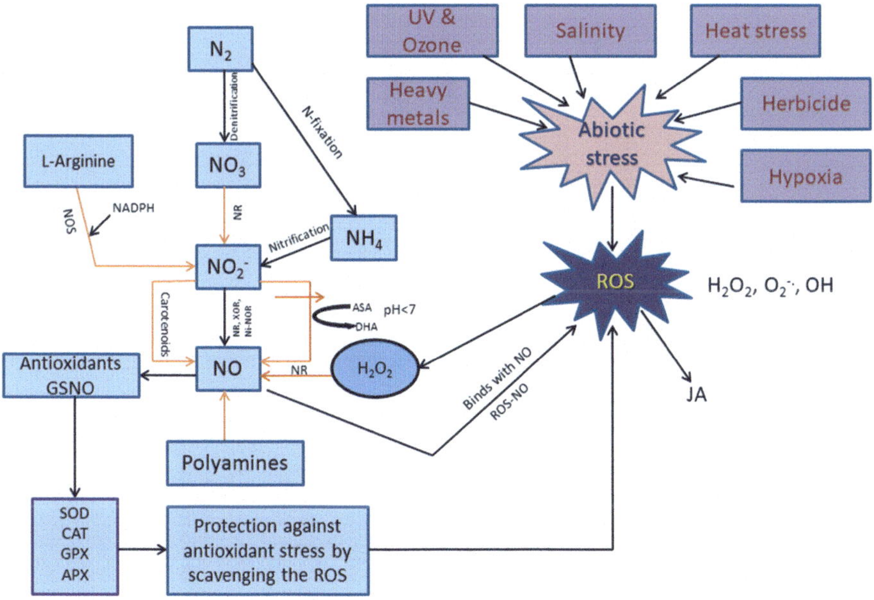

Fig. 13.3 Schematic diagram showing the crosstalk between NO and RS and their interregulatory mechanism

formation illustrates its potential antioxidant role. On the other hand, excess NO can result in nitrosative stress, so a favorable balance of ROS/NO is important.

Heavy metal(loid)s contamination affects the overall performance of plant growth across all agroclimatic zones worldwide. Nitric oxide plays a crucial role in resistance against heavy metal(loid) stress in plants (Table 13.1). The detoxification and antioxidative properties of NO is also found in soybean cell cultures under cadmium (Cd) and Cu stress (Singh et al. 2008). Hu et al. (2007) also reported that NO improved wheat seeds germination and alleviated oxidative stress against Cu toxicity by enhancing the activity of SOD and CAT and correspondingly decreasing in the lipoxygenase activity and lipid peroxidation. Nitric oxide was reported to have the ability to reduce Cu-induced toxicity. Nitric oxide is most likely mediated through the modulation in the activities of antioxidant enzymes (CAT, POD, and AXP) involved in H2O2 detoxification and in the maintenance of cellular redox couples (GR) and contents of molecular antioxidants particularly nonprotein thiol, ascorbate, and its redox status (Tewari et al. 2008). Nitric oxide protects plants against Al^{+3}-induced oxidative stress, and increased root elongation was correlated with a decrease in Al^{+3} accumulation in root apices (Wang and Yang 2005). Sun et al. (2015) reported that nitrate reductase (NR)-mediated early NO burst could be responsible in protecting roots of wheat against Al toxicity.

Table 13.1 Impact of nitric oxide on various metals/metalloids stress

Stress	NO donor	Plants	Stress effect on			Impact of NO on abiotic stress-induced changes in antioxidant properties						References
			H2O2	NO	Antioxidant	H2O2	MDA	SOD	CAT	APX	GR	
0.25, 0.5 m M As	0.25 mM SNP	*Triticum aestivum*	+		±	↓	↓		↑	↔	↑	Hasanuzzaman and Fujita (2013)
0.025 mM As	0.1 mM SNP	*Festuca arundinacea*	+	+	+	↓	↓	↑	↑	↑		Jin et al. (2010)
0.0250.5 mM As	0.05 mM SNP	*Oryza sativa*	+	_	+	↓	↓	↓	↓	↓		Singh et al. (2008)
0.05 mM Cu	0.1 mM SNP	*Panax ginseng*	+	+	_	↓	↓	↑	↑	↑	↑	Tewari et al. (2008)
0.025 mM Cd	5 mM SNP	*Brassica juncea*	+	+	+		↓	↓	↓	↓		Verma et al. (2013)
5 mM Cd	0.1 mM SNP	*Oryza sativa*	+		+	↓	↓	↓	↓	↓	↓	Hsu and Kao (2007)
0.5 mM Cd	0.1 mM SNP	*Helianthus annuus*			±		↓	↓	↓		↓	Laspina et al. (2005)

Arrows indicate up-, down-, or no regulation. +, −, ±, and = indicate up-, down-, differential, or no regulation

5 Role of Polyamines Against Metals/Metalloids Stress

Polyamines (PAs) are low-molecular-weight, aliphatic polycations compound, which are found in all the livings organisms (Chen et al. 2018; Kumar et al. 2019). Three abundant PAs, i.e., putrescine (Put), spermine (Spm), and spermidine (Spm), play a vital role in stress tolerance. In plants, PAs are biosynthesized mainly via two enzymes activity, i.e., ornithine decarboxylase (ODC) and arginine decarboxylase (ADC) (Vera-Sirera et al. 2010; Pegg and Casero 2011; Gupta et al. 2013) (Fig. 13.2). Stress-derived changes in the levels of cellular PA provide clues on their possible implication in plants stress amelioration. Several studies show that PAs could ameliorate the metals/metalloids stress toxicity by (i) maintaining integrity of plant cell membrane, (ii) reversal of growth inhibition due to abiotic stress, (iii) modulation in the expression of osmotically responsive genes, (iv) reduction of superoxide radical and H2O2 levels, (v) reduction in the accumulation of Na^+ and Cl^- ions in different organs, and (vi) increasing activities of antioxidant enzymes (Tang and Newton 2005; Ndayiragije and Lutts 2006; Wang et al. 2007; Yiu et al. 2009; Zhang et al. 2009; Hussain et al. 2011; Alcázar et al. 2010; Kumar et al. 2019). To prove the role of PAs against the abiotic stress, the isolated ethyl methanesulfonate (EMS) mutants of *Arabidopsis* with reduced ADC activity exhibited reduced accumulation of PAs and less tolerance compared with the wild type of plants (Watson et al. 1998; Kasinathan and Wingler 2004). Similarly, the ADC2 knockout mutant exhibited no obvious phenotypic change under normal growth conditions and was more sensitive to metal stress, which can be partially reversed by uptake of Put (Urano et al. 2009). This suggests that ADC2 is a key gene and Put derived from the ADC pathway in providing stress tolerance. In another report, mutant of acl5/spms was found to be sensitive against abiotic stress such as drought, salinity, and metalloids which recovered upon the application of Spm (Yamaguchi et al. 2006, 2007). These studies revealed that the PAs are important molecule involved in stress management in the plants.

6 Role of Glutaredoxins Against Metals/Metalloids Stress

Glutaredoxins (GRXs) are universal, oxidoreductase enzymes with molecular mass between 10 and 15 kDa (Fomenko and Gladyshev 2002). Glutaredoxin/glutathione/glutathione reductase (GRX/GSH/GR) is mainly involved in maintaining the redox state of the cell. Glutaredoxins regulate the activity of proteins or enzyme through reversible glutathionylation or reduction of disulfide bonds (-S-S-) in order to maintain conformational integrity throughout the oxidative stress (Rouhier et al. 2008; Meyer et al. 2009). Glutaredoxins uses GSH and NADPH as a cofactor, generated through NADPH-dependent GSH reductase (GR) system (Rouhier et al. 2008; Zaffagnini et al. 2008) (Fig. 13.4).

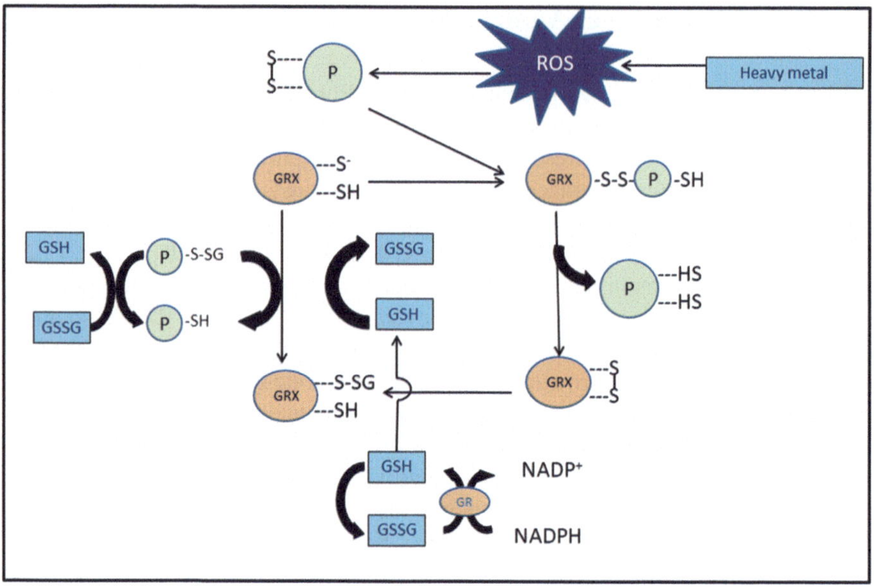

Fig. 13.4 Schematic model of redox activity of glutaredoxin under oxidative stress. (*P* protein, *GSSG* oxidized glutathione, *GSH* reduced glutathione, *GRX* glutaredoxin)

In plants, GRXs consist of a large family of genes, and others remain to be explored on the basis of mode of function under normal and metals/metalloids stress conditions (Jozefczak et al. 2012). A study reported that GRX gene (*PvGRX5*) from As-tolerant fern *Pteris vittata* when integrated in *E. coli* exhibited tolerance against 10 mM As(V) treatments (Sundaram et al. 2008, 2009; Sundaram and Rathinasabapathi 2010). Tomato (*Lycopersicon esculentum*) plants silenced with *SlGRX1* gene showed susceptibility toward metal(loid) stress, whereas, overexpressed gene in *Arabidopsis* provides tolerance against these stresses (Guo et al. 2010). Rice GRX (*OsGrx*) protects glutamine synthetase toward metal stresses (Lee et al. 2002). The role of GRXs in the mitigation of metalloids is well recognized in the prokaryotic as well as mammalian system (Ning et al. 2018). However, in plants, the GRXs are mostly reported against cadmium stress (Li 2014). A study reported that exposure of arsenic enhanced the expression of four GRX genes (*Os01g26912, Os02g40500, Os02g40500*) in rice plants, indicating GRXs are directly/indirectly involved in the amelioration of As-induced stress (Rai et al. 2011).

7 Role of Glyoxalase System Against Metals/Metalloids Stress

The glyoxalase system is an important pathway of cellular metabolism, which involves in the methylglyoxal (MG) detoxification. Various stresses such as heavy metal(loid)s, salinity, drought, and high temperature drastically increase the rate of glycolysis, which leads to the MG formation (Hoque et al. 2016). Removal of phosphoryl group by β-elimination from the unstable triose phosphates emphasizes the formation of MG. Methylglyoxal is a genotoxic agent, and at higher concentration, it directly disrupts the structure of the proteins through the formation of advanced glycation end products (AGEs). It also damages DNA structure and, consequently, enhances cell death (Borysiuk et al. 2018).

In plants, MG is eliminated by the two-glyoxalase pathway containing glyoxalase I (Gly I) and glyoxalase II (Gly II) (Sankaranarayanan et al. 2017) (Fig. 13.5). In condition where reduced glutathione (GSH) is abundant, MG is converted into hemithioacetal (HTA), where Gly-I enzyme converts HTA into S-D-lactoyl-glutathione (SLG). Further, the toxic SLG is detoxified through Gly-II, converting SLG to nontoxic D-lactate, where the GSH is also freed by the compound and gets back to the normal cellular metabolism (Yadav et al. 2008). There are several studies showing the increase in activities of Gly-I and Gly-II, conferring tolerance against heavy metal(loid)s in plants. Overexpression of the glyoxalase pathway in transgenic tobacco and rice plants has been found to prevent an increase of ROS and MG under stress conditions by maintaining GSH homeostasis and antioxidant enzyme levels (Yadav et al. 2005; Singla-Pareek et al. 2006, 2008). The specific activity of glyoxalase I (Gly I) was increased along with the other antioxidant enzyme activities in *Vigna radiata* plants in response to Cd stress (1 mM, 48 h) (Hossain et al. 2010).

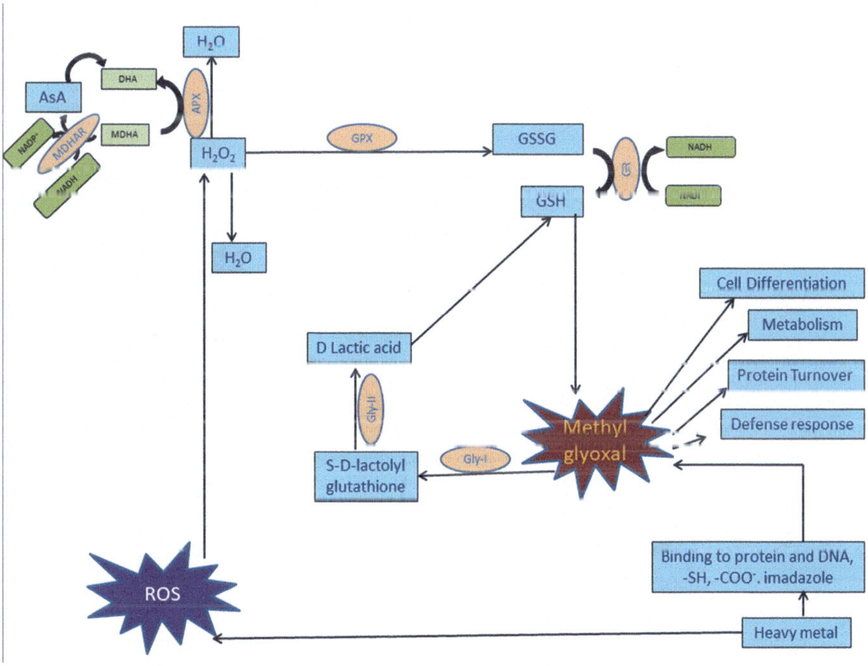

Fig. 13.5 Schematic diagram showing pathways of glyoxalase in detoxification of methylglyoxal generated against oxidative stress

8 Dehydrins

Dehydrins (DHNs), also known as the group 2 or D-11 family of late embryogenesis abundant (LEA) proteins, are widely distributed in the plant kingdom which accumulates under environmental stresses, viz., drought, low temperature, and salinity. Dehydrins are hydrophilic, thermostable, stress-responsive proteins and contain high-charged amino acids belonging to late embryogenesis abundant (LEA) family (Hasanuzzaman et al. 2015). Dehydrins contain conserved lysine-rich motifs known as K-segments consisting of 15-amino acid motif sequence present in 1–11 copies near to the C-terminus of dehydrin (EKKGIMDKIKEKLPG) (Fig. 13.6).

Apart from the K-segment, dehydrins can possess other conserved sequential motifs: the tyrosine-rich Y-segment (consensus (V/T)D(E/Q)YGNP) near to the N-terminus and the serine-rich S-segment (Kosová et al. 2010). S-segment is formed by a stretch of 4–10 phosphorylable serine-rich tract which is proposed to be related to the nuclear transport of dehydrins. Dehydrins form an amphiphilic helix which interacts with the membranes or proteins to prevent inactivation under dehydration (Liu et al. 2017a, b). According to the number and combination of these three conserved domains, a "YnSnKn" nomenclature scheme within the dehydrin family has been developed (Close 1996). Recently, antioxidative activity has also been attrib-

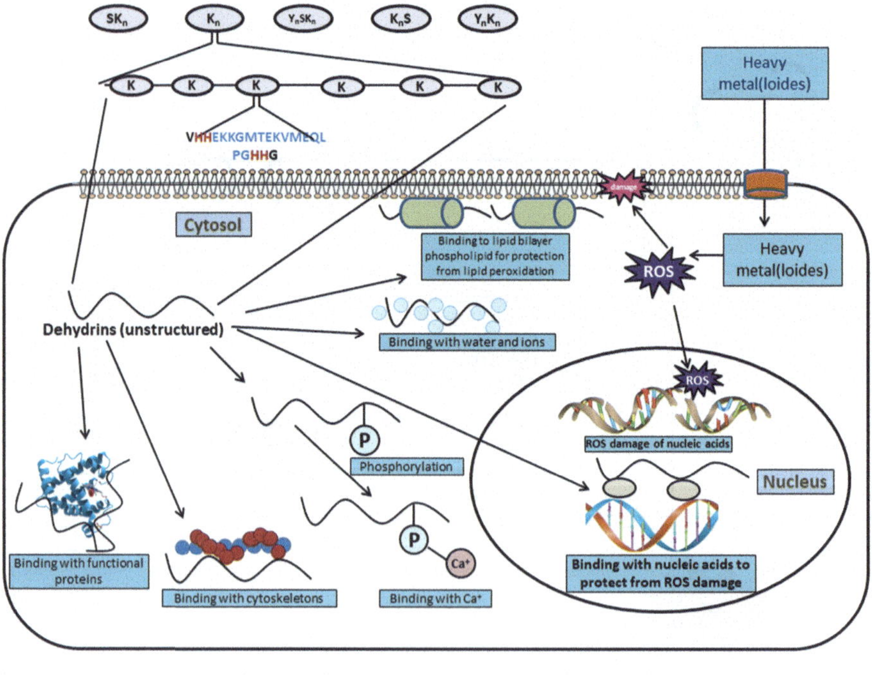

Fig. 13.6 Generalized schematic illustration of various functions of dehydrins in plant cell

uted to dehydrins, since the environmental stresses which induce the accumulation of dehydrins also produce ROS, inducing oxidative stress in plants (Sun and Lin 2010). The localization of dehydrins in subcellular components also strengthens the role of dehydrins playing role in antioxidative activity. Dehydrins are reported to occur in (i) nucleus to possibly stabilize chromosome against ROS (Carjuzaa et al. 2008; Egerton-Warburton et al. 1997; Hara et al. 2009), (ii) in vicinity of plasma membrane to protect the lipid peroxidation (Danyluk et al. 1998), and (iii) in mitochondria and chloroplast (Borovskii et al. 2000; Mueller et al. 2003) to protect the cellular machinery from H2O2 being generated from PS-II (Salin 1991). Dehydrins are also known as responsive to ABA (RAB) proteins, synergistically elevating with ABA. Various stress related to dehydration (e.g., osmotic stress, drought, salinity, and heat) enhances the level of DHNs (Fig. 13.6). The limited studies indicate the role of DHNs in the amelioration of metals/metalloids stress. In a study, the expression of two DHN genes of *Brassica juncea*, i.e., *Bj*DHN2 and *Bj*DHN3, was found to enhance with heavy metal(loid)s accumulation. The overexpression of *Bj*DHN2 and *Bj*DHN3 in transgenic plants exhibits heavy metal(loid)s tolerance (Xu et al. 2008). Similarly, the overexpression of *DREB* transcription factor (plays a crucial role in abiotic stress tolerance) in plants exhibited the enhanced expression of DHNs such as RD29A, COR15a, and RD29B, which also enhanced the tolerance against heavy metal(loid)s stress (Ban et al. 2011; Yu et al. 2018). On the other hand, DHNs can stabilize the free transition metal ions by chelating them with motifs, reducing oxidative stress (Svensson et al. 2002). A study using metal ion affinity chromatography (IMAC) shows CuCOR15 dehydrin has the capability to chelate and immobilize Fe^{3+}, Co^{2+}, Ni^{2+}, Cu^{2+}, and Zn^{2+} ions. The chelation capacity of dehydrin indicates it could be an in vivo tool for metal tolerance in plants. In tobacco plants, the over expression of maize dehydrin gene, i.e., *Zm*DHN13, correlated with the tolerance to oxidative stress (Liu et al. 2017a, b). Dehydrins are reported to accumulate against heavy metal(loid)s. Expression of a bean dehydrin gene *Pv*SR3 was upregulated by H2O2 and heavy metal(loid)s, which are known to catalyze the production of toxic ROS. There are also report of radical scavenging ability of a citrus dehydrin (CuCOR19) owing to high content of amino acid glycine (Gly), histidine (His), and lysine (Lys)—residues which are susceptible to oxidative modification (Dean et al. 1997). Catalytic metals, such as Fe and Cu, mainly exist as cofactors in many enzymes and proteins in plants under normal growth conditions. These metals can be released as free ions when plants are exposed to environmental stresses, producing highly toxic radicals via the Haber-Weiss or the Fenton reactions (Mittler 2002). Owing to the high His content, having strong metal-binding domains like His-His pairs and His-X3His motifs, plant dehydrins can be suggested to be ion-sequestering antioxidative proteins.

9 Heavy Metal(loid)s Chelation: Metallothioneins and Phytochelatins

Chelation of the toxic/excess metals in the cytosol by high-affinity ligands is a unique and efficient strategy for heavy metal(loid) detoxification and tolerance. In addition to organic acids and amino acids, two classes of peptides, viz., metallothioneins (MTs) and phytochelatins (PCs), are predominant in heavy metal(loid) detoxification by chelation. Phytochelatins are family of metal-ligating peptides synthesized non-translationally, having the general structure of (γ-Glu-Cys)n-Gly where n = 2–11. The PCs are synthesized using GSH as substrate and PC synthase enzyme, which is induced by the presence of metals (Cobbet 2000; Goldsborough 2000). The reported metals which are ligated with the PCs and sequestered in vacuoles are Cd, Zn, and As. Ligated metal-PC complex are transported across the tonoplast membrane into the vacuole through ABC and CAX transporters. Metallothioneins are low-molecular-mass cysteine-rich polypeptides encoded by a family of genes which can bind with heavy metal(loid)s and play a role in their intracellular sequestration. MT were discovered 57 years ago as a Cd-binding protein in horse kidney. MT genes and proteins have been reported from many prokaryotic (bacteria) and eukaryotic organisms (plants, invertebrates, and vertebrates) which are induced in the presence of metals and also play a role in its detoxification and intracellular metal ions homeostasis (Robinson et al. 1993; Cobbett and Goldsbrough 2002; Roychowdhury et al. 2018, 2019). The two metal-binding, cysteine-rich domains in MTs give these metalloproteins their characteristic dumbbell conformation (Fig. 13.7).

Class I MTs are widespread in vertebrates containing 20 highly conserved cysteine, while class II MTs include all those from plants and fungi as well as from non-vertebrate animals which do not follow any strict arrangement of Cys residues. However, due to lack of scientific consensus, phytochelatins are considered as class III metallothioneins. Following the discovery of PCs as a Cd-binding ligand in plants, MT proteins were discovered as Zn-binding proteins from wheat embryos in plants (Lane et al. 1987). This provided the evidence that in plants, in addition to PCs, MTs also play a role in metal(loid) binding and homeostasis (Fig. 13.8). MTs are involved in Cu tolerance and homeostasis in plants: Some plant MTs are functional copper-binding proteins, and expression of some MT genes is induced by copper.

10 Heavy Metal(loid) Transporters

Plants require six macronutrients (Ca, K, Mg, N, P, S) and eight micronutrients (B, Cl, Cu, Fe, Mo, Mn, Ni, Zn), and absence of any one of these elements does not allow transition through a full life cycle. For example, Fe is a key component of heme proteins for a range of enzymes (e.g., cytochromes, catalase, and Fe-S pro-

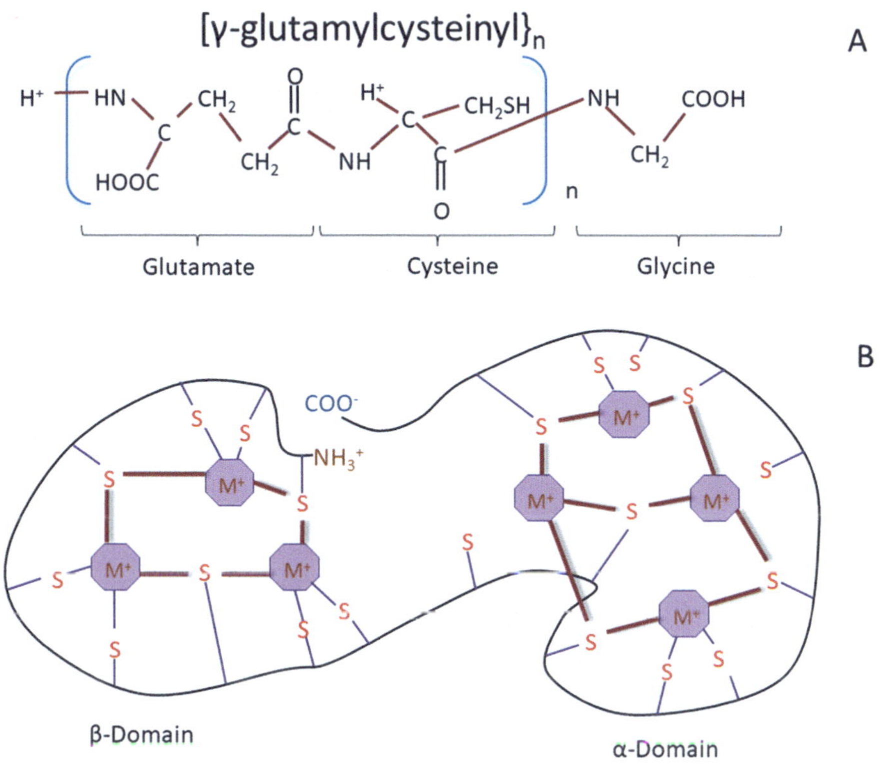

Fig. 13.7 Schematic structure of (**a**) phytochelatin and (**b**) metallothionein proteins that have been classified based on the arrangement of their cysteine (Cys) residues

teins such as ferredoxin), Cu is key cofactor for certain electron transfer proteins in photosynthesis (e.g., plastocyanin) and respiration (e.g.. cytochrome C oxidase) and lignification (laccase), while Mn is less redox active but is also involved in photosynthesis (e.g., O2 evolution). Zn is non-redox active but has a key structural and/or catalytic role in many proteins and enzymes. Although essential, these metals can be toxic when present in excess as they can produce ROS through Fenton-type reaction.

$$\mathbf{Fe^{2+} + H_2O_2 \ \ intermediate\ complex\ \ Fe^{3+} + OH^{\bullet} + OH^{-}}$$

Cation/s transport across the plant plasma membrane is governed by an electrochemical gradient of protons generated due to the pumping of protons out of the cell by H^+-ATPases, creating an electrochemical potential (ECP) across the plasma membrane (PM). Using this ECP, several other transporters can efflux and influx cations across the PM. Modern genetic and molecular techniques such as sequence comparison and functional complementation of yeast mutants have identified a wide range of transition metal transporter gene families in plants (Fig. 13.9). These

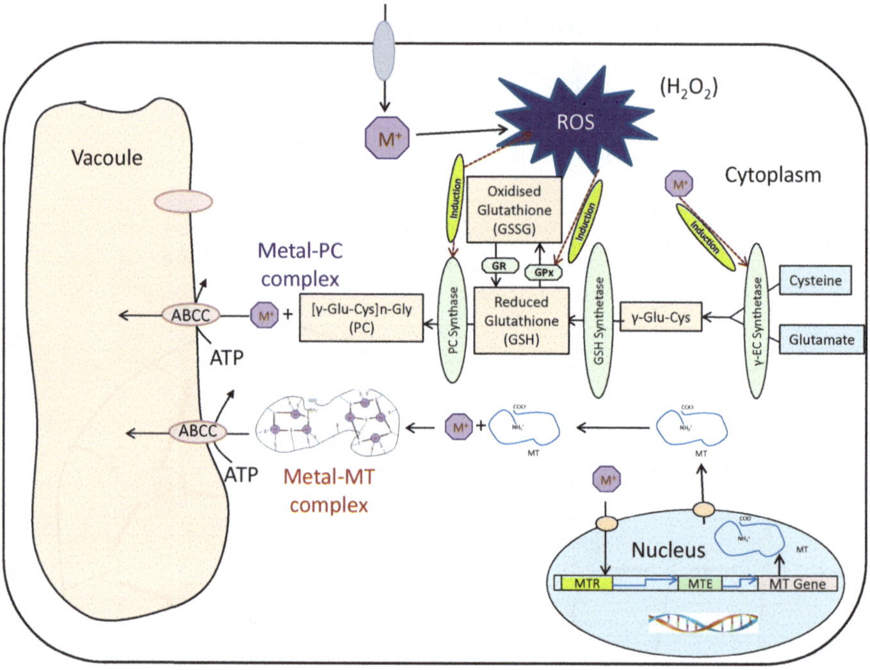

Fig. 13.8 Schematic diagram showing pathway of phytochelatin and metallothionein induction by heavy metal(loid) [M^+] in a plant cell

include the heavy metal(loid) (or CPx-type) ATPases, natural resistance-associated macrophage proteins (Nramps), cation diffusion facilitators (CDFs), ZIP family (ZRT, IRT-like protein), the cation antiporters, YSL (yellow-stripe-like) transporter, NAS (nicotianamine synthase), SAMS (S-adenosylmethionine synthetase), FER (ferritin Fe (III) binding), CDF (cation diffusion facilitator), and IREG (iron-regulated transporter) family (Hall and Williams 2003; Guerinot 2000; Williams et al. 2000; Talke et al. 2006; van de Mortel et al. 2006; Kramer et al. 2007; Memon and Schroder 2009; Maestri et al. 2010).

In plants, ZIPs have been identified in both dicots and monocots, such as 18 ZIP genes in *Arabidopsis* and 16 from rice (Ivanov and Bauer 2017). In *Arabidopsis, IRT1*is expressed against Fe deficiency and plays a role in Mn/Zn transport. A few studies have also implied that ZIPs may be involved in Cd transport. Among the transport of major trace cations, the uptake of Fe is of particular interest as there are two different strategies between non-graminaceous plants (strategy I) and graminaceous plants (strategy II) Briat and Lobreaux (1997) (Fig. 13.10).

The strategy I plants rely on ferric reductase oxidase (FRO) to reduce the ferric chelates at the root PM and then consequently on the iron-regulated transporter (IRT) to absorb the generated Fe^{2+} ions across the root PM. On the contrary, strategy II, used by graminaceous plants (e.g., barley, maize, and rice), depends on the bio-

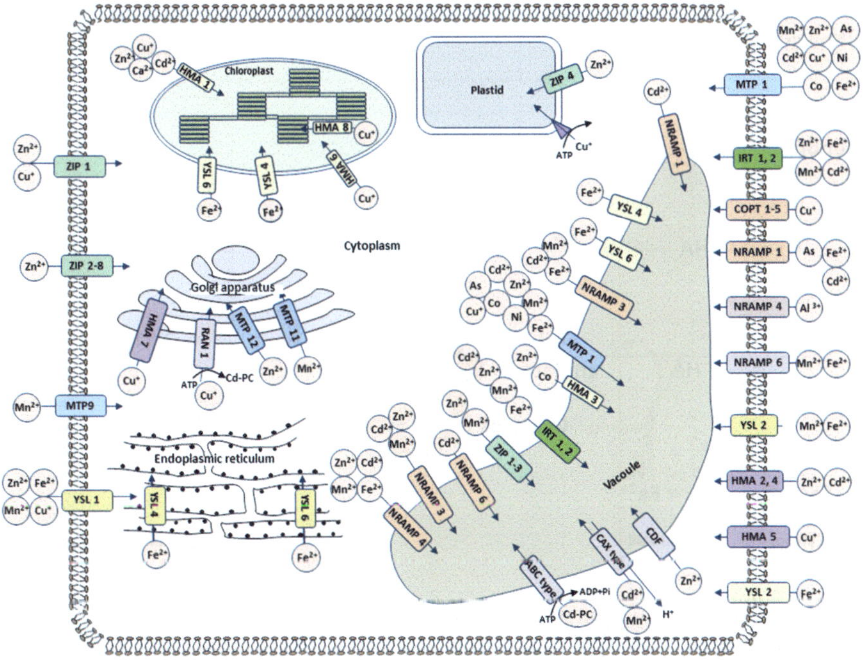

Fig. 13.9 Schematic diagram of putative locations of different metal transporters in the plant cell. (Modified from Clemens 2001; Hall and Williams 2003; Ding et al. 2017)

synthesis and secretion of phytosiderophores (PS) such as mugineic acids (MAs) which mobilize Fe^{3+} from the soil matrix. The secreted MAs into the rhizosphere solubilize Fe^{3+} from the soil matrix, and the resulting Fe-PS complexes are taken up through yellow stripe (YS) and yellow stripe-like (YSL) transporters. Graminaceous plants, despite using strategy II, can also have ferrous transporters (IRT1, NRAMP) allowing the absorption of Fe^{2+} in addition to its strategy II-based Fe^{3+}-MA uptake. Iron-regulated transporter (IRT1) is a member of a small gene family encoding a predicted protein of 339 amino acids, with eight transmembrane domains. Four histidine-glycine repeats constitute potential metal-binding domains between TM domains three and four. Five ZIP transporter genes have been reported for rice *Os*IRT1, *Os*IRT2, *Os*ZIP1, *Os*ZIP3, and *Os*ZIP4. *Os*IRT1 and *Os*IRT2 are ferrous iron transporters. *Os*ZIP1 is primarily associated with metal uptake and *Os*ZIP3 with overall Zn homeostasis in the cell, especially in leaves. COPT family transporter is known for the uptake of Cu in plants. They work as homo- or heterotrimer of a three-TM domain forming a channel for specific Cu^{+} transport. The FRO present in the root PM facilitates the reduction of the unavailable Cu^{2+} to Cu^{+}, consequently uptaken by COPT transporters. Similar to the reduction of Cu^{2+} to Cu^{+}, Mn^{3+} and Mn^{4+} need to be reduced to Mn^{2+} to be absorbed by IRT1 present in root PM. In addition, the NRAMP transporters can also transport Mn. In rice plants, both

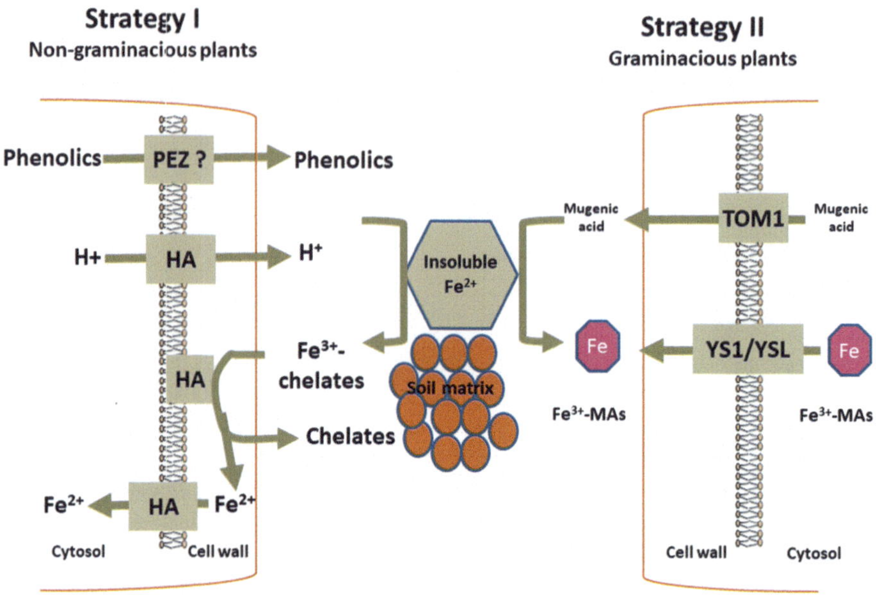

Fig. 13.10 Schematic diagram showing strategies for Fe uptake in graminaceous and non-graminaceous plants

NRAMP1 and NRAMP 5 are reported to transport Mn. NRAMP transporters are a monomeric protein spanning 11 TM domains which transports a wide range of metals, i.e., Fe^{2+}, Mn^{2+}, Co^{2+}, and Zn^{2+}. Furthermore, Cu^{+} transport is also mediated by members of the zinc-regulated transporter and iron-regulated transporter (ZRT-IRT)-like proteins (ZIP). These are a ubiquitous family of divalent metals transporters (primarily Fe^{2+}, Zn^{2+}, Ni^{2+}, Cd, and Mn^{2}) acting as homodimer where each monomer has eight TM domains. The metal specificity of the ZIP transporters is conferred by conserved histidine-rich cytosolic loops between three and four TM domains. Zn, Mn^{2+}, and Cu^{+} uptake also takes place through YS and YSL family (e.g., *Zm*YS1 and *Hv*YSL2) are also involved in the transport of Zn-PS, Cu-PS, and Mn-PS complexes. YSL transporters (yellow stripe-like proteins) are phenotypes identified in maize containing a range of 11–16 TM regions and function across a H^{+} symporter. Members of the YSL family can only be found in plants, although YSL belong to the larger OPT (oligopeptide transporter) family that is also present in fungi. YSL transporters do not use free metals as substrate but a complex of metals with nicotianamine (NA) or its derivatives. IRT1 also facilitates the transport of Ni as Ni-PS complexes (Pinto and Ferreira 2015). P1b-ATPases are a clade of the P-type superfamily of ATPases (which also includes the Na^{+}/K^{+}-ATPase or the H^{+}-ATPase which is a monomer with several TM domains from six to eight) where the last two cytosolic loops are enlarged and comprised the ATPase function domain (proximity to C-terminus) that drives transport of the metal. These transporters frequently have cytosolic metal-binding domains in N- and/or C-terminal with a regulatory function. P1b-ATPases are involved in long-distance Cu^{+} and Zn^{2+} transport in plants, as well as metal transport into organelle. CDF transporters (cation diffusion facilitator) is a homodimer facilitating the transport of divalent metals such as Fe^{2+}, Zn^{2+}, or Mn^{2+}, coupled to a H^{+} antiporter. The monomer has six transmembrane domains with a His-rich region in the cytosol between the fourth and the fifth TM domain. There are three metal-binding domains in the protein (site I in the transmembrane region, site II in the membrane-cytosol interface, and site III in the C-terminal domain), but only I and III seem to be directly involved in transport.

11 Role of Phytohormones Against Metals/Metalloids Stress

Recent evidence suggests that phytohormones not only regulate the normal function of the plants but are also the key molecules in the plants which are elevated during the wide range of stresses. Several phytohormones such as abscisic acid (ABA), auxin, brassinosteroids, ethylene, gibberellic acid (GA), and cytokinins (CKs) have been reported to play a role during metal stress (Zhu et al. 2012; Masood et al. 2016; Bücker-Neto 2017).

Abscisic acid (ABA) plays a crucial role in plants, involving seed development and dormancy (Finkelstein 2013) and imparting tolerance to plants against adverse environmental conditions, and its signaling pathway is a central regulator of abiotic stress response in plants (Danquah et al. 2014). The metals Ni, Zn, and Cd reduce water potential and stomatal conductance in the plants where elevated level of ABA recovers the inhibition (Rucińska-Sobkowiak 2016). Abscisic acid is reported to play a role in increasing the growth of roots of aquatic plants *Typha latifolia* and *Phragmites australis* (Fediuc et al. 2005), in potato tubers (Stroinski et al. 2010), and in rice plants (Kim et al. 2014) against Cd exposure. Similarly, increase in ABA levels increased in wheat seeds when exposed to Hg, Cd, and Cu (Munzuro et al. 2008). The expression of genes involved in ABA synthesis, viz., 9-cis-epoxycarotenoid dioxygenase 2 (*Os*NCED2 and *Os*NCED3), in rice plant has been found to increase in coincidence with upregulation of four ABA signaling genes against As and Va stress (Huang et al. 2012). Although the mechanism underlying the ABA signaling pathway in response to heavy metal(loid)s exposure remains largely unexplored, the pathway involves pyrabactin resistance1 (PYR1)/PYR1-like (PYL)/regulatory components of ABA receptors (RCAR), PP2C, and SNF1-related protein kinase 2 (*Sn*RK2) (Fig. 13.11).

Abscisic acid is also known to induce transient mitogen-activated protein kinase (MAPK) kinase activity (Burnett et al. 2010). Elevated accumulation of Cd and Cu enhances MAPK activity and subsequently the level of ABA in plants to counter the stress (Yeh et al. 2004; Liu et al. 2018; Hsu and Kao 2003). Hence, there exists interrelationship between ABA and MAPK in countering heavy metal(loid)-induced stress. During stress, ABA also regulates the different genes expression involved in the late embryogenesis abundant (LEA) proteins, dehydrins, and stress-responsive proteins (Vaseva et al. 2010).

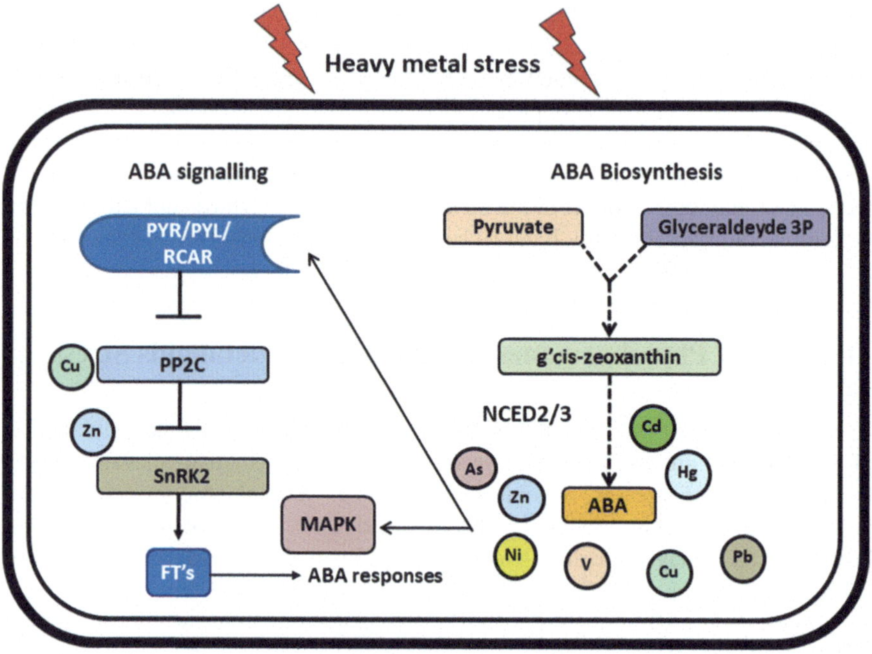

Fig. 13.11 Schematic diagram of abscisic acid and heavy metal(loid)s interaction in plants. (Modified from: Bücker-Neto et al. 2017)

The involvement of ethylene (ET) has been reported in response to abiotic stress induced by several elicitors including heavy metal(loid)s (Maksymiec 2007; DalCorso et al. 2010; Khan et al. 2015). Ethylene is synthesized from methionine, which is converted to S-adenosylmethionine (SAM) by SAM synthetase. SAM, as a substrate, is capable of forming 1-aminocyclopropane-1-carboxylic acid (ACC) by ACC synthase (ACS) (Fig. 13.12). This is the rate-limiting step in the ethylene biosynthesis pathway and ultimately releases 5′-methylthioadenosine (MTA), which is recycled again to methionine via the "Yang cycle." However, in presence of O_2, ACC is degraded by ACC oxidase (ACO) forming ethylene, CO_2 and cyanide in plants (Yang and Hoffman 1984). The enzymes ACS and ACO are the two fundamental components of ET biosynthesis. Stress induced by heavy metal(loid) increases the expression of genes encoding these two proteins, resulting in higher ET production (Schellingen et al. 2014; Khan et al. 2015). ACS2 and ACS6 are phosphorylated by MAPKs, which in turn increase their stability and activity. The transcription factor WRKY33 protein binds directly to the W-boxes in the promoters of ACS2 and ACS6 genes in vivo; hence, WRKY33 activates ACS2 and ACS6 expression downstream of the MPK3/MPK6 cascade (Li et al. 2012). EIN2 gene is an important gene ethylene signaling pathway and acts as a transducer of ethylene and stress responses (Alonso et al. 1999). The transcript levels of EIN2 increase against Pb treatment in *Arabidopsis* seedlings (Cao et al. 2009).

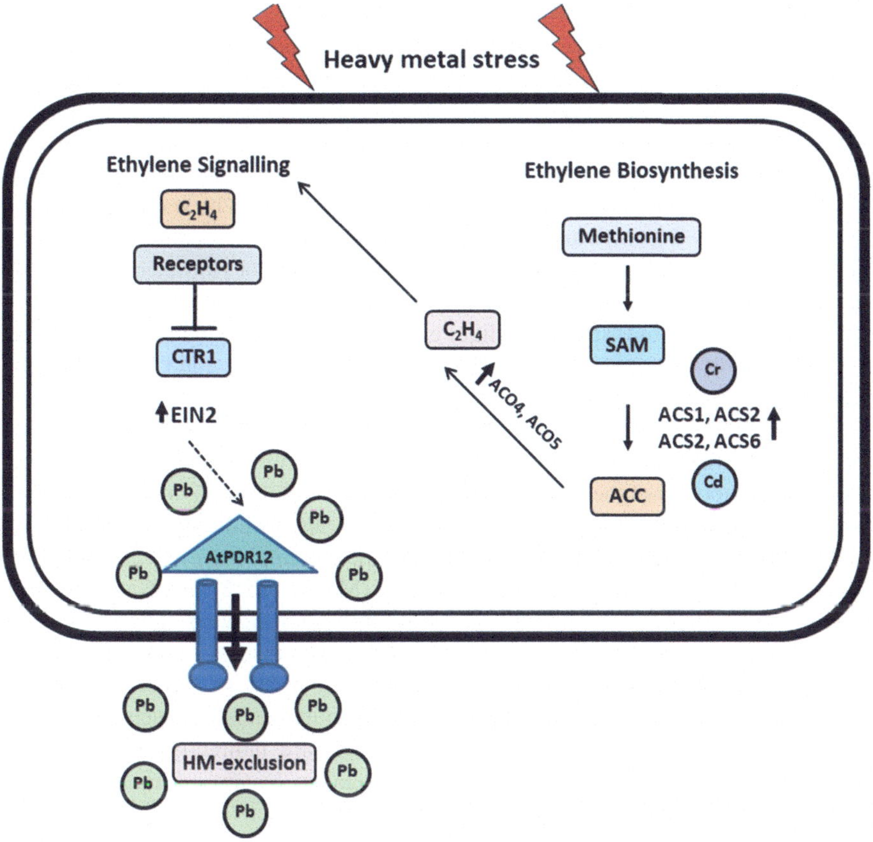

Fig. 13.12 Schematic diagram illustrating the ethylene signaling against heavy metal(loid) stress. (Modified from: Bücker Neto et al. 2017)

A study suggests that the Cd tolerance is related to the induction of ET biosynthesis (Cao et al. 2014). The *Arabidopsis thaliana* ein2-1 mutants (a gene involved in the ethylene biosynthesis) exposed to Al significantly enhance antioxidants enzymes, i.e., SOD and CAT activities, with respect to the WT plants. The study indicates that the elevated level of ET can ameliorate Al toxicity by regulating antioxidant enzymes in the plants (Zhang et al. 2014). Rice plants treated with Cr showed the enhanced expression of genes 1-aminocyclopropane-1-carboxylic acid synthase and 1-aminocyclopropane-1-carboxylic acid oxidase are involved in the ET biosynthesis, highlighting the role of ET in Cr tolerance (Steffens 2014; Trinh et al. 2014). The accumulation of Pb exhibited enhanced transcript levels of EIN2 in *Arabidopsis* seedlings, coinciding with the heavy metal(loid) tolerance (Cao et al. 2009). Another study also reported that *Arabidopsis* etr1-1 and ein3-3 mutants exhibited less sensitivity toward Li accumulation (Bueso et al. 2007).

The gibberellic acids (GAs) are the tetracyclic diterpenoid carboxylic acid compounds, in which the only GA1 functions as phytohormone (Sponsel and Hedden 2010).

The GAs regulate the seed germination, leaf expansion, development of fruits, and stem elongation (Salazar-Cerezo et al. 2018). A study reported that the GAs enhance the tolerance against heavy metal(loid)s in wheat by enhancing *Ta*MYB73 gene expression (He et al. 2011). The accumulation of Cd reported inducing iron-regulated transporter 1 (IRT1), which facilitates the uptake of Cd. The elevated level of GA against Cd toxicity was found to suppress the IRT1, thereby reducing the toxic manifestation in *Arabidopsis thaliana* (Zhu et al. 2012). The GA was also found to reverse the reduced growth and chlorophyll content in wheat seedlings exposed to Ni (Siddiqui et al. 2011). On the other hand, the GA is not restricted only to the regulation of genes expression and chlorophyll content; it also enhances the antioxidants against metal(loid) stress (Siddiqui et al. 2011). Several studies show that the elevated level of GA coincides with the enhanced peroxidase and CAT activity in plants (Wojtania et al. 2016).

Among the phytohormones, ABA, auxin, GA, ET, and cytokinins (CKs) are closely concurrent to the nitrogen metabolism. Cytokinins are the N6-prenylated derivative compound of the adenine, which regulates cell division, nutrient metabolism, and nodulation (Savelieva et al. 2018). CKs are mainly synthesized in the seeds, fruits, and roots (Matsuo et al. 2012). The accumulation of CKs has been reported to mitigate heavy metal(loid)s toxicity (Manara 2012). In plants, elevated accumulation of CKs was able to increase transpiration, which can stimulate photosynthesis rate, coinciding with reversal of plant biomass against heavy metal(loid)s stress (Cortleven and Schmülling 2015). During As toxicity, a CK receptor, cytokinin response-1 (CRE1), was downregulated and simultaneously induced the group 1 sulfate transporters. This indicates that the CKs are indirectly regulating the sulfur-related defense mechanism in the plants against As stress (Srivastava et al. 2009). In another study, depletion of CK was found to activate thiol compounds, which are directly involved in the defense mechanism against As stress in *Arabidopsis thaliana* (Mohan et al. 2016). It is reported that the Cd treatment inhibits the growth rate and net photosynthetic rate in plants. The adenosine-5′-phosphosulfate reductase (APR), an enzyme involved in the sulfate assimilation, was also reported to enhance with the elevated level of CKs in *Arabidopsis*, under the metal stress (Fu et al. 2018). The sulfur is an essential nutrient of the plants, which is also involved in the synthesis of defense mechanism compounds, such as cysteine, GSH, and GRX, against a wide range of stresses (Mallick et al. 2013). The studies indicate that the CKs may regulate genes expression, sulfur metabolism, and level of RNS to counter the metals/metalloids stress.

Auxin chemically known as indole-3-acetic acid (IAA) is an important phytohormone in several physiological aspects of plant growth and signaling in the stress tolerance mechanism. The expression of IAA in plants is tightly regulated by the degradation of AUX/IAA transcriptional repressors that are associated to ARFs (auxin-responsive factors) and thus inhibits the transcription of auxin-responsive genes ultimately regulating the plant physiology (Vanneste and Friml 2009) (Fig. 13.13). Evidence suggests that auxin reverses the inhibited shoot and root growth during heavy metal(loid) stress (Wang et al. 2015). Generally, the higher accumulation of metal(loids) such as As and Cd reduces root growth, which coin-

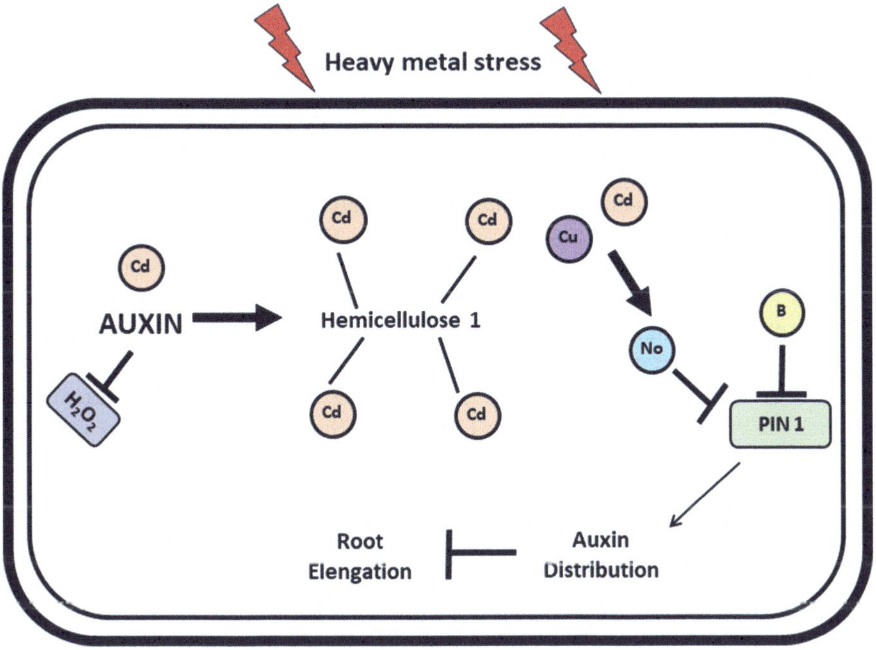

Fig. 13.13 Schematic diagram illustrating the interactions between auxin and heavy metal(loid)s in plants. (Modified from: Bücker-Neto et al. 2017)

cides with the reduced level of auxin (Ronzan et al. 2018). It is reported that the hemicellulose-1 pays a crucial role in heavy metal(loid) stress tolerance. A recent study shows that the elevated level of auxin under Cd stress enhances the level of hemicelluloses, which restricts the heavy metal(loid)s movement within the root (Zhu et al. 2013). The accumulation of heavy metal(loid) enhances the level of auxin regulation via PINFORMED1 (PIN1) protein (an auxin efflux carrier for meristematic root growth under normal conditions) in plants (Bücker-Neto et al. 2017) and is involved in the hormone distribution under heavy metal(loid) stress. In *Arabidopsis thaliana*, the mutant lines of PIN1 (*pin1* DR5rev::GFP) exhibited no auxin activity in both the meristem and elongation zones compared to the normal lines (DR5rev::GFP) against Cu stress, indicating that PIN1 is involved in hormone distribution under heavy metal(loid) stress (Yuan et al. 2013). Evidences suggest that there exists interrelationship between ET and auxin. Ethylene induces the expression of genes involved in auxin biosynthesis and stimulates the auxin transport toward the elongation zone, thereby regulating root elongation (Ruzicka et al. 2007). Under boron (B) deficiency, *Arabidopsis* auxin, ethylene, and ROS participate together in signaling the reduction in root elongation (Camacho- Cristóbal et al. 2015). Generally, heavy metal(loid) stress leads to decrease in endogenous auxins levels. The level of three auxins (IAA, NAA, and IBA) was reported to reduce in *Brassica juncea* against As (Srivastava et al. 2013) and barley root tip against Cd (Zelinová et al. 2015). Nitric oxide (NO) also represses auxin transport to roots, which hinders its elongation (Yuan and Huang 2016).

Brassinosteroids (BRs) regulate several plant growth aspects, i.e., cell expansion and elongation, photomorphogenesis, flowering, male fertility, seed germination, vascular differentiation, plant architecture, stomata formation, and senescence, and it acts during abiotic stress. Phytotoxicity induced by heavy metal(loid)s is consequent upon ROS production in plants. Alternatively, antioxidant enzyme activities are also regulated by BRs (Cao et al. 2005); hence, BR-induced stress tolerance can be attributed to increased expression of antioxidant genes (Xia et al. 2009) (Fig. 13.14). Exogenous application of 24-epiBL has been found to ameliorate Ni stress in *Brassica juncea* by enhancing the activity of antioxidant enzymes (Kanwar et al. 2013), in wheat against Ni toxicity (Yusuf et al. 2011), and in *Raphanus sativus* and *Vigna radiate* (Yusuf et al. 2012), by elevation of different antioxidant enzyme activity. These reports suggest that BRs definitely play a role against amelioration of abiotic stress in general by elevating the antioxidant activity.

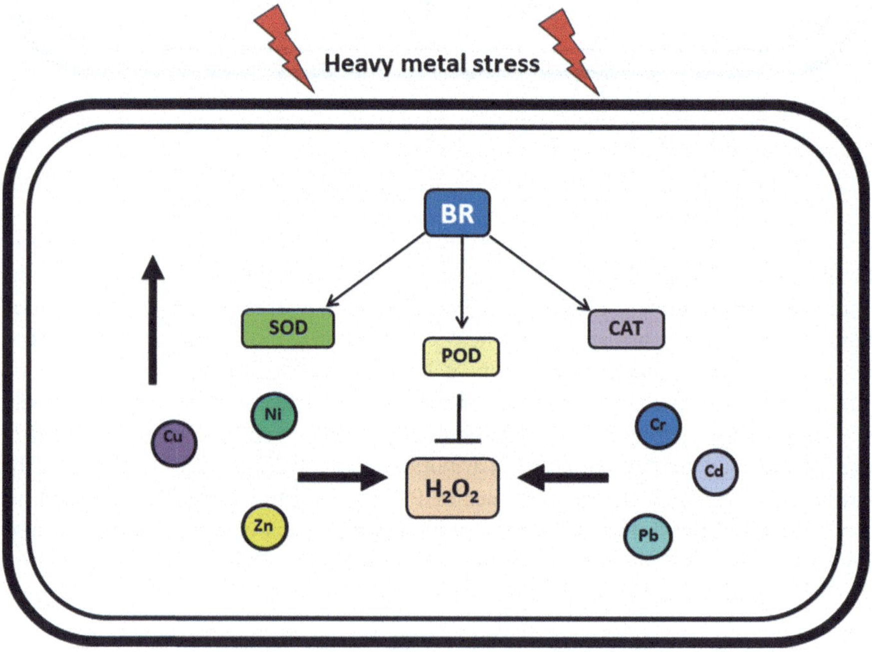

Fig. 13.14 Diagram showing the brassinosteroids-induced stress tolerance in plants, under heavy metal(loid) exposure. (Modified from: Bücker-Neto et al. 2017)

12 Conclusion

The changing climate and demands of the population explosion misbalance the metals/metalloids in the environment, which consequently emphasizes the metals uptake in the crops. The uptake of metals/metalloids above optimum range leads to the toxic manifestation in the crops. Against metals/metalloids toxicity, a crop activates their mitigation strategies to provide tolerance. In the different mitigation strategies, the antioxidants system mainly counteracts the generation of ROS in the crop plants. Simultaneously, the NO also supports the mitigation strategies by nitrosylation process to activate antioxidants system proteins. Similarly, in the plants, phytohormones regulate the different signaling cascades, which involve in the different mitigation strategies. However, the GABA shunt pathway protects plants by synthesizing the TCA cycle intermediate, which inhibits during stress. Similarly, the PAs mainly protect plants by the preservation of bimolecular structure integrity and modulation in the expression of osmotically responsive genes. However, during oxidative stress, the GRXs are involved in the maintaining redox state of the cell. Simultaneously, the glyoxalase system detoxifies the MG, which synthesize during stress. On the other hand, DHNs stabilize the free transition metal ions by chelating them with their motifs. The regulation of the stress-responsive phytohormones, i.e., auxin, cytokinin, brassinosteroids, abscisic acid, and ethylene, can help not only in enabling crop plant to be resilient toward climate-induced stress, i.e., drought and high temperature, but also in overcoming the heavy metal(loid)-induced stress by regulating its uptake. Along with these mitigation strategies, specific agronomic practices using precise use of phytochemicals can help in growing crop plants in future climate change scenario.

Acknowledgment The authors are thankful to Dilip K. Chakraborty toward his skillful help in drawings, the figures, and illustrations.

References

Abercrombie JM, Halfhill MD, Ranjan P, Rao MR, Saxton AM, Yuan JS, Stewart CN (2008) Transcriptional responses of *Arabidopsis thaliana* plants to as (V) stress. BMC Plant Biol 8:87

Alcázar R, Planas J, Saxena T, Zarza X, Bortolotti C, Cuevas J, Bitrián M, Tiburcio AF, Altabella T (2010) Putrescine accumulation confers drought tolerance in transgenic *Arabidopsis* plants over-expressing the homologous Arginine decarboxylase 2 gene. Plant Physiol Biochem 48:547–552

Al-Whaibi MH (2011) Plant heat shock proteins: a mini review. J King Saud Univ Sci 23(2):139–150

Alonso JM, Hirayama T, Roman G, Nourizadeh S, Ecker JR (1999) EIN2, a bifunctional transducer of ethylene and stress responses in Arabidopsis. Science 284(5423):2148–2152

Ban Q, Liu G, Wang Y (2011) A DREB gene from Limonium bicolor mediates molecular and physiological responses to copper stress in transgenic tobacco. J Plant Physiol 168:449–458

Basu A, Roychowdhury R, Bhattacharyya SS, Tah J (2012) Estimation of major heavy metals (Fe, Cu and Zn) in the fruit part of *Cucumis sativus* L. World J Sci Technol 2(7):01–03

Burnett EC, Desikan R, Moser RC, Neill SJ (2010) ABA activation of an MBP kinase in Pisum sativum epidermal peels correlates with stomatal responses to ABA. J Exp Bot 51(343):197–205
Beligni MV, Lamattina L (1999) Nitric oxide protects against cellular damage produced by methylviologen herbicides in potato plants. Nitric Oxide 3:199–208
Borovskii GB, Stupnikova IV, Antipina AA, Downs CA, Voinikov VK (2000) Accumulation of dehydrin-like proteins in the mitochondria of cold-treated plants. J Plant Physiol 156:797–800
Borysiuk K, Ostaszewska-Bugajska M, Vaultier MN, Hasenfratz-Sauder MP, Szal B (2018) Enhanced formation of methylglyoxal-derived advanced glycation end products in *Arabidopsis* under ammonium nutrition. Front Plant Sci 9:667
Bouche N, Fromm H (2004) GABA in plants: just a metabolite? Trends Plant Sci 9:110–115
Bouché N, Fait A, Bouchez D, Møller SG, Fromm H (2003) Mitochondrial succinic- semialdehyde dehydrogenase of the γ-aminobutyrate shunt is required to restrict levels of reactive oxygen intermediates in plants. Proc Natl Acad Sci 100:6843–6848
Bown AW, Shelp BJ (2016) Plant GABA: not just a metabolite. Trends Plant Sci 21:811–813
Briat JF, Lobreaux S (1997) Iron transport and storage in plants. Trends Plant Sci 2:187–193
Bücker-Neto L, Paiva ALS, Machado RD, Arenhart RA, Margis-Pinheiro M (2017) Interactions between plant hormones and heavy metal(loid)s responses. Genet Mol Biol 40(1):373–386
Bueso E, Alejandro S, Carbonell P, Perez-Amador MA, Fayos J, Bellés JM, Rodriguez PL, Serrano R (2007) The lithium tolerance of the *Arabidopsis* cat2 mutant reveals a cross-talk between oxidative stress and ethylene. Plant J 52:1052–1065
Camacho-Cristóbal JJ, Martín-Rejano EM, Herrera-Rodríguez MB, Navarro-Gochicoa MT, Rexach J, González-Fontes A (2015) Boron deficiency inhibits root cell elongation via an ethylene/auxin/ROS-dependent pathway in Arabidopsis seedlings. Journal of experimental botany 66(13):3831–3840
Cao S, Chen Z, Liu G, Jiang L, Yuan H, Ren G, Bian X, Jian H, Ma X (2009) The *Arabidopsis* ethylene-insensitive 2 gene is required for lead resistance. Plant Physiol Biochem 47:308–312
Cao F, Chen F, Sun H, Zhang G, Chen ZH, Wu F (2014) Genome-wide transcriptome and functional analysis of two contrasting genotypes reveals key genes for cadmium tolerance in barley. BMC Genomics 15:611
Carjuzaa P, Castellión M, Distéfano AJ, del Vas MS (2008) Detection and subcellular localization of dehydrin-like proteins in quinoa (*Chenopodium quinoa* Willd.) embryos. Protoplasma 233:149–156
Chakraborty S, Pattanayak A, Mandal S, Das M, Roychowdhury R (2014) An overview of climate change: causes, trends and implications. In: Roychowdhury R (ed) Crop improvement in the era of climate change. IK International Publishing House, New Delhi, pp 1–29
Chen D, Shao Q, Yin L, Younis A, Zheng B (2018) Polyamine function in plants: metabolism, regulation on development, and roles in abiotic stress responses. Front Plant Sci 9:1945. https://doi.org/10.3389/fpls.2018.01945
Close TJ (1996) Dehydrins: emergence of a biochemical role of a family of plant dehydration proteins. Physiol Plant 97(4):795–803.
Clemens S (2001) Molecular mechanisms of plant metal tolerance and homeostasis. Planta 212:475–486
Cobbet CS (2000) Phytochelatin biosynthesis and function in heavy metal detoxification. Curr Opinions Curr Biol 3:211–216
Cobbett C, Goldsbrough P (2002) Phytochelatins and metallothioneins: roles in heavy metal detoxification and homeostasis. Annu Rev Plant Biol 53:159–182
Cao S, Xu Q, Cao Y, Qian K, An K, Zhu Y, Binzeng H, Zhao H, Kuai B (2005) Loss-of-function mutations in DET2 gene lead to an enhanced resistance to oxidative stress in Arabidopsis. Physiol Plant 123(1):57–66.
Cortleven A, Schmülling T (2015) Regulation of chloroplast development and function by cytokinin. J Exp Bot 66:4999–5013
Danquah A, de Zelicourt A, Colcombet J, Hirt H (2014) The role of ABA and MAPK signaling pathways in plant abiotic stress responses. Biotechnol Adv 32:40–52

Danyluk J, Perron A, Houde M, Lamin A, Flower B, Benhamou N, Sarhan F (1998) Accumulation of an acidic dehydrin in the vicinity of the plasma membrane during cold acclimation of wheat. Plant Cell 10:623–638

DalCorso G, Farinati S, Furini A. Regulatory networks of cadmium stress in plants (2010) Plant Signal Behav 5(6):663–667

Das K, Roychoudhury A (2014) Reactive oxygen species (ROS) and response of antioxidants as ROS-scavengers during environmental stress in plants. Front Environ Sci 2:53

Dean RT, Fu S, Stocker R, Davis MJ (1997) Biochemistry and pathology of radical- mediated protein oxidation. Biochem J 324:1–18

Ding J, Ji C, Cai H (2017) Membrane transporter families of metal microelements make plants grow better and healthier. Asian J Plant Sci Res 7:1–27

Egerton-Warburton LM, Balsamo RA, Close TJ (1997) Temporal accumulation and ultrastructural localization of dehydrins in *Zea mays*. Physiol Plant 101:545–555

Finkelstein R (2013) Abscisic acid synthesis and response. The *Arabidopsis* book/Am Soc Plant Biol 11:e0166. https://doi.org/10.1199/tab.0166

Fomenko DE, Gladyshev VN (2002) CxxS: fold-independent redox motif revealed by genome-wide searches for thiol/disulfide oxidoreductase function. Protein Sci 11:2285–2296

Foyer CH, Noctor G (2011) Ascorbate and glutathione: the heart of the redox hub. Plant Physiol 155:2–18

Foyer CH, Shigeoka S (2011) Understanding oxidative stress and antioxidant functions to enhance photosynthesis. Plant Physiol 155:93–100

Fritioff Å, Kautsky L, Greger M (2005) Influence of temperature and salinity on heavy metal uptake by submersed plants. Environ Pollut 133:265–274

Fediuc E, Lips SH, Erdei L. O-acetylserine (thiol) lyase activity in Phragmites and Typha plants under cadmium and NaCl stress conditions and the involvement of ABA in the stress response (2005) J plant physiol 162(8):865–872

Fu Y, Tang J, Yao GF, Huang ZQ, Li YH, Han Z, Chen XY, Hu LY, Hu KD, Zhang H (2018) Central role of adenosine 5′-Phosphosulfate Reductase in the control of plant hydrogen Sulfide metabolism. Front Plant Sci 9:1404. https://doi.org/10.3389/fpls.2018.01404

Ghasemzadeh A, Ghasemzadeh N (2011) Flavonoids and phenolic acids: role and biochemical activity in plants and human. J Med Plant Res 5:6697–6703

Goldsborough P (2000) Metal tolerance in plants: the role of phytochelatins and metallothioneins. In: Terry N, Banuelos G (eds) Phytoremediation of contaminated soil and water. CRC Press LLC, Boca Raton, pp 221–233

González-Guerrero M, Escudero V, Saéz Á, Tejada-Jiménez M (2016) Transition metal transport in plants and associated endosymbionts: arbuscular mycorrhizal fungi and rhizobia. Front Plant Sci 7:1088

Guerinot ML (2000) The ZIP family of metal transporters. Biochim Biophys Acta 1465:190–198

Guo Y, Huang C, Xie Y, Song F, Zhou X (2010) A tomato glutaredoxin gene SlGRX1 regulates plant responses to oxidative, drought and salt stresses. Planta 232:1499–1509

Gupta K, Dey A, Gupta B (2013) Plant polyamines in abiotic stress responses. Acta Physiol Plant 35:2015–2036

Hall JL, Williams LE (2003) Transition metal transporters in plants. J Exp Bot 54:2601–2613

Hara M, Shinoda Y, Tanaka Y, Kuboi T (2009) DNA binding of citrus dehydrin promoted by zinc ion. Plant Cell Environ 32:532–541

Hasanuzzaman M, Nahar K, Alam M, Roychowdhury R, Fujita M (2013) Physiological, biochemical, and molecular mechanisms of heat stress tolerance in plants. Int J Mol Sci 14(5):9643–9684

Hasanuzzaman M, Fujita M (2013) Exogenous sodium nitroprusside alleviates arsenic-induced oxidative stress in wheat (Triticum aestivum L.) seedlings by enhancing antioxidant defense and glyoxalase system. Ecotoxicology 22(3):584–596

Hasanuzzaman M, Roychowdhury R, Karmakar J, Dey N, Nahar K, Fujita M (2015) Recent advances in biotechnology and genomic approaches for abiotic stress tolerance in crop plants. In: Devarajan T, Jeyabalan S (eds) Genomics and proteomics: concepts, technologies and applications. Apple Academic Press, pp 333–366 Apple Academic Press, Oakville, Canada.

He Y, Li W, Lv J, Jia Y, Wang M, Xia G (2011) Ectopic expression of a wheat MYB transcription factor gene, TaMYB73, improves salinity stress tolerance in *Arabidopsis thaliana*. J Exp Bot 63:1511–1522

Hoque TS, Hossain MA, Mostofa MG, Burritt DJ, Fujita M, Tran LSP (2016) Methylglyoxal: an emerging signaling molecule in plant abiotic stress responses and tolerance. Front Plant Sci 7:1341

Hossain MA, Hasanuzzaman M, Fujita M (2010) Up-regulation of antioxidant and glyoxalase systems by exogenous glycine betaine and proline in mung bean confer tolerance to cadmium stress. Physiol Mol Biol Plants 16(3):259–272

Hsu YT, Kao CH (2003) Role of abscisic acid in cadmium tolerance of rice (Oryza sativa L.) seedlings. Plant, Cell Environ 26(6):867–874

Hu KD, Hu L, Li YH, Zhang FQ, Zhang H (2007) Protective roles of nitric oxide on germination and antioxidant metabolism in wheat seeds under copper stress. Plant Growth Regul 53:173–183

Huang TL, Nguyen QTT, Fu SF, Lin CY, Chen YC, Huang HJ (2012) Transcriptomic changes and signalling pathways induced by arsenic stress in rice roots. Plant Mol Biol 80:587–608

Hussain SS, Ali M, Ahmad M, Siddique KH (2011) Polyamines: natural and engineered abiotic and biotic stress tolerance in plants. Biotechnol Adv 29:300–311

Ivanov R, Bauer P (2017) Sequence and coexpression analysis of iron-regulated ZIP transporter genes reveals crossing points between iron acquisition strategies in green algae and land plants. Plant Soil 418(1–2):61–73

Jin JW, Xu YF, Huang YF (2010) Protective effect of nitric oxide against arsenic-induced oxidative damage in tall fescue leaves. Afr J Biotech 9(11):1619–1627

Jozefczak M, Remans T, Vangronsveld J, Cuypers A (2012) Glutathione is a key player in metal-induced oxidative stress defenses. Int J Mol Sci 13:3145–3175

Kasinathan V, Wingler A (2004) Effect of reduced arginine decarboxylase activity on salt tolerance and on polyamine formation during salt stress in *Arabidopsis thaliana*. Physiol Plant 121:101–107

Kanwar MK, Bhardwaj R, Chowdhary SP, Arora P, Sharma P, Kumar S (2013) Isolation and characterization of 24-Epibrassinolide from Brassica juncea L. and its effects on growth, Ni ion uptake, antioxidant defense of Brassica plants and in vitro cytotoxicity. Acta Physiol Plant 35(4):1351–1362

Keiluweit M, Bougoure JJ, Nico PS, Pett-Ridge J, Weber PK, Kleber M (2015) Mineral protection of soil carbon counteracted by root exudates. Nat Clim Chang 5:588

Kosová K, Prášil IT, Vítámvás P (2010) Role of Dehydrins in plant stress response. In: Handbook of plant and crop stress, pp 240–265. https://doi.org/10.1201/b10329-13

Kramer U, Talke I, Hanikenne M (2007) Transition metal transport. FEBS Lett 581:2263–2272

Kumar N, Dubey AK, Upadhyay AK, Gautam A, Ranjan R, Srikishna S, Sahu N, Behera SK, Mallick S (2017) GABA accretion reduces Lsi-1 and Lsi-2 gene expressions and modulates physiological responses in Oryza sativa to provide tolerance towards arsenic. Sci Rep 7:8786

Khan MI, Nazir F, Asgher M, Per TS, Khan NA (2015) Selenium and sulfur influence ethylene formation and alleviate cadmium-induced oxidative stress by improving proline and glutathione production in wheat. J Plant Physiol 173:9–18

Kim YH, Khan AL, Kim DH, Lee SY, Kim KM, Waqas M, Jung HY, Shin JH, Kim JG, Lee IJ (2014) Silicon mitigates heavy metal stress by regulating P-type heavy metal ATPases, Oryza sativa low silicon genes, and endogenous phytohormones. BMC Plant Biol 14(1):13

Kumar N, Gautam A, Dubey AK, Ranjan R, Pandey A, Kumari B, Singh G, Mandotra S, Chauhan PS, Srikrishna S, Dutta V (2019) GABA mediated reduction of arsenite toxicity in rice seedling through modulation of fatty acids, stress responsive amino acids and polyamines biosynthesis. Ecotoxicol Environ Saf 173:15–27

Laspina NV, Groppa MD, Tomaro ML, Benavides MP (2005) Nitric oxide protects sunflower leaves against Cd-induced oxidative stress. Plant Sci 169(2):323–330

Lane BG, Kajioka R, Kennedy TD (1987) The wheat germ Ec protein is a zinc containing metallothionein. Biochem Cell Biol 65:1001–1005

Lee KO, Lee JR, Yoo JY, Jang HH, Moon JC, Jung BG, Chi YH, Park SK, Lee SS, Lim CO, Yun DJ (2002) GSH-dependent peroxidase activity of the rice (*Oryza sativa*) glutaredoxin, a thioltransferase. Biochem Biophys Res Commun 296:1152–1156

Li S (2014) Redox modulation matters: emerging functions for glutaredoxins in plant development and stress responses. Plan Theory 3:559–582

Li G, Meng X, Wang R, Mao G, Han L, Liu Y, Zhang S (2012) Dual-level regulation of ACC synthase activity by MPK3/MPK6 cascade and its downstream WRKY transcription factor during ethylene induction in Arabidopsis. PLoS Genet 8(6)

Liu Y, Song Q, Li D, Yang X, Li D (2017a) Multifunctional roles of plant dehydrins in response to environmental stresses. Front Plant Sci 8:1018

Liu Y, Wang L, Zhang T, Yang X, Li D (2017b) Functional characterization of KS-type dehydrin ZmDHN13 and its related conserved domains under oxidative stress. Sci Rep 7:7361

Liu J, Wang J, Lee S, Wen R (2018) Copper-caused oxidative stress triggers the activation of antioxidant enzymes via ZmMPK3 in maize leaves. PLoS One 13:e0203612

Maestri E, Marmiroli M, Visioli G, Marmiroli N (2010) Metal tolerance and hyperaccumulation: costs and trade-offs between trade and environment. Environ Exp Bot 68:1–13

Mallick S, Kumar N, Singh AP, Sinam G, Yadav RN, Sinha S (2013) Role of sulfate in detoxification of arsenate-induced toxicity in *Zea mays* L.(SRHM 445): nutrient status and antioxidants. J Plant Interact 8:140–154

Masood A, Khan MI, Fatma M, Asgher M, Per TS, Khan NA (2016) Involvement of ethylene in gibberellic acid-induced sulfur assimilation, photosynthetic responses, and alleviation of cadmium stress in mustard. Plant Physiol. Biochem 104:1–10

Manara A (2012) "Plant responses to heavy metal toxicity," in Plants and Heavy Metals,, A. Furini, Ed., SpringerBriefs in Molecular Science, pp. 27–53, Springer, Dordrecht, Netherlands.

Marschner H (2011) Marschner's mineral nutrition of higher plants. Academic press United States of America

Matsuo S, Kikuchi K, Fukuda M, Honda I, Imanishi S (2012) Roles and regulation of cytokinins in tomato fruit development. J Exp Bot 63:5569–5579

Memon AR, Schroder P (2009) Implications of metal accumulation mechanisms to phytoremediation. Environ Sci Pollut Res 16:162–175

Meyer Y, Buchanan BB, Vignols F, Reichheld JP (2009) Thioredoxins and glutaredoxins: unifying elements in redox biology. Annu Rev Genet 43:335–367

Mittler R (2002) Oxidative stress, antioxidants and stress tolerance. Trends Plant Sci 7:405–410

Mohan TC, Castrillo G, Navarro C, Zarco-Fernández S, Ramireddy E, Mateo C, Zamarreño AM, Paz-Ares J, Muñoz R, García-Mina JM, Hernández LE (2016) Cytokinin determines thiol-mediated arsenic tolerance and accumulation. Plant Physiol 171:1418–1426

Maksymiec W (2007) Signaling responses in plants to heavy metal stress. Acta Physiol Plant 29(3):177

Mueller JK, Heckathorn SA, Fernando D (2003) Identification of a chloroplast dehydrin in leaves of mature plants. Int J Plant Sci 164:535–542

Ndayiragije A, Lutts S (2006) Do exogenous polyamines have an impact on the response of a salt sensitive rice cultivar to NaCl? J Plant Physiol 163:506–516

Neill SJ, Desikan R, Clarke A, Hurst RD, Hancock JT (2002) Hydrogen peroxide and nitric oxide as signalling molecules in plants. J Exp Bot 53:1237–1247

Ning X, Sun Y, Wang C, Zhang W, Sun M, Hu H, Liu J, Yang L (2018) A Rice CPYC-type Glutaredoxin *Os*GRX20 in protection against bacterial blight, methyl Viologen and salt stresses. Front Plant Sci 9:111

Noyes PD, McElwee MK, Miller HD, Clark BW, Van Tiem LA, Walcott KC, Erwin KN, Levin ED (2009) The toxicology of climate change: environmental contaminants in a warming world. Environ Int 35:971–986

Parankusam S, Adimulam SS, Bhatnagar-Mathur P, Sharma KK (2017) Nitric oxide (NO) in plant heat stress tolerance: current knowledge and perspectives. Front Plant Sci 8:1582

Pegg AE, Casero Jr. RA (2011) Polyamines: Methods and Protocols, Methods in Molecular Biology, Osaka, Japan vol. 720, Humana Press, pp 3–35. https://doi.org/10.1007/978-1-61779-034-8_1

Pinto E, Ferreira IMPLVO (2015) Cation transporters/channels in plants: tools for nutrient biofortification. J Plant Physiol 179:64–82

Qaderi MM, Kurepin LV, Reid DM (2012) Effects of temperature and watering regime on growth, gas exchange and abscisic acid content of canola (*Brassica napus*) seedlings. Environ Exp Bot 75:107–113

Rai A, Tripathi P, Dwivedi S, Dubey S, Shri M, Kumar S, Tripathi PK, Dave R, Kumar A, Singh R, Adhikari B (2011) Arsenic tolerances in rice (*Oryza sativa*) have a predominant role in transcriptional regulation of a set of genes including Sulphur assimilation pathway and antioxidant system. Chemosphere 82:986–995

Ramesh SA, Tyerman SD, Xu B, Bose J, Kaur S, Conn V, Domingos P, Ullah S, Wege S, Shabala S, Feijó JA (2015) GABA signalling modulates plant growth by directly regulating the activity of plant-specific anion transporters. Nat Commun 6:7879

Robinson NJ, Tommey AM, Kuske C, Jackson PJ (1993) Plant metallothioneins. Biochem J 295:1–10

Rockel P, Strube F, Rockel A, Wildt J, Kaiser WM (2002) Regulation of nitric oxide (NO) production by plant nitrate reductase *in vivo* and *in vitro*. J Exp Bot 53:103–110

Ronzan M, Piacentini D, Fattorini L, Della RF, Eiche E, Riemann M, Altamura MM, Falasca G (2018) Cadmium and arsenic affect root development in *Oryza sativa* L. negatively interacting with auxin. Environ Exp Bot 151:64–75

Rouhier N, San Koh C, Gelhaye E, Corbier C, Favier F, Didierjean C, Jacquot JP (2008) Redox based anti-oxidant systems in plants: biochemical and structural analyses. Biochim Biophys Acta Gen Subj 1780:1249–1260

Roychowdhury R (2014) Crop improvement in the era of climate change. IK International Publishing House, New Delhi, p 496

Roychowdhury R, Tah J (2011) Differential response by different parts of *Solanum melongena* L. for heavy metal accumulation. Plant Sci Feed 1(6):80–83

Roychowdhury R, Khan MH, Choudhury S (2018) Arsenic in rice: an overview on stress implications, tolerance and mitigation strategies. In: Hasanuzzaman M, Nahar K, Fujita M (eds) Plants under metal and metalloid stress. Springer, Singapore, pp 401–415

Roychowdhury R, Khan MH, Choudhury S (2019) Physiological and molecular responses for metalloid stress in rice – a comprehensive overview. In: Hasanuzzaman M, Fujita M, Nahar K, Biswas J (eds) Advances in rice research for abiotic stress tolerance. Woodhead Publishing, Elsevier, pp 341–369. United Kingdom and United States

Růžička K, Ljung K, Vanneste S, Podhorská R, Beeckman T, Friml J, Benková E (2007) Ethylene regulates root growth through effects on auxin biosynthesis and transport-dependent auxin distribution. Plant Cell 19(7):2197–2212

Rucińska-Sobkowiak R (2016) Water relations in plants subjected to heavy metal stresses. Acta Physiol Plant 38:257

Salazar-Cerezo S, Martínez-Montiel N, García-Sánchez J, Pérez-y-Terrón R, Martínez-Contreras RD (2018) Gibberellin biosynthesis and metabolism: a convergent route for plants, fungi and bacteria. Microbiol Res 208:85–98

Salin ML (1991) Chloroplast and mitochondrial mechanisms for protection against oxygen toxicity. Free Radic Res Commun 12:851–858

Sankaranarayanan S, Jamshed M, Kumar A, Skori L, Scandola S, Wang T, Spiegel D, Samuel M (2017) Glyoxalase goes green: the expanding roles of glyoxalase in plants. Int J Mol Sci 18:898

Savelieva EM, Oslovsky VE, Karlov DS, Kurochkin NN, Getman IA, Lomin SN, Sidorov GV, Mikhailov SN, Osolodkin DI, Romanov GA (2018) Cytokinin activity of N 6- benzylade-

nine derivatives assayed by interaction with the receptors in planta, in vitro, and in silico. Phytochemistry 149:161–177
Srivastava S, Srivastava AK, Suprasanna P, D'souza SF (2013) Identification and profiling of arsenic stress-induced microRNAs in Brassica juncea. J Exp Bot 64(1):303–315
Schellingen K, Van Der Straeten D, Vandenbussche F, Prinsen E, Remans T, Vangronsveld J, Cuypers A (2014) Cadmium-induced ethylene production and responses in Arabidopsis thaliana rely on ACS2 and ACS6 gene expression. BMC Plant Biol 14(1):214
Schmöger ME, Oven M, Grill E (2000) Detoxification of arsenic by phytochelatins in plants. Plant Physiol 122:793–802
Seneviratne SI, Nicholls N, Easterling D, Goodess CM, Kanae S, Kossin J, Luo Y, Marengo J, McInnes K, Rahimi M, Reichstein M, Sorteberg A, Vera C, Zhang X (2012) Changes in climate extremes and their impacts on the natural physical environment. In: Field CB, Barros V, Stocker TF, Qin D, Dokken DJ, Ebi KL, Mastrandrea MD, Mach KJ, Plattner G-K, Allen SK, Tignor M, Midgley PM (eds) Managing the risks of extreme events and disasters to advance climate change adaptation. A Special Report of Working Groups I and II of the Intergovernmental Panel on Climate Change (IPCC). Cambridge University Press, Cambridge/New York, pp 109–230
Siddiqui MH, Al-Whaibi MH, Basalah MO (2011) Interactive effect of calcium and gibberellin on nickel tolerance in relation to antioxidant systems in *Triticum aestivum* L. Protoplasma 248:503–511
Singh S, Mosley LM (2003) Trace metal levels in drinking water on Viti Levu, Fiji Islands. S Pac J Nat Appl Sci 21:31–34
Singh HP, Batish DR, Kaur G, Arora K, Kohli RK (2008) Nitric oxide (as sodium nitroprusside) supplementation ameliorates cd toxicity in hydroponically grown wheat roots. Environ Exp Bot 63:158–167
Singla-Pareek SL, Yadav SK, Pareek A, Reddy MK, Sopory SK (2006) Transgenic tobacco overexpressing glyoxalase pathway enzymes grow and set viable seeds in zinc- spiked soils. Plant Physiol 140:613–623
Singla-Pareek SL, Yadav SK, Pareek A, Reddy MK, Sopory SK (2008) Enhancing salt tolerance in a crop plant by overexpression of glyoxalase II. Transgenic Res 17:171–180
Song FL, Gan RY, Zhang Y, Xiao Q, Kuang L, Li HB (2010) Total phenolic contents and antioxidant capacities of selected Chinese medicinal plants. Int J Mol Sci 11:2362–2372
Sponsel VM, Hedden P (2010) Gibberellin biosynthesis and inactivation. In: Plant hormones. Springer, Dordrecht, pp 63–94
Srivastava S, Srivastava AK, Suprasanna P, D'souza SF (2009) Comparative biochemical and transcriptional profiling of two contrasting varieties of *Brassica juncea* L. in response to arsenic exposure reveals mechanisms of stress perception and tolerance. J Exp Bot 60:3419–3431
Stroiński A, Chadzinikolau T, Giżewska K, Zielezińska M (2010) ABA or cadmium induced phytochelatin synthesis in potato tubers. Biol Plantarum 54(1):117–120
Steffens B (2014) The role of ethylene and ROS in salinity, heavy metal, and flooding responses in rice. Front Plant Sci 5:685
Steward FC (1949) γ-Aminobutyric acid: a constituent of the potato tuber? Science 110:439–440
Stöhr C, Strube F, Marx G, Ullrich WR, Rockel P (2001) A plasma membrane-bound enzyme of tobacco roots catalyses the formation of nitric oxide from nitrite. Planta 212:835–841
Sun X, Lin H-H (2010) Role of plant dehydrins in antioxidation mechanisms Biologia 65(5):755–759 DOI: 10.2478/s11756-010-0090-y
Sun C, Liu L, Yu Y, Liu W, Lu L, Jin C, Lin X (2015) Nitric oxide alleviates aluminum-induced oxidative damage through regulating the ascorbate-glutathione cycle in roots of wheat. J Integr Plant Biol 57:550–561
Sundaram S, Rathinasabapathi B (2010) Transgenic expression of fern *Pteris vittata* glutaredoxin PvGrx5 in *Arabidopsis thaliana* increases plant tolerance to high temperature stress and reduces oxidative damage to proteins. Planta 231:361

Sundaram S, Rathinasabapathi B, Ma LQ, Rosen BP (2008) An arsenate-activated glutaredoxin from the arsenic hyperaccumulator fern *Pteris vittata* L. regulates intracellular arsenite. J Biol Chem 283:6095–6101

Sundaram S, Wu S, Ma LQ, Rathinasabapathi B (2009) Expression of a *Pteris vittata* glutaredoxin PvGRX5 in transgenic *Arabidopsis thaliana* increases plant arsenic tolerance and decreases arsenic accumulation in the leaves. Plant Cell Environ 32:851–858

Suzuki M, Takahashi M, Tsukamoto T, Watanabe S, Matsuhashi S, Yazaki J, Kishimoto N, Kikuchi S, Nakanishi H, Mori S, Nishizawa NK (2006) Biosynthesis and secretion of mugineic acid family phytosiderophores in zinc-deficient barley. Plant J 48:85–97

Svensson J, Ismail AM, Palva ET, Close TJ (2002) Dehydrins. In: Storey KB, Storey JM (eds) Sensing signaling and cell adaptation. Elsevier, Amsterdam/Boston

Talke I, Hanikenne M, Krämer U (2006) Zn dependent global transcriptional control, transcriptional de-regulation and higher gene copy number for genes in metal homeostasis of the hyperaccumulator Arabidopsis halleri. Plant Physiol 142:148–167

Tang W, Newton RJ (2005) Polyamines reduce salt-induced oxidative damage by increasing the activities of antioxidant enzymes and decreasing lipid peroxidation in *Virginia pine*. Plant Growth Regul 46:31–43

Tewari RK, Kim S, Hahn EJ, Paek KY (2008) Involvement of nitric oxide-induced NADPH oxidase in adventitious root growth and antioxidant defense in Panax ginseng. Plant Biotechnol Rep 2:113–122

Trinh NN, Huang TL, Chi WC, Fu SF, Chen CC, Huang HJ (2014) Chromium stress response effect on signal transduction and expression of signaling genes in rice. Physiol Plant 150:205–224

Tripathi RD, Tripathi P, Dwivedi S, Dubey S, Chakrabarty D (2012) Arsenomics: omics of arsenic metabolism in plants. Front Physiol 3:275

Tun NN, Santa-Catarina C, Begum T, Silveira V, Handro W, Floh EI, Scherer GF (2006) Polyamines induce rapid biosynthesis of nitric oxide (NO) in Arabidopsis thaliana seedlings. Plant Cell Physiol 47:346–354

Urano K, Maruyama K, Ogata Y, Morishita Y, Takeda M, Sakurai N, Suzuki H, Saito K, Shibata D, Kobayashi M, Yamaguchi-Shinozaki K (2009) Characterization of the ABA-regulated global responses to dehydration in Arabidopsis by metabolomics. Plant J 57:1065–1078

van de Mortel JE, Villanueva LA, Schat H, Kwekkeboom J, Coughlan S, Moerland PD, Ver Loren van Themaat E, Koornneef M, Aarts MGM (2006) Large expression differences in genes for iron and Zn homeostasis, stress response, and lignin biosynthesis distinguish roots of Arabidopsis thaliana and the related metal hyperaccumulator Thlaspi caerulescens. Plant Physiol 142:1127–1147

Vaseva II, Grigorova BS, Simova-Stoilova LP, Demirevska KN, Feller U (2010) Abscisic acid and late embryogenesis abundant protein profile changes in winter wheat under progressive drought stress. Plant Biol 12:698–707

Verma K, Mehta SK, Shekhawat GS (2013) Nitric oxide (NO) counteracts cadmium induced cytotoxic processes mediated by reactive oxygen species (ROS) in Brassica juncea: cross-talk between ROS, NO and antioxidant responses. Biometals 26(2):255–269

Vera-Sirera F, Minguet EG, Singh SK, Ljung K, Tuominen H, Blázquez MA, Carbonell J (2010) Role of polyamines in plant vascular development. Plant Physiol Biochem 48:534–539

Vanneste S, Friml J (2009) Auxin: a trigger for change in plant development. Cell 136(6):1005–1016

Vranová E, Langebartels C, Van Montagu M, Inzé D, Van Camp W (2000) Oxidative stress, heat shock and drought differentially affect expression of a tobacco protein phosphatase 2C. J Exp Bot 51:1763–1764

Wang YS, Yang ZM (2005) Nitric oxide reduces aluminum toxicity by preventing oxidative stress in the roots of Cassia tora L. Plant Cell Physiol 46:1915–1923

Wang X, Shi G, Xu Q, Hu J (2007) Exogenous polyamines enhance copper tolerance of *Nymphoides peltatum*. J Plant Physiol 164:1062–1070

Wang R, Wang J, Zhao L, Yang S, Song Y (2015) Impact of heavy metal stresses on the growth and auxin homeostasis of Arabidopsis seedlings. Biometals 28(1):123–132

Watson MB, Emory KK, Piatak RM, Malmberg RL (1998) Arginine decarboxylase (polyamine synthesis) mutants of Arabidopsis thaliana exhibit altered root growth. Plant J 13:231–239

Wijngaard RR, van der Perk M, der Grift B, de Nijs TCM, Bierkens MFP (2017) The impact of climate change on metal transport in a lowland catchment. Water Air Soil Pollut 228:107

Williams LE, Pittman JK, Hall JL (2000) Emerging mechanisms for heavy metal transport in plants. Biochim Biophys Acta Biomembr 1465:104–126

Wojtania A, Skrzypek E, Gabryszewska E (2016) Morphological and biochemical responses to gibberellic acid in Magnolia×'Spectrum' in vitro. Acta Biologica Cracoviensia s Botanica 58:103–111

Xu J, Zhang YX, Wei W, Han L, Guan ZQ, Wang Z, Chai TY (2008) BjDHNs confer heavy-metal tolerance in plants. Mol Biotechnol 38:91–98

Xia XJ, Huang LF, Zhou YH, Mao WH, Shi K, Wu JX, Asami T, Chen Z, Yu JQ (2009) Brassinosteroids promote photosynthesis and growth by enhancing activation of Rubisco and expression of photosynthetic genes in Cucumis sativus. Planta 230(6):1185

Yadav SK, Singla-Pareek SL, Ray M, Reddy MK, Sopory SK (2005) Transgenic tobacco plants overexpressing glyoxalase enzymes resist an increase in methylglyoxal and maintain higher reduced glutathione levels under salinity stress. FEBS Lett 579:6265–6271

Yadav SK, Singla-Pareek SL, Sopory SK (2008) An overview on the role of methylglyoxal and glyoxalases in plants. Drug Metabol Drug Interact 23:51–68

Yamaguchi K, Takahashi Y, Berberich T, Imai A, Miyazaki A, Takahashi T, Michael A, Kusano T (2006) The polyamine spermine protects against high salt stress in *Arabidopsis thaliana*. FEBS Lett 580:6783–6788

Yamaguchi K, Takahashi Y, Berberich T, Imai A, Takahashi T, Michael AJ, Kusano T (2007) A protective role for the polyamine spermine against drought stress in *Arabidopsis*. Biochem Biophys Res Commun 352:486–490

Yeh CM, Hsiao LJ, Huang HJ (2004) Cadmium activates a mitogen-activated protein kinase gene and MBP kinases in rice. Plant Cell Physiol 45:1306–1312

Yin H, Li Y, Xiao J, Xu Z, Cheng X, Liu Q (2013) Enhanced root exudation stimulates soil nitrogen transformations in a subalpine coniferous forest under experimental warming. Glob Chang Biol 19:2158–2167

Yiu JC, Liu CW, Fang DYT, Lai YS (2009) Waterlogging tolerance of welsh onion (*Allium fistulosum* L.) enhanced by exogenous spermidine and spermine. Plant Physiol Biochem 47:710–716

Yu L, Longqing L, Qiang Z, Yiming Y, Heling W, Ruijun W, Jihui Z (2013) Influence of temperature on the heavy metal accumulation of five vegetable species in semiarid area of Northwest China. Chem Ecol 29:353–365

Yang SF, Hoffman NE. Ethylene biosynthesis and its regulation in higher plants (1984) Annu Rev Plant Physiol 35(1):155–189

Yuan HM, Huang X (2016) Inhibition of root meristem growth by cadmium involves nitric oxide-mediated repression of auxin accumulation and signalling in Arabidopsis. Plant, Cell Environ 39(1):120–135

Yuan HM, Xu HH, Liu WC, Lu YT (2013) Copper regulates primary root elongation through PIN1-mediated auxin redistribution. Plant and Cell Physiol 54(5):766–778

Yusuf M, Fariduddin Q, Hayat S, Hasan SA, Ahmad A (2011) Protective response of 28 homo brassinolide in cultivars of Triticum aestivum with different levels of nickel. Arch Environ Cont Toxicol 60(1):68–76

Yusuf M, Fariduddin Q, Ahmad A (2012) 24-Epibrassinolide modulates growth, nodulation, antioxidant system, and osmolyte in tolerant and sensitive varieties of Vigna radiata under different levels of nickel: a shotgun approach. Plant Physiol Biochem 57:143–53

Yu Z, Wang X, Zhang L (2018) Structural and functional dynamics of dehydrins: a plant protector protein under abiotic stress. Int J Mol Sci 19:3420

Zaffagnini M, Michelet L, Massot V, Trost P, Lemaire SD (2008) Biochemical characterization of glutaredoxins from *Chlamydomonas reinhardtii* reveals the unique properties of a chloroplastic CGFS-type glutaredoxin. J Biol Chem 283:8868–8876

Zelinová V, Alemayehu A, Bočová B, Huttová J, Tamás L (2015) Cadmium-induced reactive oxygen species generation, changes in morphogenic responses and activity of some enzymes in barley root tip are regulated by auxin. Biologia 70(3):356–364

Zhang W, Jiang B, Li W, Song H, Yu Y, Chen J (2009) Polyamines enhance chilling tolerance of cucumber (*Cucumis sativus* L.) through modulating antioxidative system. Sci Hortic 122:200–208

Zhang Z, Liao H, Lucas WJ (2014) Molecular mechanisms underlying phosphate sensing, signaling, and adaptation in plants. J Integ Plant Biol 56(3):192–220

Zhu XF, Jiang T, Wang ZW, Lei GJ, Shi YZ, Li GX, Zheng SJ (2012) Gibberellic acid alleviates cadmium toxicity by reducing nitric oxide accumulation and expression of IRT1 in *Arabidopsis thaliana*. J Hazard Mater 239:302–307

Zhu XF, Wang ZW, Dong F, Lei GJ, Shi YZ, Li GX, Zheng SJ (2013) Exogenous auxin alleviates cadmium toxicity in *Arabidopsis thaliana* by stimulating synthesis of hemicellulose 1 and increasing the cadmium fixation capacity of root cell walls. J Hazard Mater 263:398–403

Chapter 14
Arsenic-Induced Stress and Mitigation Strategies in Plants

Iti Sharma

Abstract Arsenic (As) is a toxic metalloid and inorganic arsenic is carcinogenic to humans and animals. About 50 ppb (μg/L) has been the standard for arsenic in drinking water in the USA since 1942. In the 1960s, published data from Taiwan indicated that arsenic in drinking water could cause skin cancer. In 2001, USEPA reduced the limit from 50 to 10 μg/L. According to WHO, the newly recommended guideline value for drinking water is 10 μg/L (Australia 7 μg/L). Arsenic enters in the plant system via contaminated irrigation water or soil. The toxic effect of arsenic in plants causes many metabolic disorders and often leads to death of the plants. Plants are equipped with multilayer defense weapons to encounter arsenic toxicity inside the system. Arsenic is analogous to phosphate and competes with phosphate molecules at transporter site. Once it enters in the plant cell, phytochelatins are ready for sequestration and accumulation of arsenic in vacuoles. Plants have enzymatic and nonenzymatic defense systems for arsenic-mediated oxidative damage. The chapter presents the latest research and findings for interaction between plant cell and arsenic.

Keywords Arsenic · Plants · Transporter · Defense · Methylation · Antioxidants · Phytochelatins

Abbreviations

AR	Arsenate reductase
DMA	Dimethyl arsenic acid
GR	Glutathione reductase
Grx	Glutaredoxin-linked arsenate reductase
MMA	Monomethylarsonic acid
TMAsO	Trimethylarsine oxide
Trx	Thioredoxin-coupled arsenate reductase

I. Sharma (✉)
Birla Institute of Technology and Science (BITS), Pilani, Rajasthan, India

R. Roychowdhury et al. (eds.), *Sustainable Agriculture in the Era of Climate Change*, https://doi.org/10.1007/978-3-030-45669-6_14

1 Introduction

Arsenic (As) contamination in soil, air, and water is a rising apprehension around the world. Arsenic is a crystalline toxic and carcinogenic metalloid from 5A group of a transition element. It exhibits various oxidation states such as As^{+3} and As^{+5} due to $4s^24p^3$ outer electronic configuration (Matera and Le Hecho 2001).

It is a major element of the earth's crust and exposed to the environment by mining activities and groundwater in rivers (Drewniak and Sklodowska 2013). Sulfide ores of gold, iron, and copper contain arsenic, for example, realgar (As4S4), arsenopyrite (FeAsS), and orpiment (As2S3) are common arsenic minerals (Hoang et al. 2010). The solubility of arsenic in water arises many health concerns especially in Southeast Asia (Abbas et al. 2018). The US Environmental Protection Agency (USEPA 2001) and WHO set 10 $\mu g\ L^{-1}$ as the permissible limit of arsenic in drinking water. Natural sources as volcanic discharge, weathering of rocks, and hot springs and anthropogenic sources such as the use of pesticide, herbicide, wood preservative, mining, combustion, and industrial effluents are also considered as major factors for increasing arsenic contamination in soil, irrigation water, and food chain. Freshwater resources have arsenic concentration ranging from 0.15 to 0.45 $\mu g\ L^{-1}$, where $<2\ \mu g\ L^{-1}$ arsenic has been reported in marine water (Ng 2005). Generally, soil arsenic ranges from 0.2 to 41 mg Kg^{-1}, but the continuous use of pesticide and herbicide causes arsenic pollution 400–900 mg Kg^{-1} in field soil (Kabata-Pendias and Pendias 2001).

2 Interaction Between Arsenic and Plants

The plants are exposed to arsenic through contaminated soil and water. The inorganic forms such as arsenite and arsenate are more toxic than monomethyl arsine, dimethyl arsine, arsenobetaine, and arsenocholine. Microbes can methylate arsenic in the form of monomethylarsonic acid (MMA), dimethyl arsenic acid (DMA), and trimethylarsine oxide (TMAsO), and all these are bioavailable for plants (Ye et al. 2012). The MMA and DMA are basic components of herbicide and pesticide used in agricultural practices (Williams et al. 2007). Methylated arsenic forms are also observed in root and shoot of maize, *Holcus lanatus*, *Trifolium pretense*, and rice (Zavala et al. 2008; Roychowdhury et al. 2018, 2019). The uptake of DMA may be correlated with phosphate deficiency in the nearby area (Abbas and Meharg 2008). Another speculated report shows arsenate in epidermis and arsenite in ground tissue, endodermis, pericycle, xylem, and phloem in the root of wheat and rice exposed to arsenate (Kopittke et al. 2013).

3 Bioavailability of Arsenic to Plants

Arsenic forms complex molecules with ferrous and aluminum in the soil. Arsenate is present in immobilized complex form than arsenite within the aerobic environment. The anaerobic conditions support the reduction of arsenate to arsenite; therefore, higher influx rate of arsenite is reported by iron plaques formed in roots of rice crop grown in water-logging fields (Yamaguchi et al. 2014). The solubility and bioavailability areas are affected by soil pH, speciation, and leaching; therefore, a limited amount of arsenic is available to plants. The arsenite (As^{+3}) is a more toxic and soluble form than arsenate (As^{+5}). According to a study, arsenite adsorbed on soil particle at pH 7.0 and arsenate at pH 4.0 (Pierce and Moore 1982). The reduction reactions in rhizospheric microorganism play a significant role in the bioavailability of arsenic to plant roots (Nearing et al. 2014). It has been reported by various researchers that root and shoot extract has 95% arsenite in plants grown in arsenate-rich medium (Carey et al. 2010). The genes involved in arsenic uptake are identified by researchers in different plant species.

4 Arsenic Uptake Mechanism in Plants

The uptake mechanism for arsenate and arsenite is different as arsenate enters via phosphate transporter system whereas arsenite through nodulin 26-like intrinsic protein (NIP) aquaglyceroporin or natural channels (Zhao et al. 2009). The arsenite is analogous to silicon due to the same tetrahedral and Pka value; therefore, silicon transporter Lsi1 (*NIP2;1*) also supports AsIII influx in rice roots (Ma et al. 2008). Figure 14.1 summarizes the newly identified arsenite (As^{+3}) transporters and their functions in plant cells. The similar structure of arsenate and phosphate activates phosphate transports to facilitate the transport of arsenate (Sharma and Travlos 2012). Recently traced phosphate transporters are AtPHT1;4, AtPHT1;7, OsPHT1, and PvPht1;3 reported in *Arabidopsis*, *Oryza sativa*, and *Pteris vittata* (LeBlanc et al. 2013; Kamiya et al. 2013; DiTusa et al. 2016; Roychowdhury et al. 2018, 2019). Monomethylarsonic acid (MMA), Dimethyl arsenic acid (DMA) and trimethylarsine oxide (TMAsO) are taken up by aquaglyceroporins and aquaporin *Lsi1* and *NIP2;1* in rice roots. The translocation rate of DMA is faster than inorganic arsenic in roots and shoots of rice (Carey et al. 2011).

5 Phytochemistry of Arsenic Toxicity

Arsenic is considered as an important toxic and stress factor in plants which leads to metabolic disorders during the transition of oxidation states. The toxicity appears in a symptom such as wilting of leaves, slow root growth and shoot growth, leaf

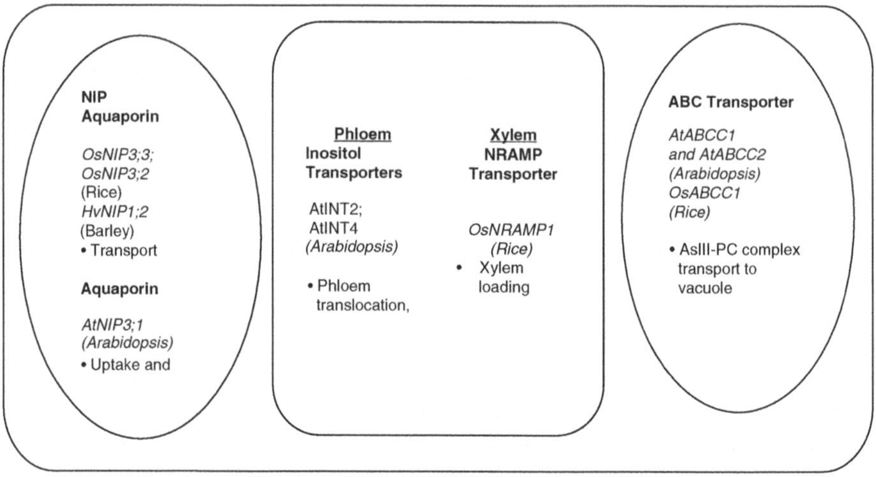

Fig. 14.1 Arsenite (As^{+3}) transporters and their functions in plant cells

necrosis, violet leaf color, and ultimately plant loss (Adriano 1986). Most of the arsenite is chelated with phytochelatins, but some amount of arsenic can affect the activity of enzymes and proteins due to affinity toward sulfhydryl groups (Oremland and Stolz 2003). Arsenate remains neutral, but methylated arsenic (DMA) can form dimethylarsinothioyl glutathione (DMAS-GS) with thiol groups residing in sulfur-rich plant cells of *Brassica oleracea*. Arsenate is analogous to phosphate; therefore, it can disrupt oxidative phosphorylation and production of ATP (Oremland and Stolz 2003). The replacement of phosphate with arsenate in ATP is known as aminolysis (Meharg and MacNair 1994). Fowler (1977) also reported that arsenate can replace phosphate in DNA molecule and consequently damage the nucleic acid of the organism.

6 Defense Against Arsenic in Plants

The plant copes with potential stress factors by developing stress resistance. This implies all morphological and physiological measures required to inhibit the stress (Sharma and Anita 2016). The following strategies can be differentiated according to the requirement:

1. Tolerance toward the stress factor without suffering great damage
2. Defense against the stress factor by suitable protective mechanisms
3. Reversion of the stress effect by repairing the damage which has occurred

Plants have genetically tolerant traits against high levels of arsenic. Potatoes, cabbage tomatoes, carrots, tobacco, rye, Sudan grass, and grapes are highly tolerant, whereas strawberries, corn, beet, and squash are moderately tolerant. Similarly, onion, cucumbers, and legumes show low tolerance against arsenic (Su et al. 2010). According to various studies, the conversion mechanism is less efficient in higher plants; therefore, numerous species of plants have inorganic soluble forms of arsenic accumulated in plant cells. Some specific plant species are reported as hyperaccumulator which can accumulate more than 1000 mg Kg^{-1} arsenic in their cells. *Agrostis tenuis*, *Agrostis stolonifera*, *Agrostis canina*, *Jasione montana*, *Calluna vulgaris*, *Holcus lanatus*, *Ceratophyllum demersum*, *Cheilanthes sinuata*, *Adiantum raddianum*, *Polystichum acrostichoides*, *Actiniopteris radiata*, *Pellaea rotundifolia*, *Nephrolepis cordifolia* (Eisler 1994; Srivastava et al. 2010), *Pteris vittata* (Ma et al. 2001), and *Pityrogramma calomelanos* are well-known hyperaccumulators of arsenic. *Pseudotsuga menziesii* is hyperaccumulator gymnosperm (Haug et al. 2004), whereas rice is considered as hyperaccumulator monocot angiosperm as 722 μg g^{-1} arsenic is accumulated in edible grains cultivated near mining areas (Okkenhaug et al. 2012). *Hordeum vulgare and Triticum aestivum* also store a small amount of arsenic within plant cell (Su et al. 2010).

6.1 First Line of Defense

6.1.1 Methylation

Once arsenic enters in the lower plant cell, the activated defense system leads to methylation and biotransformation for the conversion of inorganic arsenic into non-toxic organic forms such as monomethyl arsenic, dimethyl arsenic, arsenobetaine, and arsenocholine. The arsenic is methylated by Challenger's pathway, in fungi and microbes, but the pathway is still not clear in higher plant.

Algal cells have the gene for S-adenosylmethionine methyltransferase (ArsM) and ArsM proteins for methylation of arsenic (Qin et al. 2006). However, researchers identified AS3MT enzyme which is able to transfer thiol group from GSH to arsenite and leads to the formation of MMA and DMA (Hayakawa et al. 2005). The ArsM enzyme is also identified in *Cyanidioschyzon merolae* and *Prochlorococcus* for methylation of arsenic in higher plants (Ajees et al. 2012; Saunders and Rocap 2016).

6.1.2 Phytochelatins

The higher plants have distinct defense mechanism for detoxification and accumulation of arsenic at first checkpoint. The higher plants have thiol-rich, heavy-metal-binding polypeptide called phytochelatins for holding and accumulation of inorganic arsenic molecules immediately after the entry in to the cells (Zenk 1996; Roychowdhury et al. 2018, 2019). The arsenic binding with phytochelatins is also reported in *Rauvolfia serpentina*, *Arabidopsis* sp., and *Silene vulgaris* (Kamiya and Fujirawa 2011).

Their general makeup is two or more γ-glutamylcysteine units that repeat and have glycine as the terminal residue. The glutathione is a nonprotein thiol donor or precursor for phytochelatins. The glutathione produces phytochelatins in the presence of transpeptidase phytochelatin synthetase enzyme (Pawlik-Skwronska 2001). The synthesized phytochelatins are able to bind some metals in the cytosol, and the phytochelatin–metal complex is transported to the plant vacuole (Rauser 1990). It has been proven in X-ray absorption spectroscopy reports that both root and shoot cells of *Brassica juncea* and *Arabidopsis thaliana* have arsenite-tris-thiol complex, which is similar to As(III)-tris-glutathione (Castillo-Michel et al. 2011). The researchers suggest that the thiol group is contributed by glutathione and phytochelatins (Pickering et al. 2006). Zhao et al. (2003) reported that only 4.5% of total arsenic concentration is complexed as glutathione–arsenite–phytochelatin complex with phytochelatins in *Pteris vittata*. Similarly, only 1% arsenic is present as phytochelatin complex in arsenic hyperaccumulator, *P. cretica* (Raab et al. 2004). Kamiya and Fujirawa (2011) reported that the PC-deficient mutant (cad1-3) of *Arabidopsis* is highly susceptible to arsenate than control plants. Sequestration of As–PC complex in plant vacuoles and efflux depends on the phytochelatin synthetase and specific transporter located on the tonoplast such as PvACR3, AtABCC1, AtABCC2, OsABCC1, and ABCC transporters (Song et al. 2010). Arsenic accumulation and simultaneously high efflux rate of arsenite to surrounding medium also balanced the arsenic stress in the plant cell (Chen et al. 2016).

The latest report suggests that increased sulfur supply has a correlation with decreased arsenic concentration in rice plants (Dixit et al. 2015; Zhang et al. 2016; Nishida et al. 2016). Similarly, Zhang et al. (2016) also notified that sulfur-mediated downregulation of phosphate transporters and aquaporin gene and upregulation of ABC transporter and phytochelatin synthetase gene are important strategies for mitigation of arsenic in the plant cell (He et al. 2016; Yang et al. 2016). It has been reported that phytochelatins are able to complex and capture arsenic only in a limited range of concentration. Therefore, phytochelatins may not be the main source of detoxification of arsenite in arsenic hyperaccumulator.

6.1.3 Arsenate Reductase (AR)

Arsenate is analogous to phosphate and subsequently leads to replacement reaction and metabolic disorders. The enzyme arsenate reductase (Arsc) is able to reduce arsenate into arsenite using glutathione as an electron donor. The resulting arsenite is available for detoxification through phytochelatins. The arsenate reductase shows some variation in the active site of catalytic cysteine residue and reduction mechanism. Therefore, researchers have isolated and characterized various classes of arsenate reductase, such as thioredoxin-coupled arsenate reductase (Trx) and glutaredoxin-linked arsenate reductase (Grx). The arsenate reductase is evident in both prokaryotic and eukaryotic organisms (Mukhopadhyay et al. 2000). Table 14.1 shows the genes cloned and characterized by AR in different plant species. In plants

Table 14.1 The genes cloned and characterized for AR in different plant species

Genes	Plant	References
ATQ1 (AR)	Mouse-ear cress (*Arabidopsis thaliana*)	Chao et al. (2014) and Sanchez-Bermejo et al. (2014)
ACR3	Mouse-ear cress (*Arabidopsis thaliana*)	Ali et al. (2012)
ScACR3	Rice (*Oryza sativa*)	Duan et al. (2011)
AtACR2	Mouse-ear cress (*Arabidopsis thaliana*)	Duan et al. (2007)
OsACR2	Rice (*Oryza sativa*)	Duan et al. (2007)
HlAsr	Velvet grass (*Holcus lanatus*)	Dhankher et al. (2006)
PvACR2	Chinese brake (*Pteris vittata*)	Ellis et al. (2006)

also arsenate reductase is isolated from the root extract of hyperaccumulator *Pteris vittata*.

6.2 Second Line of Defense

Another form of arsenic toxicity is oxidative stress due to the production of reactive oxygen species (ROS) during the conversion of oxidation state. Some ROS such as superoxide anions, hydrogen peroxide (H_2O_2), and hydroxyl radicals cause damage to cell components (Gill and Tuteja 2010). These ROS are also developed during the methylation of arsenite within plant cells (Zaman and Pardini 1996). Plants which are exposed to severe stress show increased susceptibility to photo-inhibition with subsequent development of chlorosis (Cenkci et al. 2010). Although generation of ROS is a common phenomenon of electron transport system in plant cells, addition of liberated free electron during conversion of As^{+5} to As^{+3} (Sharma 2012), electron leakage, or inhibition of enzymes related to electron transport system are other ways of ROS generation in plant cells exposed to arsenic (Talukdar 2013). Sometimes hydrogen peroxide and superoxide anions react with each other during Haber–Weiss and Fenton reaction and consequently form highly damaging ROS called hydroxyl radical. The hydroxyl radical immediately attacks DNA and protein and initiates lipid peroxidation of membranes (Fenton 1899).

These reactive oxygen species are highly reactive and destructive for many biological reactions. Plants have specified enzymatic and nonenzymatic components to detoxify arsenic-induced reactive oxygen species. In response to the creation of ROS, plants synthesize enzymatic and nonenzymatic antioxidants. The nonenzymatic antioxidant molecules are L-ascorbic acid, reduced glutathione (GSH), α-tocopherols, and carotenoids, whereas antioxidant enzymes are superoxide dismutase (EC 1.15.1.1), ascorbate peroxidase (EC 1.11.1.11), peroxidase (EC 1.11.1.7), catalase (EC 1.11.1.6), glutathione reductase (EC 1.6.4.2), and arsenate reductase (give the EC1.20.2.1) (Sharma 2013).

6.2.1 Enzymatic Antioxidants

The singlet oxygen is the first excited state of oxygen produced in photosystem II (PSII) electron transfer reaction of the chloroplast. It is highly reactive and destructive to nearby biomolecules (Garg and Singla 2011). The superoxide anion (O_2^-) is produced during Mehler reaction and photooxidation reaction in chloroplast, mitochondria, glyoxysome, and peroxisome (Alscher et al. 2002). The enzyme superoxide dismutase (SOD)-mediated reduction of molecular oxygen produced superoxide radical. The consumption of oxygen takes place in cytochrome oxidase of mitochondria or reduction of singlet oxygen in flavoprotein region of NADH dehydrogenase during respiration. The superoxide may also develop through transfer of electron between ubiquinone to cytochrome C_1 region of the respiratory chain (Mubarak et al. 2016). Superoxide dismutase is associated with various metal cofactors: CuSOD and ZnSOD are located in cytosol, peroxisome, plastid, and root nodules; MnSOD is located in the mitochondria; and FeSOD is located in the plastids. Because SOD can degrade superoxide anions, it can play a very important role in the defense of cells upon stress (Fridovich 1995).

The dismutation of superoxide radical by superoxide dismutase (SOD) enzyme results in the generation of comparatively less toxic hydrogen peroxide. The catalase is a tetrameric heme-containing enzyme found in the peroxisomes, glyoxysome, cytosol, mitochondria, and root nodules to convert hydrogen peroxide into water and molecular oxygen in peroxisome (Sharma and Ahmad 2014).

$$O_2 + H_2O_2 \rightarrow H_2O + 1/2O_2$$

High catalase activity is reported in arsenic-tolerant and hyperaccumulator plants such as *Zea mays* L. and *P. vittata* (Mylona et al. 1998). There are several forms, or isozymes, of CAT like CAT-1 and CAT-2. These isozymes may respond differently under the same conditions (Horváth et al. 2002). The reaction occurs in the presence of ascorbate peroxidases (APX), and ascorbate acts as an electron donor for peroxidation reaction. Hydrogen peroxide converts in water and mono-dehydro-ascorbic acid within the chloroplast. A vast literature is available on arsenic-mediated oxidative damage and role of enzymatic antioxidant battery for defense against stress factor (Mylona et al. 1998; Hartley-Whitaker et al. 2001; Singh et al. 2006).

6.2.2 Nonenzymatic Antioxidant Molecules

As an antioxidant, ascorbic acid is located in apoplast cytosol and vacuole. Ascorbic acid can manage ROS through the direct elimination of superoxide anions, hydrogen peroxide, and hydroxyl radicals (Sharma 2012). It also acts as a secondary antioxidant due to the maintenance of the reduction of another plant antioxidant α-tocopherol (Conklin 2001) or indirectly through the action of ascorbate peroxidase (Asada 1992). Glutathione is composed of the amino acids, glutamate, cysteine, and glycine. It is, directly and indirectly, demanding in defense and detoxification within the plant system. Glutathione also acts as a precursor for phytochelatin (PC) synthesis (Kneer and Zenk 1992; Zenk 1996; Pawlik-Skwronska 2001). It also binds with reactive oxygen species in the presence of enzyme glutathione *S*-transferases (GSTs). A cyclic pathway of glutathione reductase (GR) regulated glutathione to glutathione disulfide (GSSG) and again glutathione during detoxification of ROS in plant species (Hameed et al. 2014; Roychowdhury et al. 2018, 2019). Glutathione reductase is the enzyme that, in conjunction with NADPH, catalyzes the reduction of GSSG to GSH.

$$GSSG + NADPH - GSH + NADP^{+}$$

Glutathione reductase has been detected in bacteria, yeast, plants, and animals. It is essentially responsible for maintaining the GSH levels in the plant cell exposed to arsenic. Thus, the advancement of the experimental studies shows various avenues of research in the field of arsenic toxicity and tolerance in plant species. Some knowledge gaps and future prospects are still present like transporter for DMA and MMA, which are still unknown. The role of sulfur and silicon supply for mitigation of arsenic toxicity in field practices required more research.

7 Conclusion

Arsenic is an important stress factor which can damage cellular metabolism and can lead to death of the plant. It is a transition element found in the inorganic form of arsenate and arsenite. It can hinder the biological process due to the interaction and inhibition of thiol groups of enzymes. During the conversion of their oxidative state, arsenic produces reactive oxygen species due to an imbalance between generation and elimination of ROS that leads to rapid oxidation reaction in plant cells. Oxidative stress causes damage to cell membrane lipid bilayer, fragmentation of peptide chain, and altered electrical charge, DNA damage and mutation, base degradation, and single-strand breakage. To minimize the damaging effect of ROS, plants have enzymatic and nonenzymatic defense systems including metal binding phytochelatins. The biochemistry and toxicology of arsenic are still complicated in higher plants.

References

Abbas MHH, Meharg AA (2008) Arsenate, arsenite and dimethyl arsenic acid (DMA) uptake and tolerance in maize (*Zea mays* L.). Plant Soil 304:277–289

Abbas G, Murtaza B, Bibi I, Shahid M, Niazi NK, Khan MI, Amjad M, Hussain M, Natasha (2018) Arsenic uptake, toxicity, detoxification, and speciation in plants: physiological, biochemical, and molecular aspects. Int J Environ Res Public Health 15:59. https://doi.org/10.3390/ijerph15010059

Adriano DC (1986) Trace elements in the terrestrial environment. Springer, New York
Ajees AA, Marapakala K, Packianathan C, Sankaran B, Rosen BP (2012) Structure of an As(III) S-adenosylmethionine methyltransferase: insights into the mechanism of arsenic biotransformation. Biochemistry 51:5476–5485
Ali W, Isner JC, Isayenkov SV, Liu W, Zhao FJ, Maathuis FJM (2012) Heterologous expression of the yeast arsenite efflux system ACR3 improves *Arabidopsis thaliana* tolerance to arsenic stress. New Phytol 194:716–723
Alscher RG, Erturk N, Heath LS (2002) Role of superoxide dismutases (SODs) in controlling oxidative stress in plants. J Exp Bot 53:1331–1341
Asada K (1992) Ascorbate peroxidase – a hydrogen peroxide-scavenging enzyme in plants. Physiol Plant 85(2):235–241
Carey AM, Scheckel KG, Lombi E, Newville M, Choi Y, Norton GJ, Charnock JM, Feldmann J, Price AH, Meharg AA (2010) Grain unloading of arsenic species in rice. Plant Physiol 152:309–319
Carey AM, Norton GJ, Deacon C, Scheckel KG, Lombi E, Punshon T, Guerinot ML, Lanzirotti A, Newville M, Choi Y, Price AH, Meharg AA (2011) Phloem transport of arsenic species from flag leaf to grain during grain filling. New Phytol 192:87–98
Castillo-Michel H, Hernandez-Viezcas J, Dokken KM, Marcus MA, Peralta-Videa JR, Gardea-Torresdey JL (2011) Localization and speciation of arsenic in soil and desert plant *Parkinsonia florida* using mu XRF and mu XANES. Environ Sci Technol 45:7848–7785
Cenkci S, Cigerci IH, Yıldız M, Özay C, Bozdag A, Terzi H (2010) Lead contamination reduces chlorophyll biosynthesis and genomic template stability in *Brassica rapa* L. Environ Exp Bot 67:467–473
Chao DY, Chen Y, Chen J, Shi S, Chen Z, Wang C et al (2014) Genome-wide association mapping identifies a new arsenate reductase enzyme critical for limiting arsenic accumulation in plants. PLoS Biol 12:e1002009
Chen Y, Fu JW, Han YH, Rathinasabapathi B, Ma LQ (2016) High As exposure induced substantial arsenite efflux in As-hyperaccumulator *Pteris vittata*. Chemosphere 144:2189–2194
Conklin PL (2001) Recent advances in the role and biosynthesis of ascorbic acid in plants. Plant Cell and Environment24: 383–394
Dhankher OP, Rosen BP, McKinney EC, Meagher RB (2006) Hyperaccumulation of arsenic in the shoots of Arabidopsis silenced for arsenate reductase (ACR2). Proc Natl Acad Sci U S A 103:5413–5418
DiTusa SF, Fontenot EB, Wallace RW, Silvers MA, Steele NT, Elnagar AH, Dearman KM, Smith AP (2016) A member of the phosphate transporter 1 (Pht1) family from the arsenic-hyperaccumulating fern *Pteris vittata* is a high-affinity arsenate transporter. New Phytol 209:762–772
Dixit G, Singh AP, Kumar A, Singh PK, Kumar S, Dwivedi S et al (2015) Sulfur mediated reduction of arsenic toxicity involves efficient thiol metabolism and the antioxidant defense system in rice. J Hazard Mater 298:241–251
Drewniak L, Sklodowska A (2013) Arsenic-transforming microbes and their role in biomining processes. Environ Sci Pollut Res 20:7728–7739
Duan GL, Zhou Y, Tong YP, Mukhopadhyay R, Rosen BP, Zhu YG (2007) A CDC25 homologue from rice functions as an arsenate reductase. New Phytol 174:311–321
Duan G, Kamiya T, Ishikawa S, Arao T, Fujiwara T (2011) Expressing *ScACR3* in rice enhanced arsenite efflux and reduced arsenic accumulation in rice grains. Plant Cell Physiol 53:154–163
Eisler R (1994) Part 11: human health and ecosystem effects. In: Nriagu JO (ed) Arsenic in the environment. New York, Wiley
Ellis DR, Gumaelius L, Indriolo E, Pickering IJ, Banks JA, Salt DE (2006) A novel arsenate reductase from the arsenic hyperaccumulating fern *Pteris vittata*. Plant Physiol 141:1544–1554
Fenton HJH (1899) The oxidation of polyhydric alcohols in presence of iron. Proc Chem Soc 25:224
Fowler BA (1977) Toxicology of environmental arsenic. In RA Goyer, MA Mehlman, eds, Toxicology of Trace Elements. Wiley & Sons, New York, pp 79–122

Fridovich I (1995) Superoxide radical and superoxide dismutase. Annu Rev Biochem 64:97–112
Garg N, Singla P (2011) Arsenic toxicity in crop plants: physiological effects and tolerance mechanisms. Environ Chem Lett 9:303–321
Gill SS, Tuteja N (2010) Reactive oxygen species and antioxidant machinery in abiotic stress tolerance in crop plants. Plant Physiol Biochem 48:909–930
Hameed A, Sharma I, Kumar A, Azooz MM, Lone HA, Ahmad P (2014) Glutathione metabolism in plants under environmental stress. In: Ahmad P (ed) Oxidative damage to plants: antioxidant networks and signaling. Academic Press, Elsevier USA pp 183–193
Hartley-Whitaker J, Ainsworth G, Meharg A (2001) Copper-and-arsenic induced oxidative stress in Holcus lanatus L. Cloned with differential sensitivity. Plant Cell Environ 24:713–722
Haug CM, Reimer KJ, Cullen WR (2004) Arsenic uptake by the Douglas-fir (*Pseudotsuga menziesie*). Appl Organomet Chem 18:626–630
Hayakawa T, Kobayashi Y, Cui X, Hirano S (2005) A new metabolic pathway of arsenite: arsenic-glutathione complexes are substrates for human arsenic methyltransferase Cyt19. Arch Toxicol 79:183–191
He Z, Yan H, Chen Y, Shen H, Xu W, Zhang H et al (2016) An aquaporin PvTIP4;1 from *Pteris vittata* may mediate arsenite uptake. New Phytol 209:746–761
Hoang TH, Ju-Yong K, Sunbaek B, Kyoung-Woong K (2010) Source and fate of As in the environment. Geosy Eng 13:35–42
Horváth E, Janda T, Szalai G, Páldi E (2002) In vitro salicylic acid inhibition of catalase activity in maize: differences between the isozymes and a possible role in the induction of chilling tolerance. Plant Sci 163:1129–1135
Kabata-Pendias A, Pendias H (2001) Trace elements in soils and plants, 3rd edn. CRC Press, Boca Raton
Kamiya T, Islam MR, Duan GL, Uraguchi S, Fujiwara T (2013) Phosphate deficiency signaling pathway is a target of arsenate and phosphate transporter *OsPT1* is involved in As accumulation in shoots of rice. Soil Sci Plant Nutr 59:580–590
Kamiya T, Fujiwara T (2011) A novel allele of the Arabidopsis phytochelatin synthase 1 gene conferring high sensitivity to arsenic and antimony. Soil Sci Plant Nutr 57(2):272–278
Kneer R, Zenk MH (1992) Phytochelatins protect plant enzymes from heavy metal poisoning. Phytochemistry 31:2663–2667
Kopittke PM, de Jonge MD, Wang P, McKenna BA, Lombi E, Paterson DJ, Howard DL, James SA et al (2013) Laterally resolved speciation of arsenic in roots of wheat and rice using fluorescence-XANES imaging. New Phytol 201:1251–1262
LeBlanc MS, McKinney EC, Meagher RB, Smith AP (2013) Hijacking membrane transporters for arsenic phytoextraction. J Biotechnol 163:1–9
Ma LQ, Komar KM, Tu C, Zhang WH, Cai Y, Kennelley ED (2001) A fern that hyperaccumulates arsenic. Nature 409:579
Ma JF, Yamaji N, Mitani N, Xu XY, Su YH, McGrath SP et al (2008) Transporters of arsenite in rice and their role in arsenic accumulation in rice grain. Proc Natl Acad Sci U S A 105:9931–9935
Matera V, Le Hecho I (2001) In: Selim HM, Spark DL (eds) Heavy metal release in soils. Lewis publishers, Boca Rotan
Meharg AA, MacNair MR (1994) Relationship between plants that differ in their tolerance to arsenate. Plant Soil 162:99–106
Mubarak H, Mirza N, Chai L Y, Yang Z H, Yong W, Tang C J, Mahmood Q, Pervez A, Farooq U, Fahad S (2016) Biochemical and metabolic changes in arsenic contaminated *Boehmeria nivea* L. BioMed Res Int 2016:1123 1128
Mukhopadhyay R, Shi J, Rosen BP (2000) Purification and characterization of Acr2p, the *Saccharomyces cerevisiae* arsenate reductase. J Biol Chem 275:21149–21157
Mylona PV, Polidoros AN, Scandalios JG (1998) Modulation of antioxidant responses by arsenic in maize. Free Rad Bio Med 25:576–585
Nearing MM, Koch I, Reimer KJ (2014) Complementary arsenic speciation methods: a review. Spectrochim Acta B At Spectrosc 99:150–162

Ng JC (2005) ENVIRONMENTAL contamination of arsenic and its toxicological impact on humans. Environ Chem 2:146–160

Nishida S, Duan G, Ohkama-Ohtsu N, Uraguchi S, Fujiwara T (2016) Enhanced arsenic sensitivity with excess phytochelatin accumulation in shoots of a SULTR1; 2 knockout mutant of *Arabidopsis thaliana* (L.) Heynh. Soil Sci Plant Nutr 62:367–372

Okkenhaug G, Zhu YG, He J, Li X, Luo L, Mulder J (2012) Antimony (Sb) and arsenic (As) in Sb mining impacted paddy soil from Xikuangshan, China: differences in mechanisms controlling soil sequestration and uptake in rice. Environ Sci Technol 46:3155–3162

Oremland RS, Stolz JF (2003) The ecology of arsenic. Science 300:939–944

Pawlik-Skwronska B (2001) Phytochelatin production in freshwater algae *Stigeoclonium* in response to heavy metals contained in mining water; effects of some environmental factors. Aquat Toxicol 52:241–249

Pickering IJ, Gumaelius L, Harris HH, Prince RC, Hirsch G, Banks JA, Salt DE, George GN (2006) Localizing the biochemical transformations of arsenate in a hyperaccumulating fern. Environ Sci Technol 40:5010–5014

Pierce ML, Moore CB (1982) Adsorption of arsenite and arsenate on amorphous iron oxyhydroxide. Water Res 16:1247–1253

Qin J, Rosen BP, Zhang Y, Wang GJ, Franke S, Rensing C (2006) Arsenic detoxification and evolution of trimethylarsine gas by a microbial arsenite Sadenosylmethionine methyltransferase. Proc Natl Acad Sci U S A 103:2075–2080

Raab A, Meharg AA, Jaspars M, Genney DR, Feldmann J (2004) Arsenic-glutathione complexes: their stability in solution and during separation by different HPLC modes. J Atomic Absorp Spec 19:183–190

Rauser WE (1990) Phytochelatins. Annu Rev Biochem 59:61–86

Roychowdhury R, Khan MH, Choudhury S (2018) Arsenic in rice: an overview on stress implications, tolerance and mitigation strategies. In: Hasanuzzaman M, Nahar K, Fujita M (eds) Plants under metal and metalloid stress. Springer, Singapore, pp 401–415

Roychowdhury R, Khan MH, Choudhury S (2019) Physiological and molecular responses for metalloid stress in rice a comprehensive overview. In: Hasanuzzaman M, Fujita M, Nahar K, Biswas J (eds) Advances in rice research for abiotic stress tolerance. Woodhead Publishing, Elsevier UK, pp 341–369

Sanchez-Bermejo E, Castrillo G, del Llano B, Navarrol C, Zarco-Fernandez S, Martinez-Herrera DJ et al (2014) Natural variation in arsenate tolerance identifies an arsenate reductase in *Arabidopsis thaliana*. Nat Commun 5:4617

Saunders JK, Rocap G (2016) Genomic potential for arsenic efflux and methylation varies among global *Prochlorococcus* populations. The ISME J 10:197–209

Sharma I (2012) Arsenic induced oxidative stress in plants. Biologia 67:447–453

Sharma I (2013) Arsenic stress in plants: an inside story. In: Khalid RH, Ahmad P, Munir O (eds) Crop improvement new approaches and modern techniques. Springer, New York/Dordrecht/Heidelberg/London, pp 379–400

Sharma I, Ahmad P (2014) Catalase: a versatile antioxidant in plants. In: Ahmad P (ed) Oxidative damage to plants: antioxidant networks and signaling. Academic Press, Elsevier USA pp 131–143

Sharma I, Anita (2016) Survey of vegetation grown near copper mine. Int J Res Med Sci Basic Sci 2(11):20–29

Sharma I, Travlos IS (2012) Phosphate supply as a promoter of tolerance to arsenic in pearl millet. Int J Plant Prod 6(4):443–456

Singh N, Ma LQ, Shrivastava M, Rathinasapathi B (2006) Metabolic adaptation to arsenic-induced oxidative stress in *Pteris vittata* L and *Pteris ensiformis* L. Plant Sci 170:274–282

Song WY, Park J, Mendoza-Cózatl DG, Suter-Grotemeyer M, Shim D, Hörtensteiner S et al (2010) Arsenic tolerance in Arabidopsis is mediated by two ABCC-type phytochelatin transporters. Proc Natl Acad Sci U S A 107:21187–21192

Srivastava M, Santos J, Srivastava P, Ma LQ (2010) Comparison of arsenic accumulation in 18 fern species and four *Pteris vittata* accessions. Bioresour Technol 101:2691–2699

Su YH, McGrath SP, Zhao FJ (2010) Rice is more efficient in arsenite uptake and translocation than wheat and barley. Plant Soil 328:27–34

Talukdar D (2013) Arsenic-induced changes in growth and antioxidant metabolism of fenugreek. Russ J Plant Physiol 60:652–660

United States Environmental Protection Agency (USEPA) (2001) National primary drinking water regulations: arsenic and clarifications to source contaminants monitoring. Fed Regist 66:6975–6706

Williams PN, Raab A, Feldmann J, Meharg AA (2007) Market basket survey shows elevated levels of As in South Central US processed rice compared to California: consequences for human dietary exposure. Environ Sci Technol 41:2178–2183

Yamaguchi N, Ohkura T, Takahashi Y, Maejima Y, Arao T (2014) Arsenic distribution and speciation near rice roots influenced by iron plaques and redox conditions of the soil matrix. Environ Sci Technol 48:1549–1556

Yang J, Gao MX, Hu H, Ding XM, Lin HW, Wang L et al (2016) OsCLT1, a CRT-like transporter 1, is required for glutathione homeostasis and arsenic tolerance in rice. New Phytol 211:658–670

Ye J, Rensing C, Rosen BP, Zhu YG (2012) Arsenic biomethylation by photosynthetic organisms. Trends Plant Sci 17:155–162

Zaman K, Pardini RS (1996) An overview of the relationship between oxidative stress and mercury and arsenic. Toxic Sub Mech 15(3):151–181

Zavala YJ, Gerads R, Gürleyük H, Duxbury JM (2008) Arsenic in rice: II. Arsenic speciation in USA grain and implications for human health. Environ Sci Technol 42:3861–3866

Zenk MH (1996) Heavy metal detoxification in higher plants: a review. Gene 179:21–30

Zhang J, Zhao CY, Liu J, Song R, Du YX, Li JZ et al (2016) Influence of sulfur on transcription of genes involved in arsenic accumulation in rice grains. Plant Mol Biol Rep 34:556–565

Zhao FJ, Wang JR, Barker JHA, Schat H, Bleeker PM, McGrath SP (2003) The role of phytochelatins in arsenic tolerance in the hyperaccumulator *Pteris vittata*. New Phytol 159:403–410

Zhao FJ, Ma JF, Meharg AA, McGrath SP (2009) Arsenic uptake and metabolism in plants. New Phytol 181:777–794

Chapter 15
Arsenic Tolerance and Signaling Mechanisms in Plants

Sudhakar Srivastava and Vaishali Yadav

Abstract Arsenic (As) contamination is now prevalent in large areas of a number of countries. Arsenic stress during vegetative and reproductive phases of plant growth affects the growth, flowering, and fruit/grain development due to a number of physiological, biochemical, and molecular effects. The metabolism of carbon, nitrogen, and sulfur is affected along with changes in photosynthetic and respiratory processes. The avoidance of As toxicity requires fine-tuning of several processes starting from uptake and transport of As to As detoxification and prevention of stress manifestation. Thus, the regulatory processes promoting concerted action of several pathways become an important determinant of As stress tolerance of a plant. These include various phytohormones (auxins, jasmonates, ethylene, abscisic acid, salicylic acid, etc.), transcription factors, microRNAs, reactive oxygen and reactive nitrogen species, and kinases. It is also important to note that an early perception of arsenic entry into plants and signaling of the information to distant parts (leaves) ensure that plants remain prepared to tackle the As stress. Although the research on signaling aspects of arsenic stress is yet in the primary stage, the information available is of potential application and future research. The present work describes current knowledge on perception and signaling of As stress in plants.

Keywords Arsenic · MicroRNAs · Nitric oxide · Phytohormones · Reactive oxygen species · Transcription factors

1 Introduction

Arsenic (As) is a highly toxic metalloid having an array of toxic manifestations and carcinogenicity on various organisms, including humans (Duker et al. 2005). Arsenic is present in nature due to its use in agricultural and industrial sectors as well as due to natural biogeochemical factors leading to a large amount of As being

S. Srivastava (✉) · V. Yadav
Plant Stress Biology Laboratory, Institute of Environment and Sustainable Development, Banaras Hindu University, Varanasi, India
e-mail: sudhakar.iesd@bhu.ac.in

R. Roychowdhury et al. (eds.), *Sustainable Agriculture in the Era of Climate Change*, https://doi.org/10.1007/978-3-030-45669-6_15

released in the environment (Shukla et al. 2017). Arsenic is posing serious health concerns specifically in West Bengal in India and Bangladesh. Exposure to As for a long time leads to the development of some serious symptoms such as skin lesions and different types of cancers (Kumar et al. 2015). There are many forms of As present in nature. These can be categorized into inorganic [arsenite: As(III) and arsenate: As(V)] and organic [monomethylarsonic acid: MMA, dimethylarsinic acid: DMA]. It is known that mono-, di-, and tri-methylated forms of As are produced in the soil by microbial processes (Mestrot et al. 2011). The entry of toxic As species in plants also occurs via competitive uptake through transporters of essential metals by the mechanism of ionic mimicry (Verbruggen et al. 2009; Ye et al. 2010). The influx of As(III) occurs primarily via nodulin 26-like intrinsic proteins (NIPs) class of aquaglyceroporins (Verbruggen et al. 2009; Bienert et al. 2008; Roychowdhury et al. 2018, 2019), whereas that of As(V) is driven by phosphate transporters (Wu et al. 2011).

The plants suffer toxicity when the As accumulation passes a certain level called threshold level (Srivastava et al. 2007). Arsenic exposure to plants affects several physiological, biochemical, and morphological processes which include germination, growth, and biomass production (Finnegan and Chen 2012). Arsenic toxicity impacts photosynthesis by reducing chlorophyll content in the leaves and by altering photosynthetic efficiency, electron transport rate, transpiration, etc. (Srivastava et al. 2013a). Arsenic, although a non-redox-active metalloid, also induces the production of reactive oxygen species (ROS) by inter-conversion of one ionic form to another and by overconsumption of reduced glutathione (GSH) in the process (Mylona et al. 1998). The oxidative stress conditions in the presence of As cause lipid peroxidation of membranes, damage proteins and DNA, and induce toxicity (Srivastava et al. 2011). Arsenic exposure also causes genotoxic effects like mitotic index depression, increased chromosomal aberrations like a break, fragments, c-metaphase, and multipolar arrangements (Gupta et al. 2018). By translocation to the shoot, the plant growth can be severely inhibited by As by arresting cellular expansion and biomass accumulation as well as by compromising plant reproductive capacity resulting in loss of yield and fruit production (Garg and Singla 2011).

In the presence of As, it becomes important for the plants to prepare and be equipped with mechanisms to tolerate the stress so as to survive and propagate. Plants tolerate As stress by an array of metabolic processes such as binding of As to thiol (-SH) containing ligands, glutathione (GSH), and phytochelatins (PCs) followed by As–thiol sequestration in vacuoles (Srivastava et al. 2007; Song et al. 2014; Tang et al. 2019), reduction of As(V) to As(III) (Chao et al. 2014; Shi et al. 2016), decrease in uptake of As by altered expression of transporters (Srivastava et al. 2014), protection against oxidative stress via activation of antioxidant enzymes and molecules (Srivastava et al. 2007; Roychowdhury et al. 2018, 2019), anatomical and morphological changes allowing plants to bind As to cell walls and to maximize nutrient uptake and translocation (root phenotype), etc. There are a large number of metabolites such as cysteine, glutathione, proline, α-tocopherol, γ-aminobutyric acid (GABA), and carotenoids which have a role in As tolerance (Pathare et al. 2013; Singh et al. 2016). The physiological and biochemical experiments to omics approach like transcriptomics, metabolomics, and proteomics have been employed to unravel the response networks for As tolerance in plants (Yu et al. 2012; Dixit et al. 2015; Srivastava et al. 2015; Roychowdhury et al. 2018, 2019). It is clear that a complex interplay of various mechanisms and their concerted response is important for tackling As stress. This necessitates the involvement of a number of regulatory and signaling processes to operate for the early perception and signaling of As stress and for coordinating various physiological, biochemical, molecular, and metabolic responses (Pathare et al. 2013). This chapter presents the knowledge gained in the field of As stress perception and signaling in plants.

2 Perception and Signaling of Arsenic Stress

The perception and signaling involve sensing the entry of As in plants and transfer this information among various plant cells and tissues from receptor systems to effector ones. The timely perception of the stress allows plants to respond appropriately to the incoming stress. Apart from the perception of the stress, the signal must reach to all cells, tissues, and organs so as to prepare them for stress that might reach to them sooner or later. In the case of As stress, a few studies have been conducted that provide valuable insights into the perception and signaling of As stress. Some of the well-known signaling mechanisms constitute mitogen-activated protein kinases (MAPKs), reactive oxygen species (ROS) and reactive nitrogen species (RNS), hormones [abscisic acid (ABA), jasmonic acid (JA) and other jasmonates, salicylic acid (SA), ethylene, brassinosteroids, etc.], microRNAs, and transcription factors. In the following sections, the available knowledge is presented to showcase the aspects of perception and signaling of As stress in plants (Roychowdhury et al. 2018, 2019).

3 Phytohormones

Phytohormones regulate crucial physiological responses of plants during growth, such as cell division, cell differentiation, organ formation, leaf senescence, flowering, and grain development, and also in response to various abiotic and biotic stresses (Shukla et al. 2017). Auxins play the role of key regulators of phenotypic changes that directly affect responses of plants to the stresses (Nibau et al. 2008). The auxin accumulation takes place at the root tip, and the plant apical tissues are a place of auxin synthesis. Therefore, auxins are transported downward via phloem toward the tip of the root (Benkova et al. 2003). The exogenous supply of an auxin, indole-3-acetic acid (IAA), was found to improve the growth of *Brassica juncea* and enhance tolerance of plants under As stress. The effects were found to be mediated through changes in the expression of microRNAs like miR167, miR319, and miR854 (Srivastava et al. 2013b). Selenium (Se) is known to ameliorate As toxicity in rice plants (Chauhan et al. 2017). The exogenous supply of auxin along with Se further augments As tolerance of rice plants in comparison to that achieved through Se alone supply (Pandey and Gupta 2015). In *A. thaliana*, mutants of auxin transporters *aux1*, *pin1*, and *pin2* have been found to be more sensitive to As(III) as compared to WT plants. Further, the supply of auxin transport inhibitors also compromised plants' tolerance to As(III) of WT plants, while exogenous IAA supply improved As(III) tolerance of *aux1* mutants (Krishnamurthy and Rathinasabapathi 2013).

Jasmonates play a vital role in mediating several processes involved in the overall development of plants and abiotic and biotic stress response. Some of the developmental processes include senescence, tuber formation, pollen maturation, root growth, and interactions with other organisms (Carvalhais et al. 2015). The activation of jasmonic acid (JA) signaling pathway happens due to the conjugation of JA to the amino acids, i.e., L-isoleucine (Ile), which produces an active hormone called as (3R, 7S)-jasmonoyl-L-isoleucine (JA-Ile). The F-box protein Coronatine-Insensitive 1 (COI1) is a co-receptor of JA-Ile, which binds with JA-Ile, for ubiquitin-mediated degradation of JAZs (jasmonate ZIM-domain) proteins. This releases transcription factors, i.e., MYC2, which activate or repress gene expression associated with JA (Carvalhais et al. 2015). JAs can regulate sulfur metabolism and the level of GSH and PCs to ameliorate the toxic effects of metals (Maksymiec et al. 2007). Jasmonates are suggested to play the primary role in sensing the entry of As in plants via an indirect perception of As-induced stress to sulfur metabolism (Srivastava et al. 2009). The significant changes in the level of JA and methyl jasmonate (MeJA) were observed in As-stressed *B. juncea*. Further, the exogenous supply of JA was found to improve the growth of *B. juncea* plants as well as tolerance under As stress (Srivastava et al. 2013b). An involvement of JA in the responses to As stress in rice and *B. juncea* was also suggested when the transcriptome analysis showed significant upregulation of genes of JA biosynthesis and JAZs (Yu et al. 2012; Srivastava et al. 2015).

Abscisic acid (ABA) is a derivative of carotenoid and is an essential phytohormone, which plays a vital role against various kinds of stresses like low temperature, heat stress, drought, salinity, and metals (Vishwakarma et al. 2017). In transcriptomic analyses in rice and *B. juncea* under As(III) and/or As(V) stress, a change in ABA metabolism genes was noticed (Chakrabarty et al. 2009; Srivastava et al. 2015). It has been found that there is an increase in ABA synthesis under As stress in *B. juncea*. An evidence of interaction between ABA and miR159 was also found to have an impact on As stress response (Srivastava et al. 2013b). Salicylic acid (SA), a phenolic compound, is synthesized from phenylalanine in cytoplasm and chorismate in chloroplast by two different pathways (Vlot et al. 2009). The main functions of SA include seed germination, respiration, and thermal tolerance. It plays a vital role in plant response to a variety of abiotic stresses (Vlot et al. 2009). A reversion in the plant growth and oxidative stress upon the exogenous application of SA in the rice has been found under As stress. It could also reduce As translocation to shoots through changes in As transporter genes (Singh et al. 2015). SA also acts as a direct scavenger of ROS and affects oxidative metabolism and increases antioxidant potential (Singh et al. 2015). An interplay of SA and nitric oxide (NO) in the regulation of As stress is also demonstrated. In rice plants exposed to As(III), a decline in the level of NO and SA was noticed. Upon exogenous supply of SA, the level of SA, as well as NO, was found to increase through the increase in the activity of nitrate reductase. The supplementation of SA was found to reduce As concentrations while increasing that of nutrient elements like Zn, Mn, and Fe in rice shoots (Singh et al. 2017). Ethylene is the only plant hormone present in a gaseous state. This hormone is involved in processes like fruit ripening and abscission and root hair formation (Lewis et al. 2011). An increase in ethylene levels under metal stress has been found, and the role of ethylene in the regulation of metal stress responses of plants is demonstrated (Masood et al. 2012; Khan and Khan 2014). However, the role of ethylene in As stress responses of plants is not yet elucidated. Nonetheless, ethylene biosynthesis and ethylene-responsive genes have been observed to upregulated during transcriptome analysis of rice plants upon As exposure (Chakrabarty et al. 2009; Yu et al. 2012).

Cytokinins are derivatives of adenine and are involved in cell division, nutrient metabolism, nodulation, and circadian rhythms (Argueso et al. 2010). Cytokinins act as negative regulators of lateral root formation via interplay with auxins (Laplaze et al. 2007). Srivastava et al. (2009) found that in the presence of As stress, *B. juncea* shows downregulation of cytokinin response 1 (CRE1) gene, which is a cytokinin receptor. The negative effect on cytokinin reception led to an increase in the expression of sulfate transporters. Mohan et al. (2016) developed transgenic *Arabidopsis thaliana* lines having a constitutive expression of cytokinin oxidase/dehydrogenase 1 (CKX1) gene to achieve cytokinin depletion. It was found that cytokinin-depleted lines became As(V) tolerant through efficient suppression of As(V) transporter and PHT1;1 and also due to increase in the expression of As(V) reductase gene as well as due to increase in the level of thiol compounds (GSH and PCs). Gibberellins are a large family of plant growth hormones associated mainly with seed germination, leaf expansion, and floral development (Daviere and Achard 2013). The role of gibberellins in metal stresses like cadmium, nickel, and lead is revealed (Meng et al. 2009; Sharaf et al. 2009). However, the role of gibberellins in As stress responses of plants is suggested only due to observed changes in the expression of gibberellin synthesis and response genes (Chakrabarty et al. 2009). The above discussion implicates that coordinated action of several phytohormones operates under As stress to fine-tune plants' responses. However, more experiments are needed in the future to delineate the in-depth mechanisms.

4 Reactive Oxygen and Nitrogen Species (ROS and RNS)

The oxygen-dependent life on Earth also entails the continuous production and dismutation of reactive oxygen species (ROS) due to the transfer of electrons or excess energy to oxygen. The process normally occurs at an optimal rate, and the role of ROS as an essential component of growth and development is known (Mittler 2002).

The ROS include singlet oxygen ($^1O_2$), hydrogen peroxide (H_2O_2), superoxide radical ($O_2\cdot-$), and hydroxyl radical (·OH) (Foyer and Noctor 2000). However, in the presence of metals including As, the rate of ROS production may surpass that of ROS dismutation leading to an excess of oxygen molecules wandering to react with biomolecules. The excessive ROS production under As stress causes oxidative stress and damages membranes, proteins, and DNA (Srivastava et al. 2011). However, ROS stimulation under As stress can also act as a signaling component. In fact, an increase in the activity of prooxidant enzymes like ascorbate oxidase and NADPH oxidase, involved in ROS generation, has been observed in response to As stress (Srivastava et al. 2011). Research on NADPH oxidase mutant lines of *A. thaliana* demonstrated the role of another prooxidant enzyme, glycolate oxidase, in As-induced stress responses (Gupta et al. 2013). Further, ROS are known to interact with other signaling components like NO and phytohormones (Shukla et al. 2017; Singh et al. 2017).

Reactive nitrogen species constitute NO and compounds derived from NO reaction with biomolecules like S-nitrosothiols (Shukla et al. 2017). NO plays a critical part in abiotic and biotic stress responses of plants due to its ability to interact with several biomolecules, proteins, ROS, and phytohormones (Mengel et al. 2013; Grun et al. 2016). NO interacts with biomolecules in three types of mechanisms: S-nitrosylation, metal nitrosylation, and tyrosine nitration (Lamotte et al. 2014). S-nitrosylation is the reaction of NO with thiol-containing molecules and proteins. Metal nitrosylation is the binding of NO to transition metals to form metalloproteins. In tyrosine nitration, a nitro group is added to tyrosine residues. Nitric oxide supplementation has been found to play a vital role in the reduction of As accumulation in rice (Singh et al. 2015). Involvement of NO and SA interaction in As stress responses of rice was also demonstrated (Singh et al. 2017). As induced, nitro-oxidative stress occurs due to the effect of As on the activity of glutathione reductase and *S*-nitrosoglutathione reductase. This, in turn, decreases GSH and GSNO content (Leterrier et al. 2012). NO-mediated protection against As toxicity has also been noticed in *Luffa acutangula* (Singh et al. 2013), *Hordeum vulgare* (Shukla et al. 2015), and *Pistia stratiotes* (Farnese et al. 2017). NO supplementation improved plant physiology and maintained the organellar structure. In rice plants, NO supplementation has been found to reduce As-induced oxidative stress and improve thiol metabolism and expression of genes (Praveen and Gupta 2018). In *B. juncea*, exogenous supply of sodium nitroprusside (NO donor) was found to reduce As accumulation and improve growth and mineral nutrient status of plants. The supply of NO was also found to affect nitrogen metabolism and the expression of auxin transporter genes (Praveen et al. 2019).

5 MicroRNAs (miRNAs)

Posttranscriptional regulation of mRNAs by other small RNAs is an important component of gene expression regulation (Axtell 2013; Bej and Basak 2014). MicroRNAs (miRNAs) are highly conserved small RNAs of 20–24 nucleotides that regulate the expression of genes by targeting specific mRNAs. The miRNAs either cleave the target mRNA or decrease translation of mRNA into protein (Xie et al. 2015; Stepien et al. 2017; Srivastava and Pathare 2017). These miRNAs regulate several processes of growth and development of plants, metabolic pathways, and enzymatic reactions. In the case of abiotic and biotic stresses, including metals, the role of miRNAs has been revealed and has emerged as a crucial component (Srivastava and Pathare 2017). A few As-responsive miRNAs like miR168, miR169, miR319, miR416, miR444, miR820, miR1430, miR1437, etc. were computationally identified by Tuli et al. (2010). In another study comparing high and low As-accumulating varieties of rice, miRNA profiling identified a number of As(III)- and As(V)-responsive miRNAs. These miRNAs either demonstrated a common response to both As species and/or in both varieties or showed variety and As(III)-/As(V)-specific response (Sharma et al. 2015). The miR164, miR171, miR395, miR529, miR820, and miR1432 showed downregulation in both As(III) and As(V) stress, whereas miR408, miR1861, miR2102, and miR2907 showed upregulation. miR396 and miR528 showed a variable response in As(III) and As(V) stress, while miR399 and miR1846 demonstrated rice variety-specific response (Sharma et al. 2015). In *B. juncea*, As(V) stress-induced changes in the miRNA profile were evaluated for variable time points of 1 h and 4 h. The study identified a total of 69 miRNAs of 18 miRNA families that showed changes in expression pattern in time- and organ-dependent manner (Srivastava et al. 2013b). The upregulated miRNAs, miR426, miR472, and miR390, at 1 h depicted downregulation at 4 h, while the reverse trend was true for miR395. The time-dependent (6 h, 24 h, 72 h) and organ-dependent (root, shoot) real-time RT PCR analysis of few miRNAs (miR156, miR159, miR162, miR167, miR838, and miR854) further confirmed that expression of miRNAs changes dynamically and differentially in various tissues in response to As(V) stress in *B. juncea* (Srivastava et al. 2013b). The response of miRNAs also varies from plant to plant as seen in *O. sativa* and *B. juncea* (Liu and Zhang 2012, Srivastava et al. 2013b) where, for example, miR164 and miR172 were upregulated in *B. juncea* but showed downregulation in *O. sativa*. Further, the varying concentration of As(III) from low to high also has an influence on the miRNA profile of rice plants (Yu et al. 2012). In elemental interactions of As and selenium (Se) also, the role of miRNAs has been demonstrated Pandey et al. (2015). It is evident that miRNAs constitute a dynamic component of regulatory processes involved in As stress responses of plants, and the profile of miRNAs is an As concentration, As species, plant species, plant organ, and time-dependent process. Through dynamic changes, miRNAs affect crucial processes of metal detoxification (sulfur metabolism), metal uptake and distribution (transporters), metal stress (antioxidant mechanisms), and metal stress response (transcription factors, hormones)

(Curaba et al. 2014; Srivastava and Pathare 2017). In sulfur metabolism, miR395 regulates sulfur transporter (SULTR2;1) and ATP sulfurylase (APS), while miR838 and miR854 affect SULTR2;1 and SULTR2;2 and serine acetyl transferase (SAT) (Liang et al. 2010, Srivastava et al. 2013b). The regulation of Cu/Zn superoxide dismutase, the important enzyme of antioxidant defense dismutating superoxide ion to hydrogen peroxide, is achieved via miR398 (Bouche 2010). The level and signaling of phytohormones are also regulated by miRNAs, for example, miR160 and miR167 for auxin and miR319, miR168, and miR169 for jasmonates (Schommer et al. 2008).

6 Transcription Factors (TFs) and Kinases

The transcription process of mRNA synthesis and hence the gene expression are regulated by a number of transcription factors (TFs) (Anumalla et al. 2016). The role of several TFs is tuned in response to stress so as to achieve desirable expression of the required set of genes. A number of transcriptome studies have revealed the involvement of several families of TFs in As stress responses in plants like rice and *B. juncea*. The important TFs include WRKY, Myb, NAM, ATAF, NF-YA, AP2/EREBP, GATA, heat shock TFs, basis helix–loop–helix, homeobox, C2H2 zinc fingers, DOF zinc finger, etc. (Chakrabarty et al. 2009, Yu et al. 2012, Srivastava et al. 2015, Rai et al. 2015). The role of WRKY TF, WRKY6, was experimentally demonstrated by Castrillo et al. (2013) who found that As(V) stress-induced suppression of As(V)/phosphate transporter, PHT1;1, was regulated by WRKY6 in *A. thaliana*. Recently, the role of another WRKY TF, WRKY28, was studied in rice. WRKY28 was found to be an As(V)-inducible TF, and it regulated As accumulation in rice seedlings through effects on jasmonates (Wang et al. 2018). Mitogen-activated protein kinases (MAPKs) are a group of proteins, which are evolutionarily conserved and play important roles in signal transduction. MAPKs function in a cascade. Generally, the signals are relayed through MAPK kinase kinases (MAPKKKs), MAPK kinases (MAPKKs), and MAPK. In most cases, a single MAPK module transmits more than one signal in plants (Rodriguez et al. 2010). In the case of As, the role of MAPK3, MAPK4, and MKK4 is demonstrated in rice (Rao et al. 2011).

7 Conclusion and Future Perspective

The research on arsenic has mostly focused on arsenic chemistry in field and management practices to regulate arsenic levels in rice plant. The research on arsenic–plant interactions demonstrates involvement of various hormones, microRNAs, and kinases in the process of arsenic stress perception, signaling, and tolerance. However, the present picture mostly shows isolated changes in various signaling processes. The future research needs to bring about exact mechanisms of integration of various signaling molecules, hormones, and proteins so as construct a holistic picture. There is a need to reveal how arsenic infiltration is perceived by the plants. Hypothetically, arsenic entry might be perceived through interaction with a sensor protein, and this needs to be elucidated in coming years. The regulatory mechanisms involved in transporters' expression in roots, shoot, flag leaves, and seeds need to be understood so as to enable development of low-grain arsenic-accumulating rice varieties. The interconversion of arsenic species from one form to another constitutes an important mechanism of regulation of arsenic stress, its concentration in tissues, and its transport within plant. The process itself needs to be delineated in more detail. In addition, the aim should be to elucidate regulatory processes involved in arsenic metabolism. In conclusion, the research needs to focus on enhancing our knowledge about mechanistic details of arsenic–plant interactions.

References

Anumalla M, Roychowdhury R, Geda CK, Bharathkumar S, Goutam KD, Mohandev TSS (2016) Mechanism of stress signal transduction and involvement of stress inducible transcription factors and genes in response to abiotic stresses in plant. Int J Recent Sci Res 7(8):12754–12771

Argueso CT, Raines T, Kieber JJ (2010) Cytokinin signalling and transcriptional networks. Curr Opin Plant Biol 13:533–539

Axtell MJ (2013) Classification and comparison of small RNAs from plants. Annu Rev Plant Biol 64:137–159

Bej S, Basak J (2014) MicroRNAs: the potential biomarkers in plant stress response. American J Plant Sci 5(5):748

Benkova E, Michiniewicz M, Sauer M, Teichmann T, Seifertova D, Jurgens G, Friml J (2003) Local efflux dependent auxin gradients as a common module for plant organ formation. Cell 115:591–602

Bienert GP, Thorsen M, Schussler MD, Nilsson HR, Wagner A, Tamas MJ, Jahn TP (2008) A subgroup of plant aquaporins facilitates the bi-directional diffusion of $As(OH)_3$ and $Sb(OH)_3$ across membranes. BMC Biol 6:26

Bouche N (2010) New insights into miR398 functions in Arabidopsis. Plant Signal Behav 5:684–686

Carvalhais LC, Dennis PG, Badri DV, Kidd BN, Vivanco JM, Schenk PM (2015) Linking jasmonic acid signaling, root exudates, and rhizosphere microbiomes. Plant Microbe Interact 28:1049–1058

Castrillo G, Sánchez-Bermejo E, de Lorenzo L, Crevillén P, Fraile-Escanciano A, Mohan TC (2013) WRKY6 transcription factor restricts arsenate uptake and transposon activation in *Arabidopsis*. Plant Cell 25:2944–2957

Chakrabarty D, Trivedi PK, Mishra P, Tiwari M, Shri M, Shukla D, Kumar S, Rai A, Pandey A, Nigam D, Tripathi RD, Tuli R (2009) Comparative transcriptome analysis of arsenate and arsenite stresses in rice seedlings. Chemosphere 74:688–702

Chao DY, Chen Y, Chen JG, Shi SL, Chen ZR, Wang CC, Danku JM, Zhao FJ, Salt DE (2014) Genome-wide association mapping identifies a new arsenate reductase enzyme critical for limiting arsenic accumulation in plants. PLoS Biol 12:e1002009

Chauhan R, Awasthi S, Tripathi P, Mishra S, Dwivedi S, Niranjan A, Mallick S, Tripathi P, Pande V, Tripathi RD (2017) Selenite modulates the level of phenolics and nutrient element to alleviate the toxicity of arsenite in rice (*Oryza sativa* L.). Ecotox Environ Safety 138:47–55

Curaba J, Singh MB, Bhalla PL (2014) miRNAs in the crosstalk between phytohormones signalling pathways. J Exp Bot 65:1425–1438
Daviere JM, Achard P (2013) Gibberellin signaling in plants. Development 140:1147–1151
Dixit G, Singh AP, Kumar A, Dwivedi S, Deeba F, Kumar S, Suman S, Adhikari B, Shukla Y, Trivedi PK, Pandey V, Tripathi RD (2015) Sulfur alleviates arsenic toxicity by reducing its accumulation and modulating proteome, amino acids and thiol metabolism in rice leaves. Sci Rep 5:16205
Duker AA, Carranza EJM, Hale M (2005) Arsenic geochemistry and health. Environ Int 31:631–641
Farnese FS, Oliveira JA, Paiva EAS, Menezes-Silva PE, da Silva AA, Campos FV, Ribeiro C (2017) The involvement of nitric oxide in integration of plant physiological and ultrastructural adjustments in response to arsenic. Front Plant Sci 8:516
Finnegan PM, Chen W (2012) Arsenic toxicity: the effect on plant metabolism. Front Plant Sci 3:182
Foyer CH, Noctor G (2000) Oxygen processing in photosynthesis: regulation and signalling. New Phytol 146:359–388
Garg N, Singla P (2011) Arsenic toxicity in crop plants: physiological effects and tolerance mechanisms. Environ Chem Lett 9:303–321
Grun S, Lindermayr C, Sell S, Durner J (2016) Nitric oxide and gene regulation in plants. J Exp Bot 57:507–516
Gupta DK, Inouhe M, Rodriguez-Serrano M, Romero-Puertas MC, Sandalio LM (2013) Oxidative stress and arsenic toxicity: role of NADPH oxidase. Chemosphere 90:1987–1996
Gupta G, Mishra K, Srivastava S, Kumar A (2018) Cytotoxic assessment of chromium and arsenic using chromosomal behavior of root meristem in *Allium cepa* L. Bull Environ Contam Toxicol 100:803–808
Khan MIR, Khan NA (2014) Ethylene reverses photosynthetic inhibition by nickel and zinc in mustard through changes in PS II activity, photosynthetic nitrogen use efficiency, and antioxidant metabolism. Protoplasma 251:1007–1019
Krishnamurthy A, Rathinasabapathi B (2013) Auxin and its transport play a role in plant tolerance to arsenite-induced oxidative stress in *Arabidopsis thaliana*. Plant Cell Environ 36:1838–1849
Kumar S, Dubey RS, Tripathi RD, Chakrabarty D, Trivedi PK (2015) Omics and biotechnology of arsenic stress and detoxification in plants: current updates and prospective. Environ Int 74:221–230
Lamotte O, Bertoldo JB, Besson-Bard A, Rosnoblet C, Aimé S, Hichami S, Terenzi H, Wendehenne D (2014) Protein S-nitrosylation: specificity and identification strategies in plants. Front Chem 2:114
Laplaze L, Benkova E, Casimiro I, Maes L, Vanneste S, Swarup R, Weijers D et al (2007) Cytokinins act directly on lateral root founder cells to initiate root initiation. Plant Cell 19:3889–3900
Leterrier M, Airaki M, Palma JM, Chaki M, Barroso JB, Corpas FJ (2012) Arsenic triggers the nitric oxide (NO) and S-nitrosoglutathione (GSNO) metabolism in Arabidopsis. Environ Pollut 166:136–143
Lewis DR, Negi S, Sukumar P, Muday GK (2011) Ethylene inhibits lateral root development, increases IAA transport and expression of PIN3 and PIN7 auxin efflux carriers. Development 138:3485–3495
Liang G, Yang F, Yu D (2010) MicroRNA395 mediates regulation of sulfate accumulation and allocation in *Arabidopsis thaliana*. Plant J 62:1046–1057
Liu Q, Zhang H (2012) Molecular identification and analysis of arsenite stress-responsive miRNAs in rice. J Agric Food Chem 60:6524–6536
Maksymiec W, Wójcik M, Krupa Z (2007) Variation in oxidative stress and photochemical activity in *Arabidopsis thaliana* leaves subjected to cadmium and excess copper in the presence or absence of jasmonate and ascorbate. Chemosphere 66:421–427
Masood A, Iqbal N, Khan NA (2012) Role of ethylene in alleviation of cadmium-induced photosynthetic capacity inhibition by Sulphur in mustard. Plant Cell Environ 35:524–533

Meng H, Hua S, Shamsi IH, Jilani G, Li Y, Jiang L (2009) Cadmium-induced stress on the seed germination and seedling growth of *Brassica napus* and its alleviation through exogenous plant growth regulators. Plant Growth Regul 58:47–59

Mengel A, Chaki M, Shekariesfahlan A, Lidermayr C (2013) Effect of nitric oxide on gene transcription – S-nitrosylation of nuclear proteins. Front Plant Sci 4:293

Mestrot A, Feldmann J, Krupp EM, Hossain MS, Roman-Ross G, Meharg AA (2011) Field fluxes and speciation of arsines emanating from soils. Enviorn Sci Technol 45:1798–1804

Mittler R (2002) Oxidative stress, antioxidants and stress tolerance. Trends Plant Sci 7:405–410

Mohan TC, Castrillo G, Navarro C, Zarco-Fernandez S, Ramireddy E, Mateo C et al (2016) Cytokinin determines thiol-mediated arsenic tolerance and accumulation. Plant Physiol 171:1418–1426

Mylona PV, Polidoros AN, Scandalios JG (1998) Modulation of antioxidant responses by arsenic in maize. Free Radic Biol Med 25:576–585

Nibau C, Gibbs DJ, Coates JC (2008) Branching out in new directions: the control of root architecture by lateral root formation. New Phytol 179:595–614

Pandey C, Gupta M (2015) Selenium and auxin mitigates as stress in rice by combining the role of stress indicators, modulators and genotoxicity assay. J Hazard Mater 287:384–391

Pandey C, Raghuram B, Sinha AK, Gupta M (2015) miRNA plays a role in the antagonistic effect of selenium on arsenic stress in rice seedlings. Metallomics 7:857–866

Pathare V, Srivastava S, Suprasanna P (2013) Evaluation of effects of arsenic on carbon, nitrogen, and sulfur metabolism in two contrasting varieties of *Brassica juncea*. Acta Physiol Plant 35:3377–3389

Praveen A, Gupta M (2018) Nitric oxide confronts arsenic stimulated oxidative stress and root architecture through distinct gene expression of auxin transporters, nutrient related genes and modulates biochemical responses in *Oryza sativa* L. Environ Pollut 240:950–962

Praveen A, Pandey A, Gupta M (2019) Nitric oxide alters nitrogen metabolism and PIN gene expressions by playing protective role in arsenic challenged *Brassica juncea* L. Ecotox Environ Safety 176:95–107

Rai A, Bhardwaj A, Misra P, Bag SK, Adhikari B, Tripathi RD, Trivedi PK, Chakrabarty D (2015) Comparative transcriptional profiling of contrasting rice genotypes shows expression differences during arsenic stress. Plant Genome 8:1–14

Rao KP, Vani G, Kumar K, Wankhede DP, Misra M, Gupta M, Sinha AK (2011) Arsenic stress activates MAP kinase in rice roots and leaves. Arch Biochem Biophys 506:73–82

Rodriguez MC, Petersen M, Mundy J (2010) Mitogen-activated protein kinase signaling in plants. Annu Rev Plant Biol 61:621–649

Roychowdhury R, Khan MH, Choudhury S (2018) Arsenic in rice: an overview on stress implications, tolerance and mitigation strategies. In: Hasanuzzaman M, Nahar K, Fujita M (eds) Plants under metal and metalloid stress. Springer, Singapore, pp 401–415

Roychowdhury R, Khan MH, Choudhury S (2019) Physiological and molecular responses for metalloid stress in rice – a comprehensive overview. In: Hasanuzzaman M, Fujita M, Nahar K, Biswas J (eds) Advances in rice research for abiotic stress tolerance. Woodhead Publishing/Elsevier, USA, pp 341–369

Schommer C, Palatnik JF, Aggarwal P, Chetelat A, Cubas P, Farmer EE, Nath U, Weigel D (2008) Control of jasmonate biosynthesis and senescence by miR319 targets. PLoS Biol 6:e230

Sharaf AEMM, Farghal II, Sofy MR (2009) Role of gibberellic acid in abolishing the detrimental effects of Cd and Pb on the broad bean and lupin plants. Res J Agric Biol Sci 5:668–673

Sharma N, Tripathi A, Sanan-Mishra N (2015) Profiling the expression domains of a rice specific miroRNAunder stress. Front Plant Sci 6:333

Shi S, Wang T, Chen Z, Tang Z, Wu Z, Salt DE, Chao DY, Zhao FJ (2016) OsHAC1;1 and OsHAC1;2 function as arsenate reductases and regulate arsenic accumulation. Plant Physiol 172:1708–1719

Shukla P, Singh S, Dubey P, Singh A, Singh AK (2015) Nitric oxide mediated amelioration of arsenic toxicity which alters the alternative oxidase (Aox1) gene expression in *Hordeum vulgare* L. Ecotox Environ Safety 120:59–65

Shukla A, Srivastava S, Suprasanna P (2017) Genomics of metal stress-mediated signaling and plant adaptive responses in reference to phytohormones. Curr Genomics 18:512–522

Singh VP, Srivastava PK, Prasad SM (2013) Nitric oxide alleviates arsenic-induced toxic effects in ridged *Luffa* seedlings. Plant Physiol Biochem 71:155–163

Singh AP, Dixit G, Kumar A, Mishra S, Singh PK, Dwivedi S (2015) Nitric oxide alleviated arsenic toxicity by modulation of antioxidants and thiol metabolism in rice (*Oryza sativa* L.). Front Plant Sci 6:1272

Singh S, Parihar P, Singh R, Singh VP, Prasad SM (2016) Heavy metal tolerance in plants: role of transcriptomics, proteomics, metabolomics, and ionomics. Front Plant Sci 6:1143

Singh AP, Dixit G, Kumar A, Mishra S, Kumar N, Dixit S (2017) A protective role for nitric oxide and salicylic acid for arsenite phytotoxicity in rice (*Oryza sativa* L.). Plant Physiol Biochem 115:163–173

Song WY, Yamaki T, Yamaji N, Ko D, Jung KH, Fujii-Kashino M, An G, Martinoia E, Lee Y, Ma F (2014) A rice ABC transporter, OsABCC1, reduces arsenic accumulation in the grain. Proc Natl Acad Sci U S A 111:15699–15704

Srivastava S, Pathare V (2017) Role of microRNAs in arsenic stress tolerance of plants. Intl J Plant Environ 3:9–15

Srivastava S, Mishra S, Tripathi RD, Dwivedi S, Trivedi PK, Tandon PK (2007) Phytochelatins and antioxidant systems respond differentially during arsenite and arsenate stress in *Hydrilla verticillata* (L.f.) Royle. Environ Sci Technol 41:2930–2936

Srivastava S, Srivastava AK, Suprasanna P, D'Souza SF (2009) Comparative biochemical and transcriptional profiling of two contrasting varieties of *Brassica juncea* L. in response to arsenic exposure reveals mechanisms of stress perception and tolerance. J Exp Bot 60:3419–3431

Srivastava S, Suprasanna P, D'Souza SF (2011) Redox state and energetic equilibrium determine the magnitude of stress in *Hydrilla verticillata* upon exposure to arsenate. Protoplasma 248:805–815

Srivastava S, Srivastava AK, Singh B, Suprasanna P, D'Souza SF (2013a) The effect of arsenic on pigment composition and photosynthesis in *Hydrilla verticillata* (L.f.) Royle. Biol Plant 57:385–389

Srivastava S, Srivastava AK, Suprasanna P, D'souza SF (2013b) Identification and profiling of arsenic stress-induced microRNAs in *Brassica juncea*. J Exp Bot 64:303–315

Srivastava AK, Srivastava S, Mishra S, D'Souza SF, Suprasanna P (2014) Identification of redox-regulated components of arsenate (AsV) tolerance through thiourea supplementation in rice. Metallomics 9:1718–1730

Srivastava S, Srivastava AK, Sablok G, Deshpande T, Suprasanna P (2015) Transcriptomics profiling of Indian mustard (*Brassica juncea*) under arsenate stress identifies key candidate genes and regulatory pathways. Front Plant Sci 6:646

Stepien A, Knop K, Dolata J, Taube M, Bajczyk M, Barciszewska-Pacak M, Pacak A, Jarmolowski A, Szweykowska-Kulinska Z (2017) Posttranscriptional coordination of splicing and miRNA biogenesis in plants. WIREs RNA 8:e1403

Tang Z, Chen Y, Miller AJ, Zhao FJ (2019) The C-type ATP-binding cassette transporter OsABCC7 is involved in the root-to-shoot translocation of arsenic in rice. Plant Cell Physiol (in press) 60:1525. https://doi.org/10.1093/pcp/pcz054

Tuli R, Chakrabarty D, Trivedi PK, Tripathi RD (2010) Recent advances in arsenic accumulation and metabolism in rice. Mol Breed 26:307–323

Verbruggen N, Hermans C, Schat H (2009) Mechanisms to cope with arsenic and cadmium excess in plants. Curr Opin Plant Biol 12:364–372

Vishwakarma K, Upadhyay N, Kumar N, Yadav G, Singh J, Mishra RK et al (2017) Abscisic acid signaling and abiotic stress tolerance in plants: a review on current knowledge and future prospects. Front Plant Sci 8:161

Vlot AC, Dempsey DMA, Klessig DF (2009) Salicylic acid, a multifaceted hormone to combat disease. Annu Rev Phytopathol 47:177–206

Wang P, Xu X, Tang Z, Zhang W, Huang X-Y, Zhao F-J (2018) OsWRKY28 regualtes phosphate and arsenate accumulation, root system architecture and fertility in rice. Front Plant Sci 9:1330

Wu Z, Ren H, McGrath SP, Wu P, Zhao FJ (2011) Investigating the contribution of the phosphate transport pathway to arsenic accumulation in rice. Plant Physiol 157:1498–1508

Xie M, Zhang S, Yu B (2015) microRNA biogenesis, degradation and activity in plants. Cell Mol Life Sci 72:87–99

Ye WL, Wood BA, Stoud JL, Andralojc PJ, Raab A, McGrath SP, Feldmann J, Zhao FJ (2010) Arsenic speciation in phloem and xylem exudates of castor bean. Plant Physiol 154:1505–1513

Yu LJ, Luo YF, Liao B, Xie LJ, Chen L, Xiao S, Li JT, Hu SN, Shu WS (2012) Comparative transcriptome analysis of transporters, phytohormone and lipid metabolism pathways in response to arsenic stress in rice (*Oryza sativa*). New Phytol 195:97–112

Chapter 16
Heme Oxygenase 1(HO1): An Enzyme of Plant System and Its Role Against Various Abiotic Stresses

Khushbu Verma and Afroz Alam

Abstract Heme oxygenase 1 (HO1) is a well-known protector against oxidative stress in an animal system. In plant system, recently it's also well explored against different stresses. Heme oxygenase (HO) scavenges free heme released from heme proteins with the production of ferrous iron (Fe^{2+}), biliverdin-IXα (BV-IXα), and carbon monoxide (CO). HO illustrates functional diversity and engages in several functions. It regulates positively in antioxidant defense mechanism against various types of abiotic stress. Among all forms of HO, HO1 has been well explored against abiotic stresses as metal stress, salt stress, and drought stress; however, the differing roles of the other three HOs are still less identified. HO1 defects lead to reduced chlorophyll content. It delays gibberellin-triggered programmed cell death of aleurone layer due to drought stress by interacting with nitric oxide. HO1 along with nitric oxide and iron shows the long-distance signaling response to salt stress. Overall, the study helps to explore highlights on the recent studies related to plant HO research involving its antioxidant role against abiotic stresses.

Keywords Abiotic stress · Antioxidant defense mechanism · Drought stress · Heme oxygenase 1 (HO1) · Salt stress

1 Introduction

The role of heme oxygenase (HO, EC 1.14.99.3) and its location were firstly observed in an animal system. It binds to heme and forms a 1:1 complex (Kikuchi et al. 2005). In case of mammal system, heme oxygenase is a membrane-bound

K. Verma
Faculty of Agriculture & Veterinary Science, Jayoti Vidyapeeth Women's University, Jaipur, Rajasthan, India

A. Alam (✉)
Department of Bioscience and Biotechnology, Banasthali Vidyapith, Banasthali, Rajasthan, India

R. Roychowdhury et al. (eds.), *Sustainable Agriculture in the Era of Climate Change*, https://doi.org/10.1007/978-3-030-45669-6_16

enzyme along with C-terminal hydrophobic tail. Two isoforms HO1 and HO2 of heme oxygenase have been observed (Yoshida et al. 1991). Davis et al. (2001) and Emborg et al. (2006) have reported that HOs consist of a small gene family with four members in the plant system. It has been categorized into two classes, HO1-like genes (including HO3 and HO4 of *Arabidopsis*) and HO2 genes, on the basis of amino acid sequence alignments in HO proteins. However, in cyanobacteria, algae, and higher plants, HO1 is soluble in nature (Unno et al. 2007). The report has revealed that HO1 is a single dense sphere consisting of mostly α-helix. Heme oxygenase has been recognized as a catalyst in plant system for the oxidation of heme to biliverdin IXα (BV), CO, and Fe^{+2} (Noriega et al. 2004; Shekhawat and Verma 2010; Shekhawat et al. 2011). In mammals, this BV IXα converts bilirubin IX α through bilirubin reductase enzyme (Ortiz de Montellano and Wilks 2001). Unno et al. (2007) have suggested that substrate heme adjusts in a sandwich-like structure between proximal and distal helix. The proximal end contains His 25 which is the proximal heme ligand involved in the binding of heme. Proximal and distal helices of substrate-free form of HO move freely with their conformation changes followed by heme-binding site opening. The fundamental structural folds of HO1 are conserved with different heme-binding pockets (Linly et al. 2006). In animal system, structure and size of HO1 are about comparable with few differences. Muramoto et al. (1999) have reported HO protein of molecular mass of 32.6 KDa having 55 amino acid chloroplast transit peptides in *Arabidopsis thaliana*. The main objective of this chapter is to highlight and be enlightened to the diversification in the regulatory role of HO1 along with its other members of the HO family in the plant cell. The main focus of this study is about the cross-talk between HO1 and other antioxidant enzymes or no enzymatic molecules.

2 Location of HO Gene Family in Plant Cell

The location of heme oxygenase 1 (HO1) in plant system has an interesting study after finding its presence in the plant system. The location of HO1 confirms more possible findings in regulatory roles in plant system. Due to the major regulatory role of HO1 in the phytochrome synthesis, it was suggested that the location of heme oxygenase is in the chloroplast. This fact was later on confirmed as well. It has been reported that HO1, among members of HO family, is located mainly in the stroma of chloroplast by using green fluorescent protein (GFP) reporter and immunoblot (Muramoto et al. 1999). Shekhawat and Verma (2010) have reported that HO1, a metalloprotein, catalyzes oxidative degradation of heme and protects against abiotic stress. HO1 catalyzes the heme degradation into biliverdin (BV IXα), carbon monoxide (CO), and ferrous ions (Fe^{+2}) (Fig. 16.1). Expression of HO1 responds to extensive cellular oxidative stress. Dixit et al. (2014) have suggested that oxidative stress elevates the level of reactive oxygen species (ROS), which causes the induction of HO1 expression. ROS mostly arises in the mitochondrion, connected with the leakage of partially reduced oxygen from the ETS under a normal and stressed

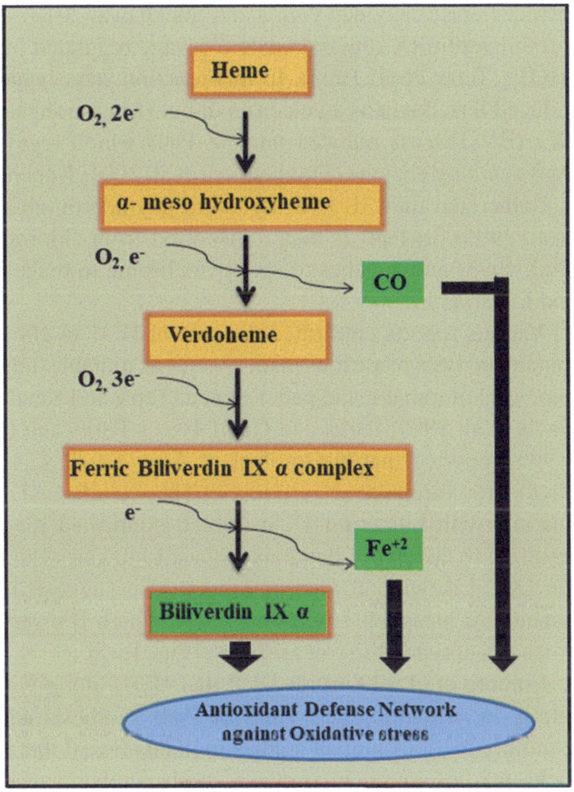

Fig. 16.1 Role of heme oxygenase 1 (HO1) in the degradation of heme and in antioxidant defense mechanism

condition. So, there is a possibility of co-localization of HO1 in the mitochondrion. Possible mitochondrial localization of HO1 might be on its inner membranous space because of its regulatory role in the synthesis of electron donor species. Subcellular localization of HO1 in chloroplast with mitochondrion has been confirmed in leaves of *Glycine max* L. (Dixit et al. 2014).

3 Regulatory Role of HO1 in Chromophore and Chlorophyll Synthesis

This is a well-known fact that active phytochrome leads to growth and photomorphogenic responses to red (R) and far-red light (FR) (Smith 1995; Quail 2002). In higher plants, HO1 has been discovered as the regulatory member in phytochrome–chromophore metabolism pathway. Holo-phytochrome protein is made up of PHY polypeptides and PφB chromophore. In case of PφB chromophore synthetic pathway, 5-aminolevulinic acid is the first precursor which is further converted into protoporphyrin IX followed by the formation of heme in the presence of ferrochelatase enzyme (Verma and Shekhawat 2013). Further, heme is transformed into biliverdin IXα by oxidative cleavage regulated by heme oxygenase 1 enzyme (HO1) (Terry et al. 1993). In this reaction step, heme oxygenase 1 makes use of reduced ferredoxin as an electron donor (Muramoto et al. 2002). Further, biliverdin IXα (BV IXα) is reduced into 3Z-PφB which regulated a ferredoxin-dependent PφB synthase enzyme (Frankenberg et al. 2001; Kohchi et al. 2001). Lastly, 3Z-PφB is isomerized into 3E-PφB by using phytochromobilin isomerase enzyme (Terry et al. 1995) 3E-PφB is then transported from chloroplast to the cytoplasm (Terry et al. 1993) and attaches to apophytochrome to make holo-phytochrome (Lagarias and Lagarias 1989).

Various reports confirmed the role of HO1 in chromophore synthesis by using mutant analysis of various plant species as mutant study of hy1 of *Arabidopsis thaliana*, yg-2 of tomato, and pcd-1 of pea (Terry and Kendrick 1999; Weller et al. 1996; Davis et al. 1999; Parks and Quail 1991). Parks and Quail (1991) and Davis et al. (1999) reported that hy1 mutant of *Arabidopsis* was not able to make PφB- and phytochrome-regulated responses. They confirmed it further by feeding these mutants with biliverdin IX α, which confirmed the negative effect of mutation resulting to lacking of heme oxygenase1 enzyme which converts heme into biliverdin IXα. Likewise, few other reports also suggested that mutant au and yg-2 of tomato are incapable to produce PφB, which is a resultant in plants with yellow-green phenotype (Terry et al. 2001) (Fig. 16.2).

Emborg et al. (2006) revealed that HO1, along with other members of the HO family in *Arabidopsis*, is vital for PφB synthesis and involved in the holo-phys assembly. Various mutant studies in plants reveal that HO is a key enzyme for phytochrome chromophore synthesis with involvement in the chlorophyll synthesis. hy1 and hy2 mutant of *Arabidopsis*, dark-grown au and yg-2 mutant of tomato, and pcd1 and pcd2 mutant of pea showed less Pchlide (precursor of chlorophyll) compared to their respective wild type. Beale and Weinstein (1991) and Cornah et al. (2003) revealed in their studies that there is a sharing pathway in between chlorophyll and phytochrome chromophore synthesis. In case of chlorophyll synthesis, Pchlide is formed by metal addition into protoporphyrin IX. In chlorophyll synthesis pathway, two steps are regulatory and rate-limiting. First is 5-aminolevulinic acid (5-ALA) formation from glutamate and second is metal insertion into protoporphyrin IX (Beale and Weinstein 1991; Cornah et al. 2003). Further, this Phclide forms chlorophyllide followed by chlorophyll synthesis. Terry and Kendrick (1999) revealed the indirect regulatory role of HO1 in chlorophyll synthesis by the mutant study of a yg-2 mutant of tomato. They observed that heme accumulated in mutant due to the inactivation of HO 1 enzyme which gives the origin of feedback inhibition for synthesis of 5-ALA. In spite of the plant system, feedback inhibition for 5-ALA synthesis through accumulated heme was reported in other systems also including cyanobacteria, green algae, and higher plants (Beale and Weinstein 1991; Castelfranco and Zeng 1991). However, Terry and Kendrick (1996, 1999) recommended in their report that inhibition of Pchlide synthesis is not connected to the inhibition of phytochrome. Some reports strongly support this opinion in which it

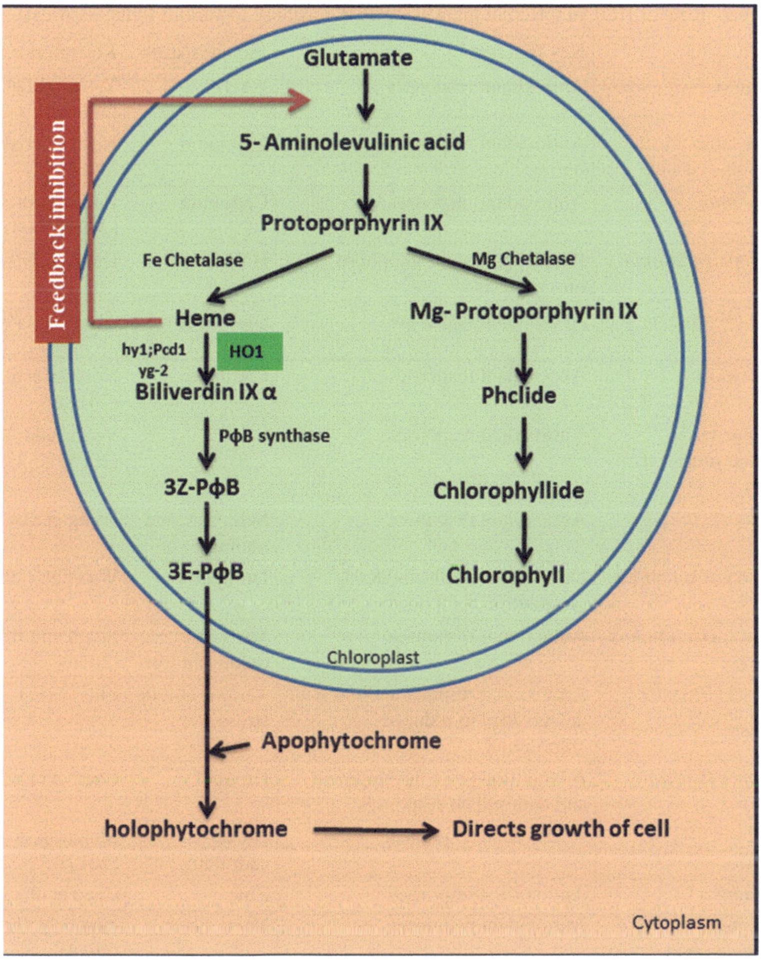

Fig. 16.2 Role of heme oxygenase 1 (HO1) in chromophore synthesis and indirect regulation of chlorophyll synthesis

was revealed that pale yellow-green phenotype had shown during their growth in several phytochrome mutant plants (Chory et al. 1989; Weller et al. 1996) (Table 16.1).

On the other hand, a severe reduction in Pchlide synthesis was reported in au and yg-2 mutant plants (Terry et al. 2001) which recommended that phytochrome also contributed to reduced Pchlide level in plant tissue (Terry et al. 2001). Hence, HO is a key enzyme for phytochrome chromophore synthesis and has an indirect regulatory role for chlorophyll synthesis with feedback inhibition of 5-ALA (Verma and Shekhawat 2013).

Table 16.1 Role of HO1 in different plant systems in response to various inducers/stress

Plants	Key function	Inducers/stress	Key references
Soybean (*Glycine max*)	Antioxidant responses	UV-B	Yannarelli et al. (2006)
Glycine max	Antioxidant responses	Cadmium	Noriega et al. (2004)
Glycine max	Antioxidant responses	Cadmium	Balestrasse et al. (2005, 2008)
Arabidopsis thaliana	Phytochrome–chromophore biosynthesis	HY1 locus	Davis (1999)
Alfalfa (*Medicago sativa*)	Antioxidant responses	Cadmium	Cui et al. (2011)
Glycine max	Antioxidant responses	Cadmium	Noriega et al. (2012)
Black mustard (*Brassica juncea* (L.) Czern.)	Antioxidant responses	Salt	Verma and Alam (2015)
Arabidopsis thaliana	Antioxidant responses	Salt, iron, and mercury	Wang et al. (2016)
Rice (*Oryza sativa* L.)	Delays gibberellin-induced programmed cell death	Drought	Wu et al. (2016)
Sunflower (*Helianthus annuus*)	Antioxidant responses	Salt/nitric oxide and iron	Singh and Bhatla (2016)
Rapeseed (*Brassica napus*)	Antioxidant responses along with defects lead to reduced chlorophyll	Various abiotic stress	Zhu (2017)
Cucumber (*Cucumis sativus*)	Adventitious root development and antioxidant responses	Drought	Chen et al. (2017)
Mung bean (*Vigna radiata*)	Antioxidant responses	Nickel–cadmium	Mahawar et al. (2017)
Oryza sativa L.	Antioxidant responses	Zinc	Chen et al. (2018)
Black mustard (*Brassica nigra*)	Antioxidant responses	Nano-silver	Amooaghaie et al. (2018)

4 Role of HO1 in Various Abiotic Stresses as an Antioxidant Enzyme

Heme oxygenase (HO, EC 1.14.99.3) enzyme is mainly responsible for stereospecific cleavage of heme and the heme oxidation to biliverdin IXα (BV), CO, and Fe^{+2} (Shekhawat and Verma 2010; Shekhawat et al. 2011). Many reports have supported the fact of HO's participation in plant system for cell protection mechanism against oxidative stress. In the response to stress conditions, HO has been acknowledged as a significant enzyme for ROS scavenging in both animals and plants (Ryter and Tyrrell 2000; Baranano et al. 2002; Noriega et al. 2004; Balestrasse et al. 2005, 2008). HO1 is directly not involved in these mechanisms. It shows involvement with its byproducts BV and BR as a component of the antioxidant defense mechanism in animals and plants (Stocker 1990; Ryter and Tyrrell 2000; Gonzales et al. 2002; Noriega et al. 2003, 2004; Balestrasse et al. 2008).

A lot of reports support HO1 as an antioxidant with all its byproducts of HO reaction. Balestrasse et al. (2005, 2008) suggested that HO1 was the main component of the antioxidant defense network in *Glycine max* L. plant against cadmium stress. In *Glycine max* L. under cadmium and salt stress, the level of BV increased because of the enhancement in the expression of HO1 in response to overcoming oxidative stress (Balestrasse et al. 2005, 2008; Zilli et al. 2009). Yannarelli et al. (2006) reported an elevation of HO1 activity against abiotic stress and enhancement of HO1 mRNA in *Glycine max* L. against UV-B stress. However, there are also some controversial reports. Various reports have revealed about the induction of HO1 by several factors such as its own substrate heme, NaCl, heavy metals, cadmium toxicity, UV-A and UV-B radiation, hyperoxia, and hypoxia (Maines and Kappas 1976; Maines 1988; Keyse and Tyrrell 1989; Tomaro et al. 1991; Noriega et al. 2004; Yannarelli et al. 2006; Balestrasse et al. 2008; Zilli et al. 2009) (Table 16.1).

According to previous reports, HOs have their contribution in only developmental biology like biosynthesis of phytochrome chromophore in plants (Muramoto et al. 2002; Terry et al. 2002; Kohchi et al. 2005; Emborg et al. 2006; Gohya et al. 2006). Analysis of the regulatory role of HO1 in Chinese cabbage, *Medicago sativa*, and wheat has also revealed that HO1 plays a significant role against various abiotic stresses in plant system (Cui et al. 2011; Fu et al. 2011; Jin et al. 2011; Xu et al. 2011). Noriega et al. (2012) suggested that HO1 is concerned in defense exerted by jasmonic acid against Cd stress in *Glycine max* roots.

In the past, the regulatory role of heme oxygenase 1 (HO1) under-reported as of the Antioxidant Defense Responses of *Brassica juncea* (L.) Czern. against Salt Stress Condition with treatment days dependant analysis and concentration-dependent analysis (Verma and Alam 2015; Verma et al. 2015). Wang (2016) revealed about characterization of AtHO1 promoter in reaction to salt stress, iron deficiency, and excess mercury exposure. Recently, Mahawar et al. (2017) has reported about positive role of HO1 in ROS quenching against Cd- and Ni-induced cytotoxicity in *Vigna radiata* L. Chen et al. (2018) has reported about the involvement of hemin through the HO1 system (along with its byproducts) against zinc stress in *Oryza sativa* L. with less zinc accumulation. Amooaghaie et al. (2018) have revealed about the role of HO1 activity and endogenous NO homeostasis which modulates the antioxidant defense system of *Brassica nigra* against stress induced due to nano-silver toxicity.

5 Cross-Talk Between HO1 Enzyme and Other Antioxidant Enzymes

In fact, for Pchlide, HO1 has a regulator role in antioxidant defense mechanism along with other antioxidants. Various reports revealed a pattern of expression of other antioxidant enzymes along with the elevated expression of HO1 (Balestrasse et al. 2005, 2008; Verma and Alam 2015; Verma et al. 2015). Balestrasse et al. (2008) who reported with their data propose that stimulation of CAT and HO1 occurred in nodules of soybean plants in the response to cell protection against oxidative damage due to cadmium stress. In *Brassica juncea* L. Czern., ascorbate peroxidase (APX), peroxidase (POD), and glutathione reductase (GR) expressed their highest activity at 150 mM NaCl concentration along with the upregulated expression of HO1 at the biochemical and molecular level (Verma et al. 2015). In *Vigna radiata*, HO1 activity was found to be highest against Cd stress and Ni stress with the increased activity of other antioxidant enzymes like CAT and APX (Mahawar et al. 2017). Chen et al. (2017) also reported about the possibility of CO byproduct of heme degradation by HO1; in H_2-induced adventitious root development under drought-induced stress, there is enhancement of leaf chlorophyll content and activation of antioxidant enzymes against oxidative stress. These reports suggest that HO1 has direct and indirect involvement through its products against oxidative stress and it plays as antioxidant enzyme.

6 Role of HO1 in Long-Distance Signaling Along with Other Molecules Against Abiotic Stress

Some reports have suggested and supported that HO1 has a role in long-distance signaling with other enzymatic and nonenzymatic molecules. As in past research, it has been identified that HO1 has indirect regulation in the synthesis of chlorophyll (Verma and Shekhawat 2013). This opinion was strongly supported by further research in *Brassica napus* by comparative sequencing analysis of *BnaC07.HO1* gene in wild and mutant species (Zhu et al. 2017).

On the other hand, few reports indicate about the role of HO1 along with nitric oxide and iron in the long-distance signaling response to salt stress (Singh and Bhatla 2016). It was suggested by this report that NO certainly alters HO1 activity in sunflower seedling cotyledons and there is a possible connection between endogenous NO, NaCl stress, and iron homeostasis by alteration of HO1 activity. HO1 also has a role in delaying gibberellin-triggered programmed cell death of aleurone layer due to drought stress by interacting with nitric oxide. It was suggested in evidence that GA regulates the expression of HO1 in germinating rice aleurone layers under drought stress condition (Wu et al. 2016). Santa-Cruz et al. (2017) also suggested that low methylation of HO1 promoter enhances the antioxidant activity against environmental stress. Their study has reported that some stress-related transcription factors are involved in HO1 upregulation against UV-B radiation and suggested a signaling link between TFs and HO1 expression. On the other hand, it's a well-known fact that accumulated proline plays a significant responsibility in the response of tolerance of plants against various abiotic stress. Gaseous signaling molecules such as a nitric oxide (NO) and carbon monoxide (CO), which are byproducts generated via HO1-regulated heme degradation process and hydrogen sulfide (H_2S), are involved in cell signaling. Somewhere, all these molecules regulate the accumulation of proline under abiotic-induced stress conditions (He and He 2017).

7 Conclusion

In plant system, heme oxygenase 1 (HO1) is a key enzyme for avoiding the negative and harmful effects of various abiotic stresses. By degradation of free heme, heme oxygenase (HO) generates byproducts such as ferrous iron (Fe^{2+}), biliverdin-IXα (BV-IXα), and carbon monoxide (CO). HO1, the member of HO family, shows diversification in regulatory role with its byproducts (BV-IXα, CO, and Fe^{2+}) and engages in various steps of antioxidant defense. In spite of the antioxidant role of HO1, it shows a role in chromophore synthesis, indirect regulation of chlorophyll, delayed programmed cell death, and lateral root development under abiotic stress. CO byproduct of HO1 also plays a role in H_2-induced adventitious root development under drought-induced stress by elevating the chlorophyll content and activation of antioxidant enzymes. Upregulation of HO1 activity has a connection with transcription factors to avoid the adverse effect of UV-B-induced stress. CO along with NO and H_2S has a signaling role in an enhancement of proline concentration under abiotic stress condition. This review concludes that HO1 has various regulatory roles in plant system as an antioxidant as well as a regulatory role in various signaling pathways and in developmental biology in regulating the chromophore, chlorophyll synthesis, and adventitious root development.

References

Amooaghaie R, Tabatabaei F, Ahadi A (2018) Alterations in *HO-1* expression, heme oxygenase activity and endogenous NO homeostasis modulate antioxidant responses of *Brassica nigra* against nano silver toxicity. J Plant Physiol 228:75–84

Balestrasse KB, Noriega GO, Batlle A, Tomaro ML (2005) Involvement of heme oxygenase as antioxidant defense in soybean nodules. Free Radic Res 39:145–151

Balestrasse KB, Yannarelli GG, Noriega GO, Batlle A, Tomaro ML (2008) Heme oxygenase and catalase gene expression in nodules and roots of soybean plants subjected to cadmium stress. Biometals 21:433–441

Baranano DE, Rao M, Ferris CD, Snyder SH (2002) Biliverdin reductase: a major physiological cytoprotectant. Proc Natl Acad Sci U S A 99:16093–16098

Beale SI, Weinstein JD (1991) Biochemistry and regulation of photosynthetic pigment formation in plants and algae. In: Jordan PM (ed) Biosynthesis of tetrapyrroles. Elsvier, Amsterdam, pp 155–235

Castelfranco PA, Zeng XH (1991) Regulation of 5-aminolevulinic acid synthesis in developing chloroplasts an endogenous inhibitor from the thylakoid membranes. Plant Physiol 97:1–6

Chen Y, Wang M, Hu L et al (2017) Carbon monoxide is involved in hydrogen gas-induced adventitious root development in cucumber under simulated drought stress. Orig Res 8. https://doi.org/10.3389/fpls.2017.00128

Chen Q, Gong C, Ju X et al (2018) Hemin through the Heme oxygenase 1/ferrous Iron, carbon monoxide system involved in zinc tolerance in *Oryza Sativa* L. J Plant Growth Regul 37:947–957

Chory J, Peto CA, Ashbaugh M, Sagnaich R, Pratt L, Ausubel F (1989) Different roles for phytochrome in etiolated and green plants deduced from characterization of Arabidopsis thaliana mutants. Plant Cell 1:867–880

Cornah JE, Terry MJ, Smith AG (2003) Green or red: what stops the traffic in the tetrapyrrole pathway? Trends Plant Sci 8:224–230

Cui W, Fu G, Wu H, Shen W (2011) Cadmium- induced heme oxygenase-1 gene expression is associated with the depletion of glutathione in the roots of *Medicago sativa*. Biometals 24:93–103

Davis SJ, Kurepa J, Vierstra RD (1999) The *Arabidopsis thaliana* HY1 locus required for phytochrome-chromophore biosynthesis, encodes a protein related to heme oxygenases. Proc Natl Acad Sci U S A 96:6541–6546

Davis SJ, Bhoo SH, Durski AM, Walker JM, Vierstra RD (2001) The heme oxygenase family required for phytochrome chromophore biosynthesis is necessary for proper photomorphogenesis in higher plants. Plant Physiol 126:656–669

Dixit S, Verma K, Shekhawat GS (2014) Subcellular localization of Hemeoxygenase-1 in mitochondria and the possible implication as its role as an antioxidant in soybean plant (*Glycine max*). Protoplasma 251:671–675

Emborg TJ, Walker JM, Noh B, Vierstra RD (2006) Multiple heme oxygenase family members contribute to the biosynthesis of the phytochrome chromophore in Arabidopsis. Plant Physiol 140:856–868

Frankenberg N, Mukougawa K, Kohchi T, Lagarias JC (2001) Functional genome analysis of the HY2 family of ferredoxin-dependent bilin reductase from oxygenic photosynthetic organisms. Plant Cell 13:965–978

Fu G, Zhang L, Cui W, Wang Y, Shen W, Ren Y (2011) Induction of heme oxygenase-1 with β-CD-hemin complex mitigates cadmium–induced oxidative damage in the roots of *Medicago sativa*. Plant Soil 345:271–285

Gohya T, Zhang X, Yoshida T, Migita CT (2006) Spectroscopic characterization of a higher plant heme oxygenase isoform-1 from *Glycine max* (soybean) – coordination structure of the heme complex and catabolism of heme. FEBS J 273:5384–5399

Gonzales S, Erario MA, Tomaro ML (2002) Heme oxygenase-1 induction and dependent increase in ferritin. A protective antioxidant stratagem in hemintreated rat brain. Dev Neurosci 24:61–168

He H, He L-F (2017) Regulation of gaseous signaling molecules on proline metabolism in plants. Plant Cell Rep 37:387–392

Jin Q-J, Yuan X-X, Cui W-T, Han B, Feng J-F, Xu S, Shen W-B (2011) Isolation and characterization of a heme oxygenase-1 gene from Chinese cabbage. Mol Biotechnol 50:8–17

Keyse SM, Tyrrell AM (1989) Heme oxygenase is the major 32 kDa stress protein induced in human skin fibroblasts of UV-A radiation, hydrogen peroxide and sodium arsenite. Proc Natl Acad Sci U S A 86:99–103

Kikuchi G, Yoshida T, Noguchi M (2005) Heme oxygenase and heme degradation. Biochem Bioph Res Co 338:558–567

Kohchi T, Mukougawa K, Frankenberg N, Masuda M, Yokota A, Lagarias JC (2001) The Arabidopsis HY2 gene encodes phytochromobilin synthase, a ferredoxin- dependent biliverdin reductase. Plant Cell 13:425–436

Kohchi T, Kataoka H, Linley PJ (2005) Biosynthesis of chromophores for phytochrome and related photoreceptors. Plant Biotechnology 22:409–413

Lagarias JC, Lagarias DM (1989) Self assembly of synthetic phytochrome holoprotein in vitro. Proc Natl Acad Sci U S A 86:5778–5780

Linly PJ, Landsberger M, Kohchi T, Cooper JB, Terry MJ (2006) The molecular basis of heme oxygenase deficiency in the pcd1 mutant of pea. FBES J 273:2594–2606

Mahawar L, Kumar R, Shekhawat GS (2017) Evaluation of heme oxygenase 1 (HO 1) in cd and Ni induced cytotoxicity and crosstalk with ROS quenching enzymes in two to four leaf stage seedlings of Vigna radiate. Protoplasma 255:527. https://doi.org/10.1007/s00709-017-1166-0

Maines MD (1988) Heme oxygenase: function, multiplicity, regulatory mechanism and clinical applications. FASEB J 2:2557–2568

Maines MD, Kappas A (1976) Studies on the mechanism of induction of heme oxygenase by cobalt and other metal ions. Biochem J 154:125–131

Muramoto T, Kohchi T, Yokota A, Goodman HM (1999) The Arabidopsis photomorphogenic mutant hy1 is deficient in phytochrome chromophore biosynthesis as result of a mutation in a plastid heme oxygenase. Plant Cell 11:335–348

Muramoto T, Tsurui N, Terry MJ, Yokota A, Kohchi T (2002) Expression and biochemical properties of a ferredoxin- dependent heme oxygenase required for phytochrome chromophore synthesis. Plant Physiol 130:1958–1966

Noriega GO, Tomaro ML, Batlle AMC (2003) Bilirubin is highly effective in preventing in vivo d-aminolevulinic acid-induced oxidative cell damage. Biochim Biophys Acta 1638:173–178

Noriega GO, Balestrasse KB, Batlle A, Tomaro ML (2004) Heme oxygenase exerts a protective role against oxidative stress in soybean leaves. Biochem Biophys Res Commun 323:1003–1008

Noriega G, Cruz DS, Batlle A, Tomaro M, Balestresse K (2012) Heme oxygenase is involved in the protection exerted by jasmoic acid against cadmium stress in soybean roots. J Plant Growth Regul 31:79–89

Ortiz de Montellano PR, Wilks A (2001) Heme oxygenase structure and mechanism. Adv Inog Chem 51:359–407

Parks BM, Quail PH (1991) Phytochrome deficient hy1 and hy2 long hypocotyl mutants of Arabidopsis are defective in phytochrome chromophore biosynthesis. Plant Cell 3:1177–1186

Quail PH (2002) Phytochrome photosensory signaling networks. Nat Rev Mo Cell Biol 3:85–93

Ryter SW, Tyrrell RM (2000) The heme synthesis and degradation pathways: role in oxidant sensitivity. Heme oxygenase has both pro and antioxidant properties. Free Radic Biol Med 28:289–309

Santa-Cruz D, Pacienza N, Zilli C, Pagano E, Balestrasse K, Yannarelli G (2017) Heme oxygenase up-regulation under ultraviolet-B radiation is not epigenetically restricted and involves specific stress-related transcriptions factors. Redox Biol 12:549–557

Shekhawat GS, Verma K (2010) Haem oxygenase (HO): an overlooked enzyme of plant metabolism and defence. J Exp Bot 61:2255–2270

Shekhawat GS, Dixit S, Verma K, Nasybullina EI, Kosmachevskaya OV, Topunov AF (2011) Heme oxygenase: enzyme with functional diversity. J Stress Physiol Biochem 7:88–94

Singh N, Bhatla SC (2016) Nitric oxide and iron modulate heme oxygenase activity as a long distance signaling response to salt stress in sunflower seedling cotyledons. Nitric Oxide 53:54–64

Smith H (1995) Physiological and ecological functions within the phytochrome family. Annu Rev Plant Physiol 46:269–315

Stocker R (1990) Induction of haem oxygenase as a defence against oxidative stress. Free Radic Res Commun 9:101–112

Terry MJ, Kendrick RE (1996) The aurea and yellow-green-2 mutants of tomato are deficient in phytochrome chromophore synthesis. J Biol Chem 271:21681–21686

Terry MJ, Kendrick RE (1999) Feedback inhibition of chlorophyll synthesis in the phytochrome chromophore- deficient aurea and yellow-green-2 mutants of tomato. Plant Physiol 119:143–115

Terry MJ, Wahleithner JA, Lagarias JC (1993) Biosynthesis of the plant photoreceptor phyto chrome arch. Biochem Biophys 306:1–15

Terry MJ, Mcdowell MT, Lagarias JC (1995) (3Z)- and (3E)- Phytochromobilin are intermediates in the biosynthesis of the phytochrome chromophore. J Biol Chem 270:11111–11118

Terry MJ, Ryberg M, Raitt CE, Page AM (2001) Altered etioplast development in phytochrome chromophore- deficient mutants. Planta 214:314–325

Terry MJ, Linley PJ, Kohchi T (2002) Making light of it: the role of plant heme oxygenases in phytochrome chromophore synthesis, Biochem Soc Trans 30:604–609

Tomaro ML, Frydman J, Frydman RB (1991) Heme oxygenase induction by CoCl2, co-protoporphyrin IX, phenylhydrazine, and diamide: evidence for oxidative stress involvement. Arch Biochem Biophys 286:610–617

Unno M, Matsui T, Ikeda-Saito M (2007) Structure and catalytic mechanism of heme oxygenase. Nat Prod Rep 24:553–570

Verma K, Alam A (2015) Role of Heme Oxygenase1 in alteration of antioxidant defense responses of *Brassica juncea* (L.) Czern. under stress condition. Int J Sci Res Environ Sci 3(3):0119–0128

Verma K, Shekhawat GS (2013) Phytochrome-chromophore biosynthesis and chloroplast development: possible role and regulation of HO. In: Moliere A, Vigneron E (eds) New developments in chromophore research. Nova Science Publishers, Hauppauge, pp 267–280

Verma K, Dixit S, Shekhawat GS, Alam A (2015) Antioxidant activity of heme oxygenase 1 in *Brassica juncea* (L.) Czern. (Indian mustard) under salt stress. Turk J Biol 39:540–549

Wang FQ, Yang J, Dai C, Wu MZ, Zhang YH, Shen WB (2016) Characterization of *Arabidopsis thaliana heme oxygenase 1* promoter in response to salinity, iron deficiency, and mercury exposure. Biol Plant 61(1):35–47

Weller JL, Terry MJ, Rameau C, Reid JB, Kendrick RE (1996) The phytochrome-deficient *pcd1* mutant of pea is unable to convert heme to biliverdin IXα. Plant Cell 8:55–67

Wu H, Zheng Y, Liu J, Zhang H, Chen H (2016) Heme oxygenase-1 delays gibberellin induced programmed cell death of rice aleurone layers subjected to drought stress by interacting with nitric oxide. Front Plant Sci. https://doi.org/10.3389/fpls.2015.01267

Xu DK, Jin QJ, Xie YJ, Liu YH, Lin YT, Shen WP (2011) Characterization of a wheat heme oxygenase-1 gene and its responses to different abiotic stresses. Int J Mol Sci 12:7692–7707

Yannarelli GG, Noriega GO, Batlle A, Tomaro ML (2006) Heme oxygenase up regulation in ultraviolet-B irradiated soybean plants involves reactive oxygen species. Planta 224:1154–1162

Yoshida T, Ishikawa K, Sato M (1991) Degradation of heme by a soluble peptide of heme oxygenase obtained from rat liver microsomes by mild trypsinization. Eur J Biochem 199:729–733

Zhu L (2017) Heme oxygenase 1 defects lead to reduced chlorophyll in *Brassica napus*. Plant Mol Biol. https://doi.org/10.1007/s11103-017-0583-y

Zilli CG, Santa-Cruz DM, Yannarelli CG, Noriega GO, Tomaro ML, Balestrasse KB (2009) Heme oxygenase contributes to alleviate salinity damage in *Glycine max* leaves. Int J Biochem Cell Biol 2009:1–9

Chapter 17
Role of Plant Growth-Promoting Rhizobacteria (PGPR) for Crop Stress Management

Ashutosh Kabiraj, Krishnendu Majhi, Urmi Halder, Moitri Let, and Rajib Bandopadhyay

Abstract Crops under both abiotic and biotic stress are the major constraints on productivity. A number of factors like physical disorders, disease susceptibility, toxicity, hormonal imbalance, and nutritional deficiency interfere with the growth and development of plant under stress condition. Under these circumstances, rhizoremediation with the help of the plant growth-promoting rhizobacteria can mitigate stress-induced adverse effects on crop productivity. Plant growth-promoting rhizobacteria and their associated molecules play dual role by affecting both nutrition and resistance concomitantly through overlapping mechanisms. These free-living plant growth-promoting rhizobacteria actively colonize plant roots, exerting beneficial effects using their own metabolism or by directly affecting the plant metabolism. Rhizobial symbiosis has great agricultural importance in terms of improving soil fertility and crop productivity due to their synergistic as well as antagonistic interactions with other microbes in the soil environment. Plant growth-promoting rhizobacteria trigger elicitors, produce siderophores which deprive iron nutrition, and also induce cell wall-degrading extracellular enzymes as defense responses against plant pathogens. PGPR have the ability to induce the secretion of phytohormones, volatile compounds, antibiotics, and toxins which play an important role in plant growth. Rhizobacteria trigger N-acyl homoserine lactones (AHLs) like autoinducer molecules to regulate the gene expression as a part of quorum sensing. Other than these, plant growth-promoting rhizobacteria stimulate endogenous hormones of hosts to enhance stress tolerance. The mutualistic symbiosis triggers NOD factors and NOP effectors, while nonsymbiotic bacterial molecules enhance plant nutrient acquisition and growth. Here in this chapter, we have discussed and reviewed comprehensively the effectivity and mechanisms of plant growth-promoting rhizobacteria for enhancing crop productivity under different stress conditions.

A. Kabiraj · K. Majhi · U. Halder · M. Let · R. Bandopadhyay (✉)
Department of Botany (UGC-Centre for Advanced Study), The University of Burdwan, Burdwan, West Bengal, India

R. Roychowdhury et al. (eds.), *Sustainable Agriculture in the Era of Climate Change*, https://doi.org/10.1007/978-3-030-45669-6_17

Keywords Crop stress · Defense response · Mutual symbiosis · PGPR · Rhizoremediation

1 Introduction

Environmental change with rapidly increasing population throughout the world is becoming a big challenge to feed all the people. Within 2020, the population of the world will be about 8 billion (Glick 2012) and it would be 9 billion in 2050 (Vejan et al. 2016). About 50% of grain yield is required to be increased in most important crops like rice, wheat, and maize to feed all the people in 2050 (Shrivastava and Kumar 2015). Rapid changes to the environment mainly due to excessive use of chemical fertilizers and hazardous material in the field are decreasing crop productivity in one hand and increasing pollution on the other hand (Chakraborty et al. 2014; Roychowdhury 2014). Plant–microbes interactions draw a lot of attention to many scientists from time to time throughout the world. Deep relationship and interactions between plants, soil, soil microfauna, and microorganisms take place at rhizosphere (Antoun and Prevost 2005). Rhizospheric microorganisms that live in the rhizosphere of plants may or may not invade the plant root for shelter and make a symbiotic association with plants by providing some essential elements as well as protection. Rhizobacteria are involved in the promotion of plant growth and development, known as plant growth-promoting rhizobacteria, simply, PGPR. A bacterium can be called as PGPR when it shows three important characters or at least two characters as stimulation, fast colonization, and beneficial activity to plants on growth (Bhattacharyya and Jha 2012). Despite huge numbers of microbes present in the rhizosphere, only 7–15% microbial cells occupy the root surface (Gray and Smith 2005), and only 1–2% bacteria are responsible for the beneficial activity to the plants as PGPR (Beneduzi et al. 2012). There are two basic types of PGPR: intracellular PGPR (iPGPR), which makes root nodule and resides in it; another one is extracellular PGPR (ePGPR), which cannot make nodule and resides outside of the root. iPGPR can fix nitrogen symbiotically in the root nodule of the host; on the other hand, ePGPR helps by providing protection to plant, forming siderophores, increasing phytohormone production, enhancing the resistant potential to plants, etc. (Gray and Smith 2005). Basically, *Rhizobia* and *Frankia* are not called as PGPR (Antoun and Prevost 2005), but in this study, we will also focus on the *Rhizobia* as they have a direct effect on growth of the plants by nitrogen fixation.

The concept of nitrogen cycle was first inaugurated by Reyset in 1856 by describing the release of nitrogen from organic matter. Berthelot in 1885 was able to demonstrate chemical nitrogen fixing by lightning strike. Although biological nitrogen fixation by the microorganisms was first introduced by Jodinin in 1862. At the end of nineteenth to early twentieth century nitrogen-fixing microbes got much attention, and interest is increasing day to day (Elmerich 2007).

Being sessile, plant has to deal with different types of environmental hazards like biotic (pathogenic fungi, bacteria, virus, nematodes, etc.) as well as abiotic (extreme temperature, salt stress, flood, drought, high wind, etc.). Plants overcome these situations by modulating their mode of gene expression (Yang et al. 2009; Roychowdhury et al. 2014; Hasanuzzaman et al. 2015; Anumalla et al. 2016).

Acceleration of nutrient availability, assimilation of nutrients, suppression of disease-causing microorganisms, and enhancing growth and metabolisms are beneficial activities for plant which are commonly performed by PGPR (Perez-Montano et al. 2014). For example, PGPR and other plant beneficial microorganisms help plants to overcome the stress conditions by exhorting induced systemic tolerance (IST) (Yang et al. 2009).

In stress conditions, PGPR induces many stress-tolerating genes, proteins, enzymes, etc. In drought stress, transcription of *ERD15* gene takes place in *Arabidopsis thaliana* by the activity of PGPR *Paenibacillus polymyxa* (Timmusk and Wagner 1999). *Achromobacter piechaudii* ARV8 is another example of PGPR, which produces 1-aminocyclopropane-1-carboxylate (ACC) deaminase in drought stress condition, inhibits the function of ethylene (responsible for the reduction of root and shoot length) in pepper (*Capsicum annuum* L.) and tomato (*Solanum lycopersicum*) plants (Mayak et al. 2004). *HIGH-AFFINITY* K^+ *TRANSPORTER 1* (*HKT1*) is a transporter protein expressed in the *Arabidopsis* responsible for the Na^+ import to the root system; in presence of PGPR, its expression is decreased in salt stress condition (Yang et al. 2009).

According to the experiment of Guo and Chi 2014, presence of PGPR influenced Cd accumulation in root and it is balanced in rhizospheric region of *Lolium multiflorum* Lam. but in case of *Glycine max* L. Cd accumulation showed significantly decreases in both root as well as shoot.

However, in this review, we point out recent knowledge on PGPR, a very brief history, and its impact on the plant growth and development in stress conditions, like salt, drought, and heavy metal.

2 Impact of Environmental Stresses on Crop Productivity

2.1 *Adaptation of Defense System Under Stress Conditions*

Some plants are not able to take action against the pathogenic microorganisms. Usually, some physical and chemical materials are the basic weapons for the plant defense. Pathogenic microorganisms induce systemic acquired resistance (SAR) in plants. SAR is associated with the pathogen-related (PR) protein and salicylic acid. On the other hand, induced systemic resistance (ISR) is induced by PGPR. ISR is associated with ethylene and jasmonic acid production. PGPR induce the production of oxidative enzymes like peroxidase, superoxide dismutase, etc. which give protection to plant from different pathogens like bacteria, fungi, virus, etc. (Kumar and Verma 2019). Wheat plants treated with PGPR *Dietzianatrono limnaea* STR1, supplemented with 150 mM salt concentration, showed better dry weight and length against control condition due to the overexpression of ABA-responsive gene (*ABARE*) and *TaOPR1* gene in root and shoot system (Bharti et al. 2016). Some Rhizobacteria produce exopolysaccharides that accumulate Na^+ ions and relieve plants from salt stress (Arora et al. 2013). Physiological, biochemical, and morphological adaptations with different beneficial activities of PGPR help to induce the defense system in plants to overcome other abiotic stresses.

2.2 Nitrogen Fixation Under Stress Conditions

The eukaryotic organisms are not able to fix molecular nitrogen into their cells. Plants used to uptake nitrogenous compounds such as nitrate, ammonia, etc. from the environment through their root system. Here, the role of microorganisms is noticeable; many beneficial free-living and symbiotic microbes are well documented as nitrogen fixers in plants. But in stress condition, nitrogen fixation is also hampered dramatically. So, the basic metabolisms in plants get partially or fully arrested because of the low level of nitrogen-containing compounds resulting reduction of growth and development. Nitrogen fixation is a very energy-consuming procedure. PGPR is not too good as nitrogen fixers (Martínez-Viveros et al. 2010), but it helps plants directly or indirectly to overcome the stressed condition. Symbiotic association by *Rhizobium* sp. with nodule formation is restricted in the legume plants only, so nonleguminous plants are dependent on rhizospheric bacteria (Martínez-Viveros et al. 2010). The well-known nitrogen-fixing nonsymbiotic bacteria are as *Azoarcus* sp. (Reinhold-Hurek et al. 1993), *Burkholderia* sp. (Santos et al. 2001); *Azospirillum* sp. (Bashan and de-Bashan 2010), etc. *nif* gene is responsible for the biological nitrogen fixation, and this gene is present also in the PGPR (Gupta et al. 2015). *Pseudomonas stutzeri* A1501 is an ACC deaminase-producing strain containing *acdS* gene; besides ACC deaminase production in salt stress, it also regulates the function of nitrogenase, an important nitrogen-fixing enzyme, and increases crop yield in rice plant (Han et al. 2015).

3 Plant Growth-Promoting Rhizobacteria (PGPR)

3.1 History of PGPR

The term PGPR was first used by Joseph W. Kloepper in late 1970s (Vessey 2003) and defined by Kloepper and Schroth (Kloepper 1978). From the last decade of the nineteenth century, nitrogen-fixing bacteria act as a PGPR, and its molecular mechanism (Bhattacharyya and Jha 2012) has become an interesting topic to the scientists. At the very beginning, PGPR studies were restricted on beneficial activity regarding biological control of plant diseases only (Antoun and Prevost 2005).

The common plants which make symbiotic associations with the rhizobia are soybean (*Glycine max*), alfalfa (*Medicago sativa*), bean (*Phaseolus vulgaris*), pea (*Pisum sativum*), clover (*Trifolium* sp.), peanut (*Arachis hypogaea*), acacia (*Acacia* sp.), lentil (*Lens culinaris*), vetch (*Coronilla* sp.), birdsfoot trefoil (*Lotus corniculatus*), chickpea (*Cicer arietinum*), etc. (Gray and Smith 2005). Rhizobial species that can associate with these plants are as *Bradyrhizobium japonicum* E109 (Cassan et al. 2009), *Bradyrhizobium japonicum* 532C, etc. Many bacterial strains like *Aeromonas hydrophila* P73, *Pseudomonas fluorescens* 31-12, *Serratia liquefaciens*2-68, *Pseudomonas putida* G11-32, etc. (Zhang et al. 1996), *Rhizobium* sp. (Nyoki and Ndakidemi 2018), and *Sphingomonas* sp. LK11 (Bilal et al. 2018) make relationships with soybean (*Glycine max*). *Ensifer meliloti* (Cedeño-García et al. 2018; Kang et al. 2018), *Rhizobium radiobacter*, *Rhizobium rosettiformans* (Kang et al. 2018); *Sinorhizobium meliloti*, *Achromobacter spanium*, *Serratia plymuthica* (Aroua et al. 2018), *Sinorhizobium meliloti* with *Paenibacillus mucilaginosus* (Ju et al. 2019), etc. can make symbiotic relationship with alfalfa (*Medicago sativa*) plants. Bean (*Phaseolus vulgaris*) plants are associated with the common rhizobacteria as *Azospirillum brasilense* (Malinich and Bauer 2018) and *Rhizobium tropici* (Nogales et al. 2002) along with *Rhizobium* sp. (Ormeño-Orrillo et al. 2012), *Rhizobium tropici*, *R. etli*, *R. gallicum*, *R. leguminosarum* bv. *phaseoli*, *R. giardinii* (Fernandez-Aunión et al. 2010), etc. One of the common legume plant peas (*Pisum sativum*) is associated with rhizobacteria as *Streptomyces lydicus* WYEC108 (Tokala et al. 2002), *Bacillus thuringeinsis*-KR1, along with *R. leguminosarum* (Mishra et al. 2009), etc. Co-inoculation of PGPR with the *Rhizobium* is also extensively studied in case of the pea plant.

Many species of *Bacillus* and *Pseudomonas* belong to free-living PGPR, i.e., ePGPR (Beneduzi et al. 2012). Other bacterial species such as *Enterobacter*, *Klebsiella*, *Azotobacter*, *Variovorax*, *Azospirillum*, *Serratia*, *Burkholderia* (Nadeem et al. 2014; Vejan et al. 2016), etc. are also reported as ePGPR. These bacteria are involved directly or indirectly with the plant growth and development.

3.2 PGPR to Mitigating Stress

Different types of abiotic stresses like salinity, drought, heavy metal, water logging, temperature, water contamination, air pollutant, etc. and biotic stresses like pathogenicity, weeds, parasites, etc. are present in the environment (Saleem et al. 2007). By interactions with plants, PGPR help them to mitigate both abiotic and biotic stresses. Some examples of PGPR that mitigate stresses like salt by *Achromobacter piechaudii*, (Mayak et al. 1999) and *Variovorax paradoxus*5C-2 against drought on pea plants (Dodd et al. 2004); *Pseudomonas putida* UW4 to tomato plants (Grichko and Glick 2001); *Burkholderia phytofirmans* relieves potato plants in temperature (Bensalim et al. 1998); *Pseudomonas fluorescens* can reduce pathogenicity stress over *Chamaecytisus proliferus* plant (Donate Correa et al. 2005); *Kluyvera ascorbata* SUD165 has an effect on *Brassica napus* in heavy metal stress

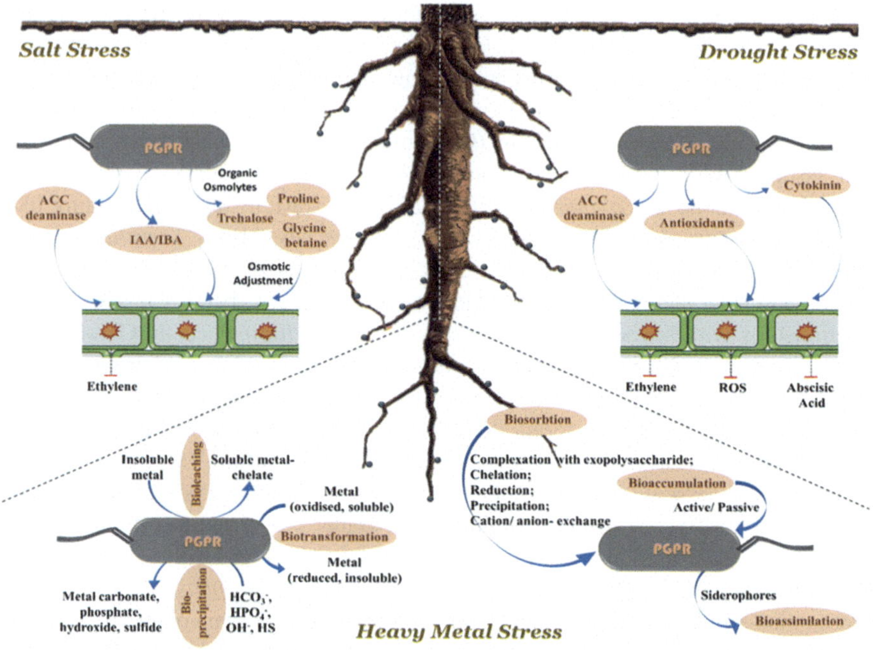

Fig. 17.1 Activity of PGPR to mitigate different stress conditions on plants

(Burd et al. 1998). ACC deaminase is the very common enzyme, produced by PGPR, which helps plant to mitigate all abovementioned stresses. Here, in this review, we just discuss the most important plant biotic and abiotic stresses (salt, drought, and heavy metal) and their mitigation strategies (Fig. 17.1).

3.3 Potential Synergistic As Well As Antagonistic Effects of PGPR

There are many biotic factors like bacteria, fungi, nematodes, parasites, etc. that directly and indirectly interact with plants. These organisms may involve in beneficial activities with plants in one hand, while in another, they can produce a detrimental effect on normal metabolism.

PGPR have the ability to interact with these factors and reduce their pathological activities by the production of antibiotic, siderophore, HCN, etc. (Gupta et al. 2015). For example, antibiotics are produced as amphisin, tropolone, oomycin A, phenazine, pyrrolnitrin, tensin, 2, 4-diacetylphloroglucinol (DAPG), pyoluteorin, and cyclic lipopeptides by different species of *Pseudomonas* (Loper and Gross 2007). From the study of Srivastava et al. 2010 it is revealed that *Trichoderma*, fluorescent *Pseudomonas*, and *Glomus* have a synergistic effect on the *Fusarium* wilting disease of potato. Alizadeh et al. in 2013 reported synergistic effects of *Pseudomonas* sp. Ps14 and fungus *Trichodermaharzianum*Tr6 on *Cucumis sativus* to express some defense-related genes. In studies on cotton, *Gossypium hirsutum*, two different bacterial strains, *Azospirillum* sp. AZ204 and *Pseudomonas fluorescens* Pf1, showed growth promotion against normal conditions (Marimuthu et al. 2013). Also *Pseudomonas aeruginosa* PHU094, *Trichoderma harzianum* THU0816, and *Mesorhizobium* sp. RL091 have the capability to activate the phenylpropanoid pathway (Singh et al. 2014).

Some microorganisms have the ability to inhibit the growth of other microorganisms by secretion of some toxic chemicals like antibiotics. PGPR also produce some chemicals and destroy the growth of many pathogenic microbes (Siddiqui and Singh 2005) like synthesis of hydrolytic enzymes (protease, lipase, glucanase, etc.), competition for nutrient, regulation of ethylene production and siderophore and antibiotic production, etc. (Beneduzi et al. 2012). Siderophores or iron-chelating chemicals are a good weapon for the rhizospheric microbes. More than 100 types of siderophores produced by microbes are discovered until now. PGPR can produce siderophores, attract iron ions, and accumulate iron for their metabolic activity. So, pathogenic bacteria are deprived of iron and ultimately die. For example, siderophore pseudobactin is produced by *Pseudomonas putida* B10 that can inhibit the growth of *Fusarium oxysporum* (Kloepper et al. 1980). Bacteriocins are the chemicals produced by bacteria that are antagonistic to the same group of bacteria. *E. coli*, a gram-negative bacterium, produces bacteriocin and colicin, which is antagonistic to many gram-negative bacteria (Beneduzi et al. 2012). Chitinase and beta-glucanase are two important enzymes produced by PGPR that can inhibit the growth of fungi (Vejan et al. 2016). Induced systemic resistance (ISR) and systemic acquired resistance (SAR) synergistically affect against the biotic stress in presence of PGPR. Species of *Pseudomonas* are mainly responsible for the stimulation of these responses (Fig. 17.2).

4 Rhizoremediation to Mitigate Stress-Induced Adverse Effects on Crop Productivity

4.1 Effects of PGPR on Salty Crops

About 20% agricultural lands and 50% crop (about 5.2 billion hectares of fertile land, Numan et al. 2018) are under salt stress in the world (Paul and Lade 2014). When electrical conductivity of a saturated paste soil extract is ECc $\geq$ 4dS/m, it is known as saline soil (Forni et al. 2017). There are five different classes of soil salinity, such as nonsaline, slightly saline, moderately saline, strongly saline, and very strongly saline (Paul and Lade 2014). Among all, less number of salt-tolerable plants can grow in very strongly saline class. Soil salinity can inhibit many process in plant including protein synthesis, lipid metabolism, photosynthesis, etc.

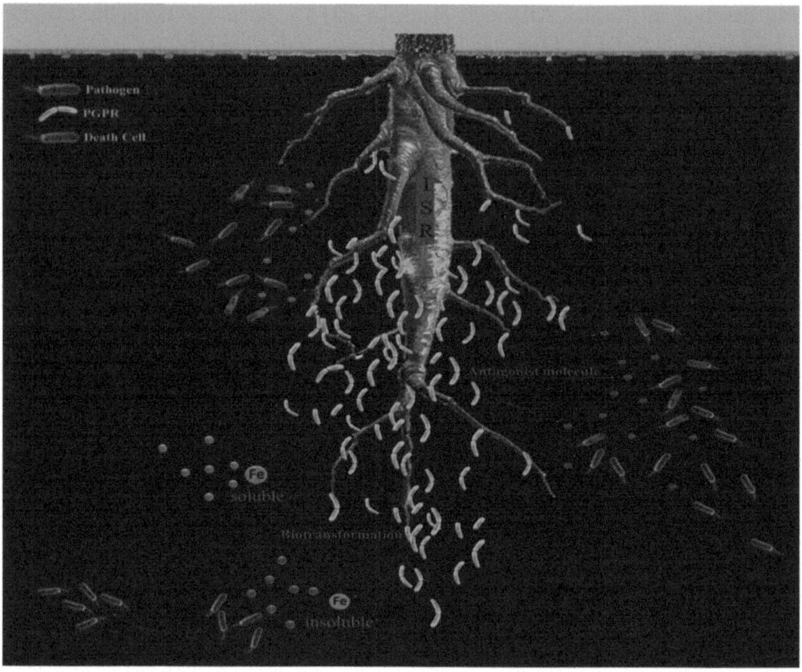

Fig. 17.2 PGPR-induced defense system against pathogenic attack on plants

Usually, salts induce ROS production, such as superoxide radicals (O^{2-}), hydroxyl radicals (OH), and hydrogen peroxide (H_2O_2), are responsible for DNA damage, protein degradation, and lipid peroxidations of membranes. It also hampers seed set and crop yield and reduces flowering in different plants like wheat, barley, rice, cotton, etc. (Numan et al. 2018). Accumulation of sodium and chlorine ions in the soil can reduce the availability of other important essential elements; cause high osmotic potential; affect ion transport, DNA damage, cell viability (Jha and Subramanian 2014), plant morphology, and root and shoot growth; etc.

To overcome the salt stress, plants upregulate different enzyme production such as superoxide dismutase (SOD), ascorbate peroxidase (APX), and proline catalase (CAT), essential to scavenge and detoxify the effects of ROS (Noreen et al. 2010). Na^+/H^+ anti-transporter plays a crucial role in sodium accumulation inside the vacuole of plant cell. Sometimes Na^+ is transported and accumulated inside the older cells of plant, and ultimately these cells sacrifice themselves. SOS response genes (SOS1, SOS2, and SOS3) are expressed under tight regulations of salt stress in plants (Numan et al. 2018). In salinity stress, plants upregulate ABA production in root and shoot system (Cramer and Quarrie 2002; Kang et al. 2014a, b). Nitric oxide regulates the synthesis of H^+ ATPase actively and forces to Na^+/H^+ exchange by H^+ gradient formation which leads to the homeostasis of Na^+ and K^+ ultimately (Zhang et al. 2008a, b). Genes are expressed by plant to mitigate salt stress as *P5CS mod* in tobacco (Hong et al. 2000), *BADH1* (betaine aldehyde dehydrogenase) in tomato

(Zhang et al. 2001), *DcHsp17.7* in carrot (Song and Ahn 2011), *SOS1* in *Brassica* (Chakraborty et al. 2012), etc.

In the presence of PGPR, root length, shoot length, and dry weight of the rice are increased in both salty and normal conditions (Jha and Subramanian 2014). Studies have shown that different hormones, siderophores, HCN productions, phosphate solubilizations, etc. (Sarkar et al. 2018) have been accelerated in plants in the presence of PGPR. Auxins like indole-3-acetic acid (IAA), indole-3-butyric acid (IBA), etc. and its precursor may be produced by the bacteria through its metabolism. IAA induces to produce a huge number of lateral roots, increase the length of the hypocotyl (Zhao et al. 2001) and shoot to root ratio, and reduce root elongation (Loper and Schroth 1986). Cytokinin and gibberellins, produced by the bacteria, are also directly involved to mitigate salt stress in plants by promoting its growth. GA1, GA19, GA20, and GA44 gibberellins are produced by different bacteria (Numan et al. 2018). Volatile organic compounds (VOCs) are produced by PGPR for plant stress management. Ryu et al. in 2004 reported that 2, 3-butanediol and acetoin are two VOCs produced by bacteria able to promote growth in *Arabidopsis thaliana*. Nonsymbiotic nitrogen fixation and organic and inorganic phosphate solubilizations are also done by these specific types of bacteria in salt stress. Downregulating the expression of K^+ ion transporter (HKT1) and upregulating the same gene in shoot may provide protection to plants from salt stress causing less accumulation of Na^+ (Zhang et al. 2008a, b). Activities of caspase-like protease, superoxide dismutase, lipid peroxidation, etc. are reduced in the presence of PGPR in rice plants (Jha and Subramanian 2014). Transcription factors such as *TaMYB* and *TaWRKY* are modulated by PGPR *Dietzianatrono limnaea* STR1 to activate the genes of wheat plants which are actively involved in salt elimination from the cytosol by proper expressions of transporter genes. *TaST*, a salt stress-induced gene; *TaNHX1*, *TaHAK*, and *TaHKT1*, ion transporter genes; and *APX*, *MnSOD*, *CAT*, *POD*, *GPX*b, and *GR* antioxidant proteins are also expressed on wheat plant in presence of PGPR *D. natronolimnaea* STR1 (Bharti et al. 2016) (Fig. 17.3).

Sulla carsona is a species of Leguminosae used as cattle food in salty regions of the world. Presence of PGPR-like *Acinetobacter* sp. Br3, *Pseudomonas putida* Br18, and *Curtobacterium* sp. Br20 along with *Sulla carcona* shows increases biomass, more chlorophyll content and antioxidant property. (Hmaeid et al. 2019). Other plant growth-regulating rhizobacteria and their impacts on plant against salt stress are discussed in Table 17.1.

4.2 *Effects of PGPR on Thirsty Crops*

In drought stress, plant faces very detrimental effect on the crop production (Vinocur and Altman 2005). Damage of photosynthetic apparatus and change of chlorophyll content in the plants (Ortiz et al. 2015) are major issues in this condition. Due to less amount of water, concentrations and viscosity of the cells increased dramatically;

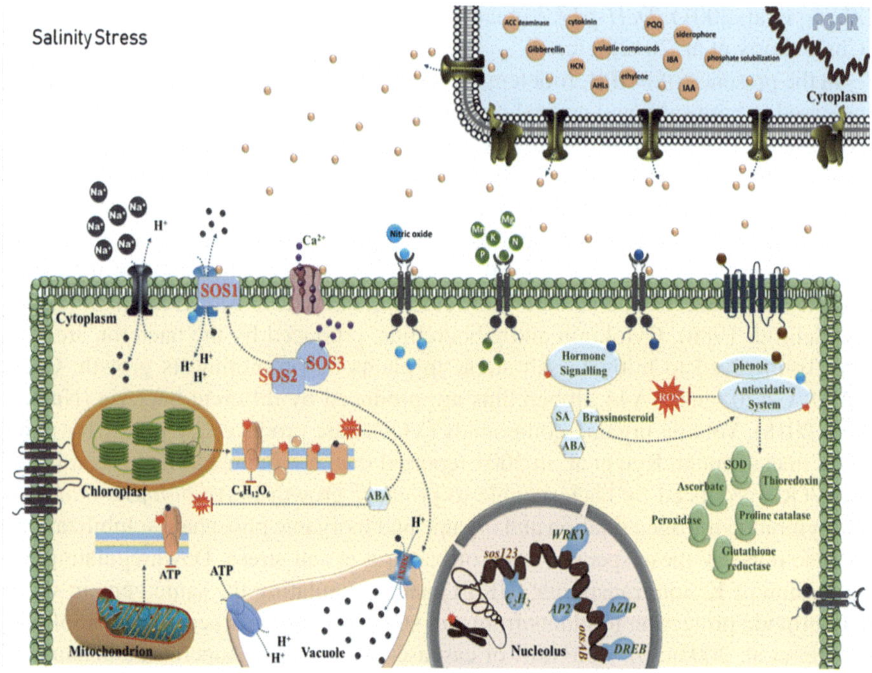

Fig. 17.3 Rhizoremediation to mitigate salt-induced adverse effects on plant cell

proteins or enzymes which are mixed in the cytoplasm may easily come near to each other and may deteriorate (Hoekstra et al. 2001).

According to Farooq et al. (2009), dry and fresh weight of the root and shoot is remarkably reduced and root length is increased in drought stress. Nutrition uptake and transpiration in plant are also hampered in water scarcity. Ions like SO_2^{4-}, NO_3^-, etc. are not assimilated due to the unavailability of energy (Grossman and Takahashi 2001) in drought condition which negatively affects plant growth and development.

PGPR produce phytohormone (abscisic acid, gibberellin, indole acetic acid, cytokinin); important enzymes like ACC deaminase, which is responsible for reduction of ethylene in root system; exopolysaccharide, which increases systemic tolerance in plant; etc. (Yang et al. 2009; Dimkpa et al. 2009; Timmusk and Nevo 2011; Kim et al. 2012; Timmusk et al. 2014). In 2007, Arkhipova et al. reported the effect of *Bacillus* sp. on lettuce plant and concluded the effect of bacterial cytokinin on plant growth in drying soil. ACC deaminase-producing bacteria *Variovorax paradoxus* showed better growth and development of pea plant in the drying soil by deactivating ethylene production (Belimov et al. 2009). ACC deaminase metabolizes ACC into ammonia and α-ketobutyrate (Saleem et al. 2007). In pepper plants, *Bacillus licheniformis* K11 can upregulate several genes like *Cadhn*, *VA*, *sHSP*, and *CaPR-10* and their respective proteins such as dehydrin-like protein, vacuolar H^+-ATPase, small heat shock protein, pathogenesis-related protein 10, etc.

Table 17.1 PGPR activity against salt stress

PGPR	Crop	Mode of applications	Remarks	References
Pseudomonas simiae AU	*Glycine max* (soybean)	Treating soybean (*Glycine max*) seeds with 20-microliter bacterial culture and 10 $mmoll^{-1}$ NaCl concentration	*Pseudomonas simiae* can produce volatile substances responsible for enhancement of soybean seedling growth at 10 mmol/L NaCl concentration	Vaishnav et al. (2015)
Bacillus subtilis EY2, *Bacillus atrophaeus* EY6, *Bacillus sphaericus* GC subgroup B EY30, *Staphylococcus kloosii* EY37, *Kocuria erythromyxa* EY43	*Fragaria ananassa* (strawberry)	Incubate strawberry roots in bacterial suspension (10^8 CFU/ml) along with 10% NaCl concentration	Increasing the ratio of different essential elements like Fe, K, N, Ca, etc. in leaves and root when treated with the bacteria against controls. It also increase plant growth, chlorophyll content, etc.	Karlidag et al. (2013)
Pseudomonas mendocina	*Lactuca sativa* (lettuce)	Inoculation of bacterial suspension in plant	Have a great impact on the growth of lettuce plant against salinity stress but less effective than chemical fertilizers	Kohler et al. (2010)
Azospirillum sp.	*Triticum aestivum* (wheat)	Bacterial inoculums contain 10^7CFUs per ml. Salt stress was given by using sodium chloride	Increase the content of the chlorophyll a, chlorophyll b, chlorophyll ab, and also proline accumulation	Zarea et al. (2012)
Alcaligenes sp., *Bacillus* sp., *Ochrobactrum* sp.	*Oryza sativa* (rice)	Rice seeds were incubated for 1 h at room temperature with sterile (0.3 M) $MgSO_4$ solution and bacterial suspension	Fresh weight of root increases 311.48%, 281%, and 260% against controls; shoot length and chlorophyll content are also enhanced	Bal et al (2013)

(continued)

Table 17.1 (continued)

PGPR	Crop	Mode of applications	Remarks	References
Pseudomonas putida KT2440, *Novosphingobium* sp. HR1a	*Citrus macrophylla* (alemow)	50 cm long plants were treated with both the PGPR strains supplemented with 60 mM and 90 mM NaCl solutions.	Accumulation of IAA is increased in leaves, Proline and Chloride accumulation is reduced in roots.	Vives-Peris et al. (2018)
Pseudomonas fluorescence	*Rosmarinus officinalis* (rosemary)	Treating the plants with bacterial suspension (109 cfu/ml) and different concentration of NaCl as 2.5 g/l, 5 g/l, 7.5 g/l, 10 g/l, and 12 g/l.	By increasing the concentration of NaCl more than 10 g/L, the uninoculated plant shows a decrease in the essential oil content, but in the case of inoculated plants with the bacteria, it showed constant essential oil content	Bidgoli et al. (2019)
Klebsiella sp. SBP-8	*Triticum aestivum* (wheat)	Treating sterilized wheat seeds with 1×10^8 CFU ml^{-1} bacterial solution and different concentrations of salts as 150 mM, 175 mM, and 200 mM.	In response to the salt stress, the bacterium can produce different antioxidants, long-chain alkenes and fatty alcohols which give the potentiality to tolerate abiotic stress	Singh and Jha (2017)

(Lim and Kim 2013). *Pseudomonas putida* MTCC5279 showed regulation of several important genes like *DREB1A*, *NAC*, *LEA*, *DHN*, etc. (Tiwari et al. 2016).

The activity of phosphatase and accumulation of the proline in roots and leaves, respectively, increased in lettuce plants due to the interaction of *Pseudomonas mendocina*; also activities of peroxidase and catalase are enhanced, and superoxidase dismutase is reduced (Kohler et al. 2008). Some examples of PGPR and their potentiality against drought stress are shown in Table 17.2.

Table 17.2 PGPR activity against drought stress

PGPR	Crops	Remarks	References
Achromobacter piechaudii ARV8	*Capsicum annuum* (pepper); *Solanum lycopersicum* (tomato)	ACC deaminase synthesis and IST stimulation	Mayak et al. (2004)
Azospirillum sp.	*Triticum aestivum* (wheat)	IAA production	Dimkpa et al. (2009)
Azospirillum brasilense	*Lycopersicon esculentum* (tomato)	Produces nitric acid responsible for IAA biosynthesis and stimulates adventitious root development	Molina-Favero et al. (2008)
Bacillus thuringiensis	*Lavandula dentate* (French lavender)	Downregulation of glutathione reductase and ascorbate peroxidase activity and induction of physiological, nutritional activities	Armada et al. (2014)
Bacillus 23-B, *Pseudomonas* 6-P, *Mesorhizobium ciceri*	*Cicer arietinum* (chickpea)	Higher proline content, enhance root and shoot length, fresh weight, germination, etc.	Sharma et al. (2013)
Bacillus subtilis, *Bacillus licheniformis*	*Glycine max* (soybean)	Increase in dry weight, oil and protein content, grain yield, leaf area index, etc.	Mondani et al. (2019)
Pseudomonas putida H-2-3	*Glycine max* (soybean)	Increase leaf length, chlorophyll content	Kang et al. (2014a, b)
Phyllobacterium brassicacearum STM196	*Arabidopsis* sp. (thale cress)	Better tolerance by changing transpiration, ABA content, photosynthesis	Bresson et al. (2013)

4.3 *Effects of PGPR on Crops Under Heavy Metal Stress*

The environment is becoming polluted beyond our expectation because of progressive industrialization and urbanization. Besides other factors, excessive accumulation of heavy metals in soil, water, and air causes loss of soil fertility, which affects metabolic pathways of plants, ecosystem functioning, and health issues to humans and animals.

Heavy metals (>5 g/cm^3) are classified into three types, namely, precious metals, radionuclides, and toxic metals. Precious metals [palladium (Pd), platinum (Pt), silver (Ag), gold (Au), ruthenium (Ru), etc.] are less reactive with high economic value; radionuclides [uranium (U), thorium (Th), radium (Ra), americium (Am), etc.] contain unstable nucleus and emit harmful rays like alpha, beta, and gamma; and toxic metals [mercury (Hg), chromium (Cr), lead (Pb), zinc (Zn), copper (Cu), nickel (Ni), cadmium (Cd), arsenic (As), cobalt (Co), tin (Sn), selenium (Se), etc.] are known for its potential toxicity mainly in environmental contexts. The nature and concentration of elements determine the toxicity of heavy metals. A little amount of heavy metals (copper, cobalt, molybdenum, etc.) are required for the metabolic pathways of organisms, but if the amount is high, then it could be harmful to the organisms (Roychowdhury and Tah 2011; Basu et al. 2012; Roychowdhury et al. 2018, 2019).

Heavy metals are nonbiodegradable in nature and they are hard to remove. There are many processes to mitigate heavy metals from the environment like ultrafiltration, immobilization, coagulation, electrodialysis, soil washing, chemical precipitation, stabilization, ion exchange, etc., but these processes are too expensive because it requires many chemical reagents and high energy sources (Gupta et al. 2016; Selatnia et al. 2004). The most efficient methods of removing heavy metals from environment are use of microbes, which have the ability to degrade heavy metals by means of its intrinsic properties or to convert it into toxic to nontoxic form (Gupta et al. 2016; Ledin 2000).

PGPR developed many mechanisms and play an important role in extraction process of heavy metal. These mechanisms include

(i) Biotransformation or mineralization – alteration of highly toxic metals into low or nontoxic forms (Gupta and Diwan 2017).
(ii) Metals bind with metal-binding proteins and peptides – metal-binding proteins like metallothioneins and phytochelatins (Mejare and Bulow 2001) and peptides composed of metal-binding amino acids (mainly cysteine and histidine residues).
(iii) Methylation, volatilization, and demethylation processes mediated by microorganisms to remove toxic metals (Ullah et al. 2015).
(iv) Extrusion metals that are extruded out from the bacterial cells through plasmid or chromosomal mediated methods (Tak et al. 2013).
(v) Exclusion metal ions change in the position of target sites (Tak et al. 2013).

Overexpression of GSH synthetase in the cytosol of Indian mustard (*Brassica juncea*) by *E. coli gshII* gene enhanced accumulation and tolerance of Cd (Mosa et al. 2016). Mercuric ion reductase (encoded by *merA* gene) and organomercurial lyase (encoded by *merB* gene) present on bacterial cell help to convert the toxic form of mercury into less toxic forms (Meagher 2000; Dhankher et al. 2012) (Fig. 17.4).

Bacillus sp. SC2b isolated from *Sedum plumbizincicola* tolerate high concentration of Cd, Zn, and Pb (Ma et al. 2015). *Microbacterium oxydans* AY509223 isolated from *Alyssum murale* mobilized high concentration of Ni present in Ni-contaminated soil (Abou-Shanab et al. 2006). Some Cr-resistant PGPR play an important role to harbor the tolerance capacity of heavy metals like Cu, Pb, Zn, and Cd (Ma et al. 2015). A detailed account on heavy metal tolerance is given in Table 17.3.

5 Genetic Engineering Approaches of PGPR

As bacteria thrive in always changing environments, genetic material of microbes changes in many ways to overcome different types of stresses. Rhizospheric bacteria, *Pseudomonas putida* VTw33, shows *ars* operon with *arsH* gene (Chang et al. 2018);

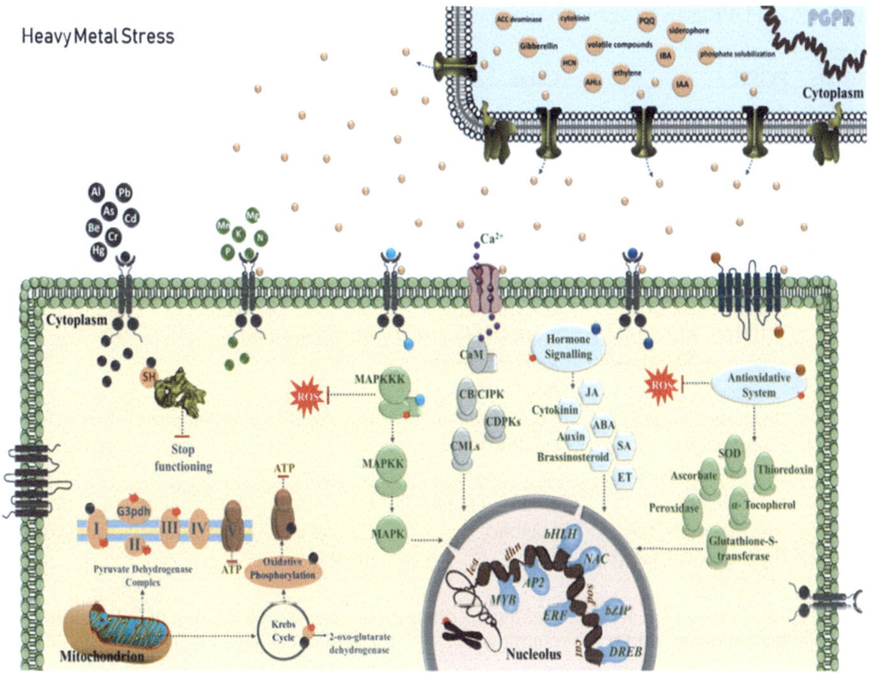

Fig. 17.4 Rhizoremediation to mitigate heavy metal-induced adverse effects on plant cell

the same operon with *ars R* gene (Ramanathan et al. 1998) can bioremediate the arsenic metalloid stress. Mercury causes many severe diseases in human and is toxic to plants. Mercury-tolerating gene, mercury reductase (*merA*), is present in some PGPR. This gene was incorporated in plants by genetic engineering and showed that transgenic plants were able to phytoremediate mercury (Karenlampi et al. 2000). The report says that many bacterial strains as *Pseudomonas putida*, *Ralstonia eutropha*, *E. coli*, *Mycobacterium marianum*, etc. are genetically modified and applied on the heavy metal-contaminated field (Sarma and Prasad 2019). Not only that, genetically modified plant species may be applied with the PGPR, and its synergistic activity is also very promising approach. Genetically engineered *Arabidopsis thaliana* can remove cadmium and lead when inoculated with Rhizobacteria (Bhattacharyya and Jha 2012).

6 Conclusion and Future Perspective

Food scarcity along with population burst throughout the world is becoming a very common problem, and we need to develop sustainable agriculture to feed all the people. Due to climate change and anthropogenic activities, plants are facing different biotic and abiotic stresses throughout their life. Besides this, we are using a

Table 17.3 PGPR activity against heavy metal stress

Heavy metals	PGPR	Crop	Remarks	References
Cr, Cd	*Azotobacter* sp.	*Lepidium sativum* (peppergrass)	Stimulate the growth and increases the biomass of the plant	Sobariu et al. (2017)
Zn, Cd, Pb, Cu	*Phyllobacterium myrsinacearum* RC6b	*Sedum alfredii* (*Sedum*); *Medicago sativa* L. (alfalfa)	Hyperaccumulation of heavy metals, shoot biomass increased	Liu et al. (2015)
Pb	*Kluyvera ascorbate* SUD165, *Kluyvera ascorbate* SUD165/26	*Brassica napus* (canola); *Brassica juncea* (Indian mustard)	Increased dry weight of Indian mustard	Burd et al. (2000)
Pb, Cu, Cd	*Bradyrhizobium* sp., *Pseudomonas* sp., *Ochrobactrumcytisi*	*Lupinus luteus* (yellow lupin)	Increased in plant biomass	Dary et al. (2010)
Cd	*Pseudomonas fluorescens ACC9*, *Pseudomonas tolaasii ACC23*	*Brassica napus* (canola)	Increased accumulation of Cd and also increases plant biomass	Dell'Amico et al. (2008)
Cd	*Micrococcus* sp. *MU1*, *Klebsiella* sp. *BAM1*	*Helianthus annuus* (sunflower)	Increased mobilization of Cd in affected soil and also takes out Cd ions from an aqueous solution	Prapagdee et al. (2013)
Ni	*Kluyvera ascorbata* SUD165, *Kluyvera ascorbata* SUD165/26	*Lycopersicon esculentum* (tomato); *Brassica juncea* (Indian mustard)	Increased dry weights and length of the plants	Burd et al. (2000)
Zn, Pb	*Kluyvera ascorbate* SUD165, *Kluyvera ascorbate* SUD165/26	*Brassica napus* (canola); *Brassica juncea* (Indian mustard)	Increased chlorophyll level in canola plants	Burd et al. (2000)
			Increased dry weight of Indian mustard	

tremendous amount of chemical fertilizers, chemical pesticides, and herbicides for increased crop production, leads to loss of agronomic fields and productivity from day to day. Stresses are different types; among them, drought stress, salt stress, and heavy metal stress are more detrimental and cause the main loss in agriculture. Plants' own defense system is not enough to overcome these detrimental stressed conditions. Plant growth-promoting bacteria (PGPR) play a central role to mitigate stress by physiological, biochemical, and molecular modification of plant responses on stress. Synergistic effects of PGPR are also extensively studiedand it revealed that more than one microorganism gave the better result against the individual one. Bacteria with multiple functions against separate stresses may be a very useful tool for trace management and improvement of the crop. Among different bacterial strains, *Pseudomonas* sp. is much common in rhizosphere and potent bacteria for the member of synergistic activity. PGPR help in nitrogen fixation by their *nif* genes, which produce siderophore and antibiotics to inhibit the growth of other microorganisms, and through ACC deaminase activity ethylene content is reduced which leads to continuous growth and development of plants under stress conditions. Heavy metal-accumulating bacteria can accumulate different heavy metals like cadmium, lead, mercury, copper, arsenic, etc. Not only that, genetic engineering approaches help to insert desired microbial genes to the microorganisms and plants that can express microbial proteins which help to bioremediate heavy metal from the environment. The activity of eukaryotic organisms, i.e., fungi, in association with the prokaryotic bacteria to mitigate several stresses in plants is an interesting topic, and getting much more attention to fight against the common use of chemical fertilizers.

Acknowledgments Authors are thankful to UGC Centre for Advanced Studies, Department of Botany, The University of Burdwan, for pursuing the research work. AK is thankful to DHESTBT (WB-DBT) for providing the research fund [Memo no. 30 (Sanc.)-BT/ST/P/S&T/2G-48/2017]. KM and ML are grateful to UGC-JRF for supporting and proving fund to continue research work. U.H. is thankful to SRF (state-funded) for the finance assistance [Fc (Sc.)/RS/SF/BOT./2017-18/22].

References

Abou-Shanab RA, Angle JS, Chaney RL (2006) Bacterial inoculants affecting nickel uptake by *Alyssum murale* from low, moderate and high Ni soils. Soil Biol Biochem 38(9):2882–2889

Alizadeh H, Behboudi K, Ahmadzadeh M et al (2013) Induced systemic resistance in cucumber and *Arabidopsis thaliana* by the combination of *Trichoderma harzianum* Tr6 and *Pseudomonas* sp. Ps14. Biol Control 65(1):14–23

Antoun H, Prévost D (2005) Ecology of plant growth promoting rhizobacteria. In: Siddiqui ZA (ed) PGPR: biocontrol and biofertilization. Springer, Dordrecht, pp 1–38

Anumalla M, Roychowdhury R, Geda CK, Bharathkumar S, Goutam KD, TSS M (2016) Mechanism of stress signal transduction and involvement of stress inducible transcription factors and genes in response to abiotic stresses in plant. Int J Recent Sci Res 7(8):12754–12771

Arkhipova TN, Prinsen E, Veselov SU et al (2007) Cytokinin producing bacteria enhance plant growth in drying soil. Plant Soil 292(1–2):305–315

Armada E, Roldán A, Azcon R (2014) Differential activity of autochthonous bacteria in controlling drought stress in native *Lavandula* and *Salvia* plants species under drought conditions in natural arid soil. Microb Ecol 67(2):410–420

Aroua I, Abid G, Souissi F et al (2018) Identification of two pesticide-tolerant bacteria isolated from Medicago sativa nodule useful for organic soil phytostabilization. Int Microbiol 22(1):111–120

Arora NK, Tewari S, Singh R (2013) Multifaceted plant-associated microbes and their mechanisms diminish the concept of direct and indirect PGPRs. In: Plant microbe symbiosis: Fundamentals and advances, Springer, New Delhi, p 411–449

Bal HB, Nayak L, Das S et al (2013) Isolation of ACC deaminase producing PGPR from rice rhizosphere and evaluating their plant growth promoting activity under salt stress. Plant Soil 366(1–2):93–105

Bashan Y, de-Bashan LE (2010) How the plant growth-promoting bacterium *Azospirillum* promotes plant growth-a critical assessment. Adv Agron 108:77–136

Basu A, Roychowdhury R, Bhattacharyya SS, Tah J (2012) Estimation of major heavy metals (Fe, Cu and Zn) in the fruit part of *Cucumis sativus* L. World J Sci Technol 2(7):01–03
Belimov AA, Dodd IC, Hontzeas N et al (2009) Rhizosphere bacteria containing 1-aminocyclopro pane-1-carboxylate deaminase increase yield of plants grown in drying soil via both local and systemic hormone signaling. New Phytol 181(2):413–423
Beneduzi A, Ambrosini A, Passaglia LM (2012) Plant growth-promoting rhizobacteria (PGPR): theirpotential as antagonists and biocontrol agents. Genet Mol Biol 35(4):1044–1051
Bensalim S, Nowak J, Asiedu SK (1998) A plant growth promoting rhizobacterium and temperature effects on performance of 18 clones of potato. Am J Potato Res 75(3):145–152
Bharti N, Pandey SS, Barnawal D et al (2016) Plant growth promoting rhizobacteria *Dietzianatronolimnaea* modulates the expression of stress responsive genes providing protection of wheat from salinity stress. Sci Rep 6:34768
Bhattacharyya PN, Jha DK (2012) Plant growth-promoting rhizobacteria (PGPR): emergence in agriculture. World J Microbiol Biotechnol 28(4):1327–1350
Bidgoli RD, Azarnezhad N, Akhbari M et al (2019) Salinity stress and PGPR effects on essential oil changes in *Rosmarinus officinalis* L. Agric Food Secur 8(1):2
Bilal S, Khan AL, Shahzad R et al (2018) Mechanisms of Cr (VI) resistance by endophytic *Sphingomonas* sp. LK11 and its Cr (VI) phytotoxic mitigating effects in soybean (*Glycine max* L.). Ecotoxicol Environ Saf 164:648–658
Bresson J, Varoquaux F, Bontpart T et al (2013) The PGPR strain *Phyllobacterium brassicacearum* STM196 induces a reproductive delay and physiological changes that result in improved drought tolerance in *Arabidopsis*. New Phytol 200(2):558–569
Burd GI, Dixon DG, Glick BR (1998) A plant growth-promoting bacterium that decreases nickel toxicity in seedlings. Appl Environ Microbiol 64(10):3663–3668
Burd GI, Dixon DG, Glick BR (2000) Plant growth-promoting bacteria that decrease heavy metal toxicity in plants. Can J Microbiol 46(3):237–245
Cassan F, Perrig D, Sgroy V et al (2009) *Azospirillum brasilense* Az39 and *Bradyrhizobium japonicum* E109, inoculated singly or in combination, promote seed germination and early seedling growth in corn (*Zea mays* L.) and soybean (*Glycine max* L.). Eur J Soil Biol 45(1):28–35
Cedeño-García GA, Gerding M, Moraga G et al (2018) Plant growth promoting rhizobacteria with ACC deaminase activity isolated from Mediterranean dryland areas in Chile: effects on early nodulation in alfalfa. Chilean J Agric Res 78(3):360–369
Chakraborty K, Sairam RK, Bhattacharya RC (2012) Differential expression of salt overly sensitive pathway genes determines salinity stress tolerance in *Brassica* genotypes. Plant Physiol Biochem 51:90–101
Chakraborty S, Pattanayak A, Mandal S, Das M, Roychowdhury R (2014) An overview of climate change: causes, trends and implications. In: Roychowdhury R (ed) Crop improvement in the era of climate change. IK International Publishing House, New Delhi, pp 1–29
Chang JS, Yoon IH, Kim KW (2018) Arsenic biotransformation potential of microbial *arsH*responses in the biogeochemical cycling of arsenic-contaminated groundwater. Chemosphere 191:729–737
Cramer GR, Quarrie SA (2002) Corrigendum to: abscisic acid is correlated with the leaf growth inhibition of four genotypes of maize differing in their response to salinity. Funct Plant Biol 29(1):111–115
Dary M, Chamber-Pérez MA, Palomares AJ et al (2010) "*In situ*" phytostabilisation of heavy metal polluted soils using *Lupinus luteus* inoculated with metal resistant plant-growth promoting rhizobacteria. J Hazard Mater 177(1–3):323–330
Dell'Amico E, Cavalca L, Andreoni V (2008) Improvement of *Brassica napus* growth under cadmium stress by cadmium-resistant rhizobacteria. Soil Biol Biochem 40(1):74–84
Dhankher OP, Pilon-Smits EA, Meagher RB et al (2012) Biotechnological approaches for phytoremediation. In: Plant biotechnology and agriculture. Academic, San Diego, pp 309–328
Dimkpa C, Weinand T, Asch F (2009) Plant–rhizobacteria interactions alleviate abiotic stress conditions. Plant Cell Environ 32(12):1682–1694

Dodd IC, Belimov AA, Sobeih WY et al (2004) Will modifying plant ethylene status improve plant productivity in water-limited environments. In: Proceedings for the 4th international crop science congress, Brisbane, Australia 26

Donate-Correa J, León-Barrios M, Pérez-Galdona R (2005) Screening for plant growth-promoting rhizobacteria in *Chamaecytisus proliferus* (tagasaste), a forage tree-shrub legume endemic to the Canary Islands. Plant Soil 266(1–2):261–272

Elmerich C (2007) Historical perspective: from bacterization to endophytes. In: Associative and endophytic nitrogen-fixing bacteria and cyanobacterial associations. Springer, Dordrecht, pp 1–20

Estrada-De Los Santos P, Bustillos-Cristales R, Caballero-Mellado J (2001) *Burkholderia*, a genus rich in plant-associated nitrogen fixers with wide environmental and geographic distribution. Appl Environ Microbiol 67(6):2790–2798

Farooq M, Wahid A, Kobayashi N et al (2009) Plant drought stress: effects, mechanisms and management. In: Sustainable agriculture. Springer, Dordrecht, pp 153–188

Fernandez-Aunión C, Hamouda TB, Iglesias-Guerra F et al (2010) Biosynthesis of compatible solutes in rhizobial strains isolated from *Phaseolus vulgaris* nodules in Tunisian fields. BMC Microbiol 10(1):192

Forni C, Duca D, Glick BR (2017) Mechanisms of plant response to salt and drought stress and their alteration by rhizobacteria. Plant Soil 410(1–2):335–356

Glick BR (2012) Plant growth-promoting bacteria: mechanisms and applications. Scientifica 2012:1–15

Gray EJ, Smith DL (2005) Intracellular and extracellular PGPR: commonalities and distinctions in the plant bacterium signaling processes. Soil Biol Biochem 37(3):395–412

Grichko VP, Glick BR (2001) Amelioration of flooding stress by ACC deaminase-containing plant growth-promoting bacteria. Plant Physiol Biochem 39(1):11–17

Grossman A, Takahashi H (2001) Macronutrient utilization by photosynthetic eukaryotes and the fabric of interactions. Annu Rev Plant Biol 52(1):163–210

Guo J, Chi J (2014) Effect of Cd-tolerant plant growth-promoting rhizobium on plant growth and Cd uptake by *Lolium multiflorum* Lam. and *Glycine max* (L.) Merr. in Cd-contaminated soil. Plant Soil 375(1–2):205–214

Gupta P, Diwan B (2017) Bacterial exopolysaccharide mediated heavy metal removal: a review on biosynthesis, mechanism and remediation strategies. Biotechnol Rep 13:58–71

Gupta G, Parihar SS, Ahirwar NK et al (2015) Plant growth promoting rhizobacteria (PGPR): current and future prospects for development of sustainable agriculture. J Microb Biochem Technol 7(2):096–102

Gupta A, Joia J, Sood A et al (2016) Microbes as potential tool for remediation of heavy metals: a review. J Microb Biochem Technol 8(4):364–372

Han Y, Wang R, Yang Z et al (2015) 1-Aminocyclopropane-1-carboxylate deaminase from *Pseudomonas stutzeri* A1501 facilitates the growth of rice in the presence of salt or heavy metals. J Microbiol Biotechnol 25(7):1119–1128

Hasanuzzaman M, Roychowdhury R, Karmakar J, Dey N, Nahar K, Fujita M (2015) Recent advances in biotechnology and genomic approaches for abiotic stress tolerance in crop plants. In: Devarajan T, Jeyabalan S (eds) Genomics and proteomics: concepts, technologies and applications. Apple Academic Press, Canada, pp 333–366

Hmaeid N, Wali M, Mahmoud OM et al (2019) Efficient rhizobacteria promote growth and alleviate NaCl-induced stress in the plant species *Sulla carnosa*. Appl Soil Ecol 133:104–113

Hoekstra FA, Golovina EA, Buitink J (2001) Mechanisms of plant desiccation tolerance. Trends Plant Sci 6(9):431–438

Hong Z, Lakkineni K, Zhang Z et al (2000) Removal of feedback inhibition of Δ1-pyrroline-5-carboxylate synthetase results in increased proline accumulation and protection of plants from osmotic stress. Plant Physiol 122(4):1129–1136

Jha Y, Subramanian RB (2014) PGPR regulate caspase-like activity, programmed cell death, and antioxidant enzyme activity in paddy under salinity. Physiol Mol Biol Plants 20(2):201–207

Ju W, Liu L, Fang L et al (2019) Impact of co-inoculation with plant-growth-promoting rhizobacteria and rhizobium on the biochemical responses of alfalfa-soil system in copper contaminated soil. Ecotoxicol Environ Saf 167:218–226

Kang SM, Khan AL, Waqas M et al (2014a) Plant growth-promoting rhizobacteria reduce adverse effects of salinity and osmotic stress by regulating phytohormones and antioxidants in *Cucumis sativus*. J Plant Interact 9(1):673–682

Kang SM, Radhakrishnan R, Khan AL et al (2014b) Gibberellin secreting rhizobacterium, *Pseudomonas putida* H-2-3 modulates the hormonal and stress physiology of soybean to improve the plant growth under saline and drought conditions. Plant Physiol Biochem 84:115–124

Kang W, Shi S, Xu L (2018) Diversity and symbiotic divergence of endophytic and non-endophytic rhizobia of *Medicago sativa*. Ann Microbiol 68(5):247–260

Kärenlampi S, Schat H, Vangronsveld J et al (2000) Genetic engineering in the improvement of plants for phytoremediation of metal polluted soils. Environ Pollut 107(2):225–231

Karlidag H, Yildirim E, Turan M et al (2013) Plant growth-promoting rhizobacteria mitigate deleterious effects of salt stress on strawberry plants (Fragaria× ananassa). HortScience 48(5):563–567

Kim YC, Glick BR, Bashan Y et al (2012) Enhancement of plant drought tolerance by microbes. In: Plant responses to drought stress. Springer, Berlin, pp 383–413

Kloepper JW (1978) Plant growth-promoting rhizobacteria on radishes. In Proceedings of of the 4th International conference on plant pathogenic bacteria, Station de Pathologie Vegetale et Phytobacteriologie, vol 2. INRA, Angers, France, pp 879–882

Kloepper JW, Leong J, Teintze M et al (1980) *Pseudomonas* siderophores: a mechanism explaining disease-suppressive soils. Curr Microbiol 4(5):317–320

Kohler J, Hernández JA, Caravaca F et al (2008) Plant-growth-promoting rhizobacteria and arbuscular mycorrhizal fungi modify alleviation biochemical mechanisms in water-stressed plants. Funct Plant Biol 35(2):141–151

Kohler J, Caravaca F, Roldán A (2010) An AM fungus and a PGPR intensify the adverse effects of salinity on the stability of rhizosphere soil aggregates of *Lactuca sativa*. Soil Biol Biochem 42(3):429–434

Kumar A, Verma JP (2019) The role of microbes to improve crop productivity and soil health. In: Ecological wisdom inspired restoration engineering. Springer, Cham, pp 249–265

Ledin M (2000) Accumulation of metals by microorganisms—processes and importance for soil systems. Earth-Sci Rev 51(1–4):1–31

Lim JH, Kim SD (2013) Induction of drought stress resistance by multi-functional PGPR *Bacillus licheniformis* K11. Plant Pathol J 29(2):201–208

Liu ZF, Ge HG, Li C et al (2015) Enhanced phytoextraction of heavy metals from contaminated soil by plant co-cropping associated with PGPR. Water Air Soil Pollut 226(3):1–10

Loper JE, Gross H (2007) Genomic analysis of antifungal metabolite production by *Pseudomonas fluorescens* Pf-5. Eur J Plant Pathol 119:265–278

Loper JE, Schroth MN (1986) Influence of bacterial sources of indole-3-acetic acid on root elongation of sugar beet. Phytopathology 76(4):386–389

Ma Y, Oliveira RS, Wu L et al (2015) Inoculation with metal-mobilizing plant-growth-promoting rhizobacterium *Bacillus* sp. SC2b and its role in rhizoremediation. J Toxicol Environ Health 78(13–14):931–944

Malinich EA, Bauer CE (2018) The plant growth promoting bacterium *Azospirillum brasilense* is vertically transmitted in *Phaseolus vulgaris* (common bean). Symbiosis 76(2):97–108

Marimuthu S, Ramamoorthy V, Samiyappan R et al (2013) Intercropping system with combined application of *Azospirillum* and *Pseudomonas fluorescens* reduces root rot incidence caused by *Rhizoctonia bataticola* and increases seed cotton yield. J Phytopathol 161(6):405–411

Martínez-Viveros O, Jorquera MA, Crowley DE et al (2010) Mechanisms and practical considerations involved in plant growth promotion by rhizobacteria. J Soil Sci Plant 10(3):293–319

Mayak S, Tirosh T, Glick BR (1999) Effect of wild-type and mutant plant growth-promoting rhizobacteria on the rooting of mung bean cuttings. J Plant Growth Regul 18(2):49–53

Mayak S, Tirosh T, Glick BR (2004) Plant growth-promoting bacteria that confer resistance to water stress in tomatoes and peppers. Plant Sci 166(2):525–530

Meagher RB (2000) Phytoremediation of toxic elemental and organic pollutants. Curr Opin Plant Biol 3(2):153–162

Mejáre M, Bülow L (2001) Metal-binding proteins and peptides in bioremediation and phytoremediation of heavy metals. Trends Biotechnol 19(2):67–73

Mishra PK, Mishra S, Selvakumar G et al (2009) Coinoculation of *Bacillus thuringeinsis*-KR1 with *Rhizobium leguminosarum* enhances plant growth and nodulation of pea (*Pisum sativum* L.) and lentil (*Lens culinaris* L.). World J Microb Biot 25(5):753–761

Molina-Favero C, Creus CM, Simontacchi M et al (2008) Aerobic nitric oxide production by *Azospirillum brasilense* Sp245 and its influence on root architecture in tomato. Mol Plant-Microbe Interact 21(7):1001–1009

Mondani F, Khani K, Honarmand SJ et al (2019) Evaluating effects of plant growth-promoting rhizobacteria on the radiation use efficiency and yield of soybean (*Glycine max*) under water deficit stress condition. Agric Water Manage 213:707–713

Mosa KA, Saadoun I, Kumar K et al (2016) Potential biotechnological strategies for the cleanup of heavy metals and metalloids. Front Plant Sci 7:303

Nadeem SM, Ahmad M, Zahir ZA et al (2014) The role of mycorrhizae and plant growth promoting rhizobacteria (PGPR) in improving crop productivity under stressful environments. Biotechnol Adv 32(2):429–448

Nogales J, Campos R, Ben Abdelkhalek H et al (2002) *Rhizobium tropici* genes involved in free-living salt tolerance are required for the establishment of efficient nitrogen-fixing symbiosis with *Phaseolus vulgaris*. Mol Plant-Microbe Interact 15(3):225–232

Noreen Z, Ashraf M, Akram NA (2010) Salt-induced regulation of some key antioxidant enzymes and physio-biochemical phenomena in five diverse cultivars of turnip (*Brassica rapa* L.). J Agron Crop Sci 196(4):273–285

Numan M, Bashir S, Khan Y et al (2018) Plant growth promoting bacteria as an alternative strategy for salt tolerance in plants: a review. Microbiol Res 209:21–32

Nyoki D, Ndakidemi PA (2018) Root length, nodulation and biological nitrogen fixation of rhizobium inoculated soybean (*Glycine max* [L.] Merr.) grown under maize (*Zea mays* L.) intercropping systems and P and K fertilization. Adv Biores 9(1):173–180

Ormeño-Orrillo E, Menna P, Almeida LG et al (2012) Genomic basis of broad host range and environmental adaptability of *Rhizobium tropici* CIAT 899 and *Rhizobium* sp. PRF 81 which are used in inoculants for common bean (*Phaseolus vulgaris* L.). BMC Genomics 13(1):735

Ortiz N, Armada E, Duque E et al (2015) Contribution of arbuscular mycorrhizal fungi and/or bacteria to enhancing plant drought tolerance under natural soil conditions: effectiveness of autochthonous or allochthonous strains. J Plant Physiol 174:87–96

Paul D, Lade H (2014) Plant-growth-promoting rhizobacteria to improve crop growth in saline soils: a review. Agron Sustain Dev 34(4):737–752

Pérez-Montaño F, Alías-Villegas C, Bellogín RA et al (2014) Plant growth promotion in cereal and leguminous agricultural important plants: from microorganism capacities to crop production. Microbiol Res 169(5–6):325–336

Prapagdee B, Chanprasert M, Mongkolsuk S (2013) Bioaugmentation with cadmium-resistant plant growth-promoting rhizobacteria to assist cadmium phytoextraction by *Helianthus annuus*. Chemosphere 92(6):659–666

Ramanathan S, Shi W, Rosen BP et al (1998) Bacteria-based chemiluminescence sensing system using β-galactosidase under the control of the ArsR regulatory protein of the ars operon. Anal Chim Acta 369(3):189–195

Reinhold-Hurek B, Hurek T, Gillis M et al (1993) *Azoarcus* gen. nov., nitrogen-fixing proteobacteria associated with roots of Kallar grass (*Leptochloafusca* (L.) Kunth), and description of two species, *Azoarcusindigens* sp. nov. and *Azoarcuscommunis* sp. nov. Int J Syst Evol Microbiol 43(3):574–584

Roychowdhury R (2014) Crop improvement in the era of climate change. IK International Publishing House, New Delhi, p 496

Roychowdhury R, Tah J (2011) Differential response by different parts of *Solanum melongena* L. for heavy metal accumulation. Plant Sci Feed 1(6):80–83

Roychowdhury R, Taoutaou A, Hakeem KR, Gawwad MR, Tah J (2014) Molecular marker-assisted technologies for crop improvement. In: Roychowdhury R (ed) Crop improvement in the era of climate change. IK International Publishing House, New Delhi, pp 241–258

Roychowdhury R, Khan MH, Choudhury S (2018) Arsenic in rice: an overview on stress implications, tolerance and mitigation strategies. In: Hasanuzzaman M, Nahar K, Fujita M (eds) Plants under metal and metalloid stress. Springer, Singapore, pp 401–415

Roychowdhury R, Khan MH, Choudhury S (2019) Physiological and molecular responses for metalloid stress in rice – a comprehensive overview. In: Hasanuzzaman M, Fujita M, Nahar K, Biswas J (eds) Advances in rice research for abiotic stress tolerance. Woodhead Publishing/ Elsevier, USA, pp 341–369

Ryu CM, Farag MA, Hu CH et al (2004) Bacterial volatiles induce systemic resistance in *Arabidopsis*. Plant Physiol 134(3):1017–1026

Saleem M, Arshad M, Hussain S et al (2007) Perspective of plant growth promoting rhizobacteria (PGPR) containing ACC deaminase in stress agriculture. J Ind Microbiol Biotechnol 34(10):635–648

Sarkar A, Ghosh PK, Pramanik K et al (2018) A halotolerant *Enterobacter* sp. displaying ACC deaminase activity promotes rice seedling growth under salt stress. Microbiol Res 169(1):20–32

Sarma H, Prasad MN (2019) Metabolic engineering of Rhizobacteria associated with plants for remediation of toxic metals and metalloids. In: Transgenic plant technology for remediation of toxic metals and metalloids. Academic, London, pp 299–318

Selatnia A, Boukazoula A, Kechid N et al (2004) Biosorption of lead (II) from aqueous solution by a bacterial dead *Streptomyces rimosus* biomass. Biochem Eng 19(2):127–135

Sharma P, Khanna V, Kumari P (2013) Efficacy of aminocyclopropane-1-carboxylic acid (ACC)-deaminase-producing rhizobacteria in ameliorating water stress in chickpea under axenic conditions. Afr J Microbiol Res 7(50):5749–5757

Shrivastava P, Kumar R (2015) Soil salinity: a serious environmental issue and plant growth promoting bacteria as one of the tools for its alleviation. Saudi J Biol Sci 22(2):123–131

Siddiqui ZA, Singh LP (2005) Effects of fly ash, *Pseudomonas striata* and *Rhizobium* on the reproduction of nematode Meloidogyne incognita and on the growth and transpiration of pea. J Environ Biol 26(1):117–122

Singh RP, Jha PN (2017) Analysis of fatty acid composition of PGPR *Klebsiella* sp. SBP-8 and its role in ameliorating salt stress in wheat. Symbiosis 73(3):213–222

Singh A, Jain A, Sarma BK et al (2014) Rhizosphere competent microbial consortium mediates rapid changes in phenolic profiles in chickpea during *Sclerotium rolfsii* infection. Microbiol Res 169(5–6):353–360

Sobariu DL, Fertu DI, Diaconu M et al (2017) Rhizobacteria and plant symbiosis in heavy metal uptake and its implications for soil bioremediation. New Biotechnol 39:125–134

Song NH, Ahn YJ (2011) DcHsp17. 7, a small heat shock protein in carrot, is tissue-specifically expressed under salt stress and confers tolerance to salinity. New Biotechnol 28(6):698–704

Srivastava R, Khalid A, Singh US et al (2010) Evaluation of arbuscular mycorrhizal fungus, fluorescent *Pseudomonas* and *Trichoderma harzianum* formulation against *Fusarium oxysporum* f. sp. *Lycopersici* for the management of tomato wilt. Biol Control 53(1):24–31

Tak HI, Ahmad F, Babalola OO (2013) Advances in the application of plant growth-promoting rhizobacteria in phytoremediation of heavy metals. Rev Environ Contam Toxicol 223(Springer):33–52

Timmusk S, Nevo E (2011) Plant root associated biofilms: perspectives for natural product mining. In: Bacteria in agrobiology: plant nutrient management. Springer, Heidelberg, pp 285–300

Timmusk S, Wagner EG (1999) The plant-growth-promoting rhizobacterium *Paenibacillus-polymyxa* induces changes in *Arabidopsis thaliana* gene expression: a possible connection between biotic and abiotic stress responses. Mol Plant-Microbe Interact 12(11):951–959

Timmusk S, El-Daim IA, Copolovici L et al (2014) Drought-tolerance of wheat improved by rhizosphere bacteria from harsh environments: enhanced biomass production and reduced emissions of stress volatiles. PLoS One 9(5):e96086

Tiwari S, Lata C, Chauhan PS et al (2016) *Pseudomonas putida* attunes morphophysiological, biochemical and molecular responses in *Cicer arietinum* L. during drought stress and recovery. Plant Physiol Biochem 99:108–117

Tokala RK, Strap JL, Jung CM et al (2002) Novel plant-microbe rhizosphere interaction involving *Streptomyces lydicus* WYEC108 and the pea plant (*Pisum sativum*). Appl Environ Microbiol 68(5):2161–2171

Ullah A, Heng S, Munis MF et al (2015) Phytoremediation of heavy metals assisted by plant growth promoting (PGP) bacteria: a review. Environ Exp Bot 117:28–40

Vaishnav A, Kumari S, Jain S et al (2015) Putative bacterial volatile-mediated growth in soybean (*Glycine max* L. Merrill) and expression of induced proteins under salt stress. J Appl Microbiol 119(2):539–551

Vejan P, Abdullah R, Khadiran T et al (2016) Role of plant growth promoting rhizobacteria in agricultural sustainability-a review. Molecules 21(5):1–17

Vessey JK (2003) Plant growth promoting rhizobacteria as biofertilizers. Plant Soil 255(2):571–586

Vinocur B, Altman A (2005) Recent advances in engineering plant tolerance to abiotic stress: achievements and limitations. Curr Opin Biotechnol 16(2):123–132

Vives-Peris V, Gómez-Cadenas A, Pérez-Clemente RM (2018) Salt stress alleviation in citrus plants by plant growth-promoting rhizobacteria *Pseudomonas putida* and *Novosphingobium* sp. Plant Cell Rep 37(11):1557–1569

Yang J, Kloepper JW, Ryu CM (2009) Rhizosphere bacteria help plants tolerate abiotic stress. Trends Plant Sci 14(1):1–4

Zarea MJ, Hajinia S, Karimi N et al (2012) Effect of *Piriformospora indica* and *Azospirillum* strains from saline or non-saline soil on mitigation of the effects of NaCl. Soil Biol Biochem 45:139–146

Zhang F, Dashti N, Hynes RK et al (1996) Plant growth promoting rhizobacteria and soybean [*Glycine max* (L.) Merr.] nodulation and nitrogen fixation at suboptimal root zone temperatures. Ann Bot 77(5):453–460

Zhang HX, Hodson JN, Williams JP et al (2001) Engineering salt-tolerant *Brassica* plants: characterization of yield and seed oil quality in transgenic plants with increased vacuolar sodium accumulation. Proc Natl Acad Sci 98(22):12832–12836

Zhang H, Kim MS, Sun Y et al (2008a) Soil bacteria confer plant salt tolerance by tissue-specific regulation of the sodium transporter HKT1. Mol Plant-Microbe Interact 21(6):737–744

Zhang H, Xie X, Kim MS et al (2008b) Soil bacteria augment *Arabidopsis* photosynthesis by decreasing glucose sensing and abscisic acid levels in planta. Plant J 56(2):264–273

Zhao Y, Christensen SK, Fankhauser C et al (2001) A role for flavin monooxygenase-like enzymes in auxin biosynthesis. Science 291(5502):306–309

Chapter 18
Plant Tissue Culture and Crop Improvement

Kiranmai Chadipiralla, Pachipala Gayathri, Vemula Rajani, and Pichili Vijaya Bhaskar Reddy

Abstract Due to the drastically increasing population associated with limited natural and traditionally cultivated crops, novel methods are required to address this concern and thereby increase crop productivity. With the advent of various scientific technologies, such as plant tissue culture, crop improvement through various in vitro protocols involving genetic manipulation has come to the forefront. Plant tissue culture is an advanced in vitro protocol, through which regeneration of organs, tissues, or plant cells can be obtained on an artificially prepared nutrient medium. Employing various novel gene transfer methods, the preferred characteristic traits from one plant can be passed on to another plant simply by introducing the gene responsible for that particular character. In plant tissue culture, several techniques like protoplast fusion, anther culture, and embryo transfer have been used to produce new genetically variant cops. Tissue culture helps in mass multiplication and clonal propagation of plants from any tiny part of the plant tissue. In recent years, this technique has been vastly used for conservation of germplasm as well as in the commercialization of various crops. Cell culture techniques are playing a major role in enhancing crop improvement potential by producing somaclonal and gametoclonal variants. Several varieties of crops including but not limited to vegetable (resistant to pest and disease-free), fruits (seedless, pink fleshed), ornamental plants, and sporeless mushrooms have been developed. This was possible only due to the diverse advanced protocols that are rapidly expanding in the field of tissue culture. Keeping in view of the importance and impact of tissue culture in enhancing the quality and yield of crops, this chapter is focused on reviewing methodologies employed in plant tissue culture along with the challenges that lie ahead.

Keywords Plant tissue culture · Micropropagation · Protoplast fusion · Anther culture · *Agrobacterium tumefaciens*

K. Chadipiralla (✉) · P. Gayathri · V. Rajani
Department of Biotechnology, Vikrama Simhapuri University, Nellore, Andhra Pradesh, India

P. V. B. Reddy (✉)
Department of Life Science and Bioinformatics, Assam University, Diphu Campus, Diphu, Assam, India

R. Roychowdhury et al. (eds.), *Sustainable Agriculture in the Era of Climate Change*, https://doi.org/10.1007/978-3-030-45669-6_18

1 Introduction – Crop Improvement

As human beings are depending on food for their existence, crop improvement has taken a center stage in order to enhance the yield so as to cater to the needs of the rapidly growing populations (Brown and Thorpe 2011; Chakraborty et al. 2014; Roychowdhury et al. 2014). The field of crop improvement sprouted up almost 10,000 years ago. Even in the ancient world, primitive people were on the search for a constant change through the transition from hunting to the cultivation of different crops. Although the cop improvement was existing for the past thousands of years, only two techniques have been used for crop improvement. According to Bogorad, a plant molecular biologist, at Harvard University, there are two basic techniques to improve crops in ancient days – selection and breeding.

1.1 Selection

Selection is the primary step that helps in identifying the salient characteristic features that may be used for the passing of traits from one generation to another. The farmers select plants which possess advantageous characters like those that produce the size of the fruit, color, taste, pest resistance, drought resistance, etc.

1.2 Breeding

Breeding refers to the reproductive method of crossing two parental plants with the desired traits to produce an offspring generation acquiring the advantageous qualities from both the parents. The conventional process of plant breeding could be a success or a failure because early plant breeders didn't know the methods of genetic transmission of traits. If by chance any valuable traits are transferred to the offspring, such breeds could be maintained. By following general breeding techniques (the physical basis of inheritance), one does not know the actual mechanism that lies behind the development of new breeds by crossing two plants (Roychowdhury et al. 2014). After the early period of 1900, Johann Gregor Mendel revealed the key aspects of breeding by his amazing experiments in the 1860s. Employing common pea plant (*Pisum sativum*) in his garden in Austria, Mendel identified that hereditary information is stored in discrete units which are now called genes. Although the importance of these experiments was not immediately recognized, his findings were helpful in revolutionizing this field after his death. In general, methods of crop improvement involved plant introduction, acclimatization, mutation breeding, tissue culture, genetic engineering, etc.

1.3 Biological Revolution

Upon understanding the importance of Mendel's findings, the breeders were able to apply the concepts to develop the novel breeds with more precision. In the year 1930, there was a successful production of hybrid corn which helped in doubling the crop yield within a limited period. Rapid increase in the output was only possible and can be attributed to Mendel's laws.

2 Tissue Culture Protocols Employed in Crop Improvement

Plant tissue culture has been well recognized and extensively used in in vitro cultivation technology that is popularly used to regenerate and cultivate plant parts, plant, seeds etc., on a sterile known composition of nutrient media (Gamborg and Philips 2010; Reddy et al. 2013) Plant tissue culture is an ultimate source that has recently been revolutionized to a state where any kind of plant material could be used for regeneration in artificial favorable conditions by applying various in vitro techniques such as micropropagation. These in vitro methods are intensely being used to improve a variety of crops with expected characters or traits.

2.1 Somatic Embryogenesis

One artificial method of developing a plant or its embryo is somatic embryogenesis. A whole plant/plant embryo is acquired from any somatic cell that is otherwise not usually implicated in the embryo development process (Chadipiralla et al. 2018). Such embryos are not actively involved in embryo development. There is no seed coat or endosperm around the somatic embryo.

Somatic embryos can be differentiated into a bipolar structure containing root and shoot areas. Somatic embryos are almost similar to zygotic embryos and can also be matured and germinated as well (Steward et al. 1958; Kato and Takeuchi 1963; William and O'Rourke 1974; Nomura and Komamine 1985). Somatic embryogenesis in carrot through such in vitro technique has been earlier reported in case of carrot. A typical protocol representing the steps involved in somatic embryogenesis is shown in (Fig. 18.1). The importance of somatic embryogenesis has been reiterated by combining efficient traits from different plants and successfully obtaining the desired genotype (Sharp et al. 1980). Generation of multiple subsequent embryos from within the principal/primary somatic embryos is called proliferative embryogenesis. Such secondary embryos result in the production of vegetative propagates in bulk and also in a limited period of time.

Development of such secondary embryos further plays an important role and helps in recovering a huge number of plantlets either by genetic transformation,

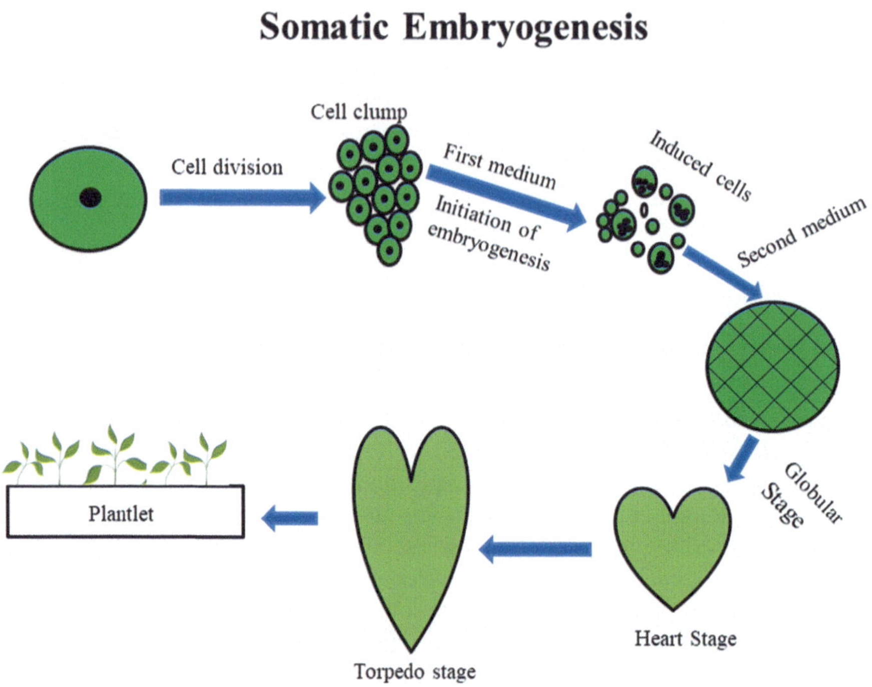

Fig. 18.1 Picture representing the different events occurring during somatic embryogenesis

induced mutation, or clonal propagation. These embryogenetic cells are regenerated into plants by exposing embryogenic cells to microprojectile bombardment or by other genetic transformation protocols (Chen and Beversdorf 1994). Thus, for the introduction of a mutation through in vitro conditions through γ- radiation, secondary embryogenesis can be used by mutation induction through altered starch composition (Joseph et al. 2004).

2.2 *Anther Culture and Embryo Rescue*

2.2.1 Anther Culture

Anther culture is one of the best techniques to improve crops by collecting anthers under aseptic conditions from immature flowers in which the developing anthers are at a decisive stage and normally are grown on an artificial nutrient-containing substratum (Manipal and Shekhawat 2010). Such microspores developed in the cultured anther are generated into callus or embryoids, which further result in plantlets (Fig. 18.2).

In the year 1921, Berger has identified the existence of haploids in Datura plant. Tulekar has worked on the proliferation of haploid callus from the matured pollen

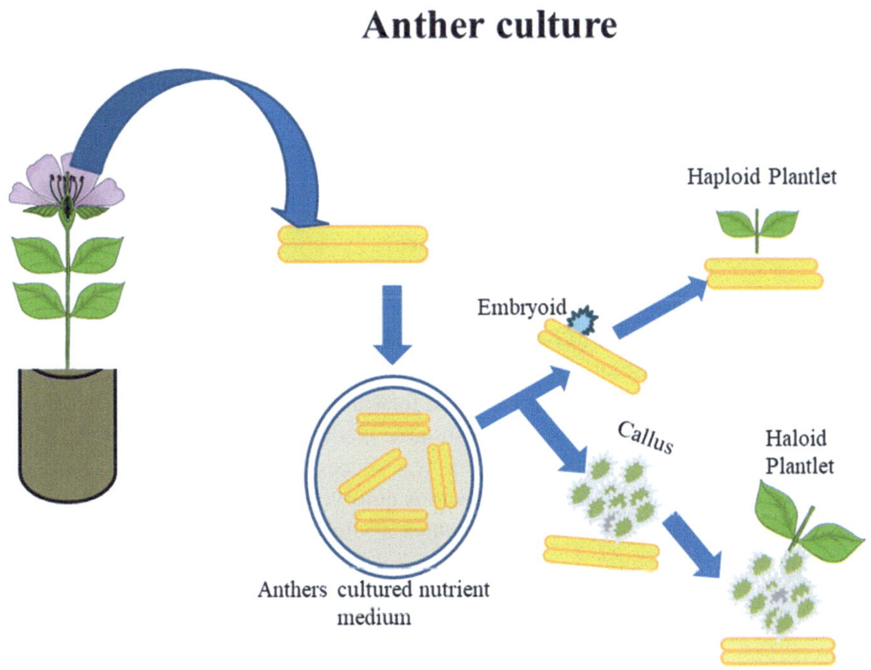

Fig. 18.2 Representation of process of development of a haploid plant through anther culture

grain in *Ginkgo biloba* plant in 1953. Bourgin and Hitch developed the full-fledged haploid plants from *Nicotiana tobacum*. There are two methods for anther culture that include direct and indirect culture. Embryogenesis is the direct method where anther behaves like a zygote and synthesizes embryoids and later develop into haploid plantlets. Organogenesis is an indirect method where anther undergoes repeated cell division and forms the callus which later develops into haploid plantlets.

Applications of anther culture:

(1) It can be used to improve the crops by producing haploid plants.
(2) Improve crops of cereals, vegetables, and seasonal crops like watermelon, asparagus, cabbage, broccoli, etc.
(3) By applying anther culture, it successfully improved cabbage crop and pak choi, etc.
(4) This technique could be used for developing haploid cultures for the purpose of developing various horticulture plants.
(5) The anther culture-derived embryos or haploids are known to be highly stable and hence become a good alternative in cryogenic studies and conservation of germplasm.
(6) In vitro generation of the anther culture-derived haploids is rapid and hence save a significant amount of time as compared to the conventional plant breeding techniques. While it takes several years to develop a homozygous or an

isozygous plant through traditional inbreeding techniques, anther culture could help to generate the same within a short span of 1 year.

(7) The success of anther culture lies in the aspect of bringing in gene variability. Hence this technique of mass multiplication and propagation could be vital in inducing genetic variability and thereby improving the quality.

2.2.2 Embryo Rescue or Embryo Culture

Nurturing and blooming of an immature embryo into a viable plant is called "embryo rescue," which has proved to be a successful and the oldest in vitro technique discovered by "Charles Bonnet" in the eighteenth century. Following the demonstration of this technique, many scientists came forward and started placing the embryos in different nutrient media in search of suitable media continuously for 14 years. In the year 1904, the in vitro embryo rescue was successfully performed by Hanning, who has described the failures of precocious embryos which produce small, weak, and inviable plantlets. Embryo rescue plays a vital role in crop improvement by nurturing the weak embryo, consequently helping in the survival of them. The remote and general hybridization techniques fail to undergo general sexual reproduction; thus, embryo rescue can support in evading this problem. Through the embryo rescue technique, the artificial medium is used as a replacement of the endosperm, hence allowing the embryo for its development. A representative protocol for the embryo–endosperm transplantation used in the embryo rescue is shown in Fig. 18.3.

One significant application of embryo rescue is that it helps in generating interspecific and intergeneric hybrids. Embryo abortion is one reason for the occurrence of interspecific incompatibility. Though the fertilization occurs in embryo even in small, shrunken seeds by following wide hybridization, it fails to develop into a seed. In such a scenario, embryo rescue is the best method to overcome the above problem. This technique is also used to recover maternal haploids, which are developed as a consequence of interspecific hybridization through chromosomal elimination. Embryo rescue has been proved to be quite successful in breaking the dormancy of seed and further helps in overcoming the immaturity of seed. Embryo rescue helps in increasing the probability of transfer of genetic variations in the process of plant breeding thereby improving the quality of the offsprings (Balla and Brozik 1993). This technique helps in the termination of seed abortion. Embryo rescuing is a successful in vitro technique that helps in the preservation of the invaluable germplasm of several crops through cryopreservation, and samples could be revived at a later point of time.

Improved Crops by Embryo Rescue/Embryo Culture

Cain et al. (1983) has used embryo rescue and developed seedless grapes for the first time, and subsequently Aguero et al. (1995) further modified the embryo rescue protocol in order to improve the crop variety (Singh and Shetty 2011). Prasad et al. (1996) has elucidated the role of embryo rescue in production of seedless lime.

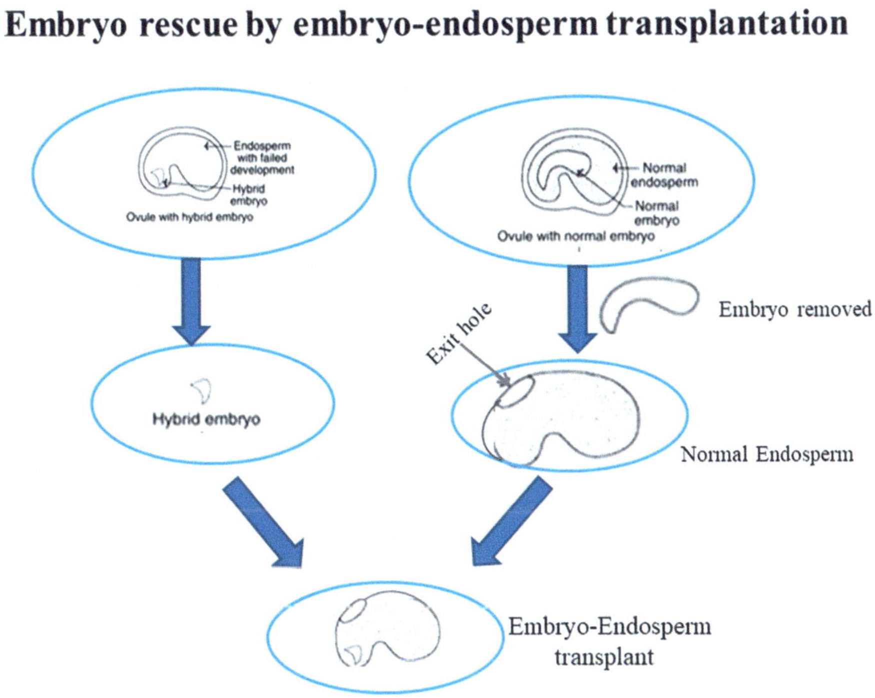

Fig. 18.3 A pictorial representation of the embryo–endosperm transplantation protocol used in the process of embryo rescue

Haploidization/monoploidization is also one of the best techniques in plant biotechnology to improve crops and their traits. Embryo culture is largely used for the production of monoploids. Kasha and Kao has developed *Hordeum bulbosum* monoploids in the year 1970. Zenkteler and Zenkteler (2016) developed haploid plants of lettuce. It is highly useful in testing seed viability. Embryo rescue can improve germination. Burgos and Ledbetter (1993) applied this tool and successfully germinated apricot seed whose embryo was inhibited to survive (Fig. 18.3).

2.3 *Protoplast Culture and Fusion*

2.3.1 Protoplast Culture

Protoplast Isolation

Owing to the sexual incompatibility barriers, it is impossible to produce full hybrids by crossing two distant species with desired traits. It has proved as a serious handicap in crop improvement. Protoplast culture is the best method to overcome this problem by isolating protoplasts from plant cells. A cell in which the cell wall has been removed to expose the cell membrane is termed as a protoplast. Gentle isolation of intact viable and undamaged protoplast in a short time is the key to the success of this protocol. There are two methods to isolate protoplasts that include mechanical method and enzymatic method. Mechanical method involves the removal of the cell wall from a plasmolyzed cell by incubating the cell in a hypertonic sucrose solution. The cell walls of a plasmolyzed cell are carefully dissected out without damaging the shrunken protoplasm. Unaltered and intact protoplasts are then released by inducing osmotic swelling upon treatment with hypoosmotic sucrose solution. Klercker was the first to use mechanical isolation method in the year 1892 from *Stratiotes aloides* by plasmolysis (Fig. 18.4). This work was extended later for the isolation of protoplasts from onion bulbs. Unlike osmotic shock in mechanical methods, the enzymatic method of protoplast isolation involves the use of digestive enzymes to dissolve the cell wall consequently exposing the inner plasma membrane. Enzymatic protoplast isolation has been demonstrated by Cocking in the year 1960 using cellulase enzyme to dissolve the cell wall of the tomato roots and release the protoplasts. After going through various modifications, this technique is being used to isolate protoplasts even from higher plants. The isolation of protoplast by the enzymatic method may involve two methods. It could be a one-step method (direct) or two-step (sequential) method. In the direct method, the segments of the leaf are properly incubated in a solution containing enzymes such as pectinase and cellulase that degrade the complex cell wall components.

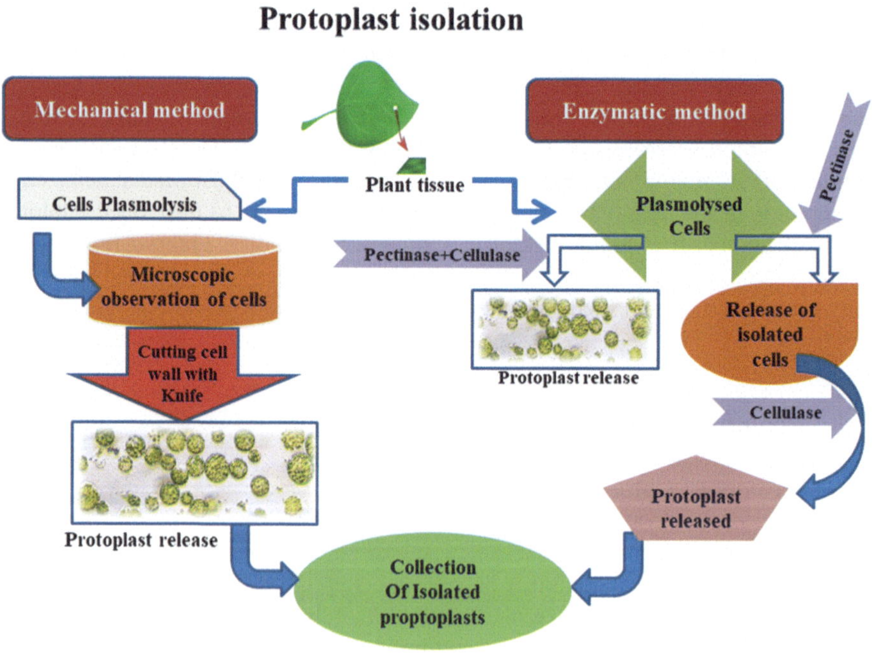

Fig. 18.4 A typical representation of the steps involved in the process of isolation of collection of protoplasts

The clean protoplasts are then isolated by centrifuging and washing with sorbitol solution. While in the direct method, both enzymes are simultaneously added, sequential method involves the use of two enzymes one after the other. In the sequential method, the plant tissue is first incubated with a pectinase enzyme that helps in degrading the tissue lamella. The cell suspension that is devoid of the lamella is now subjected to cellulase enzyme that aids in releasing the undamaged free protoplast. After isolation by following several steps, purification is required for viable protoplasts to remove the undigested material.

Culturing the Protoplast

Following purification, the protoplasts are tested for their viability, and then the positive samples are then cultured at a known concentration on artificial media. Different methods are there to culture protoplasts.

Hanging Drop Method

This method of protoplast culture was discovered by Kao in 1970. In this method protoplasts at known density were suspended in the medium at a low-light intensity and incubated in an inverted position. This technique is usually performed in such cases wherein the normal growth of tissue is being inhibited by the plane substrate on which the tissue is growing. This method helps to overcome such growth inhibition by reducing the surface area–volume ratio and also by minimizing the evaporation rate. The medium is normally placed on a glass slide which is kept inverted. This makes the sample suspended on the inverted slide which does not allow the spreading of tissue sample due to the influential gravity beneath and also due to the effective surface tension.

Agar Plating Method

This method was developed by Bergmann in the year 1960. This method has been in use since its ease and convenience. A very small volume of the isolated pure protoplast is made into a suspension in the medium and is immediately mixed in the melted agar placed in a culture dish. While the temperature of the media is regulated at 4 °C at the preparation stage, it is later incubated in a contamination-free environment at a temperature of about 27–30 °C. This method is advantageous to handle a large number of protocols at a time; planting ability can also be determined easily. Clumping of the protoplasts when placed in culture media is a recurring problem that could be overcome by this agar plating method. In order to overcome any drawbacks of this method, advanced research techniques such as protoplast fusion were developed to improve varieties of crops.

Microculture Technique

In the microculture technique, a tiny droplet of culture medium containing either single or multiple protoplasts is usually placed on a microscopic slide. Two sides of the microscopic glass slide are covered with two cover slips. The third cover slip is placed over these two cover slips for shielding the protoplast suspension. Jones et al. (1960) established this technique. Later in the year 1965, Vasil and Hildebrad practically applied this technique on isolated cells of tobacco to demonstrate the totipotency to rejuvenate. These findings were further taken to the next stage of cultivating a complete a mature tobacco plant from isolated protoplasts of tobacco (Takebe et al. 1971).

Multidrop Array (MDA) Technique
This is an advanced method of hanging drop technique that was developed by Potrykus et al. (1979) where a minute quantity of a plant tissue sample is used for a mechanical screening and detection of a variety of nutritional molecules as well as other hormones.

2.3.2 Protoplast Fusion

After isolation and purification, protoplasts undergo viability test and are later cultivated by any one of the above methods. A genetic modification in which plants from two distinct species of plants are fused together is called protoplast fusion to synthesize new hybrid plants possessing both the parental characteristics, which are called somatic hybrids (Fig. 18.5). Generally, somatic hybrids are produced either by two different varieties or by the same species (e.g., somatic hybrids between flowering potato and nonflowering potato plants) or between two different species (e.g., production of *Triticale* somatically hybridized wheat and rye *Secale*).

The process of protoplast fusion is induced by the addition of polyethylene glycol (PEG). By this process, it is possible to fuse multiple protoplasts at a time, and further, there is no need for plant hormones for regeneration. Protoplast itself behaves like germinating spore. Although Ca^{++} and $NaNO_3$ are also used at alkaline pH, the results vary depending on the type of the organism where the fusion is being initiated. Moreover, few reports revealed high pH exhibited toxicity on the protoplast.

Improved Crops by Protoplast Fusion
Using somatic fusion, resistant variety of potato which is resistant to potato leaf roll disease was developed. The somatic fusion of *Solanum tuberosum*, a wild variety, with nontuber-bearing potato plant *Solanum brevidens* resistant to viral disease resulted in the development of a resistant variety of crop plant *Solanum tuberosum*.

2.4 *Transformation*

Crop improvement by the transgenic method through transformation technology is another core research tool that has emerged as a great application in this field. Once Griffith (1928) and Avery et al. (1944) demonstrated transforming principle, the road was laid for initiating the gene transformation experiments in plants. Prior to these studies, scientists working on plant gene transformation techniques had germinated seeds which were generally exposed to the entire genome, but the majority of successful transformation techniques were however not that much flexible under aseptic conditions (Kleinhof and Behki 1977).

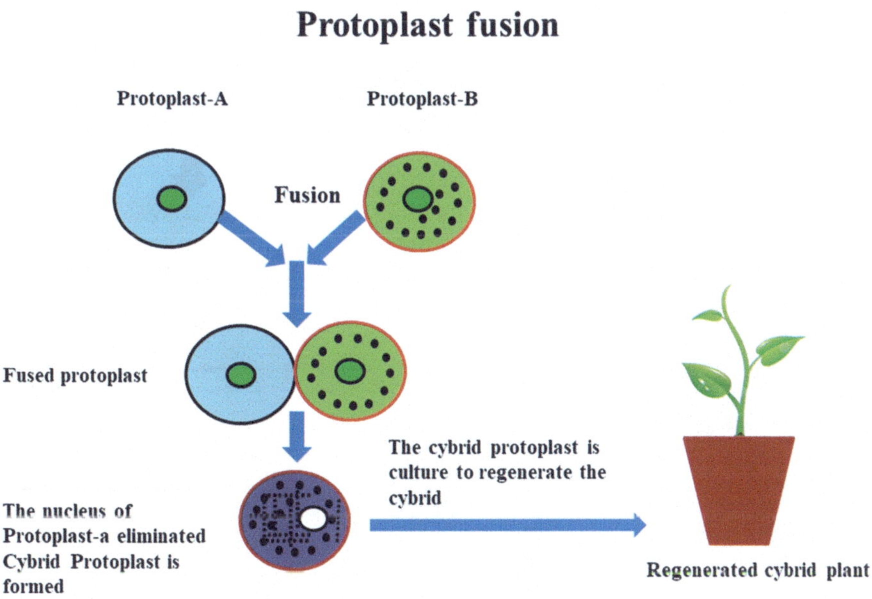

Fig. 18.5 Depiction of the protoplast fusion from two different sources

In the year 1977, Montoya et al. had successfully invented a novel gene transformation protocol in plants by *Agrobacterium tumefaciens*, which led to the development of crown gall in higher plants. This technique was a wonderful method to induce tumor formation, which can be the saddle for developing gene transformation in dicotyledonous plants. Gene transformation can be performed through two methods, namely, direct or indirect methods (Fig. 18.6).

In case of the direct method, there are several techniques to transfer gene of interest in plants. The list includes imbibitions, gene gun, liposome method, osmotic method, microinjection, infiltration, shoot apex, and silicon-mediated transformation (SCMT) (Rao et al. 2009). Drawback with direct method of DNA uptake is that the incorporation DNA pattern is random and uncertain. The effective result and the success rate of gene transformation depend on the selection procedure, shoot regeneration, and embryogenesis (Sobhanian et al. 2012). In independent gene transfer methods (*Agrobacterium*-based but not dependent on tissue culture), there is a difficulty in regeneration and production of genetically modified nuclear or plastid genome in many transgenic plants like rice, corn, and tea (Mayers et al. 2010). To overcome these problems with independent gene transformation and direct gene transformation, advanced research findings suggest tissue culture based *Agrobacterium*-mediated transformation (TCBAT) method, which is an indirect method for effective and enhanced success rate. The last three decades have resulted in great experimental and scientific development in terms of crop improvement deploying *Agrobacterium*-mediated transformation. By following general conventional methods and cross-breeding, it is extremely not possible to develop crops

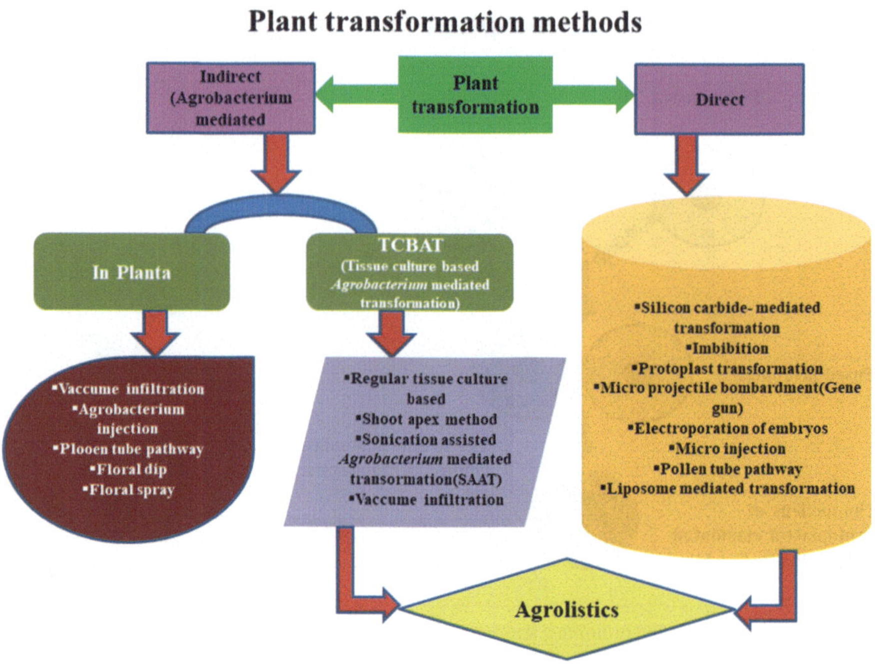

Fig. 18.6 A flowchart showing the details of the various methods used in plant transformation

with desirable traits, whereas with gene transformation, it is absolutely credible to get desirable traits and also getting varieties of crops with the gene of interest within the short time. This method has come as a boon to overcome the problems and obstacles of conventional and cross-breeding in crop improvement. More recently, agrolistics, a combination of direct and *Agrobacterium*-mediated transformation, is also being used and playing an effective role in gene transfer.

2.4.1 Improved Crops by Indirect Gene Transformation – In Planta Transformation

Although the technology in the field of tissue culture and regeneration methodologies has been well developed, the regeneration in a variety of crops for the purpose of gene transfer has not been completely put to use to its full capacity and hence has been the major limitation (Kesavareddy et al. 2018). Therefore, novel methods that can help in the transfer of genetic traits by avoiding the plant tissue culture protocols may be required (Table 18.1). Such a requirement helped in the development of in planta method of gene transformation (Kalbande and Patil 2016). This method involves the implantation of the DNA directly into the whole plant in vivo. Cocultivation of *Agrobacterium* or microprojectile bombardment methods is used to lodge the foreign DNA into the actively dividing apical meristematic tissue of the plant (Chee and Slightom 1995; Brich 1997). Similar studies have also been designed to transfer the DNA directly in the ovary (Zhou et al. 1983).

2.4.2 Tissue Culture–Based Agrobacterium-Mediated Transformation

The success of the *Agrobacterium*-mediated transformation-based tissue culture technology depends on the assessment and standardization of various factors (Kapildev et al. 2016). The factors that are vital primarily include the selection and dosage of the plant regulators and the bacterial strain that needs to be used for transformation (Table 18.2). In addition to these factors, we have to follow the proper nutrient medium to be used, the genotype of plant, temperature, light, antibiotics, target tissue, and standardization of methods in the selection of transgenic cells (Palla and Pijut 2015). Apart from the setting of these factors, another important component that is required to be present in the reaction mixture for effective and enhanced transfer of T-DNA is acetosyringone. However, the degree of transfer can be improved by the supplementation of different surfactants and also by subjecting the sample to sonication and also by applying the vacuum filtration (Tohidfar and Mohsenpour 2010).

In order to cater to the needs of the increasing populations and provide them with sufficient food, there was a great deal of urgency to focus on increasing the yield of the edible crops. In this regard, one great option that researchers and breeders too could look at crop improvement was through gene transfer methods.

Using the gene transfer methods, Toriyama et al. (1988) used electroporation for the transgenic rice production of rice. However, few studies used electroporation along with PEG (polyethylene glycol) in the recovery of fertile transgenic plants, namely, japonica and indica rice. For the production of improved transgenic rice (*Oryza sativa* L.), microprojectile bombardment was adopted for the successful transformation of immature rice embryos (Cao et al. 1992; Li et al. 1993). For the first time, Fujimoto et al. (1993) fused *Cry* gene, responsible to produce δ-endotoxin from *Bacillus thuringiensis*, the rice strain, which is a pest-resistant variety as compared to its wild type. Imran et al. (2012) worked on wheat crop and developed a drought-tolerant media for wheat callus induction with PEG 6000, which has induced osmotic stress. Such a stress induction showed a significant effect on callus morphology and helped in screening drought-tolerant cell lines. Setter (2003) together worked on three crops simultaneously (wheat, barley, and oats) and improved waterlogging tolerant germplasm.

2.5 Somaclonal Variation

It is quite common that changes occur due to chromosomal rearrangement after the production of a plant by in vitro propagation. Such variations are very common in plants which regenerated from callus. The changes could be either phenotypic or

Table 18.1 Depiction of the various in planta gene transformation methods applied in crop improvement

Plant species	Gene method of transformation	GOL/ SMG/RG	Trait	Year of work done	References
Indica rice (*Oryza sativa* L.)	*Agrobacterium* transformation	LEA protein gene Emv 2	Increased capacity to efficiently survive in dry and drought conditions	2008	Rajesh et al.
Indica rice (*Oryza sativa* L.)	*Agrobacterium* transformation	Cry1C- Coding sequence of hpt and gus	Insect resistance	2011	Ignacimuthu and Raveendar
Rice (*Oryza sativa* L.)	*Agrobacterium* transformation	Gus	Regulation and expression of α- amylase gene	1992	Ming-Tsair Chan et al.
Indica rice (*Oryza sativa* L.)	Vacuum infiltration through piercing	Bar /gfp	Controlling rice salt tolerance	2009	Lin et al.
Soyabean (*Glycine aestivum* L.)	Pollen tube	Gus	Improving soya bean seed composition	2009	Liu et al.
Wheat (*Triticum aestivum* L.)	Floral dip	Npt-II	To develop kanamycin-resistant plants	2009	Zale et al.
Wheat (*Triticum aestivum* L.)	*Agrobacterium* transformation	Gus, Npt -II	Rapid production of transgenic plants via *A. tumefaciens*	1997	Ming Cheng et al.
Maize (*Zea mays* L.)	Floral dip	Hpt-II, gfp	Resistance to hygromycin	2012	Mu et al.
Cotton (*Gossypium hirsutum* L.)	Pollen tube	Npt-II	High-frequency regeneration	2010	Tanveer Khan et al.
Onion (*Allium cepa* L	*Agrobacterium* transformation	Gfp	To develop transient transformation method (to enhance transformation efficiency)	2014	Xu et al.
Okra (*Abelmoschus esculentus* L.	Vacuum infiltration/ SAAT	Bar, hpt-II, gus	Disease resistance	2015	Manickavasaga M et al.
Sugarcane (*Saccharum officinarum* L.)	Vacuum infiltration/ SAAT	Bar, hpt-II, gus	To enhance transformation efficiency for the transformation of economically important genes	2015	Mayavan et al.

(continued)

Table 18.1 (continued)

Plant species	Gene method of transformation	GOL/ SMG/RG	Trait	Year of work done	References
Tomato (*Solanum lycopersicum* L.)	*Agrobacterium* transformation	Hpt-II	To improve the ability to withstand cold stress	2015	Shah et al.
Tomato (*Solanum lycopersicum* L.	*Agrobacterium* transformation	rolB and Gus	Resistance against fungal pathogens	2014	Waheed A et al.

GOI genes of interest, *SMG* selection marker genes, *SAAT* sonication-assisted *Agrobacterium*-mediated transformation, *RG* reporter genes, *Hpt* Hypoxanthine osphor ribosyl-transferase, *gus* β-Glucuronidase, *Bar* Basta tolerance, *gfp* Green Fluorescence Protein, *NptII* Neomycin Phosphotransferase II, *LEA 2* Late Embryogenesis Abundant Protein

genotypic. Commonly generated somaclonal changes include a change either in the number of chromosomes (polyploidy and aneuploidy) or in the structure of the chromosome (transposons) and sometimes in DNA sequences (mutations). At times the epigenetic-related events such as gene methylation and gene amplification also are exhibited. The process of generation of increased variability in plants that are obtained through in vitro cell culture protocols or from adventitious shoots is called somaclonal variations (Fig. 18.7). Such variations could be caused by genetic mutations under in vitro conditions or due to the exposure to any chemicals. Whatever the reason could be, in the majority of the cases, these variations that surface out help in the development or generation of new cultivars and hence seen as advantageous. The advantages in the newly developed cultivar may be in terms of disease resistance, drought resistance, pesticide resistant, requiring less water content, desired color development in ornamental plants, etc.

2.5.1 Improved Crops by Somaclonal Variation

Through the application of somaclonal variation technique, several herbicide-resistant crop varieties have been developed. An array of tissue culture techniques that have been developed have invaluably helped in the selection process consequently helping to generate the herbicide-resistant crops (Chaleff 1988; and Chaleff and Ray 1984; Hughes 1983; Maliga et al. 1987). Similarly, the selection of traits in the process of developing herbicide-resistant crops has been obtained through protoplast (rape seed), callus cultures (maize), suspension cultures (maize), and microspores (rapeseed) (Anderson and Georgeson 1989; Tuberosa and Lucchese 1990; Swasson et al. 1988; Parker et al. 1990). Studies of Tashiro et al. have identified and demonstrated the occurrence of morphological changes in ginger because of somaclonal variation (Tashiro et al. 1995).

Table 18.2 Tissue culture–based *Agrobacterium*-mediated gene transformation in different plants

Plant species	Additive/treatment	GOL/SMG/RG	Trait	Year of work done	References
Cucumber (*Cucumis sativus L.*)	Acetosyringone/vacuum infiltration	Act, npt-II, gfp	To enhance transformation efficiency	2013	Nanasato et al.
Chickpea (*Cicer arietinum L.*)	Particle gun bombardment	Cry I A(C)	To induce insecticidal activity	2007	Shivani I et al.
Black gram (*Vigna mungo L. Hepper)*	Acetosyringone	Npt-II,uidA	To establish the protocol to use for transferring new traits	2015	Sainger et al.
Broccoli (*Brassica oleracea var. ltalica*)	Acetosyringone	athsp101, npt-II,uc	To optimize the *Agrobacterium*-based gene transfer to produce transgenic broccoli	2015	Ravanfar and Abdul
Indian ginseng (*Withania somnifera L.*)	Acetosyringone/SAAT/vacuum infiltration	npt-II, gusA	Establishment of stable transformation protocol via *Agrobacterium tumefaciens*	2015	Sivanadhan et al.
Cucumber (*Cucumis sativus L.*)	Acetosyringone	Npt-II	To enhance the efficiency of *Agrobacterium* infection in the proximal region of explants	2016	Baskaran et al.
Chrysanthemum (*Chrysanthemum indicum L.*)	Acetosyringone	rsmyb1, bar	To transfer anthocyanin regulatory gene to modify flower color	2016	Naing et al.
Jerusalem artichoke (*Helianthus tuberosus L.*)		hpt-II gus	To develop an efficient regeneration and transformation system of *Helianthus tuberosus* L.	2016	Kim et al.

GOI genes of interest, *SMG* selection marker genes, *SAAT* sonication-assisted *Agrobacterium*-mediated transformation, *RG* reporter genes

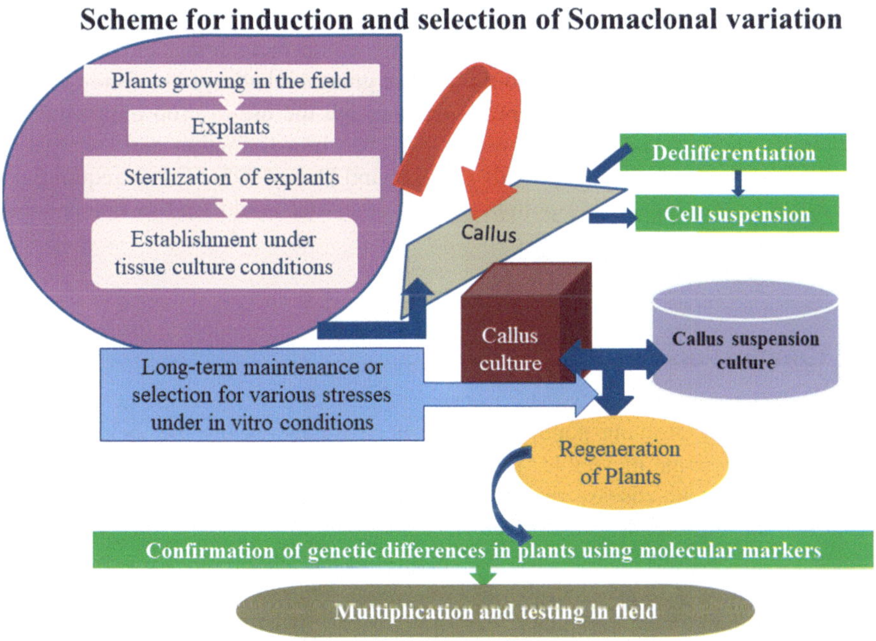

Fig. 18.7 A schematic representation of the steps involved in the induction and selection process of somaclonal variation

3 Future Challenges of Plant Tissue Culture in Crop Improvement

The relevance of plant tissue culture in the field of crop improvement is rapidly growing and hence increased the demand for such in vitro techniques (Smith and Drew 1990). The two major reasons that are attracting scientists toward tissue culture and crop improvement, the first one being the need to meet the food supply for the worldwide population and the second is to play a major role in producing bioactive products commercially that is expected to reap the business profits worldwide. It seems that the future belongs to the adaptation of tissue culture techniques in various fields of science along with tissue culture-dependent crop improvement. Major considerations of these application cloud are the preservation of plant genetic resources and the transformation of plants by the consideration of human needs in restoring genetic transformation to provide sufficient and safety foods. It is estimated that the world's population could be around 8 billion by 2020. Hence, it is necessary to support crop development methods and also develop crops with better traits such as disease resistance, withstand various stress conditions, survive with minimal water, etc. (García-Gonzáles et al. 2010).

The challenge of plant tissue culture-based *Agrobacterium*-dependent transformation is to create transgenic plants in maintaining constant as well as risk-free transgenic plants. Another challenge is to overcome unwanted mutations in somaclonal variation. It has to reduce food scarcity while maintaining environmental integrity and also control global pollution. The greatest challenge the future lies in developing crops that require no pesticides through the use of crop improvement techniques such as tissue culture-dependent gene transformation and eventually result in the pesticide-free crop fields (Dekker and Duke 1995). This is required to prevent the soil from getting polluted which otherwise could lead to tremendous loss to the frequent crop failures.

4 Conclusions

With the rapidly growing populations, the need for increasing the yield of food products has become inevitable. To address this issue, improved crops with better traits such as yield, disease resistance, stress tolerant, etc. are to be developed. Although the traditional protocols of plant breeding have partially succeeded, the output results are not able to meet the requirements of the current times. and the problems will only increase in the years to come when the agriculturally utilized lands will be significantly reduced. Further, the drastically changing climatic condition across the globe is another factor that will augment the problem. Hence, the in vitro plant tissue culture techniques have become an integral part of the advanced biotechnology methods in addressing the problem. As discussed in the chapter, the developments that have surfaced in recent years in the field of plant tissue culture will hopefully meet demands of the future generations. Overall, the plant tissue culture may help in developing a healthy as well as greener world for the generations to come.

References

Aguero C, Riquelme C, Tizio R (1995) Embryo rescue from seedless grapevine (*Vitis vinifera* L.) treated with growth-retardants. Vitis 34:73–76

Anderson PC, Georgeson M (1989) Herbicide-tolerant mutants of corn. Genome 31(2):994–999

Avery OT et al (1944) Studies on the chemical nature of the substance inducing transformation of Pneumococcal types. J Exp Med 79:137–157

Balla I, Brozik S (1993) Embryo culture of sweet cherry hybrids. In: II International cherry symposium, pp 385–386

Baskaran P, Soós V, Balázs E, Van Staden J (2016) Shoot apical meristem injection: a novel and efficient method to obtain transformed cucumber plants. S Afr J Bot 103:210–215

Brich RG (1997) Plant transformation: problems and strategies for practical application. AnnRevPlant PhysiolPlant MolBiol 48:297–326

Brown DC, Thorpe TA (2011) Crop improvement through tissue culture. World J Microbiol Biotechnol 11:409–415

Burgos L, Ledbetter CA (1993) Improved efficiency in apricot breeding: effects of embryo development and nutrient media on *in vitro* germination and seedling establishment. Plant Cell Tissue Organ Cult 35(3):217–222

Cain DW, Emershad RL, Tarailo RE (1983) In-ovulo embryo culture and seedling development of seeded and seedless grapes (*Vitis vinifera* L.). Vitis 22:9–14

Cao J, Duan X, McElroy D, Wu R (1992) Regeneration of herbicide resistant transgenic rice plants following micro projectile-mediated transformation of suspension culture cells. Plant Cell Rep 11:586–591

Chadipiralla K, Pachipala G, Allam US, Reddy PVB (2018) Recent advances in biotechnology (set of 3 volumes) vol 3, pp 137–154. isbn: 978-81-8329-936-7

Chakraborty S, Pattanayak A, Mandal S, Das M, Roychowdhury R (2014) An overview of climate change: causes, trends and implications. In: Roychowdhury R (ed) Crop improvement in the era of climate change. IK International Publishing House, New Delhi, pp 1–29

Chaleff RS (1988) A second mutant enhances resistance of a tobacco mutant of sulfonylurea herbicides. Theor Appl Genet 76:177–182

Chaleff RS, Ray TB (1984) Herbicide resistant mutants from tobacco cell cultures. Science 223:1148–1151

Chan MT, Lee TM, Chang HH (1992) Plant cell physiology. Oxford J 33(5):577–588

Chen JL, Beversdorf WD (1994) A combined use of microprojectile bombardment and DNA imbibition enhances transformation frequency of canola (*Brassica napus* L.) r. Theor Appl Genet 88:187–192

Chee PP, Slightom JL (1995) Transformation of soybean (*Glycine max)* via A*grobacterium tumefaciens* and analysis of transformed plants. In: Gartland KMA, Davey MR (eds) *Agrobacterium* Protocols, Methods Mol Bio (vol. 44). Springer, Totowa, NJ, pp 101–119

Cheng M, Fry JE, Pang S, Zhou H, Hironaka CM, Ducan DR, Conner TW, Monsanto YW (1997) Genetic transformation of wheat mediated by *Agrobacterium tumefacience*. Plant Physiol 115:971–980

Dekker J, Duke SO (1995). Advances in agronomy

Fujimoto H, Itoh K, Yamamoto M, Kyozuka J, Shimamoto K (1993) Insect resistant rice generated by introduction of a modified β-endotoxin gene of *Bacillus thuringiensis*. Biotechnology 11:1151–1155

Gamborg OL, Philips GC (2010) Plant cell tissue culture and organ culture fundamental methods. Springer, Berlin

García-Gonzáles R, Catolica MD, Quiroz K, Carrasco B, Caligari P (2010) Plant tissue culture: current status, opportunities and challenges. Cienciae Investigacion Agaria 37(3):5–30

Griffith F (1928) The significance of pneumococcal types. J Hyg 27:113–159

Hughes K (1983) Selection for herbicide resistance. In: Evans DA, Sharp WR et al (eds) Handbook of plant cell culture. MacMillan, New York, p 442

Ignacimuthu S, Raveendar S (2011) Agrobacterium mediated transformation of Indica rice (*Oryza sativa* L.) for insect resistance. Natl Acad Agric Sci (South Korea) 179(2):277–286

Imran M, Razzaq A, Ishfaq AH, Kaleem S, Khan AA, Qayyum A, Ahmad M (2012) Interaction of callus selection media and stress duration for *in vitro* selection of drought tolerant callus of wheat. Afr J Biotechnol 11(17):4000–4006

Jones LE, Hildebrandt AC, Riker AJ, Wu JH (1960) Growth of somatic tobacco cells in microculture. Am J Bot 47(6):468–475

Joseph R, Yeoh H-H, Loh C-S (2004) Induced mutations in cassava using somatic embryos and the identification of mutant plants with altered starch yield and composition. Plant Cell Rep 23:91–98

Kalbande BB, Patil AS (2016) Plant tissue culture independent *Agrobacterium tumefaciens* mediated in-planta transformation strategy for upland cotton (*Gossypium hirsutum*). J Genet Eng Biotechnol 14:9–18

Kapildev G, Chinnathambi A, Sivanandhan G, Rajesh M, Vasudevan V, Mayavan S, Arun M, Jeyaraj M, Alharbi SA, Selvaraj N, Ganapathi A (2016) High-efficient *Agrobacterium*-mediated in planta transformation in black gram (*Vigna mungo* (L.) Hepper). Acta Physiol Plant 38:205

Kasha KJ, Kao KN (1970) High frequency haploid production in barley (Hordeum vulgare L.). Nature 225(5235):874–876

Kato H, Takeuchi M (1963) Morphogenesis *In Vitro* starting from single cells of carrot root. Plant Cell Physiol 4:243–245

Keshavareddy G, Kumar ARV, Ramu VS (2018) Methods of plant transformation- a review. Int J Curr Microbiol App Sci 7:2319–7706

Khan T, Reddy VS, Leelavathi S (2010) High-frequency regeneration via somatic embryogenesis of an elite recalcitrant cotton genotype (*Gossypium hirsutum* L.) and efficient agrobacterium – mediated transformation. Plant Cell Tissue Org Cult 101:323–330

Kim MJ, An DJ, Moon KB, Cho HS, Min SR, Sohn JH, Jeon JH, Kim HS (2016) Highly efficient plant regeneration and agrobaçterium-mediated transformation of *Helianthus tuberosus* L. Ind Crop Prod 83:670–679

Kleinhof A, Behki R (1977) Prospects for plant genome modification by nonconventional methods. Annu Rev Genet 11:79–101

Li L, Qu R, de Kochko A, Fauquet C, Beachy RN (1993) An improved rice transformation system using the biolistic method. Plant Cell Rep 12:250–255

Lin J, Zhou B, Yang Y, Mei J, Zhao X, Guo X, Huang X, Tang D, Liu X (2009) Piercing and vacuum infiltration of the mature embryo: a simplified method for agrobacterium-mediated transformation of Indica rice. Plant Cell Rep 28:1065–1074

Liu J, Su Q, An L, Yang A (2009) Transfer of a minimal linear marker-free and vector-free smGFP cassette into soybean via ovary-drip transformation. Biotechnol Lett 31:295–303

Maliga P, Fejes E, Steinback K, Menczel L (1987) Cell culture approaches for obtainingherbicide-resistant chloroplasts in crop plants. ACS Symp Ser 334:115

Manickavasagam M, Subramanyam K, Ishwarya R, Elayaraja D, Ganapathi (2015) Assessment of factors influencing the tissue culture-independent agrobacterium-mediated in planta genetic transformation of okra (*Abelmoschus esculentus* (L.) Moench). Plant Cell Tissue Organ Cult 123:309–320

Manipal S, Shekhawat V (2010) Plant biotechnology – *in-vitro* principles, techniques and applications. MJP Publishers, Chennai, pp 105–124

Mayavan S, Subramanyam K, Jaganath B, Sathish D, Manickavasagam M, Ganapathi A (2015) Agrobacterium mediated in planta genetic transformation of sugarcane sets. Plant Cell Rep 34:1835–1848

Mayers B, Zaltsman A, Lacroix B, Kozlovsky SV, Krichevsky A (2010) Nuclear and plastid genetic engineering of plants:omparision of opportunities and challenges. Biotechnol Adv 28:747–756

Montoya AL, Chilton MD, Gordon MP, Sciaky D, Nester EW (1977) Octopine and nopaline metabolism in Agrobacterium tumefaciens and crown gall tumor cells: role of plasmid genes. J Bacteriol 129:101–107

Mu G, Chang N, Xiang K, Sheng Y, Zhang Z, Pan G (2012) Genetic transformation of maize female inflorescence flowering floral dip method mediated by agrobacterium. Biotechnology 11:178–183

Naing AH, Ai TN, Jeon SM, Lim SH, Kim CK (2016) An efficient protocol for agrobacterium-mediated genetic transformation of recalcitrant *Chrysanthemum* cultivar Shinma. Acta Physiol Plant 38:1–9

Nanasato Y, Konagaya K, Okuzaki A, Tsuda M, Tabei Y (2013) Improvement of agrobacterium-mediated transformation of cucumber (*Cucumis sativus* L.) by combination of vacuum infiltration and co-cultivation on filter paper wicks. Plant Biotechnol Rep 7:267–276

Nomura K, Komamine A (1985) Identification and isolation of single cells that produce somatic embryos at a high frequency in a carrot suspension culture. Plant Physiol 79:988–991

Palla KJ, Pijut PM (2015) Agrobacterium-mediated genetic transformation of *Fraxinus americana* hypocotyls. Plant Cell Tissue Organ Cult 120:631–641

Parker WB, Somers DA, Wyse DI, Keith RA, Burton JD, Gronwald JW, Gengebach BG (1990) Selection and characterization of sethoxydim- tolerant maiz tissue cultures. Plant Physiol 92:1220

Potrykus CT, Harms H, Lorz (1979) Multiple –drop –array (MDM)technique for the large-scale testing of culture media variations in hanging microdrop culture of single cell system. I: The technique. Plant Sci Lett 14(3):231–235

Prasad MBNV, Sahijram L, A Rekha (1996) Role of embryo culture techniques in the improvement of seedless lime fruit quality. National symposium on horticulture biotechnology, Bangalore, India, Oct 28–30

Rajesh S, Krishnaveni S, Sudhakar D, Raveendran M, Sivakumar P, Gnanam R, Manickam A (2008) Agrobacterium mediated transformation of Indica Rice (*Oryza sativa* L.), IR64 with Mungbean LEA protein gene for water-stress tolerance. Am J Plant Physiol 3:101–110

Rao AQ, Bakhsh A, Kiani S, Shahzad K, Shahid AA, Husnain T, Riazuddin S (2009) The myth of plant transformation. Biotechnol Adv 27:753–763

Ravanfar SA, Abdul AM (2015) Shoot tip regeneration and optimization of *Agrobacterium tumefaciens*-mediated transformation of broccoli (*Brassica oleracea* var. italica) cv. Green marvel. Plant Biotechnol Rep 9:27–36

Reddy BS, Karmakar J, Roychowdhury R, Dey N (2013) Optimization of callus induction and callus multiplication in rice (*Oryza sativa* L.) landraces. Res Plant Biol 3(5):41–44

Roychowdhury R, Taoutaou A, Hakeem KR, Gawwad MR, Tah J (2014) Molecular marker-assisted technologies for crop improvement. In: Roychowdhury R (ed) Crop improvement in the era of climate change. IK International Publishing House, New Delhi, pp 241–258

Sainger M, Chaudhary D, Dahiya S, Jaiwal R, Jaiwal PK (2015) Development of an efficient in vitro plant regeneration system amenable to agrobacterium-mediated transformation of a recalcitrant grain legume blackgram (*Vigna mungo* L. Hepper). Physiol Mol Biol Plants 21:505–517

Setter TLI (2003) Waters review of prospects for germplasm improvement for water logging tolerance in wheat, barley and oats. Plant Soil 253(1):1–34

Shah HS, Ali S, Jan SA, Din JU, Ali GM (2015) Piercing and incubation method of in planta transformation producing stable transgenic plants by over expressing DREB1A gene in tomato (*Solanum lycopersicum* mill). Plant Cell Tissue Org Cult 120:1139–1157

Sharp WR, Sondahl MR, Caldas LS, Maraffa SB (1980) The physiology of in vitro asexual embryogenesis. Hortic Rev 2:268–310

Shivani I, Hari SM, Susan E (2007) Genetic transformation of Chick pea (*Cicer arietium* L.) with insecticidal crystalprotein gene using particle gun bombardment. Plant cell rep 26(6):755

Singh G, Shetty S (2011) Impact of tissue culture on agriculture in India. Invited Rev Biotechnol Bioinf Bioeng 3:279–288

Sivanandhan G, Kapil Dev G, Theboral J, Selvaraj N, Ganapathi A, Manickavasagam M (2015) Sonication, vacuum infiltration and thiol compounds enhance the agrobacterium-mediated transformation frequency of *Withania somnifera* (L) Dunal. PLoS One 10:e0124693

Smith MK, Drew RA (1990) New current applications of tissue culture in plant propagation and improvement. Aust J Plant Physiol 17:267–289

Sobhanian N, Habashy AA, Farshad FE, Tohidfar M (2012) Optimizing regeneration and reporter gene (gus) transformation of alfalfa (Medicago sativa). Ann Biol Res 3:2419–2427

Steward FC, Mapes MO, Mears K (1958) Growth and organized development of cultured cells. Organization in cultures grown from freely suspended cells. Am J Bot 45:705–708

Swasson EB, Coumans MP, Brown GL, Patel JD, Beversdorf (1988) The characterization of herbi cide tolerant plants in *Brassica napus* L. after *in vitro* selection of microspores and protoplasts. Plant Cell Rep 7(2):83–87

Takebe I, Labib G, Melchers G (1971) Regeneration of whole plants from isolated mesophyll protoplasts of tobacco. Naturwissenschaften 58:318–320

Tashiro Y, Onimaru H, Shigyo M, Isshiki S, Miyazaki S (1995) Isozyme mutation induced by treatment of cultured shoot tips with alkylating agent in ginger cultivars (Zinger officinale Rose.). Bull Fac Agric Saga Univ 79:29–35

Tohidfar M, Mohsenpour M (2010) Effective factor in cotton (*Gossypium* Spp) transformation using agrobacterium. J Agric Biotechnol 2:1–24

Toriyama K, Arimoto Y, Uchimiya H, Hinata K (1988) Transgenic rice plants after direct gene transfer into protoplasts. Biotechnology 6:1072–1074

Tuberosa R, Lucchese C (1990) Selection of maize cell lines tolerant to the non-selective herbicide Basta. Chim Oggi 8:43–46

Waheed A, Jhsan H, Mohammad TW, Kiran KS, Mysore BM (2014) *Agrobacterium*-mediated transformation of tomato with rolB gene results in enhancement of fruit quality and foliar resistance against fungal pathogens. PLoS One 9(5):e96979

Williams KC, O'Rourke PK (1974) Decorticated safflower meal as protein supplement in diets fed either restrictively or *ad libitum* to barrow and gilt pigs over 45 kg live weight. Aust J Exp Agric Husb 14(66):12–16

Xu K, Huang X, Wu M, Wang Y, Chang Y, Liu K, Zhang J, Zhang Y, Zhang F, Yi L, Li T, Wang R, Tan G, Li C (2014) A rapid highly efficient and economical method of *Agrobacterium*-mediated in planta transient transformation in living onion epidermis. PLoS One 9:e83556

Zale JM, Agarwal S, Loar S, Steber CM (2009) Evidence for stable transformation of wheat by floral dip in agrobacterium tumefaciens. Plant Cell Rep 28:903–913

Zenkteler E, Zenkteler M (2016) Development of haploid embryos and plants of *Lactuca sativa* induced by distant pollination with *Helianthus annuus* and *H. tuberosus*. Euphytica 208:439–451

Zhou GY, Weng J, Zeng Y, Huang J, Qian S, Liu G (1983) Introduction of exogenous DNA into cotton embryos. Methods Enzymol 101:433–481

Chapter 19
Abiotic and Biotic Stress Research in Plants: A Gizmatic Approach of Modern Omics Technologies

Nilofer Sheikh, Dina Barman, and Kaushik Bhattacharjee

Abstract Plants are the basic source of food for humans, birds, and animals because of which they occupy the primary position in the food chain. Being the immovable organisms, plants have to withstand adverse environmental and biological conditions as a result of which they are subjected to various types of abiotic and biotic stresses. These stresses affect the growth and reproduction of plants and also cause loss in the yield of crops. Plants have the natural ability to tolerate these stresses by modifying various molecular, cellular, or physiological mechanisms within them and by understanding these mechanisms; a stress-tolerant plant can be developed. Conventional approaches are inadequate to understand these complex mechanisms, so advanced molecular and system biology tools have been used to overcome these problems. One of the noticeable achievements that laid the foundation for development and improvement of stress tolerance in plants is the omics technologies such as genomics, transcriptomics, metabolomics, proteogenomics, etc. which use large scale data of biological origin such as genes, proteins, etc. along with heavy data mining or bioinformatics component. Recent developments in the field of omics technologies hold an immense potential to reshape and improve plants for stress tolerance by modifying their gene regulation or other regulatory networks. The modern omics technologies have been successfully applied for improvement of abiotic or biotic stress-tolerant plants. The present review gives a brief description of the various omics technologies implied for plant stresses (biotic and abiotic) and also discusses the future prospects related to the use of these technologies.

N. Sheikh
Department of Botany, University of Science and Technology, Meghalaya, Baridua, India

D. Barman
Department of Botany, Gauhati University, Guwahati, India

K. Bhattacharjee (✉)
Division of Life Sciences, Institute of Advanced Study in Science and Technology, Guwahati, India

R. Roychowdhury et al. (eds.), *Sustainable Agriculture in the Era of Climate Change*, https://doi.org/10.1007/978-3-030-45669-6_19

Keywords Abiotic stress, Biotic stress, Crop improvement, Omics technologies, Genomics, Transcriptomics, Metabolomics, Proteogenomics

1 Background

With growing global warming and abnormal climate change, plants are encountered with a number of abiotic and biotic stresses which ultimately affect their metabolism, growth, yield, and development (Pandey et al. 2017). These stresses may be natural or human-induced (Kumar and Verma 2018). The stress conditions result in hormonal imbalance, ion toxicity, susceptibility to disease, and nutrient mobilization which lead to affect the plant growth and development (Kumar and Verma 2018). Drought, salinity, heat, cold, chilling, heavy metal, freezing, nutrient, high-light intensity, and ozone are some of the abiotic stresses which affect growth and development of plants (Cavanagh et al. 2008). In addition to these abiotic stresses, plants also face the threat of various kind of biotic stresses which occur as a result of interaction between plant and other living organisms such as infection by pathogens (including bacteria, fungi, viruses, and nematodes) or attack by herbivore pests (Atkinson and Urwin 2012). Sometimes plants face multiple stresses at the same time where one stress condition enhances or reduces the effect of another stress condition. As a consequence of different abiotic stresses, the strength of occurrences of detrimental effects of biotic stresses on plants is increased and/or decreased (Pandey et al. 2017).

In response to various stresses, plants express different stress-coping mechanisms based on type and strength of the stress, for example, accumulation of proline during drought and salinity, formation of heat shock proteins, performance of photosynthesis, antioxidants, responses to pathogen, hormone signaling, osmolyte synthesis, etc. in order to detect changes of environmental conditions, prevent damage, and ensure their survival (Atkinson and Urwin 2012; Prasch and Sonnewald 2013). The response of plants to individual stress is found to be different from multiple stresses (Atkinson and Urwin 2012). The mechanism underlying the tolerance of plants to different kind of stresses is still unknown, and further studies are required in this context. "Omics" tools, including genomics, transcriptomics, proteomics, metabolomics, and phenomics, with the association of bioinformatics and computational approaches, enlightened us to understand the mechanism underlying the tolerance of plants and to produce stress-tolerant plants. Understanding the importance, this chapter provides an update on the recent findings related to various omics tools used in different kind of biotic and abiotic stresses and also lays down future prospect for various researches for improvement of stress tolerance in plants.

2 Stress and Plant Response

Mostly plants are exposed to multiple stresses at the same time which are unfavorable for their growth and development, for instance, during summer season plants face concurrently heat stress, drought stress, and high-light stress. These multiple stresses interact with one another so that the effect of one stress depends on another. Plants may consider the multiple stresses as entirely new stress and produce a tailored response to the new stress; however, sometimes the individual stresses would usually elicit opposing reactions. For instance, *Arabidopsis* confers resistance to osmotic, salt, and heat stress on expressing pigeon pea proline-rich protein; however, plants affected by these stresses are of stunted size (Priyanka et al. 2010).

The simultaneous occurrence of biotic and abiotic stresses is crucial where the effects of an abiotic stress weaken the plant defenses, and thus plants become more susceptible to pathogen. Sharma et al. (2007) observed that on exposing the high temperature stress in wheat, the plant becomes more prone to bacterial, viral, fungal, and nematode pathogens. However, sometimes it was also observed that abiotic stress interacts positively with pathogen stress. For example, on exposure to drought stress, plants may close stomata to reduce transpiration which has positive effect on resistance to pathogen (Beattie 2011). Similarly, viral infection can provide protection from drought stress (Xu et al. 2008). It is due to the fact that pathogenicity actively interferes with plant-water relations (Beattie 2011). The combination of stresses, i.e., abiotic-biotic or abiotic-abiotic, also neutralizes the effect of each other which leads to overall neutral or positive impact on plants. For example, in response to individual drought and ozone stresses in *Medicago truncatula* (alfalfa), the growth becomes affected; however, when both the stresses act in combination, the plant becomes tolerant to these stresses (Pandey et al. 2017). Hence, the mechanisms underlying plant responses to multiple simultaneous stresses are crucial. However, omics strategy makes us understand the exact mechanism behind such response which also helps researcher to develop broad-spectrum stress-tolerant crops.

The plants mainly respond to stresses by means of various crosstalks among the signal pathways (Takahashi et al. 2004). One of the earliest signals is reactive oxygen species (ROS) and reactive nitrogen species (RNS) which are involved in many biotic and abiotic stress responses and thus modify enzyme activity and gene regulation (Molassiotis and Fotopoulos 2011). Hormone such as abscisic acid (ABA) is also considered as important regulators of plant responses to abiotic stress, while defense against different biotic stresses is specified by antagonism between the salicylic acid (SA) and jasmonic acid (JA)/ethylene signaling pathways. However, ABA acts both synergistically and antagonistically to biotic stresses and thus generates a complex network of interacting pathways with crosstalk at different levels (Yasuda et al. 2008). Transcription factors (TFs) are also of prime importance for conferring multiple stress tolerance response as they control a wide range of downstream events (Xu et al. 2011a).

3 Omics Technologies to Understand the Stress Tolerance and Plant Response

The Sanskrit word "om" depicts fullness, and the term omics presumably has gotten from the Sanskrit "om." Along these lines, omics infers totality. As of now, the post-fix omics is employed to portray different advanced fields of science, for example, genomics, proteomics, metabolomics, and so forth, to depict studies involving the sum total of genes, proteins, and metabolites, respectively, within an organism or a cell. With the appearance of new advances and gained learning, the quantity of fields in omics and their applications in differing fields is quickly expanding in the post-genomics period. Modern researches use both molecular and systems biology for study of plant stress or the improvement of plant stress responses (Table 19.1). They provide new insights and open new horizons for understanding stresses and stress responses as well as their improvement and resistance to stresses (Duque et al. 2013). The applications of few of the omics approaches in studying plant stresses are discuss below in detail (Fig. 19.1).

3.1 Genomics

Genomics deals with the identification of gene sequences, intragenic sequences, gene structures, and annotations (Duque et al. 2013). The study of genes involves various processes such as DNA extraction, amplification using PCR techniques, DNA sequencing, as well as sequence assembly (Ramegowda et al. 2014). Functional genomics approaches help in identifying the functional genes involved in abiotic and biotic stress responses in plants and also unravel the interactions between genes involved in regulatory networks (Ramegowda et al. 2014). Functional genomics approaches involve sequencing- or hybridization-based methodologies. The sequencing-based approaches use expressed sequence tags (ESTs) sequencing and serial analysis of gene expression (SAGE) where ESTs have been used to explore the discovery of expressed gene and genome annotation and SAGE can measure the expression of gene. Array-based techniques are one of the widely used hybridization-based methodologies where target DNA is allowed to hybridize with cDNA or oligonucleotide probes attached to a surface to assess expression. This technique is widely used to disclose stress responses in different crops and industrially and agriculturally important plant species (Akpinar et al. 2013). Researchers have identified different stress-responsive genes in different variety of plants with functional genomics approaches (Arpat et al. 2004; Micheletto et al. 2007). Jeong et al. (2010) identified rice NAC-domain gene, *OsNAC10*, with functional genomics approach. They observed that when this gene is under the control of constitutive promoter *GOS2*, *RCc3* showed enhanced tolerance to drought condition.

In addition to functional genomics approaches, structural genomics with next-generation sequencing (NGS) platforms such as Roche 454 GS FLX Titanium or

Table 19.1 Different omics tools and technologies used

Omics tools	Technologies used
Genomics	Next-generation sequencing (NGS) – Roche Platforms (GS Jr, GS FLX+), Ilumina (MiniSeq, MiSeq, HiSeq, and NovaSeq), SOLID and Ion Torrent, PacBio (RSII and Sequel), OxNano (MiniON and GridION/Prome-thION); bead-based flow cytometric method; expressed sequence tag (EST) analysis
Proteomics	Protein extraction and purification; 1-D and 2-D PAGE; reversed-phase high-performance liquid chromatography (RPLC); MALDI-TOF-MS/MS; ESI-TOF-MS/MS; X-ray crystallography; nuclear magnetic resonance (NMR) spectroscopy; MS/MS-based isotope-coded affinity tags (ICAT) and isobaric tag for relative and absolute quantitation (iTRAQ); multidimensional protein identification technology (MudPIT)
Transcriptomics	RNA-Seq; expressed sequence tags (ESTs); cDNA-AFLP; RFLP-coupled domain-directed differential Display; microarray; next-generation sequencing (NGS) – Roche Platforms (GS Jr, GS FLX+), Ilumina (MiniSeq, MiSeq, HiSeq, and NovaSeq), SOLID and Ion Torrent, PacBio (RSII and Sequel), OxNano (MiniON and GridION/Prome-thION)
Metabolomics	Thin-layer chromatography (TLC); high-performance thin-layer chromatography (HPTLC); liquid chromatography-mass spectrometry (LC-MS), high-performance liquid chromatography (HPLC); capillary electrophoresis-mass spectrometry (CE-MS); gas chromatography-mass spectrometry (GC-MS); nuclear magnetic resonance (NMR); Fourier-transform infrared (FT-IR) spectroscopy
Plant glycomics	Chromatographic techniques – LC, HPLC; tandem mass spectrometry; MALDI-mass spectrometry – MALDI-TOF-MS/MS; ESI-mass spectrometry – LC-ESI-MS/MS; NMR spectroscopy – gCOSY, TOCOSY; Fourier-transform ion cyclotron mass spectrometry (FT-ICR-MS); Microarray – carbohydrate microarray, Neoglycolipid (NGL)-based oligosaccharide microarray, lectin microarray, glycogene microarray
Plant lipidomics	Liquid chromatography-mass spectrometry (LC-MS); ion trap mass spectrometer, triple quadrupole, MALDI-TOF MS, Fourier transform mass spectrometer (FT-MS); direct-infusion ESI-based MS
Phenomics	Technologies used in genomics, transcriptomics, metabolomics
Metatranscriptomics	Technologies used in genomics and transcriptomics
Cytogenomics and mutagenomics	Technologies used in genomics; TILLING (Targeting Induced Local Lesions IN Genomes)
Plant miRNomics	Technologies used in genomics and transcriptomics
Plant secretomics	Technologies used in genomics, proteomics, and metabolomics
Signalomics	Technologies used in genomics, proteomics, and metabolomics
Thiolomics	Technologies used in genomics, proteomics, and metabolomics
Transplastomics and chloroplastomics	Technologies used in genomics, proteomics, and transcriptomics
Plant mitochodriomics	Technologies used in genomics, proteomics, and metabolomics
Micromorphomics	Technologies used in genomics, proteomics, and metabolomics
Microbiomics in plants	Technologies used in genomics
Cryobionomics	Technologies used in genomics, proteomics, and metabolomics

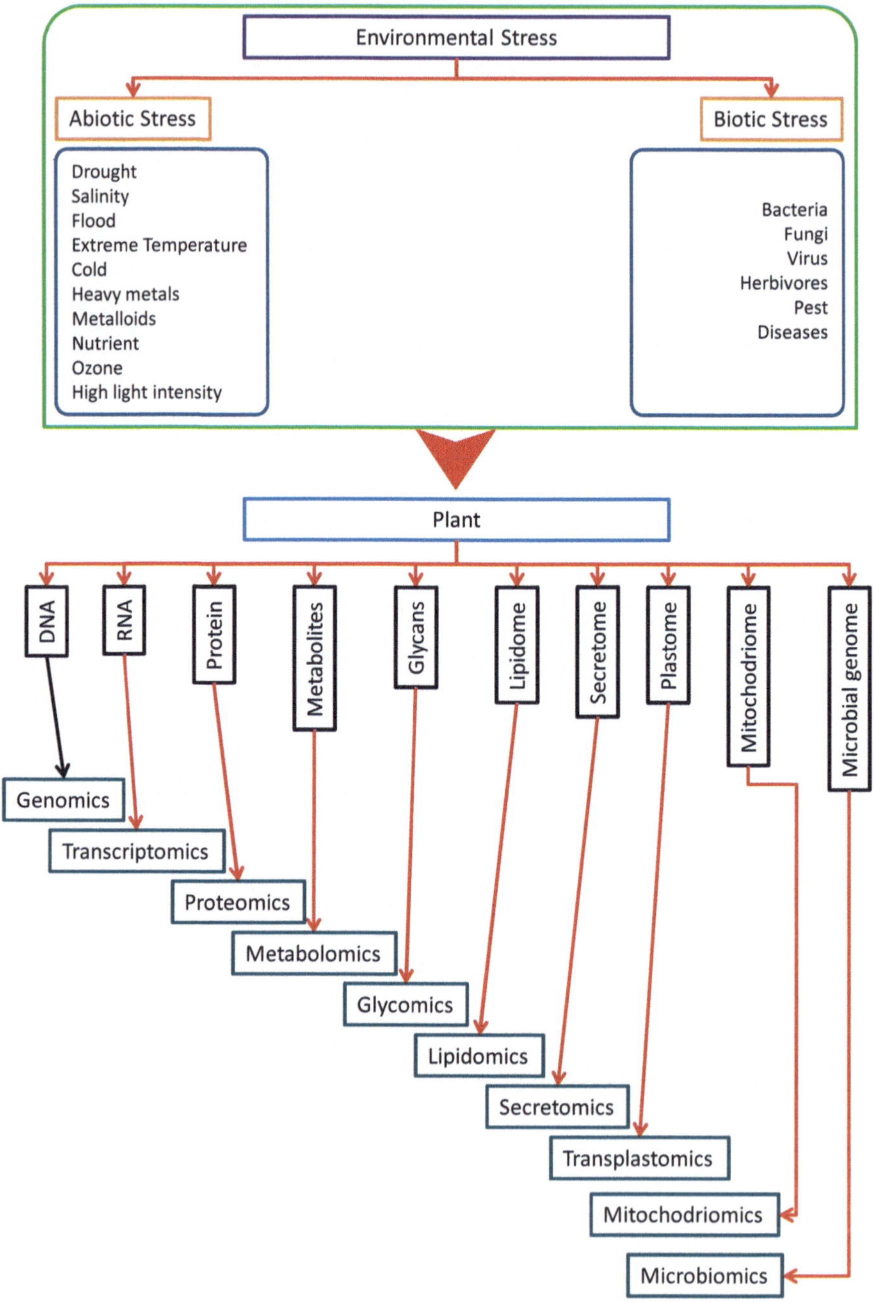

Fig. 19.1 Schematic diagram representing multipurpose application of omics approaches

Illumina Solexa Genome Analyzer and comparative genomics can also be used to get detailed information of genomic features of plants. The NGS-mediated shotgun sequences have been used to develop molecular markers to detect polymorphisms at the DNA sequence level which help plant breeders to create new breeds that can tolerate several biotic and abiotic stresses (Akpinar et al. 2013). Genomics in combination with other omics technologies such as transcriptomics and proteomics help plant breeders to create new stress-tolerant breeds (Agarwal et al. 2014) and to understand the regulation and expression of those genes (Yanik et al. 2013; Zadraznik et al. 2013).

3.2 *Transcriptomics*

Transcriptomics is the study of the complete set of RNA transcripts that are produced by the genome, under specific set of conditions by a cell using high-throughput methods. Comparison of transcriptomes permits the identification of genes that are differentially expressed in particular plant, or in response to different stresses. The transcriptomics technology has provided a valuable insight to diverse plant stress response and tolerance. It provides better understanding of abiotic stress candidate genes and plant-pathogen relationship. Various technologies such as hybridization-based approaches, sequence-based approaches, and RNA sequences are used for transcriptomic studies (Tan et al. 2009). Transcriptomics analysis of broccoli plants treated with N6-benzylaminopurine showed that genes encode for molecular chaperones and stress is upregulated (Liu et al. 2013). Transcriptomics analysis of oak plants showed upregulation of genes associated with plant defense when subjected to long-term mild drought (Spieb et al. 2012). On exposure to salt stress, transcriptomics analysis of plants assisted in the identification of important transcripts and related associations between various physiological processes (Maathius 2006). Exposure to both cold and salinity stresses resulted in downregulation of most of the photosynthetic genes and upregulation of genes encoding transcription factor (Evers et al. 2012). Vashisth et al. (2018) identified stress-responsive genes by transcriptome sequencing of *Artemisia annua* which help to tolerate different abiotic stresses including salt, cold, drought, and waterlogging.

3.3 *Proteomics*

Protein is regarded as the building block of all living organism. The term proteome implies to the overall proteins in a cell, tissue, or life form. Proteomics is the investigation of the whole protein complement of a cell, tissue, or living being underneath a particular, designate set of conditions. Proteomics encompass the identity, biochemical properties, and functional roles, and how quantities, modifications, and structures change during development and in response to internal and external stimuli these proteins. Various types of proteins respond to various stresses at pre- and post-transcriptional and translational levels. These stress inducible proteins could be comprehensively useful for developing stress tolerance in plant. Thus the proteomics study offers an avenue for the discovery of new proteins and pathways associated with crop physiological and stress responses. Few proteomic studies have been conducted in plants under various abiotic stresses including salt stress (Nam et al. 2012), drought stress (Castillejo et al. 2008), and heat stress (Rollins et al. 2013). Plant biologists also used proteomics approach to study the response of *Arabidopsis thaliana*, *Nicotiana tobaccum*, and *Agrostis stolonifera* in response to saline stress (Jiang et al. 2007; Razavizadeh et al. 2009; Xu et al. 2011b). Besides this, some economically important agricultural crop plants have also been investigated under saline stress with proteomics approach, e.g., canola (Bandehagh et al. 2011), sugar beet (Wakeel et al. 2011), potato (Aghaei et al. 2008), and cucumber (Du et al. 2010). Wang et al. (2014) reported 53 differentially expressed protein in *Kandelia candel* where expressions of some of the proteins are upregulated on subjecting to salt stress. Zhao et al. (2008) detected proteins involved in signal transduction while studying the guard cell proteome under drought stress in *Arabidopsis thaliana*. Passamani et al. (2017) used proteomics approach to study the response of salt stress in micro-propagated shoots of two sugarcane cultivars. It was observed that proteins which are involved in calcium-dependent protein kinase, photosystem I, phospholipase D, and glyceraldehyde-3-phosphate dehydrogenase were abundant in salt-tolerant ones. Proteomics also provide information of interactions between crop-pathogen, bacterial pathogens, and elicitors (Casasoli et al. 2008; Margaria and Palmano 2011).

3.4 Metabolomics

Metabolomics is a technology to determine and quantify metabolites involved in different life processes (Deshmukh et al. 2014). Since plants can synthesize a wide range of metabolites to adapt different stress condition, identification and quantification of these metabolites provide better understanding of stress biology in plants (Badjakov et al. 2012). The metabolic fingerprinting can be performed by different techniques such as nuclear magnetic resonance (NMR), MS, Fourier-transform ion cyclotron resonance mass spectrometry, or Fourier-transform infrared (FT-IR) spectroscopy. The identification and quantification of the metabolites can be done by NMR, GC-MS, liquid chromatography-mass spectrometry (LC-MS), capillary electrophoresis-mass spectrometry (CE-MS), and FT-IR spectroscopy (Bagati et al. 2018).

In response to salinity stress, researchers utilized metabolomic approach to know the metabolic effect of salinity in a variety of crop and related plant species including sea lavender (*Limonium latifolium*), rice (Hirai et al. 2004), tomato (Lenz et al. 2011), and grapevine (Cramer et al. 2011). Kaplan et al. (2004) found a set of metabolites in *Arabidopsis* in response to heat and cold stress. Morsy et al. (2007)

observed the accumulation of galactose and raffinose in chilling-tolerant genotype and declination of level of sugar in the chilling-sensitive genotype of rice. In response to drought stress, metabolic regulation in photosynthesis and accumulation of osmolytes occur. *Arabidopsis* can respond to drought stress on accumulating metabolites including amino acids such as proline, raffinose family oligosaccharides, γ-amino butyrate (GABA), and metabolites of tricarboxylic acid (TCA) cycle (Urano et al. 2009). The GC-MS profiling with microarray analysis was conducted by Gong et al. (2005) to study and compare the metabolic profile due to stress competence in the extremophile *Thellungiella halophila* with *Arabidopsis* which showed variation of metabolites between *Arabidopsis* and *Thellungiella* under salt and osmotic stresses. Nikiforova et al. (2005) studied the metabolite profiling of *Arabidopsis* due to sulfur stress. Similarly, Hernandez et al. (2007) identified a set of metabolites such as amino acid, polyols, and sugar which were increased due to phosphorus stress. Various studies on heavy metal-induced stress in plants were also done using NMR-based metabolites fingerprinting on *Silene cucubalus* cell cultures by Bailey et al. (2003).

3.5 Phenomics

Phenomics is the acquisition of multidimensional phenotypic data in an organism (Houle et al. 2010). It extends our knowledge of stress tolerance mechanisms in crop plants. The linking of phenomics to different omics-technologies such as genomics, transcriptomics, and metabolomics helps to analyze the performance of plants in different stress conditions and to fill the gap between response of gene and complex targeted traits for crop improvement (Singh et al. 2018). Plants are sensitive to high heat/cold temperature, and plants can tolerate these stresses on changing physiological and biochemical traits which can be studied by phenotyping (Singh et al. 2018). Phenomics can also be used to find salinity stress behavior in plants which occurs in two phases, i.e., osmotic phase and tissue-tolerant phase. It was observed that osmotic-sensitive plants are more prone to reduction of growth rate in comparison to osmotic-tolerant plants on exposure to NaCl. This reduction of growth rate can be calculated by daily imaging the plant or by infrared thermography (Sirault et al. 2009). James and Sirault (2012) also used infrared thermography to screen a large number of wheat genotypes in response to salinity tolerance. In tissue tolerant phase, in response to NaCl stress, sensitive plants are not able to compartmentalize Na^+ away from cytosol resulting in early senescence in leaves which leads to alter the leaf color. These color changes are visible by light imaging techniques with a Scanalyzer 3D Discovery Platform (Rajendran et al. 2009). Similarly phenomics study with different phenotypic systems such as visible, infrared, and hyperspectral imaging is also able to identify plant's response to drought conditions (Iyer-Pascuzzi et al. 2010).

Plants are more prone to various pathogens resulting in loss of yield. Pathogens result in increasing chlorophyll fluorescence at early stage of infection. These symptoms are generally monitored visually which is time-consuming; however, the high-throughput phenotyping detects the symptoms on monitoring chlorophyll fluorescence more easily even before its detection by human eyes (Yang et al. 2012). Researchers have also developed different high-throughput phenotyping systems such as hyperspectral imaging system, ground-based real-time remote-sensing system, remote pheromone trap monitoring system, etc. to detect plant pathogenic diseases at an early stage (Moshou et al. 2005; Fukatsu et al. 2012; Yang et al. 2012).

3.6 Meta-transcriptomics

Metatranscriptomics deals with the expression of the mRNA within microbial community to identify genes or genetic pathways (Jiang et al. 2016). With this approach, it is possible to identify induced systemic resistance offered by microbial strains in plants in response to various pathogens. For example, *Pseudomonas fluorescens* and *Bacillus* sp. induce resistance to tomato leaf curl virus and tobacco mosaic virus, respectively (Wang et al. 2009; Sangeetha et al. 2010). It also provides information of bacterial gene expression profiles in control/wild and treated plant species/mutant plant species with various agents to adapt defense response (Jones and Dangl 2006). Metatranscriptomics study also provided information of stress tolerance in *Salix* sp. grown in polluted land where Rubisco activase transcripts were downregulated and thiamine thiazole synthase and CP12 were uniformly upregulated (Brereton et al. 2016). The mRNA-based metatranscriptomics can reveal changes of rhizosphere microbial flora in different stress conditions. Hayden et al. (2018) used metatranscriptomics of wheat rhizospheric microflora in disease suppressive and non-suppressive soil to *Rhizoctonia solani* AG8. The results showed that in suppressive soil the polyketide cyclase, terpenoid biosynthesis backbone (dxs) and cold shock proteins (csp) genes get expressed. Whereas in non-suppressive soil, antibiotic genes, genes involved in detoxifying reactive oxygen species (ROS) and superoxide radicals genes get expressed.

3.7 Cytogenomics and Mutagenomics

The plant cytogenomics and mutagenomics are important breakthrough in history of genetics and plant breeding. With the aid of advanced technologies like DNA-based-specific fluorescence banding, GISH, and FISH-guided chromosome painting to the classical plant cytogenetics, the identification, localization, and mapping of chromosome-specific markers in plants are greatly facilitated, which are of high importance in plant breeding, molecular systematics, species identification, detection of hybrid nature, detection of alien chromosomes, chromosomal aberrations, and analysis of somaclonal variations and diversity (Chaudhary et al. 2011). On the other hand, mutational approach offers a powerful tool to study the genetic and molecular mechanism protecting plants against diverse types of biotic and abiotic stresses. The induction of mutation in plants results in alteration of the susceptible alleles of resistant genes and often leads to improvement of yield, morphological traits and increased tolerance to biotic and abiotic stresses (Kharkwal and Shu 2009). For example the induction of mutation by γ-rays, X-rays, NaN_3, EMS, NMU, N-ethyl nitroso urea, colchicine, transposon, etc. in cereals (rice, wheat, maize, barley)/millets, grains, and vegetables results in the plants becoming more tolerant to both biotic and abiotic stresses (Sinjushin and Talukdar 2015). This induced mutation can be easily identified by genomic-based technologies on comparing with wild and normal type of plants (Sinjushin and Talukdar 2015).

3.8 Plant miRNomics

In plants, microRNAs or miRNAs are short (22 bp approx.), endogenously expressed and non-translated RNAs which are regulating the expression of messenger RNAs by targeting transcripts for cleavage or translational repression and sometimes considered as negative gene regulatory molecules. They are considered as conserved regions and play an important role in plant growth and development. Additionally miRNAs also occupy an imperative part in plant stress responses (Kruszka et al. 2012). The expression profiles of most of the miRNAs changed in response to stress, and this stress-responsive miRNAs control the attenuated plant growth and development under stress condition (Mendoza-Soto et al. 2012). Different types of miRNA are involved in various type of abiotic and biotic stress response. Among abiotic stress response, miR169 in *Arabidopsis* were modified to adapt to drought and cold (Li et al. 2008; Zhou et al. 2008). Similarly, in response to salt stress, the expressions of different salt-responsive miRNAs are altered, and some of the miRNAs respond to salt stress by regulating transcription factors (Ding et al. 2009). In response to biotic stress such as plant disease development due to viral infection, miRNAs are involved in silencing of virus-induced gene. Though the exact mechanism behind it is not clear, it is presumed that plants may utilize general RNA silencing machinery to destroy viral RNAs (Hohn and Vazquez 2011). Plants also resist bacterial infection by miRNA-guided regulations of multiple plant hormone pathways. In response to infection of *Pseudomonas syringae*, miR393a gets expressed in plants, and auxin receptors AFB1 are downregulated which result in repression of auxin perception and signaling and thus protecting plants against pathogenic bacteria (Navarro et al. 2006). There are some miRNAs which can respond to multiple biotic and abiotic stresses, for example, miR156 can respond to drought, salt, cold, heat, ABA, oxidative, hypoxia, and UV B (Kruszka et al. 2012; Sunkar et al. 2012).

3.9 Plant Glycomics

Glycomics describes studies designed to define the complete repository of sugars (glycans) and the interaction of these with other macromolecules that a cell or an organism produces under specified conditions of time, location, and environment. Since the sugars play key role in many biological processes such as signaling, stress responses, and immunity, glycomics study is beneficial to study stress responses of plants. Plant glycomic research has significant applications in biopharming of proteins and enzymes (Yadav et al. 2015). The genomics and proteomics don't provide information of biosynthesis of glycan; hence the study of glycomics is lagging behind; however advancement in technology aided the progress of glycomics (Yadav et al. 2015). The covalent linkage of glycan to protein and lipid is considered as the most structurally diverse modification which provides both structural and functional diversity. Attaching glycan to asparagine (N) residues of proteins in plant cell causes variations in the endoplasmic reticulum and the Golgi apparatus, and the *N*-glycan modifications in the Golgi apparatus result in complex *N*-glycans. This complex *N*-glycans protect the growth of root in response to salt/osmotic stress. They also regulate biosynthesis of cell wall and proliferation of cell (von Schaewen et al. 2008). Plant lectins with fungicidal and insecticidal activities are used to form transgenic plants which have the capacity to enhance resistance of plant to pests and phytopathogens (Melnykova et al. 2013). Introducing lectin-like kinases genes in plants genome results in the enhancement of plants defense against various environmental stresses (Melnykova et al. 2013).

3.10 Plant Lipidomics

Lipidomics has transpired as a unique field that allows the comprehensive analysis of cellular lipidome including chemical structures and the quantitative composition with respect to cell signaling, membrane architecture, transcriptional and translational modulation, and cell-cell and cell-protein interactions. In plants, lipids and lipid-based derivatives are essential cellular constituents that play a key role in storage of carbon energy, cell compartmentalization, protection against pathogens, and developmental processes.

Plants can alter membrane lipidome and membrane lipid remodeling to neutralize abiotic stresses. Different researchers utilized lipidomics approach to know the response of plants in connection with different biotic and abiotic stresses. Chen et al. (2013) observed that plant model such as *Arabidopsis* can alter membrane lipid composition in response to cold stress for adaptation and survival in that condition. Similarly lipidomics study of drought-tolerant and drought-sensitive thyme plants (*Thymus serpyllum* L. and *Thymus vulgaris* L.) showed that sensitive plants can decrease the level of galactolipids and phospholipids, while tolerant plants showed increasing level of lipids involved in signaling (Moradi et al. 2017). Similarly, in response to laminar shear stress in *Taxus cuspidata*, phosphatidic acid (PA) content was found to be increased with enhanced activation of phospholipase D (PLD) and phospholipase C (PLC) in comparison to control cells which demonstrated that phospholipids and related phospholipases have imperative roles in mechanotransduction of *T. cuspidata* cells in response to shear stress (Han and Yuan 2009). Liu et al. (2017) performed lipidomics analysis to unravel the effect of nitrogen fertilization on lipid metabolism in *Camellia sinensis* L. and concluded that application of nitrogen fertilizer can balance the lipid metabolism and leads to formation of improved quality of tea with flavor/aroma origin compounds. Lipids also act as signaling molecules that can trigger genes against stress response, for example, ERECTA; proton antiporters TNHX1 and a proton pyrophosphatase TVP1 can protect *Arabidopsis* against salinity and drought stress (Han and Yuan 2009).

There are mainly two different techniques used in the identification and quantification of lipids, i.e., mass spectrometry (MS)-based techniques and non-MS-based techniques. Mass spectrometry (MS)-based techniques involve the use of ion trap mass spectrometer, triple quadrupole, MALDI-TOF MS, Fourier-transform mass spectrometer (FTMS), and direct-infusion ESI-based MS. Non-MS-based techniques involve nuclear magnetic resonance and LC-MRM platforms.

3.11 Plant Secretomics

Plant secretomics is an emerging field of proteomics which deals with secreted protein of plants that have significant importance in the formation of cell wall structure, cell-to-cell interaction, appropriate response to environmental stimuli, and defense against pathogens (Kamoun 2009). The secreted proteins in the apoplast or the extracellular space mediate major defense responses (Grant and Lamb 2006). Thus, plant secretome might play a key role in the early recognition and defense against pathogen attack and improves our insight of defense mechanism during plant-pathogen interactions. Researchers have identified different secreted proteins in response to fungal pathogens (Okushima et al. 2000; Chivasa et al. 2005; Oh et al. 2005). Plants can also produce different elicitor-responsive proteins in response to various pathogens. Thus whole protein extracts from plants augment our understanding of plant-pathogen interactions and defense signaling (Lee et al. 2006; Rampitsch et al. 2006). On attack of pathogens, plants produce salicylic acid which stimulates the expression or secretion of pathogenesis-related proteins which offer the plants to acquire resistance. In plant secretome analyses of *Capsicum annuum* infected with *Phytophthora capsici*, 75 secretory proteins were identified, most of which were stress-related proteins (Yeom et al. 2011). These stress-mediated pathogenesis-related proteins in apoplast are also effective in restraining the growth of pathogens in plants (Pechanova et al. 2010).

3.12 Signalomics

Plant growth regulators (PGRs) play important roles in the growth and development. They also act as signaling molecules in various metabolic processes in plants. For example, roots sense the onset of water stress and can communicate the need to close leaf stomata by altering abscisic acid (ABA) levels in the shoot. Day length and temperature regulate synthesis and transport of gibberellins, which promote stem elongation and stolon formation. Epinasty is characterized by downward curvature of leaves on the adaxial surface of the petiole which occurs in response to stress. This movement reduces foliar absorption of light, retards transpirational water loss, and reduces drought-induced wilting (Abeles et al. 1992). Recent studies in tomato and other crops suggest the possibility of alternative signals involved in root-based communication of drought stress. Holbrook et al. (2002) used two tomato mutants deficient in ABA synthesis to investigate the role of root-derived ABA. Additional experiments using split roots (partial drying) and grafting coupled with maintenance of turgor pressure (pressure chamber) indicated the presence of a root signal that was independent of the root genotype. Similar results were obtained by Fambrini et al. (1995), who used sunflower mutants deficient in ABA and partial root drying of tomato (Sobeih et al. 2004). These studies suggest the presence of an alternative root-based chemical signal capable of altering leaf ABA concentration and availability, facilitating stomatal responses.

3.13 Thiolomics

Thiol (sulfur) is an important constituent, for synthesizing of protein for growing plants is determined with the subsequent uptake and distribution of sulfate in response to demand and environmental factors. Cysteine (cys) is the first molecule in plant metabolism that contains both sulfur and nitrogen. The biosynthesis of cysteine is of utmost importance for the synthesis of a number of essential metabolites in plant. Ruiz and Blumwald (2002) reported that the biosynthesis of cys and glutathione (GSH) was greatly increased in *Brassica napus* when exposed to saline conditions. In gray poplar hybrid, the GSH contents increased in response to NaCl stress in leaves but not in roots (Herschbach et al. 2010). The roles of GSH in modulating plant growth and development have been explored in grass pea and in lentil genotypes under water stress (Talukdar 2013). The pivotal roles played by thiol-cascade in conferring tolerance to heavy metal stress in plants were demonstrated by Harada et al. (2001) who engineered tobacco plants expressing a rice CS gene in order to make it tolerant to toxic levels of cadmium. Similarly, a combined study of transcript, enzymatic, and metabolic profiling in the moss, *Physcomitrella patens*, revealed vital involvement of sulfate assimilatory genes under Cd stress (Rother et al. 2006). The central role played by the GSH and phytochelatins (PCs) in the detoxification of the metalloid indicates a critical importance for S-metabolism in determining plant survival in As-contaminated soils (Ahsan et al. 2008).

3.14 Transplastomics and Chloroplastomics

Plastids are semi-autonomous, endosymbiotic organelles of prokaryotic origin. They contain circular double-stranded DNA and have retained their own nucleic acid and protein synthesis machinery (Wani et al. 2015). The plastid genome, termed as plastome, is present in several identical copies in each plastid. Transplastomics is the omics science related to study of plastid transformation. Plastid transformation represents an alternative to existing nuclear transgenic technologies for improving crop quality and productivity under adverse growth conditions. Transplastomic plants which are resistant to insects due to the presence of crystal proteins of *Bacillus thuringiensis* (Bt) are considered as safe biological insecticides which have immense role in agriculture (Romeis et al. 2006). Since 1994, several transgenic crops expressing Bt crystal proteins, e.g., Cry1Ab in maize and Cry1Ac in cotton, have been commercialized and grown worldwide on millions of hectares and have significantly decreased the use of insecticides (Kota et al. 1999). A novel and non-Bt-type insect resistance strategy has recently been demonstrated by expressing long double-stranded RNA (dsRNA) targeting in order to activate RNA interference that disrupts expression of the target gene in the insects (Zhang et al. 2015). Plastid transformation is also regarded as a potential tool to increase disease resistance to phytopathogenic bacteria and fungi due to high concentrations of the target protein accumulating in a single compartment and released only during hypersensitive reaction, e.g., msi-99 transgene, which encodes a magainin 2 analog antimicrobial peptide, that was first successfully introduced to the plastid genome and was proven to be efficient against different bacteria and fungi (DeGray et al. 2001). Similarly, the introduction of a single agglutinin gene of the *Pinellia ternata* herb resulted in high levels of protein expression in leaf chloroplasts and was effective against broad-spectrum resistance against various pests including aphids, flies, lepidopteran insects, and bacterial and viral pathogens (Jin et al. 2012). The expression of the chloroperoxidase-encoding gene of *Pseudomonas pyrrocinia* in transplastomic plants conferred a similar level of fungal resistance in vitro and in plants (Ruhlman et al. 2014). Hence, transplastomic plants conferred a broad spectrum of resistance not only against different pests and diseases but also against abiotic (salt, osmotic, and oxidative) stresses.

The chloroplast is the most remarkable organelle of plant cells which is the site of a myriad of different biochemical reactions. The knowledge of chloroplast transcriptomes largely leads to the emergence of a very interesting field, directly related to the improvement of crops by photosynthesis/Rubisco engineering. Besides lowering plant nitrogen demand by modulating Rubisco levels, efforts have been made for the improvement of some of its enzymatic traits such as CO_2 affinity, CO_2/O_2

specificity, etc. in tobacco (Zhu et al. 2010). Chloroplast transcriptome of *A. thaliana* has been explored during seed development, stratification, germination, and early seedling development (Demarsy et al. 2012). The thylakoid proteome of sugar beet was explored under iron deficiency conditions, while carbon metabolism-related proteins displayed greater abundance under this stress conditions (Andaluz et al. 2006). Furthermore, the proteome of rice leaves under cold stress was reported to contain 60 unique proteins, whose functions are related to protein synthesis and folding, cell wall synthesis, protein degradation, energy production, and signal transduction. These proteomes of leaves are the most affected organelles by cold stress and possibly can be regarded as mediators of plant responses to cold stress (Cui et al. 2005).

3.15 Plant Mitochodriomics

Mitochondria, the powerhouse of the cell, are an important organelle responsible for production of required amount of ATP for other cellular functions. The mitochondrial genome size of animals is about 16.5 kbp in length whereas between 200 and 2000 kbp in plants. With the discovery of mitochondria in 1840, various progresses have been made in understanding the role of mitochondria in the regulation of energy metabolism of the cell. But the advent of mitochondrial omics technology has given an edge to understand the various regulation mechanism of the cell. In many plants, proline (amino acid) accumulates in response to water stress which later degenerates (Kiyosue et al. 1996). Thus the mitochondria play an important role in recovery of water stress by degrading accumulated proline (Mani et al. 2002).

Plants which are subjected to abiotic stress face disruption of cellular homeostasis. Fratianni et al. (2001) when isolated mitochondria from potato cells acclimated to water stress were found to have an increased capacity of several transporters like dicarboxylate, nucleotide, and K channel which could possibly be related to oxidative stress and energy management. In plants, mitochondrial potassium channels exhibiting high conductivity were discovered and have been suggested to play a major role in energy dissipation and prevention of oxidative stress (Ruy et al. 2004). Durum wheat is a drought-tolerant cereal for which mitochondrial biology in relation to water stress has been extensively studied (Pastore et al. 2007). In addition, Pastore et al. (2007) suggested mitochondrial acclimation to water deficit by applying osmotic or salt stress to seedlings which resulted in the activation of UCP and the potassium channel. Pastore et al. (2003) suggested that the Mal/OAA shuttle was the main route for reoxidation of cytosolic NADH accumulating in stress conditions in durum wheat mitochondria. Most higher plant seeds are desiccation-tolerant, and researches have suggested that mitochondria have a unique property to withstand desiccation with the existence of protective mechanisms, e.g., maize and pea mitochondria isolated from imbibing seeds were found to oxidize succinate and external NADH oxidation at high rates, while TCA cycle activities developed later during germination (Logan et al. 2001; Benamar et al. 2003).

3.16 Micromorphomics

Micromorphomics is the micromorphology-based analysis which is applied to all living organisms which have a great role in stress physiology. In response to stress, plants can adapt, avoid, or may overcome these stresses by means of various physiological and biochemical mechanisms. Plant architecture thus plays an important role in stress physiology. The study of micromorphomics includes measurements from all levels of plant organization like molecules, organelles, cells, tissues, organs, whole plants, and populations which are often necessary to analyze in response to stress. In response to drought and salt stress, plants can roll the leaf extensively (Srivastava 2001), and in response to heavy metals content, necrosis along veins and leaf chlorosis occur in plant (Andre et al. 2006). "Stomatal clustering" is also recognized as a new micromorphological marker for environmental adaptation in terrestrial plants (Gan et al. 2010). Azmat et al. (2009) stated that increased stomatal density coupled with decreased stomatal size is considered as a good adaptive feature of plants in response to heavy metal toxicity and polluted environment. The larger number of small stomata ensures sufficient flow of carbon dioxide for photosynthesis, keeping transpiration to a minimum.

Roots are also considered a sensor organ which is influenced by different stress like drought, salinity, heat, etc. Production of ramified root system in response to drought was reported in many crop plants like rice, wheat, soybean, maize, and sunflower (Tahir et al. 2002; Jaleel et al. 2009). Increased root area is also observed in *Arabidopsis thaliana* when the plant is exposed to abiotic stress (Olmos et al. 2006). Elongation of root and decrease in root diameter in response to drought in many wild species were also reported. Small diameter root maximizes absorptive surfaces, thus increasing rates of water and nutrient uptake (Reader et al. 1993).

A flower is a modified stem which ensures maximal reproductive success. The reproductive parts in flowering plants are often highly sensitive to environmental stress than the vegetative phase. For example, in saffron fresh weight of flower stigma (stigma yield) initially increases with salinity, but in extreme saline condition, stigma yields rapidly decrease (Torbaghan et al. 2011). High temperature also affects the stigma position of flower affecting the pollen capturing capacity (Talukdar 2015). Similarly, exposure to high temperature resulted in male sterility in barley because of non-viable pollen and failure of anther dehiscence (Talukdar 2015).

3.17 Microbiomics in Plants

Plant microbiome shows a close symbiotic relationship between plants and their associated microorganisms. The structure of the plant microbiome is determined by different factors, including type of plant species, plant age, soil properties, nutrient status, and climatic conditions (Berg and Smalla 2009). As an example, the associated microbiome of medicinal plants *Matricaria chamomilla* and the African nightshade, *Solanum distichum*, showed that despite their being grown in direct proximity to one another, their structural (16S rRNA genes) as well as functional (diazotrophic community, *nifH* gene encoding the nitrogenase reductase subunit) colonization profiles revealed a high degree of plant specificity (Koberl et al. 2011). Plant-associated microbes interaction helps the growth and development of the plant and also facilitates the plant to promote stress resistance. Pandey et al. (2016) have demonstrated that application of *Trichoderma harzianum* induces the rice genotypes to mitigate stress response by upregulating the genes for aquaporin, dehydrin, and malondialdehyde. Similarly, *Pseudomonas* sp. and *Acinetobacter* sp. augment production of IAA and ACC deaminase to grow barley and oats in salt-affected soil (Chang et al. 2014).

3.18 Cryobionomics

Cryopreservation is the long-term storage of biological samples in liquid nitrogen at ultralow temperatures (−196 °C) which ensures storage of nonorthodox seeds, the germplasm of vegetatively propagated species and biotechnologically important plant cell lines, etc. (Martinez-Montero and Harding 2015). However, during the process of cryopreservation, there is a possibility of cryo-injury and cryogenic/non-cryogenic stress factors to cells/tissues; hence it is important to ensure of any destabilizing effects before they are reintroduced into natural environments which can be easily confirmed by cryobionomics approach (Martinez-Montero and Harding 2015). Generally cold acclimatization is necessary before the exposure to liquid nitrogen. It was observed that genes respond to adapt to the extreme stresses as imposed by cryopreservation which can be studied by transcriptomic approach (Volk 2010). In cryopreservation, it is necessary to dehydrate the tissue with sugars such as sucrose which also acts as osmotic cryoprotectant. It was observed from sucrose-mediated osmotic stress on the banana meristem results that most of the proteins were significantly up- or downregulated (Carpentier et al. 2007). Hence, in this case it is necessary to maintain the osmoprotective level of intracellular sucrose and augment the expression of some energy-conserving glycolysis genes and the conservation of the cell. Similarly it was observed that proteins from *Vanilla planifolia*-dissected apices responded to PVS3 cryoprotective treatments where some were up- or downregulated proteins (González-Arnao et al. 2011).

3.19 Bioinformatics

All the major omics technologies are highly dependent on this omics science, i.e., bioinformatics and computational tools. Various omics approaches provide high-throughput and large-scale properties which require proper handling, analyses, visualization, and storage. Hence all omics research is tightly bound with strong bioinformatics and computational tools that perform the various analyses (El-Metwally et al. 2014). Computational tools and informatics analyses along with associated methods and algorithms in biology help to analyze biological data in an accurate, fast, and error-free manner (Orozco et al. 2013). Several bioinformatics tasks include genome sequence assembly, sequence alignment, gene prediction, DNA-protein and protein-protein interaction, interaction and regulatory networks analysis, etc. Special type of visualization is required from the huge amount of data analysis and processing by genome sequencer and mass spectrometers. Few of the examples of data visualization are integrated genome tool, PRIDE inspector and ConPath, 3omics, Peppy, etc. The large amount of data and results obtained from various bioinformatics analysis is required to be available and accessible to the scientific community. Several types of databases are available online for deposition and storing the biological data such as plant genomic information (Yu et al. 2013), plant transcriptomics information (Priya and Jain 2013), and plant proteomics information (Cheng et al. 2014).

3.20 Multi-omics Approach

Combination of different omics approaches such as metabolomics, transcriptomics, and proteomics can help to link metabolites with their producer genes, as well as allow the mapping of these metabolites/genes into metabolic pathways (Kumar et al. 2014), and the detailed output of plant metabolic pathways can be harnessed to the plant-specific metabolic pathway database. Similarly, phenomics which is the high-throughput analysis of plant physiology (Furbank and Tester 2011) can also be integrated with all other omics to identify the candidate genes involved in abiotic stresses. Identified genes by all the abovementioned omics technologies can be then functionally characterized using functional genomics approaches like gene over expression or silencing. The integrated approach of transcriptomics and metabolomics can reveal the connections between genes and metabolites (Urano et al. 2009). The combination of metabolomics with other omics techniques facilitates to correlate the metabolite levels and expression level of genes/proteins (Srivastava et al. 2013). Cramer et al. (2007) used integrated approach of metabolomics and transcriptomics to assess response of grapevine to water and salinity stress. Integrated approaches of transcriptomics and proteomics analysis also revealed that nearly five sulfate transporter genes in rice (Norton et al. 2008; Chakrabarty et al. 2009), three in *Brassica* (Srivastava et al. 2009), and at least one transporter in *Arabidopsis* (Sung et al. 2009) are upregulated in roots due to exposure to heavy metal like Arsenic.

4 Conclusions and Future Outlook

Omics technology plays an important role in biotic and abiotic stress responses in plants. It plays significant role in crop quality management under stress condition. With the use of genomics, transcriptomics, proteomics, metabolomics, etc. in plant breeding, plant breeder can improve the consistency and predictability by reducing the time and expense of producing better quality food crops that are resistant to various stresses. It also helps to understand the mechanism involved in plant-microbes and plant-pests' interactions and the various responses involved in these interactions. Thus, the use of various omics approaches has created a platform for further research by establishing networks of interaction between genes, proteins, and metabolites that are involved in stress response mechanism.

References

Abeles F, Morgan P, Saltveit M (1992) Ethylene in plant biology, 2nd edn. Academic, San Diego

Agarwal P, Parida SK, Mahto A et al (2014) Expanding frontiers in plant transcriptomics in aid of functional genomics and molecular breeding. Biotechnol J 9:1480–1492

Aghaei K, Ehsanpour AA, Komatsu S (2008) Proteome analysis of potato under salt stress. J Proteome Res 7:4858–4868

Ahsan N, Lee DG, Alam I et al (2008) Comparative proteomic study of arsenic-induced differentially expressed proteins in rice roots reveals glutathione plays a central role during as stress. Proteomics 8:3561–3576

Akpinar BA, Lucas SJ, Budak H (2013) Genomics approaches for crop improvement against abiotic stress. Sci World J 2013:361921

Andaluz S, López-Millán A, De Las Rivas J et al (2006) Proteomic profiles of thylakoid membranes and changes in response to iron deficiency. Photosynth Res 89:141–155

Andre O, Vollenweider P, Günthardt-George MS (2006) Foliage response to heavy metal contamination in Sycamore Maple (*Acer pseudoplatanus* L.). Snow Landsc Res 80(3):275–288

Arpat A, Waugh M, Sullivan JP et al (2004) Functional genomics of cell elongation in developing cotton fibers. Plant Mol Biol 54:911–929

Atkinson NJ, Urwin PE (2012) The interaction of plant biotic and abiotic stresses: from genes to the field. J Exp Bot 63(10):3523–3544

Azmat R, Haider S, Nasreen H et al (2009) A viable alternative mechanism in adapting the plants to heavy metal environment. Pak J Bot 416:2729–2738

Badjakov I, Kondakova V, Atanassov A (2012) Metabolomics: current view on fruit quality in relation to human health. In: Benkeblia N (ed) Sustainable agriculture and new biotechnologies. CRC Press, Boca Raton, pp 303–319

Bagati S, Mahajan R, Nazir M et al (2018) "Omics": a gateway towards abiotic stress tolerance. In: Zargar SM, Zargar MY (eds) Abiotic stress-mediated sensing and signaling in plants: an omics perspective. Springer, Singapore, pp 1–46

Bailey NJ, Oven M, Holmes E et al (2003) Metabolomic analysis of the consequences of cadmium exposure in *Silene cucubalus* cell cultures via ^{1}H NMR spectroscopy and chemometrics. Phytochemistry 62:851–858

Bandehagh A, Salekdeh GH, Toorchi M et al (2011) Comparative proteomic analysis of canola leaves under salinity stress. Proteomics 11:1965–1975

Beattie GA (2011) Water relations in the interaction of foliar bacterial pathogens with plants. Annu Rev Phytopathol 49:533–555

Benamar A, Tallon C, Macherel D (2003) Membrane integrity and oxidative properties of mitochondria isolated from imbibing pea seeds after priming or accelerated ageing. Seed Sci Res 13:35–45

Berg G, Smalla K (2009) Plant species and soil type cooperatively shape the structure and function of microbial communities in the rhizosphere. FEMS Microbiol Ecol 68:1–13

Brereton NJ, Gonzalez E, Marleau J et al (2016) Comparative transcriptomic approaches exploring contamination stress tolerance in *Salix* sp. Reveal the importance for a metaorganismal *de Novo* assembly approach for nonmodel plants. Plant Physiol 171(1):3–24

Carpentier SC, Witters E, Laukens K et al (2007) Banana (*Musa* spp.) as a model to study the meristem proteome: acclimation to osmotic stress. Proteomics 7:92–105

Casasoli M, Spadoni S, Lilley KS et al (2008) Identification by 2-D DIGE of apoplastic proteins regulated by oligogalacturonides in *Arabidopsis thaliana*. Proteomics 8:1042–1054

Castillejo MA, Maldonado AM, Ogueta S et al (2008) Proteomic analysis of responses to drought stress in sunflower (*Helianthus annuus*) leaves by 2DE gel electrophoresis and mass spectrometry. Proteomics J 1:59–71

Cavanagh C, Morell M, Mackay I et al (2008) From mutations to MAGIC: resources for gene discovery, validation and delivery in crop plants. Curr Opin Plant Biol 11:215–221

Chakrabarty D, Trivedi PK, Misra P et al (2009) Comparative transcriptome analysis of arsenate and arsenite stresses in rice seedlings. Chemosphere 74:688–702

Chang P, Gerhardt KE, Huang XD et al (2014) Plant growth promoting bacteria facilitate the growth of barley and oats in salt impacted soil: implications for phytoremediation of saline soils. Int J Phytoremediation 16:1133–1147

Chaudhary HK, Sood VK, Tayeng T et al (2011) Molecular cytogenetics in physical mapping of genomes and alien introgressions. In: Pratap A, Kumar J (eds) Biology and breeding of food legumes. CABI, Oxfordshire, pp 131–146

Chen D, Yan X, Xu J et al (2013) Lipidomic profiling and discovery of lipid biomarkers in *Stephanodiscus* sp. under cold stress. Metabolomics 9(5):949–959

Cheng H, Deng W, Wang Y et al (2014) dbPPT: a comprehensive database of protein phosphorylation in plants. Database 2014:bau121

Chivasa S, Simon WJ, Yu XL et al (2005) Pathogen elicitor–induced changes in the maize extracellular matrix proteome. Proteomics 5:4894–4904

Cramer GR, Ergul A, Grimplet J et al (2007) Water and salinity stress in grapevines: early and late changes in transcript and metabolite profiles. Funct Integr Genomics 7:111–134

Cramer G, Urano K, Delrot S et al (2011) Effect of abiotic stress on plant: a systems biology perspective. BMC Plant Biol 11:163

Cui S, Huang F, Wang J et al (2005) A proteomic analysis of cold stress responses in rice seedlings. Proteomics 5:3162–3172

DeGray G, Rajasekaran K, Smith F et al (2001) Expression of an antimicrobial peptide via the chloroplast genome to control phytopathogenic bacteria and fungi. Plant Physiol 127:852–862

Demarsy E, Buhr F, Lambert E et al (2012) Characterization of the plastid-specific germination and seedling establishment transcriptional programme. J Exp Bot 63:925–939

Deshmukh R, Sonah H, Patil G et al (2014) Integrating omic approaches for abiotic stress tolerance in soybean. Front Plant Sci 5:1–12

Ding D, Zhang L, Wang H et al (2009) Differential expression of miRNAs in response to salt stress in maize roots. Ann Bot 103:29–38

Du CX, Fan HF, Guo SR et al (2010) Proteomic analysis of cucumber seedling roots subjected to salt stress. Phytochemistry 71:1450–1459

Duque AS, de Almeida AM, da Silva AB et al (2013) Abiotic stress responses in plants: unraveling the complexity of genes and networks to survive. In: Vahdati K, Leslie C (eds) Abiotic stress-plant responses and applications in agriculture, pp 49–101

El-Metwally S, Ouda OM, Helmy M (2014) Next generation sequencing technologies and challenges in sequence assembly. Springer

Evers D, Legay S, Lamoreux D et al (2012) Towards a synthetic view of potato cold and salt stress response by transcriptomic and proteomic analyses. Plant Mol Biol 78:503–514

Fambrini M, Vernieri P, Toncelli ML et al (1995) Characterization of a wilty sunflower (*Helianthus annuus* L.) mutant. III. Phenotypic interaction in reciprocal grafts from wilty mutant and wild-type plants. J Exp Bot 46:525–530

Fratianni A, Pastore D, Pallotta ML et al (2001) Increase of membrane permeability of mitochondria isolated from water stress adapted potato cells. Biosci Rep 21:81–91

Fukatsu T, Watanabe T, Hu HM et al (2012) Field monitoring support system for the occurrence of *Leptocorisa chinensis* Dallas (Hemiptera: Alydidae) using synthetic attractants field servers, and image analysis. Comput Electron Agric 80:8–16

Furbank RT, Tester M (2011) Phenomics-technologies to relieve the phenotyping bottleneck. Trends Plant Sci 16:635–644

Gan Y, Zhou L, Shen ZJ et al (2010) Stomatal clustering, a new marker for environmental perception and adaptation in terrestrial plants. Bot Stud 51:325–336

Gong Q, Li P, Ma S et al (2005) Salinity stress adaptation competence in the extremophile *Thellungiella halophila* in comparison with its relative *Arabidopsis thaliana*. Plant J 44:826–839

González-Arnao MT, Durán-Sánchez B, Jiménez-Francisco B et al (2011) Cryopreservation and proteomic analysis of vanilla (*V. planifolia* A.) apices treated with osmoprotectants. Acta Hortic 908:67–72

Grant M, Lamb C (2006) Systemic immunity. Curr Opin Plant Biol 9:414–420

Han PP, Yuan YJ (2009) Lipidomic analysis reveals activation of phospholipid signaling in mechanotransduction of *Taxus* cuspidate cells in response to shear stress. FASEB J 23(2):623–630

Harada E, Coi Y-E, Tsuchisaka A et al (2001) Transgenic tobacco plants expressing a rice cysteine synthase gene are tolerant to toxic levels of cadmium. J Plant Physiol 158:655–661

Hayden HL, Savin KW, Wadeson J et al (2018) Comparative metatranscriptomics of wheat rhizosphere microbiomes in disease suppressive and non-suppressive soils for *Rhizoctonia solani* AG8. Front Microbiol 9:859

Hernandez G, Ramirez M, Valdes-Lopez O et al (2007) Phosphorus stress in common bean: root transcript and metabolic responses. Plant Physiol 144:752–767

Herschbach C, Teuber M, Eiblmeier M et al (2010) Changes in sulphur metabolism of grey poplar (*Populus* x canescens) leaves during salt stress: a metabolic link to photorespiration. Tree Physiol 30:1161–1173

Hirai MY, Yano M, Goodinowe DB et al (2004) Integration of transcriptomics and metabolomics for understanding of global responses to nutritional stresses in *Arabidopsis thaliana*. Proc Natl Acad Sci U S A 101(27):10205–10210

Hohn T, Vazquez F (2011) RNA silencing pathways of plants: silencing and its suppression by plant DNA viruses. Biochim Biophys Acta 1809(11–12):588–600

Holbrook NM, Shashidhar VR, James RA et al (2002) Stomatal control in tomato with ABA-deficient roots: response of grafted plants to soil drying. J Exp Bot 53:1503–1514

Houle D, Govindaraju DR, Omholt S (2010) Phenomics: the next challenge. Nat Rev Genet 11:855–866

Iyer-Pascuzzi AS, Symonova O, Mileyko Y et al (2010) Imaging and analysis platform for automatic phenotyping and trait ranking of plant root systems. Plant Physiol 152:1148–1157

Jaleel CA, Manivannan P, Wahid A et al (2009) Drought stress in plants: a review on morphological characteristics and pigments composition. Int J Agric Biol 11:100–105

James RA, Sirault XR (2012) Infrared thermography in plant phenotyping for salinity tolerance. Methods Mol Biol 913:173–189

Jeong JS, Kim YS, Baek KH et al (2010) Root-specific expression of OsNAC10 improves drought tolerance and grain yield in rice under field drought conditions. Plant Physiol 153(1):185–197

Jiang Y, Yang B, Harris NS et al (2007) Comparative proteomic analysis of NaCl stress-responsive proteins in *Arabidopsis* roots. J Exp Bot 58:3591–3607

Jiang Y, Xiong X, Danska J et al (2016) Metatranscriptomic analysis of diverse microbial communities reveals core metabolic pathways and microbiome-specific functionality. Microbiome 4:2

Jin S, Zhang X, Daniell H (2012) *Pinellia ternata* agglutinin expression in chloroplasts confers broad spectrum resistance against aphid, whitefly, Lepidopteran insects, bacterial and viral pathogens. Plant Biotechnol J 10:313–327

Jones JDG, Dangl JL (2006) The plant immune system. Nature 444:323–329

Kamoun S (2009) The secretome of plant-associated fungi and oomycetes. In: Deising VH (ed) Plant relationships, The Mycota, 2nd edn. Springer, Berlin/Heidelberg, pp 173–180

Kaplan F, Kopka J, Haskell DW et al (2004) Exploring the temperature stress metabolome of *Arabidopsis*. Plant Physiol 136:4159–4168

Kharkwal MC, Shu QY (2009) The role of induced mutations in world food security. In: Shu QY (ed) Induced plant mutations in the genomics era. Food and Agriculture Organization of the United Nations, Rome, pp 33–38

Kiyosue T, Yoshiba Y, Yamaguchi-Shinozaki K et al (1996) A nuclear gene encoding mitochondrial proline dehydrogenase, an enzyme involved in proline metabolism, is upregulated by proline but downregulated by dehydration in *Arabidopsis*. Plant Cell 8:1323–1335

Koberl M, Müller H, Ramadan EM et al (2011) Desert farming benefits from microbial potential in arid soils and promotes diversity and plant health. PLoS One 6:e24452

Kota M, Daniell H, Varma S et al (1999) Overexpression of the *Bacillus thuringiensis* (Bt) Cry2Aa2 protein in chloroplasts confers resistance to plants against susceptible and Bt-resistant insects. Proc Natl Acad Sci U S A 96:1840–1845

Kruszka K, Pieczynski M, Windels D (2012) Role of microRNAs and other sRNAs of plants in their changing environments. J Plant Physiol 169(16):1664–1672

Kumar A, Verma JP (2018) Does plant-microbe interaction confer stress tolerance in plants: a review? Microbiol Res 207:41–52

Kumar R, Khurana A, Sharma AK (2014) Role of plant hormones and their interplay in development and ripening of flesh fruits. J Exp Bot 65:4561–4575

Lee J, Bricker TM, Lefevre M et al (2006) Proteomic and genetic approaches to identifying defense-related proteins in rice challenged with the fungal pathogen *Rhizoctonia solani*. Mol Plant Pathol 7:405–416

Lenz T, Fischer JJ, Dreger M (2011) Probing small molecule-protein interactions: a new perspective for functional proteomics. J Proteome 75(1):100–115

Li WX, Oono Y, Zhu J et al (2008) The *Arabidopsis* NFYA5 transcription factor is regulated transcriptionally and post- transcriptionally to promote drought resistance. Plant Cell 20:2238–2251

Liu MS, Li HC, Lai YM et al (2013) Proteomics and transcriptomics of broccoli subjected to exogenously supplied and transgenic senescence-induced cytokinin for amelioration of postharvest yellowing. J Proteome 93:133–144

Liu MY, Burgos A, Ma L et al (2017) Lipidomics analysis unravels the effect of nitrogen fertilization on lipid metabolism in tea plant (*Camellia sinensis* L.). BMC Plant Biol 17:165

Logan DC, Millar AH, Sweetlove LJ et al (2001) Mitochondrial biogenesis during germination in maize embryos. Plant Physiol 125:662–672

Maathius FJM (2006) The role of monovalent cation transporters in plant responses to salinity. J Exp Bot 57:1137–1147

Mani S, van de Cotte B, Van Montagu M et al (2002) Altered levels of proline dehydrogenase cause hypersensitivity to proline and its analogs in *Arabidopsis*. Plant Physiol 128:73–83

Margaria P, Palmano S (2011) Response of the *Vitis vinifera* L. Cv. 'Nebbiolo' proteome to Flavescence dore'e phytoplasma infection. Proteomics 11:212–224

Martinez-Montero ME, Harding K (2015) Cryobionomics: evaluating the concept in plant cryopreservation. In: Barh D, Khan MS, Davies E (eds) PlantOmics: the omics of plant science. Springer, pp 655–682

Melnykova NM, Mykhalkiv LM, Mamenko PM, Kots YS (2013) The areas of application for plant lectins. Biopolym Cell 29(5):357–366

Mendoza-Soto AB, Sánchez F, Hernández G (2012) MicroRNAs as regulators in plant metal toxicity response. Front Plant Sci 3:105

Micheletto S, Rodriguez-Uribe L, Hernandez R et al (2007) Comparative transcript profiling in roots of *Phaseolus acutifolius* and *P. vulgaris* under water deficit stress. Plant Sci 73:510–520

Molassiotis A, Fotopoulos V (2011) Oxidative and nitrosative signaling in plant. Two branches in the same tree? Plant Signal Behav 6:210–214

Moradi P, Mahdavi A, Khoshkam M et al (2017) Lipidomics unravels the role of leaf lipids in thyme plant response to drought stress. Int J Mol Sci 18:2067

Morsy MR, Jouve L, Hausman JF et al (2007) Alteration of oxidative and carbohydrate metabolism under abiotic stress in two rice (*Oryza sativa* L.) genotypes contrasting in chilling tolerance. J Plant Physiol 164:157–167

Moshou D, Bravo C, Oberti R et al (2005) Plant disease detection based on data fusion of hyperspectral and multispectral fluorescence imaging using Kohonen maps. Real-Time Image 11:75–83

Nam MH, Huh SM, Kim KM et al (2012) Comparative proteomic analysis of early salt stress-responsive proteins in roots of SnRK2 transgenic rice. Proteome Sci 10:25

Navarro L, Dunoyer P, Jay F et al (2006) A plant miRNA contributes to antibacterial resistance by repressing auxin signaling. Science 312:436

Nikiforova VJ, Kopka J, Tolstikov V et al (2005) Systems rebalancing of metabolism in response to sulfur deprivation, as revealed by metabolome analysis of *Arabidopsis* plants. Plant Physiol 138:304–318

Norton GJ, Lou-Hing DE, Meharg AA et al (2008) Rice-arsenate interactions in hydroponics: whole genome transcriptional analysis. J Exp Bot 59:2267–2276

Oh IS, Park AR, Bae MS et al (2005) Secretome analysis reveals an Arabidopsis lipase involved in defense against *Alternaria brassicicola*. Plant Cell 17:2832–2847

Okushima Y, Koizumi N, Kusano T et al (2000) Secreted proteins of tobacco cultured BY2 cells: identification of a new member of pathogenesis,-related proteins. Plant Mol Biol 42:479–488

Olmos E, Kiddle G, Pellny TK et al (2006) Modulation of plant morphology, root architecture, and cell structure by low vitamin C in *Arabidopsis thaliana*. J Exp Bot 57(8):1645–1655

Orozco A, Morera J, Jiménez S et al (2013) A review of bioinformatics training applied to research in molecular medicine, agriculture and biodiversity in Costa Rica and Central America. Brief Bioinform 14:661–670

Pandey V, Ansari MW, Tula S et al (2016) Dose-dependent response of *Trichoderma harzianum* in improving drought tolerance in rice genotypes. Planta 243:1251–1264

Pandey P, Irulappan V, Bagavathiannan MV et al (2017) Impact of combined abiotic and biotic stresses on plant growth and avenues for crop improvement by exploiting Physio-morphological traits. Front Plant Sci 8:537

Passamani LZ, Barbosa RR, Reis RS et al (2017) Salt stress induces changes in the proteomic profile of micro-propagated sugarcane shoots. PLoS One 12(4):e0176076

Pastore D, Di PS, Passarella S (2003) Isolated durum wheat and potato cell mitochondria oxidize externally added NADH mostly via the malate/oxaloacetate shuttle with a rate that depends on the carrier-mediated transport. Plant Physiol 133:2029–2039

Pastore D, Trono D, Laus MN et al (2007) Possible plant mitochondria involvement in cell adaptation to drought stress. A case study: durum wheat mitochondria. J Exp Bot 58:195–210

Pechanova O, Hsu CY, Adams JP et al (2010) Apoplast proteome reveals that extracellular matrix contributes to multi-stress response in poplar. BMC Genomics 11:674

Prasch CM, Sonnewald U (2013) Simultaneous application of heat, drought, and virus to *Arabidopsis* plants reveals significant shifts in signaling networks. Plant Physiol 162:1849–1866

Priya P, Jain M (2013) RiceSRTFDB: a database of rice transcription factors containing comprehensive expression, cis-regulatory element and mutant information to facilitate gene function analysis. Database (Oxford) 2013:bat027

Priyanka B, Sekhar K, Reddy VD et al (2010) Expression of pigeon pea hybrid-proline-rich protein encoding gene (CcHyPRP) in yeast and Arabidopsis affords multiple abiotic stress tolerance. Plant Biotechnol J 8:76–87

Rajendran K, Tester M, Roy SJ (2009) Quantifying the three main components of salinity tolerance in cereals. Plant Cell Environ 32:237–249

Ramegowda V, Mysore KS, Senthil-Kumar M (2014) Virus-induced gene silencing is a versatile tool for unraveling the functional relevance of multiple abiotic-stress-responsive genes in crop plants. Front Plant Sci 5:323

Rampitsch C, Bykova NV, McCallum B et al (2006) Analysis of the wheat and *Puccinia triticina* (leaf rust) proteomes during a susceptible host-pathogen interaction. Proteomics 6:1897–1907

Razavizadeh R, Ehsanpour AA, Ahsan N et al (2009) Proteome analysis of tobacco leaves under salt stress. Peptides 30:1651–1659

Reader RJ, Jalili A, Grime JP et al (1993) A comparative-study of plasticity in seedling rooting depth in drying soil. J Ecol 81:543–550

Rollins JA, Habte E, Templer SE et al (2013) Leaf proteome alterations in the context of physiological and morphological responses to drought and heat stress in barley (*Hordeum vulgare* L.). J Exp Bot 64:3201–3212

Romeis J, Meissle M, Bigler F (2006) Transgenic crops expressing *Bacillus thuringiensis* toxins and biological control. Nat Biotechnol 24:63–71

Rother M, Krauss GJ, Grass G et al (2006) Sulphate assimilation under Cd stress in *Physcomitrella patens*-combined transcript, enzyme and metabolite profiling. Plant Cell Environ 29:1801–1811

Ruhlman T, Rajasekaran K, Cary JW (2014) Expression of chloroperoxidase from *Pseudomonas pyrrocinia* in tobacco plastids for fungal resistance. Plant Sci 228:98–106

Ruiz JM, Blumwald E (2002) Salinity-induced glutathione synthesis in *Brassica napus*. Planta 214:965–969

Ruy F, Vercesi AE, Andrade PBM et al (2004) A highly active ATP-insensitive K^+ import pathway in plant mitochondria. J Bioenerg Biomembr 36:195–202

Sangeetha G, Thangavelu R, Rani SU et al (2010) Induction of systemic resistance by mixtures of antagonist bacteria for the management of crown rot complex on banana. Acta Physiol Plant 32:1177–1187

Sharma RC, Duveiller E, Ortiz-Ferrara G (2007) Progress and challenge towards reducing wheat spot blotch threat in the Eastern Gangetic Plains of South Asia: is climate change already taking its toll? Field Crop Res 103:109–118

Singh B, Mishra S, Bohra A et al (2018) Crop Phenomics for abiotic stress tolerance in crop plants. In: Wani SH (ed) Biochemical, physiological and molecular avenues for combating abiotic stress tolerance in plants. Elsevier Inc, pp 277–296

Sinjushin A, Talukdar D (2015) Cytogenomics and mutagenomics in plant functional biology and breeding. In: Barh D, Khan MS, Davie E (eds) Phenomics: technologies and applications in plant and agriculture. Springer, pp 113–156

Sirault XRR, James RA, Furbank RT (2009) A new screening method for osmotic component of salinity tolerance in cereals using infrared thermography. Funct Plant Biol 36:970–977

Sobeih WY, Dodd IC, Bacon MA et al (2004) Long distance signals regulating stomatal conductance and leaf growth in tomato (*Lycopersicon esculentum*) plants subjected to partial rootzone drying. J Exp Bot 55:2353–5363

Spieb N, Oufir M, Matusikova I et al (2012) Ecophysiological and transcriptomic responses of oak (*Quercus robur*) to long-term drought exposure and rewatering. Environ Exp Bot 77:117–126

Srivastava LM (2001) Plant growth and development. Academic, San Diego/London

Srivastava S, Srivastava AK, Suprasanna P et al (2009) Comparative biochemical and transcriptional profiling of two contrasting varieties of *Brassica juncea* L. in response to arsenic exposure reveals mechanisms of stress perception and tolerance. J Exp Bot 60:3419–3431

Srivastava V, Obudulu O, Bygdell J et al (2013) OnPLS integration of transcriptomic, proteomic and metabolomic data shows multi-level oxidative stress responses in the cambium of transgenic hipI- superoxide dismutase *Populus* plants. BMC Genomics 14:893

Sung DY, Kim TH, Komives EA et al (2009) ARS5 is a component of the 26S proteasome complex, and negatively regulates thiol biosynthesis and arsenic tolerance in *Arabidopsis*. Plant J 59:802–812

Sunkar R, Li YF, Jagadeeswaran G (2012) Functions of microRNAs in plant stress responses. Trends Plant Sci 17(4):196–203

Tahir MHN, Imran M, Hussain MK (2002) Evaluation of sunflower (*Helianthus annuus* L.) inbred lines for drought tolerance. Int J Agric Biol 3:398–400

Takahashi S, Seki M, Ishida J et al (2004) Monitoring the expression profiles of genes induced by hyperosmotic, high salinity and oxidative stress and abscisic acid treatment in *Arabidopsis* cell culture using a full-length cDNA microarray. Plant Mol Biol 56(1):29–55

Talukdar D (2013) Selenium priming selectively ameliorates weed–induced phytotoxicity by modulating antioxidant defense components in lentil (*Lens culinaris* Medik.) and grass pea (*Lathyrus sativus* L.). Ann Rev Res Biol 3(3):195–212

Talukdar T (2015) Micromorphomics: a morphological dissection to unveil environmental stress. In: Barh D, Khan MS, Davies E (eds) Plant omics: the omics of plant science. Springer, pp 615–632

Tan KC, Ipcho SVS, Trengove RD et al (2009) Assessing the impact of transcriptomics, proteomics and metabolomics on fungal phytopathology. Mol Plant Pathol 10:703–715

Torbaghan ME, Torbaghan ME, Ahmadi M (2011) The effect of salt stress on flower yield and growth parameters of saffron (*Crocus sativus* L) in greenhouse condition. Int Res J Agric Sci Soil Sci 1(10):421–427

Urano K, Maruyama K, Ogata Y et al (2009) Characterization of the ABA-regulated global responses to dehydration in *Arabidopsis* by metabolomics. Plant J 7:1065–1078

Vashisth D, Kumar R, Rastogi S et al (2018) Transcriptome changes induced by abiotic stresses in *Artemisia annua*. Sci Rep 8:3423

Volk GM (2010) Application of functional genomics and proteomics to plant cryopreservation. Curr Genomics 11:24–29

Von Schaewen A, Frank J, Koiwa H (2008) Role of complex N-glycans in plant stress tolerance. Plant Signal Behav 3(10):871–873

Wakeel A, Asif AR, Pitann B et al (2011) Proteome analysis of sugar beet (*Beta vulgaris* L.) elucidates constitutive adaptation during the first phase of salt stress. J Plant Physiol 168:519–526

Wang SA, Wu HJ, Qiao JQ et al (2009) Molecular mechanism of plant growth promotion and induced systemic resistance to tobacco mosaic virus by *Bacillus* spp. J Microbiol Biotechnol 19:1250–1258

Wang L, Liu X, Liang M et al (2014) Proteomic analysis of salt-responsive proteins in the leaves of mangrove *Kandelia candel* during short-term stress. PLoS One 9:e83141

Wani SH, Sah SK, Sagi L et al (2015) Transplastomic plants for innovations in agriculture. A review. Agron Sustain Dev 35:1391

Xu P, Chen F, Mannas JP, Feldman T et al (2008) Virus infection improves drought tolerance. New Phytol 180:911–921

Xu ZS, Chen M, Li LC, Ma YZ (2011a) Functions and application of the AP2/ERF transcription factor family in crop improvement. J Integr Plant Biol 53:570–585

Xu C, Jiang Z, Huang B (2011b) Nitrogen deficiency-induced protein changes in immature and mature leaves of creeping bentgrass. J Am Soc Hortic Sci 136:399–407

Yadav S, Yadav DK, Yadav N, Khurana SMP (2015) Plant glycomics: advances and applications. In: Barh D, Khan MS, Davie E (eds) Phenomics: technologies and applications in plant and agriculture. Springer, pp 299–329

Yang Y, Chai RY, He Y (2012) Early detection of rice blast (*Pyricularia*) at seedling stage in Nipponbare rice variety using near-infrared hyper-spectral image. Afr J Biotechnol 11:6809–6817

Yanik H, Turktas M, Dundar E et al (2013) Genome-wide identification of alternate bearing-associated microRNAs (miRNAs) in olive (*Olea europaea* L.). BMC Plant Biol 13:10

Yasuda M, Ishikawa A, Jikumaru Y et al (2008) Antagonistic interaction between systemic acquired resistance and the abscisic acid-mediated abiotic stress response in *Arabidopsis*. Plant Cell 20:1678–1692

Yeom SI, Baek HK, Oh SK et al (2011) Use of a secretion trap screen in pepper following *Phytophthora capsici* infection reveals novel functions of secreted plant proteins in modulating cell death. Mol Plant-Microbe Interact 24:671–684

Yu J, Zhao M, Wang X et al (2013) Bolbase: a comprehensive genomics database for *Brassica oleracea*. BMC Genomics 14:664

Zadraznik T, Hollung K, Egge-Jacobsen W et al (2013) Differential proteomic analysis of drought stress response in leaves of common bean (*Phaseolus vulgaris* L.). J Proteome 78:254–272

Zhang J, Khan SA, Hasse C et al (2015) Full crop protection from an insect pest by expression of long double-stranded RNAs in plastids. Science 347:991–994

Zhao Z, Zhang W, Stanley BA et al (2008) Functional proteomics of *Arabidopsis thaliana* guard cells uncovers new stomatal signaling pathways. Plant Cell 20:3210–3226

Zhou X, Wang G, Sutoh K et al (2008) Identification of cold-inducible micro RNAs in plants by transcriptome analysis. Biochim Biophys Acta 1779(11):780–788

Zhu G, Kurek I, Liu L (2010) Engineering photosynthetic enzymes involved in CO_2-assimilation by gene shuffling. In: Rebeiz CA, Benning C, Bohnert HJ et al (eds) The chloroplast: Basics and application, Advances in photosynthesis and respiration. Springer, Dordrecht, pp 307–322

Chapter 20
Involvement of Microbes in Different Abiotic Stress Environments of Cropping Lands

Muhammad Zulqurnain Haider, Farah Saeed, Aqsa Ali, Qasim Ali, Noman Habib, Muhammad Tariq Javed, Muhammad Azeem, Naeem Iqbal, Shafaqat Ali, Imran Khan, Faisal Mahmood, Youcai Xiong, and Muhammad Afzaal

Abstract Global adverse changes in environment spiked up the abiotic (salinity, water scarcity, accumulation of heavy metals, extreme temperatures, etc.) and biotic stresses (increase in phytopathogens, etc.). This adverse environment is considered responsible for reduction in production of agriculture, hindering the bioremediation application, and even alteration in forest ecosystems. This alarming scenario forcing the scientists to look and call our natural allies, i.e., plant growth-promoting microbes (PGPM), for their assistance to develop new strategies and technologies which are sustainable eco-friendly alternatives to the formal agricultural practices like massive use of pesticides and fertilizers, etc. The results from the previous research on PGPM including rhizobia, plant growth-promoting bacteria (PGPB), arbuscular mycorrhizal (AM) fungi, vesicular-arbuscular mycorrhizal (VAM) fungi, and ectomycorrhizal (EM) fungi have demonstrated a huge unrevealed potential to cope with major environmental threats to agriculture production. In this regard, a collective piece of information is not still available which represents the missing or revealing pieces of the jigsaw puzzle of PGPM assistance to alleviate adverse envi-

M. Z. Haider (✉) · F. Saeed · A. Ali · Q. Ali · N. Habib · M. T. Javed · M. Azeem · N. Iqbal
Department of Botany, Government College University, Faisalabad, Pakistan
e-mail: drmzhaider@gcuf.edu.pk

S. Ali · F. Mahmood
Department of Environment Sciences, Government College University, Faisalabad, Pakistan

I. Khan
Department of Agronomy, University of Agriculture, Faisalabad, Pakistan

Y. Xiong
MOE (Ministry of Education) Engineering Research Center for Dryland Agriculture and Ecological Conservation, School of Life Sciences, Lanzhou University, Lanzhou, Gansu, China

M. Afzaal
School of Environment and Life Sciences, Salford, UK

R. Roychowdhury et al. (eds.), *Sustainable Agriculture in the Era of Climate Change*, https://doi.org/10.1007/978-3-030-45669-6_20

ronment footprints on agriculture production. The focus of this chapter is to explore the prospective of plant-microbe interactions to cope up with the adverse environment-related challenges of agriculture production through reviewing the progress in knowledge of mechanism underlying these interactions for sustainable development in agriculture.

Keywords Mycorrhizal Fungi · Plant Growth Promoting Rhizobacteria (PGPR) · Plants · Drought · Salt stress · Heavy metal stress · Temperature stress

Abbreviations

SA	Sustainable agriculture
MF	Mycorrhizal fungi
PGPB	Plant growth-promoting bacteria
PGPM	Plant growth-promoting microbes
ECM	Ectomycorrhizal fungi
EM	Endomycorrhizal fungi
HMS	Heavy metal stress
AMF	Arbuscular mycorrhizal fungi
VAM fungi	Vesicular-arbuscular mycorrhizal fungi
PMA	Plant-microbe association
EPS	Exopolysaccharide
RBT	Rhizobitoxine
BF	Biofertilizer
BP	Biopesticides
PMI	Plant-microbe interaction
ROS	Reactive oxygen species
LPO	Lipid peroxidation
WSTM	Water stress tolerant microbes
IAA	Indole-3-acetic acid
ABA	Abscisic acid
ACC	1-aminocyclopropane-1-carboxylate
SR	Systemic resistance
P	Phosphate
C	Carbon
MN	Mycorrhizal network
PS	Phosphate solubilizing
NM	Non-mycorrhizal
PGPs	Plant Growth Promoters
HM	Heavy metals
Hg	Mercury
Pb	Lead
Cd	Cadmium

Cr	Chromimium
Co	Cobalt
Zn	Zinc
Cu	Copper
Ni	Nickel
Ar	Arsenic

1 Introduction

Global climate has been changed mainly during the twenty-first century. This change will lead to great loss of food production and its availability for the future (Battisti and Naylor 2009). The world population increase is being estimated up to 7–8.9 billion in the coming years till 2050 (Singh et al. 2011). Increase in population and sudden changes in climate reduce the soil ability to sustain the agricultural production. As it is obvious now that the traditional agricultural practices are unable to support the tremendous global population increase (Masciarelli et al. 2014). So, the alarming scenario of food scarcity making a tough challenge for crop growers along with policy makers to make it possible to grow food and feed the huge continuously increasing world population. The conditions are becoming worst especially in underdeveloped countries having larger share in huge population growth but lower in food prodction. These countries are facing the multitude problems linking with overpopulation growth like shortage of food, shelter, agricultural lands, and resources. The food production is being shortened due to shrinkage of agricultural lands (cities expansion) although the new crop varieties and modern cultivating strategies somehow try to sustain the crop production (Roychowdhury 2011, 2014; Roychowdhury et al. 2011a, b, c, d, 2012a, b, c, d, 2013a, b, c, 2014, 2018, 2019; Roychowdhury and Tah 2011a, b, c, d, e, f, g, h, 2013; Mamgain et al. 2013; Hasanuzzaman et al. 2013; Chakraborty et al. 2014; Anumalla et al. 2015; Anumalla et al. 2016; Hasanuzzaman et al. 2015).

In addition, there are other multifaceted factors which influence directly or indirectly the crop production through altering the soil texture responsible for the soil health and fertility (a base for optimum crop growth and yield). For simplicity, these factors could be categorized as abiotic and biotic stresses which are responsible for suboptimal plant growth and yield. Throughout the world, these biotic and abiotic stresses are being involved in the reduction of agricultural production up to 30% and 50%, respectively. These stresses could be arisen either through natural (sudden/successive environmental changes and development of amiable conditions for plant pathogens growth, an indirect factor) or human anthropogenic activities. Biotic stress includes the factors involved in damages of plants/crops instigated by living organism like viruses, parasitic bacteria, and pathogenic fungi; ecto- and endoparasites; insects; weeds; and other indigenous, incursive, or cultivated plants (Newton et al. 2011). These biotic factors are although involved in reduction of crop production and present in the vicinity (surroundings) of the crops/plants but could not considered as environment factors. The adverse environment factors or major abiotic stresses include temperature, drought, flooding, salinity, and heavy metal stress. These stresses are intercepted by plants as an evident by altering the morpho-physiology of plants, their biochemistry, and even gene expression patterns. The major impact of these stresses is on basic needs of plants like nutrient availability and their uptake, the soil fertility and health, soil microbial diversity, etc. (Chodak et al. 2015).

Multiple solutions are being adopted regarding the survival of plants in diverse soil conditions, in continuously changing environment, and to adapt a sustainable agriculture (SA, only viable future agriculture), mainly the utilization of plant-associated microbial communities like plant root-associated fungi (mycorrhizal fungi, MF) and bacteria (plant growth-promoting bacteria, PGPB), under different kinds of abiotic stresses. Based on the plant-microbe relations, microbes are classified as favorable, lethal, and neutral microbes.

Nowadays, the multiple biotic and abiotic stresses are being tackled by using plant growth-promoting microbes (PGPM) for assisting plants to overcome the adverse effects of water stress (drought and flooding), salt stress, and heavy metal stress (HMS). So, for the scientists and crop growers, the development of the strategies, organisms (PGMP), and sustainable methods (biofertilizer, BF) is becoming a hot topic and challenging task for maintaining the plant fitness and health to fulfill the increasing demands of a rapidly growing world population (Vimal et al. 2017).

2 PGPM-Assisted Stress Tolerance in Plants

The growth and development of plants and their survival under adverse condition are now being enforced with the help of stress tolerant PGPM including arbuscular mycorrhizal (AM fungi) /vesicular-arbuscular mycorrhizal fungi (VAM fungi) and PGPR (Nadeem et al. 2014). The whole process involved the utilization of different biochemical and molecular approaches for isolation of microbes having the capability to sustain the extreme environment, their identification, and characterization of their potential to enhance the growth and development of plants under various stresses. The inoculation of PGPM was involved in the regulation of hormonal and nutritional balance of plants through the production of plant growth regulator in the rhizosphere and hence induction of resistance to phytopathogens (Spence and Bais 2015). Some microbes can produce the various metabolites which act as toxic to the surrounding phytopathogens and hindered their growth. For example, siderophores (an iron-chelating compound synthesized by microbes in the soil) reduce the iron availability to rhizospheric plant pathogens which resulted in the slowdown in growth rate of the pathogen and ultimately an increase in plant survival rate (Złoch et al. 2016). In addition, the associations of plants and microbes (PMA) also facilitate the growth of plants by fixing atmospheric nitrogen, solubilizing the soil phosphate, and producing the plant-related hormones (Ahmad et al. 2011). The recent research on this fascination approach also shed light on some other phenomenons like nutrient mobilization by microbes including the production of exopolysaccharide (EPS), rhizobitoxine (RBT), etc. (Vardharajula et al. 2011) that are very helpful to mitigate the effects of stressful environment on plants. For example, RBT under stressed condition can inhibit the production of ethylene, hence enhancing the plant growth and development (Kumar et al. 2009).

In addition, these kinds of microbes may have the ability to produce the key enzymes such as chitinase, glucanase, and ACC deaminase (Farooq et al. 2009) which could be involved in the growth and development of plants. Most of the microbes of this kind play a multitude role during PMA like *Bacillus* and *Paenibacillus* which improve the growth and health of plants in three different ways: (i) promotes the growth by improved nutrition accusation of host plant; (ii) acts as antagonist against pathogens by stimulating the defense response mechanisms of plants; and (iii) promotes sustainable agriculture practices (Govindasamy et al. 2010). Furthermore, these PGPM are found to be potential substitutes (BF and biopesticides, BP) for inorganic fertilizers and pesticides as well as a means of saving of almost 20–25% cost of crop production. So, these types of agricultural practices can be economical for crop producers, a step toward organic foods and vegetables production in SA. Therefore, the exploitation of plant-microbe interaction (PMI) for better growth and yields of plants could be a hot research topic in the future and could be a foundation stone of paradigm shift toward sustainable agriculture to ensure safe and healthy food.

3 Drought Stress

Drought stress is considered as an unavailability of the sufficient water quantity for the optimal growth and development of plants. This condition could probably be arisen in near future, a bitter conclusion of the analysis of past and current scenario of water availability for plants and agricultural utilization. The worldwide environmental changes are probably going to expand dry spells which could modify global examples of natural issues, i.e., vegetation generation and decay (Feyen and Dankers 2009). Dry spells (drought) mediated obstruction of plants/crops can be characterized as their capacity to overcome the threatening water deficiency. Dry spell threat has remarkable complications for subterranean carbon (C) and other supplement cycles. It might have influence on soil forms and texture through changes in C distribution to roots and foliage just as C turnover in the rhizosphere. Roots and shoots are interconnected to manage the good balance during distribution of assets and biomass production (Farrar and Jones 2000). This balance may be altered under changing ecological conditions. Water deficiency due to dry spells forced plants to make internal adjustments or synchronize its metabolic processes to support their development or survival (Malinowski and Belesky 2000). These adjustments include (i) dry season evasion, which might be because of improved water take-up by a broader root framework, decreased transpiration rate due to stomatal closer

(Peñuelas et al. 2004), or water stocking in plant tissues and (ii) dry spell tolerance and recovery from dry spell which contain accumulation and translocation of various metabolites, osmotic adjustments, or partitioning in cells (Malinowski and Belesky 2000).

The size and rate of environmental changes basically forced the plants to the extent at which plants under natural conditions can resist/tolerate the environmental change and their capacity to synchronize their internal processes. The static lifestyle of plants might be the most powerful force for the adaptive nature of plants to environmental/ecological changes (Jump and Peñuelas 2005). Ecological conditions, for example, temperature, light, and water availability, are the key factors in characterizing the distribution of plants under a specific environment. In addition, the responses of plants to a changed environment may rely on their ability to form symbiotic associations with microorganisms.

Moreover, the prerequisite for sustainable agriculture development is to explore the basis of plants and microorganisms interactions (e.g., arbuscular mycorrhizal (AM) fungi and phosphate-solubilizing (PS) bacteria), demonstrating direct impact on growth and drought stress tolerance of plants (Gryndler et al. 2000; Marulanda et al. 2009). The effects of plant growth-promoting microbes (PGPM) on plants are summarized in Table 20.1.

3.1 Impressions of PGPR on Plants Under Drought Stress

Water-limited condition (drought) for a prolonged period during the growth of the plants is considered as environmental stress which is a focus of environmental and agricultural scientists which implicit alarming conditions (especially economy in addition to food) in human society (Disante et al. 2011; Mishra and Singh 2010). The water stress effects on plants are evident of plant cell size reduction, unstable membrane integrity, reactive oxygen species (ROS) production, and increased leaf senescence collectively which resulted in reduction of crop production (Tiwari et al. 2016). It also leads to influence the plant growth at physio-molecular level by enhanced ethylene biosynthesis, deterioration of chlorophyll content, and photosynthesis apparatus and inhibition of photosynthesis (Lata and Prasad 2011), as well as impaired membrane integrity, protein unstable conformation, lipid peroxidation (LPO), and ultimately cell death (Tiwari et al. 2016).

Keeping in view the drastic effects of drought, the search for water stress tolerant microbes (WSTM) could be a suitable and effective strategy to minimize its effects on plants under water limiting scenario. The WSTM are found to be beneficial for improving plant growth and development under drought because of their evolved or adapted tolerant mechanisms under low water potential such as thickening of cell walls or induction of a dormant stage, accumulation of various osmotica, and EPS production. In addition to ensuring their own survival, WSTM improve the nutrient acquisition through molting the microenvironmental condition near the plant's root zone for the continuous growth of plants. There are multiple mechanisms reported

Table 20.1 Effects of plant growth-promoting microbes (PGPM) on plants

Plant name	Micro organisms	Effects	References
Cucumber	*Pseudomonas putida*	Stimulate seed germination	Amer and Utkhede (2000)
		Increased shoot length	
		Enhanced leaf chlorophyll content	
		Higher total dry matter	
		Higher grain yield	
		Higher N content	
Cotton	*Pseudomonas putida*	Stimulate seed germination	Anjum et al. (2007)
		Increased shoot length	
		Enhanced leaf chlorophyll content	
		Higher total dry matter	
		Higher grain yield	
		Higher N content	
Orchid	*Rhizobium*, *Microbacterium*, and	Most active IAA producers	Tsavkelova et al. (2007)
	Mycobacterium		
Plants	*Bacillus subtilis*,	Increase the production of IAA	Idris et al. (2004)
	Bacillus amyloliquefaciens		
Rice seed priming	*Pseudomonas fluorescens* and *Pseudomonas aeruginosa*	Increase the IAA production	Karnwal (2009)
Rice, wheat, and maize	PGPR and PHPR	Production of auxins, cytokinins, and abscisic acids	Hayat et al. (2008)
		Increase root growth	
		Development and yield of various agricultural crops	
All crops plants	*Pseudomonas* strains	Production of IAA	Khakipour et al. (2008)
Sweet potato	*Enterobacter*, *Pseudomonas*, *Stenotrophomonas*	IAA production	Khan and Doty (2009)
Canola and mung bean	*Pseudomonas putida*	Stimulate the root growth and development	Glick et al. (2007)
Rice	*Azospirillum*	Root and shoot growth	Pedraza et al. (2009)
		Nutrient uptake	
		Grain yield	
Corn	*Azospirillum*	Root and shoot growth	Cassan et al. (2009)
		Nutrient uptake	
		Grain yield	
Canola	*Pseudomonas*	Increased the length of seedling roots	Caron et al. (1995)

(continued)

Table 20.1 (continued)

Plant name	Micro organisms	Effects	References
Soybean	*Pseudomonas*	Seed emergence	Zaidi (2003)
		Shoot, root	
		Length and dry weight	
		Numbers of nodules	

for WSTM responses under water stress. The potential mechanisms include synthesis of (i) phytohormones like auxin (IAA), cytokinin, and abscisic acid (ABA), (ii) bacterial EPS, (iii) ACC deaminase, and (iv) systemic stress tolerance induction (Porcel et al. 2014; Farooq et al. 2009).

Moreover, PGPR can synthesize plants like hormones that promote plant growth and development under unfavorable circumstances. Jiang et al. (2013) reported that the seed inoculation with these microbes resulted in the tremendously increased ABA production. Furthermore, *Azospirillum brasilense* was also found to mitigate the adverse effects of drought in *Arabidopsis thaliana* through increased ABA biosynthesis (Cohen et al. 2015).

Additionally, 1-aminocyclopropane-1- carboxylate (ACC) is a precursor of ethylene biosynthetic pathway which acts as a signaling hormone under stress conditions activating the whole stress responding cascades. Bacterial ACC deaminase in the soil hydrolyzes ACC into ammonia and alpha-ketobutyrate (Bal et al. 2013) and, hence, can hinder/inhibit the ethylene biosynthesis pathway and downstream cascade of events. The inoculation with *Pseudomonas putida*, *Proteus penneri*, *Pseudomonas aeruginosa*, and *Alcaligenes faecalis* was found to reduce the adverse effects of drought stress as evidenced by increase in chlorophyll contents, higher shoot length, and enhanced biomass in soybean (Kang et al. 2014) and maize plants (Naseem and Bano 2014). The studies on the effects of PGPR on plants under drought stress are arranged in Table 20.2.

3.2 Impressions of Fungi on Plant Under Drought Stress

Fungi have the capability to establish a symbiotic association with the roots of plants called the mycorrhizal ("myco," fungus, "rhizal," "associated with root") fungi. Most of the terrestrial plants have this fascinating association (various kinds) with different types of fungi (although host and fungal specificity is considered an obligation). During this decade, an interesting fact was revealed about fungal-plant symbiosis that the fungi are involved in long-distance communication among plants of same species and/or different species via mycorrhizal network (MN) of the fungal hyphae. It opened the new horizons of exploration and invites critical thinking in this specific area. The literature related to fungal inoculation which aid to minimize the adverse effects of drought is summarized in Table 20.3.

Table 20.2 Influences of bacterial inoculation on plants under water stress

Stress	Bacterial inoculate	Plant species	Effects	Reference
Nutrient deficiency	*Bacillus polymyxa*,	Maize (*Z. mays*)	Increase nutrient uptake ability like K, N and P	Egamberdiyeva (2007)
	Mycobacterium phlei,			
	Pseudomonas alcaligenes			
Drought	Osmotolerant bacteria (not completely characterized)	Rice (*Oryza sativa*)	Enhance the root and shoot biomass	Yuwono et al. (2005)
Drought	*Azospirillum* and	Wheat (*T. aestivum*) and maize	Mitigate the water stress by enhance the production of GBA	Creus et al. (2004, 2005)
	Azospirillum lipoferum			
Drought	*Azospirillum brasilense*	Maize (*Z. mays*)	Increase in proline contents of root and leaves	Casanovas et al. (2002)
			Increase in root growth	
			Increase in root fresh dry weight	
Drought	*Azospirillum brasilense*	Common bean (*P. vulgaris*)	Increase in root area and growth	German et al. (2000)
Drying soil	*Variovorax paradoxus*	Pea (*Pisum sativum*)	Help to alleviate the abiotic stress	Belimov et al. (2009)
Drying soil	*Bacillus*	Lettuce (*L. sativa*)	Increase the level of auxins, cytokinins, gibberellins, abscisic acid (ABA), and ethylene	Arkhipova et al. (2007)
Osmotic stress (45% PEG)	*Arthrobacter* sp.,	Pepper (*C. annuum*)	Help to mitigate the abiotic stress	Sziderics et al. (2007)
	Bacillus sp.			
Osmotic stress (20% PEG) in the dark	*Azospirillum*	Wheat (*T. aestivum*)		Pereyra et al. (2006)
			Enhanced phosphatidylcholine contents	
			Decrease in phosphatidylethanolamine content	
Osmotic stress (20% PEG)	*Azospirillum brasilense*	Wheat (*T. aestivum*)	Enhance in phosphatidylcholine	Creus et al. (1998)
			Decrease of phosphatidylethanolamine	
			Increase the phospholipid content level	

(continued)

Table 20.2 (continued)

Stress	Bacterial inoculate	Plant species	Effects	Reference
Drought	*Azospirillum lipoferum*	*Zea mays* (maize)	Enhance the total soluble sugar, proline, and free amino acid	Bano et al. (2013)
			Improve the root length, shoot fresh weight, shoot dry weight, root fresh weight, and root dry	
Drought	*Bacillus* spp.	*Zea mays* (maize)	Improved the content of proline, sugars, free amino acids	Vardharajula et al. (2011)
			Decrease electrolyte leakage	
			It also decreases the activity of antioxidants enzyme (catalase, glutathione peroxidase)	
Drought	*Pseudomonas putida H*-2–3	Soybean	Decrease the level of abscisic acid, salicylic acid	Kang et al. (2014)
			Increase the level of jasmonic acid content	
			Activation of antioxidant by decreasing the level of flavonoids, superoxide dismutase, and radical scavenging activity	
Drought	*Bacillus amyloliquefaciens* 5113	*Triticum aestivum* (Wheat)	Bacterial-mediated plant attenuated transcript level	Kasim et al. (2013)
	Azospirillum brasilense NO40		Maintains homeostasis	
Drought	*Rhizobium leguminosarum* (LR-30),	*Triticum aestivum* (Wheat)	Catalase, exopolysaccharides, and IAA produced by the rhizobia	Hussain et al. (2014)
	Mesorhizobium ciceri (CR-30 and CR-39), and		Improved the growth and biomass	
			Drought tolerance index	
	Rhizobium phaseoli (MR-2			
Drought	*Bacillus thuringiensis*	*Lavandula dentata*	IAA induced higher proline and K-content	Armada et al. (2014)
			Improved nutritional, physiological, and metabolic activities	
			Decreased glutathione reductase (GR) and ascorbate peroxidase (APX) activity	

Stress	Bacterial inoculate	Plant species	Effects	Reference
Drought	*Pseudomonas putida* MTCC5279 (RA)	*Cicer arietinum* L.	Osmolyte accumulation	Tiwari et al. (2016)
			ROS scavenging ability	
			Stress-responsive gene expressions	
Drought	*Azospirillum* sp.	*Lettuce*	Promote aerial biomass, chlorophyll, and ascorbic acid content	Fasciglione et al. (2015)
			Better overall visual quality, hue, chroma, and antioxidant capacity	
			Lower browning intensity	
Drought	*Azospirillum brasilense* sp. 245	*Arabidopsis*	Improved plants seed yield and plants' survival	Cohen et al. (2015)
			Proline levels and relative leaf water content	
			It also decreased stomatal conductance malondialdehyde and relative soil water content	
Drought	*Phyllobacterium brassicacearum* strain	*Arabidopsis*	Enhanced ABA content resulted in decreased leaf transpiration	Bresson et al. (2013)
	STM196		Delay in reproductive development	
			Increased biomass and water use efficiency	
Drought	*Pseudomonas libanensis* TR1 and	*Brassica oxyrrhina*	Increased plant growth	Ma et al. (2016a, b)
	Pseudomonas reactans Ph3R3		Leaf relative water and pigment content	
			Decreased concentrations of proline and malondialdehyde in leaves	
Drought	*Sinorhizobium medicae*	*Medicago truncatula*	Root nodulation	Staudinger et al. (2016)
			Nutrient acquisition of nutrient during drought stress	
Drought	*R. tropici* and	Bean	Mitigate some of the negative effects of drought stress	Figueiredo et al. (2008)
	Pseudomonas polymyxa			
Drought	*Azospirillum lipoferum*	Wheat and maize	Alleviation of drought stress	Creus et al. (2004)

(continued)

Table 20.2 (continued)

Stress	Bacterial inoculate	Plant species	Effects	Reference
Drought	*Bacillus* sp.	Lettuce	Increased root, shoot growth of lettuce	Vivas et al. (2003)
			N, P, and K uptake (up to 50%) in lettuce under drought stress conditions	
Drought	*Azospirillum*	Wheat	Stimulated the production of lengthy root hairs	Creus et al. (2004)
			lateral roots	
			improved the root diameter and area	
Drought	*Bacillus subtilis*,	Chickpea (*Cicer arietinum*)	Promote plant growth under nutrient and water deficiency	Khan et al. (2018a, b)
	Bacillus thuringiensis and		Enhance the mechanisms against pathogen attack	
	Bacillus megaterium			
Drought	*Bacillus subtilis* strain (UD1022)	Agricultural crops	Increase water uptake ability	Zhang et al. (2018a, b)
Drought	PGPR	Plants	Amelioration of drought stress	Khan et al. (2018a, b)
			Increase in plant growth	
Drought	*Pseudomonas chlororaphis*	*Lycopersicum esculentum (tomato)*	Enhance the production of terpenes	Brilli et al. (2019)
			Increase the production of IAA and ABA	
			Improve the antioxidant contents	
			Enhance proline and water use efficiency	

Table 20.3 Influences of fungal inoculation on plants under drought stress

Stress	Micro organisms	Plant	Effects	Reference
Drought	*Funneliformis mosseae*, *Sordariomycetes*, *Chytridiomycota*, and *Ascomycota*	Citrus	Enhance the ability to mitigate the drought stress	He et al. (2019)
			Increase the root growth	
Drought	*Penicillium resedanum (LK6)*	*Capsicum annuum*	Increase in plant biomass	Khan et al. (2013a, b)
Drought	*Grosmannia clavigera*	*Pinus banksiana*	Enhance the growth	Wood et al. (2016)
Drought	Arbuscular mycorrhizal (AM) fungi	Poa	Improve the root morphology	Moradi and Salimi (2013)
			Increase the yield	
Drought	*Vesicular-arbuscular mycorrhiza* (VAM)	Agricultural crops	Increase the plant biomass	Song (2005)
			Increase root area	
			Increase the P absorbance	
			Mitigate the drought stress	
Drought	Arbuscular mycorrhizal (AM) fungus	Wheat	Mitigate the stress	Al-Karaki and Al-Raddad (1997)
			Increase in plant growth	
			Improve the nutrient uptake ability	
Flooding	AM fungi and *Bradyrhizobium*	Dragonblood (*Pterocarpus officinalis*)	Development of adv. roots, aerenchyma	Fougnies et al. (2007)
			Hypertrophied lenticels	

3.2.1 Impression of AM Fungi on Plant Under Drought Stress

Among these fungi, arbuscular mycorrhizal (AM) fungi are plant root associates (>80% plant families, Smith and Read 2010) that often assist the plants in the key process of nutrient accusation, i.e., as phosphorus and other soil nutrients in addition to water, by significantly increasing the plant's capacity to uptake nutrient (Entry et al. 2002) probably through enhanced absorption network. In global semi-arid regions, lower water availability along with infertile soils having lowest phosphorus contents was collectively involved in decreased crop production. Generally, wheat (*Triticum durum* L.) is being grown under rainfed conditions in these regions, where drought condition could be established at any stage of their growth.

The researchers established the fact that indigenous AM fungi could assist the plants in nutrition accusation; thus, growth and development of plants could be induced with the help of these symbiotic microorganisms under abovementioned conditions (Kupulnik and Kushnir 1991; Ellis et al. 1985). The mycorrhizal-assisted improvement of nutrient uptake like P, Zn, Fe, Cu, and Mn and the growth of plants under lower soil P and deficiency of adequate water supply are well-documented (Tobar et al. 1994; Manjunath and Habte 1991; Michelsen and Rosendahl 1990; Raju et al. 1990; Nelsen 1987).

In addition, mycorrhizae development with plant could also induce the systemic resistance (SR) to drought of the host in addition to better growth and nutrient status, a well-documented fact (Ruiz-Lozano et al. 1995; Davies et al. 1992; Hardie and Leyton 1981). This induced SR may be accompanied by enhanced P accusation from P starving soils (Nelsen 1987; Huang et al. 1985). However, the P independent SR to drought of AM fungi-hosting plants is also available in the literature (Bethlenfalvay et al. 1988; Auge et al. 1986). The responses of plants to AM fungal inoculation showed remarkable variations depending upon difference among wheat cultivars (Azcon and Ocampo 1981) and genotypes under drought stress (Kupulnik and Kushnir 1991). The capabilities of fungi and plants to undergo mycorrhizal association showed specificity between symbionts representing a heritable trait (Mercy et al. 1990; Lackie et al. 1988).

Previous studies suggested that AM fungi enforced the host plant to tolerate the effects of drought conditions by adopting various mechanisms like (i) increase in water absorption ability through extension of roots via mycorrhizal extraradical hyphae (Ruth et al. 2011; Zhang et al. 2018a), (ii) balanced osmotic adjustment (Wu et al. 2017; Zhang et al. 2018b), (iii) enhanced production of antioxidants (Huang et al. 2017), (iv) amelioration of soil structure through glomalin production (Chi et al. 2018), and (v) increased expression of aquaporin coding/regulatory gene (He et al. 2019). The exploration of molecular and genetic basis of this oldest relationship could be the most fascinating research area in coming years. The recent trends in research are the utilization AM fungi with soil amendment like biochar, etc. to improve the growth and development of plants/crops under adverse environment conditions

3.2.2 Impression of VAM Fungi on Plant Under Drought Stress

VAM fungi are considered as the most common universal partners in mycorrhizal establishment. There is large number of published data available demonstrating that VAM fungi have the ability of modulation in water relations of the host plant for better growth under drought stress. Augé reviewed the literature very systematically on water relations of plant under drought during VAM symbiosis (Augé 2001). This part of the chapter may shed lights on the insights of VAM fungal-plant relationship under current scenario.

Impression of VAM Fungi on Morphology of Plants Under Drought Stress

The impression of VAM fungi on morphology of host plants under drought conditions (widely studied during the last century) can be studied through comparison of VAM-associated plants with non-mycorrhizal (NM) plants. For instance, VAM-associated *Acacia* and rose showed more leaf abscission formation under drought conditions as compared with NM plants (Henderson and Davies 1990; Osonubi et al. 1991). Similarly, in wheat plants, VAM association also reduced the leaf senescence and necrosis in comparison with NM wheat. Furthermore, in maize, VAM association enhanced the green leaf area compared with NM maize under drought stress. In addiion, VAM association in *Alfalfa* resulted in delayed leaf senescence under water deficit conditions and VAM symbiosis led to reduce the pod abortion extent in soybeans under drought conditions (Augé 2001). Moreover, VAM-associated plants have higher tendency to recover quickly from wilting (an effect of drought) following the withdrawal of drought conditions (Gemma et al. 1997). Generally, VAM symbiosis did not show any effect on plant stomatal density and guard cell size in leaves although the difference in transpiration or stomatal conductance rate was observed between VAM and NM plants (Henderson and Davies 1990).

Impression of VAM Fungi on Metabolism of Plants Under Drought Stress

In this part, the various plant metabolic processes were highlighted showing influences of VAM fungal association on the host plant.

VAM Association Mediated Water Relations of Plants Under Drought Stress

The plant water contents are very important for carrying out normal physiological processes. Under normal conditions (without stress), the VAM fungal association with the host plants showed the unchanged water contents (Bryla and Duniway 1997; Goicoechea et al. 1997). But the drought condition induced the lower water potential of leaves in a plant which is prorogued with VAM fungal interaction with host plant (El-Tohamy et al. 1999; Dixon et al. 1994). Moreover, the VAM fungal association also enhanced the ability of host plants to recover faster from the drought after withdrawal of stressed conditions (Subramanian and Charest 1997). The transpiration rate linked to stomatal conductance was also influenced by VAM fungal partnerships and reported an increase of about three times under VAM fungal partnership in various plants species like wheat, cowpea, soybeans, rose, safflower, lettuce, etc. (Augé et al. 2001) In contrast, there are also reports present which indicated that VAM fungi have not affected the stomatal conductance of the host plant, for example, citrus taxa.

VAM Association Mediated Photosynthesis of Plants Under Drought Stress

The photosynthesis process in plants is an essential metabolic process involved directly in their survival. This ubiquitous process is linked with the presence of photosynthetic units (chlorophyll contents) and stomatal conductance in leaves of the plants in addition to the water status. The VAM fungal partnership also showed positive effects on this process too, either through enhancement in chlorophyll contents of host plant or improved stomatal conductance under water-scarce conditions and/or higher photosynthetic storage rate and export to larger sink (VAM fungal hyphae act as an additional sink for photosynthates) in comparison with NM plants (Augé et al. 2001; Gemma et al. 1997; Mathur and Vyas 1995; Davies et al. 1993). Moreover, these all benefits of VAM partnership usually rely on fungal species types under drought stress. For instance, one *Glomus* specie is found to be involved in enhancing P-use photosynthetic ability of the host plants, while other *Glomus* species affected otherwise (decreased P-use photosynthesis) in comparison with NM plants (Ruiz-Lozano et al. 1995). So, it could be concluded that mycobionts showed variable influences on a plant's ubiquitous photosynthesis process.

VAM Association Mediated Nutrient Absorption in Plants Capacity Under Drought Stress

Nutrient absorption in plants has become challenging under adverse environmental conditions. Thanks to VAM fungi for enhancing the ability of the host plants to absorb much more nutrients for sustaining their growth and development under stressed conditions. VAM fungi was reported to increase the nutrient acquisition ability of host plants like in agriculture crops (sugarcane, mung bean, wheat, tomato, etc.) as well as trees (apple, orange, wild jujube, etc.) (Wu and Xia 2004). It is because of the small diameter of fungal hyphae enabling them to extract nutrients from minute soil pores which are inaccessible to the root hairs. For instance, the colonization with *Glomus* spp. or *G. caledonium* or *G. mosseae* enhanced the biomass production up to 1.5, 1.80, and 1.99 times of NM mung beans even at 12% soil water content (He et al. 1999). Similarly, drought condition reduced the root and shoot biomass of maize plants around 31% and 12% in VAM fungal partnership in comparison with NM plants around 55% and 23%, respectively (Subramanian and Charest 1997). It indicated that VAM fungal association with host plants influenced positively the growth by improving the nutrient status especially plant's P-status. The P uptake capacity of plants reduced drastically with drought but could be rescued with VAM fungi. This rescue operation is not restricted to P absorption but also extended to the absorption of other elements. For example, the absorption of Cu and Zn was found enhanced in VAM fungal-plant symbiosis in comparison with NM plants, while Mn and Boron absorption in plants was decreased within VAM fungal-plant symbiosis under drought conditions (George et al. 1994; Marschner and Dell 1994; Bethlenfalvay and Franson 1989). Moreover, the absorption of other mineral elements like N, K, Ca, Mg, Fe, Na, and Mo was little modulated with the application of VAM fungi under drought stress (Augé et al. 2001).

VAM Association Mediated Protective Adaptation in Plants Under Drought Stress

Plants' protective strategies against stress condition include osmotic adjustment with accumulation of osmotica like soluble carbohydrates, amino acids, imino acids, proline, enzyme protectants, etc. These measurements of these biochemical parameters were an indication of plants' health status. The variations in the values of these indicators among VAM fungal-plant association and NM plants could suggest the levels of sustainability of plants under drought stress. For instance, the lower sugar content in plants representing their avoiding behavior to drought stress or lower level presence of drought-induced VAM fungal strain. In VAM fungal-associated tobacco plants, the soluble carbohydrate accumulation remained significantly lower under drought in comparison with NM plants (Schellenbaum et al. 1998) indicating lower level of drought-induced strains. Similar behavior was also reported in other plant species, i.e., rose and pepper under drought stress (Auge et al. 1992; Augé et al. 2001; Davies et al. 1993). Moreover, the drought-induced accumulation of amino acid and imino acid was also reported under VAM fungal-plant partnership (Schellenbaum et al. 1998; Subramanian and Charest 1995) contrary to the other reports (Subramanian and Charest 1995; Auge et al. 1992). The accumulation of other osmolytes like proline and free polyamines was also found in consistence with sugar contents among VAM-colonized and NM plants under drought conditions. The activities of various enzymes like superoxide dismutase (lettuces), acid phosphatase (*Alfalfa*), glutamate-ammonia ligase (maize), and glutamine and glutamate reductase (soybean) were found enhanced during the VAM fungal-plant partnership under drought conditions (Ruiz-Lozano et al. 1996; Goicoechea et al. 1997; Ruiz-Lozano 2003). It indicates that that VAM fungal-plant partnership harnessed the plant's capacity to minimize the adverse effects of drought stress.

Nitrate reductase activity in leaves and roots was also increased by VAM symbiosis in numerous studies (Panwar 1993; Azcón et al. 1996; Azcón and Tobar 1998). Where reported, total protein concentrations have been consistently higher in VAM than NM plants during drought, which were considered by the authors to be a beneficial VAM effect (Subramanian and Charest 1995; Ruiz-Lozano et al. 1996; Augé et al. 2001).

4 Salt Stress (Salinity)

The soil salinity is a major threat to agriculture worldwide and continuously increasing in magnitude in crop cultivating regions globally (Huang et al. 2008; Cirillo et al. 2018). This adverse environmental factor is very badly affecting the performance of plants as well as the global economy. The soils having higher salt concentrations (saline soils) comprise of 1/20th of global land and 1/5th of irrigated lands, which account for 1/20th of the global total and 1/5th of irrigated lands, causing around \$27.3 billion loss in agricultural production annually (Qadir et al. 2014). The salinity, generally considered as the sodium (Na+) salts excess in the soil, is becoming more severe with the addition of drought which is lethal to growth and development of plants and an open challenge to agricultural crop production and natural ecosystems' sustainability (Daliakopoulos et al. 2016). Increase in concentration of different cations/anions like Na^+, Cl^-, K^+, and Ca^{2+} in soil from normal values also considered as salinity (Shrivastava and Kumar 2015). Generally, enhancement in these ions is due to low rainfall and higher rate of weathering. The overall responses to salinity are comprised of processes like sensing and signal perception, ion homeostasis maintenance, metabolism alteration, regulation of stomata, photosynthesis, cell division and cell expansions, modifications in architecture, and morphology and phenology of plants by altered resource allocation plane (Munns and Tester 2008; Hairmansis et al. 2014). On the other hand, the growth and functioning of microorganisms is also affected by salinity mainly through changing the osmotic potential. In the following, the role of PGPM in amelioration of plant growth and development is described.

4.1 Impression of PGPR on Plant Under Salt Stress

A large number of literatures have been cited revealing the role of PGPs (plant growth-promoting organisms) and endophyte microbes in increasing the salt tolerance in plants (Barassi et al. 2006). PGPs and endophytic microbes directly affect the plant growth by different mechanisms. PGPM improved the plant growth and production under stressful conditions through different modes. This includes the production of different phytohormones (auxin, gibberellins, cytokinin, etc.), improvement in nitrogen fixation, and improved nutrient uptake and nutrient mobilization as well as the production of siderophores (Hayat et al. 2010). All these mechanisms lead to change in nutrient architecture and enhanced nutrient uptake under salt stress (Egamberdieva and Kucharova 2007). The other major indirect method for the improvement in crop growth is the reduction in pathogenic microorganisms and increase in beneficial root-colonizing bacteria that produce ACC deaminase while lowering ethylene biosynthesis (Damodaran et al. 2013). The other method to improve the plant production is the use of microbial biofilms. For instance, biofilms of PGPB promote the growth and development of plants under salt stress (Kasim et al. 2016). As an example, lettuce seeds inoculated with *Azospirillum* showed better growth and production under salt stress in comparison with non-inoculated seeds (Barassi et al. 2006). In another experiment, chili pepper plants inoculated with *Pseudomonas stutzeri* increase the salt tolerance and improved the plant growth under stressful conditions (Bacilio et al. 2016).

Moreover, the recent trend in PGMP utilization for improvement of plant production is to study the combine effects of PGPR and fungi. For instance, the inoculation of bacteria in combination with fungi also found very helpful to improving the plant growth as in maize co-inoculation of *R. intraradices* and *Massilia* sp.

RK4 in combination with AM fungi improved growth and salt stress tolerance (Krishnamoorthy et al. 2016). The influences of bacterial and fungal inoculations on plants under salt stress present in literature are tabulated in Tables 20.4 and 20.5.

4.1.1 Mechanism of PGPM-Assisted Salinity Stress Tolerance

It has been studied that the inoculation of PGPB inhibited the ethylene production and improved the root nodulation under salt stress. PGPB improve the plant growth by the accumulation of different osmolytes in cytoplasm that maintains the plant cellular osmotic potential (Vardharajula et al. 2011). Two rhizospheric bacteria *Bacillus pumilus* and *Bacillus subtilis* isolated from saline soil showed induced IAA production under salt stress when inoculated (Damodaran et al. 2013). In another study, it has been found that the inoculation with *Rhizobium* and *Pseudomonas* improved the salt tolerance in maize (Bano and Fatima 2009). This improvement is due to the decrease in LPO, accumulation of proline, as well as the selective improvement in nutrient uptake. In rice, inoculation with *P. pseudoalcaligenes* and *Bacillus pumilus* enhances salinity tolerance also with the improvement in GB production (Jha et al. 2011). *Acinetobacter* spp. and *Pseudomonas* sp. induce the IAA production in barley and oat and improve the plant growth (Chang et al. 2014). Inoculation with *Pseudomonas pseudoalcaligenes* and *Bacillus pumilus* improves the antioxidant activity and decreases the lipid peroxidation in rice under salt stress. Exogenous use of PGPR improves the plant by inducing the root nodulation and decreasing the disease susceptibility in cotton (Egamberdieva et al. 2015). In wheat, inoculation with *Azospirillum* strains improves the plant fresh and dry biomasses under salt stress (Nia et al. 2012). *Azospirillum* inoculation improves the fruit quality and fruit storage in lettuce (Fasciglione et al. 2015). *P. fluorescens*, *P. aeruginosa*, and *P. stutzeri* induce the production of phytohormones and induce the salt tolerance in tomato (Bal et al. 2013; Tank and Saraf 2010).

5 Heavy Metal Stresses

The plants under natural conditions must cope with various stresses. Among all these threatening conditions, heavy metals (HM) like mercury (Hg), lead (Pb), cadmium (Cd), chromium (Cr), cobalt (Co), zinc (Zn), copper (CU), nickel (Ni), arsenic (Ar), etc. are being accumulated in soils over a long period through human activities like industrial waste, application of chemical fertilizers, smelting, and untreated domestic sewage disposal (Aydinalp and Marinova 2009). The human's anthropogenic activities are responsible for accumulation of these metals in soil surface at rhizosphere zone and leaching down to the groundwater (Gupta and Ali 2012; Dağhan and Öztürk 2015; Hakeem et al. 2015; Öztürk et al. 2015a, b; Basheer 2018a). All the HM are considered as nonbiodegradable i.e. could not be removed from environment via natural means. While, some HM are reported as immobile i.e.

Table 20.4 Influences of bacterial inoculation on plants under salt stress

Bacterial inoculate	Plant species	Effects	Reference
Azospirillum brasilense	Pea (*Phaseolus vulgaris*)	Increase exudation of plant flavonoids compared to *Rhizobium* alone, implying an inoculation of flavonoid genes in the presence of *Azospirillum*	Dardanelli et al. (2008)
		Also alleviate the salt stress	
Pseudomonas syringae	Maize (*Zea mays*)	ACC deaminase activity	Nadeem et al. (2007)
Pseudomonas fluorescens		Increase the uptake of P and K	
Enterobacter aerogenes			
Pseudomonas fluorescens	Groundnut (*Arachis hypogaea*)	Reduce the entire ethylene level	Saravanakumar and Samiyappan (2007)
		Reduce the stress activity	
Azospirillum	Lettuce (*Lactuca sativa*)	Increase in germination and growth rate	Barassi et al. (2006)
Achromobacter piechaudii	Tomato (*Lycopersicon esculentum*)	Reduce the entire ethylene level	Mayak et al. (2004)
		Reduce the stress activity	
Aeromonas hydrophila/*caviae*	Wheat (*Triticum aestivum*)	Reduce the Na^+ uptake activity	Ashraf et al. (2004)
Bacillus insolitus			
Bacillus sp.			
Azospirillum	Maize (*Z. mays*)	Enhance the ability of Na^+, K^+, and Ca^{2+}	Hamdia et al. (2004)
Azospirillum brasilense	Chickpeas (*Cicer arietinum*), faba beans (*Vicia faba* L.)	Enhance in nodulation	Hamaoui et al. (2001)
		Increase in root and shoot growth	
		Crop yield	
Cyanobacteria	Rice	Alleviated the adverse effect of salinity on the plant growth and development	Rodríguez et al. (2006)
Pseudomonas aurantiaca and *Pseudomonas extremorientalis*	Medicinal plant *Silybum marianum*	Alleviate salt stress	Egamberdieva et al. (2013)
Pseudomonas strains	Melon	Root and shoot stimulation under saline soil conditions	Egamberdieva et al. (2013)
Azospirillum brasilense	Wheat	Alleviation of salt stress	Nabti et al. (2010)

(continued)

Table 20.4 (continued)

Bacterial inoculate	Plant species	Effects	Reference
Pseudomonas strains	Goat's rue plants	Promoted enlargement of root system	Egamberdieva (2012)
		Enhancing nutrient uptake, nodulation, and growth of goat's rue plants grown either in salt	
Pseudomonas sp.	Basil	Alleviated the effects of salinity on plant growth	Golpayegani and Tilebeni (2011)
Bacillus lentus			
Pseudomonas putida	Wheat	Stimulate seed germination	Egamberdieva and Kucharova (2009)
		Increased shoot length	
		Leaf chlorophyll content	
		Total dry matter	
		Grain yield	
		N content	
Azotobacter sp.	Wheat	Enhance the production of IAA	Thakuria et al. (2004) and Egamberdieva et al. (2008)
Azospirillum brasilense			
Bacillus circulans			
Bacillus magaterium			
Pseudomonas pickettii			
Bacillus amyloliquefaciens			
Cellulomonas sp.			
Mycoplana bullata			
Mycobacterium phlei			
Pseudomonas agglomerans			
Pseudomonas extremorientalis, *Pseudomonas putida* and *Rahnella aquatilis*			

remained fixed at their accumulating place and others considered as mobile indicating that their ability to be absorbed by roots of plants through diffusion, endocytosis or through specific membrane transporters (Öztürk 1989; Ali and Jain 2004; Ashraf et al. 2010; Ali et al. 2018; Basheer 2018b; Burakova et al. 2018). Moreover, some of the HM are among the trace micronutrients like Zn, Cu, and Ni because of their role as cofactors of various enzymes. On the other hand, some HM like Cd and Pb are constituents of pesticide, and their accumulation beyond certain limits makes them toxic (Gough 1979; Sharma and Ali 2011; Ali et al. 2017). The influences of bacterial or fungal inoculation on plants under heavy metal stress present in the literature are summarized in Tables 20.6 and 20.7. It is a very interesting fact that plants have a bioaccumulating nature which means they can uptake HM from soil and water whether they required or not for their growth and development (Yilmaz et al. 2006; Ali et al. 2016). The bioaccumulation rate and tolerance of plants vary

Table 20.5 Influences of fungal inoculation on plants under salt stress

Stress	Microorganisms	Plant	Effects	Reference
Salt	*Glomus mosseae* or *Paraglomus occultums*	Citrus	Enhance the nutrient uptake ability (K+, Ca2+, and Mg2+)	Wu et al. (2010)
			Reduce the Na^+ uptake	
			Increase both plant performance (leaf number, leaf area, shoot, and root dry weights)	
			Enhance leaf relative water content	
Salt	*Trichoderma harzianum*	Indian mustard	Increase the antioxidant activity	Ahmad et al. (2015)
Salt	*Piriformospora indica*	Barley (*Hordeum vulgare*)	Increase in antioxidant and enzyme activity	Baltruschat et al. (2008)
Salt	Arbuscular mycorrhizal (AM)	*Allium sativum*	Increase the antioxidant activity	Borde et al. (2010)
			Enhance plant growth	
Biotic and abiotic stress	*Colletotrichum* species	Crops	Increase the growth	Rodriguez et al. (2008)
			Mitigate the stress	
Salt	*Glomus intraradices*	Lettuce	Increase in proline content	Jahromi et al. (2008)
			Increase root growth and water relative content	
Abiotic stress	Endophytic fungi	Agricultural crops	Decrease the abiotic stress	Khan et al. (2013a, b)
			Enhance the GA production	
Salinity, drought, and heat stresses	*Penicillium resedanum* LK6	*Capsicum annuum* L.	Increase in chlorophyll content	Khan et al. (2013a, b)
			Plant fresh and dry weight	
			Increase GA content	
Salt	Arbuscular mycorrhiza (AM) fungi	*Jatropha curcas*	Enhance the plant growth	Kumar et al. (2010)
			Increase antioxidant activity	

widely among different plant species. HM stress inhibits or denatured various essential enzymes and other proteins. It also interfered with membrane integrity leading to impaired plant metabolic reactions like photosysnthesis, respiration, and ion homeostasis (Hossain et al. 2012). The exogenous application through seed priming of rice with *Bacillus subtilis* and *Bacillus megaterium* is helpful to mitigate the iron toxicity and reduce the iron concentration in leaves (Asch and Padham 2005; Terré et al. 2007).

Table 20.6 Influences of bacterial inoculation on plants under heavy metal stress

Stress	Bacteria	Plant	Effects	Reference
Iron toxicity	*Bacillus subtilis*	Rice (*O. sativa*)	Mitigate the stress	Asch and Padham (2005) and Terré et al. (2007)
	Bacillus megaterium		Improve plant growth	
	Bacillus sp.			
Arsenic toxicity	*Staphylococcus arlettae*	*Brassica juncea*	Enhance the soil fertility and nutrient ion concentration	Srivastava et al. (2013)
Pb/Zn toxicity	*Phyllobacterium myrsinacearum*	*Sedum plumbizincicola*	Resistance to 350 mg/L Cd	Ma et al. (2013)
			1000 mg/LZn	
			1200 mg/L Pb	
Zn toxicity	*Pseudomonas aeruginosa*	Wheat *(Triticum aestivum)*	Increase the protein content	Islam et al. (2014)
			Plant fresh and dry weight	
			N and P uptake	
Zn toxicity	*Enterobacter intermedius*	*Sinapis alba*	Enhance the production of auxin	Plociniczak et al. (2013)
	MH8b		Improve the P solubility	
Cd, AS, Cu, Pb, and Zn toxicity	*Pseudomonas koreensis*	*Miscanthus sinensis*	Improve the production of IAA	Babu et al. (2015)
	AGB-1			
Zn toxicity	*Pseudomonas brassicacearum*, *Rhizobium leguminosarum*	*Brassica juncea*	Act as chelating compounds	Adediran et al. (2016)
Hg toxicity	*Photobacterium* spp.	*Phragmites australis*	IAA, mercury reductase activity	Mathew et al. (2015)
Copper	Plant growth-promoting rhizobacteria and rhizobium	*Alfalfa*	Increase plant growth	Ju et al. (2019)
			Alleviate the Cu stress	
Heavy metal	*Alcaligenes faecalis*	Sorghum plant (*Sorghum bicolor*, L.)	Enhance the growth	El-Meihy et al. (2019)
	MG257493.1, *Bacillus cereus* MG257494.1 and *Alcaligenes faecalis* MG966440.1		Eliminate the heavy metal stress	

(continued)

Table 20.6 (continued)

Stress	Bacteria	Plant	Effects	Reference
Cd	*Pseudomonas aeruginosa* and	Tomato	Alleviate the Cd stress	Khanna et al. (2019a, b)
	Burkholderia gladioli		Modulate the *CHS*, *PAL*, *CS*, *SDH*, *FH*, and MS	
Cd	*Pseudomonas aeruginosa* and	*L. esculentum*	Mitigate the Cd stress	Khanna et al. (2019a, b)
	Burkholderia gladioli		Decrease toxicity	
			Enhance chlorophyll content	
Cd	*Leptomyrmex varians* and	Mustard	Enhance in germination rate	Pal et al. (2019)
	Pseudomonas putida		Increase in chlorophyll content	
			Increase in root and shoot length	
			Enhance the production of IAA	
Fe and Cu toxicity and deficiency	*Azospirillum brasilense*	Cucumber	Overcome the Fe and Cu deficiency	Marastoni et al. (2019)
			Mitigate the Cu stress	
			Enhance the root growth	
			Improve the nutrient uptake ability	

Table 20.7 Influences of fungal inoculation on plants under heavy metal stress

Stress	Micro organisms	Plant	Effects	Reference
Heavy metal	AM fungi	Agricultural crops	Alleviate the heavy metal stress	Clarholm and Skyllberg (2013)
Pb	*Amanita muscaria*	*Salix dasyclados*	Decrease the conc. of Pb	Hrynkiewicz and Baum (2013)
			Increase in plant weight	

6 Temperatures Stress

Plants due to their sessile lifestyle must withstand and synchronize their metabolisms according to the changing environment for flourishing and survival. Particularly, the temperature is an environmental factor which shows tremendous fluctuation throughout the year in addition to the day and night modulation and enforcing the plants for continuous alteration in their metabolism. Higher temperatures (even lower temperature) than the usual temperature have adverse effects on plant growth and development, thus considered as temperature stress. The extreme temperature change resulted in cell damage and the death of the whole plant. While, the mild temprature change required the osmotic adjustment in plants as their survival strategy. Plant metabolic machinery starts the biosynthesis of stress protectants like sugar alcohols (sorbitol, inositol, etc.) soluble sugars (trehalos, sucrose, etc.), and nitrogenous compounds (proline, Gb (glycine betaines)) having lower weights (Janská et al. 2010). Temperature change is affecting the plant roots and has influence on the enlargement of roots. Increase in light intensity at certain points enhances the development of plants but after increase in optimum light intensity causes serious damage to roots (Erickson 1959).

6.1 Microbes Under Temperature Stress

Microbes are also helping the plants to overcome the harmful effects of temperature stress. For example, Zhang et al. (1997) studied different bacterial influence and their effects on beans under extreme temperature condition and noticed the changes in root anatomy, physiology, and growth of plants. The authors reported that these bacteria played an important role to sustain the morphology and biochemical attributes of the plant and improve the plant root growth and helped to overcome the effects of temperature stress. Similarly, Bensalim et al. (1998) studied the effects of *Burkholderia phytofirmans* PsJN on potatoes grown in different temperatures (20 °C day, 15 °C night; 33 °C day, 25 °C night). Authors found that the bacterial inoculation showed larger number of tuber formation at high temperature, and increase in root and shoot length, fresh and dry weight of the plant through mitigating the adverse effects of temperature stress. Furthermore, in another study, the decrease in root and shoot weight was reported at freezing temperature along with damaged plant membrane, while the bacterial exogenous application of *B. phytofirmans* PsJN on grapevine increases the rate of post-chilling recovery. Moreover, the grapevine plants with decreased antioxidant activity at such cold temperature was rescued with inoculation of *B. phytofirmans* demostrating the enhancement in the carbohydrate, proline, phenolics, starch and chlorophyll contents (Barka et al. 2006).

On the other hand, AM fungi are known to have enforcing ability to develop the tolerance in plant against abiotic stresses and in addition to ameliorate the nutrient status of the host plants (Hajiboland et al. 2018; Porcel et al. 2012). Only very few studies have been found reporting the role of AM fungi against cold stress tolerance in plants, might be due to the lower inbuilt potential of AM fungi to sustain cold temperatures (Liu et al. 2004). Generally, fungi are very sensitive to temperature changes; even the seasonal fluctuations in temperature reduced the fungal growth and rate of colonization of host root (Smith and Read 2010). In various cold-susceptible plant species like maize, rice, and cucumber, the AM fungal association was reported to ameliorate the cold stress tolerance potential of the plants. The mechanism of induced cold tolerance in host plants might be due to their enhanced ability of osmotic adjustments. Because AM fungi enforced the host plant with higher water absorption which is required to maintain the stomatal conductance and essential for photosynthesis resulting in soluble sugar and proline accumulation. Moreover, the leaf chlorophyll contents, rate of photosynthesis, and antioxidant levels in AM fungal symbiotic plant were found to be enhanced under lower temperature stress (Zhu et al. 2010; Liu et al. 2013; Chen et al. 2013). It was also reported that lower temperature stress induced the production of phenolic compounds and their deposition into plant cell walls (Akula and Ravishankar 2011). So, AM fungal colonization of host root enhanced the phenolic metabolism in plants which might be responsible for AM fungal-assisted cold tolerance (Chen et al. 2013). Influences of bacterial inoculation on plants under temperature stress are summarized in Table 20.8.

Table 20.8 Influences of bacterial inoculation on plants under temperature stress

Stress	Type Bacterial inoculate	Plant species	Effects	Reference
Temperature	*Burkholderia phytofirmans*	Grapevine (*Vitis vinifera*)	63% enhance the tuberization	Barka et al. (2006)
			Increase carbohydrate, proline, phenolics, starch, and chlorophyll content	
Temperature	*Bacillus phytofirmans*	Potato (*Solanum tuberosum*)	Increase plant biomass	Bensalim et al. (1998)
			Root shoot length	
			Improve tuber formation	
Temperature	*Aeromonas hydrophila*	Soy bean (*Glycine max*)	Improve the root anatomy	Zhang et al. (1997)
	Serratia liquefaciens			
	Serratia proteamaculans		Mitigate the temp stress	
Heat	*Bacillus amyloliquefaciens*	Wheat	Reduced regeneration of reactive oxygen species	El-Daim et al. (2014)
	Azospirillum brasilense		Preactivation of heat shock transcription factors	
			Changes in metabolome	
Heat and drought	*Curvularia protuberata*	*Dichanthelium lanuginosum* *Solanum lycopersicum*	Colonization of roots	de Zelicourt et al. (2013)
Heat and salt	*Pseudomonas* putida KT2440	Citrus	Increase in root growth1.	Vives-Peris et al. (2018)
	Novosphingobium sp. HR1a			

7 Conclusions

Although, the different abiotic stresses are threatening the agriculture production all over the world and making the situation worst, the research scientists are adopting integrative (basic to applied) approaches to overcome or even minimizing this threatening scenario. Indeed, they all are involved in revealing the physiological, molecular, cellular, and biochemical mechanisms governing the world of plant-microbe interactions under natural as well as under the influence of adverse environmental conditions. The collected knowledge will certainly pave the way forward to develop new strategies and techniques for PGPM-based solution to get rid/minimize the threat of adverse environmental conditions to sustainable agriculture production and to rejuvenate the forest ecosystems for the better tomorrow.

References

Adediran GA, Ngwenya BT, Mosselman JFW, Heal K (2016) Bacteria–zinc co-localization implicates enhanced synthesis of cysteine-ric peptides in zinc detoxification when *Brassica juncea* is inoculated with *Rhizobium leguminosarum*. New Phytol 209:280–293

Ahmad M, Zahir ZA, Asghar HN, Asghar M (2011) Inducing salt tolerance in mung bean through coinoculation with rhizobia and plant-growth-promoting *rhizobacteria* containing 1-aminocyc lopropane-1-carboxylate deaminase. Can J Microbiol 57(7):578–589

Ahmad P, Hashem A, Abd-Allah EF, Alqarawi AA, John R, Egamberdieva D, Gucel S (2015) Role of *Trichoderma harzianum* in mitigating NaCl stress in Indian mustard (*Brassica juncea* L) through antioxidative defense system. Front plant Sci 6:868

Akula R, Ravishankar GA (2011) Influence of abiotic stress signals on secondary metabolites in plants. Plant Signal Behav 6:1720–1731

Ali I, Jain CK (2004) Advances in arsenic speciation techniques. Int J Environ Anal Chem 84(12):947–964

Ali I, Alothman ZA, Al-Warthan A (2016) Sorption, kinetics and thermodynamics studies of atrazine herbicide removal from water using iron nano-composite material. Int J Environ Sci Technol 13(2):733–742

Ali I, Alothman ZA, Alwarthan A (2017) Supra molecular mechanism of the removal of 17-β-estradiol endocrine disturbing pollutant from water on functionalized iron nanoparticles. J Mol Liq 241:123–129

Ali I, Alharbi OM, Alothman ZA, Badjah AY, Alwarthan A (2018) Artificial neural network modelling of amido black dye sorption on iron composite nano material: kinetics and thermodynamics studies. J Mol Liq 250:1–8

Al-Karaki GN, Al-Raddad A (1997) Effects of arbuscular mycorrhizal fungi and drought stress on growth and nutrient uptake of two wheat genotypes differing in drought resistance. Mycorrhiza 7(2):83–88

Amer GA, Utkhede RS (2000) Development of formulations of biological agents for management of root rot of lettuce and cucumber. Can J Microbiol 46(9):809–816

Anjum MA, Sajjad MR, Akhtar N, Qureshi MA, Iqbal A, Mahmud-ul-Hasan JAR (2007) Response of cotton to plant growth promoting rhizobacteria (PGPR) inoculation under different levels of nitrogen. J Agric Res 45:135–143

Anumalla M, Roychowdhury R, Geda CK, Mazid M, Rathoure AK (2015) Utilization of plant genetic resources and diversity analysis tools for sustainable crop improvement with special emphasis on rice. Int J Adv Res 3(3):1155–1175

Anumalla M, Roychowdhury R, Geda CK, Bharathkumar S, Goutam KD, Mohandev TSS (2016) Mechanism of stress signal transduction and involvement of stress inducible transcription factors and genes in response to abiotic stresses in plant. Int J Rec Sci Res 7(8):12754–12771

Arkhipova TN, Prinsen E, Veselov SU, Martinenko EV, Melentiev AI, Kudoyarova GR (2007) Cytokinin producing bacteria enhance plant growth in drying soil. Plant Soil 292:305–315

Armada E, Roldán A, Azcon R (2014) Differential activity of autochthonous bacteria in controlling drought stress in native *Lavandula* and *Salvia* plants species under drought conditions in the natural arid soil. Microb Ecol 67:410–420

Asch F, Padham JL (2005) Root associated bacteria suppress symptoms of iron toxicity in lowland rice. The Global Food and Product Chain-Dynamics Innovations Conflicts Strategies 276

Ashraf M, Hasnain S, Berge O, Mahmood T (2004) Inoculating wheat seedlings with exopolysaccharide-producing bacteria restricts sodium uptake and stimulates plant growth under salt stress. Biol Fertil Soil 40:157–162

Ashraf M, Ozturk M, Ahmad MSA (2010) Toxins and their phytoremediation. In: Plant adaptation and phytoremediation. Springer, Dordrecht, pp 1–32

Augé RM (2001) Water relations, drought and vesicular-arbuscular mycorrhizal symbiosis. Mycorrhiza 11:3–42

Augé RM, Stodola AJ, Tims JE, Saxton AM (2001) Moisture retention properties of a mycorrhizal soil. Plant and Soil, 230(1):87–97.

Auge RM, Schekel KA, Wample RL (1986) Osmotic adjustment in leaves of VA mycorrhizal and non-mycorrhizal rose plants in response to drought stress. Plant Physiol 82:765–770

Auge RM, Foster JC, Loescher WH, Stodola AJW (1992) Symplastic molality of free amino acids and sugars in *Rosa* roots with regard to VAM and drought symbiosis. Symbiosis 12:1–17

Aydinalp C, Marinova S (2009) The effects of heavy metals on seed germination and plant growth on *Alfalfa* plant (*Medicago sativa*). Bulg J Agric Sci 15(4):347–350

Azcon R, Ocampo JA (1981) Factors affecting vesicular-arbuscular infection and mycorrhizal dependency of thirteen wheat cultivars. New Phytol 87:677–685

Azcón R, Tobar RM (1998) Activity of nitrate reductase and glutamine synthetase in shoot and root of mycorrhizal *Allium cepa*. Effect of drought stress. Plant Sci 133:1–8

Azcón R, Gomez M, Tobar R (1996) Physiological and nutritional responses by *Lactuca sativa* to nitrogen sources and mycorrhizal fungi under drought. Biol Fertil Soils 22:156–161

Babu AG, Shea PJ, Sudhakar D, Jung IB, Oh BT (2015) Potential use of *Pseudomonas koreensis* AGB-1 in association with *Miscanthus sinensis* to remediate heavy metal (loid)-contaminated mining site soil. J Environ Manag 151:160–166

Bacilio M, Moreno M, Bashan Y (2016) Mitigation of negative effects of progressive soil salinity gradients by application of humic acids and inoculation with *Pseudomonas stutzeri* in a salt-tolerant and a salt-susceptible pepper. Appl Soil Ecol 107:394–404

Bal HB, Nayak L, Das S, Adhya TK (2013) Isolation of ACC deaminase producing PGPR from rice rhizosphere and evaluating their plant growth promoting activity under salt stress. Plant Soil 366:93–105

Baltruschat H, Fodor J, Harrach BD, Niemczyk E, Barna B, Gullner G, Zuccaro A (2008) Salt tolerance of barley induced by the root endophyte Piriformospora indica is associated with a strong increase in antioxidants. New Phytol 180(2):501–510

Bano A, Fatima M (2009) Salt tolerance in *Zea mays* (L.) following inoculation with *Rhizobium* and *Pseudomonas*. Biol Fertil Soil 45:405–413

Bano QUDSIA, Ilyas N, Bano A, Zafar NADIA, Akram ABIDA, Hassan F (2013) Effect of *Azospirillum* inoculation on maize (*Zea mays* L.) under drought stress. Pak J Bot 45:13–20

Barassi CA, Ayrault G, Creus CM, Sueldo RJ, Sobrero MT (2006) Seed inoculation with *Azospirillum mitigates* NaCl effects on lettuce. Sci Hortic 109:8–14

Barka EA, Nowak J, Clement C (2006) Enhancement of chilling resistance of inoculated grapevine plantlets with a plant growth-promoting rhizobacterium, *Burkholderia phytofirmans* strain PsJN. Appl Environ Microbiol 72:7246–7252

Basheer AA (2018a) Chemical chiral pollution: impact on the society and science and need of the regulations in the 21st century. Chirality 30(4):402–406

Basheer AA (2018b) New generation nano-adsorbents for the removal of emerging contaminants in water. J Mol Liq 261:583–593

Battisti DS, Naylor RL (2009) Historical warnings of future food insecurity with unprecedented seasonal heat. Science 323:240–244

Belimov AA, Dodd IC, Hontzeas N, Theobald JC, SafronovaVI DWJ (2009) Rhizosphere bacteria containing 1-aminocyclopropane-1-carboxylate deaminase increase yield of plants grown in drying soil via both local and systemic hormone signalling. New Phytol 181:413–423

Bensalim S, Nowak J, Asiedu SK (1998) A plant growth promoting rhizobacterium and temperature effects on performance of 18 clones of potato. Am J Potato Res 75:145–152

Bethlenfalvay GJ, Franson RL (1989) Manganese toxicity alleviated by mycorrhizae in soybean. J Plant Nutr 12:953–972

Bethlenfalvay GJ, Brown MS, Ames RN, Thomas RS (1988) Effects of drought on host and endophyte development in mycorrhizal soybeans in relation to water use and phosphate uptake. Physiol Plant 72:565–571

Borde M, Dudhane M, Jite PK (2010) AM fungi influences the photosynthetic activity, growth and antioxidant enzymes in *Allium sativum* L. under salinity condition. Not Sci Biol 2(4):64–71

Bresson J, Varoquaux F, Bontpart T, Touraine B, Vile D (2013) The PGPR strain *Phyllobacterium brassicacearum* STM196 induces a reproductive delay and physiological changes that result in improved drought tolerance in *Arabidopsis*. New Phytol 200:558–569

Brilli F, Pollastri S, Raio A, Baraldi R, Neri L, Bartolini P, Balestrini R (2019) Root colonization by Pseudomonas chlororaphis primes tomato (*Lycopersicum esculentum*) plants for enhanced tolerance to water stress. J Plant Physiol 232:82–93

Bryla DR, Duniway JM (1997) Effects of mycorrhizal infection on drought tolerance and recovery in safflower and wheat. Plant Soil 197:95–103

Burakova EA, Dyachkova TP, Rukhov AV, Tugolukov EN, Galunin EV, Tkachev AG, Ali I (2018) Novel and economic method of carbon nanotubes synthesis on a nickel magnesium oxide catalyst using microwave radiation. J Mol Liq 253:340–346

Caron M, Patten CL, Ghosh S (1995) Effects of plant growth promoting rhizobacteria Pseudomonas putidaGR-122 on the physiology of canola roots. Proc Plant Growth Regul Soc Am 7:18–20

Casanovas EM, Barassi CA, Sueldo RJ (2002) *Azospirillum* inoculation mitigates water stress effects in maize seedlings. Cereal Res Commun 30:343–350

Cassan F, Perrig D, SgroyV MO, Penna C, Luna V (2009) *Azospirillum brasilense* Az39 and *Bradyrhizobium japonicum* E109, inoculated singly or in combination, promote seed germination and early seedling growth in corn (*Zea mays* L.) and soybean (*Glycine max* L.). Eur J Soil Biol 45:28–35

Chakraborty S, Pattanayak A, Mandal S, Das M, Roychowdhury R (2014) An overview of climate change: causes, trends and implications. In: Roychowdhury R (ed) Crop improvement in the era of climate change. IK International Publishing House, New Delhi, pp 1–29

Chang P, Gerhardt KE, Huang XD, Yu XM, Glick BR, Gerwing PD, Greenberg BM (2014) Plant growth-promoting bacteria facilitate the growth of barley and oats in salt-impacted soil: implications for phytoremediation of saline soils. Int J Phytoremedation 16:1133–1147

Chen S, Jin W, Liu A, Zhang S, Liu D, Wang F, Lin X, He C (2013) Arbuscular mycorrhizal fungi (AMF) increase growth and secondary metabolism in cucumber subjected to low temperature stress. Sci Hortic 160:222–229

Chi GG, Srivastava AK, Wu QS (2018) Exogenous easily extractable glomalin related soil protein improves drought tolerance of trifoliate orange. Arch Agron Soil Sci 64(10):1341–1350

Chodak M, Gołębiewski M, Morawska-Płoskonka J, Kuduk K, Niklińska M (2015) Soil chemical properties affect the reaction of forest soil bacteria to drought and rewetting stress. Ann Microbiol 65:1627–1637

Cirillo V, Masin R, Maggio A, Zanin G (2018) Crop-weed interactions in saline environments. Eur J Agron 99:51–61

Clarholm M, Skyllberg U (2013) Translocation of metals by trees and fungi regulates pH, soil organic matter turnover and nitrogen availability in acidic forest soils. Soil Biol Biochem 63:142e153

Cohen AC, Bottini R, Pontin M, Berli FJ, Moreno D, Boccanlandro H, Piccoli PN (2015) *Azospirillum brasilense* ameliorates the response of *Arabidopsis thaliana* to drought mainly via enhancement of ABA levels. Physiol Plant 153:79–90

Creus CM, Sueldo RJ, Barassi CA (1998) Water relations in *Azospirillum*-inoculated wheat seedlings under osmotic stress. Can J Bot 76:238–244

Creus CM, Sueldo RJ, Barassi CA (2004) Water relations and yield in *Azospirillum*-inoculated wheat exposed to drought in the field. Can J Bot 82:273–281

Creus CM, Graziano M, Casanovas EM, Pereyra MA, Simontacchi M, Puntarulo S, Barassi CA, Lamattina L (2005) Nitric oxide is involved in the *Azospirillum brasilense*-induced lateral root formation in tomato. Planta 221:297–303

de Zelicourt A, Al-Yousif M, Hirt H (2013) Rhizosphere microbes as essential partners for plant stress tolerance. Mol Plant 6:242–245

Dağhan H, Öztürk M (2015) Soil pollution in Turkey and remediation methods. In: Soil remediation and plants: prospects and challenges, pp 287–312

Daliakopoulos I, Tsanis I, Koutroulis A, Kourgialas N, Varouchakis A, Karatzas G et al (2016) The threat of soil salinity: a European scale review. Sci Total Environ 573:727–739

Damodaran T, Sah V, Rai RB, Sharma DK, Mishra VK, Jha SK, Kannan R (2013) Isolation of salt tolerant endophytic and rhizospheric bacteria by natural selection and screening for promising plant growth-promoting rhizobacteria (PGPR) and growth *vigour* in tomato under sodic environment. Afr J Microbiol Res 7:5082–5089

Dardanelli MS, Fernández de Córdoba FJ, Rosario Espuny M, Rodríguez Carvajal MA, ME SD, Gil Serrano AM, Okon Y, Megías M (2008) Effect of *Azospirillum brasilense* co inoculated with *Rhizobium* on *Phaseolus vulgaris* flavonoids and Nod factor production under salt stress. Soil Biol Biochem 40:2713–2721

Davies FT Jr, Potter JR, Linderman RG (1992) Mycorrhiza and repeated drought exposure affect drought resistance and extraradical hyphae development of pepper plants independent of plant size and nutrient content. J Plant Physiol 139:289–294

Davies FT, Potter JR, Linderman RG (1993) Drought resistance of mycorrhizal pepper plants independent of leaf P-concentration – response in gas exchange and water relations. Physiol Plant 87:45–53

Disante KB, Fuentes D, Cortina J (2011) Response to the drought of Zn-stressed *Quercus suber* L. seedlings. Environ Exp Bot 70:96–103

Dixon RK, Rao MV, Garg VK (1994) Water relations and gas exchange of mycorrhizal *Leucaena leucocephala* seedlings. J Trop For Sci 6:542–552

Egamberdieva D (2012) *Pseudomonas chlororaphis*: a salt-tolerant bacterial inoculant for plant growth stimulation under saline soil conditions. Acta Physiol Plant 34:751–756

Egamberdieva D, Kucharova Z (2009) Selection for rot colonizing bacteria stimulating wheat growth in saline soils. Biol Fertil Soil 45:563–571

Egamberdieva D, Kamilova F, Validov S, Gafurova L, Kucharova Z, Lugtenberg B (2008) High incidence of plant growth-stimulating bacteria associated with the rhizosphere of wheat grown on salinated soil in Uzbekistan. Environ Microbiol 10:1–9

Egamberdieva D, Berg G, Lindström K, Räsänen LA (2013) Alleviation of salt stress of symbiotic *Galega officinalis* L. (Goat's Rue) by co-inoculation of rhizobium with root colonizing *Pseudomonas*. Plant Soil 369(1–2):453–465

Egamberdieva D, Jabborova D, Hashem A (2015) Pseudomonas induces salinity tolerance in cotton (*Gossypium hirsutum*) and resistance to *Fusarium* root rot through the modulation of indole-3-acetic acid. Saudi J Soil Sci 22:773–779

Egamberdiyeva D, Kucharova Z (2007) Managing fungal diseases of tomato and wheat by potential biocontrol agents in salinated soils of Uzbekistan. In "Best practice in Disease, pest and weed Management; The State of the Art". Edited by: D.V. Alford, F. Feldmann, J. Hasler and A Von Tiedemann, 64–65 pp.

Egamberdiyeva D (2007) The effect of plant growth promoting bacteria on growth and nutrient uptake of maize in two different soils. Appl Soil Ecol 36:184–189

El-Daim IAA, Bejai S, Meijer J (2014) Improved heat stress tolerance of wheat seedlings by bacterial seed treatment. Plant Soil 379:337–350

Ellis JR, Larsen HJ, Boosalis MG (1985) Drought resistance of wheat plants inoculated with vesicular-arbuscular mycorrhizae. Plant Soil 86:369–378

El-Meihy RM, Abou-Aly HE, Youssef AM, Tewfike TA, El-Alkshar EA (2019) Efficiency of heavy metals-tolerant plant growth promoting bacteria for alleviating heavy metals toxicity on sorghum. Environ Exp Bot

El-Tohamy W, Schnitzler WH, El-Behairy U, El-Beltagy MS (1999) Effect of VA mycorrhiza on improving drought and chilling tolerance of bean plants. J Appl Bot 73:178–183

Entry JA, Rygiewicz PT, Watrud LS, Donnelly PK (2002) Influence of adverse soil conditions on the formation and function of *Arbuscular mycorrhizas*. Adv Environ Res 7:123–138

Erickson RO (1959) Integration of plant growth processes. Am Nat 93:225–235

Farooq M, Wahid A, Kobayashi N, Fujita D, Basra SM (2009) Plant drought stress: effects, mechanisms, and management. In: Sustainable agriculture. Springer, Dordrecht, pp 153–188

Farrar JF, Jones DL (2000) The control of carbon acquisition by roots. New Phytol 147:43–53

Fasciglione G, Casanovas EM, Quillehauquy V, Yommi AK, Goñi MG, Roura SI, Barassi CA (2015) *Azospirillum* inoculation effects on growth, product quality and storage life of lettuce plants grown under salt stress. Sci Hortic 195:154–162

Feyen L, Dankers R (2009) Impact of global warming on streamflow drought in Europe. Geophys Res Atmos 114

Figueiredo MV, Burity HA, Martínez CR, Chanway CP (2008) Alleviation of drought stress in the common bean (*Phaseolus vulgaris* L.) by co-inoculation with *Paenibacillus polymyxa* and *Rhizobium tropici*. Appl Soil Ecol 40(1):182–188

Fougnies L, Renciot S, Muller F, Plenchette C, Prin Y, De Faria SM, Bâ AM (2007) Arbuscular mycorrhizal colonization and nodulation improve flooding tolerance in *Pterocarpus officinalis* Jacq. Seedlings. Mycorrhiza 17(3):159–166

Gemma JN, Koske RE, Roberts EM, Jackson N, de Antonis KM (1997) Mycorrhizal fungi improve drought resistance in creeping bent grass. J Turfgrass Sci 73:15–29

George E, Römheld V, Marschner H (1994) Contribution of mycorrhizal fungi to micronutrient uptake by plants. In: Manthey JA, Crowley DE, Luster DG (eds) Biochem of metal micro in the rhizo Lewis, London, pp 93–109

German MA, Burdman S, Okon Y, Kigel J (2000) Effects of *Azospirillum brasilense* on root morphology of common bean (*Phaseolus vulgaris* L.) under different water regimes. Biol Fertil Soils 32:259–264

Glick BR, Todorovic B, Czarny J, Cheng Z, Duan J, McConkey B (2007) Promotion of plant growth by bacterial ACC deaminase. Critical Rev Plant Sci 26:227–242

Goicoechea N, Antolin MC, Sánchez-Díaz M (1997) Influence of arbuscular mycorrhizae and rhizobium on nutrient content and water relations in drought-stressed *Alfalfa*. Plant Soil 192:261–268

Golpayegani A, Tilebeni HG (2011) Effect of biological fertilizers on biochemical and physiological parameters of basil (*Ocimum basilicm* L.) medicine plant. Am Eurasian J Agric Environ Sci 11(3):411–416

Gough LP (1979) Element concentrations toxic to plants, animals, and man. US Geological Survey Washington DC: 1466

Govindasamy V, Senthilkumar M, Magheshwaran V, Kumar U, Bose P, Sharma V, Annapurna K (2010) *Bacillus* and *Paenibacillus* spp.: potential PGPR for sustainable agriculture. In: Plant growth and health promoting bacteria, Microbiology monographs 18. Springer, Berlin/ Heidelberg

Gryndler M, Hršelová H, Stříteská D (2000) Effect of soil bacteria on hyphal growth of the arbuscular mycorrhizal fungus *Glomus claroideum*. Folia Microbiol 45(6):545–551

Gupta VK, Ali I (2012) Environmental water: advances in treatment, remediation and recycling. Newnes

Hairmansis A, Berger B, Tester M, Roy SJ (2014) Image-based phenotyping for non-destructive screening of different salinity tolerance traits in rice. Rice 7:16

Hajiboland R, Moradtalab N, Aliasgharzad N, Eshaghi Z, Feizy J (2018) Silicon influences growth and mycorrhizal responsiveness in strawberry plants. Physiol Mol Biol Plants 24:1103–1115

Hakeem K, Sabir M, Ozturk M, Mermut AR (eds) (2015) Soil remediation and plants: prospects and challenges. Academic

Hamdia ABE, Shaddad MAK, Doaa MM (2004) Mechanisms of salt tolerance and interactive effects of *Azospirillum brasilense* inoculation on maize cultivars grown under salt stress conditions. Plant Growth Regul 44:165–174

Hardie K, Leyton L (1981) The influence of vesicular-arbuscular mycorrhizae on growth and water relations of red clover. I. In phosphate deficient soil. New Phytol 89:599–608

Hasanuzzaman M, Nahar K, Alam M, Roychowdhury R, Fujita M (2013) Physiological, biochemical, and molecular mechanisms of heat stress tolerance in plants. Int J Mol Sci 14(5):9643–9684

Hasanuzzaman M, Roychowdhury R, Karmakar J, Dey N, Nahar K, Fujita M (2015) Recent advances in biotechnology and genomic approaches for abiotic stress tolerance in crop plants. In: Devarajan T, Jeyabalan S (eds) Genomics and proteomics: concepts, technologies and applications. Apple Academic Press, pp 333–366

Hamaoui B, Abbadi JM, Burdman S, Rashid A, Sarig S, Okon Y (2001) Effects of inoculation with Azospirillum brasilense on chickpeas (Cicer arietinum) and faba beans (Vicia faba) under different growth conditions. Agronomie, 21(6–7):553–560.

Hayat R, Ali S, Siddique MT, Chatha TH (2008) Biological nitrogen fixation of summer legumes and their residual effects on subsequent rainfed wheat yield. Pak J Bot 40:711–722

Hayat R, Ali S, Amara U, Khalid R, Ahmed I (2010) Soil beneficial bacteria and their role in plant growth promotion: a review. Ann Microbiol 60:579–598

He X, Zhao J, Li S (1999) Effects of water stress and VA mycorrhizal fungi on the growth of mung bean. Acta Agric Nucleatae Sin 14(5):290–294

He JD, Wu QS, Zou YN (2019) Effects of mycorrhiza and drought stress on the diversity of fungal community in soils and roots of trifoliate orange. Biotech 18(1):32–41

Henderson JC, Davies FT (1990) Drought acclimation and the morphology of mycorrhizal *Rosa hybrida* L. cv Ferdy is independent of leaf elemental content. New Phytol 115:503–510

Hossain MA, Piyatida P, da Silva JAT, Fujita M (2012) Molecular mechanism of heavy metal toxicity and tolerance in plants: central role of glutathione in detoxification of reactive oxygen species and methylglyoxal and in heavy metal chelation. J Bot 2012:1–37

Hrynkiewicz K, Baum C (2013) Selection of ectomycorrhizal willow genotype in phytoextraction of heavy metals. Environ Technol 34(2):225–230

Huang RS, Smith WK, Yost RS (1985) Influence of vesicular arbuscular mycorrhiza on growth, water relations and leaf orientation in *Leucaena leucocephala* (Lam.) de Wit. New Phytol 99:229–243

Huang S, Spielmeyer W, Lagudah ES, Munns R (2008) Comparative mapping of HKT genes in wheat, barley, and rice, key determinants of Na+ transport, and salt tolerance. J Exp Bot 59(4):927–937

Huang YM, Zou YN, Wu QS (2017) Alleviation of drought stress by mycorrhizas is related to increased root H_2O_2 efflux in trifoliate orange. Sci Rep 7:42335

Hussain MB, Zahir ZA, Asghar HN, Asgher M (2014) Can catalase and exopolysaccharides producing rhizobia ameliorate drought stress in wheat? Int J Agric Biol 16:3–13

Idris EES, Bochow H, Ross H, Boriss F (2004) Use of *Bacillus subtilis* as biocontrol agent. 6. Phytohormone action of culture filtrate prepared from plant growth promoting *Bacillus amyloliquefaciens* FZB24, FZB42, FZB45 and *Bacillus subtilis* FZB37. J Plant Dis Prot 111:583–597

Islam F, Yasmeen T, Ali Q, Ali S, Arif MS, Hussain S et al (2014) Influence of *Pseudomonas aeruginosa* as PGPR on oxidative stress tolerance in wheat under Zn stress. Ecotoxicol Environ Saf 104:285–293

Jahromi F, Aroca R, Porcel R, Ruiz-Lozano JM (2008) Influence of salinity on the in vitro development of *Glomus intraradices* and on the in vivo physiological and molecular responses of mycorrhizal lettuce plants. Microb Ecol 55(1):45

Janská A, Maršík P, Zelenková S, Ovesná J (2010) Cold stress and acclimation-what is important for metabolic adjustment? Plant Biol 12:395–405

Jha Y, Subramanian RB, Patel S (2011) The combination of endophytic and rhizospheric plant growth promoting rhizobacteria in *Oryza sativa* shows higher accumulation of osmo protectant against saline stress. Acta Physiol Plant 33:797–802

Jiang S, Zhang D, Wang L, Pan J, Liu Y, Kong X, Li D (2013) A maize calcium-dependent protein kinase gene, ZmCPK4, positively regulated abscisic acid signaling and enhanced drought stress tolerance in transgenic *Arabidopsis*. Plant Physiol Biochem 71:112–120

Ju W, Liu L, Fang L, Cui Y, Duan C, Wu H (2019) Impact of co-inoculation with plant-growth-promoting rhizobacteria and rhizobium on the biochemical responses of alfalfa-soil system in copper contaminated soil. Ecotoxicol Environ Saf 167:218–226

Jump AS, Peñuelas J (2005) Running to stand still: adaptation and the response to plants to rapid climate change. Ecol Lett 8:1010–1020

Kang SM, Radhakrishnan R, Khan AL, Kim MJ, Park JM, Kim BR, Lee IJ (2014) Gibberellin secreting rhizobacterium, *Pseudomonas putida* H-2-3modulates the hormonal and stress physiology of soybean to improve the plant growth under saline and drought conditions. Plant Physiol Biochem 84:115–124

Karnwal A (2009) Production of indole acetic acid by fluorescent *Pseudomonas* in the presence of l-tryptophan and rice root exudates. J Plant Pathol 91:61–63

Kasim WA, Osman ME, Omar MN, El-Daim IAA, Bejai S, Meijer J (2013) Control of drought stress in wheat using plant-growth-promoting bacteria. J Plant Growth Regul 32:122–130

Kasim WA, Gaafar RM, Abou Ali RM, Omar MN, Hewait HM (2016) Effect of biofilm forming plant growth promoting rhizobacteria on salinity tolerance in barley. Ann Agric Sci 61:217–227

Khakipour N, Khavazi K, Mojallali H, Pazira E, Asadirahmani H (2008) Production of auxin hormone by fluorescent pseudomonads. Am Eurasian J Agric Environ Sci 4:687–692

Khan Z, Doty SL (2009) Characterization of bacterial endophytes of sweet potato plants. Plant Soil 322:197–207

Khan AL, Waqas M, Hamayun M, Al-Harrasi A, Al-Rawahi A, Lee IJ (2013a) Co-synergism of endophyte *Penicillium resedanum* LK6 with salicylic acid helped *Capsicum annuum* in biomass recovery and osmotic stress mitigation. BMC Microbiol 13(1):51

Khan AL, Waqas M, Hamayun M, Al-Harrasi A, Al-Rawahi A, Lee IJ (2013b) Co-synergism of endophyte Penicillium resedanum LK6 with salicylic acid helped *Capsicum annuum* in biomass recovery and osmotic stress mitigation. BMC Microbiol 13(1):51

Khan N, Bano A, Zandi P (2018a) Effects of exogenously applied plant growth regulators in combination with PGPR on the physiology and root growth of chickpea (*Cicer arietinum*) and their role in drought tolerance. J Plant Interact 13(1):239–247

Khan N, Bano A, Shahid MA, Nasim W, Babar MA (2018b) Interaction between PGPR and PGR for water conservation and plant growth attributes under drought condition. Biologia 73(11):1083–1098

Khanna K, Jamwal VL, Gandhi SG, Ohri P, Bhardwaj R (2019a) Metal resistant PGPR lowered Cd uptake and expression of metal transporter genes with improved growth and photosynthetic pigments in *Lycopersicon esculentum* under metal toxicity. Sci Rep 9(1):5855

Khanna K, Jamwal VL, Sharma A, Gandhi SG, Ohri P, Bhardwaj R, Ahmad P (2019b) Supplementation with plant growth promoting rhizobacteria (PGPR) alleviates cadmium toxicity in *Solanum lycopersicum* by modulating the expression of secondary metabolites. Chemosphere 230:628–639

Krishnamoorthy R, Kim K, Subramanian P, Senthilkumar M, Anandham R, Sa T (2016) The tolerance of maize to salinity in coastal reclamation soil. Agric Ecosyst Environ 231:233–239

Kumar KV, Srivastava S, Singh N, Behl HM (2009) Role of metal resistant plant growth promoting bacteria in ameliorating fly ash to the growth of *Brassica juncea*. Hazard Mater 170(1):51–57

Kumar A, Sharma S, Mishra S (2010) Influence of arbuscular mycorrhizal (AM) fungi and salinity on seedling growth, solute accumulation, and mycorrhizal dependency of *Jatropha curcas* L. J Plant Growth Regul 29(3):297–306

Kupulnik Y, Kushnir U (1991) Growth dependency of wild, primitive and modern cultivated wheat lines on vesicular-arbuscular mycorrhizae fungi. Euphytica 56:27–36

Lackie SM, Bowley SR, Peterson RL (1988) Comparison of colonization among half-sib families of *Medicago sativa* L. by *Glomus versiforme* (Daniels and Trappe) Berch. New Phytol 108:477–482

Lata C, Prasad M (2011) The role of DREBs in the regulation of abiotic stress responses in plants. Exp Bot 62:4731–4748

Liu A, Wang B, Hamel C (2004) Arbuscular mycorrhiza colonization and development at suboptimal root zone temperature. Mycorrhiza 14:93–101

Liu ZL, Li YJ, Hou HY, Zhu XC, Rai V, He XY, Tian CJ (2013) Differences in the arbuscular mycorrhizal fungi-improved rice resistance to low temperature at two N levels: aspects of N and C metabolism on the plant side. Plant Physiol Biochem 71:87–95

Ma Y, Rajkumar M, Luo Y, Freitas H (2013) Phytoextraction of heavy metal polluted soils using *Sedum plumbizincicola inoculated* with metal mobilizing *Phyllobacterium myrsinacearum* RC6b. Chemosphere 93:1386–1392

Ma Y, Rajkumar M, Zhang C, Freitas H (2016a) Inoculation of *Brassica oxyrrhina* with plant growth promoting bacteria for the improvement of heavy metal phytoremediation under drought conditions. J Hazard Mater 320:36–44

Ma Y, Rajkumar M, Zhang C, Freitas H (2016b) The beneficial role of bacterial endophytes in heavy metal phytoremediation. J Environ Manag 174:14–25

Malinowski DP, Belesky DP (2000) Adaptations of endophyte-infected cool season grasses to environmental stresses: mechanisms of drought and mineral stress tolerance. Crop Sci 40:923–940

Mamgain A, Roychowdhury R, Tah J (2013) *Alternaria* pathogenicity and its strategic controls. Res J Biol 1:1–9

Manjunath A, Habte M (1991) Relationship between mycorrhizal dependency and rate variables associated with phosphorus uptake, utilization and growth. Commun Soil Sci Plant Anal 22:1423–1437

Marastoni L, Pii Y, Maver M, Valentinuzzi F, Cesco S, Mimmo T (2019) Role of *Azospirillum brasilense* in triggering different Fe chelate reductase enzymes in cucumber plants subjected to both nutrient deficiency and toxicity. Plant Physiol Biochem 136:118–126

Marschner H, Dell B (1994) Nutrient uptake in mycorrhizal symbiosis. Plant Soil 159:89–102

Marulanda A, Barea JM, Azcón R (2009) Stimulation of plant growth and drought tolerance by native microorganisms (AM fungi and bacteria) from dry environments: mechanisms related to bacterial effectiveness. J Plant Growth Regul 28(2):115–124

Masciarelli O, Llanes A, Luna V (2014) A new PGPR co-inoculated with *Bradyrhizobium japonicum* enhances soybean nodulation. Microbiol Res 169:609–615

Mathew DC, Ho YN, Gicana RG, Mathe GM, Chien MC, Huang C (2015) A rhizosphere-associated symbiont, *Photobacterium* spp. strain MELD1, and its targeted synergistic activity for phytoprotection against mercury. PLoS One 10:e0121178

Mathur N, Vyas A (1995) Influence of VAM on net photosynthesis and transpiration of *Ziziphus mauritiana*. J Plant Physiol 147:328–330

Mayak S, Tirosh T, Glick BR (2004) Plant growth-promoting bacteria confer resistance in tomato plants to salt stress. Plant Physiol Biochem 42:565–572

Mercy MA, Shivanshanker G, Bagyaraj DJ (1990) Mycorrhizal colonization in cowpea is host dependent and heritable. Plant Soil 121:292–294

Michelsen A, Rosendahl S (1990) The effect of VA mycorrhizal fungi, phosphorus and drought stress on the growth of *Acacia nilotica* and *Leucaena leucocephala* seedlings. Plant Soil 124:7–13

Mishra AK, Singh VP (2010) A review of drought concepts. J Hydrol 39:202–216

Moradi S, Salimi S (2013) Effects of arbuscular mycorrhizal fungi on root morphological properties of Poa in drought stress conditions. Int J Agric Crop Sci 5(6):591

Munns R, Tester M (2008) Mechanisms of salinity tolerance. Annu Rev Plant Biol 59:651–681

Nabti E, Sahnoune M, Ghoul M, Fischer D, Hofmann A, Rothballer M, Schmid M, Hartmann A (2010) Restoration of growth of durum wheat (*Triticum durum* var. waha) under saline conditions due to inoculation with the rhizosphere bacterium *Azospirillum brasilense* NH and extracts of the marine alga *Ulva lactuca*. J Plant Growth Regul 29:6–22

Nadeem SM, Zahir ZA, Naveed M, Arshad M (2007) Preliminary investigations on inducing salt tolerance in maize through inoculation with rhizobacteria containing ACC deaminase activity. Can J Microbiol 53:1141–1149

Nadeem SM, Ahmad M, Zahir ZA, Javaid A, Ashraf M (2014) The role of mycorrhizae and plant growth promoting rhizobacteria (PGPR) in improving crop productivity under stressful environments. Biotechnol Adv 32(2):429–448

Naseem H, Bano A (2014) Role of plant growth-promoting rhizobacteria and their exopolysaccharide in drought tolerance of maize. Plant Interact 9:689–701

Nelsen CE (1987) The water relations of vesicular-arbuscular mycorrhizal systems. In: Safir GR (ed) Ecophysiology of VA mycorrhizal plants. CRC Press, Boca Raton, pp 71–91

Newton AC, Johnson SN, Gregory PJ (2011) Implications of climate change for diseases, crop yields and food security. Euphytica 179:3–18

Nia SH, Zarea MJ, Rejali F, Varma A (2012) Yield and yield components of wheat as affected by salinity and inoculation with *Azospirillum* strains from saline or non-saline soil. J Saudi Soc Agric Sci 11:113–121

Osonubi O, Mulongoy K, Awotoye OO, Atayese MO, Okali DUU (1991) Effects of ectomycorrhizal and vesicular-arbuscular mycorrhizal fungi on drought tolerance of four leguminous woody seedlings. Plant Soil 136:131–143

Öztürk MA (ed) (1989) Plants and pollutants in developed and developing countries. Ege University, Botany Department, Izmir

Öztürk MA, Ashraf M, Aksoy A, Ahmad MSA (eds) (2015a) Phytoremediation for green energy. Springer

Öztürk MA, Ashraf M, Aksoy A, Ahmad MSA, Hakeem KR (eds) (2015b) Plants, pollutants and remediation. Springer

Pal AK, Mandal S, Sengupta C (2019) Exploitation of IAA producing PGPR on mustard (*Brassica nigra* L.) seedling growth under cadmium stress condition in comparison with exogenous IAA application. Plant Sci Today 6(1):22–30

Panwar JDS (1993) Response of VAM and *Azospirillum* inoculation to water status and grain yield in wheat under water stress conditions. Indian J Plant Physiol 36:41–43

Pedraza RO, Bellone CH, de Bellone S, Sorte PMB, Teixeira KRD (2009) *Azospirillum* inoculation and nitrogen fertilization effect on grain yield and on the diversity of endophytic bacteria in the phyllosphere of rice rain fed crop. Eur J Soil Biol 45:36–43

Peñuelas J, Gordon C, Llorens L, Nielsen T, Tietema A, Beier C, Bruna P, Emmett B, Estiarte M, Gorissen A (2004) Nonintrusive field experiments show different plant responses to warming and drought among sites, seasons, and species in a north–south European gradient. Ecosystems 7:598–612

Pereyra MA, Zalazar CA, Barassi CA (2006) Root phospholipids in *Azospirillum*-inoculated wheat seedlings exposed to water stress. Plant Physiol Biochem 44:873–879

Plociniczak T, Sinkkonen A, Romantschuk M, Piotrowska-seget Z (2013) Characterization of *Enterobacter intermedius* MH8b and its use for the enhancement of heavy metal uptake by *Sinapis alba* L. Appl Soil Ecol 63:1–7

Porcel R, Aroca R, Ruiz-Lozano JM (2012) Salinity stress alleviation using arbuscular mycorrhizal fungi. A review. Agron Sustain Dev 32:181–200

Porcel R, Zamarreño ÁM, García-Mina JM, Aroca R (2014) Involvement of plant endogenous ABA in *Bacillus megaterium* PGPR activity in tomato plants. BMC Plant Biol 14:36

Qadir M, Quillerou E, Nangia V, Murtaza G, Singh M, Thomas RJ, Drechse P, Noble AD (2014) Economics of salt-induced land degradation and restoration. Nat Res Forum 38:282–295

Raju PS, Clark RB, Ellis JR, Duncan RR, Maranville JW (1990) Benefit and cost analysis and phosphorus mycorrhizal fungi colonizations with sorghum (*Sorghum bicolor*) genotypes grown at varied phosphorus levels. Plant Soil 124:199–204

Rodríguez AA, Stella AM, Storni MM, Zulpa G, Zaccaro MC (2006) Effects of cyanobacterial extracellular products and gibberellic acid on salinity tolerance in *Oryza sativa* L. Saline Syst 2:1–4

Rodriguez RJ, Henson J, Van Volkenburgh E, Hoy M, Wright L, Beckwith F, Redman RS (2008) Stress tolerance in plants via habitat-adapted symbiosis. ISME J 2(4):404

Roychowdhury R (2011) Effect of chemical mutagens on carnation (*Dianthus caryophyllus* L.): a mutation breeding approach. LAP Lambert Academic Publishing

Roychowdhury R (2014) Crop improvement in the era of climate change. IK International Publishing House, New Delhi, p 496

Roychowdhury R, Tah J (2011a) Assessment of chemical mutagenic effects in mutation breeding programme for M_1 generation of Carnation (*Dianthus caryophyllus*). Res Plant Biol 1(4):23–32

Roychowdhury R, Tah J (2011b) Chemical mutagenic action on seed germination and related agro-metrical traits in M_1 Dianthus generation. Curr Bot 2(8):19–23

Roychowdhury R, Tah J (2011c) Mutation breeding in *Dianthus caryophyllus* for economic traits. Electron J Plant Breed 2(2):282–286

Roychowdhury R, Tah J (2011d) Evaluation of genetic parameters for agro-metrical characters in carnation genotypes. Afr Crop Sci J 19(3):183–188

Roychowdhury R, Tah J (2011e) Genetic variability for different quantitative traits in *Dianthus caryophyllus* L. during mutation breeding. Int J Sci Nat 2(4):778–781

Roychowdhury R, Tah J (2011f) Germination behaviors in M_2 generation of *Dianthus* after chemical mutagenesis. Int J Adv Sci Tech Res 2(1):448–454

Roychowdhury R, Tah J (2011g) Genetic variability study for yield and associated quantitative characters in mutant genotypes of *Dianthus caryophyllus* L. Int J Biosci 1(5):38–44

Roychowdhury R, Tah J (2011h) Differential response by different parts of *Solanum melongena* L. for heavy metal accumulation. Plant Sci Feed 1(6):80–83

Roychowdhury R, Tah J (2013) Mutagenesis – a potential approach for crop improvement. In: Hakeem KR, Ahmad P, Ozturk M (eds) Crop improvement – new approaches and modern techniques. Springer, Boston, pp 149–187

Roychowdhury R, Sultana P, Tah J (2011a) Morphological architecture of foliar stomata in M_2 Carnation (*Dianthus caryophyllus* L.) genotypes using Scanning Electron Microscopy (SEM). Electron J Plant Breed 2(4):583–588

Roychowdhury R, Bandopadhyay A, Dalal T, Tah J (2011b) Biometrical analysis for some agro-economic characters in M_1 generation of *Dianthus caryophyllus*. Plant Arch 11(2):989–994

Roychowdhury R, Tah J, Dalal T, Bandyopadhyay A (2011c) Selection response and correlation studies for metrical traits in mutant Carnation (*Dianthus caryophyllus* L.) genotypes. Cont J Agric Sci 5(3):06–14

Roychowdhury R, Roy S, Tah J (2011d) Estimation of heritable components of variation and character selection in eggplant (*Solanum melongena* L.) for mutation breeding programme. Cont J Biol Sci 4(2):31–36

Roychowdhury R, Alam MJ, Bishnu S, Dalal T, Tah J (2012a) Comparative study for effects of chemical mutagenesis on seed germination, survivability and pollen sterility in M_1 and M_2 generations of Dianthus. Plant Breed Seed Sci 65(1):29–38

Roychowdhury R, Datta S, Gupta P, Tah J (2012b) Analysis of genetic parameters on mutant populations of mung bean (*Vigna radiata* L.) after ethyl methane sulphonate treatment. Not Sci Biol 4(1):137–143

Roychowdhury R, Mamgain A, Ray S, Tah J (2012c) Effect of gibberellic acid, kinetin and indole 3-acetic acid on seed germination performance of *Dianthus caryophyllus* (Carnation). Agric Conspec Sci 77(3):157–160

Roychowdhury R, Karmakar J, Dey N (2012d) PCR-compatible genomic DNA isolation from different tissues of rice (*Oryza sativa*) for SSR fingerprinting. Eurasian J BioSci 6(1):85–90

Roychowdhury R, Karmakar J, Karmakar J, Adak MK, Dey N (2013a) Physio-biochemical and microsatellite based profiling of lowland rice (*Oryza sativa* L.) landraces for osmotic stress tolerance. Am J Plant Sci 4(12):52

Roychowdhury R, Abdel Gawwad MR, Banerjee U, Bishnu S, Tah J (2013b) Status, trends and prospects of organic farming in India: a review. J Plant Biol Res 2:38–48

Roychowdhury R, Banherjee U, Slofkova S, Tah J (2013c) Organic farming for crop improvement and sustainable agriculture in the era of climate change. OnLine J Biol Sci 13(2):50–65

Roychowdhury R, Taoutaou A, Hakeem KR, Gawwad MR, Tah J (2014) Molecular marker-assisted technologies for crop improvement. In: Roychowdhury R (ed) Crop improvement in the era of climate change. IK International Publishing House, New Delhi, pp 241–258

Roychowdhury R, Khan MH, Choudhury S (2018) Arsenic in rice: an overview on stress implications, tolerance and mitigation strategies. In: Hasanuzzaman M, Nahar K, Fujita M (eds) Plants under metal and metalloid stress. Springer, Singapore, pp 401–415

Roychowdhury R, Khan MH, Choudhury S (2019) Physiological and molecular responses for metalloid stress in rice – a comprehensive overview. In: Hasanuzzaman M, Fujita M, Nahar K, Biswas J (eds) Advances in rice research for abiotic stress tolerance. Woodhead Publishing/ Elsevier, pp 341–369

Ruiz-Lozano JM (2003) Arbuscular mycorrhizal symbiosis and alleviation of osmotic stress new perspectives for molecular studies. Mycorrhiza 13:309–317

Ruiz-Lozano JM, Azon R, Gomez M (1995) Effects of arbuscular-mycorrhizal *Glomus* species on drought tolerance: physiological and nutritional plant responses. Appl Environ Microbiol 61:456–460

Ruiz-Lozano JM, Azcón R, Palma JM (1996) Superoxide dismutase activity in arbuscular mycorrhizal *Lactuca sativa* plants subjected to drought stress. New Phytol 134:327–333

Ruth B, Khalvati M, Schmidhalter U (2011) Quantification of mycorrhizal water uptake via high-resolution on-line water content sensors. Plant Soil 342(1–2):459–468

Saravanakumar D, Samiyappan R (2007) ACC deaminase from *Pseudomonas fluorescens* mediated saline resistance in groundnut (*Arachis hypogaea*) plants. J Appl Microbiol 102:1283–1292

Schellenbaum L, Müller J, Boller T, Wiemken A, Schüepp H (1998) Effects of drought on non-mycorrhizal and mycorrhizal maize: changes in the pools of non-structural carbohydrates, in the activities of invertase and trehalase, and in the pools of amino acids and imino acids. New Phytol 138:59–66

Sharma S, Ali I (2011) Adsorption of Rhodamine B dye from aqueous solution onto acid activated mango (*Mangifera indica*) leaf powder: equilibrium, kinetic and thermodynamic studies. J Toxicol Environ Health Sci 3(10):286–297

Shrivastava P, Kumar R (2015) Soil salinity: a serious environmental issue and plant growth promoting bacteria as one of the tools for its alleviation. Saudi J BiolSci 22:123–131

Singh JS, Pandey VC, Singh DP (2011) Efficient soil microorganisms: a new dimension for sustainable agriculture and environmental development. Agric Ecosyst Environ 140:339–353

Smith SE, Read DJ (2010) Mycorrhizal symbiosis. Academic, London

Song H (2005) Effects of VAM on host plant in the condition of drought stress and its mechanisms. Electron J Biol 1(3):44–48

Spence C, Bais H (2015) The role of plant growth regulators as chemical signals in plant–microbe interactions: a double-edged sword. Curr Opin Plant Biol 27:52–58

Srivastava S, Verma PC, Chaudhr V, Singh N, Abhilash PC, Kumar KV (2013) Influence of inoculation of arsenic-resistant *Staphylococcus arlettae* on growth and arsenic uptake in *Brassica juncea* (L.) Czern.Var. R-46. J Hazard Mater 262:1039–1047

Staudinger C, Mehmeti-Tershani V, Gil-Quintana E, Gonzalez EM, Hofhansl F, Bachmann G, Wienkoop S (2016) Evidence for a rhizobia-induced drought stress response strategy in *Medicago truncatula*. J Proteome 136:202–213

Subramanian KS, Charest C (1995) Influence of arbuscular mycorrhizae on the metabolism of maize under drought stress. Mycorrhiza 5:273–278

Subramanian KS, Charest C (1997) Nutritional, growth, and reproductive responses of maize (*Zea mays* L.) to arbuscular mycorrhizal inoculation during and after drought stress at tasseling. Mycorrhiza 7:25–32

Sziderics AH, Rasche F, Trognitz F, Sessitsch A, Wilhelm E (2007) Bacterial endophytes contribute to abiotic stress adaptation in pepper plants (*Capsicum annuum* L.). Can J Microbiol 53:1195–1202

Tank N, Saraf M (2010) Salinity-resistant plant growth promoting rhizobacteria ameliorates sodium chloride stress on tomato plants. J Plant Interact 5:51–58

Terré S, Asch F, Padham J, Sikora RA, Becker M (2007) Influence of root zone bacteria on root iron plaque formation in rice subjected to iron toxicity. In: Tielkes E (ed) Utilisation of diversity in land use systems: sustainable and organic approaches to meet human needs. Tropentag, Witzenhausen, p 446

Thakuria D, Talukdar NC, Goswami C, Hazarika S, Boro RC, Khan MR (2004) Characterization and screening of bacteria from the rhizosphere of rice grown in acidic soils of Assam. Curr Sci 86:978–985

Tiwari S, Lata C, Chauhan PS, Nautiyal CS (2016) *Pseudomonas putida* attunes morphophysiological, biochemical and molecular responses in *Cicer arietinum* L. during drought stress and recovery. Plant Physiol Biochem 99:108–117

Tobar R, Azcón R, Barea JM (1994) Improved nitrogen uptake and transport from 15N-labelled nitrate by external hyphae of arbuscular mycorrhiza under water-stressed conditions. New Phytol 126(1):119–122

Tsavkelova EA, Cherdyntseva TA, Klimova SY, Shestakov AI, Botina SG, Netrusov AI (2007) Orchid-associated bacteria produce indole-3-aceticacid, promote seed germination, and increase their microbial yield in response to exogenous auxin. Arch Microbiol 188:655–664

Vardharajula S, Zulfikar Ali S, Grover M, Reddy G, Bandi V (2011) Drought-tolerant plant growth promoting *Bacillus* spp.: effect on growth, osmolytes, and antioxidant status of maize under drought stress. Plant Interact 6(1):1–14

Vimal SR, Singh JS, Arora NK, Singh S (2017) Soil-plant-microbe interactions in stressed agriculture management: a review. Pedosphere 27(2):177–192

Vivas A, Marulanda A, Ruiz-Lozano JM, Barea JM, Azcón R (2003) Influence of a *Bacillus* sp. on physiological activities of two arbuscular mycorrhizal fungi and on plant responses to PEG-induced drought stress. Mycorrhiza 13(5):249–256

Vives-Peris V, Molina L, Segura A, Gómez-Cadenas A, Pérez-Clemente RM (2018) Root exudates from citrus plants subjected to abiotic stress conditions have a positive effect on rhizobacteria. J Plant Physiol 228:208–217

Wood JL, Tang C, Franks AE (2016) Microbial associated plant growth and heavy metal accumulation to improve phytoextraction of contaminated soils. Soil Biol Biochem 103:131–137

Wu QS, Xia RX (2004) The relation between vesicular-arbuscular mycorrhizae and water metabolism in plants. Chin Agric Sci Bull 20(1):188–192

Wu QS, Zou YN, Liu W, Ye XF, Zai HF, Zhao LJ (2010) Alleviation of salt stress in citrus seedlings inoculated with mycorrhiza: changes in leaf antioxidant defense systems. Plant Soil Environ 56(10):470–475

Wu HH, Zou YN, Rahman MM, Ni QD, Wu QS (2017) Mycorrhizas alter sucrose and proline metabolism in trifoliate orange exposed to drought stress. Sci Rep 7:42389

Yilmaz R, Sakcali S, Yarci C, Aksoy A, Ozturk M (2006) Use of *Aesculus hippocastanum* L. as a biomonitor of heavy metal pollution. Pak J Bot 38(5):1519–1527
Yuwono T, Handayani D, Soedarsono J (2005) The role of osmotolerant rhizobacteria in rice growth under different drought conditions. Aust J Agric Res 56:715–721
Zaidi SFA (2003) Biocontrol of *Fusarium oxysporum* by plant growth promoting rhizobacteria in soybean. Ann Agric Res 24:676–678
Zhang F, Dashti N, Hynes RK, Smith DL (1997) Plant growth promoting rhizobacteria and soybean [*Glycine max* (L.) Merr] growth and physiology at suboptimal root zone temperatures. Ann Bot 79:243–249
Zhang F, Jia-Dong HE, Qiu-Dan NI, Qiang-Sheng WU, Ying-Ning ZOU (2018a) Enhancement of drought tolerance in trifoliate orange by mycorrhiza: changes in root sucrose and proline metabolisms. Not Bot Hortic Agrobot Cluj-Napoca 46(1):270–276
Zhang F, Zou YN, Wu QS (2018b) Quantitative estimation of water uptake by mycorrhizal extraradical hyphae in citrus under drought stress. Sci Hortic 229:132–136
Zhu XC, Song FB, Xu HW (2010) Arbuscular mycorrhizae improves low temperature stress in maize via alterations in host water status and photosynthesis. Plant Soil 331:129–137
Złoch M, Thiem D, Gadzała-Kopciuch R, Hrynkiewicz K (2016) Synthesis of siderophores by plant-associated metallotolerant bacteria under exposure to Cd^{2+}. Chemosphere 156:312–313

Chapter 21
Molecular Insight of Plant–Pathogen Interaction

Anil Kumar, Chunoti Changwal, Alkesh Hada, and Prashant Kumar Singh

Abstract Plants are constantly exposed to a broad range of potential pathogens and pests and therefore have developed a highly sophisticated immune system to perceive and prevent such threats. Plants use pattern recognition receptors (PRRs) as well as nucleotide-binding and leucine-rich repeat to monitor nonself and damaged-self and altered-self patterns as signs of potential danger. PRRs of plants stimulate local and systemic immunity after pathogen-/microbe-associated molecular patterns (P/MAMPs) have been recognized. Then, a cascade of response begins that, through the identification of PAMPs, results in plant immunity known as PAMP-triggered immunity (PTI). Successful pathogens bypass PTI by expressing a suite of effector protein that destroys or interferes with host defense system. Recent developments have revealed novel PRRs, their respective ligands, and pathways affecting PRR activity and signaling. To quickly identify microbial patterns and their cognate PRRs, innovative approaches have emerged. Here, in this chapter, we illustrate known PAMPs and effectors of various pathogens such as bacteria, fungi, nematode, and viruses recognized by plant receptors reported till date. We have also discussed different approaches for identifying PAMPs, effectors, and PRRs. Eventually, we highlight the evolving biotechnological potential of the use of PRRs to enhance the wide spectrum and possibly long-lasting resistance of disease in plants.

A. Kumar (✉)
Department of Entomology, Nematology and Chemistry Units, Agricultural Research Organization (ARO) - Volcani Center, Rishon LeZion, Israel

C. Changwal
Division of Plant Physiology, ICAR-Indian Agricultural Research Institute, New Delhi, India

A. Hada
Department of Biotechnology, National College (Autonomous), Tirruchirappalli, India

P. K. Singh
Institute of Plant Sciences, Agricultural Research Organization (ARO) – The Volcani Center, Rishon LeZion, Israel

Department of Biotechnology, Pachhunga University College, Mizoram Central University, Aizawl, Mizoram, India

R. Roychowdhury et al. (eds.), *Sustainable Agriculture in the Era of Climate Change*, https://doi.org/10.1007/978-3-030-45669-6_21

Keywords Pathogen-associated molecular patterns · Effectors · Crop improvement · Genetic engineering · Root knot nematode · Bacteria · PAMP-triggered immunity · Pattern recognition receptors

Abbreviations

Avr	Avirulence proteins
BAK1	Brassinosteroid Insensitive-Associated Kinase 1
BIK1	Botrytis-Induced Kinase 1
CaMV	*Cauliflower mosaic virus*
CDK	Calcium-dependent
CEBiP	Chitin elicitor-binding protein
CERK1	Chitin elicitor kinase-1 receptor
CP	Coat protein
CWDEs	Cell wall-degrading enzymes
EFR	Elongation factor receptor
EGF	Epidermal growth factor
EIX	Ethylene-inducing xylanase
EMS	Ethyl methanesulfonate
ET	Ethylene
ETI	Effector-triggered immunity
ETS	Effector-triggered susceptibility
FCs	Feeding cells
FLS2	Flagellin-sensitive 2
HR	Hypersensitive response
JA	Jasmonic acid
LCD	Localized cell death
LPS	Lipopolysaccharide
LRRs	Leucine-rich repeat
LysM	Lysin motif
MAMPs	Microbe-associated molecular patterns
NAMPs	Nematode-associated molecular pattern
NB	Nucleotide-binding domains
NB-LRR	Nucleotide-binding leucine-rich repeat
NIS1	Necrosis-inducing secreted protein 1
NLP	Nep1-like protein
PAMPs	Pathogen-associated molecular patterns
PGNs	Peptidoglycans
PPNs	Plant parasite nematodes
PPRs	Pattern recognition receptors
PTI	PAMP-triggered immunity
PVX	*Potato virus X*
RBOHD	Respiratory burst oxidase homolog D
RKNA	Root knot nematodes

RKs	Plasma-membrane-localized receptor kinases
RLPs	Receptor-like proteins
ROS	Reactive oxygen species
SA	Salicylic acid
ssRNA	Single-stranded RNA
T3SS	Type III secretion system
TMV	*Tobacco mosaic virus*
TuMV	*Turnip mosaic virus*
TYLCV	*Tomato yellow leaf curl virus*
WMV	*Watermelon mosaic virus*; *Cucumber mosaic virus*

1 Introduction

Plants are used as food sources by a multitude of different pathogens, including fungi, bacteria, nematodes, and viruses. Major crop diseases induced by plant pathogens pose a serious risk to worldwide food security (Pennisi 2010; Jones 2013). The coevolution of plant pathogens and their hosts has led to extremely evolved strategies for microbial invasion and mechanisms for plant counter-defense. The association interactions among plants and pathogens are very complex, while the latter flourish within the plants (Vorholt 2012; Bulgarelli et al. 2013). For many years, a significant number of studies have been performed to understand how plants and pathogens recognize each other in order to establish a fruitful relationship. To defend themselves against a surplus of pathogens, plant immunity is usually cell-autonomous, since each cell can activate an immune response. Nonetheless, systemic signals emerge after a local infection that control the spreading of the immune reaction and which can develop an increased level of immune system across the plants. Unlike animals, plants lack an adaptive immune response and are completely dependent to innate immune systems and that need to be extremely rapid, reliable to survive attacks. Perhaps the most important step in the response to disease is the introduction of pathogens into host cells (Imam et al. 2016). Each pathogen has a unique mechanism for evading the plant cells, such as fungi using a specialized structure, called hyphae; bacteria pass into the plant cells via trichomes, stomata lenticels, as well as other openings; and nematodes reach the plant cell through root and root caps and form nematode feeding site, while viruses can reach the plant cells only through physical injury (Layne 1967; Fox et al. 1971; Getz 1983; Holbein et al. 2016).

When the pathogen strikes the initial defense barriers, two main components of the immune response of the plant are generated (Dodds and Rathjen 2010). First, receptor proteins termed pattern recognition receptors (PRRs), present on external surface of the plant cell, recognize conserved microbial elicitors called pathogen-associated molecular patterns (PAMPs); activation of PRRs contributes to PAMP-triggered immunity (PTI). The first layer of defense is mediated by surface-localized pattern recognition receptors (PRRs), leading to PRR-triggered immunity (PTI). PAMPs are recognized by pattern recognition receptors (PRRs), surface receptor proteins of the host, and are conserved in pathogenic and nonpathogenic microbes. PRRs may be classified as plasma-membrane-localized receptor kinases (RKs) or receptor-like proteins (RLPs; Zipfel 2014). Early associations between PAMPs and PRRs outcome in proper defenses by triggering multicomponent and multilayered responses. Defense establishment is induced by several pathways that may contain influx of Ca^2+, reactive oxygen species (ROS) production, and kinase activation, i.e., calcium-dependent (CDK) and mitogen-activated, protein phosphorylation, and various plant hormone synthesis including ethylene (ET), jasmonic acid (JA), salicylic acid (SA), and acting as signal events/molecules (Pieterse et al. 2009). To develop an efficient parasitism, pathogens need to generate an evolved protein, the effector to avoid the activation of PRR-mediated signaling. The intracellular sensitive perception of pathogens and the recognition of pathogen-associated molecular patterns, such as lipopolysaccharides and flagellin, lead to the activation of the plant basal defense (or resistance), which is the first defense response, and trigger a generic mechanism consisting of plant cell wall thickening, papilla deposition, apoplast acidification, and signal transduction and transcription of defense genes. This generic basal defense mechanism has been observed in several incompatible plant–pathogen interactions and is believed to corroborate the observation that most plants are resistant to invasion by most pathogens. Therefore, successful pathogens must evolve mechanisms to interfere with or suppress basal defense to take over the host and develop disease.

The second layer of defense includes the detection of microbe virulence molecules called effectors by intracellular receptors of plant cell; this recognition causes effector-triggered immunity (ETI). Events of recognition are primarily facilitated by a class of receptor proteins containing nucleotide-binding (NB) domains and leucine-rich repeat (LRRs). NB-LRR proteins are encoded by the plant resistance (R) gene and are immune to a variety of pathogens such as bacteria, fungi, viruses, oomycetes, and insects (Dodds and Rathjen 2010). In some cases, pathogens can evade the host cell using the effector and interfere or suppress PTI signaling, resulting in the effector-triggered susceptibility (ETS). In general, PTI and ETI give rise to comparable reactions, although ETI is qualitatively stronger and quicker and often includes a type of localized cell death (LCD) termed as hypersensitive response (HR) that prevents pathogen propagation (Jones and Dangl 2006). Approximately 20% of all genes in a single crop species respond to pathogen identification with coordinated alterations in the transcript level (Eulgem and Somssich 2007).

2 Strategies for Identification and Characterization of PAMPs and PPR

Plant-recognized PAMPs have been identified from all kingdoms of life, but most of the currently known PAMPs are from bacteria, fungi, or oomycetes. Following methods have been used to identify PAMPs and PPRs.

2.1 Sequence and Genome Analysis

Though forward genetics is established and a powerful strategy to the identification of new PRRs, its potential is rather limited in the discovery of PAMPs. In fact, genetic mapping requires strains that harbor genetic variations in PAMPs-encoding genes that lead in differential eliciting activity. However, since several PAMPs might be required for essential biological functions, mutants with loss or modification of these PAMPs are probable to be unsuitable in natural environments. Moreover, forward genetic techniques may be considered under laboratory conditions that are beneficial to the growth of these mutants. For instance, the gene AvrStb6 encodes the secreted protein of *Zymoseptoria tritici*, which has been identified by genetic mapping (Zhong et al. 2017). Comparative population genomics were also used to define the Ave1 apoplastic elicitor secreted by a *Verticillium dahliae* fungus (De Jonge et al. 2012) in combination with expression profiling, although comparative transcriptome sequencing was used to detect the *Avr5* fungal protein generated by *Cladosporium fulvum* (Mesarich et al. 2014). Furthermore, through sequence homology search of gene orthology, the variation in primary protein sequence of PAMPs has led in the identification of variants of bacterial flagellin (Felix et al. 1999), oomycete elicitin (Derevnina et al. 2016), and fungal cerato platanin (Pazzagli et al. 2014). Recently, in silico analyses have been largely adopted to identify features of apoplastic elicitors such as transcription pattern, presence of a signal peptide, absence of transmembrane domains, etc. By analyzing gene transcription profiles, the protein Gr VAP1 from nematodes (Lozano-Torres et al. 2012) and several fungal enzymes capable of inducing a cell death reaction were also identified (Huang et al. 2012; Fang et al. 2016). Evolutionary signatures are under powerful selective pressure to develop and avoid identification by PRRs because apoplastic elicitors reveal the pathogen existence and cause plant defenses (Brunner and Nürnberger 2012). Most PAMPs, however, support key biological functions, thereby constraining recurrent mutations. However, under powerful negative selection in functional locations, amino acids with relaxed selective pressure demonstrate powerful positive selection to avoid PRR recognition. Thus, methodology based on bioinformatics offers excellent ability to identify novel PAMPs.

2.2 Forward Genetic Approaches

Among the most efficient methods to identifying PRRs is forward genetics. This approach needs two parents with varying phenotypic characteristics, namely, individuals with and without responsiveness to a specific PAMP. After generating mapping populations, the genetic loci that influence these traits are determined by genetic mapping using next generation sequencing. It is possible to identify parents with differential phenotypes from plant breeding programs, mutagenized populations, or natural ecotype specimens. Moreover, natural variation is a great source of potential phenotype diversity regarding PAMP perception and responsiveness. Using forward genetic approaches, many studies have identified plasma membrane-localized receptor kinases (RKs) as being involved in crop resistance to various pathogens. For instances, Xa-21 from *Oryza longistaminata* was identified as an LRR-RK conferring resistance to Xoo using positional cloning (Khush et al. 1990; Song et al. 1995). A large collection of *A. thaliana* natural accessions were used to investigate variations in PAMP responsiveness. Furthermore, ethyl methanesulfonate (EMS)-mutagenized population of the *Lotus japonicus* and *Arabidopsis thaliana* were used to identify bacterial elicitor flagellin and exopolysaccharide receptor EPR3, respectively (Gómez-Gómez and Boller 2000; Kawaharada et al. 2015).

2.3 Reverse Genetic Approaches

For the systematic functional analysis of RKs and receptor-like proteins (RLPs), collections of *A. thaliana* lines with T-DNA insertion mutants were generated with particular focus for those whose expression is triggered by PAMP infection. A collection of T-DNA insertion mutants in 44 genes encoding LRR-RLPs (Albert et al. 2015) has also identified AtRLP23, the PRR for Nep1-like protein (NLP), and 169 of 216 LRR-RK mutants have been screened to identify XPS1 as the PRR for bacterial xup25 (Mott et al. 2016). The expression of dominant negative types of RKs could be another strong strategy for evaluating their function as probable PRRs for a reverse genetic approach. To sum up, PTI is an efficient immunity layer that can be utilized to provide long-lasting plant protection. As PRRs can recognize a broad variety of microbial elicitors, the development and engineering of unique PRRs will allow new breeding approaches to leverage the broad variety of elicitor perceptions to improve the resistance of plant diseases.

3 Plant–Bacteria Interactions

Plant pathogenic bacteria have become accountable for economic damage to agriculture and pose a risk to worldwide food production. The relationship between the bacteria and their host is quite vibrant. They are mainly extracellular pathogens which reproduce between host cells in the intercellular spaces. In higher crops, bacteria cause several diseases by invading through injuries and natural openings like microscopic pores in the leaf epidermis (important for gas exchange) and stomata. The multiple diseases caused by the bacteria in higher crops are galls, halo blight cankers, and fire blight leaf spots. They have diverse tactics for infection and are classified into biotrophic and necrotrophic pathogens. Generally, bacteria rely on different secretion pathways to penetrate plant defenses and establish efficient colonization of the host plant (Mehta et al. 2008). Six secretion systems (types I–VI) were identified in bacteria, which are mostly differentiated by their constituent proteins (Costa et al. 2015). Type III secretion system (T3SS) is the primary secretion system used by pathogenic bacteria during infection. Such a system allows bacteria to specifically inject proteins, called effectors, into the host cell and suppress cellular defense. T3SS is valuable for pathogenicity and is conserved of many gram-negative bacteria (Galán and Collmer 1999; Cornelis and Van Gijsegem 2000). The best characterized T3SS effectors are called avirulence (Avr) proteins, which were reported in many plant pathogens (Staskawicz et al. 1984; Keen 1990; Lahaye and Bonas 2001).

Typically, plants protect themselves against pathogen invasions; PTI is one of the first defense responses and triggered by plant pattern recognition receptors (PRRs) upon recognition of PAMPs (Zipfel 2014; Table 21.1). PAMPs are conserved microbial molecules that are necessary for pathogen survival, including flagellin, bacterial elongation factor (EF)-Tu, peptidoglycan, chitin, or cell wall-derived molecules. PTI responses can be defeated by type III effector proteins interfering with PTI responses and hence promoting bacterial virulence. This so-called effector-triggered susceptibility is counteracted by a second row of plant defense reactions, called ETI, which is activated by plant resistance products (R) when individual effector proteins are recognized (Cui et al. 2015). Among the well-studied plant PRRs are flagellin receptor FLS2 (flagellin-sensitive 2), EFR (EF-Tu receptor), and the XA21 (Gómez-Gómez and Boller 2000; Klink et al. 2009; Wu et al. 2014). In the case of *Arabidopsis* receptor kinase FLS2 (Flagellin Sensing 2), which recognizes directly

Table 21.1 Plant receptor interacts with its ligand during plant–bacteria interactions

PRR	Ligand	Family	References
FLS2 (*Arabidopsis thaliana*)	flg22	LRR-RK	Gómez-Gómez and Boller (2000); Zhang et al. (2017)
FLS3 (tomato)	flgII-28	LRR-RK	Hind et al. (2016)
EFR (*Arabidopsis thaliana*)	elf18	LRR-RK	Kunze et al. (2004); Zipfel et al. (2006)
CORE (tomato)	Cold-shock protein (csp22)	LRR-RLK	Wang et al. (2016)
CSPR (*Nicotina benthamiana*)	Cold-shock protein (csp22)	LRR-RLP	Saur et al. (2016)
XA21 (rice)	RaxX(RAx21)	LRR-RK	Pruitt et al. (2015)
EPR3 (*Lotus japonicas*)	Extracellular polysaccharides L	LysM-RLK	Kawaharada et al. (2015)
ReMAX/RLP1 (*Arabidopsis thaliana*)	eMAX	LRR-RLP	Jehle et al. (2013)
LORE (*Arabidopsis thaliana*)	Lipopolysaccharide (LPS)	Lectin-RK	Ranf et al. (2015)
XPS1 (*Arabidopsis thaliana*)	xup25	LRR-RK	Mott et al. (2016)
RLP23 (*Arabidopsis thaliana*)	Nep1-like protein (nlp20)	LRR-RLP	Albert et al. (2015)
LYM1/LYM3 (*Arabidopsis thaliana*)	LysM-RLP	Peptidoglycan	Willmann et al. (2011)

to bacterial flagellin, Flg22 (N-terminal 22-amino-acid epitope), and forms a signaling complex, this PAMP recognition was very well characterized (Li et al. 2005; Dunning et al. 2007; Forsyth et al. 2010). In specific, FLS2 is needed for bacterial resistance to leaf surface inoculation, offering proof that connects PRR to pathogen resistance (Zipfel 2014). The crystal structure of the FLS2-BAK1 ectodomains associated with flg22 shows that BAK1 functions as a co-receptor (Sun et al. 2013). FLS2 interacts with Brassinosteroid Insensitive-Associated Kinase 1 (BAK1) after elicitation and forms a complex of proteins, and protein phosphorylation takes place (Schulze et al. 2010). It is also noted that Botrytis-Induced Kinase 1 (BIK1) pathogen-mediated upregulation effectively regulates the complex FLS2-BAK1. In vitro study showed that both FLS2 and BAK1 are phosphorylated by BIK1. BIK1 also phosphorylated NADPH oxidase RBOHD (respiratory burst oxidase homolog D), leading to the production of ROS (Kadota et al. 2014). In addition to BIK1, FLS2-mediated signaling events involve heterotrimeric G proteins. The Gα subunit XLG2 in *Arabidopsis* suppresses proteasome-dependent degradation of BIK1 (Liang et al. 2016). BIK1 interacts with and phosphorylates XLG2 that dissociates from BIK1 after flg22 perception and interacts with RBOHD, indicating involvement in ROS burst regulation (Liang et al. 2016).

Similarly, the elongation factor receptor (EFR) requires BAK1 for PTI upon detection of elf18, a peptide obtained from the bacterial elongation factor EF-Tu, the most abundant and conserved bacterial proteins. Elf18 perception appears to be limited to the Brassicaceae family (Boller and Felix 2009); however the implementation of EFR alone is adequate to impart elf18 responsiveness and bacterial resistance in other plant families (Zipfel et al. 2006). PTI induction is facilitated by the fact that both FLS2 and EFR belong to the RLK superfamily with the intracellular kinase domain of the extracellular LRR domain (Bonardi et al. 2012). Like FLS2, EFR also undergoes autophosphorylation, indicating that both FLS2 and EFR carry active kinase domains (Xiang et al. 2008). In most of the reports, BAK1 positively regulates the plant immunity, but in some cases, it was also found to be a negative regulator (for more detail read review – Noman et al. 2019). For instance, *Arabidopsis* BAK1 mutant lines exhibited resistance to *Pseudomonas syringae* through overproduction of salicylic acid (SA) compared to control plant (Veronese et al. 2006). On the other side, the same mutant plants are highly susceptible to *Botrytis cinerea*. Since many studies illustrate deficiencies in FLS2-mediated immunity, the apparent function of BAK1 in plant–pathogen interactions is difficult to determine. In order to determine the precise function of PRRs and various molecular regulators during plant defense, comprehensive studies are therefore needed.

The *Xa21* gene in rice (*Oryza sativa*) was the first PPR to be identified as receptor-like kinase (RLK), which imparts resistance to (diverse strain of bacteria) *Xanthomonas oryzae* pv. *oryzae* (Xoo). Similar to FLS2 and EFR, Xa21 contains non-RD kinase, the existence of which has been suggested to be associated with the role of innate immunity across kingdoms (Dardick and Ronald 2006). A series of research proved convincingly that the elicitor inducing Xa21-mediated resistance is Ax21, a sulfated protein secreted by the Xoo type I secretion system. A synthetic peptide of tyrosine-sulfated 17-amino acid corresponding to the Ax21 N-terminus is entirely involved in the generation of Xa21-mediated resistance. Cross-linking studies proposed that Ax21 interacts with XA21. Ax21 is conserved in most *Xanthomonas* species, and tyrosine sulfation is needed to be recognized by XA21 (Shen et al. 2002; Lee et al. 2009). These results are conceptually important as they show that a protein earlier categorized as a R protein is actually a PRR that mediates powerful PTI resistance.

An inventory of PAMPS and PRRs in plants has been expanded (Table 21.1; Boutrot and Zipfel 2017). PAMP identification specificities often show plant lineage-specific or intraspecies differences, potentially representing plant adaptation to the diversification of MAMPs in microbial communities. Natural variation trials in tomatoes disclosed class XII LRR-RLK FLS3 as the flgII-28 receptor, showing that plants use distinct PRRs to identify distinct patterns of the same microbial protein (Hind et al. 2016). Intriguingly, flgII-28 alleles have an impact on bacterial virulence in *A. thaliana* in a way that depends on the FLS2 (Clarke et al. 2013). It is fascinating to evaluate how flgII-28 modifications influence the recognition of flg22 and/or FLS2-triggered immunity (Saijo et al. 2018). Peptidoglycans (PGNs) constitute cell walls of most bacteria, and their perception induces immune responses in plants, insects, and mammals (Felix and Boller 2003; Guan and Mariuzza 2007; Gust et al. 2007). PGN perception happens through LysM-containing proteins that are common of almost all kingdoms of life (Zhang et al. 2009). In *Arabidopsis*, LysM-RLPs AtLYM1 and AtLYM3 identify PGNs and are likely to behave in a LysM-RLK CERK1 complex during bacterial resistance (Willmann et al. 2011). In particular, LYM1/LYM3 may bind both complex and soluble PGNs to Gram-negative or Gram-positive bacteria, whereas CERK1 does not.

Lipopolysaccharide (LPS) is the major outer membrane component of Gram negative bacteria, which protects the bacteria buts act as MAMPs for mammal and plant hosts (Ranf 2016; Kagan 2017). Convincing proof for the perception of the plant LPS was acquired with the finding of *A. thaliana* lectin S-domain RLK LORE specifically needed for LPS identification by *Pseudomonas* and *Xanthomonas* species (Ranf et al. 2015). LORE-mediated LPS recognition is limited to the Brassicaceae family; however, the implementation of LORE is adequate to transfer the specificity of LPS recognition to *Nicotiana benthamiana*. This refers to the critical position of LORE in LPS perception, even though immediate LPS–LORE binding remains to be proved. Comparative genomics approaches will be useful for MAMP–PRR identification. For example, this approach identified six new MAMPs in *P. syringae*, such as xanthine/uracil permease (xup25 epitope). The xup25 directly recognized by class XII LRR-RLK XPS1 confers bacterial resistance (Mott et al. 2016). In tobacco, cold-shock proteins (csp15/csp22 epitope) are identified as immunogenic (Felix and Boller, 2003). Natural variation experiments in tomatoes resulted to class XII LRR-RK CORE (close to EFR and Xa21) being identified as a high-affinity receptor for csp22 (Wang et al. 2016). CORE remains in the Solanaceae family, although its implementation gives complete sensitivity to csp22 and, in particular, increased resistance to Pst in *A. thaliana*.

4 Plant–Fungus Interactions

Fungi are important plant pathogens that cause greater yield losses compared to bacteria or viruses with numerous diseases on economically important crops (Tournas 2005). Pathogenic fungi can exhibit biotrophic, necrotrophic, or saprophytic (rely on dead and decaying organic matter for nutrition) kind of lifestyle (Mendgen and Hahn 2002; Glazebrook 2005; Howlett 2006). Biotrophic fungi, to obtain the nutrients from living host cells, form a structure like haustoria or hyphae throughout their life (Heath 1997; Schulze-Lefert 2004). The haustorial membrane contains ATPases which acidifies the extrahaustorial space and hence creates a pH gradient, which drives sugar- and amino acid-H+ symport catalyzed by transporters of haustorial membrane (Struck et al. 1996). The secretion of CWDEs is also very limited in biotrophic fungi (Oliver and Ipcho 2004; Schulze-Lefert 2004). It is to be noted that, in obligate biotrophs of CWDEs, secretion does not result in maceration of tissue. The infection process in biotrophic fungi is aided by appressoria formation, secretion of CWDEs, and pathogen effector molecules. Initially, fungal necrotrophs infect the host cells, and this includes different phases of conidial association followed by germination, lesion formation, and finally softening and sporulation of the tissue (Prins et al. 2000). After infection, the low molecular weight or peptide toxin production, haustoria, appressoria, hyphae formation, toxic secondary metabolites production, phytohormones and ROS formation, and cell wall-degrading enzyme's (CWDEs) secretion help in penetration (Mendgen and Hahn 2002; Oliver and Ipcho 2004; Daub and Chung 2009; Horbach et al. 2011; Stergiopoulos et al. 2013). Fungi also suppress the defense of the host cell, thus abetting their own propagation (Prins et al. 2000). Plant pathogens secrete effectors molecules that enable them to interact with their hosts and influence the outcome of this interaction. The recognition of fungal pathogens by host plants is based on PAMPs (Table 21.2). PAMPs are generally conserved molecules in the body of fungal pathogen, for instance, xylanase and chitin are well-known PAMPs of fungi (Ron and Avni 2004; Kaku et al. 2006; Miya et al. 2007). These PAMPs are recognized by their PRRs, leading to PTI reaction (Bent and Mackey 2007; Hückelhoven 2007; Zipfel 2009). To neutralize PTI, fungi produce effector proteins which suppress the immune response of the host and alter the physiology of the host cells for their multiplication and infection (Göhre and Robatzek 2008; Boller and He 2009; Stergiopoulos and de Wit 2009; Stergiopoulos et al. 2010; Koeck et al. 2011). Multiple bacterial effectors target BAK1/SERK3 and BIK1 immune kinases which are RLKs and receptor-like cytoplasmic kinases (RLCKs), respectively. Recently, it has been observed that a broadly conserved fungal effector, necrosis-inducing secreted protein 1 (NIS1) of Ascomycota and Basidiomycota is able to suppress PTI by targeting BAK1 and BIK1, just like bacteria (Irieda et al. 2019). Chitin is an essential component of the filamentous fungi cell wall. Two types of lysin motif (LysM) proteins, CERK1 (chitin elicitor kinase-1 receptor) and CEBiP (chitin elicitor binding protein), have been recognized as the chitin cell surface receptor-like protein (Wan et al. 2008). Gene silencing of these genes has shown that these LysM

Table 21.2 Plant receptor interacts with its ligand during plant–fungus interactions

PRR	Type	Ligand	References
CERK1 (*Arabidopsis thaliana*)	LysM-RLK	Chitin	Miya et al. (2007); Wan et al. (2008)
LYK5 (*Arabidopsis thaliana*)	LysM-RLK	Chitin	Cao et al. (2014)
I-3 (*Solanum pennellii*)	LEC-RLK	Avr3/Six1	Catanzariti et al. (2015)
RLP30 (*Arabidopsis thaliana*)	LRR-RLP	SCFE1	Zhang et al. (2013)
Eix1/Eix2 (*Solanum pennellii*)	LRR-RLP	Ethylene-inducing xylanase	Ron and Avni (2004)
Cf-2 (*Solanum pimpinellifolium*)	LRR-RLP	Avr2, gr-VAP1	Dixon et al. (1996); Rooney et al. (2005)
RLP42/RBPG1 (*Arabidopsis thaliana*)	LRR-RLP	Endopolygalacturonases	Zhang et al. (2013)
Cf-4 (*Solanum hirsutum*)	LRR-RLP	Avr4	Burke and Kadonaga (1997)
Cf-5 (*Solanum lycopersicum*)	LRR-RLP	Avr5	Dixon et al. (1998)
Cf-9 (*Solanum* pimpinellifolium)	LRR-RLP	Avr9	Jones et al. (1994); Luderer et al. (2001)
Hcr9-4E (*Solanum hirsutum*)	LRR-RLP	Avr4E	Burke and Kadonaga (1997); Weinberger et al. (1999)
LepR3/RLM (*Brassica napus*)	LRR-RLP	AvrLm1, AvrLm2	Larkan et al. (2013); Larkan et al. (2015)
I (*Solanum pimpinellifolium*)	LRR-RLP	Avr1/Six4	Catanzariti et al. (2017)
CEBiP (*Oryza sativa*)	LysM-RLP	Chitin	Kaku et al. (2006), Shimizu et al. (2010); Hayafune et al. (2014)
LYP4, LYP6 (*Oryza sativa*)	LysM-RLP	Peptidoglycan, chitin	Liu et al. (2012)
LYM1, LYM3 (*Arabidopsis thaliana*)	LysM-RLP	Peptidoglycan	Willmann et al. (2011)
LYM2 (*Arabidopsis thaliana*)	LysM-RLP	Chitin	Faulkner et al. (2013)

proteins in rice are required for chitin perception and signaling, while CEBiP-type molecules of *Arabidopsis* are not involved in chitin signaling, suggesting the distinction between the chitin receptor systems in these model crops. The recognition of chitin requires LysM-RLK CERK1 of host plant (Shinya et al. 2015). In the case of rice, OsCERK1 may not bind chitin directly but works in a receptor complex with the chitin-binding LysM-RLP OsCEBiP (Yamaji et al. 2010; Hayafune et al. 2014). Furthermore, OsLYP4 and OsLYP6, CEBiP homologs also bind chitin and contribute to chitin responses and disease resistance, also appear to function with OsCERK1 (Liu et al. 2012; Kouzai et al. 2014). Several reports showed that CERK1 directly binds chitin in *Arabidopsis thaliana* and thus mediates chitin-triggered immunity (Wan et al. 2008; Petutschnig et al. 2010). It has also been shown that peptidoglycan perception in *Arabidopsis* requires lysin motif protein LYM1, LYM2, and LYM3 that are CEBiP homologs and they are dispensable for CERK1-dependent chitin recognition (Wang et al. 2012; Faulkner et al. 2013), while LYM1 and LYM3 are required for PGN recognition together with CERK1 (Willmann et al. 2011). Most of the research has suggested that AtCERK1 is the major chitin receptor and mediates chitin-induced signaling, while recent studies revealed that AtLYK5 is the primary receptor to mediate CERK1 homodimerization and phosphorylation in response to chitin (Cao et al. 2014). Double mutant of LYK5/LYK4 makes plants chitin insensitive. Two separate systems of perception in *Arabidopsis*, including LYK5/LYK4-CERK1 or LYM2, are used for different chitin responses.

Ethylene-inducing xylanase (EIX) is recognized as a potent elicitor in tomato through the LRR-RLPs Eix2 and Eix1 (Ron and Avni 2004). Eix2 triggers defense responses in the plant after exposure to EIX, including Eix1 upregulation, where the EIX binding in turn attenuates Eix2 signaling by operating with BAK1 (Bar et al. 2010). Fungal endopolygalacturonases (PGs) also display elicitor activity (Poinssot et al. 2003). Studies of natural variation demonstrated that LRR-RLP42 (RBGP1) of *Arabidopsis* recognizes several *Botrytis cinerea* PGs and mediates PG-induced *Hyaloperonospora arabidopsidis* (Hpa) resistance which is dependent on SOBIR1 (Zhang et al. 2014). Despite extensive studies of disastrous oomycete pathogens including *Phytophthora infestans*, knowledge of oomycete MAMP receptors in plants is still limited (Raaymakers and Van Den Ackerveken 2016). Oomycete MAMPs include cell wall components, β-1,3- and β-1,6-glucans, glucan-chitosaccharides, and cellulose-binding elicitor lectin (Khatib et al. 2004; Nars et al. 2013). Natural variations and requirement of BAK1 for CBEL responsiveness have been described in *A. thaliana* (Larroque et al. 2013). MAMPs are also provided by the transglutaminase GP42 (Nürnberger et al. 1994, 1995; Hahlbrock et al. 1995) and eicosapolyenoic acids of *Phytophthora mycelia* (Bostock et al. 1981). Cognate PRRs have yet to be identified for these MAMPs.

Various species of fungi such as *Phytophthora* secrete many proteins with proven effector activity (Kamoun 2006). Some of the effectors, known as avirulence (Avr) proteins, show specific gene-for-gene interactions with host resistance (R) proteins (Flor 1942). Avr protein consist of two conserved motifs, viz., dEER (aspartate, glutamate, and arginine) and RxLR (arginine, any amino acid, leucine, and arginine) (Dodds et al. 2009; Tyler 2009). Several reports have shown the significance of these motifs; the mutational analysis in both of the motifs prevents the transfer of Avr1b and Avr3a from fungi (*Phytophthora sojae* and *P. infestans*) into the host cell; hence these motifs are crucial for delivering the avr protein across the host plasma membrane (Whisson et al. 2007; Dou et al. 2008). These effector molecules suppress the host defense system after entering the host cell and cause cell death (Bos et al. 2006; Dou et al. 2008). In biotrophs, effectors are needed during penetration to downregulate PTI induced by fungal interaction with the host plant.

5 Plant–Nematode Interactions

Plant parasite nematodes (PPNs) are obligate biotrophs feeding on plant's root and complete its life cycle in ~3–10 weeks. They outbreak most economically important crops, causing significant yield losses worldwide (Abad et al. 2008). The most harmful PPNs are root-knot nematodes (RKNs, *Meloidogyne* spp.) as well as cyst nematodes (CNs, *Globodera* spp. and *Heterodera* spp.) and are endoparasitic in nature. They establish a close relationship with their host plants and form specialized multinucleate feeding cells (FCs) from which they extract water and nutrient to sustain life. The molecular mechanism underlying the interaction between nematodes and plants remains understudied, despite significant progress in recent years. Like many plant pathogens, these nematodes also overcome host defenses for its own advantage (Mitchum et al. 2013; Mantelin et al. 2015). In order to fight these pests, scientists concentrated on host-specific resistance, in which one or more genes in the parent plant impart heritable resistance. (Barbary et al. 2015; Mitchum 2016; Saucet et al. 2016). Basal defense responses, PTI, were also revealed to be essential during plant–nematode associations (Holbein et al. 2016). Numerous studies had exhibited that PTI and ETI responses were produced by plants upon nematode infection (Hamamouch et al. 2011; Kandoth et al. 2011; Siddique et al. 2014; Manosalva et al. 2015; Sidonskaya et al. 2016). The first comprehensive study on nematode-associated molecular pattern (NAMPs) illustrated the perception of nematode-produced ascarosides by plant cells, activating the expression of gene associated with MAMP-triggered immunity (Manosalva et al. 2015). Ascarosides act as pheromones and are highly conserved among nematodes (Choe et al. 2012). Ascr#18 is the most abundant among root-knot and cyst nematodes and is responsible for increased resistance in potato, tomato, barley, and *Arabidopsis* against various pathogens, including virus, bacteria fungi, and oomycete and nematodes (Manosalva et al. 2015). Above evidences show that the nematode-secreted ascarosides are likely to be NAMPS perceived by plants for the induction of basal defense mechanisms. However, there is still no real proof that plant cell surface receptors recognize ascarosides. In addition, a study contended that the protein nature of potential NAMPs contributes to the induction of *Arabidopsis* basal immunity (Mendy et al. 2017). To prove this, NemaWater was prepared to incubate the plant parasitic nematodes J2s (*Meloidogyne incognita* and *Heterodera schachtii*) in water for 24 hours. Then the roots of *Arabidopsis* were treated with NemaWater, and apoplastic ROS bursts were formed that are characteristic of induced plant basal immunity. This study also showed that PTI responses were mediated by NILR1, which was directly induced after infection with nematode (Mendy et al. 2017). Thus, for perceiving a protein of nematode and inducing plant basal immunity against nematode infection, this was the first report on NILR1 as the PRR. However, further studies are still needed to understand the NILR1 downstream signaling.

5.1 *Effectors of the Plant Parasitic Nematodes Suppress Plant Defense*

Most of the plant parasitic nematodes (PPNs) have a unique structure known as a stylet (hollow, protrusible mouth spear) used to puncture plant cell walls, facilitate the migration of PPNs into plant tissues, and finally inject esophageal secretions (effector protein) into the host cells to initiate nematode feeding site (Davis et al. 2000). These stylet-secreted effector proteins help to modulate cellular mechanisms that promote the growth and development of PPNs. Majority of nematodes secreted effector protein, also linked to suppression host defenses (Table 21.3), although they are not NAMPs. Various effectors have been characterized which stressed the plant to activate defense responses (Smant and Jones 2011). For instance, *Arabidopsis* plants overexpressing nematode effector, Hs10A06, are more susceptible to cyst nematode, bacterial, and viral pathogens and upregulate smaller amounts of pathogenesis-related (PR) proteins. Hs10A06 acts on signaling molecule, salicylic acid, and the antioxidant machinery, thus shielding nematodes against plant defense responses (Hewezi et al. 2010). Similarly, the Hs4F01 annexin-like effector interacts with an oxidoreductase of the 2OG-Fe (II) oxygenase family to inhibit the triggering of host defense (Patel et al. 2010). One more CN effector, Hg30C02, interacts to a potential PR protein of plant, β-1, 3-endoglucanase, and is involved in defense suppression (Hamamouch et al. 2012). Moreover, first calreticulin of *M. incognita*, Mi-CRT has been revealed to have a direct effect on PTI suppression (Jaouannet et al. 2013).

The cyst nematode effector's repertoire includes a large family of secreted effectors having a SPRY domain, named SPRYSECs (SP1a and Ryanodine receptor (SPRY) (Rehman et al. 2009; Rosso et al. 2009). The *G. pallida* SPRYSEC (GpRBP-1) has been exhibited to be the avirulence target of the GPA-2 NB-LRR-resistant protein (Sacco et al. 2009). Remarkably, the GrSPRYSEC-19 protein has been shown to suppress the ETI facilitated by numerous NB-LRR-resistant proteins, including GPA-2 (Postma et al. 2012). One more class of A Gr-VAP1 effector protein is considered as PAMP because it activates basal innate immunity in potato roots upon *G. rostochiensis* infestation. Nematode effector proteins can modulate auxin-based signaling and defense responses. For example, to control syncytia formation, 19C07 *H. schachtii* effector binds to LAX3 auxin transporter of *Arabidopsis* (Lee et al. 2011).

There is a long list of the nematode effectors which suppress the plant immunity (Table 21.3), for example, MiSGCR, *Meloidogyne incognita* effector, suppresses plant cell death caused by oomycete or bacterial triggers of plant defense (Nguyen et al. 2018). Similarly, HgGLAND4 and HaGland5 downregulate the defense-related gene expression (Barnes et al. 2018; Yang et al. 2019). During interaction, the effector Mi-CRT, of *Meloidogyne incognita* is secreted in the apoplasm and suppresses plant basal defenses (Jaouannet et al. 2013). A recently identified *M. incognita* effector, RHA1B, suppresses effector-triggered immunity in potato. This effector also blocks the expression of *Acre31* and *WRKY22* genes (a marker of

Table 21.3 Nematode effectors that suppress plant immunity to establish feeding cells

Nematode effector	Nematode	Predict function	Plant target and function	References
Cyst nematode, *Globodera* spp.				
GpRBP-1	*Globodera pallida*	SPRYSEC	NB-LRR-resistant protein potato GPA-2	Sacco et al. (2009)
GpRHA1B	*Globodera pallida*	Ubiquitin ligase	Unknown, suppress effector-triggered immunity	Kud et al. (2019)
GrVAP1	*Globodera rostochiensis*	Venom allergen protein	Papain-like cysteine protease Rcr3pim	Lozano-Torres et al. (2012)
GrSPRYSEC-19	*Globodera rostochiensis*	SPRYSEC	NB-LRR protein tomato SW5F	Rehman et al. (2009)
Cyst nematode, *Heterodera spp.*				
*Ha*Gland5	*Heterodera avenae*	–	Pyruvate dehydrogenase subunit, suppress the plant defense responses	Yang et al. (2019)
*HaVAP1*and 2	*Heterodera avenae*	Venom allergen protein	CYPRO4-like protein (HvCLP), suppress the cell death induced by BAX	Luo et al. (2019)
HgGLAND4	*Heterodera glycines*	DNA-binding protein	Lipid transfer protein AtLTP, suppresses plant defenses	Barnes et al. (2018)
HgGLAND18	*Heterodera glycines*	Known	Suppression of both basal and hypersensitive cell death immune responses	Noon et al. (2016)
Hg30C02	*Heterodera glycines*	Unknown	β-1,3-endoglucanase	Hamamouch et al. (2012)
Hs19C07	*Heterodera schachtii*	Unknown	Plasma membrane auxin influx transporter AtLAX3, involved in defense suppression	Lee et al. (2011)
Hs4F01	*Heterodera schachtii*	Annexin-like	Oxidoreductase of the 2OG-Fe(II) oxygenase family, prevent the triggering of host defense	Patel et al. (2010)
Root knot nematode, *Meloidogyne* spp.				
MiSGCR1	*Meloidogyne incognita*	Small glycine- and cysteine-rich protein	Triggers of plant defense	Nguyen et al. (2018)
Mi-CRT	*Meloidogyne incognita*	Calreticulin	Suppression of plant basal defenses	Jaouannet et al. (2013)
MjTTL5	*Meloidogyne javanica*	Unknown	Ferredoxin: Thioredoxin reductase catalytic subunit AtFTRc, suppress plant immune response	Lin et al. (2016)

(continued)

Table 21.3 (continued)

Nematode effector	Nematode	Predict function	Plant target and function	References
Misp12	*Meloidogyne incognita*	Unknown	Suppression of plant defense response	Xie et al. (2016)
Mh265	*Meloidogyne hapla*	Unknown	Suppression of the host defenses	Gleason et al. (2017)
MgGPP	*Meloidogyne graminicola*	Unknown	Suppression of host defense responses	Chen et al. (2017)
Mg16820	*Meloidogyne graminicola*	Unknown	Suppression of host defense responses	Naalden et al. (2018)
Mg01965	*Meloidogyne graminicola*	Unknown	Suppression of host innate immune responses	Zhuo et al. (2019)

PTI), which is mediated by flg22, but for this, E3 activity of RHA1B is not required. RHA1B thus facilitates nematode parasitism not only by triggering degradation of NB-LRR immune receptors to block ETI signaling but also by suppressing PTI signaling through a mechanism that is still unknown to be independent of E3.

6 Plant–Virus Interactions

Plants are stubbornly prone to viruses globally as they hack the host system for their growth. It is believed that these cause more than one-third of crop diseases (Boualem et al. 2016). Although the population and distribution of viruses are somehow restricted by the management of insect vectors through the implementation of multiple chemicals, these chemical treatments cannot immediately prevent infections of plant viruses. The development program of disease-resistant crops through plant breeding is therefore crucial as the use of chemicals has a more adverse effect on animals and the atmosphere (Islam et al. 2017). In order to initiate infection, an extensive and competitive protein–protein, protein–lipid, and protein–nucleic acid interaction network is established between the plant and the virus. Plant viruses are parasites that are simple, obligatory, and intracellular and have comparatively small size of the genetic material. Viruses are pathogens based on nucleic acids and contain a unique protein called capsid, which is made up small units of capsomers. For illustration, a large number of identified plant viruses are positive single-stranded RNA (+ssRNA) viruses, generally encoding no more than a dozen proteins. They carry their genetic information as single-stranded RNA (ssRNA), double-stranded RNA (dsRNA), or DNA and lack the parts needed for their autonomous survival, so for their life cycle, they depend solely on host systems. These interactions result in the formation of complexes of viral replication that make new offspring of viral RNA in host cells. Remarkable progress has been achieved in the understanding of plant–virus interrelationship over the previous century, which filled the gap between plant–inborn virus immunity. In contrast to animal counterparts with dedicated defense cells, crops depend on each cell's ability to detect and protect against the invaders.

To defend themselves, plants have developed multilayered monitoring against plant viruses such as innate immunity, RNA silencing, translation repression, protein degradation mediated by ubiquitination and autophagy, and other dominant gene-mediated resistance defenses. Although viruses are comparatively easy genetic structures, molecular resistance systems and susceptibility to viral diseases are not completely known till date. Consequently, an efficient effector-triggered immunity (ETI) will maintain the crops' health; nevertheless, an inadequate ETI will contribute to the development of disease, i.e., the plant susceptibility. Models of total resistance of most pathogens do not match well with viral resistance due to the parasitic intracellular nature of the virus. For example, PRRs serve as a major defense component when a plasma membrane receptor perceives a fungal or bacterial MAMP (Tena et al. 2011) but cannot play a part in the fight against plant viruses as they are previously considered non-PAMP-coding pathogens. Moreover, several studies suggest that PTI also plays an essential role in both incompatible and compatible plant–virus relationship. Recently, a research suggests that the existence of BRI1-associated kinase 1 (BAK1, a plant PRR coreceptor kinas SERK3) regulator is essential for plants' antiviral defense. Mutants of *Arabidopsis* BAK-1 display increased susceptibility during compatible relationships to three distinct RNA viruses. In addition, crude viral samples (unpurified virions) trigger numerous responses to PTI markers that are dependent on BAK1. Overall, this argues that plant antiviral resistance results from BAK1-dependent PTI (Kørner et al. 2013). Similarly, L-type lectin receptor kinases (LecRKs) were recognized as hypothetical immune receptors in *Arabidopsis*. Gene silencing of CaLecRK S.5, a transmembrane RK of pepper (Capsicum annuum L), leads to increased vulnerability of this plant to several different pathogens, such as *Tobacco mosaic virus* and *Pepper mild mottle virus*, bacterium (*Xanthomonas campestris* pv. Vesicatoria), and oomycete pathogen (*Phytophthora capsici*) (Woo et al. 2016).

Another study revealed that double-stranded RNAs (dsRNAs) are conserved PAMP with virus replication (Alexopoulou et al. 2001), which induces PTI reaction when treating *Arabidopsis* with dsRNAs, which depends on the somatic embryogenesis receptor-like kinase 1 (SERK1) co-receptor but autonomous of the RNA silencing pathway (Niehl et al. 2016). These results therefore indicate that dsRNAs are authentic PAMPs in crops that activate a SERK1 signaling cascade and a unique dsRNA receptor. Likewise, the *Potato virus X* (PVX) coat protein (CP) and Triple Gene Block 1 proteins are sensed by plant cells and subsequently trigger the activation of PTI response (Perraki et al. 2018). Thus, unknown PAMPs obtained from PVX CP proteins may trigger PTI-like reactions. According to the theory, general three-dimensional folding was discovered in virus CPs with comparable materials, such as flexuous filamentous rods (Zamora et al. 2017). Moreover, at current, viral PAMPs, plant PRRs perceiving them, and cellular signals are still unknown. In the last two decades, several viral resistance factors as well as viral factors have been identified in different plant species (Table 21.4).

Table 21.4 Representative host factors against plant viruses with antiviral activity

Host factor (host name)	Cellular function	Virus factor and virus	References
Virus movement			
ESC1 (*Arabidopsis thaliana*)	Mechanosensitive ion channel	NA-CMV, (*Turnip mosaic virus*)	Zhang et al. (2019)
KELP (*Nicotiana benthamiana*)	Transcription coactivator	P30 (*Turnip mosaic virus*)	Sasaki et al. (2009)
BTR1 (*Arabidopsis thaliana*)	mRNA binding	Genomic RNA (*Turnip mosaic virus*)	Fujisaki and Ishikawa (2008)
Rsv3 (*Glycine max*)	NA	CI (*Soybean mosaic virus*)	Zhang et al. (2009)
Ny1 (*Solanum tuberosum*)	NA	NA (*Potato virus Y*)	Szajko et al. (2008); Lukan et al. (2018)
Virus formation and accumulation			
NBR1 (*Arabidopsis thaliana*)	Autophagy cargo receptor	CP, virions, HC-pro– (*Cauliflower mosaic virus*, *Turnip mosaic virus*, and *Watermelon mosaic virus*)	Hafrén et al. (2017, 2018)
PUS4- (*Nicotiana benthamiana*)	Pseudouridine synthase	Genomic RNA (*Brome mosaic virus*)	Zhu et al. (2007)
CYR1 (*Vigna mungo*)	NA	CP (*Mungbean yellow mosaic virus*)	Maiti et al. (2012)
RFP1 (*Nicotiana tabacum*)	Ubiquitination	BC1 (*Tomato yellow leaf curl virus*)	Shen et al. (2016)
Viral RNA translation			
NIK1- *Arabidopsis thaliana*	Receptor-like kinase	NSP (*Cabbage leaf curl virus*)	Zorzatto et al. (2015)
APUM5- *Arabidopsis thaliana*	mRNA binding	mRNA (*Cytomegalovirus*, *Turnip mosaic virus*)	Huh et al. (2013)
Accumulation or activity of the replication proteins			
Beclin1 (*Nicotiana benthamiana*)	Autophagy	NIb (*Turnip mosaic virus*)	Li et al. (2018)
Tm1- (*Solanum lycopersicum*)	NA	130 K (*Turnip mosaic virus*)	Ishibashi et al. (2007)
TARF- (*Nicotiana tabacum*)	Ubiquitination	126 K (*Tobacco mosaic virus*)	Yamaji et al. (2010)
Rsp5p- (yeast)	Ubiquitination	P92 (*Tomato bushy stunt virus*)	Barajas et al. (2009)
PVR4 (Capsicum *annum*)	NA	NIb (*Pepino mosaic virus*, *Potato virus Y*)	Kim et al. (2015)
mRNA stability			
DCP1, DCP2, XRN4, PARN (*Nicotiana benthamiana*, and *Arabidopsis thaliana*)	mRNA decay	NA (*Turnip mosaic virus*)	Li and Wang (2018)
XRN4 (yeast and *Nicotiana benthamiana*)	mRNA decay	mRNA (*Tomato bushy stunt virus*)	Jaag and Nagy (2009)
DCP1 (*Arabidopsis thaliana*)	mRNA decay	mRNA (*Tobacco rattle virus*)	Ma et al. (2015)

6.1 Effector of Plant Pathogenic Viruses Recognition

In 1984, the first genes of bacterial and viral avirulence (avr) were cloned. Though in the same year, viruses and bacterial avr genes were physically isolated, the problems associated with their characterization after discovery were very different, and these differences have had a major influence on the paradigm of host–microbes relationships over the past three decades. Bacterial avr proteins have eventually been shown to suppress host defenses, contributing to their classification as effectors, while study on viral avr proteins has focused on their involvement in the viral infection cycle instead of on host defenses. The ability of a pathogen to infect a plant depends upon a complex series of interactions between the two participants. The result of these interactions, whether disease or immunity, is affected by pathogenic and host proteins, small molecules, or even small RNAs, which can be used by the pathogen to promote infection or by the host to block pathogen infection (Hogenhout et al. 2009). Effectors were identified as pathogenic proteins and small molecules, which are the structure and function of the host cell, and have become a major attraction for studies (Michelmore et al. 2017). The success of viruses as pathogen relies on their potential inside host cells to reproduce. They accomplish this by actively reprogramming the metabolism of host cells to promote the process of infection and enable viruses to escape or suppress mechanisms of host defense. Recently, study on plant–pathogen interactions has stated that an understanding of effectors may provide useful insights into the identification of host resistance breeding material and contribute to basic improvements in plant biology (Michelmore et al. 2017). Effectors are pathogen-secreted polypeptides or small molecules that increase infection and influence plant immunity (Toruño et al. 2016). Secretion can require elaborate pathogen-owned machinery for delivering effectors throughout the plant cell plasma membrane and then into the cytoplasm. The effectors are transferred to their locations of action once inside the cytoplasm and can modulate many processes within the host cell, including cell death induction, hormone response, plant gene expression, and vesicular transportation pathway. Moreover, the literature includes different views on whether plant virus proteins should be considered pathogenic effectors (Kang et al. 2005; Mandadi and Scholthof 2013; de Ronde et al. 2014), which could be due to the fact that many pathogenic effectors suppress host defenses typically linked with either SA or JA pathways, which were not normally associated with virus proteins until now. However, it has now been observed that pathogen effectors can do much more than suppress plant defenses, as they can interfere with plant hormones and cause huge modifications in the host gene transcripts and are capable of cell-to-cell motion (Mitchum et al. 2013; Toruño et al. 2016). Several of these functions are also characteristic of the proteins of the plant virus. In addition, microarray studies of *Arabidopsis* infected with virus have revealed that infections cause major changes in the defense and stress response genes (Whitham et al. 2003). Therefore, virus proteins have much more in common than is usually acknowledged with effectors of other pathogens. For instance, a multifunctional plant virus protein *Cauliflower mosaic virus* (CaMV) P6 shows its similarities with other well-established pathogen effectors.

First-time CaMV, a P6 protein expressing transgene, elicited virus like symptoms in tobacco (Baughman, Jacobs and Howell 1988). The main role of P6 in eliciting plant defenses and signs of disease in resistant and susceptible plants was found shortly after the scanning of its sequence on CaMV genome (Bonneville et al. 1989). The P6 proteins of CaMV and the related *Caulimovirus* expressing P6 became the first virus genes shown through recombinant DNA techniques to be the avirulent (avr) gene responsible for activating HR in *D. stramonium* and *N. edwardsonii* (Daubert et al. 1984; Schoelz et al. 1986). later on, the function of the Avr factor eventually demonstrated by agroinfiltration experiments, in specific plant and virus strain combinations (Palanichelvam et al. 2000), though the function of P6 protein throughout the CaMV infection cycle remained unknown. The subsequent function *Figwort mosaic virus* has been shown to reprogram eukaryotic ribosomes in such a way that they can reinitiate the translation of downstream cistrons into polycistronic *Caulimovirus* 35S RN (Bonneville et al. 1989; Gowda et al. 1989; Scholthof et al. 1992; Zijlstra and Hohn 1992; De Tapia et al. 1993). P6 was named as a translational transactivator (TAV)/viroplasmin protein to acknowledge its core role in CaMV 35S genomic RNA translation. The CaMV P6 protein description mainly highlighted its function in translation till 2004.

Starting in 2004, a series of articles showed that CaMV-infected *Arabidopsis* plants had significant changes in host basal defenses. One study showed that transgenic *Arabidopsis* crops expressing P6 are insensitive to ethylene, as shown in the triple reaction test (Geri et al. 2004). At around the same period, it was shown that CaMV infections of sensitive crops stimulate ROS as well as marker genes for the defense mechanisms of ethylene and SA, although it was concluded that these pathways could not limit CaMV infections (Love et al. 2005, 2007). An assessment of P6-transgenic crops eventually shows that the expression of P6 suppresses the oxidative burst, gene expression linked with the SA defense pathway, and SA-dependent autophagy (Love et al. 2012). Furthermore, P6 enhanced transcription of JA defense pathway genes (Love et al. 2012; Zvereva et al. 2016). Thus, the P6 effector fits even the most stringent definition of a pathogen effector as it suppresses host basal defenses. Taken together, these reports show that, as with other types of pathogens, CAMV infections produce a response similar to PTI in their hosts and that over the lifetime of the infected plant, CAMV P6 is capable of suppressing host PTI. In the case of viruses, the number of effectors is restricted by the relatively small size of a virus genome, but much remains to be studied about the function of the virus effector. In essence, any new function characterized by a plant virus protein may be analogous to the discovery of a completely new effector present in a bacterial, fungal, oomycete, or nematode pathogen. The objective should be to fully comprehend how pathogenic effectors reprogram their hosts to promote infection, irrespective of the pathogen.

7 Conclusions

Every year, 10–20% of worldwide crop production is dropped owing to plant diseases, influencing food safety (Oerke 2006). In comparison to conventional chemical control methods, genetic techniques are required to enhance resistance to plant disease and the use of PRRs that provide broad-spectrum resistance to disease (Dangl et al. 2013). To improve crop productivity, plant biologists have been targeting disease resistance over the past decades. Nevertheless, current molecular observations have revealed various components of plant immunity that enable specific pathogens to be identified and tackled. Identifying various microbial patterns and their cognate receptors, pattern recognition receptor (PRR), in model plants gives an insight into the complexity of plant immune system recognition capabilities. Tremendous progress has been made in identification of diverse and huge number of patterns that are recognized by individual plant species. A significant quantity of immune receptors is limited to individual plant genera only. This may have two consequences: firstly, transferring single or multiple receptors (PPR pyramiding) in crops to consciously boost plant immunity without putting firm selective pressure on pathogens; secondly, plant species beyond models could become valuable assets for unique pattern recognition specificities. Further studies involving the comprehensive integration of genomics with biochemical methods may broaden our knowledge of PRR- and NB-LRR-facilitated signaling mechanisms, as it will open up innovative opportunities for creating disease resistance in plants.

References

Abad P, Gouzy J, Aury J-M et al (2008) Genome sequence of the metazoan plant-parasitic nematode *Meloidogyne incognita*. Nat Biotechnol 26:909–915

Albert I, Böhm H, Albert M et al (2015) An RLP23-SOBIR1-BAK1 complex mediates NLP-triggered immunity. Nat Plants 5:151402015

Alexopoulou L, Holt AC, Medzhitov R et al (2001) Recognition of double-stranded RNA and activation of NF-κB by Toll-like receptor 3. Nature 413:732–738

Bar M, Sharfman M, Ron M et al (2010) BAK1 is required for the attenuation of ethylene-inducing xylanase (Eix)-induced defense responses by the decoy receptor LeEix1. Plant J 63:791–800

Barajas D, Li Z, Nagy PD (2009) The Nedd4-type Rsp5p ubiquitin ligase inhibits tombusvirus replication by regulating degradation of the p92 replication protein and decreasing the activity of the tombusvirus replicase. J Virol 83:11751–11764

Barbary A, Djian-Caporalino C, Palloix A et al (2015) Host genetic resistance to root-knot nematodes, *Meloidogyne* spp., in Solanaceae: from genes to the field. Pest Manag Sci 71:1591–1598

Barnes SN, Wram CL, Mitchum MG et al (2018) The plant-parasitic cyst nematode effector GLAND4 is a DNA-binding protein. Mol Plant Pathol 19:2263–2276

Baughman GA, Jacobs JD, Howell SH (1988) Cauliflower mosaic virus gene VI produces a symptomatic phenotype in transgenic tobacco plants. Proc Natl Acad Sci 85:733–737

Bent AF, Mackey D (2007) Elicitors, effectors, and R genes: the new paradigm and a lifetime supply of questions. Annu Rev Phytopathol 45:399–436

Boller T, Felix G (2009) A renaissance of elicitors: perception of microbe-associated molecular patterns and danger signals by pattern-recognition receptors. Annu Rev Plant Biol 60:379–406

Boller T, He SY (2009) Innate immunity in plants: an arms race between pattern recognition receptors in plants and effectors in microbial pathogens. Science 8:742–744

Bonardi V, Cherkis K, Nishimura MT et al (2012) A new eye on NLR proteins: focused on clarity or diffused by complexity? Curr Opin Immunol 24:41–50

Bonneville JM, Sanfaçon H, Fütterer J et al (1989) Posttranscriptional trans-activation in cauliflower mosaic virus. Cell 22:1135–1143

Bos JIB, Kanneganti TD, Young C et al (2006) The C-terminal half of Phytophthora infestans RXLR effector AVR3a is sufficient to trigger R3a-mediated hypersensitivity and suppress INF1-induced cell death in *Nicotiana benthamiana*. Plant J 48:165–176

Bostock RM, Kuc JA, Laine RA (1981) Eicosapentaenoic and arachidonic acids from *Phytophthora infestans* elicit fungitoxic sesquiterpenes in the potato. Science 3:67–69

Boualem A, Dogimont C, Bendahmane A (2016) The battle for survival between viruses and their host plants. Curr Opin Virol 17:32–38

Boutrot F, Zipfel C (2017) Function, discovery, and exploitation of plant pattern recognition receptors for broad-spectrum disease resistance. Annu Rev Phytopathol 4:257–286

Brunner F, Nürnberger T (2012) Identification of immunogenic microbial patterns takes the fast lane. Proc Natl Acad Sci U S A 109:4029–4030

Bulgarelli D, Schlaeppi K, Spaepen S et al (2013) Structure and functions of the bacterial microbiota of plants. Annu Rev Plant Biol 64:807–838

Burke TW, Kadonaga JT (1997) The downstream core promoter element, DPE, is conserved from Drosophila to humans and is recognized by TAF(II)60 of *Drosophila*. Genes Dev 11:3020–3031

Cao Y, Liang Y, Tanaka K et al (2014) The kinase LYK5 is a major chitin receptor in *Arabidopsis* and forms a chitin-induced complex with related kinase CERK1. elife 23:3

Catanzariti AM, Lim GTT, Jones DA (2015) The tomato I-3 gene: a novel gene for resistance to Fusarium wilt disease. New Phytol 207:106–118

Catanzariti AM, Do HTT, Bru P et al (2017) The tomato I gene for Fusarium wilt resistance encodes an atypical leucine-rich repeat receptor-like protein whose function is nevertheless dependent on SOBIR1 and SERK3/BAK1. Plant J 89:1195–1209

Chen J, Lin B, Huang Q et al (2017) A novel *Meloidogyne graminicola* effector, MgGPP, is secreted into host cells and undergoes glycosylation in concert with proteolysis to suppress plant defenses and promote parasitism. PLoS Pathog 12:e1006301

Choe A, Von Reuss SH, Kogan D et al (2012) Ascaroside signaling is widely conserved among nematodes. Curr Biol 8:772–780

Clarke CR, Chinchilla D, Hind SR et al (2013) Allelic variation in two distinct *Pseudomonas syringae* flagellin epitopes modulates the strength of plant immune responses but not bacterial motility. New Phytol 200:847–860

Cornelis GR, Van Gijsegem F (2000) Assembly and function of type iii secretory system. Annu Rev Microbiol 54:735–774

Costa TRD, Felisberto-Rodrigues C, Meir A et al (2015) Secretion systems in gram-negative bacteria: structural and mechanistic insights. Nat Rev Microbiol 13:343–359

Cui H, Tsuda K, Parker JE (2015) Effector-triggered immunity: from pathogen perception to robust defense. Annu Rev Plant Biol 66:487–511

Dangl JL, Horvath DM, Staskawicz BJ (2013) Pivoting the plant immune system from dissection to deployment. Science 341:746–751

Dardick C, Ronald P (2006) Plant and animal pathogen recognition receptors signal through non-RD kinases. PLoS Pathog 2:e2

Daub ME, Chung K-R (2009) Photoactivated perylenequinone toxins in plant pathogenesis. The Mycota 15:197–206

Daubert SD, Schoelz J, Debao L et al (1984) Expression of disease symptoms in cauliflower mosaic virus genomic hybrids. J Mol Appl Genet 2:537–547

Davis EL, Hussey RS, Baum TJ et al (2000) Nematode parasitism genes. Annu Rev Phytopathol 38:365–396

De Jonge R, Van Esse HP, Maruthachalam K et al (2012) Tomato immune receptor Ve1 recognizes effector of multiple fungal pathogens uncovered by genome and RNA sequencing. Proc Natl Acad Sci U S A 27:5110–5115

de Ronde D, Butterbach P, Kormelink R (2014) Dominant resistance against plant viruses. Front Plant Sci 5:307

De Tapia M, Himmelbach A, Hohn T (1993) Molecular dissection of the cauliflower mosaic virus translation transactivator. EMBO J 12:3305–3314

Derevnina L, Dagdas YF, De la Concepcion JC et al (2016) Nine things to know about elicitins. New Phytol 212:888–895

Dixon MS, Hatzixanthis K, Jones DA et al (1998) The tomato disease resistance gene and six homologs show pronounced allelic variation in leucine-rich repeat copy number. Plant Cell 10:1915–1925

Dixon MS, Jones DA, Keddie JS et al (1996) The tomato Cf-2 disease resistance locus comprises two functional genes encoding leucine-rich repeat proteins. Cell 9:451–459

Dodds PN, Rathjen JP (2010) Plant immunity: towards an integrated view of plant pathogen interactions. Nat Rev Genet 11:539–534

Dodds PN, Rafiqi M, Gan PHP et al (2009) Effectors of biotrophic fungi and oomycetes: pathogenicity factors and triggers of host resistance. New Phytol 183:993–1000

Dou D, Kale SD, Wang X et al (2008) Conserved C-terminal motifs required for avirulence and suppression of cell death by *Phytophthora sojae* effector Avr1b. Plant Cell 20:1118–1113

Dunning FM, Sun W, Jansen KL et al (2007) Identification and mutational analysis of Arabidopsis FLS2 leucine-rich repeat domain residues that contribute to flagellin perception. Plant Cell 19:3297 3313

Eulgem T, Somssich IE (2007) Networks of WRKY transcription factors in defense signaling. Curr Opin Plant Biol 10:366–371

Fang A, Han Y, Zhang N et al (2016) Identification and characterization of plant cell death-inducing secreted proteins from Ustilaginoidea virens. Mol Plant-Microbe Interact 29:405–416

Faulkner C, Petutschnig E, Benitez Alfonso Y et al (2013) LYM2-dependent chitin perception limits molecular flux via plasmodesmata. Proc Natl Acad Sci U S A 110:9166–9170

Felix G, Boller T (2003) Molecular sensing of bacteria in plants. the highly conserved RNA-binding motif RNP-1 of bacterial cold shock proteins is recognized as an elicitor signal in tobacco. J Biol Chem 21:6201–6208

Felix G, Duran JD, Volko S et al (1999) Plants have a sensitive perception system for the most conserved domain of bacterial flagellin. Plant J 18:265–276

Flor HH (1942) Inheritance of pathogenicity in Melampsora lini. Phytopathology 32:653–669

Forsyth A, Mansfield JW, Grabov N et al (2010) Genetic dissection of basal resistance to Pseudomonas syringae pv. Phaseolicola in accessions of Arabidopsis. Mol Plant-Microbe Interact 23:1545–1552

Fox RTV, Manners JG, Myers A (1971) Ultrastructure of entry and spread of Erwinia carotovora var. atroseptica into potato tubers. Potato Res 14:61–73

Fujisaki K, Ishikawa M (2008) Identification of an *Arabidopsis thaliana* protein that binds to tomato mosaic virus genomic RNA and inhibits its multiplication. Virology 25:402–411

Galán JE, Collmer A (1999) Type III secretion machines: bacterial devices for protein delivery into host cells. Science 21:1322–1328

Geri C, Love AJ, Cecchini E et al (2004) *Arabidopsis* mutants that suppress the phenotype induced by transgene-mediated expression of cauliflower mosaic virus (CaMV) gene VI are less susceptible to CaMV-infection and show reduced ethylene sensitivity. Plant Mol Biol 56:111 124

Getz S (1983) Scanning electron microscopy of infection sites and lesion development on tomato fruit infected with pseudomonas syringae pv. Tomato. Phytopathology 73:39–43

Glazebrook J (2005) Contrasting mechanisms of defense against biotrophic and necrotrophic pathogens. Annu Rev Phytopathol 43:205–227

Gleason C, Polzin F, Habash SS et al (2017) Identification of two Meloidogyne hapla genes and an investigation of their roles in the plant-nematode interaction. Mol Plant-Microbe Interact 30:101–112

Göhre V, Robatzek S (2008) Breaking the barriers: microbial effector molecules subvert plant immunity. Annu Rev Phytopathol 46:189–215

Gómez-Gómez L, Boller T (2000) FLS2: an LRR receptor-like kinase involved in the perception of the bacterial elicitor flagellin in *Arabidopsis*. Mol Cell 5:1003–1011

Gowda S, Wu FC, Scholthof HB et al (1989) Gene VI of figwort mosaic virus (caulimovirus group) functions in posttranscriptional expression of genes on the full-length RNA transcript. Proc Natl Acad Sci U S A 86:9203–9207

Guan R, Mariuzza RA (2007) Peptidoglycan recognition proteins of the innate immune system. Trends Microbiol 15:127–134

Gust AA, Biswas R, Lenz HD et al (2007) Bacteria-derived peptidoglycans constitute pathogen-associated molecular patterns triggering innate immunity in *Arabidopsis*. J Biol Chem 2:32338–32348

Hafrén A, Macia JL, Love AJ et al (2017) Selective autophagy limits cauliflower mosaic virus infection by NBR1-mediated targeting of viral capsid protein and particles. Proc Natl Acad Sci U S A 114:E2026–E2035

Hafrén A, Üstün S, Hochmuth A et al (2018) Turnip mosaic virus counteracts selective autophagy of the viral silencing suppressor HCpro. Plant Physiol 176:649–662

Hahlbrock K, Scheel D, Logemann E et al (1995) Oligopeptide elicitor-mediated defense gene activation in cultured parsley cells. Proc Natl Acad Sci U S A 9:4150–4157

Hamamouch N, Li C, Seo PJ et al (2011) Expression of *Arabidopsis* pathogenesis-related genes during nematode infection. Mol Plant Pathol 12:355–364

Hamamouch N, Li C, Hewezi T et al (2012) The interaction of the novel 30C02 cyst nematode effector protein with a plant β-1,3-endoglucanase may suppress host defence to promote parasitism. J Exp Bot 63:3683–3695

Hayafune M, Berisio R, Marchetti R et al (2014) Chitin-induced activation of immune signaling by the rice receptor CEBiP relies on a unique sandwich-type dimerization. Proc Natl Acad Sci U S A 111:E404–E413

Heath MC (1997) Signalling between pathogenic rust fungi and resistant or susceptible host plants. Ann Bot 80:713–720

Hewezi T, Howe PJ, Maier TR et al (2010) Arabidopsis spermidine synthase is targeted by an effector protein of the cyst nematode *Heterodera schachtii*. Plant Physiol 152:968–984

Hind SR, Strickler SR, Boyle PC et al (2016) Tomato receptor flagellin-sensing 3 binds flgII-28 and activates the plant immune system. Nat Plants 06:16136

Hogenhout SA, Van Der Hoorn RAL, Terauchi R et al (2009) Emerging concepts in effector biology of plant-associated organisms. Mol Plant-Microbe Interact 22:115–122

Holbein J, Grundler FMW, Siddique S (2016) Plant basal resistance to nematodes: an update. J Exp Bot 67:2049–2061

Horbach R, Navarro-Quesada AR, Knogge W et al (2011) When and how to kill a plant cell: infection strategies of plant pathogenic fungi. J Plant Physiol 1:51–62

Howlett BJ (2006) Secondary metabolite toxins and nutrition of plant pathogenic fungi. Curr Opin Plant Biol 9:371–375

Huang CF, Yamaji N, Chen Z et al (2012) A tonoplast-localized half-size ABC transporter is required for internal detoxification of aluminum in rice. Plant J 69:857–867

Hückelhoven R (2007) Cell wall–associated mechanisms of disease resistance and susceptibility. Annu Rev Phytopathol 45:101–127

Huh SU, Kim MJ, Paek KH (2013) *Arabidopsis* Pumilio protein APUM5 suppresses cucumber mosaic virus infection via direct binding of viral RNAs. Proc Natl Acad Sci U S A 8:779–784

Imam J, Singh PK, Shukla P (2016) Plant microbe interactions in post genomic era: perspectives and applications. Front Microbiol 26:1488

Irieda H, Inoue Y, Mori M et al (2019) Conserved fungal effector suppresses PAMP-triggered immunity by targeting plant immune kinases. Proc Natl Acad Sci U S A 116:496–505
Ishibashi K, Masuda K, Naito S et al (2007) An inhibitor of viral RNA replication is encoded by a plant resistance gene. Proc Natl Acad Sci U S A 21:13833–13838
Islam W, Zhang J, Adnan M et al (2017) Plant virus ecology: a glimpse of recent accomplishments. Appl Ecol Environ Res 15:691–705
Jaag HM, Nagy PD (2009) Silencing of Nicotiana benthamiana Xrn4p exoribonuclease promotes tombusvirus RNA accumulation and recombination. Virology 10:344–352
Jaouannet M, Magliano M, Arguel MJ et al (2013) The root-knot nematode calreticulin mi-CRT is a key effector in plant defense suppression. Mol Plant-Microbe Interact 26:97–105
Jehle AK, Lipschis M, Albert M et al (2013) The receptor-like protein ReMAX of Arabidopsis detects the microbe-associated molecular pattern eMax from Xanthomonas. Plant Cell 25:2330–2340
Jones N (2013) Planetary disasters: it could happen one night. Nature 10:154–156
Jones JDG, Dangl JL (2006) The plant immune system. Nature 444:323–329
Jones DA, Thomas CM, Hammond-Kosack KE et al (1994) Isolation of the tomato Cf-9 gene for resistance to Cladosporium fulvum by transposon tagging. Science 4:789–793
Kadota Y, Sklenar J, Derbyshire P et al (2014) Direct regulation of the NADPH oxidase RBOHD by the PRR-associated kinase BIK1 during plant immunity. Mol Cell 10:43–55
Kagan JC (2017) Lipopolysaccharide detection across the kingdoms of life. Trends Immunol 38:696–704
Kaku H, Nishizawa Y, Ishii-Minami N et al (2006) Plant cells recognize chitin fragments for defense signaling through a plasma membrane receptor. Proc Natl Acad Sci U S A 103:11086–11091
Kamoun S (2006) A catalogue of the effector secretome of plant pathogenic oomycetes. Annu Rev Phytopathol 44:41–60
Kandoth PK, Ithal N, Recknor J et al (2011) The soybean Rhg1 locus for resistance to the soybean cyst nematode *Heterodera glycines* regulates the expression of a large number of stress- and defense-related genes in degenerating feeding cells. Plant Physiol 155:1960–1975
Kang B-C, Yeam I, Jahn MM (2005) Genetics of plant virus resistance. Annu Rev Phytopathol 43:581–621
Kawaharada Y, Kelly S, Nielsen MW et al (2015) Receptor-mediated exopolysaccharide perception controls bacterial infection. Nature 523:308–312
Keen N (1990) Gene-for-gene complementarity in plant-pathogen interactions. Annu Rev Genet 24:447–463
Khatib M, Lafitte C, Esquerré-Tugayé MT et al (2004) The CBEL elicitor of Phytophthora parasitica var. nicotianae activates defence in *Arabidopsis thaliana* via three different signalling pathways. New Phytol 162:501–510
Khush G, Bacalangco E, Ogawa T (1990) A new gene for resistance to bacterial blight from *O. longistaminata*. Rice Genet Newsl 23:558–560
Kim SB, Lee HY, Seo S et al (2015) RNA-dependent RNA polymerase (NIb) of the potyviruses is an avirulence factor for the broad-spectrum resistance gene Pvr4 in *Capsicum annuum* cv. CM334. PLoS One 11:e0119639
Klink VP, Kim KH, Martins V et al (2009) A correlation between host-mediated expression of parasite genes as tandem inverted repeats and abrogation of development of female *Heterodera glycines* cyst formation during infection of Glycine max. Planta 230:53–71
Koeck M, Hardham AR, Dodds PN (2011) The role of effectors of biotrophic and hemibiotrophic fungi in infection. Cell Microbiol 13:1849–1857
Kørner CJ, Klauser D, Niehl A et al (2013) The immunity regulator BAK1 contributes to resistance against diverse RNA viruses. Mol Plant-Microbe Interact 26:1271–1280
Kouzai Y, Nakajima K, Hayafune M et al (2014) CEBiP is the major chitin oligomer-binding protein in rice and plays a main role in the perception of chitin oligomers. Plant Mol Biol 84:519–528

Kud J, Wang W, Gross R et al (2019) The potato cyst nematode effector RHA1B is a ubiquitin ligase and uses two distinct mechanisms to suppress plant immune signaling. PLoS Pathog 15:e1007720

Kunze G, Zipfel C, Robatzek S et al (2004) The N terminus of bacterial elongation factor Tu elicits innate immunity in Arabidopsis plants. Plant Cell 16:3496–3507

Lahaye T, Bonas U (2001) Molecular secrets of bacterial type III effector proteins. Trends Plant Sci 6:479–485

Larkan NJ, Lydiate DJ, Parkin IAP et al (2013) The Brassica napus blackleg resistance gene LepR3 encodes a receptor-like protein triggered by the Leptosphaeria maculans effector AVRLM1. New Phytol 197:595–605

Larkan NJ, Ma L, Borhan MH (2015) The Brassica napus receptor-like protein RLM2 is encoded by a second allele of the LepR3/Rlm2 blackleg resistance locus. Plant Biotechnol J 13:983–992

Larroque M, Belmas E, Martinez T et al (2013) Pathogen-associated molecular pattern-triggered immunity and resistance to the root pathogen Phytophthora parasitica in *Arabidopsis*. J Exp Bot 64:3615–3625

Layne R (1967) Foliar trichomes and their importance as infection sites for Corynebacterium michiganensé ontomato. Phytopathology 57:981–985

Lee SW, Han SW, Sririyanum M et al (2009) A type I-secreted, sulfated peptide triggers XA21-mediated innate immunity. Science 6:850–853

Lee C, Chronis D, Kenning C et al (2011) The novel cyst nematode effector protein 19C07 interacts with the arabidopsis auxin influx transporter LAX3 to control feeding site development. Plant Physiol 155:866–880

Li F, Wang A (2018) RNA decay is an antiviral defense in plants that is counteracted by viral RNA silencing suppressors. PLoS Pathog 14:e1007228

Li X, Lin H, Zhang W et al (2005) Flagellin induces innate immunity in nonhost interactions that is suppressed by *Pseudomonas syringae* effectors. Proc Natl Acad Sci U S A 102:12990–12995

Li F, Zhang C, Li Y et al (2018) Beclin1 restricts RNA virus infection in plants through suppression and degradation of the viral polymerase. Nat Commun 28:1268

Liang X, Ding P, Lian K et al (2016) *Arabidopsis* heterotrimeric G proteins regulate immunity by directly coupling to the FLS2 receptor. elife 4:e13568

Lin B, Zhuo K, Chen S et al (2016) A novel nematode effector suppresses plant immunity by activating host reactive oxygen species-scavenging system. New Phytol 209:1159–1173

Liu T, Liu Z, Song C et al (2012) Chitin-induced dimerization activates a plant immune receptor. Science 1:1160–1164

Love AJ, Yun BW, Laval V et al (2005) Cauliflower mosaic virus, a compatible pathogen of Arabidopsis, engages three distinct defense-signaling pathways and activates rapid systemic generation of reactive oxygen species. Plant Physiol 139:935–948

Love AJ, Laval V, Geri C et al (2007) Components of *Arabidopsis* defense- and ethylene-signaling pathways regulate susceptibility to cauliflower mosaic virus by restricting long-distance movement. Mol Plant-Microbe Interact 20:659–670

Love AJ, Geri C, Laird J et al (2012) Cauliflower mosaic virus protein P6 inhibits signaling responses to salicylic acid and regulates innate immunity. PLoS One 7:e47535

Lozano-Torres JL, Wilbers RHP, Gawronski P et al (2012) Dual disease resistance mediated by the immune receptor Cf-2 in tomato requires a common virulence target of a fungus and a nematode. Proc Natl Acad Sci U S A 109:10119–10124

Luderer R, Rivas S, Nürnberger T et al (2001) No evidence for binding between resistance gene product Cf-9 of tomato and avirulence gene product AVR9 of *Cladosporium fulvum*. Mol Plant-Microbe Interact 14:867–876

Lukan T, Baebler Š, Pompe-Novak M et al (2018) Cell death is not sufficient for the restriction of potato virus Y spread in hypersensitive response-conferred resistance in potato. Front Plant Sci 15:168

Luo S, Liu S, Kong L et al (2019) Two venom allergen-like proteins, HaVAP1 and HaVAP2, are involved in the parasitism of *Heterodera avenae*. Mol Plant Pathol 20:471–484

Ma X, Nicole MC, Meteignier LV et al (2015) Different roles for RNA silencing and RNA processing components in virus recovery and virus-induced gene silencing in plants. J Exp Bot 66:919–932

Maiti S, Paul S, Pal A (2012) Isolation, characterization, and structure analysis of a non-TIR-NBS-LRR encoding candidate gene from MYMIV-resistant *Vigna mungo*. Mol Biotechnol 52:217–233

Mandadi KK, Scholthof KBG (2013) Plant immune responses against viruses: how does a virus cause disease. Plant Cell 25:1489–1505

Manosalva P, Manohar M, Von Reuss SH et al (2015) Conserved nematode signalling molecules elicit plant defenses and pathogen resistance. Nat Commun 6:7795

Mantelin S, Thorpe P, Jones JT (2015) Suppression of plant defences by plant-parasitic nematodes. Elsevier Ltd:273–286

Mehta A, Brasileiro ACM, Souza DSL et al (2008) Plant-pathogen interactions: what is proteomics telling us? FEBS J 275:3731–3746

Mendgen K, Hahn M (2002) Plant infection and the establishment of fungal biotrophy. Trends Plant Sci 7:352–356

Mendy B, Wang'ombe MW, Radakovic ZS et al (2017) *Arabidopsis* leucine-rich repeat receptor–like kinase NILR1 is required for induction of innate immunity to parasitic nematodes. PLoS Pathog 13:e1006284

Mesarich CH, Griffiths SA, Van Der Burgt A et al (2014) Transcriptome sequencing uncovers the Avr5 avirulence gene of the tomato leaf mold pathogen *Cladosporium fulvum*. Mol Plant-Microbe Interact 27:846–857

Michelmore R, Coaker G, Bart R et al (2017) Foundational and translational research opportunities to improve plant health. Mol Plant-Microbe Interact 30:515–516

Mitchum MG (2016) Soybean resistance to the soybean cyst nematode *Heterodera glycines*: an update. Phytopathology 155:1960–1975

Mitchum MG, Hussey RS, Baum TJ et al (2013) Nematode effector proteins: an emerging paradigm of parasitism. New Phytol 199:879–894

Miya A, Albert P, Shinya T et al (2007) CERK1, a LysM receptor kinase, is essential for chitin elicitor signaling in *Arabidopsis*. Proc Natl Acad Sci U S A 4:19613–19618

Mott GA, Thakur S, Smakowska E et al (2016) Genomic screens identify a new phytobacterial microbe-associated molecular pattern and the cognate *Arabidopsis* receptor-like kinase that mediates its immune elicitation. Genome Biol 9:98

Naalden D, Haegeman A, de Almeida-Engler J et al (2018) The Meloidogyne graminicola effector Mg16820 is secreted in the apoplast and cytoplasm to suppress plant host defense responses. Mol Plant Pathol 19:2416–2430

Nars A, Lafitte C, Chabaud M et al (2013) Aphanomyces euteiches cell wall fractions containing novel glucan-chitosaccharides induce defense genes and nuclear calcium oscillations in the plant host *Medicago truncatula*. PLoS One 23:e75039

Nguyen CN, Perfus-Barbeoch L, Quentin M et al (2018) A root-knot nematode small glycine and cysteine-rich secreted effector, MiSGCR1, is involved in plant parasitism. New Phytol 217:687–699

Niehl A, Wyrsch I, Boller T et al (2016) Double-stranded RNAs induce a pattern-triggered immune signaling pathway in plants. New Phytol 211:1008–1019

Noman A, Aqeel M, Lou Y (2019) PRRs and NB-LRRs: from signal perception to activation of plant innate immunity. Int J Mol Sci 20:1882

Noon JB, Qi M, Sill DN et al (2016) A plasmodium like virulence effector of the soybean cyst nematode suppresses plant innate immunity. New Phytol 212:444–460

Nürnberger T, Nennstiel D, Jabs T et al (1994) High affinity binding of a fungal oligopeptide elicitor to parsley plasma membranes triggers multiple defense responses. Cell 12:449–460

Nürnberger T, Nennstiel D, Hahlbrock K et al (1995) Covalent cross-linking of the Phytophthora megasperma oligopeptide elicitor to its receptor in parsley membranes. Proc Natl Acad Sci U S A 14:2338–2342

Oerke EC (2006) Crop losses to pests. J Agric Sci 144:31–43
Oliver RP, Ipcho SVS (2004) *Arabidops*is pathology breathes new life into the necrotrophs-vs.-biotrophs classification of fungal pathogens. Mol Plant Pathol 1:347–352
Palanichelvam K, Cole AB, Shababi M et al (2000) Agroinfiltration of cauliflower mosaic virus gene VI elicits hypersensitive response in Nicotiana species. Mol Plant-Microbe Interact 13:1275–1279
Patel N, Hamamouch N, Li C et al (2010) A nematode effector protein similar to annexins in host plants. J Exp Bot 61:235–248
Pazzagli L, Seidl-Seiboth V, Barsottini M et al (2014) Cerato-platanins: elicitors and effectors. Plant Sci 228:79–87
Pennisi E (2010) Armed and dangerous. Science 12:804–805
Perraki A, DeFalco TA, Derbyshire P et al (2018) Phosphocode-dependent functional dichotomy of a common co-receptor in plant signalling. Nature 561:248–252
Petutschnig EK, Jones AME, Serazetdinova L et al (2010) The Lysin motif receptor-like kinase (LysM-RLK) CERK1 is a major chitin-binding protein in *Arabidopsis thaliana* and subject to chitin-induced phosphorylation. J Biol Chem 10:28902–28911
Pieterse CMJ, Leon-Reyes A, Van Der Ent S et al (2009) Networking by small-molecule hormones in plant immunity. Nat Chem Biol 5:308–316
Poinssot B, Vandelle E, Bentéjac M et al (2003) The endopolygalacturonase 1 from Botrytis cinerea activates grapevine defense reactions unrelated to its enzymatic activity. Mol Plant-Microbe Interact 16:553–564
Postma WJ, Slootweg EJ, Rehman S et al (2012) The effector SPRYSEC-19 of Globodera rostochiensis suppresses CC-NB-LRR-mediated disease resistance in plants. Plant Physiol 160:944–954
Prins TW, Tudzynski P, von Tiedemann A et al (2000) Infection strategies of *Botrytis cinerea* and related necrotrophic pathogens. Fungal Pathol:33–64
Pruitt RN, Schwessinger B, Joe A et al (2015) The rice immune receptor XA21 recognizes a tyrosine-sulfated protein from a gram-negative bacterium. Sci Adv 1:e1500245
Raaymakers TM, Van Den Ackerveken G (2016) Extracellular recognition of oomycetes during biotrophic infection of plants. Front Plant Sci 21(7):906
Ranf S (2016) Immune sensing of lipopolysaccharide in plants and animals: same but different. PLoS Pathog 12:e1005596
Ranf S, Gisch N, Schäffer M et al (2015) A lectin S-domain receptor kinase mediates lipopolysaccharide sensing in *Arabidopsis thaliana*. Nat Immunol 16:426–433
Rehman S, Postma W, Tytgat T et al (2009) A secreted SPRY domain-containing protein (SPRYSEC) from the plant-parasitic nematode *Globodera rostochiensis* interacts with a CC-NB-LRR protein from a susceptible tomato. Mol Plant-Microbe Interact 22:330–340
Ron M, Avni A (2004) The receptor for the fungal elicitor ethylene-inducing xylanase is a member of a resistance-like gene family in tomato. Plant Cell 16:1604–1615
Rooney HCE, Van't Klooster JW, Van Der Hoorn RAL et al (2005) Cladosporium Avr2 inhibits tomato Rcr3 protease required for Cf-2-dependent disease resistance. Science 17:1783–1786
Rosso MN, Jones JT, Abad P (2009) RNAi and functional genomics in plant parasitic nematodes. Annu Rev Phytopathol 47:207–232
Sacco MA, Koropacka K, Grenier E et al (2009) The cyst nematode SPRYSEC protein RBP-1 elicits Gpa2- and RanGAP2-dependent plant cell death. PLoS Pathog 5:e1000564
Saijo Y, Loo EP, Yasuda S (2018) Pattern recognition receptors and signaling in plant–microbe interactions. Plant J 93:592–613
Sasaki N, Ogata T, Deguchi M et al (2009) Over-expression of putative transcriptional coactivator KELP interferes with tomato mosaic virus cell-to-cell movement. Mol Plant Pathol 10:161–173
Saucet SB, Van Ghelder C, Abad P et al (2016) Resistance to root-knot nematodes Meloidogyne spp. in woody plants. New Phytol 211:41–56
Saur IML, Kadota Y, Sklenar J et al (2016) NbCSPR underlies age-dependent immune responses to bacterial cold shock protein in *Nicotiana benthamiana*. Proc Natl Acad Sci U S A 22:3389–3394

Schoelz J, Shepherd RJ, Daubert S (1986) Region VI of cauliflower mosaic virus encodes a host range determinant. Mol Cell Biol 6:2632–2637
Scholthof HB, Gowda S, Wu FC et al (1992) The full-length transcript of a caulimovirus is a polycistronic mRNA whose genes are trans activated by the product of gene VI. J Virol 66:3131–3139
Schulze B, Mentzel T, Jehle TAK et al (2010) Rapid Heteromerization and Phosphorylation of Ligand-activated Plant Transmembrane Receptors and Their Associated Kinase BAK1. J Biol Chem 285:9444–9451
Schulze-Lefert P (2004) Knocking on the heaven's wall: pathogenesis of and resistance to biotrophic fungi at the cell wall. Curr Opin Plant Biol 7:377–383
Shen Y, Sharma P, DaSilva FG et al (2002) The Xanthomonas oryzae pv. Oryzae raxP and raxQ genes encode an ATP sulphurylase and adenosine-5′-phosphosulphate kinase that are required for AvrXa21 avirulence activity. Mol Microbiol 44:37–48
Shen Q, Hu T, Bao M et al (2016) Tobacco RING E3 ligase NtRFP1 mediates ubiquitination and proteasomal degradation of a geminivirus-encoded βC1. Mol Plant 6:911–925
Shimizu T, Nakano T, Takamizawa D et al (2010) Two LysM receptor molecules, CEBiP and OsCERK1, cooperatively regulate chitin elicitor signaling in rice. Plant J 64:204–214
Shinya T, Nakagawa T, Kaku H et al (2015) Chitin-mediated plant-fungal interactions: catching, hiding and handshaking. Curr Opin Plant Biol 26:64–71
Siddique S, Matera C, Radakovic ZS et al (2014) Host-pathogen interactions: parasitic worms stimulate host NADPH oxidases to produce reactive oxygen species that limit plant cell death and promote infection. Sci Signal 8:ra33
Sidonskaya E, Schweighofer A, Shubchynskyy V et al (2016) Plant resistance against the parasitic nematode Heterodera schachtii is mediated by MPK3 and MPK6 kinases, which are controlled by the MAPK phosphatase AP2C1 in *Arabidopsis*. J Exp Bot 67:107–118
Smant G, Jones J (2011) Suppression of plant defences by nematodes. Genomics and Molecular Genetics of Plant-Nematode Interactions:273–286
Song WY, Wang GL, Chen LL et al (1995) A receptor kinase-like protein encoded by the rice disease resistance gene, Xa21. Science 15:1804–1806
Staskawicz BJ, Dahlbeck D, Keen NT (1984) Cloned avirulence gene of Pseudomonas syringae pv. Glycinea determines race-specific incompatibility on *Glycine max* (L.) Merr. Proc Natl Acad Sci US A 81:6024–6028
Stergiopoulos I, de Wit PJGM (2009) Fungal effector proteins. Annu Rev Phytopathol 47:233–263
Stergiopoulos I, Van Den Burg HA, Ökmen B et al (2010) Tomato Cf resistance proteins mediate recognition of cognate homologous effectors from fungi pathogenic on dicots and monocots. Proc Natl Acad Sci U S A 20:7610–7615
Stergiopoulos I, Collemare J, Mehrabi R et al (2013) Phytotoxic secondary metabolites and peptides produced by plant pathogenic Dothideomycete fungi. FEMS Microbiol Rev 37:67–93
Struck C, Hahn M, Mendgen K (1996) Plasma membrane H+-ATPase activity in spores, germ tubes, and haustoria of the rust fungus Uromyces viciae-fabae. Fungal Genet Biol 20:30–35
Sun Y, Li L, Macho AP et al (2013) Structural basis for flg22-induced activation of the Arabidopsis FLS2-BAK1 immune complex. Science 1:624–628
Szajko K, Chrzanowska M, Witek K et al (2008) The novel gene Ny-1 on potato chromosome IX confers hypersensitive resistance to potato virus Y and is an alternative to Ry genes in potato breeding for PVY resistance. Theor Appl Genet 116:297–303
Tena G, Boudsocq M, Sheen J (2011) Protein kinase signaling networks in plant innate immunity. Curr Opin Plant Biol 14:519–529
Toruño TY, Stergiopoulos I, Coaker G (2016) Plant-pathogen effectors: cellular probes interfering with plant defenses in spatial and temporal manners. Annu Rev Phytopathol 4:419–441
Tournas VH (2005) Spoilage of vegetable crops by bacteria and fungi and related health hazards. Crit Rev Microbiol 31:33–44
Tyler BM (2009) Entering and breaking: virulence effector proteins of oomycete plant pathogens. Cell Microbiol 11:13–20

Veronese P, Nakagami H, Bluhm B et al (2006) The membrane-anchored botrytis-induced kinase1 plays distinct roles in *Arabidopsis* resistance to necrotrophic and biotrophic pathogens. Plant Cell 18:257–273

Vorholt JA (2012) Microbial life in the phyllosphere. Nat Rev Microbiol 10:828–840

Wan J, Zhang XC, Neece D et al (2008) A LysM receptor-like kinase plays a critical role in chitin signaling and fungal resistance in *Arabidopsis*. Plant Cell 20:471–481

Wang E, Schornack S, Marsh JF et al (2012) A common signaling process that promotes mycorrhizal and oomycete colonization of plants. Curr Biol 4:2242–2246

Wang L, Albert M, Einig E et al (2016) The pattern-recognition receptor CORE of Solanaceae detects bacterial cold-shock protein. Nat Plants 28:16185

Weinberger F, Friedlander M, Hoppe HG (1999) Oligoagars elicit a physiological response in *Gracilaria conferta* (Rhodophyta). J Phycol 35:747–755

Whisson SC, Boevink PC, Moleleki L et al (2007) A translocation signal for delivery of oomycete effector proteins into host plant cells. Nature 1:115–118

Whitham SA, Quan S, Chang HS et al (2003) Diverse RNA viruses elicit the expression of common sets of genes in susceptible Arabidopsis thaliana plants. Plant J 33:271–283

Willmann R, Lajunen HM, Erbs G et al (2011) *Arabidopsis* lysin-motif proteins LYM1 LYM3 CERK1 mediate bacterial peptidoglycan sensing and immunity to bacterial infection. Proc Natl Acad Sci U S A 6:19824–19829

Woo JY, Jeong KJ, Kim YJ et al (2016) CaLecRK-S.5, a pepper L-type lectin receptor kinase gene, confers broad-spectrum resistance by activating priming. J Exp Bot 67:5725–5741

Wu S, Shan L, He P (2014) Microbial signature-triggered plant defense responses and early signaling mechanisms. Plant Sci 228:118–126

Xiang T, Zong N, Zou Y et al (2008) Pseudomonas syringae effector AvrPto blocks innate immunity by targeting receptor kinases. Curr Biol 8:74–80

Xie J, Li S, Mo C et al (2016) A novel *Meloidogyne incognita* effector *misp12* suppresses plant defense response at latter stages of nematode parasitism. Front Plant Sci 7:964

Yamaji Y, Hamada K, Yoshinuma T et al (2010) Inhibitory effect on the tobacco mosaic virus infection by a plant RING finger protein. Virus Res 153:50–57

Yang S, Pan L, Chen Y et al (2019) *Heterodera avenae* GLAND5 effector interacts with pyruvate dehydrogenase subunit of plant to promote nematode parasitism. Front Microbiol 4:1241

Zamora M, Méndez-López E, Agirrezabala X et al (2017) Potyvirus virion structure shows conserved protein fold and RNA binding site in ssRNA viruses. Sci Adv 3:eaao2182

Zhang C, Hajimorad MR, Eggenberger AL et al (2009) Cytoplasmic inclusion cistron of soybean mosaic virus serves as a virulence determinant on Rsv3-genotype soybean and a symptom determinant. Virology 1:240–248

Zhang W, Fraiture M, Kolb D et al (2013) *Arabidopsis* receptor-like protein30 and receptor-like kinase suppressor OF BIR1-1/EVERSHED mediate innate immunity to necrotrophic fungi. Plant Cell 25:4227–4241

Zhang L, Kars I, Essenstam B et al (2014) Fungal endopolygalacturonases are recognized as microbe-associated molecular patterns by the arabidopsis receptor-like protein responsiveness to botrytis polygalacturonases. Plant Physiol 164:352–364

Zhang X, Valdés-López O, Arellano C et al (2017) Genetic dissection of the maize (*Zea mays* L.) MAMP response. Theor Appl Genet 130:1155–1168

Zhang Z, Tong X, Liu SY et al (2019) Genetic analysis of a piezo-like protein suppressing systemic movement of plant viruses in Arabidopsis thaliana. Sci Rep 9:3187

Zhong Z, Marcel TC, Hartmann FE et al (2017) A small secreted protein in Zymoseptoria tritici is responsible for avirulence on wheat cultivars carrying the Stb6 resistance gene. New Phytol 214:619–631

Zhu J, Gopinath K, Murali A et al (2007) RNA-binding proteins that inhibit RNA virus infection. Proc Natl Acad Sci U S A 104:3129–3134

Zhuo K, Naalden D, Nowak S et al (2019) A *Meloidogyne graminicola* C-type lectin, Mg01965, is secreted into the host apoplast to suppress plant defence and promote parasitism. Mol Plant Pathol 20:346–355

Zijlstra C, Hohn T (1992) Cauliflower mosaic virus gene VI controls translation from dicistronic expression units in transgenic Arabidopsis plants. Plant Cell 4:1471–1484

Zipfel C (2009) Early molecular events in PAMP-triggered immunity. Curr Opin Plant Biol 12:414–420

Zipfel C (2014) Plant pattern-recognition receptors. Trends Immunol 35:345–351

Zipfel C, Kunze G, Chinchilla D et al (2006) Perception of the bacterial PAMP EF-Tu by the receptor EFR restricts agrobacterium-mediated transformation. Cell 19:749–760

Zorzatto C, MacHado JPB, Lopes KVG et al (2015) NIK1-mediated translation suppression functions as a plant antiviral immunity mechanism. Nature 520:679–682

Zvereva AS, Golyaev V, Turco S et al (2016) Viral protein suppresses oxidative burst and salicylic acid-dependent autophagy and facilitates bacterial growth on virus-infected plants. New Phytol 211:1020–1034

Chapter 22
Role of miRNAs in Abiotic and Biotic Stress Management in Crop Plants

Narayanan Nair Anjali and Kalluvettankuzhy Krishnannair Sabu

Abstract MicroRNAs (miRNAs) are endogenous, 18–25 nucleotide in length, and noncoding RNAs found in animals, plants, and in some viruses. They have been recognized as an important class of gene expression regulators and are involved in a multitude of biological processes in plants, including organ differentiation and development, metabolism, and response to biotic and abiotic stress factors, to mention a few. The method of action of miRNA is by binding to a ribonucleoprotein complex which is incorporated with a protein of Argonaute family. This RNA-induced silencing complex (RISC) targets the mRNA and inhibits its expression by degrading mRNA or by translational repression. It has been reported that posttranscriptional regulation mediated by miRNAs has important roles in adverse stress conditions. Some miRNAs respond to a number of abiotic and biotic stresses, and some others are stress specific. Not only characterization of miRNAs but also identification of their targets is crucial in this field of research. The participation of miRNAs in a species-specific manner to biotic and abiotic stress responses reveals the functional diversification of miRNAs among plant species and highlights the importance of conducting miRNA studies in diverse plant species. MicroRNA-based genetic modification technology is a very powerful technique to develop high-yielding crops which can resist biotic and abiotic stress factors.

Keywords miRNA · Abiotic stress · Biotic stress · NGS · Bioinformatics

1 Introduction

Abiotic and biotic stresses adversely affect the growth, development, and yield of the plant (Roychowdhury et al. 2013, 2014; Roychowdhury 2014; Chakraborty et al. 2014; Hasanuzzaman et al. 2015). Plants execute some complex processes to detect and resist these environmental stresses like drought, salinity, extreme

N. N. Anjali · K. K. Sabu (✉)
Jawaharlal Nehru Tropical Botanic Garden and Research Institute (JNTBGRI), Thiruvananthapuram, Kerala, India
e-mail: sabu@jntbgri.res.in

R. Roychowdhury et al. (eds.), *Sustainable Agriculture in the Era of Climate Change*, https://doi.org/10.1007/978-3-030-45669-6_22

temperatures, heavy metals, and the damage caused by living organisms such as bacteria, fungi, viruses, and parasites (Hasanuzzaman et al. 2013; Mamgain et al. 2013; Roychowdhury and Tah 2013; Roychowdhury et al. 2013a, b). This complex process involves the activation of interconnected system of gene regulations, thereby changing the level of expression of genes in a transcriptional or posttranscriptional manner (Anumalla et al. 2016). Positive regulators of stress tolerance to plants are upregulated, and negative regulators are downregulated by the intricate mechanisms of plants (Shriram et al. 2016). It is very important to understand the mechanisms behind plant response to these stresses and to use this information gained in modern plant breeding programs (Anumalla et al. 2015; Dresselhaus and Hückelhoven 2018; Roychowdhury et al. 2018, 2019).

MicroRNAs (miRNAs) are endogenous, 18–25 nucleotide in length, and non-coding RNAs found in animals, plants, and in some viruses. They have been recognized as an important class of gene expression regulators (Iki 2017) and are involved in a multitude of biological processes in plants, including organ differentiation and development, metabolism, defense against pathogens, response to environmental, and biotic and abiotic stress factors to mention a few (Bej and Basak 2014; Shweta et al. 2015; Yoshikawa 2017). It has been reported that posttranscriptional regulation mediated by miRNAs has important roles in adverse stress conditions (Bari et al. 2006). Some miRNAs respond to a number of abiotic stresses, and some others are stress-specific. miRNA expression levels also differ with environmental and biotic stresses. Not only characterization of miRNAs but also identification of their targets is crucial in this field of research (Covarrubias and Reyes 2010). The first report of miRNA relation with stress was miR398 which showed downregulation upon oxidative stress in *Arabidopsis*. It targets cytosolic CSD1 and chloroplastic CSD2 which are two closely related Cu/Zn superoxide dismutase (CSD) coding genes (Sunkar et al. 2006). The participation of miRNAs in a species-specific manner to stress responses reveals the functional diversification of miRNAs among plant species and highlights the importance of conducting miRNA studies in diverse plant species. MicroRNA-based genetic modification technology is a very powerful technique to develop high-yielding crops which can resist biotic and abiotic stress factors. Several traits in a number of crops have been improved by using the miRNA-based technology.

2 Biogenesis and Mechanism of Action of miRNAs in Plants

In plants, more than 100 miRNA genes (MIR) are present in the intergenic regions throughout the genome. miRNA genes are transcribed by RNA polymerase II (pol II) to generate the primary transcripts (pri-miRNAs). Transcriptional coactivator mediator helps in the recruitment of pol II to MIR promoter. Pri-miRNAs are stabilized by the addition of 5′ 7-methylguanosine cap and 3′ polyadenylate tail. Because of the structural peculiarities of the pri-miRNA stem loop, an initial dicer-like (DCL) cleavage occurs at the base of the stem. Splicing and further processing of pri-miRNAs are carried out by the interactive functions of Hyponastic Leaves 1 (HYL1), Dawdle (DDL), Tough (TGH), and Serrate (SE) proteins as well as the cap-binding proteins CBP20 and CBP80 which result into a 70 nt precursor miRNA (pre-miRNA) having a short stem plus a 2-nucleotide 3′ overhang. Pri-miRNAs and pre-miRNAs are sequentially processed by dicer-like I (DCL I) to yield miRNA/miRNA∗ duplexes. These steps took place in the nucleus and transported to cytoplasm by HASTY (HST1) (Li et al. 2015). The duplexes are stabilized through end methylation catalyzed by Hua Enhancer 1 HEN1. The duplex is separated, and usually one strand is selected as the mature miRNA, whereas the other strand is degraded. Only the mature miRNA is integrated into the Argonaute (AGO1)-containing RNA-induced silencing complex (RISC) and accumulated with Decapping enzymes (DCP1 and DCP2) (Fig. 22.1). The miRNA-RISC targets complementary messenger RNA (mRNA) (Zhang et al. 2014; Tang et al. 2015). They regulate gene expression either by cleaving mRNA or inhibiting the translational process at the posttranscriptional level (Yang et al. 2015). Degradation of the mRNA occurs in a perfect match where specific region of the miRNA known as "seed region," i.e., six nucleotides of the miRNA (2-7 nt from the 5′end), binds to the complementary sequences located in the mRNA 3′ untranslated region (3'UTR). Repression occurs in the case of incomplete complementarity where the mRNA is not destroyed but protein translation is prevented. Sometimes, they can also involve in methylation at the transcriptional level (Brodersen et al. 2008). A single miRNA can target many mRNAs, and a mRNA can be targeted by multiple miRNAs (Zandkarimi et al. 2015; Kelly et al. 2015).

3 Abiotic Stress-Responsive miRNAs

Abiotic stress is one of the serious components that is responsible for huge yield loss in crop plants. Plants are exposed to different abiotic stresses like drought, salinity, extreme temperatures, and heavy metals during various developmental stages of growth (Singh et al. 2017). These abiotic stresses cause dreadful reactions on crops and result in substantial yield loss. miRNAs are known to play a key role in cellular responses under these stresses.

3.1 miRNAs in Response to Drought Stress

Drought is one of the major abiotic stresses that affects crop growth and productivity (Tambo and Abdoulaye 2011). It has been reported that miRNAs play important roles in response to drought stress. Using microarray, upregulation of miR393 and miR169 was detected in rice under drought stress (Zhao et al. 2007). miR393 was found to be upregulated under drought stress in many other plants like *Arabidopsis*, *Medicago*, and common bean (Sunkar and Zhu 2004; Arenas-Huertero et al. 2009).

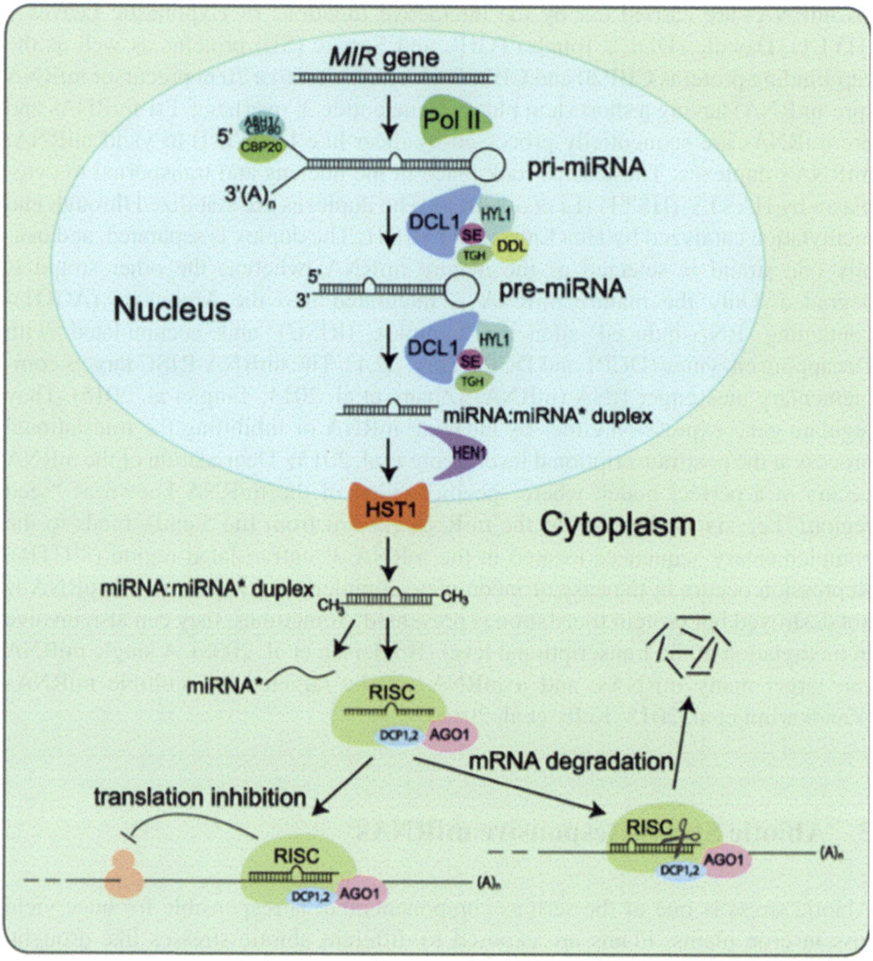

Fig. 22.1 Diagrammatic representation of miRNA biogenesis in plants. (Modified from: Yang and Li 2012)

In rice, miR393 targets OsTIR1 and OsAFB2 (Xia et al. 2012) which are auxin receptor gene homologs. Another study also reported that miR169 shows upregulation in rice (Jeong and Green 2013). Because of drought stress, a rapid increase in expression on miR169g was found in roots of rice than in shoots (Zhao et al. 2007). miR169a and miR169c which are the two members of miR169 gene family showed downregulation in response to drought in *Arabidopsis* and *Medicago truncatula* (Trindade et al. 2010). miR169 targets NFYA5 mRNA which encodes a subunit of the transcription factor named as nuclear factor Y (Li et al. 2008). miR169 showed upregulation in tomato in association with drought stress (Zhang et al. 2011a). miR169g and miR169n showed downregulation in drought-affected leaves of wheat (Stephenson et al. 2007). Zhou et al. (2010) reported that miR156, miR159, miR168, miR170, miR171, miR172, miR319, miR396, miR397, miR408, miR529, miR896, miR1030, miR1035, miR1050, miR1088, and miR1126 showed downregulation and miR159, miR319, miR395, miR474, miR845, miR851, miR854, miR896, miR901, miR903, miR1026, and miR1125 showed upregulation under drought stress in rice. miR159 was found to be upregulated under drought stress in *Arabidopsis* (Reyes and Chua 2007). It targets the transcription factor genes MYB33 and MYB101 (Sunkar et al. 2008). miR157, miR167, miR168, miR171, and miR408 were some other drought-associated miRNAs reported in *Arabidopsis* (Liu et al. 2008). miR820 was found as a novel miRNA in rice under drought stress treatment (Jeong et al. 2011). In *Populus trichocarpa* miR171, miR1446a-e, miR1444a, miR1447, and miR1450 were found to be significantly repressed by drought stress (Lu et al. 2008). By using high-throughput sequencing and microarray analysis, Li et al. (2011a, b) reported 104 upregulated and 27 downregulated miRNAs in *Populus euphratica*. miR472, miR473, and miR1444 were reported to have relation with drought stress only in *Populus* plants (Shuai et al. 2013). Twenty-five miRNAs showed upregulation and 43 showed downregulation under drought stress in maize (Li et al. 2013). Out of these 68 drought-responsive miRNAs, 34 targeted 56 mRNA genes of maize. Differential expression of many of these drought-responsive miRNA families were found under the effect of other abiotic stresses in maize (Ding et al. 2009). Members of miR319 family exhibited upregulation at the early stages of drought stress and displayed downregulation at the later stages of drought stress in maize (Li et al. 2013). miRNAs that respond to drought stress in maize targeted mRNA genes coding for MAPK (mitogen-activated protein kinase), PLD (phospholipase D), PDH (proline dehydrogenase), and POD (peroxidase) (Wei et al. 2009). miR156, one of the conserved miRNA families (Schwab et al. 2005), was identified to have response under drought stress in maize, switch grass (*Panicum virgatum*), wild emmer wheat (*Triticum dicoccoides*) (Kantar et al. 2010), *Arabidopsis*, and rice (Wu et al. 2009). miR395 and miR169 showed very high response to drought stress in tobacco (Frazier et al. 2011). pvu-miRS1, pvu-miR1514a, pvu-miR159.2, pvu-miR2118, and pvu-miR2119 were reported to be legume-specific miRNAs that accumulate upon drought stress in *Phaseolus vulgaris* (Arenas-Huertero et al. 2009). Under high water-deficit conditions, miR156 and miR162 displayed change in expression levels in *P. vulgaris* (Nageshbabu and Jyothi 2013). miR156 targets squamosal promoter-binding protein-like (SPL) genes that associate with shoot development, male fertility, anthocyanin biosynthesis, and flowering control. miR162 plays a role in miRNA biogenesis by negatively regulating Dicer-like 1 (DCL 1) gene. Out of 44 drought-responsive miRNAs discovered in cowpea, 30 were upregulated, and 14 were downregulated (Barrera-Figueroa et al. 2011). miR398 and miR408 exhibited upregulation under drought stress in *Medicago trun catula* (Trindade et al. 2010). It targets and cleaves the COX5b, CSD1, and plantacyanin genes. Stu-miR396, stu-miR156a, and stu-miR157a were reported to be induced by drought stress in potato (Hwang et al. 2011). It was found that stu-miR156a and stu-miR157a target 68 and 71 mRNA transcripts, respectively. There are reports for a phenomenon called cross-adaptation in which a plant exposed to one stress acquires resistance against other stresses (Sabehat et al. 1998).

Upregulation of miR2089 and miR2118 was observed in *M. truncatula* under water deficit condition which targets mRNA genes coding for proteins associated with disease resistance (Wang et al. 2011).

3.2 miRNAs in Response to Salinity

In *Arabidopsis* miR156, miR158, miR159, miR165, miR167, miR168, miR169, miR171, miR319, miR393, miR394, miR396, and miR397 showed downregulation, and miR398 showed upregulation under salt stress (Liu et al. 2008). miR169, which is a drought-responsive miRNA, was later reported to be induced by salt stress in rice (Liu et al. 2008). Fifty miRNAs have been reported to be expressed in soybean under salt stress (Zhang et al. 2008). Salt-responsive miRNAs like miR393, miR394, miR396, and miR156 were identified to be present in rice, maize, soybean, and *Populus* (Ding et al. 2009; Li et al. 2011b). miR398 was discovered to be associated with salt stress in *Populus tremula* (Jia et al. 2009). Ninety-eight miRNAs were found in maize which changes in their expression upon salt stress (Ding et al. 2009). Shivaprasad et al. (2011) reported about miR395 in salt stress-resistant tomato hybrids.

3.3 miRNAs in Response to Heat and Cold

miR393, miR167, miR397b, miR402, miR319c, miR165/166, miR169, miR172, miR396, miR397, and miR408 were identified as accumulated upon cold stress in *Arabidopsis* (Sunkar and Zhu 2004). Eighteen miRNAs were reported in rice which respond to cold stress. miR167 and miR319 exhibited downregulation by cold stress in rice (Lv et al. 2010). miR1425 which is a rice-specific miRNA showed reduced level of expression during cold stress (Jeong and Green 2013). It targets pentatricopeptide repeat (PPR) genes that encode for RNA-binding proteins and are involved in posttranscriptional processes. miR812q is another miRNA which was found to be upregulated during cold stress in rice (Jeong et al. 2011). miR402 was significantly expressed in all the three cases of drought, salinity, and cold stress. In rice, miR397b.2 showed response to heat stress. It was reported that 12 miRNAs responded to heat stress in wheat. The same miRNA family members showed differential regulation in response to powdery mildew infection and heat stress in wheat (Xin et al. 2010). Compared to other stress-responsive miRNAs, extreme temperature responsive miRNAs have not been explored well in plants and have been reported only in *Arabidopsis*, rice, wheat, and *Brassica* (Xin et al. 2010; Jeong et al. 2011).

3.4 miRNAs in Response to Nutrient Homeostasis

miR399 showed upregulation in response to phosphate deprivation and has been studied in many plants like *Arabidopsis*, common bean, rice, and tomato (Chiou et al. 2006; Aung et al. 2006). miR827 was also identified as miRNA associated with phosphate starvation in *Arabidopsis* (Hsieh et al. 2009) and rice (Jeong et al. 2011). miR395 showed upregulation during sulfur starvation in *Arabidopsis* (Jones-Rhoades and Bartel 2004). Nitrogen starvation-responsive miRNAs have been reported in many species like *Arabidopsis* (Liang et al. 2012), maize (Zhao et al. 2012) and rice (Nischal et al. 2012). Jeong et al. (2011) reported that miR3979, a rice-specific miRNA, was downregulated under nitrogen deprivation condition.

3.5 miRNAs in Response to Hypoxia and Oxidative Stress

Oxygen is a very essential factor required for plants. Waterlogging condition is the main reason for decreased oxygen uptake in plants, and it seriously affects the mitochondrial respiration. Low oxygen stress treatment on roots of *Arabidopsis* showed that 25 miRNAs belonging to 19 families are hypoxia responsive. A large number of these miRNAs were found to be increased under oxygen stress indicating that they are mainly involved in downregulating gene expression (Moldovan et al. 2010). miR156g, miR157d, miR158a, miR159a, miR172a, miR172b, miR391, and miR775 are some of the miRNAs found to be hypoxia-responsive in *Arabidopsis* roots. Under oxidative stress, miR398 was found to be downregulated, and superoxide dismutase (SOD) proteins were upregulated in *Arabidopsis* (Sunkar et al. 2006). miR169, miR397, miR827, miR1425, and miR528 were reported as peroxide responsive microRNAs in rice seedlings of *Arabidopsis*.

3.6 miRNAs in Response to Heavy Metals

Studies have been reported on the role of miRNAs in metal stress responses in different plants like *Arabidopsis thaliana*, *Medicago truncatula*, *Brassica napus*, *Oryza sativa*, *Nicotiana tabacum*, and *Phaseolus vulgaris* (Gielen et al. 2012). miR319, miR398, miR2118, miR159, miR1511, and miR169 are some of the miRNAs identified as manganese responsive in *Phaseolus vulgaris*. In *Arabidopsis thaliana*, miR398 was found to be copper-responsive and iron-responsive miRNA. Upon zinc exposure, miR398a and miR398b/c were documented as having elevated exposure in *Arabidopsis thaliana*. It was observed through high-throughput sequencing that miR156, miR159, miR1507, miR2651, etc. were differentially expressed in *Medicago truncatula* seedlings subjected to mercury stress. Altered expression of different miRNAs like miR171, miR819, and miR415 were observed in *Medicago truncatula* seedlings subjected to aluminum stress. Several miRNAs like miR815, miR169, and miR2121 were identified in rice in response to arsenic stress. Cadmium induced-differential expression of microRNAs like miRNA394 and miRNA400 was observed in *Brassica napus*.

4 Biotic Stress-Responsive miRNAs

The yield and quality of the crops are affected by many biotic factors like viruses, bacteria, fungi, insect pests, and nematode parasites (Jay et al. 2010). For controlling these pathogens, farmers are applying many chemicals to plants, and as a result of this, biotic factors generate resistance against these chemicals. New approaches which can develop plants resistant to pathogens and with high yield are in great demand, and microRNAs can control the expression of defense-regulated genes in response to biotic stress.

4.1 miRNAs in Response to Viral Pathogenesis

miRNA families like miR160, miR166, miR167, miR171, and miR396 were found to be upregulated in response to *Rice stripe virus* infection in rice. This indicates that miRNAs play an important role in developing viral resistance in plants by targeting viral suppressors of RNA silencing. In *Brassica napus*, miR1885 and mir158 were found to be *Turnip mosaic virus*-responsive, and in *Solanum lycopersicum*, miR482 was found to be *Cucumber mosaic virus*-responsive. It was found out in a study that miR168 was differentially expressed in *Arabidopsis* and *Nicotiana* upon *Rice stripe virus* attack. It was observed that miR159a, miR167b, and miR171a showed differential expression because of *Plum pox virus* pathogenesis in *Nicotiana benthamiana* (Simón-Mateo and García 2006). Artificial miRNAs were created from miR159 precursor of *Arabidopsis*, and it targets viral mRNA sequences encoding two gene-silencing suppressors, P69 of *Turnip yellow mosaic virus* and HC-Pro of *Turnip mosaic virus*. *Arabidopsis* plants with amiR-P69^{159} and amiR-HC-Pro^{159} constructs were found to be resistant to *Turnip yellow mosaic virus* and *Turnip mosaic virus*, respectively. *Cucumber mosaic virus* resistance was shown by transgenic tobacco plants which overexpresses an artificial miRNA (miR2b) by inhibiting the expression of silencing suppressor 2b of CMV.

4.2 miRNAs in Response to Bacterial Pathogenesis

Pathogen recognition receptors (PRRs) are present in almost all plants and animals which receive signal from pathogen-associated molecular patterns (PAMPs) like bacterial flagellin. It will activate the defense response of host against bacterial infections. It was found that microRNAs like miR159, miR393, miR319, miR160, miR165/miR166, and miR167 were upregulated, whereas miR390, miR398, and miR408 were downregulated by *P. syringae* infection in *Arabidopsis* (Li et al. 2010). The signal received from flagellin increases the resistance in *Arabidopsis* against bacterium *P. syringae*. miR393 was found to be overexpressed in case of infection with flagellin-derived peptide in *Arabidopsis* and miR393 targets and negatively regulate F-box auxin receptors transport inhibitor response 1 (TIR1), AFB2, and AFB3 coding mRNAs. Plants with elevated levels of miR393 show a decline in the level of TIR1 transcripts and enhanced bacterial resistance. Thus, it is confirmed that antibacterial resistance in plants can be achieved by miR393-mediated suppression of auxin signaling in plants. Zhang et al. (2011b) reported that miR393b was observed to be having a role against *P. syringae* pathogenesis in *Nicotiana*.

4.3 miRNAs in Response to Fungal Pathogenesis and Nematode Infections

miR408 was reported to have altered expression in wheat against *Puccinia striiformis* infection by targeting TaCLP1 gene which encodes a type of plantacyanin protein. miR396 was found to have defensive role in *Arabidopsis* against *Plectosphaerella cucumerina*, *Botrytis cinerea*, *Fusarium oxysporum*, and *Colletotrichum higginsianum* infections (Islam et al. 2018). miR1448 and miR1450 were found to be responsive to *B. dothidea* attack in *P. trichocarpa*. It was observed that miR482 has an important role in defense against *V. dahlia* infection by upregulating the disease resistance protein gene. Several miRNAs like miR161, miR164, miR167a, mir172c, miR396c, miR396a,b, and miR398 were observed to be downregulated in *Arabidopsis* against *H. schachtii* (Hewezi et al. 2008). In another study, 101 miRNAs were found to be responsive to soybean cyst nematode infection in *Glycine max*.

5 Application of Next-Generation Sequencing (NGS) and Bioinformatics for miRNA Discovery in Plants

Illumina sequencing was used to identify the microRNAs present in *Phaseolus vulgaris*. RNA was extracted from different tissues like root, stem, flower, and seedlings. Small RNA libraries were created, and 114 conserved miRNAs belonging to 33 families were discovered after sequencing (Pelaez et al. 2012). miRNAs in peanut were discovered by using the next-generation sequencing platform. Seventy-five conserved and 14 novel miRNA families were found out by the combined approaches of high-throughput sequencing and bioinformatics tools (Zhao et al. 2010). Small RNA library was created, and Solexa sequencing was performed to discover the miRNAs present in *Porphyra yezoensis*. 224 conserved miRNA families and 7 new miRNAs were detected from the small RNA reads using computational methods (Liang et al. 2010). Seventy-three miRNAs from 49 families were characterized from Asiatic cotton (Wang et al. 2012). RNA was isolated from flower and fruit of *Citrus trifoliata*, and small RNA library construction was done. Solexa sequencing generated 4,876,395 unique reads, and 63 miRNA sequences from 42 families were found to be conserved in other plant species. In addition to this, ten novel miRNAs were also discovered from *C. trifoliate* (Song et al. 2010). 323 miRNAs belonging to 276 families were identified from 11 tissues in wheat. For 124 miRNA families, 524 mRNA targets were deduced through degradome sequencing, and 5' RACE was used for confirmation. Sixty-four miRNAs were found which have crucial regulatory roles in grain development (Sun et al. 2014). Fall dormancy is a peculiar feature of *Medicago sativa* to get through the winter. Next-generation sequencing was performed to recognize the miRNAs associated with fall dormancy. Twenty-eight miRNAs were found to be responsive to this phenomenon, of which 20 were conserved and 8 were new miRNAs. Upregulation of eight conserved and downregulation of six new miRNAs in response to fall dormancy were validated using qRT PCR technique (Fan et al. 2014). Small RNA libraries constructed from leaf, stem, callus, and male and female flowers were subjected to high-throughput sequencing and 38 miRNAs were discovered which have not been reported in cassava earlier (Khatabi et al. 2016). Jiang et al. (2014) reported the miRNAs associated with pollen development in *Brassica campestris* ssp. *chinensis* by using next-generation sequencing and degradome study. RNA was isolated from the flower buds of male fertile line and male sterile line of *B. campestris* ssp. *Chinensis*, and small RNA libraries were created. Illumina sequencing was performed, and 18 miRNAs were found to be differentially expressed between the 2 libraries. Witches' broom disease is a major threat to *Paulownia fortunei* for reducing its crop productivity. A study was conducted to characterize the miRNAs present in *P. fortunei* including those miRNAs which are responsive to witches' broom disease. Small RNA libraries were created from control, and witches' broom disease infected *P. fortunei* seedlings. After deep sequencing, 95 conserved miRNAs from 18 families and 122 new miRNAs have been reported (Niu et al. 2016). RNA was isolated from seven tissues of chickpea, and libraries were constructed for Illumina sequencing. 440 miRNAs were found out which are known to be present in other plant species. Using the plant miRNA prediction tool miRDeep-P, 178 novel miRNAs were also discovered (Jain et al. 2014). Small RNA library was prepared from leaf of black pepper, and Illumina sequencing was performed. 303 conserved miRNAs were identified. Four new miRNAs and 50 precursor sequences of conserved miRNAs were also determined using black pepper transcriptome analysis. Eight miRNAs showing differential expression were validated using stem loop qRT PCR method (Asha et al. 2016).

Wang et al. (2015) reported about small RNA sequencing of non-heading Chinese cabbage by Illumina technology. Filtering of the sequence reads was done by ACGT101-miR-v3.5 package. Reads that got aligned with rRNA, tRNA, snRNA, snoRNA from Rfam, and GenBank noncoding RNA database were removed from further analysis. The remaining reads were mapped with the Chinese cabbage genome using MapMi software with default settings. miRNAs were identified from the miRBase database, and secondary structure prediction was performed using Mfold program. IDEG6 tool was utilized for finding out the differentially expressed miRNAs. Shen et al. (2010) made a study on abiotic stress-responsive miRNAs in rice. For this work, rice miRNAs were retrieved from Sanger Institute miRBase Sequence Database, and stem loop RT primer were designed for this miRNAs. Small RNA sequence reads from *Populus euphratica* were first subjected for removal of adapters, low-quality reads, poly A tail, and sequences shorter than 18 nt or longer than 30 nt. Then the clean reads were mapped against genome of *Populus* using the tool SOAP, and completely matched sequences were taken for next step analysis. Sequences which got aligned with rRNA, tRNA, snRNA, and snoRNA sequences in NCBI GenBank and Rfam 10.1 databases were discarded. Retained sequences were compared with miRNAs from miRBase database to identify the conserved miRNAs by permitting two mismatches. Secondary structures of this conserved miRNAs were identified using RNAfold program. Mireap software was employed for predicting the novel miRNAs of *Populus*. Target genes of these new miRNAs were speculated, and gene annotation was carried out using GO and KEGG databases (Si et al. 2014). Raw small RNA reads of trifoliate orange sequenced by Illumina 1 G Genome Analyzer were filtered by several computational tools, and remaining reads were used for Blastn search against miRBase 13.0 database to find out conserved miRNAs. To gather knowledge about precursor sequences, ESTs of citrus were downloaded from dbEST database available at NCBI GenBank, and sRNA sequences were mapped against these EST sequences. Through miRCat tool, secondary structures of this miRNAs were generated. Target mRNAs of these miRNAs were identified by using HarvEST C46 *Citrus* unigene database on the HarvEST Blast Search web server. Functions of target genes were elucidated by Citrus HarvEST 1.20 program (Song et al. 2010). A study was conducted by Sun et al. (2014) to find out the miRNAs and their target genes in wheat. Small RNAs sequenced using Illumina GA IIx platform were first subjected for preprocessing steps. Adapters were removed by vector strip tool available in EMBOSS package, and reads with length shorter than 18 nt and longer than 26 nt were trimmed using computational tools. Remaining reads were mapped against the wheat genome, ESTs from dbEST database, and ESTs and cDNAs from wheat genetic resources database using the tool Bowtie. Reads that matched with noncoding RNAs in Rfam database and repetitive sequences in Plant Repeat Database were eliminated. miReap software was employed for detection of miRNAs from the retained reads. Precursors of this miRNAs were identified by the information available at the International Wheat Genome Sequencing Consortium. Small RNA sequence data from *Salicornia europaea* was filtered with Beijing Genomics Institute (BGI) sRNA analysis pipeline, and SOAP tool was used for mapping the clean reads against *S. europaea* transcriptome. Reads that matched with repetitive sequences downloaded from Repeat Masker and rRNA, tRNA, snRNA, and snoRNA sequences downloaded from Rfam database were removed from further analysis. Conserved miRNAs were identified by blastn search against miRBase database, and novel miRNA prediction was done using Mireap software. Secondary structures of miRNA precursors were elucidated with Mfold program (Feng et al. 2015). Databases, software, or tools used in this study with almost same pipeline are exploited in another piece of work on miRNAs under drought condition in rice roots conducted by Bakhshi et al. (2016). For the identification of salinity-responsive miRNAs in finger millet, raw sRNA seq data was first preprocessed by removing adapter sequences and poly A tail. The clean reads were then used for Blastn search against miRBase database to identify conserved miRNAs by allowing only one mismatch. Since the whole genome of finger millet is not available, reads were mapped to closest reference genomes *Setaria italica* and *Oryza sativa* using Blast provided by Geneious 7.0 software. Secondary structures of pre-miRNAs were predicted using Quick fold tool available in Mfold software with default settings. Target genes of both conserved and novel miRNAs were identified by psRNATarget program (Selvam et al. 2015). miRBase, deepBase, PMRD, PNRD, PMTED, Plant MPSS, miRTarBase, Rfam, and ARMOUR are some of the databases previously reported in studies involved in discovery of plant miRNAs. PASmiR, isomiRex, CLC genomics Workbench, MIRFINDER, Targetfinder, mirCheck, findmiRNA, MicroInspector, RNAhybrid, CleaveLand, TAPIR, and miRDeep-P are some of the important tools for analyzing miRNAs and their targets in plants (Tripathi et al. 2015).

6 miRNA-Based Genetic Modification Technology for Developing Stress-Resistant Crops

miRNA-based genetic modification technology is a very powerful technique that can be utilized for crop productivity by developing improved cultivars with enhanced biotic and abiotic stress tolerance and increased biomass yields. Methods for manipulating miRNAs and their targets in transgenic plants include constitutive, stress-induced, or tissue-specific expression of miRNAs or their targets, RNA interference, expressing miRNA-resistant target genes, artificial target mimic, and artificial miRNAs. RNA interference (RNAi), also known as posttranscriptional gene silencing, offers defense mechanism against internal parasitic and external pathogenic nucleic acids and controls gene expression. RNAi is the process in which double-stranded RNA (dsRNA) targets specific mRNA and inhibits gene expression by degrading it. In an earlier study, an attempt was made to intensify the color of red petunia flower for making it more attractive and thereby increasing its market value. They tried to overexpress the gene coding for chalcone synthase enzyme, which is included in the pathway for the production of red pigment. It was found in some of the transgenic petunia plants that flowers become colorless. The exact reason for this phenomenon was not known at that time, and Jorgensen coined the term co-suppression for this. It can be described as a traditional form of posttranscriptional gene silencing in eukaryotes (Agrawal et al. 2003). Later, Fire et al. (1998) reported about employing RNAi technology to control gene expression in *Caenorhabditis elegans*. They studied about the design and method of introduction of dsRNA to the host organism.

dsRNA as a whole was found to be better than two separate single strands in executing the function. Single-stranded DNAs produced interference in an adequate level, but dsRNA showed very powerful silencing of gene expression. The method of action of miRNA is by binding to a ribonucleoprotein complex which is incorporated with a protein of Argonaute family. This RNA-induced silencing complex (RISC) targets the mRNA and inhibits the expression by degrading mRNA or by translational repression.

By regulating the expression of a single miRNA, plant tolerance to biotic and abiotic stress can be increased. The endogenous and artificial miRNAs can be used for crop improvement. There are several approaches for developing stress-tolerant crops by miRNA-based genetic modification technology (Fig. 22.2). If miRNA expression is positively linked with stress tolerance character, either the miRNAs should be over expressed or the expression of target genes should be knocked down by RNA interference technology. In case miRNA expression is negatively associated with stress-tolerance character, miRNA-resistant target genes have to be overexpressed, or artificial target mimics have to be introduced (Zhou and Luo 2013). Several traits in a number of crops have been improved by using the miRNA-based technology, and some of the important ones are listed in Table 22.1.

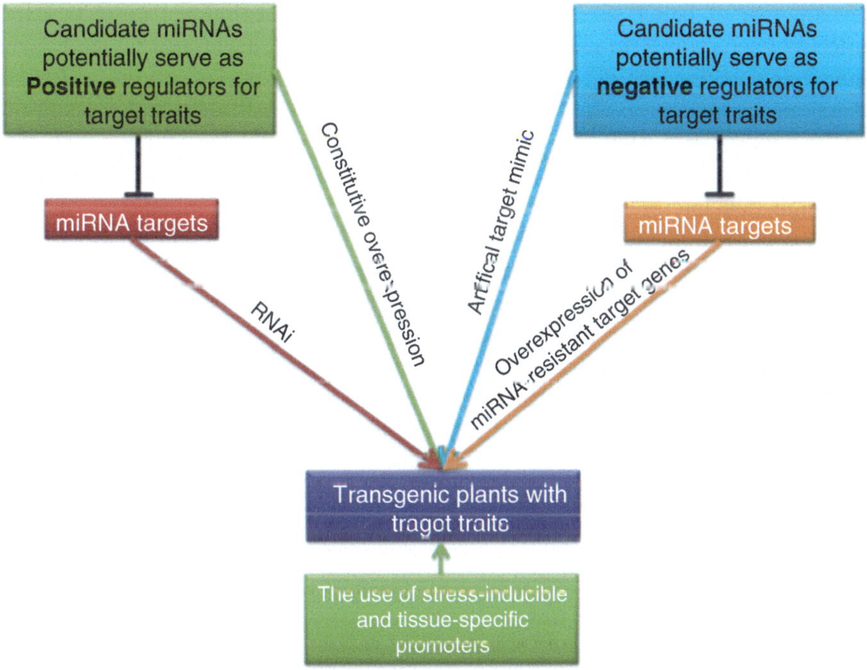

Fig. 22.2 Different approaches for developing stress tolerant crops using miRNA-based genetic modification technology

Table 22.1 High-yielding crops produced by miRNA-based genetic modification technology

Trait improved	miRNA	Target	Plant
Drought tolerance	miR169	*NFYA5*	*Arabidopsis*
	miR169	*GmNFYA3*	Soybean and *Arabidopsis*
Cold tolerance	miR319	*PCF5/PCF8*	Rice
Heat stress tolerance	miR398	*CSD1 CSD2 CCS*	*Arabidopsis*
Bacterial resistance	miR393	*TIR1*	*Arabidopsis*
Cauliflower mosaic virus (CMV) resistance	amiR171	*2b of CMV*	Tobacco
Tomato leaf curl New Delhi virus resistance	amiR-AV1–1	*AV1/AV2*	Tomato
Turnip yellow mosaic virus resistance	amiR159	*P69 of TYMV*	*Arabidopsis*

7 Conclusion and Future Prospects

Both traditional and novel methods are essential to elevate plant biotic and abiotic stress resistance levels high enough for field application in crops. Strong backup of knowledge on the mechanism behind plant response to stress will be useful for the development of better stress-tolerant varieties. Posttranscriptional regulation by miRNAs exerts a role in response of plants to stress. Conventional methods like Sanger sequencing are not much reliable in detecting species-specific miRNAs which are expressed in very small amounts. NGS provides enormous amount of data to address such needs. However, fast processing power and efficient tools are needed to manage this big data. With the advancement in computer capacity and algorithm, data interpretation becomes easier. Sometimes errors might occur during library construction and sequencing procedures which can affect the quality of the raw reads. Quality control check has to be done for the raw reads obtained from NGS platforms. It gives an idea of whether the data has any issues, and one can be conscious before starting analysis. The blend of next-generation sequencing and bioinformatics analysis has paved the way for discovery of many species-specific miRNAs of plants from several species. NGS data are now being produced in non-model organisms at a greater speed as these high-throughput studies can be performed at a reasonable budget. miRNA-based genetic modification technology represents a novel and feasible approach to improve plant stress tolerance.

References

Agrawal N, Dasaradhi PVN, Mohmmed A, Malhotra P, Bhatnagar RK, Mukherjee SK (2003) RNA interference: biology, mechanism, and applications. Microbiol Mol Biol Rev 67:657–685. https://doi.org/10.1128/MMBR.67.4.657-685.2003

Anumalla M, Roychowdhury R, Geda CK, Mazid M, Rathoure AK (2015) Utilization of plant genetic resources and diversity analysis tools for sustainable crop improvement with special emphasis on rice. Int J Adv Res 3(3):1155–1175

Anumalla M, Roychowdhury R, Geda CK, Bharathkumar S, Goutam KD, Mohandev TSS (2016) Mechanism of stress signal transduction and involvement of stress inducible transcription factors and genes in response to abiotic stresses in plant. Int J Recent Sci Res 7(8):12754–12771

Arenas-Huertero C, Perez B, Rabanal F, Blanco-Melo D, De la Rosa C, Estrada-Navarrete G, Sanchez F, Covarrubias AA, Reyes JL (2009) Conserved and novel miRNAs in the legume Phaseolus vulgaris in response to stress. Plant Mol Biol 70:385–401. https://doi.org/10.1007/s11103-009-9480-3

Asha S, Sreekumar S, Soniya EV (2016) Unravelling the complexity of microRNA-mediated gene regulation in black pepper (Piper nigrum L.) using high-throughput small RNA profiling. Plant Cell Rep 35:53–63. https://doi.org/10.1007/s00299-015-1866-x

Aung K, Lin S, Wu C, Huang Y, Su C, Chiou T (2006) pho2, a phosphate overaccumulator, is caused by a nonsense mutation in a microRNA399 target gene. Plant Physiol 141:1000–1011. https://doi.org/10.1104/pp.106.078063

Bakhshi B, Mohseni Fard E, Nikpay N, Ebrahimi MA, Bihamta MR, Mardi M, Salekdeh GH (2016) MicroRNA signatures of drought signaling in rice root. PLoS One 11:1–25. https://doi.org/10.1371/journal.pone.0156814

Bari R, Datt Pant B, Stitt M, Scheible W (2006) PHO2, MicroRNA399, and PHR1 define a phosphate-signaling pathway in plants. Plant Physiol 141:988–999. https://doi.org/10.1104/pp.106.079707

Barrera-Figueroa BE, Gao L, Diop NN, Wu Z, Ehlers JD, Roberts PA, Close TJ, Zhu J, Liu R (2011) Identification and comparative analysis of drought-associated microRNAs in two cowpea genotypes. BMC Plant Biol 11:1–11. https://doi.org/10.1186/1471-2229-11-127

Bej S, Basak J (2014) MicroRNAs: the potential biomarkers in plant stress response. Am J Plant Sci 5(5):748

Brodersen P, Sakvarelidze-Achard L, Bruun-Rasmussen M, Dunoyer P, Yamamoto YY, Sieburth L, Voinnet O (2008) Widespread translational inhibition by plant miRNAs and siRNAs. Science 320:1185–1190. https://doi.org/10.1126/science.1159151

Chakraborty S, Pattanayak A, Mandal S, Das M, Roychowdhury R (2014) An overview of climate change: causes, trends and implications. In: Roychowdhury R (ed) Crop improvement in the era of climate change. IK International Publishing House, New Delhi, pp 1–29

Chiou T, Aung K, Lin S, Wu C, Chiang S, Su C (2006) Regulation of phosphate homeostasis by MicroRNA in Arabidopsis. Plant Cell 18:412–421. https://doi.org/10.1105/tpc.105.038943

Covarrubias AA, Reyes JL (2010) Post-transcriptional gene regulation of salinity and drought responses by plant microRNAs. Plant Cell Environ 33:481–489. https://doi.org/10.1111/j.1365-3040.2009.02048.x

Ding D, Zhang L, Wang H, Liu Z, Zhang Z, Zheng Y (2009) Differential expression of miRNAs in response to salt stress in maize roots. Ann Bot 103:29–38. https://doi.org/10.1093/aob/mcn205

Dresselhaus T, Hückelhoven R (2018) Biotic and abiotic stress responses in crop plants. Agronomy 8

Fan W, Zhang S, Du H, Sun X, Shi Y, Wang C (2014) Genome-wide identification of different dormant Medicago sativa L. MicroRNAs in response to fall dormancy. PLoS One 9:e114612. https://doi.org/10.1371/journal.pone.0114612

Feng J, Wang J, Fan P, Jia W, Nie L, Jiang P, Chen X, Lv S, Wan L, Chang S, Li S, Li Y (2015) High-throughput deep sequencing reveals that microRNAs play important roles in salt tolerance of euhalophyte Salicornia europaea. BMC Plant Biol 15:1–17. https://doi.org/10.1186/s12870-015-0451-3

Fire A, Xu S, Montgomery MK, Kostas SA, Driver SE, Mello CC (1998) Potent and specific genetic interference by double-stranded RNA in Caenorhabditis elegans. Nature 391:806–811

Frazier TP, Sun G, Burklew CE, Zhang B (2011) Salt and drought stresses induce the aberrant expression of microRNA genes in tobacco. Mol Biotechnol 49:159–165. https://doi.org/10.1007/s12033-011-9387-5

Gielen H, Remans T, Vangronsveld J, Cuypers A (2012) MicroRNAs in metal stress: specific roles or secondary responses? Int J Mol Sci 13:15826–15847. https://doi.org/10.3390/ijms131215826

Hasanuzzaman M, Nahar K, Alam M, Roychowdhury R, Fujita M (2013) Physiological, biochemical, and molecular mechanisms of heat stress tolerance in plants. Int J Mol Sci 14(5):9643–9684
Hasanuzzaman M, Roychowdhury R, Karmakar J, Dey N, Nahar K, Fujita M (2015) Recent advances in biotechnology and genomic approaches for abiotic stress tolerance in crop plants. In: Devarajan T, Jeyabalan S (eds) Genomics and proteomics: concepts, technologies and applications. Apple Academic Press, Toronto, pp 333–366
Hewezi T, Howe P, Maier TR, Baum TJ (2008) Arabidopsis small RNAs and their targets during cyst nematode parasitism. Mol Plant-Microbe Interact 21:1622–1634. https://doi.org/10.1094/MPMI-21-12-1622
Hsieh T, Ibarra CA, Silva P, Zemach A, Eshed-Williams L, Fischer RL, Zilberman D (2009) Genome-wide demethylation of Arabidopsis endosperm. Science 324:1451–1454. https://doi.org/10.1126/science.1172417
Hwang E, Shin S, Kwon H (2011) Identification of microRNAs and their putative targets that respond to drought stress in Solanum tuberosum. J Korean Soc Appl Biol Chem 54:317–324. https://doi.org/10.3839/jksabc.2011.051
Iki T (2017) Messages on small RNA duplexes in plants. J Plant Res 130:7–16. https://doi.org/10.1007/s10265-016-0876-2
Islam W, Noman A, Qasim M, Wang L (2018) Plant responses to pathogen attack: small RNAs in focus. Int J Mol Sci 19. https://doi.org/10.3390/ijms19020515
Jain M, Chevala VVSN, Garg R (2014) Genome-wide discovery and differential regulation of conserved and novel microRNAs in chickpea via deep sequencing. J Exp Bot 65:5945–5958. https://doi.org/10.1093/jxb/eru333
Jay F, Renou J, Voinnet O, Navarro L (2010) Biotic stress-associated microRNAs: identification, detection, regulation, and functional analysis BT – plant microRNAs: methods and protocols. (eds: Meyers BC, Green PJ) Humana Press, Totowa, pp 183–202
Jeong D, Green PJ (2013) The role of rice microRNAs in abiotic stress responses. J Plant Biol 56:187–197. https://doi.org/10.1007/s12374-013-0213-4
Jeong D, Park S, Zhai J, Gurazada SGR, De Paoli E, Meyers BC, Green PJ (2011) Massive analysis of rice small RNAs: mechanistic implications of regulated MicroRNAs and variants for differential target RNA cleavage. Plant Cell 23:4185–4207. https://doi.org/10.1105/tpc.111.089045
Jia X, Wang W, Ren L, Chen Q, Mendu V, Willcut B, Dinkins R, Tang X, Tang G (2009) Differential and dynamic regulation of miR398 in response to ABA and salt stress in Populus tremula and Arabidopsis thaliana. Plant Mol Biol 71:51–59. https://doi.org/10.1007/s11103-009-9508-8
Jiang J, Lv M, Liang Y, Ma Z, Cao J (2014) Identification of novel and conserved miRNAs involved in pollen development in Brassica campestris ssp. chinensis by high-throughput sequencing and degradome analysis. BMC Genomics 15:146. https://doi.org/10.1186/1471-2164-15-146
Jones-Rhoades MW, Bartel DP (2004) Computational identification of plant microRNAs and their targets, including a stress-induced miRNA. Mol Cell 14:787–799. https://doi.org/10.1016/j.molcel.2004.05.027
Kantar M, Lucas SJ, Budak H (2010) miRNA expression patterns of Triticum dicoccoides in response to shock drought stress. Planta 233:471–484. https://doi.org/10.1007/s00425-010-1309-4
Kelly H, Downing T, Tuite NL, Smith T, Kerin M, Dwyer RM, Clancy E, Barry T, Reddington K (2015) Cross platform standardisation of an experimental pipeline for use in the identification of dysregulated human circulating MiRNAs. PLoS One 10:e0137389
Khatabi B, Arikit S, Xia R, Winter S, Oumar D, Mongomake K, Meyers BC, Fondong VN (2016) High-resolution identification and abundance profiling of cassava (Manihot esculenta Crantz) microRNAs. BMC Genomics 17:85. https://doi.org/10.1186/s12864-016-2391-1
Li W, Oono Y, Zhu J, He X, Wu J, Iida K, Lu X, Cui X, Jin H, Zhu J (2008) The Arabidopsis NFYA5 transcription factor is regulated transcriptionally and posttranscriptionally to promote drought resistance. Plant Cell 20:2238–2251. https://doi.org/10.1105/tpc.108.059444
Li Y, Zhang QQ, Zhang J, Wu L, Qi Y, Zhou J (2010) Identification of microRNAs involved in pathogen-associated molecular pattern-triggered plant innate immunity. Plant Physiol 152:2222–2231. https://doi.org/10.1104/pp.109.151803

Li B, Qin Y, Duan H, Yin W, Xia X (2011a) Genome-wide characterization of new and drought stress responsive microRNAs in Populus euphratica. J Exp Bot 62:3765–3779. https://doi.org/10.1093/jxb/err051

Li H, Dong Y, Yin H, Wang N, Yang J, Liu X, Wang Y, Wu J, Li X (2011b) Characterization of the stress associated microRNAs in Glycine max by deep sequencing. BMC Plant Biol 11:1–12. https://doi.org/10.1186/1471-2229-11-170

Li J, Fu F, An M, Zhou S, She Y, Li W (2013) Differential expression of microRNAs in response to drought stress in maize. J Integr Agric 12:1414–1422. https://doi.org/10.1016/S2095-3119(13)60311-1

Li W, Wang P, Li Y, Zhang K, Ding F, Nie T, Yang X, Lv Q, Zhao L (2015) Identification of MicroRNAs in response to different day lengths in soybean using high-throughput sequencing and qRT-PCR. PLoS One 10:e0132621

Liang C, Zhang X, Zou J, Xu D, Su F, Ye N (2010) Identification of miRNA from Porphyra yezoensis by high-throughput sequencing and bioinformatics analysis. PLoS One 5. https://doi.org/10.1371/journal.pone.0010698

Liang G, He H, Yu D (2012) Identification of nitrogen starvation-responsive microRNAs in Arabidopsis thaliana. PLoS One 7:e48951. https://doi.org/10.1371/journal.pone.0048951

Liu H, Tian X, Li Y, Wu C, Zheng C (2008) Microarray-based analysis of stress-regulated microRNAs in Arabidopsis thaliana. RNA 14:836–843. https://doi.org/10.1261/rna.895308

Lu S, Sun Y, Chiang VL (2008) Stress-responsive microRNAs in Populus. Plant J 55:131–151. https://doi.org/10.1111/j.1365-313X.2008.03497.x

Lv D, Bai X, Li Y, Ding X, Ge Y, Cai H, Ji W, Wu N, Zhu Y (2010) Profiling of cold-stress-responsive miRNAs in rice by microarrays. Gene 459:39–47. https://doi.org/10.1016/j.gene.2010.03.011

Mamgain A, Roychowdhury R, Tah J (2013) *Alternaria* pathogenicity and its strategic controls. Res J Biol 1:1–9

Moldovan D, Spriggs A, Yang J, Pogson BJ, Dennis ES, Wilson IW (2010) Hypoxia-responsive microRNAs and trans-acting small interfering RNAs in Arabidopsis. J Exp Bot 61:165–177. https://doi.org/10.1093/jxb/erp296

Nageshbabu R, Jyothi MN (2013) Profile of small interfering RNAs from French bean Phaseolus vulgaris under abiotic stress conditions. Int J Pharm Bio Sci 4:176–185

Nischal L, Mohsin M, Khan I, Kardam H, Wadhwa A, Abrol YP, Iqbal M, Ahmad A (2012) Identification and comparative analysis of microRNAs associated with low-N tolerance in rice genotypes. PLoS One 7:e50261. https://doi.org/10.1371/journal.pone.0050261

Niu S, Fan G, Deng M, Zhao Z, Xu E, Cao L (2016) Discovery of microRNAs and transcript targets related to witches' broom disease in Paulownia fortunei by high-throughput sequencing and degradome approach. Mol Genet Genomics 291:181–191. https://doi.org/10.1007/s00438-015-1102-y

Pelaez P, Trejo MS, Iniguez LP, Estrada Navarrete G, Covarrubias AA, Reyes JL, Sanchez F (2012) Identification and characterization of microRNAs in Phaseolus vulgaris by high-throughput sequencing. BMC Genomics 13:83

Reyes JL, Chua N (2007) ABA induction of miR159 controls transcript levels of two MYB factors during Arabidopsis seed germination. Plant J 49:592–606. https://doi.org/10.1111/j.1365-313X.2006.02980.x

Roychowdhury R (2014) Crop improvement in the era of climate change. IK International Publishing House, New Delhi, p 496

Roychowdhury R, Tah J (2013) Mutagenesis – a potential approach for crop improvement. In: Hakeem KR, Ahmad P, Ozturk M (eds) Crop improvement – new approaches and modern techniques. Springer, Boston, pp 149–187

Roychowdhury R, Karmakar J, Karmakar J, Adak MK, Dey N (2013) Physio-biochemical and microsatellite based profiling of lowland rice (*Oryza sativa* L.) landraces for osmotic stress tolerance. Am J Plant Sci 4(12):52

Roychowdhury R, Abdel Gawwad MR, Banerjee U, Bishnu S, Tah J (2013a) Status, trends and prospects of organic farming in India: a review. J Plant Biol Res 2:38–48
Roychowdhury R, Banherjee U, Slofkova S, Tah J (2013b) Organic farming for crop improvement and sustainable agriculture in the era of climate change. OnLine J Biol Sci 13(2):50–65
Roychowdhury R, Taoutaou A, Hakeem KR, Gawwad MR, Tah J (2014) Molecular marker-assisted technologies for crop improvement. In: Roychowdhury R (ed) Crop improvement in the era of climate change. IK International Publishing House, New Delhi, pp 241–258
Roychowdhury R, Khan MH, Choudhury S (2018) Arsenic in rice: an overview on stress implications, tolerance and mitigation strategies. In: Hasanuzzaman M, Nahar K, Fujita M (eds) Plants under metal and metalloid stress. Springer, Singapore, pp 401–415
Roychowdhury R, Khan MH, Choudhury S (2019) Physiological and molecular responses for metalloid stress in rice – a comprehensive overview. In: Hasanuzzaman M, Fujita M, Nahar K, Biswas J (eds) Advances in rice research for abiotic stress tolerance. Woodhead Publishing/Elsevier, pp 341–369
Sabehat A, Weiss D, Lurie S (1998) Heat-shock proteins and cross-tolerance in plants. Physiol Plant 103:437–441. https://doi.org/10.1034/j.1399-3054.1998.1030317.x
Schwab R, Palatnik JF, Riester M, Schommer C, Schmid M, Weigel D (2005) Specific effects of microRNAs on the plant transcriptome. Dev Cell 8:517–527. https://doi.org/10.1016/j.devcel.2005.01.018
Selvam J, Rahman NH, Senthil N, Boopathi NM, Raveendran M (2015) Computational identification of salinity responsive microRNAs in contrasting genotypes of finger millet (Eleusinecoracana L.). Res J Biotechnol 10:52–64
Shen J, Xie K, Xiong L (2010) Global expression profiling of rice microRNAs by one-tube stem-loop reverse transcription quantitative PCR revealed important roles of microRNAs in abiotic stress responses. Mol Genet Genomics 284:477–488. https://doi.org/10.1007/s00438-010-0581-0
Shivaprasad PV, Dunn RM, Santos BACM, Bassett A, Baulcombe D (2011) Extraordinary transgressive phenotypes of hybrid tomato are influenced by epigenetics and small silencing RNAs. EMBO J 31:257–266
Shriram V, Kumar V, Devarumath RM, Khare TS, Wani SH (2016) MicroRNAs as potential targets for abiotic stress tolerance in plants. Front Plant Sci 7:817. https://doi.org/10.3389/fpls.2016.00817
Shuai P, Liang D, Zhang Z, Yin W, Xia X (2013) Identification of drought-responsive and novel Populus trichocarpamicroRNAs by high-throughput sequencing and their targets using degradome analysis. BMC Genomics 14:233
Shweta B, Tamara B, Martin HA, Juliane F, Roman K (2015) Detecting miRNA mentions and relations in biomedical literature (version 3; referees: 2 approved, 1 approved with reservations). F1000Res 3
Si J, Zhou T, Bo W, Xu F, Wu R (2014) Genome-wide analysis of salt-responsive and novel microRNAs in Populus euphratica by deep sequencing. BMC Genet 15:1–11. https://doi.org/10.1186/1471-2156-15-S1-S6
Simón-Mateo C, García JA (2006) MicroRNA-guided processing impairs plum pox virus replication, but the virus readily evolves to escape this silencing mechanism. J Virol 80:2429–2436. https://doi.org/10.1128/JVI.80.5.2429-2436.2006
Singh I, Smita S, Mishra DC, Kumar S, Singh BK, Rai A (2017) Abiotic stress responsive miRNA-target network and related markers (SNP, SSR) in Brassica juncea. Front Plant Sci 8:1943
Song C, Wang C, Zhang C, Korir NK, Yu H, Ma Z, Fang J (2010) Deep sequencing discovery of novel and conserved microRNAs in trifoliate orange (Citrus trifoliata). BMC Genomics 11:431
Stephenson TJ, McIntyre CL, Collet C, Xue G (2007) Genome-wide identification and expression analysis of the NF-Y family of transcription factors in Triticum aestivum. Plant Mol Biol 65:77–92. https://doi.org/10.1007/s11103-007-9200-9

Sun F, Guo G, Du J, Guo W, Peng H, Ni Z, Sun Q, Yao Y (2014) Whole-genome discovery of miRNAs and their targets in wheat (Triticum aestivum L.). BMC Plant Biol 14:1–17. https://doi.org/10.1186/1471-2229-14-142

Sunkar R, Zhu J (2004) Novel and stress-regulated microRNAs and other small RNAs from Arabidopsis. Plant Cell 16:2001–2019. https://doi.org/10.1105/tpc.104.022830

Sunkar R, Kapoor A, Zhu J (2006) Posttranscriptional induction of two Cu/Zn superoxide dismutase genes in arabidopsis is mediated by downregulation of miR398 and important for oxidative stress tolerance. Plant Cell 18:2051–2065. https://doi.org/10.1105/tpc.106.041673

Sunkar R, Zhou X, Zheng Y, Zhang W, Zhu J (2008) Identification of novel and candidate miRNAs in rice by high throughput sequencing. BMC Plant Biol 8:1–17. https://doi.org/10.1186/1471-2229-8-25

Tambo JA, Abdoulaye T (2011) Climate change and agricultural technology adoption: the case of drought tolerant maize in rural Nigeria. Mitig Adapt Strateg Glob Chang 17:277–292. https://doi.org/10.1007/s11027-011-9325-7

Tang C, Yang M, Wu F, Zhao H, Pang Y, Yang R, Lu G, Yang Y (2015) Identification of miRNAs and their targets in transgenic Brassica napus and its acceptor (Westar) by high-throughput sequencing and degradome analysis. RSC Adv 5:85383–85394. https://doi.org/10.1039/C5RA14672K

Trindade I, Capitao C, Dalmay T, Fevereiro M, Santos DMD (2010) miR398 and miR408 are up-regulated in response to water deficit in Medicago truncatula. Planta 231:705–716. https://doi.org/10.1007/s00425-009-1078-0

Tripathi A, Goswami K, Sanan-Mishra N (2015) Role of bioinformatics in establishing microRNAs as modulators of abiotic stress responses: the new revolution. Front Physiol 6. https://doi.org/10.3389/fphys.2015.00286

Wang T, Chen L, Zhao M, Tian Q, Zhang W (2011) Identification of drought-responsive microRNAs in Medicago truncatula by genome-wide high-throughput sequencing. BMC Genomics 12:367. https://doi.org/10.1186/1471-2164-12-367

Wang M, Wang Q, Wang B (2012) Identification and characterization of MicroRNAs in Asiatic cotton (Gossypium arboreum L.). PLoS ONE 7. https://doi.org/10.1371/journal.pone.0033696

Wang Z, Jiang D, Zhang C, Tan H, Li Y, Lv S, Hou X, Cui X (2015) Genome-wide identification of turnip mosaic virus-responsive microRNAs in non-heading Chinese cabbage by high-throughput sequencing. Gene 571:178–187. https://doi.org/10.1016/j.gene.2015.06.047

Wei F, Zhang J, Zhou S, He R, Schaeffer M, Collura K, Kudrna D, Faga BP, Wissotski M, Golser W, Rock SM, Graves TA, Fulton RS, Coe E, Schnable PS, Schwartz DC, Ware D, Clifton SW, Wilson RK, Wing RA (2009) The physical and genetic framework of the maize B73 genome. PLoS Genet 5:e1000715

Wu L, Zhang Q, Zhou H, Ni F, Wu X, Qi Y (2009) Rice MicroRNA effector complexes and targets. Plant Cell 21:3421–3435. https://doi.org/10.1105/tpc.109.070938

Xia K, Wang R, Ou X, Fang Z, Tian C, Duan J, Wang Y, Zhang M (2012) OsTIR1 and OsAFB2 downregulation via OsmiR393 overexpression leads to more tillers, early flowering and less tolerance to salt and drought in Rice. PLoS One 7:e30039

Xin M, Wang Y, Yao Y, Xie C, Peng H, Ni Z, Sun Q (2010) Diverse set of microRNAs are responsive to powdery mildew infection and heat stress in wheat (Triticum aestivum L.). BMC Plant Biol 10:1–11. https://doi.org/10.1186/1471-2229-10-123

Yang X, Li L (2012) Analyzing the microRNA transcriptome in plants using deep sequencing data. Biology 1:297. https://doi.org/10.3390/biology1020297

Yang T, Fang L, Zhang X, Hu J, Bao S, Hao J, Li L, He Y, Jiang J, Wang F, Tian S, Zong X (2015) High-throughput development of SSR markers from pea (Pisum sativum L.) based on next generation sequencing of a purified Chinese commercial variety. PLoS ONE 10:e0139775

Yoshikawa M (2017) Pre-microRNA processing activity in nuclear extracts from Arabidopsis suspension cells. J Plant Res 130:75–82. https://doi.org/10.1007/s10265-016-0874-4

Zandkarimi H, Bedre R, Solis J, Mangu V, Baisakh N (2015) Sequencing and expression analysis of salt-responsive miRNAs and target genes in the halophyte smooth cordgrass (Spartina alternifolia Loisel). Mol Biol Rep 42:1341–1350. https://doi.org/10.1007/s11033-015-3880-z

Zhang B, Pan X, Stellwag EJ (2008) Identification of soybean microRNAs and their targets. Planta 229:161–182. https://doi.org/10.1007/s00425-008-0818-x

Zhang Q, Li J, Zhao Y, Korban SS, Han Y (2011a) Evaluation of genetic diversity in Chinese wild apple species along with apple cultivars using SSR markers. Plant Mol Biol Report 30:539–546. https://doi.org/10.1007/s11105-011-0366-6

Zhang X, Li H, Zhang J, Zhang C, Gong P, Ziaf K, Xiao F, Ye Z (2011b) Expression of artificial microRNAs in tomato confers efficient and stable virus resistance in a cell-autonomous manner. Transgenic Res 20:569–581. https://doi.org/10.1007/s11248-010-9440-3

Zhang Y, Zhu X, Chen X, Song C, Zou Z, Wang Y, Wang M, Fang W, Li X (2014) Identification and characterization of cold-responsive microRNAs in tea plant (Camellia sinensis) and their targets using high-throughput sequencing and degradome analysis. BMC Plant Biol 14:271. https://doi.org/10.1186/s12870-014-0271-x

Zhao B, Liang R, Ge L, Li W, Xiao H, Lin H, Ruan K, Jin Y (2007) Identification of drought-induced microRNAs in rice. Biochem Biophys Res Commun 354:585–590. https://doi.org/10.1016/j.bbrc.2007.01.022

Zhao C, Xia H, Frazier TP, Yao Y, Bi Y, Li A, Li M, Li C-S, Zhang B-H, Wang X-J (2010) Deep sequencing identifies novel and conserved microRNAs in peanuts (Arachis hypogaea L.). BMC Plant Biol 10

Zhao M, Tai H, Sun S, Zhang F, Xu Y, Li W (2012) Cloning and characterization of maize miRNAs involved in responses to nitrogen deficiency. PLoS One 7:e29669

Zhou M, Luo H (2013) MicroRNA-mediated gene regulation: potential applications for plant genetic engineering. Plant Mol Biol 83:59–75. https://doi.org/10.1007/s11103-013-0089-1

Zhou L, Liu Y, Liu Z, Kong D, Duan M, Luo L (2010) Genome-wide identification and analysis of drought-responsive microRNAs in Oryza sativa. J Exp Bot 61:4157–4168. https://doi.org/10.1093/jxb/erq237

Chapter 23
Recent Transgenic Approaches for Stress Tolerance in Crop Plants

Krishna Kumar Rai and Avinash Chandra Rai

Abstract Exacerbation of plant growth and productivity due to a wide range of stresses has significantly affected global food security, agricultural productivity, and quality worldwide. In order to bridge the gap between the supply and demand of the ever-increasing global population, it is indispensable to foster a new breed of stress-tolerant crops with refined traits and higher yields against several abiotic and biotic stresses. The transgenic approach of conventional breeding, owing to the limited and time-consuming success due to the complex nature of genes involved in stress tolerance, is now being widely adopted to breed crop plants with enhanced stress tolerance. Thus, identification and characterization of critical genes involved in abiotic and biotic stress tolerance are an important requisite to develop stress-tolerant crops. Genetic engineering of crop plants employs two strategies (i) either manipulating single functional gene or (ii) by editing those regulatory genes which modulate the expression of other stress-responsive genes. Genome editing using artificial nucleases such as zinc finger nucleases (ZFNs), transcription activator-like effector nucleases (TALENS), and Clustered Regulatory Interspaced Short Palindromic Repeat (CRISPR), CRISPR-associated protein 9 (Cas9), has significantly impacted basic as well as applied research including plant breeding by accelerating the editing of target genome in precise and predictable manner. Here, in this chapter, we are not going to discuss the past transgenic development approaches; mostly we will review some of the recent advancement made in the field of transgenic plants and the potential exploitation of genome-editing tools such as in conferring environmental stress tolerance in crops under field condition.

Keywords Abiotic stresses · Transgenic plants · Stress tolerance · Transcription factors · Genes

---

K. K. Rai
Department of Botany, Institute of Sciences, Banaras Hindu University, Varanasi, Uttar Pradesh, India

A. C. Rai (✉)
Institute of Plant Sciences, Agricultural Research Organization (ARO) – The Volcani Center, Rishon LeZion, Israel

R. Roychowdhury et al. (eds.), *Sustainable Agriculture in the Era of Climate Change*, https://doi.org/10.1007/978-3-030-45669-6_23

1 Introduction

The transgenic approaches have put forward astounding scope and opportunity for the improvement of several economically important horticultural crops to enhance their biotic and abiotic stress tolerance, quality and, productivity (Wang et al. 2016a, b). The genetic engineering technology offers tremendous potential by forging new genotypes for their exploitation in a breeding program, improving their nutritional quality, and enhancing their shelf life along with other agronomically important traits for improving their yield under adverse condition (Roychowdhury et al. 2013, 2014; Hasanuzzaman et al. 2013; Roychowdhury and Tah 2013; Roychowdhury 2014; Anumalla et al. 2015; Wani et al. 2016). Transgenic techniques employ myriad of functions for fostering plants with improved horticultural traits which involve isolation of a concerned gene, their ligation with desire vector to form recombinant DNAs, and their efficient transformation within the plant genome to produce a new breed of crop plants with novel function (Marco et al. 2015). Plants being sessile are usually confronted by various abiotic stresses that torment them at each level, i.e., morphological, physiological, biochemical, and molecular level that intricate severe repercussions on their growth and productivity.

Owing to its complex polygenic nature, engineering plants with enhanced abiotic stress tolerance is hard to comprehend; however, results of single-gene transfer studies showing induction of in the expression levels of genes involved in various signaling pathways are promising (Gürel et al. 2016). Transgenic plants exhibiting enhanced accumulation of proline, glycine betaine, proteins like molecular chaperones, late embryogenesis abundant (LEA) proteins, along with the increased expression of defense-related genes have been widely reported in various crop plants (Grobkinsky et al. 2016; Hasanuzzaman et al. 2015; Anumalla et al. 2016). Despite the advancements in the genetic engineering technology, earlier reports have indicated that plants engineered with transgenic approaches showed considerable accumulation of osmolytes such as proline, mannitol, and glycine betaine delineated insignificant but increased tolerance to several abiotic stresses compared to their non-transgenic counterparts (Banerjee and Roychoudhury 2016).

Nevertheless, the past recent years have witnessed the tremendous outbreak of recombinant DNA technology in major horticultural crops allowing them to combat efficiently with several of the biotic and abiotic stresses, which is the major essence and substratum of sustainable agriculture (Bakhsh and Hussain 2015). Several studies on transgenic crops have reported efficient transformation and regeneration competencies in many of the horticultural plants including cucumber, broccoli, date palm, and chili (Wani et al. 2016), and in the recent decades, several transgenic horticultural crops have been developed, and more are in pipeline. Apart from the genes involved in the antioxidant defense system and transcription factors (TFs), genes like cysteine protease inhibitors, trypsin inhibitors, and other pathogenesis-related genes have also been used for tailoring crop plants with improved tolerance to several biotic and abiotic stresses (Wang et al. 2016a, b; Roychowdhury et al. 2018, 2019). Additionally, RNA interference (RNAi) techniques have also been implicated in various horticultural crops for inducing virus resistance by expressing virus coat protein gene (Mamta and Rajam 2018).

The application of RNAi technology along with the genome-editing tool Clustered Regulatory Inter-Spread Short Palindromic Repeat/CRISPR associated protein 9 (CRISPR/Cas9) has accelerated the pace of improvement of economically important crops mainly in terms of their nutritional quality, shelf life, along with minimization of their pre- and postharvest losses (Jaganathan et al. 2018). This chapter is intended to provide the main features of some of the recent developments in the field of recombinant DNA technology for engineering crop plants against various biotic and abiotic stresses. Information about the different tools and techniques that have been employed to clone a particular gene or multiple genes and their efficient mode of transformation in the horticultural crops have also been discussed. Different types of secondary metabolites/osmolytes accumulated, scavenging of free radical generation via stimulation of antioxidant defense system, and probable role TFs in the mitigation of biotic and abiotic stresses are also critically discussed.

2 Transgenic Approach for Biotic Stress Management

The agricultural growth and productivities are adversely affected by several biotic stresses such as bacteria, fungi, and viruses and are becoming much more prominent with the exacerbation of environmental changes. Till date, conventional breeding method has gained limited success as many of the biotic and abiotic stress tolerance is governed by many genes, so for developing stress-tolerant genotypes, using transgenic approaches is the only feasible option. In this section, we have summarized recent progress made for engineering crop plants with enhanced biotic stress tolerance using a transgenic approach.

2.1 Genetic Engineering for Insect Pest Resistance

The insect pests are one of the most severe biotic stresses that intimidate crop plants throughout their life cycle. They significantly affect yield potential of several economically important agricultural crops thus posing serious threat in meeting global food demands of ever-growing global populations (Douglas 2018). The damages caused by insect pests have been grouped into two categories, viz., direct and indirect damages. Direct damages involve fouling of plant tissues, deformation and necrosis of floral organs, followed by propagation of plant pathogens, whereas indirect damages involve bruising of fruits causing a significant decrease in harvesting index leading to increased cost for their protection and production (Pappas et al. 2017). At present, insect pests cause an estimated yield loss of approximately 20% of major horticultural crops worldwide (Douglas 2018). Studies have reported that only about 0.5% of insects pose a significant threat to human health; therefore, they had been given the status "pest," which mainly belong to class Lepidopteran (Zaidi et al. 2017). Till date, neither of the conventional breeding nor molecular breeding techniques have been successful in developing cultivar tolerant to lepidopteran group of insects.

Presently, several chemicals and cultural, biological, and transgenic approaches are being deployed to effectively control the vicious attack of insect pests. Among these, the cultural control measures involve crop rotation, proper land water and harvest management, and biological control measures which involve the use of predators and parasitoids are the traditional approaches which are being used by the farmers since ancient times (Lombardo et al. 2016). Despite their slow activity and limited application in a given area at a particular time by skilled personnel, they are mostly preferred, over chemical application owing to their low-cost effectivity, easy to handle, environment-friendly and little or no adverse effect (Douglas 2018). Chemical methods making use of toxic chemicals which are being sprayed on a regular basis are one of the quicker and most effective modes of control measures compared to the rest of the methods (Jeschke 2016). However, one of the main disadvantages of the chemical method is that once they are applied higher than the recommended doses, they can affect vital signaling in plants by damaging cell surface proteins (Borel 2017). The present-day plant breeding techniques are facing significant constraint in fostering new improved genotypes due to the lack of the availability of resistant/tolerant genotypes as well as the problems imposed by linkage drag while introgressing genes from wild varieties (Dempewolf et al. 2017).

In this context, the transgenic approach has emerged as an upper-hand tool over traditional methods due to the high degree of their specificity and ability to synthesize a substantial number of insecticidal compounds (Majeed et al. 2018). In 1987, the first insect-resistant transgenic plants were developed in tomato by using Cry genes from soil bacterium *Bacillus thuringiensis* that showed improved tolerance against *Spodoptera litura* and *Heliothis virescens*. Many of the vegetable crops, such as eggplant and chili, which are widely consumed in Asian and African continents, especially in the Indian subcontinent, are much prone to lepidopteran insect (de Castro et al. 2017). Transgenic *Bt* brinjal overexpressing *cry1Ab* and *cry1Ac* has been developed and showed enhanced tolerance to lepidopteran insects (Biswas et al. 2018). Similarly, transgenic cauliflower was transformed using *cry1Ab* gene, and the transgenic cauliflower developed exhibited strong tolerance against diamondback moth (*Plutella xylostella*) (Srivastava et al. 2016). Transgenic cabbage line DTC 507 transformed using both cry1Ab and cry1B showed enhanced tolerance against diamondback moth (*Plutella xylostella*) disease (Parmar et al. 2017). Transgenic *cry1Ac* gene has also been successfully transforming in okra (*Abelmoschus esculentus*) for conferring resistance against fruit borer (*Earias vittella*) that has tremendously affected its productivity (Jan et al. 2015).

Apart from *Cry* genes, several other genes per se, trypsin inhibitor, protease inhibitors, and cysteine inhibitors, have also been transformed in various crop plants to increase their tolerance against several insect pest diseases (Tanpure et al. 2017). Parallel to this, several of these genes have been transformed in cauliflower and sweet potato and have shown enhanced resistance to *Spodoptera litura* and *Plutella xylostella* (Chen et al. 2016a, b). Genetically engineered strawberry plants overexpressing cowpea trypsin inhibitor (CpTi) gene under a strong promoter exhibited strong resilience against vine weevil and aphids (Singh et al. 2018). A further group of plant pests also include root-knot nematode (*Meloidogyne incognita*) that also adversely affects the growth and yield of many horticultural crops. Transgenic plants overexpressing several proteinase inhibitor genes especially *Cysteine proteinase* gene have been considered as the most effective strategy for controlling yield losses as they block the metabolic process of *Meloidogyne incognita* by activating the proteolytic activity of several other proteinases (Zhang et al. 2015a, b, c). Resistance against root-knot nematode has been successfully achieved by developing transgenic plants such as in brinjal, tomato, and potato and exhibited a strong reduction of 70–80% of nematode lesions compared to their non-transgenic counterparts (Seow-Neng et al. 2017).

2.2 Genetic Engineering for Disease Resistance

Diseases caused by various bacteria and fungi are the next big threat to the growth and production of horticultural crops. Due to lack of the availability of resistant/tolerant genotypes, conventional/molecular breeding has been observed to have limited but time-consuming success in fostering new improved genotypes. Here as well, genetic engineering for providing increased disease resistance has become the only viable option in the context of efficiency and time limitation. Resistance against several bacterial and fungal diseases has been done by transferring various genes like glucanase, chitinase, and pathogenesis-related (PR) genes which are capable of degrading/damaging cell wall of pathogens (Bill et al. 2016). One of the most important strategies that is used for developing disease resistance transgenic plants is the induction of systemic acquired resistance (SAR)-related genes which is often induced with the increased accumulation of endogenous salicylic acid (SA) which in turn modulates the expression of many downstream genes including pathogenesis-related genes (Chen et al. 2017). In a study, tomato *PR-5* gene has been transformed in sweet orange, and the transgenic orange developed showed enhanced tolerance to *Phytophthora citrophthora* (Takemoto and Mizuno 2016). Similarly, *Non-expressor of Pathogenesis Related (NPR-1)* gene from *Arabidopsis thaliana* has been transformed in tomato, and the developed transgenic tomato showed improved tolerance to heat stress as well as to increase tolerance to *Tomato mosaic virus* (TMV) (Bubici et al. 2017).

Transgenic tomato overexpressing *NPR1* gene also showed enhanced tolerance to bacterial wilt and fusarium wilt as well as exhibited moderate resistance against bacterial leaf spot. In this continuation, transgenic apple has also been developed with enhanced resistance against apple cedar rust caused by *V. inaequalis* and *Gymnosporangium juniperi-virginianae* and apple blight disease caused by *Erwinia amylovora* (Thakur et al. 2018). Studies have reported that overexpression of *NPR1* gene also plays a significant role in the regulation of endogenous SA level that ultimately regulates SAR-mediated response in plants. The stimulated SAR response in turn regulates the expression of several downstream genes involved in stress defense pathway, and transgenic plants developed by overexpressing *NPR1* genes have shown significant reduction in their disease severity percentage, and few of them were also able to remain unaffected up to 3–4 years of their planting under field condition (Verma et al. 2016).

Another major class of genes imparting disease resistance belong to group *chitinases*. Chitinases are the group of enzymes that catalyze the hydrolysis of chitin, an important insoluble constituent of the cell wall of various microorganisms per se, bacteria and fungi (Chen et al. 2018). *Chitinase* gene has been successfully transformed in many crops for developing a transgenic plant with improved tolerance to a variety of fungal diseases such as overexpression of tobacco *ChiC* gene in carrot, which has delineated enhanced tolerance to *Botrytis cinerea* (Su et al. 2015). Similarly, an attempt was also made to develop transgenic potato by transferring *chitinase* gene from *P. vulgaris* to confer tolerance to early blight caused by *Alternaria solani*, and the developed transgenic lines also showed enhanced tolerance to herbicides which was further confirmed by in vitro assays (Chen et al. 2018).

Researchers have also developed transgenic guava with enhanced tolerance to guava wilt disease caused by *Fusarium oxysporum* by transferring *endochitinase* gene from *Trichoderma* (Kumar et al. 2016) which is a major constraint to guava production around the globe. Technological advancement within the "OMICS" technology has led to the identification of various defense pathways associated with pathogen attack and has also facilitated the investigation of several genes involved in this defense pathway (Flexas and Gago 2018). Among such genes, the gene glucanase has also been accredited to confer resistance against fungal pathogens by stimulating the biosynthesis of an antifungal compound phytoalexins (Su et al. 2016). In a study, overexpression of *Trichoderma harzianum* β-1-3-glucanase gene under the control of strong promoter *CaMV35S* in the transgenic plants exhibited strong tolerance of about 61% to anthracnose after treating them with *Colletotrichum acutatum* (Forcelini et al. 2016). Similarly, overexpression of other non-plant antimicrobial genes such as phytoalexins, defensin, and attacin in transgenic plants has shown to confer resistance against several fungal diseases (Levy et al. 2018). For instance, transgenic capsicum annum lines overexpressing fungal gene carboxy esterase gene showed enhanced tolerance against anthracnose as carboxy esterase gene has been known to stimulate disease resistance by enhancing the expression of NPR genes which in turn activate the biosynthesis of several antimicrobial compounds in transgenic plants (Checker et al. 2018). In the past recent years, RNAi and CRISPR Cas9 technologies have been widely exploited for the development of transgenic plants with improved tolerance to several biotic and abiotic stresses by causing sequence-specific degradation of small RNAs or by switching off the expression of certain endogenous genes (Rani et al. 2016). For example, transgenic grapevine developed by silencing of susceptible gene *MLO-7* through RNAi approach showed enhanced tolerance to powdery mildew (Pessina et al. 2016).

2.3 Genetic Engineering for Virus Resistance

The plant viruses significantly affect crop growth and productivity, and till today conventional breeding has failed to generate crop plants with improving resistance/tolerance to viral diseases (Trebicki et al. 2015). Transgene technology has been promising in transferring novel stress tolerance genes in many of the crop plants for example, in fruit crop per se; papaya virus coat protein gene method has been successfully used to engineer against *Papaya ringspot virus* (PRSV) (Mishra et al. 2016). The intervention of recombinant DNA technology in the coat protein gene of the virus has been successfully transformed in papaya which has shown up to 80% more tolerance compared to non-transgenic plants (Dhekney et al. 2016). Similarly, transgenic strawberry plants overexpressing virus coat protein gene showed improved resistance to *Strawberry mild yellow-edge potexvirus* (Silva et al. 2015). In another study, transgenic chili plants transformed with the coat protein gene exhibited strong resistance against viral disease (Praveen et al. 2017). The growth, development, and productivity of *Citrullus lanatus* have been significantly affected by *Zucchini yellow mosaic virus* (ZYMV), and in order to overcome the yield-related loss, the *Citrullus lanatus* lines were transformed using virus coat protein gene, and the developed transgenic lines were completely immune to ZYMV disease (Harth et al. 2017).

Transgenic papaya plants transformed with the mutated replicase gene have also resulted in the improved resistance against PRSV disease under field condition compared to non-transgenic control plants (Fragoso et al. 2017). In this continuation, transgenic banana plants overexpressing replicase-associated gene showed enhanced tolerance to the *Banana bunchy top virus* (BBTV); on the contrary, non-transgenic banana plants showed high bunchy top symptoms (Ghag et al. 2015). Among other genome engineering technologies, RNA interference and CRISPR Cas9 have received much attention owing to their simplicity and high reproducibility. RNA interference has been reported to play a significant role in tailoring plants with enhanced virus resistance by facilitating the formation of self-complementary hairpin RNA structure under the control of *rolC* promoter, thereby controlling the systemic spread of viral disease (Leus 2018). For instance, transgenic poinsettia plants have been developed through RNA gene silencing technology which showed enhanced tolerance to *Poinsettia mosaic virus* (Noman et al. 2017). Similarly, RNA interference technology was also used to develop transgenic tomato plants expressing *AC4* gene construct, and the developed transgenic tomato plants showed enhanced resistance against *Tomato leaf curl virus* (TLCV). Correspondingly, transgenic banana plants overexpressing *siRNA* target gene showed decreased viral replication and its transformation which ultimately resulted in improved resistance against BBTV infection (Snehi et al. 2015).

3 Transgenic Approach for Abiotic Stress Management

The present-day agriculture is being continuously intimidated by several abiotic stresses such as heat drought, salinity, a cold and heavy metal, which has accounted for up to 50% decrease in crop productivity worldwide. Since conventional breeding approaches which mainly involve varietal cross and mutation breeding have resulted in impetuous conclusion, application of transgenic technology is the only viable option to engineer abiotic stress-tolerant plants by altering the expression levels of various genes of stress defense pathway. Several genes encoding stress-protective compounds, metabolites, and antioxidants have been identified in related and distantly related plants that can be exploited for engineering-sensitive plants/ varieties for multiple stress tolerance. This section mainly deals with the advancement made in the abiotic stress management in crop plants using a transgenic approach.

3.1 Genetic Engineering for Drought Stress Management

Several regulatory genes of signaling pathway provide valuable targets for tailoring plants against a wide range of abiotic stresses. Among them are transcription factors (TFs), which are capable of modulating the expression of various downstream as well as several other genes involved in the regulation of abiotic stress tolerance in plants which are the prime targets for genetic engineering for abiotic stress tolerance in plants (Marco et al. 2015). Various TFs such as member of Apetala 2/ Ethylene Responsive Factor (AP2/ERF) family TFs, MYB, MYC, WRKY, and NAC TFs can be exploited for genetic engineering of crop plants (Marco et al. 2015). In a study, transforming tomato plants with the DNA construct of *Arabidopsis* C-repeat binding factor/dehydration-responsive element binding (CBF/DREB) under constitutive expression of *nos* and *35s*, *Cauliflower mosaic virus* (*CaMV*) promoter exhibited strong tolerance to drought stress compared to non-transformed tomato plants (Kidokoro et al. 2015). Similarly, transgenic apple plants overexpressing *Oryza sativa OsMYB4* showed improved tolerance to drought stress by stimulating the expression of *MYB TFs* (Buti et al. 2018). Transforming grapevine plants with *DREB1b* gene construct had improved drought tolerance of transgenic grapevine plants compared to non-transformed plants.

Another group of TFs which has been reported to positively regulate abiotic stress tolerance in plants are WRKY TFs, which have been widely used for bioengineering stress-tolerant plants (Jiang et al. 2017). There are several studies that have implicated that overexpression of WRKY TFs is able to regulate drought stress in plants, for example, transforming *Chrysanthemum morifolium* plants with WRKY TFs was able to regulate drought stress tolerance via ABA-mediated pathway and activating the expression of various downstream genes such as *MYB2*, *DREB1A and Abscisic acid Insensitive 2* (*ABI2*). In another study, researchers have also transformed banana plants by transforming *WRKY71* gene, and the developed transgenic showed multiple abiotic stress tolerance (Zhou et al. 2015). Likewise, NAC TFs which comprises of three TFs, i.e., no apical meristem (N)/arabidopsis transcription activator factor (A)/cup-shaped cotyledon (C), thus abbreviated as NAC have also been used in the development of transgenic plants exhibiting improved abiotic stress tolerance (Hong et al. 2016). In a study, overexpression of *SNAC1, SNAC2, and OsNAC6* gene in transgenic rice plants showed enhanced salt and drought stress tolerance at the vegetative stage with improved seed setting percentage (25–40%) under field condition compared to non-transgenic counterparts (Todaka et al. 2015). Studies have also reported that overexpression of *Triticum aestivum TaNAC2* gene in *Arabidopsis* have resulted in increased tolerance to salt, drought, and cold stress. Similarly, researchers have also reported that transforming transgenic plants by TaNAC67 gene construct enhances drought tolerance by improving physiological traits, enhancing plant water status, strengthening membrane integrity, and activating several genes of stress defense pathways (Sharma et al. 2017).

Late embryogenesis abundant (LEA) proteins belonging to a group of hydrophilic proteins are also induced under drought, salt, chilling, high temperature, and desiccation stress in plants (Banerjee and Roychoudhury 2016). Studies have reported that overexpression of LEA proteins in various transgenic plants has been shown to enhance tolerance to drought and desiccation stresses by lowering membrane damage and by increasing the biosynthesis and accumulation of osmolytes proline and glycine betaine (Artur et al. 2018). Furthermore, studies have also shown that overexpressing *Arthrobacter globiformis choline oxidase* gene is also involved in imparting drought stress tolerance in transgenic potato plants by enhancing the accumulation of osmolytes proline and glycine betaine. The transgenic potato plants not only showed enhanced accumulation of osmolytes but also showed higher chlorophyll content, low lipid peroxidation, and enhanced activity of antioxidant enzymes compared to non-transformed lines (Chen et al. 2016a, b). Thus, TFs have been observed as an important player of stress defense pathway, which can be exploited for genetic engineering of crop plants with enhanced abiotic stress tolerance.

3.2 Genetic Engineering for Heat Stress Management

High-temperature stress or heat stress is one of the most important abiotic stresses that severely affects plant growth and productivity (Zhu 2016). High-temperature stress results in the generation of oxidative stress in plants by provoking membrane damage, reducing photosynthetic efficiency, increasing generation, and reducing scavenging of reactive oxygen species (ROSs) (Dwivedi et al. 2016). When the level of ROS reaches beyond a certain threshold, they cause denaturation of proteins and reduced protein homeostasis. Genetic engineering for heat tolerance in plants require transgenic plants with improved ROS scavenging ability as it has been reported that ROSs at basal level participate in many of the signaling pathways and when their level reaches beyond a certain threshold, they are effectively scavenged by enzymes of antioxidant defense pathways such as superoxide dismutase (SOD), glutathione reductase (GR), peroxidase (POX), and catalase (CAT) (Fancy et al. 2017). Therefore, transgenic plants overexpressing ROS-scavenging enzymes is one of the potential strategies to overcome heat stress-induced oxidative damages in plants. In a study, researchers have shown that overexpression of cytosolic ascorbate peroxidase *(cAPX)* enhances heat tolerance in transgenic apple plants by reducing the membrane damage and improving the photosynthetic efficiency (Zandalinas et al. 2018).

Similarly, transgenic tomato plants overexpressing *cAPX* gene have also resulted in the improved heat tolerance up to 40 °C in open field condition by decreasing the heat stress-induced oxidative injury, lipid peroxidation, and enhanced scavenging of antioxidant enzymes compared to non-transgenic tomato plants (Sadiq and Akram 2018). Likewise, an isoform of SOD, i.e., overexpression of Cu/Zn SOD from okra plants and transforming it into potato plants, has also resulted in increased heat tolerance of potato plants by stimulating several ROS-scavenging enzymes thus mediating effective quenching of ROSs in transgenic potato lines (Sadiq and Akram 2018). Among nonenzymatic methods, overexpression of genes encoding several chemical compounds such as methyl viologen, nitric oxide, and polyamines has been shown to directly involve in conferring heat stress tolerance in plants. Several studies have reported about the potential role of polyamine in imparting heat stress tolerance as in tomato plants; overexpression of S-adenosyl-I-methionine decarboxylase (SAMDC) from *Saccharomyces cerevisiae* resulted in increased production of polyamines leading to enhanced tolerance in transgenic tomato plants (Parmar et al. 2017).

Transgenic plants overexpressing heat shock proteins (HSPs) have also been known to enhance high-temperature tolerance by acting as molecular chaperones, maintaining protein homeostasis, and minimizing denaturation of proteins (Jacob et al. 2017). Most of the HSPs have been characterized from tomato and *Arabidopsis*, and studies have reported that overexpression of heat shock factor (HSF) from *Glycine max* enhanced high-temperature tolerance in soybean plants via activating the expression of several downstream genes involved in stress defense pathway (Zhang et al. 2015a, b, c). Similarly, overexpression of *VpHSF1* from *Vitis pseudoreticulata* in transgenic tobacco plants resulted in improved heat tolerance in transgenic lines compared to non-transgenic counterparts (Hu et al. 2016). The above results clearly pinpointed that HSF genes can be efficiently employed for tailoring crop plants against various abiotic stresses. However, a comprehensive understanding of efficient means of developing transgenic plants is required in order to minimize the off-target generation and negative effect on plant innate immunity.

3.3 Genetic Engineering for Salt Stress Management

Salinization of agricultural land is one of the most serious threats to the environment as well as to agricultural sustainability. Soil salinity is increasingly becoming one of the most important abiotic stresses affecting plant growth, development, and agricultural productivity worldwide (Bless et al. 2018). Currently, the world accounts for approx. 1.5 billion hectares of arable agricultural land, and so far, nearly 25%, i.e., about 77 million hectares, of the world's agricultural farmland has become unsuitable for farming due to their excessive salinization (Bless et al. 2018). Soil is generally considered as a saline soil if the electrical conductivity (EC) of the soil in the root zone is greater than 4 dS m^{-1} which corresponds to approx. 40 mM of NaCl in the soil with 15% of sodium exchangeable rate (Singh 2015). Furthermore, recent studies have reported continuous increase (10% annually) in the soil salinization due to continuous weathering of rocks, low precipitation, irrigation with saline water, and poor agronomic practices which if not controlled will lead to salinization of more than 50% of the arable agricultural land by 2050 (Singh 2018).

Like other abiotic stresses, salinity stress is also a complex regulatory network which is control by several genes (Munns and Gilliham 2015). A body of literature has documented that plants that exhibit abiotic stress tolerance like salt stress are capable of synthesizing various pathogenic-related proteins to ameliorate the adverse effect of stress conditions (Negrao et al. 2017). Among a different group of pathogenesis-related proteins, *Osmotin* proteins are the first that mark their presence upon plant encounters with climate extremes which stimulate the plant's innate immunity to counteract the negative effect of different abiotic stresses (Wan et al. 2017). In this context, a study was conducted to confirm this hypothesis where an *Osmotin* gene derived from tobacco plants was transformed in strawberry plants, and they recorded that the transgenic strawberry plants overexpressing *Osmotin* gene showed enhanced tolerance to salinity stress compared to non-transgenic plants (Sripriya et al. 2017). A similar kind of study was also performed in chili (*Capsicum annum* L.) plants where the researchers reported that transgenic chili plants overexpressing tobacco *Osmotin* gene in their T_2 generation exhibited increased tolerance to salinity stress (300 mM NaCl) directly as the result of improved photosynthesis, low lipid peroxidation, and enhanced activity of antioxidant enzymes (Ullah et al. 2018).

Enhanced susceptibility to salt stress of citrus species compared to other horticultural crops is well documented (Ahmadi et al. 2016). Researchers have also determined the efficacy of halotolerant gene 2 (*HAL2*) in conferring salt stress tolerance in crop plants via stimulating the methionine biosynthetic pathway involved in the sequestration of sodium ions (Thanananta et al. 2018). Several reports have delineated that overexpression of *HAL2* gene in several transgenic plants has resulted in increased tolerance to salt stress compared to their non-transgenic counterparts (Thanananta et al. 2018). Transgenic tomato plants with enhanced ectopic expression of *cytosolic ascorbate peroxidase* (*cAPX*) gene effectively improved salt stress tolerance in transgenic tomato by improving the scavenging of reactive oxygen species (Jiang et al. 2016). Salt stress also adversely affect the growth and productivity of potato and sweet potato (*Ipomoea batatas*) plants. However, transforming potato and sweet potato plants with *betaine aldehyde dehydrogenase* (*BADH*) gene from *Spinacia oleracea* strongly improves the tolerance of these plants to salinity stress by stimulating enhanced biosynthesis of glycine betaine that effectively decreased membrane damage, improved photosynthesis, and enhanced the activities of various enzymes involved in stress defense pathway (Wang et al. 2016a, b).

Transgenic cherry tomato lines overexpressing strawberry *D-galacturonic acid reductase (GalUR)* gene exhibited improved tolerance to 200 mM of salinity stress coupled with enhanced accumulation of ascorbic acid and increased activity of several enzymes of stress defense pathway (Lim et al. 2016). Similarly, overexpression of Na^+/H^+ antiporter gene (*TaNHX2*) in chili as well as in tomato plants increased the ability of transgenic chili and tomato plants to combat with the adverse effect of salinity stress by improving plant water contents, enhancing the accumulation of osmolytes like proline and glycine betaine, and regulating the expression of other downstream stress-responsive genes (Bulle et al. 2016).

3.4 Genetic Engineering for Metal Stress Management

The accelerated release of various metals that naturally occurs within the different layers of Earth's crust into arable agricultural land due to excessive mining and industrialization has posed a severe threat to the ecosystem, to crop productivity, and to human health (Mishra et al. 2017). However, most metals such as Cu, Fe, Zn, Ni, and Mn are of the absolute requirement by the cells. They become toxic only when they are hyperaccumulated at higher concentrations leading to the development of oxidative stress in plants (Emamverdian et al. 2015). Metals can lead to the generation of oxidative stress in plants by various processes like oxidation/reduction processes leading to the generation of ROSs, interrupting the function of various molecules for, e.g., chlorophyll molecules. In addition, accumulation of metals like Hg^+ and Cu^+ shows acute reactivity toward thiol group that instigates severe repercussion on protein structure and function (Berni et al. 2018). The conventional method of metal remediation includes soil excavation; washing and reburial have not contributed much in the remediation of metal contamination of the arable farmland. So, the use of a transgenic approach to stimulate the remediation of metals by plants is an effective alternative.

Several genes have been known to be involved in the uptake, translocation, and sequestration of metal ions, and enhanced ectopic expression of these metal-responsive genes is an effective strategy for engineering crop plants with enhanced metal stress tolerance (Marco et al. 2015). In a study, overexpression of metal chelator protein metallothioneins (MTs) in tobacco as well as in oil seed rape enhanced the tolerance of transgenic tobacco and oil seed rape plants against Cd stress (Peng et al. 2017). Similarly, transgenic cauliflower plants exhibiting ectopic expression of copper resistance-associated metallothioneins *CUP1* gene showed 15-fold higher tolerance to Cd and Cu stress compared to their non-transgenic counterparts (Ruta et al. 2018). In another study, researchers have also reported that transgenic mustard plants overexpressing *γ-glutamyl cysteine synthase* gene showed enhanced tolerance to Cd stress by stimulating the accumulation and sequestration of the ion (Yuan et al. 2015). Enhanced ectopic expression of Zn transporter protein (AtMTP1) in transgenic *Arabidopsis thaliana* plants showed improved resistance against Zn accumulation, while overexpression of *A. thaliana* calcium transporter protein CAX2 in transgenic tobacco resulted in enhanced sequestration of Ca, Cd, and Mn (Fasani et al. 2017). Another important metal transporter gene (NtCBP4) was identified from tobacco plants specifically involved in the binding of calmodulin protein. However, overexpression of this gene resulted in increased Ni and Pb accumulation in transgenic plants (Mosa et al. 2016). Parallel to this, *Arabidopsis* plants overexpressing ferric reductase (FRE1 and FRE2) gene from yeast resulted in increased uptake of iron in transgenic plants compared to non-transgenic plants. Additionally, the transgenic *Arabidopsis* plants also showed enhanced tolerance to various metals such as Fe, Cu, Mg, Mn, and Zn (Mosa et al. 2016). Correspondingly, transgenic tobacco plants overexpressing natural resistance-associated macrophage protein 1 (NRAMP1) from *A. thaliana* enhanced the resistance of transgenic plants against Fe and Cd stress. Similarly, overexpression of several other metal transporter proteins such as *Arabidopsis* IRTI metal transporter protein has shown to increase the transport of various metals such as Fe, Zn, Cd, and Mn (Agorio et al. 2017). With the advent of transgenic technology, it has now become feasible to engineer crop plants overexpressing certain metal transporter proteins to accumulate specific metals together with enhanced tolerance and improved productivity under metal-infected soil.

4 Role RNA Interference and Genome Editing Tools in the Improvement of Plant's Stress Tolerance

Plant-dependent RNA interference (RNAi) plays a significant role in improving plant tolerance against various biotic stresses (Fang and Qi 2016). RNAi in plants uses antisense-mediated silencing of important genes mediating cleavage of essential genes via expression of double-stranded RNAs (dsRNAs). Studies have well documented that the RNAi regulatory pathways are an evolutionarily conserved process which involves the generation of small interfering molecules with the help of dicer enzyme (Zhang et al. 2017). These small interfering molecules are termed as small interfering RNA (siRNA) and microRNA (miRNA) which mediate the cleavage of target RNA molecule within the RNA-induced silencing complex (RISC) containing Argonaute protein (Fang and Qi 2016). This technique has been successfully exploited in the generation of transgenic plants with enhanced tolerance to different biotic stresses. Overexpression of RNA-dependent RNA polymerase (RdRP) enzyme has been reported to enhance the accumulation of siRNA both in plants and in animals (Hunter et al. 2016). For instance, overexpression of dsRNAs in transgenic tobacco plants provided resistance against *H. armigera* (Tanpure et al. 2017). Studies have also reported that Cytochrome P450 (CYP6AE14) gene of *H. armigera* when overexpressed is able to detoxify gossypol which ultimately results in the inhibited growth of larvae (Jin et al. 2015). In another study, a cDNA library was constructed for corn rootworm by testing 200 genes to identify potential dsRNA targets. A total of 14 genes were identified of which *vacuolar ATPase (v-ATPase)* gene was identified as the potential gene for developing transgenic corn plants with enhanced tolerance to rootworm disease and exhibited minimum root damage (Khajuria et al. 2015). RNAi in conjunction with transgenic technology has been widely employed to enhance the resilience of crop plants against various stresses (Mamta and Rajam 2018). Furthermore, overexpression of *CYP6AE14* and cysteine protease gene form *Gossypium hirsutum* (*35GhCP1*) in transgenic cotton, enhance their resistance against cotton bollworm disease (Mamta and Rajam 2018). The use of RNAi and transgenic approach in the development of resistant crop plants has been summarized in Table 23.1. Overall, from the above discussion and available literatures, it has become apparent that RNA interference offers great potential in controlling the damages caused by insect pests that adversely affect plant growth and productivity; however, some refinement and modification in this technology are required in order to ameliorate the adverse effect of off-target amplification together with their evaluation/validation by conducting large-scale field trials.

The Clustered Regulatory Interspread Short Palindromic Repeats (CRISPR) CRISPR-associated protein 9 (CRISPR-Cas9) has emerged as the most sophisticated technology for genome editing (Xuefei et al. 2017). The CRISPR-Cas9 is originally derived from the adaptive immune response of archaea/bacteria which protect themselves from invading nucleic acids by catalyzing sequence-specific cleavage of nucleic acids which result in the generation of double-stranded breaks (DSBs) (Xuefei et al. 2017). Once generated, these DSBs are repaired either by nonhomologous end-joining (NHEJ) method or either by homologous recombination (HR), thus causing modification of target site within the plant genome (Table 23.2), and uses three types of engineered nuclease, viz., zinc finger nuclease (ZFN), transcription activator-like effector nuclease (TALENS), and CRISPR-Cas, for directing site-specific cleavage (Bortesi and Fischer 2015). CRISPR-Cas9 has been widely employed in a variety of agricultural crops such as in tomato, citrus, grape, etc. for activation or repression of certain target genes. Several studies have reported that CRISPR-Cas9 is extremely helpful in developing virus-resistant crops for, e.g., researchers have successfully developed virus resistance cucumber plants with nonfunctional eukaryotic translation initiation factor 4E (*eIF4E*) using CRISPR-Cas9 technology. The developed transgenic cucumber plants showed enhanced tolerance to *Cucumber vein yellowing virus* as well as to *Papaya ringspot mosaic virus* (Chandrasekaran et al. 2016a, b). In another study, transgenic grape cultivar in which ribonucleoproteins were mutated using CRISPR-Cas9 approach showed enhanced tolerance to powdery mildew disease, and transgenic apple

Table 23.1 List of recently genetically engineered crops for biotic and abiotic stress management

Crops	Biotic and abiotic stresses	Target gene	Target trait	Functional mechanism	References
Okra (*Abelmoschus esculentus*)	Insect pest	*Cry1AC*	Fruit and shoot borer resistance	Dissolution of insect midgut	Narendran et al. (2013)
Planch (*Actinidia chinensis*)	Insect pest	*sbtCry1AC*	Resistance against *Oraesia excavata*	Dissolution of insect midgut	Zhang et al. (2015a, b, c)
Daisy (*Chrysanthemum morifolium*)	Aphid resistance	*CmWRKY48*	Positive role in many biotic stresses	Inhibit excessive growth of aphids	Li et al. (2015)
Strawberry (*Fragaria ananassa*)	Fungal resistance	*b-1,3- Glucanase gene (bgn13, 1* from *Trichoderma harzianum*	Resistance to crown gall disease	Degrade fungal cell wall	Mercado et al. (2015)
Guava (*Psidium guajava*)	Fungal resistance	*Endochitinase gene* from *Trichoderma*	Prevention of conidial growth	Chitinase degrades chitin of fungal cell wall	Mishra et al. (2016)
Daisy (*Chrysanthemum Morifolium*)	Drought resistance	*CmWRKY1*	Drought stress tolerance	Positive regulator of drought stress	Fan et al. (2016)
Potato (*Solanum tuberosum*)	Heat resistance	*stnsLTP1*	Heat, drought, and salinity tolerance	Positive regulator of antioxidant defense system	Gangadhar et al. (2016)
Cherry tomato (*Solanum lycopersicum* var. cerasiforme)	Heat resistance	*D- galacturonic acid reductase gene*	Heat and drought tolerance	Regulate biosynthesis of ascorbic acid	Lim et al. (2016)
Chili pepper (*Capsicum annuum*)	Salinity tolerance	*Osmotin gene*	Salinity and drought stress	Enhanced accumulation of osmolyte proline	Bulle et al. (2016)
Bottle gourd (*Lagenaria siceraria*)	Salinity tolerance	*AVP1 gene*	Salt tolerance	Increased water content and low membrane damage	Han et al. (2015)

cultivar mutated for the same gene using CRISPR-Cas9 showed enhanced tolerance against fire blight disease (Malnoy et al. 2016). The efficacy of CRISPR-Cas9 approach has been validated in many scientific studies (Fig. 23.1); however, recently a group of researchers created a knockout mutant in watermelon using CRISPR-Cas9 that resulted in the development of seedless watermelon plants. Similarly,

Table 23.2 List of recently genetically engineered crops using genome-editing approach

Crop	Biotic stress	Genome-editing approach	Target trait	References
Tobacco (*Nicotiana benthamiana*)	*Tomato yellow leaf curl China virus*	Zinc finger nuclease	Replication-associated protein site in intragenic region	Chen et al. (2014)
Arabidopsis (*Arabidopsis thaliana*)	*Tomato yellow leaf curl virus*	Zinc finger nuclease	Replication-associated protein site in intragenic region	Mori et al. (2013)
Tobacco (*Nicotiana benthamiana*)	*Tobacco curly shoot virus*	Transcription activator like effector nuclease	Replication-associated protein site in intragenic region	Cheng et al. (2015)
Tobacco (*Nicotiana benthamiana*)	*Tomato leaf curl Yunnan virus*	Transcription activator-like effector nuclease	Replication-associated protein site in intragenic region	Ali et al. (2015)
Tobacco (*Nicotiana benthamiana*)	Tomato yellow leaf curl China beta satellite	Transcription activator like effector nuclease	Replication-associated protein site in intragenic region	Baltes et al. (2015)
Tobacco (*Nicotiana benthamiana*)	*Beet curly top virus*	Clustered regulatory inter-spread short palindromic repeat	Intragenic region, coat protein, and replication associated protein	Ji et al. (2015)
Tobacco (*Nicotiana benthamiana*) *Arabidopsis* and *Arabidopsis thaliana*	*Bean yellow dwarf virus*	Clustered regulatory inter-spread short palindromic repeat	Intragenic region, coat protein, and replication-associated protein	Ali et al. (2016)
Tobacco (*Nicotiana benthamiana*) *Arabidopsis* and *Arabidopsis thaliana*	*Beet severe curly top virus*	Clustered regulatory inter-spread short palindromic repeat	Intragenic region, coat protein, and replication-associated protein	Pyott et al. (2016)
Tobacco (*Nicotiana benthamiana*)	*Tomato yellow leaf curl Sardinian virus*	Clustered regulatory inter-spread short palindromic repeat	Elongation factor 4E	Chandrasekaran et al. (2016a, b)
Cucumber (*Cucumis sativus*)	*Papaya ring spot mosaic virus*	Clustered regulatory inter-spread short palindromic repeat	Elongation factor 4E	Chandrasekaran et al. (2016a, b)

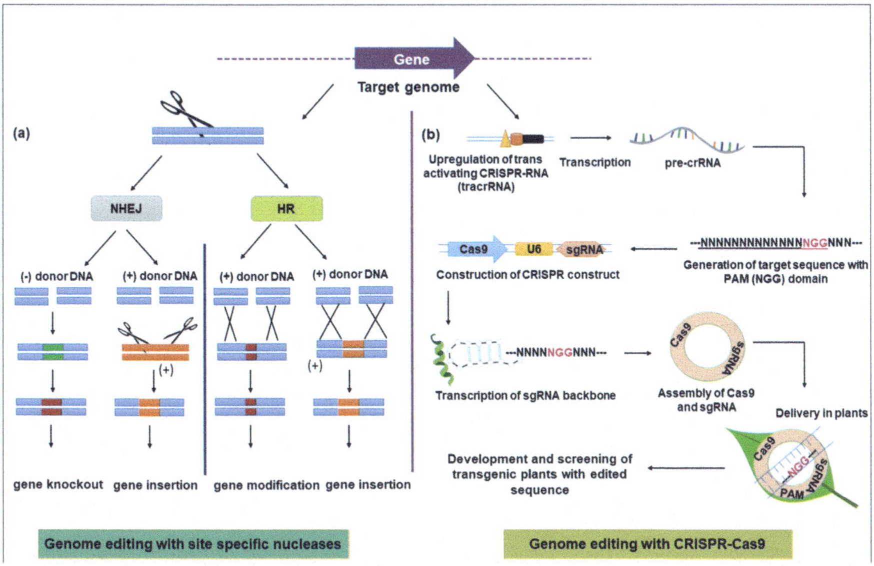

Fig. 23.1 Detailed illustration of genome editing by (**a**) site-specific nucleases where editing of target genome by nonhomologous end-joining (NHEJ) method results in the generation of gene knock-out due to random insertion and deletion of base pairs or results in the formation of gene insertion in the presence of donor DNA, which is simultaneously cleaved by the same endonuclease, whereas in case of homologous Recombination (HR), the donor DNA can be modified by being exploited for the creation of gene insertion by specifically placing nucleotide substitutions; (**b**) Clustered Regulatory Interspaced Short Palindromic Repeat (CRISPR) CRISPR associated protein 9 (Cas9) approach from bacteria involved the activation of trans-activating crRNA (tracrRNA), which then combined with pre-crRNA and following its transcription, it is then converted into single-guide RNA (sgRNA) using small RNA promoter (U6). The sgRNA and Cas9 construct are assembled and are transformed in to plants to generate transgenic plants with improved tolerance to biotic and abiotic stresses

CRISPR-Cas9 has also been utilized for the development of parthenocarpy in tomato plants by inducing mutation in *SlIAA9* gene (Ueta et al. 2017).

5 Conclusion and Future Prospective

The advent of genetic engineering/transgene technologies has revolutionized the strategies of developing crop plants against several biotic and abiotic stresses which will not only lead to the development of sustainable agriculture but also will immensely solve the problem of food security of the ever-increasing global population. In this context, advancement in the "OMICS" technology has achieved adequate commercial success in developing important horticultural crops with enhanced ability to counteract the adverse effect of biotic and abiotic stresses. Recombinant DNA technology has provided substantial mean for engineering plants with desirable traits together with genome-editing tool per se; CRISPR-Cas9 driven by RNA has overcome some of the potential barriers in developing stress tolerance cultivar. Nevertheless, the efficiency, durability, and productivity of genome engineering and editing techniques remain to be determined at large scale under field condition. Nonetheless, the coming future will allow much comprehensive analysis of various genome-engineering technologies that will further facilitate the development of crop varieties with enhanced productivity and stress tolerance.

References

Agorio A, Giraudat J, Bianchi MW, Marion J, Espagne C, Castaings L, Lelièvre F, Curie C, Thomine S, Merlot S (2017) Phosphatidylinositol 3-phosphate–binding protein AtPH1 controls the localization of the metal transporter NRAMP1 in *Arabidopsis*. Proc Natl Acad Sci USA:3354–3363

Ahmadi N, Baroiller JF, Carreras HD, Morillon R (2016) Adaptation to salinity. In: Climate change and agriculture worldwide. Springer, Dordrecht, pp 45–58

Ali Z, Abulfaraj A, Idris A, Ali S, Tashkandi M, Mahfouz MM (2015) CRISPR/Cas9-mediated viral interference in plants. Genome Biol 16:238

Ali Z, Ali S, Tashkandi M, Zaidi SS, Mahfouz MM (2016) CRISPR/Cas9-mediated immunity to geminiviruses: differential interference and evasion. Sci Rep 6:26912

Anumalla M, Roychowdhury R, Geda CK, Mazid M, Rathoure AK (2015) Utilization of plant genetic resources and diversity analysis tools for sustainable crop improvement with special emphasis on rice. Int J Adv Res 3(3):1155–1175

Anumalla M, Roychowdhury R, Geda CK, Bharathkumar S, Goutam KD, Mohandev TSS (2016) Mechanism of stress signal transduction and involvement of stress inducible transcription factors and genes in response to abiotic stresses in plant. Int J Recent Sci Res 7(8):12754–12771

Artur MA, Zhao T, Ligterink W, Schranz E, Hilhorst HW (2018) Dissecting the genomic diversification of LATE EMBRYOGENESIS ABUNDANT (LEA) protein gene families in plants. Genome Biol Evol 11(2):459–471

Bakhsh A, Hussain T (2015) Engineering crop plants against abiotic stress: current achievements and prospects. Emir J Food Agric 27:24–39

Baltes NJ, Hummel AW, Konecna E, Cegan R, Bruns AN, Bisaro DM et al (2015) Conferring resistance to geminiviruses with the CRISPR–Cas prokaryotic immune system. Nat Plants 1:15145

Banerjee A, Roychoudhury A (2016) Group II late embryogenesis abundant (LEA) proteins: structural and functional aspects in plant abiotic stress. Plant Growth Regul 79(1):1–7

Berni R, Luyckx M, Xu X, Legay S, Sergeant K, Hausman JF, Lutts S, Cai G, Guerriero G (2018) Reactive oxygen species and heavy metal stress in plants: impact on the cell wall and secondary metabolism. Environ Exp Bot. https://doi.org/10.1016/j.envexpbot.2018.10.017

Bill M, Sivakumar D, Beukes M, Korsten L (2016) Expression of pathogenesis-related (PR) genes in avocados fumigated with thyme oil vapours and control of anthracnose. Food Chem 194:938–943

Biswas R, Banerjee A, Halder U, Bandopadhyay R (2018) Transgenic research in vegetable crops with special reference to brinjal. In: Genetic engineering of horticultural crops. Academic Press, London, pp 155–167

Bless AE, Colin F, Crabit A, Devaux N, Philippon O, Follain S (2018) Landscape evolution and agricultural land salinization in coastal area: a conceptual model. Sci Total Environ 625:647–656

Borel BR (2017) When the pesticides run out. Nature 543:302–304

Bortesi L, Fischer R (2015) The CRISPR/Cas9 system for plant genome editing and beyond. Biotechnol Adv 33(1):41–52
Bubici G, Carluccio AV, Stavolone L, Cillo F (2017) Prosystemin overexpression induces transcriptional modifications of defense-related and receptor-like kinase genes and reduces the susceptibility to cucumber mosaic virus and its satellite RNAs in transgenic tomato plants. PLoS One 12(2):e0171902
Bulle M, Yarra R, Abbagani S (2016) Enhanced salinity stress tolerance in transgenic chilli pepper (*Capsicum annuum* L.) plants overexpressing the wheat antiporter (*TaNHX2*) gene. Mol Breed 36:36
Buti M, Pasquariello M, Ronga D, Milc JA, Pecchioni N, Pucciariello C, Perata P, Francia E (2018) Transcriptome profiling of short-term response to chilling stress in tolerant and sensitive *Oryza sativa* ssp. Japonica seedlings. Funct Integr Genomic 18(6):627–644
Chandrasekaran J, Brumin M, Wolf D, Leibman D, Klap C, Pearlsman M, Sherman A, Arazi T, Gal-On A (2016a) Development of broad virus resistance in non-transgenic cucumber using CRISPR/Cas9 technology. Mol Plant Pathol 17(7):1140–1153
Chandrasekaran J, Brumin M, Wolf D, Leibman D, Klap C, Pearlsman M et al (2016b) Development of broad virus resistance in non-transgenic cucumber using CRISPR/Cas9 technology. Mol Plant Pathol 17:1140–1153
Checker VG, Kushwaha HR, Kumari P, Yadav S (2018) Role of phytohormones in plant defense: signaling and cross talk. In: Molecular aspects of plant-pathogen interaction. Springer, Singapore, pp 159–184
Chen W, Qian Y, Wu X, Sun Y, Wu X, Cheng X (2014) Inhibiting replication of begomoviruses using artificial zinc finger nucleases that target viral-conserved nucleotide motif. Virus Genes 48:494–501
Chen SP, Lin IW, Chen X, Huang YH, Chang SC, Lo HS, Lu HH, Yeh KW (2016a) Sweet potato NAC transcription factor, Ib NAC 1, upregulates sporamin gene expression by binding the SWRE motif against mechanical wounding and herbivore attack. Plant J 86(3):234–248
Chen Q, Mason HS, Mor T, Sutherland A, Cardineau GA, Tacket CJ (2016b) Subunit vaccines produced using plant biotechnology. In: New generation vaccines, 4th edn. CRC Press, Boca Raton, pp 664–667
Chen S, Li X, Lavoie M, Jin Y, Xu J, Fu Z, Qian H (2017) Diclofop-methyl affects microbial rhizosphere community and induces systemic acquired resistance in rice. J Environ Sci 51:352–360
Chen J, Piao Y, Liu Y, Li X, Piao Z (2018) Genome-wide identification and expression analysis of chitinase gene family in *Brassica rapa* reveals its role in clubroot resistance. Plant Sci 270:257–267
Cheng X, Li F, Cai J, Chen W, Zhao N, Sun Y et al (2015) Artificial TALE as a convenient protein platform for engineering broad-spectrum resistance to begomoviruses. Viruses 7:4772–4782
de Castro DL, García-Gómez BI, Gómez I, Bravo A, Soberón M (2017) Identification of Bacillus thuringiensis Cry1AbMod binding proteins from *Spodoptera frugiperda*. Peptides 98:99–105
Dempewolf H, Baute G, Anderson J, Kilian B, Smith C, Guarino L (2017) Past and future use of wild relatives in crop breeding. Crop Sci 57(3):1070–1082
Dhekney SA, Kandel R, Bergey DR, Sitther V, Soorianathasundaram K, Litz RE (2016) Advances in papaya biotechnology. Biocatal Agric Biotechnol 5:133–142
Douglas AE (2018) Strategies for enhanced crop resistance to insect pests. Annu Rev Plant Biol 69:637–660
Dwivedi SL, Ceccarelli S, Blair MW, Upadhyaya HD, Are AK, Ortiz R (2016) Landrace germplasm for improving yield and abiotic stress adaptation. Trends Plant Sci 21(1):31–42
Emamverdian A, Ding Y, Mokhberdoran F, Xie Y (2015) Heavy metal stress and some mechanisms of plant defense response. Sci World J. https://doi.org/10.1155/2015/756120
Fan Q, Song A, Jiang Zhang T, Sun H, Wang Y (2016) *CmWRKY1* enhances the dehydration tolerance of chrysanthemum through the regulation of ABA-associated genes. PLoS One 11:e0150572

Fancy NN, Bahlmann AK, Loake GJ (2017) Nitric oxide function in plant abiotic stress. Plant Cell Environ 40(4):462–472

Fang X, Qi Y (2016) RNAi in plants: an Argonaute-centered view. Plant Cell 28(2):272–285

Fasani E, DalCorso G, Varotto C, Li M, Visioli G, Mattarozzi M, Furini A (2017) The MTP 1 promoters from *Arabidopsis halleri* reveal cis-regulating elements for the evolution of metal tolerance. New Phytol 214(4):1614–1630

Flexas J, Gago J (2018) A role for ecophysiology in the 'omics' era. Plant J 96(2):251–259

Forcelini BB, Seijo TE, Amiri A, Peres NA (2016) Resistance in strawberry isolates of *Colletotrichum acutatum* from Florida to quinone-outside inhibitor fungicides. Plant Dis 100(10):2050–2056

Fragoso G, Hernández M, Cervantes-Torres J, Ramírez-Aquino R, Chapula H, Villalobos N, Segura-Velázquez R, Figueroa A, Flores I, Jiménez H, Adalid L (2017) Transgenic papaya: a useful platform for oral vaccines. Planta 245(5):1037–1048

Gangadhar BH, Sajeesh K, Venkatesh J, Baskar V, Abhinandan K, Yu JW, Prasad R, Mishra RK (2016) Enhanced tolerance of transgenic potato plants over-expressing non-specific lipid transfer protein-1 (StnsLTP1) against multiple abiotic stresses. Front Plant Sci 7:1228

Ghag SB, Shekhawat UK, Hadapad AB, Ganapathi TR (2015) Stacking of host-induced gene silencing mediated resistance to banana bunchy top virus and fusarium wilt disease in transgenic banana plants. Curr Trends Biotechnol Pharm 9(3):212–221

Großkinsky DK, van der Graaff E, Roitsch T (2016) Regulation of abiotic and biotic stress responses by plant hormones. Plant Pathogen Resistance Biotechnol 9:131

Gürel F, Öztürk ZN, Uçarlı C, Rosellini D (2016) Barley genes as tools to confer abiotic stress tolerance in crops. Front Plant Sci 7:1137

Han JS, Park LI, Jeon SM, Park S, Naing AH, Kim CK (2015) Assessments of salt tolerance in a bottle gourd line expressing the Arabidopsis H?-pyrophosphatase AVP1 gene and in a watermelon plant grafted onto a transgenic bottle gourd rootstock. Plant Breed 134:233–238

Harth JE, Simmons HE, Stephenson AG (2017) Vertical infection of zucchini yellow mosaic virus via pollen transmission occurs at a lower frequency than ovule transmission. Eur J Plant Pathol 147(3):717–720

Hasanuzzaman M, Nahar K, Alam M, Roychowdhury R, Fujita M (2013) Physiological, biochemical, and molecular mechanisms of heat stress tolerance in plants. Int J Mol Sci 14(5):9643–9684

Hasanuzzaman M, Roychowdhury R, Karmakar J, Dey N, Nahar K, Fujita M (2015) Recent advances in biotechnology and genomic approaches for abiotic stress tolerance in crop plants. In: Devarajan T, Jeyabalan S (eds) Genomics and proteomics: concepts, technologies and applications. Apple Academic Press, Toronto, pp 333–366

Hong Y, Zhang H, Huang L, Li D, Song F (2016) Overexpression of a stress-responsive NAC transcription factor gene ONAC022 improves drought and salt tolerance in rice. Front Plant Sci 7:4

Hu Y, Han YT, Zhang K, Zhao FL, Li YJ, Zheng Y, Wang YJ, Wen YQ (2016) Identification and expression analysis of heat shock transcription factors in the wild Chinese grapevine (*Vitis pseudoreticulata*). Plant Physiol Biochem 99:1–10

Hunter LJ, Brockington SF, Murphy AM, Pate AE, Gruden K, MacFarlane SA, Palukaitis P, Carr JP (2016) RNA-dependent RNA polymerase 1 in potato (*Solanum tuberosum*) and its relationship to other plant RNA-dependent RNA polymerases. Sci Rep 6:23082

Jacob P, Hirt H, Bendahmane A (2017) The heat-shock protein/chaperone network and multiple stress resistance. Plant Biotechnol J 15(4):405–414

Jaganathan D, Ramasamy K, Sellamuthu G, Jayabalan S, Venkataraman G (2018) CRISPR for crop improvement: an update review. Front Plant Sci 9:985

Jan MT, Abbas N, Shad SA, Saleem MA (2015) Resistance to organophosphate, pyrethroid and biorational insecticides in populations of spotted bollworm, *Earias vittella* (Fabricius) (Lepidoptera: Noctuidae), in Pakistan. Crop Prot 78:247–252

Jeschke P (2016) Progress of modern agricultural chemistry and future prospects. Pest Manag Sci 72(3):433–455

Ji X, Zhang H, Zhang Y, Wang Y, Gao C (2015) Establishing a CRISPRCas-like immune system conferring DNA virus resistance in plants. Nat Plants 1:15144

Jiang M, Jiang JJ, He CM, Guan M (2016) Broccoli plants over-expressing a cytosolic ascorbate peroxidase gene increase resistance to downy mildew and heat stress. J Plant Pathol 1:413–420

Jiang J, Ma S, Ye N, Jiang M, Cao J, Zhang J (2017) WRKY transcription factors in plant responses to stresses. J Integr Plant Biol 59(2):86–101

Jin S, Singh ND, Li L, Zhang X, Daniell H (2015) Engineered chloroplast dsRNA silences cytochrome p450 monooxygenase, V-ATPase and chitin synthase genes in the insect gut and disrupts *Helicoverpa armigera* larval development and pupation. Plant Biotechnol J 13(3):435–446

Khajuria C, Vélez AM, Rangasamy M, Wang H, Fishilevich E, Frey ML, Carneiro NP, Gandra P, Narva KE, Siegfried BD (2015) Parental RNA interference of genes involved in embryonic development of the western corn rootworm, *Diabrotica virgifera* virgifera LeConte. Insect Biochem Mol Biol 63:54–62

Kidokoro S, Watanabe K, Ohori T, Moriwaki T, Maruyama K, Mizoi J, Myint Phyu Sin Htwe N, Fujita Y, Sekita S, Shinozaki K, Yamaguchi-Shinozaki K (2015) Soybean DREB 1/CBF-type transcription factors function in heat and drought as well as cold stress-responsive gene expression. Plant J 81(3):505–518

Kumar SA, Kumari PH, Jawahar G, Prashanth S, Suravajhala P, Katam R, Sivan P, Rao KS, Kirti PB, Kishor PK (2016) Beyond just being foot soldiers–osmotin like protein (OLP) and chitinase (Chi11) genes act as sentinels to confront salt, drought, and fungal stress tolerance in tomato. Environ Exp Bot 132:53–65

Leus L (2018) Breeding for disease resistance in ornamentals. In: Ornamental crops. Springer, Cham, pp 97–125

Levy A, Conway JM, Dangl JL, Woyke T (2018) Elucidating bacterial gene functions in the plant microbiome. Cell Host Microbe 24(4):475–485

Li P, Song A, Gao C, Jiang J, Chen S, Fang W (2015) The overexpression of a chrysanthemum WRKY transcription factor enhances aphid resistance. Plant Physiol Biochem 95:26–34

Lim MY, Jeong BR, Jung M, Harn CH (2016) Transgenic tomato plants expressing strawberry D-galacturonic acid reductase gene display enhanced tolerance to abiotic stresses. Plant Biotechnol Rep 10(2):105–116

Lombardo L, Coppola G, Zelasco S (2016) New technologies for insect-resistant and herbicide-tolerant plants. Trends Biotechnol 34(1):49–57

Majeed U, Yaqoob U, Qazi HA, Ahmad S, John R (2018) CRISPR/Cas system as an emerging technology to enhance plant viral immunity. Physiol Mol Plant Pathol 103:107–113

Malnoy M, Viola R, Jung MH, Koo OJ, Kim S, Kim JS, Velasco R, Nagamangala Kanchiswamy C (2016) DNA-free genetically edited grapevine and apple protoplast using CRISPR/Cas9 ribonucleoproteins. Front Plant Sci 7:1904

Mamta B, Rajam MV (2018) RNA interference: a promising approach for crop improvement. In: Biotechnologies of crop improvement, vol 2. Springer, Cham, pp 41–65

Marco F, Bitrián M, Carrasco P, Rajam MV, Alcázar R, Tiburcio AF (2015) Genetic engineering strategies for abiotic stress tolerance in plants. In: Plant biology and biotechnology. Springer, New Delhi, pp 579–609

Mercado JA, Barcelo M, Pliego C, Rey M, Caballero JL, MunozBlanco J, Ruano-Rosa D, Lopez-Herrera C, Santos B, RomeroMunoz F, Pliego-Alfaro F (2015) Expression of the b-1,3 glucanase gene bgn13,1 from Trichoderma harzianum in strawberry increases tolerance to crown rot diseases but interferes with plant growth. Transgenic Res 24:979–989

Mishra M, Jalil SU, Mishra RK, Kumari S, Pandey BK (2016) In vitro screening of guava plantlets transformed with endochitinase gene against *Fusarium oxysporum* f. sp. psidii. Czech J Genet Plant Breed 52:6–13

Mishra J, Singh R, Arora NK (2017) Alleviation of heavy metal stress in plants and remediation of soil by rhizosphere microorganisms. Front Microbiol 8:1706

Mori T, Takenaka K, Domoto F, Aoyama Y, Sera T (2013) Inhibition of binding of tomato yellow leaf curl virus rep to its replication origin by artificial zinc-finger protein. Mol Biotechnol 54:198–203

Mosa KA, Saadoun I, Kumar K, Helmy M, Dhankher OP (2016) Potential biotechnological strategies for the cleanup of heavy metals and metalloids. Front Plant Sci 7:303

Munns R, Gilliham M (2015) Salinity tolerance of crops–what is the cost? New Phytol 3:668–673

Narendran M, Deole SG, Harkude S, Shirale D, Nanote A, Bihani P, Parimi S, Char BR, Zehr UB (2013) Efficient genetic transformation of okra (*Abelmoschus esculentus* (L.) Moench) and generation of insect-resistant transgenic plants expressing the cry1Ac gene. Plant Cell Rep 32:1191–1198

Negrão S, Schmöckel SM, Tester M (2017) Evaluating physiological responses of plants to salinity stress. Ann Bot 119(1):1–11

Noman A, Aqeel M, Deng J, Khalid N, Sanaullah T, Shuilin H (2017) Biotechnological advancements for improving floral attributes in ornamental plants. Front Plant Sci 8:530

Pappas ML, Broekgaarden C, Broufas GD, Kant MR, Messelink GJ, Steppuhn A, Wäckers F, Van Dam NM (2017) Induced plant defences in biological control of arthropod pests: a double-edged sword. Pest Manag Sci 73(9):1780–1788

Parmar N, Singh KH, Sharma D, Singh L, Kumar P, Nanjundan J, Khan YJ, Chauhan DK, Thakur AK (2017) Genetic engineering strategies for biotic and abiotic stress tolerance and quality enhancement in horticultural crops: a comprehensive review. 3Biotech 7(4):239

Peng JS, Ding G, Meng S, Yi HY, Gong JM (2017) Enhanced metal tolerance correlates with heterotypic variation in SpMTL, a metallothionein-like protein from the hyperaccumulator *Sedum plumbizincicola*. Plant Cell Environ 40(8):1368–1378

Pessina S, Lenzi L, Perazzolli M, Dalla Costa L, Campa M, Velasco R, Malnoy M (2016) MLO genes silencing reduces susceptibility to powdery mildew in grapevine. In: X International symposium on grapevine physiology and biotechnology. IT, p 66

Praveen S, Ramesh SV, Mangrauthia SK (2017) Transgenic approaches to combat plant viruses occurring in India. In: A century of plant virology in India. Springer, Singapore, pp 783–805

Pyott DE, Sheehan E, Molnar A (2016) Engineering of CRISPR/Cas9- mediated potyvirus resistance in transgene-free *Arabidopsis* plants. Mol Plant Pathol 17:1276–1288

Rani R, Yadav P, Barbadikar KM, Baliyan N, Malhotra EV, Singh BK, Kumar A, Singh D (2016) CRISPR/Cas9: a promising way to exploit genetic variation in plants. Biotechnol Lett 38(12):1991–2006

Roychowdhury R (2014) Crop improvement in the era of climate change. IK International Publishing House, New Delhi, p 496

Roychowdhury R, Tah J (2013) Mutagenesis – a potential approach for crop improvement. In: Hakeem KR, Ahmad P, Ozturk M (eds) Crop improvement – new approaches and modern techniques. Springer, Boston, pp 149–187

Roychowdhury R, Karmakar J, Karmakar J, Adak MK, Dey N (2013) Physio-biochemical and microsatellite based profiling of lowland rice (*Oryza sativa* L.) landraces for osmotic stress tolerance. Am J Plant Sci 4(12):52

Roychowdhury R, Taoutaou A, Hakeem KR, Gawwad MR, Tah J (2014) Molecular marker-assisted technologies for crop improvement. In: Roychowdhury R (ed) Crop improvement in the era of climate change. IK International Publishing House, New Delhi, pp 241–258

Roychowdhury R, Khan MH, Choudhury S (2018) Arsenic in rice: an overview on stress implications, tolerance and mitigation strategies. In: Hasanuzzaman M, Nahar K, Fujita M (eds) Plants under metal and metalloid stress. Springer, Singapore, pp 401–415

Roychowdhury R, Khan MH, Choudhury S (2019) Physiological and molecular responses for metalloid stress in rice - a comprehensive overview. In: Hasanuzzaman M, Fujita M, Nahar K, Biswas J (eds) Advances in rice research for abiotic stress tolerance. Woodhead Publishing/ Elsevier, London, pp 341–369

Ruta LL, Banu MA, Neagoe AD, Kissen R, Bones AM, Farcasanu IC (2018) Accumulation of Ag (I) by *Saccharomyces cerevisiae* cells expressing plant metallothioneins. Cell 7(12):266

Sadiq M, Akram NA (2018) Field performance of transgenic drought-tolerant crop plants. In: Biochemical, physiological and molecular avenues for combating abiotic stress tolerance in plants, pp 83–102

Seow-Neng C, Bakar NA, Mahmood M, Chai-Ling H, Shaharuddin NA (2017) Alternative strategy in crop protection: protease inhibitors from turmeric. In: Crop improvement. Springer, Cham, pp 253–270

Sharma R, Sudan RS, Kumari S, Salgotra RK, Singh R (2017) Research article terminal heat stress-responsive genes analysis in heat susceptible HDR77 genotype of wheat (*Triticum aestivum* L.) by using semi-quantative RTPCR. Electron J Plant Breed 8(4):1124–1132

Silva KJ, Brunings A, Peres NA, Mou Z, Folta KM (2015) The Arabidopsis NPR1 gene confers broad-spectrum disease resistance in strawberry. Transgenic Res 24(4):693–704

Singh A (2015) Soil salinization and waterlogging: a threat to environment and agricultural sustainability. Ecol Indic 57:128–130

Singh A (2018) Salinization of agricultural lands due to poor drainage: a viewpoint. Ecol Indic 95:127–130

Singh S, Singh A, Kumar S, Mittal P, Singh IK (2018) Protease inhibitors: recent advancement in its usage as a potential biocontrol agent for insect pest management. Insect Sci. https://doi.org/10.1111/1744-7917.12641

Snehi SK, Raj SK, Prasad V, Singh V (2015) Recent research findings related to management strategies of Begomoviruses. J Plant Pathol Microbiol 6(6):273

Sripriya R, Parameswari C, Veluthambi K (2017) Enhancement of sheath blight tolerance in transgenic rice by combined expression of tobacco osmotin (ap24) and rice chitinase (chi11) genes. Vitro Cell Dev Biol Plant 53(1):12–21

Srivastava DK, Kumar P, Sharma S, Gaur A, Gambhir G (2016) Genetic engineering for insect resistance in economically important vegetable crops. In: Plant tissue culture: propagation, conservation and crop improvement. Springer, Singapore, pp 343–378

Su Y, Xu L, Wang S, Wang Z, Yang Y, Chen Y, Que Y (2015) Identification, phylogeny, and transcript of chitinase family genes in sugarcane. Sci Rep 5:10708

Su Y, Wang Z, Liu F, Li Z, Peng Q, Guo J, Xu L, Que Y (2016) Isolation and characterization of ScGluD2, a new sugarcane beta-1, 3-glucanase D family gene induced by *Sporisorium scitamineum*, ABA, H_2O_2, NaCl, and $CdCl_2$ stresses. Front Plant Sci 7:1348

Takemoto D, Mizuno Y (2016) Belowground and aboveground strategies of plant resistance against Phytophthora species. In: Belowground Defence Strategies in Plants. Springer, Cham, pp. 151–169

Tanpure RS, Barbole RS, Dawkar VV, Waichal YA, Joshi RS, Giri AP, Gupta VS (2017) Improved tolerance against *Helicoverpa armigera* in transgenic tomato over-expressing multi-domain proteinase inhibitor gene from *Capsicum annuum*. Physiol Mol Biol Plants 23(3):597–604

Thakur AK, Singh KH, Sharma D, Singh L, Parmar N, Nanjundan J, Khan YJ (2018) Transgenic development for biotic and abiotic stress management in horticultural crops. In: Genetic engineering of horticultural crops. Academic Press, London, pp 353–386

Thanananta N, Vuttipongchaikij S, Apisitwanich S (2018) Agrobacterium-mediated transformation of a *Eucalyptus camaldulensis*× *E. tereticornis* hybrid using peeled nodal-stem segments with yeast *HAL2* for improving salt tolerance. New For 49(3):311–327

Todaka D, Shinozaki K, Yamaguchi-Shinozaki K (2015) Recent advances in the dissection of drought-stress regulatory networks and strategies for development of drought-tolerant transgenic rice plants. Front Plant Sci 6:84

Trębicki P, Nancarrow N, Cole E, Bosque-Pérez NA, Constable FE, Freeman AJ, Rodoni B, Yen AL, Luck JE, Fitzgerald GJ (2015) Virus disease in wheat predicted to increase with a changing climate. Glob Chang Biol 21(9):3511–3519

Ueta R, Abe C, Watanabe T, Sugano SS, Ishihara R, Ezura H, Osakabe Y, Osakabe K (2017) Rapid breeding of parthenocarpic tomato plants using CRISPR/Cas9. Sci Rep 7(1):507

Ullah A, Hussain A, Shaban M, Khan AH, Alariqi M, Gul S, Jun Z, Lin S, Li J, Jin S, Munis MF (2018) Osmotin: a plant defense tool against biotic and abiotic stresses. Plant Physiol Biochem 123:149–159

Verma V, Ravindran P, Kumar PP (2016) Plant hormone-mediated regulation of stress responses. BMC Plant Biol 16(1):86

Wan Q, Hongbo S, Zhaolong X, Jia L, Dayong Z, Yihong H (2017) Salinity tolerance mechanism of osmotin and osmotin-like proteins: a promising candidate for enhancing plant salt tolerance. Curr Genomics 18(6):553–556

Wang FW, Wang ML, Guo C, Wang N, Li XW, Chen H, Dong YY, Chen XF, Wang ZM, Li HY (2016a) Cloning and characterization of a novel betaine aldehyde dehydrogenase gene from *Suaeda corniculata*. Genet Mol Res 15:15027848

Wang H, Wang H, Shao H, Tang X (2016b) Recent advances in utilizing transcription factors to improve plant abiotic stress tolerance by transgenic technology. Front Plant Sci 7:67

Wani SH, Sah SK, Hossain MA, Kumar V, Balachandran SM (2016) Transgenic approaches for abiotic stress tolerance in crop plants. In: Advances in plant breeding strategies: agronomic, abiotic and biotic stress traits. Springer, Cham, pp 345–396

Xuefei Y, Yajuan L, Yuan G, Beibei S, Lina M, Xinghai Y, Da H (2017) The application of CRISPR/Cas9 system in Gene Knock-out of *Nicotiana benthamiana*. Mol Plant Breed 1:6

Yuan H, Zhang Y, Huang S, Yang Y, Gu C (2015) Effects of exogenous glutathione and cysteine on growth, lead accumulation, and tolerance of *Iris lactea* var. chinensis. Environ Sci Pollut Res 22(4):2808–2816

Zaidi SS, Briddon RW, Mansoor S (2017) Engineering dual begomovirus-*Bemisia tabaci* resistance in plants. Trends Plant Sci 22(1):6–8

Zandalinas SI, Mittler R, Balfagón D, Arbona V, Gómez-Cadenas A (2018) Plant adaptations to the combination of drought and high temperatures. Physiol Plant 162(1):2–12

Zhang L, Davies LJ, Elling AA (2015a) A *Meloidogyne incognita* effector is imported into the nucleus and exhibits transcriptional activation activity in planta. Mol Plant Pathol 16(1):48–60

Zhang HY, Liu HM, Liu XZ (2015b) Production of transgenic kiwifruit plants harboring the SbtCry1Ac gene. Genet Mol Res 14:8483–8489

Zhang L, Zhao HK, Dong QL, Zhang YY, Wang YM, Li HY, Xing GJ, Li QY, Dong YS (2015c) Genome-wide analysis and expression profiling under heat and drought treatments of HSP70 gene family in soybean (*Glycine max* L.). Front Plant Sci 6:773

Zhang J, Khan SA, Heckel DG, Bock R (2017) Next-generation insect-resistant plants: RNAi-mediated crop protection. Trends Biotechnol 35(9):871–882

Zhou L, Wang NN, Gong SY, Lu R, Li Y, Li XB (2015) Overexpression of a cotton (*Gossypium hirsutum*) *WRKY* gene, *GhWRKY34*, in *Arabidopsis* enhances salt-tolerance of the transgenic plants. Plant Physiol Biochem 96:311–320

Zhu JK (2016) Abiotic stress signaling and responses in plants. Cell 167(2):313–324

Chapter 24
Improvement of Crop's Stress Tolerance by Gene Editing CRISPR/CAS9 System

Avinash Singh, Rajib Roychowdhury, Toolika Singh, Wenjing Wang, Deepanker Yadav, Ajay Kumar, Arpan Modi, Avinash Chandra Rai, Sandeep Ghughe, Anil Kumar, and Prashant Kumar Singh

Abstract There is an urgent need to enhance agriculture productivity to feed the world's ever-increasing population on the one hand and stresses reducing global agriculture productivity on the other hand. Hence there is only one way to boost global agriculture productivity through innovative breeding technology that can provide access to food security worldwide. Moreover, the availability of fully

---

A. Singh · T. Singh
Department of Biochemistry and Molecular Biology Section, Medical University of South Carolina, Charleston, SC, USA

R. Roychowdhury
Department of Vegetables and Field Crops, Institute of Plant Sciences, Agricultural Research Organization (ARO) – The Volcani Center, Rishon Lezion, Israel

A. Modi · A. C. Rai · S. Ghughe
Institute of Plant Sciences, Agricultural Research Organization (ARO) – The Volcani Center, Rishon LeZion, Israel

W. Wang
State Key Laboratory of Cotton Biology, Henan Key Laboratory of Plant Stress Biology School of Life Science, Henan University, Kaifeng, Henan, China

D. Yadav
Department of Botany, Guru Ghasidas Vishwavidyalaya, Bilaspur, Chhattisgarh, India

A. Kumar
Department of Entomology, Nematology and Chemistry Units, Agricultural Research Organization (ARO) - Volcani Cente, Rishon LeZion, Israel

A. Kumar
Department of Entomology, Institute of Plant Protection, Agricultural Research Organization (ARO) – The Volcani Center, Rishon LeZion, Israel

P. K. Singh (✉)
Institute of Plant Sciences, Agricultural Research Organization (ARO) – The Volcani Center, Rishon LeZion, Israel

Department of Biotechnology, Pachhunga University College, Mizoram Central University, Aizawl, Mizoram, India
e-mail: singhpk@pucollege.edu.in

R. Roychowdhury et al. (eds.), *Sustainable Agriculture in the Era of Climate Change*, https://doi.org/10.1007/978-3-030-45669-6_24

sequenced genome of various crops in combination with the advancement in genome editing technologies (GETs) has opened the doors to plant biologists to edit almost any desirable trait as well as provided a magic stick for the crop's functional genomics. First-generation GETs such as *z*inc *f*inger *n*ucleases (ZFNs), *s*ite-*s*pecific *n*ucleases (SSNs), meganucleases, and *t*ranscription *a*ctivator-*l*ike *e*ffector *n*ucleases (TALENs) enable plant biologists to target any gene of interest. However, most of these technologies are extravagant, tedious, and burdensome. With the emergence of second-generation GETs, *c*lustered *r*egularly *i*nterspaced *s*hort *p*alindromic *r*epeats/Cas9 (CRISPR/Cas9) offers an efficiently targeted modification of almost all crops and accelerating the crop improvement programs. CRISPR involves designing, cloning, and/or gene-free editing methods. CRISPR/Cas9 system includes a Cas9 protein that makes double-strand cut and a small guide RNA molecule that directs Cas9 to a specific sequence of DNA to cleave. The native DNA repair machinery of the cell generally repairs the nick and facilitates gene editing. CRISPR/Cas9-mediated genome editing (CMGE) has revolutionized agriculture by offering a tool for trait improvement, gene regulation, development of virus resistance, and the generation of mutant libraries. This chapter reviews a brief introduction about CRISPR/Cas9, future perspectives of CRISPR in plant synthetic biology and domestication, advances in CRISPR delivery systems, and specificity of editing. Besides this, there is a short discussion over challenges and opportunities for plant breeding and its fate in agriculture. The chapter also covers an inventory of CRISPR-edited crops until dates for enhancing crop stress tolerance.

Keywords CRISPR · Cas9 · Crop improvement · Biotic stress · Abiotic stress · Gene editing

1 Introduction

Extreme weather conditions, shrinking agriculturally fertile land, and cumulative biotic and abiotic stresses on the one hand while on the other hand drastically growing population urgently call for enhancing global food productivity (FAOSTAT 2016). The development of new technologies for sustainable crop production or elite crop varieties, which can cope with the changing environment, is the only way to improve crop productivity. Moreover, understanding the genetic basis of plant response to environmental stresses is a furthermost requirement of plant biologists, and in the past few decades, various genes and their regulatory network under stress have been identified (Abdelrahman et al. 2018). Plant breeding is relying on introgression of genes associated with unique agronomic traits; however, the extensive application of cultivars generated through these programs reduces the genetic diversity in some useful crops (Flint-Garcia 2018). Hence, the existence of genetic diversity in improvement programs of crops is disapprovingly essential, and a large number of new valuable traits are unceasingly hunted in wild relatives of current crops. Nonetheless, the documentation of the genes governing these characteristics and their introgression for the generation of elite varieties is a tedious and labor-intensive process. Breeders have been using random mutagenesis tool for decades in mutational breeding programs as a substitution to use the natural genetic variation and delivered >3000 commercial varieties of >230 diverse crops and plants (Oladosu et al. 2016). Hence, creating unprecedented changes in the gene pool of crops is urgently needed to develop new cultivars with environmental stress tolerance without compromising with yield. For the upgrading of crop species, gene-editing techniques have been widely used for the past three decades as classical breeding techniques are time-consuming and require several screening steps (Ma et al. 2016). Crop improvement programs begin with the dealing of transgenic approach, which comes with various shortcomings such as nonspecific as well as unstable integration of transgenes into the crop genome, human health concerns, and not environment-friendly (Stephens and Barakate 2017). Again, gene targeting (GT, hereinafter) has been utilized as a tool for plant biotechnology and breeding for genome alterations within a target locus through homologous recombination (HR, hereinafter) (Paszkowski et al. 1988); unfortunately, it possesses low efficiency and does not apply to plants with complex genomes.

Nevertheless, nowadays, with the emergence of new-generation sequencing (NGS, hereinafter) technologies, QTLs identification (Zhu et al. 2017), a reliable intrusion approach, is desirable for crop improvement programs. Gene-editing technologies (GETs, hereinafter) offer a highly specific solution to target the exact loci accurately and contrast to breeding tools, and it enables the integration of a particular gene of interest at a precise location, provides a means to alter a few nucleotides of genes, and can be replaced for modifying existing alleles with another ortholog (Kamburova et al. 2017). GETs rely on specific proteins to recognize a particular DNA sequence to produce single- or double-stranded breaks (Zhang et al. 2017a, b). It is straightforward to gain nuclease activity; however, the recognition of the particular DNA sequence is a very crucial and cumbersome process. Amino acid-based identification of DNA requires an assemblage of heavy protein molecules for a minimum 19–23 bp sequence for selectivity.

GETs have undergone advancements over time to time from 2005 and 2011 and have been broadly classified into first- and the second-generation technologies (also known as new breeding technologies) with emergence of *s*ite-*s*pecific *n*ucleases (SSNs, hereinafter), *z*inc *f*inger *n*ucleases (ZFNs, hereinafter), *t*ranscription *a*ctivator-*l*ike *e*ffector *n*ucleases (TALENs, hereinafter), and meganucleases. Although first- and second-generation editing technologies, such as ZFNs, SSNs, meganucleases, and TALENS, are still in use, these technologies have a limitation and are tedious, labor-intensive, and technically more complicated (Zhang et al. 2017a, b).

*C*lustered *r*egularly *i*nterspersed *s*hort *p*alindromic *r*epeat/Cas9 (CRISPR/Cas9, hereinafter) is the third-generation GET and is viewed as the most potent genome editing tool by the generation of DNA double-stranded breaks (DSBs, hereinafter) at particular loci (Jinek et al. 2012). These DSBs stimulate both nonhomologous end joining (NHEJ, hereinafter) (Salomon and Pucht 1998) and homology-directed recombination (HDR, hereinafter) (Puchta et al. 1996) machinery of the cell to repair the breaks. The targeted region(s) repaired by either HDR or NHEJ possess substitution and insertion/deletion (INDELS) mutations (Jinek et al. 2012). Unlike the new breeding technologies, the third-generation genome editing tool, i.e., CRISPR/Cas9, is quite simple and more comfortable in design and methodologies and is time as well as cost-effective. CRISPR-edited crops carry edited DNA for the desired trait, and such plants can be used for breeding programs to develop varieties, utilized directly without any objections related to GM crops (Jansson 2018; Waltz 2018).

2 CRISPR/Cas9: A Tool for Gene Editing

CRISPR was identified as a family of DNA sequences in prokaryotic genomes as an adaptive immune response against antiviral defense system (Rousseau et al. 2009; Barrangou 2015). The CRISPR acronym (Jansen et al. 2002) denotes repetitive sequences bordered with non-repetitive DNA and was discovered as downstream to *iap* genes of *Escherichia coli* (Ishino et al. 1987). Later it was identified that these non-repetitive sequences are homologous with plasmids- and phages-derived foreign DNA. However, Cas (CRISPR-associated protein) is an enzyme which utilizes the CRISPR sequences as a guide sequence to recognize and cleave specific DNA strands, which are complementary to the CRISPR sequence. Cas protein in combination with CRISPR forms a CRISPR/Cas tool (Zhang et al. 2014a, b). The CRISPR/Cas system has been categorized into three distinct types (types I, II, and III) and with common Cas1 and Cas2 proteins depending upon the presence of a specific signature Cas protein (Makarova et al. 2011). At particular locations within the CRISPR sequence across the genome, CRISPR/Cas can accommodate short sequences of non-native-genetic material (spacers) (Bhaya et al. 2011; Wiedenheft et al. 2012). Various Cas-encoding genes and CRISPR loci with spacers which are located in the vicinity constitute the necessary components of the CRISPR/Cas system (Bhaya et al. 2011). The CRISPR/Cas possesses RNase and/or DNase activity for either of the targets, i.e., RNA or DNA (Wiedenheft et al. 2009). Mojica et al. (2005) explained the mechanism of HDR-based genome editing, and later the technology of CRISPR/Cas9 cleavage has been identified as an auspicious GET (Jinek et al. 2012; Liu et al. 2017a, b).

CRISPR/Cas discovery has transformed the plant and animal biology researches because of its utility in genome editing. Unlike ZFNs and TALENs, CRISPR is more forthright and includes a 20-nucleotide guide RNA (fused crRNA+tracrRNA=gRNA) complementary to the DNA, which has to be edited, and a Cas9 nuclease, which cleaves three to four bases beyond *p*rotospacer *a*djacent *m*otif (PAM) to generate DSBs. These DSBs are repaired through the cell repair machinery and create gene editing (Fig. 24.1). Hence CRISPR relies on cell repair system. As stated before, the DSB repair is brought up in two ways by the enzymes: (1) The most common outcomes of CRISPR by enzyme scan stitch them back together of the dangled ends and result in either addition or deletion of one or more bases in the

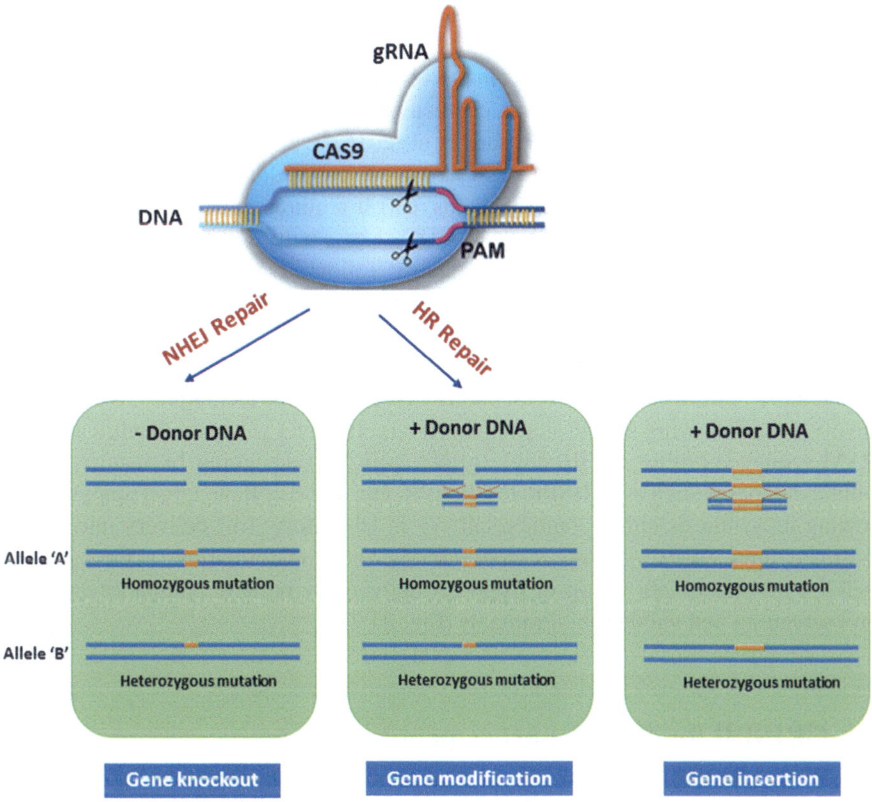

Fig. 24.1 Mechanism of CRISPR/Cas9 in genome editing mechanism. (Modified from: Ding et al. 2016; Abdelrahman et al. 2018)

targeted DNA, leading to gene function disruption known as nonhomologous end joining (NHEJ). (2) Alternatively, there are some other enzymes, which patch the breaks by using a single-stranded DNA upstream and downstream of the cut that matches to the DNA sequence. Hence, a complementary DNA strand known as homology-directed recombination (HDR/HR) is created to achieve the repair. Both the mentioned processes are a bit enigmatic, and no one has reported what happens when CRISPR enters inside the cell (Richardson et al. 2018). In human cells, it has been notified that Cas9-induced repairs occur through single-strand template repair (SSTR, hereinafter) system which calls for the inter-strand cross-link repair system known as Fanconi anemia (FA) pathway, which diverts repair toward single-strand template repair (SSTR) (Richardson et al. 2018). Nevertheless, the technical easiness of the CRISPR/Cas9 prepares to construct extremely fast with routinely achieved cloning in the laboratory in one afternoon. As a result, CRISPR/Cas9-mediated genome editing (CMGE, hereinafter) is routinely used in thousands of laboratories for genome mutagenesis purposes. Israeli scientist Ayal Hendel, Bar-Ilan University, working on nanotechnology-based CRISPR technique stated that

CRISPR allows the cutting of mutated DNA and its replacement with the healthy one. Furthermore, researchers are also hoping that by applying CMGE a reduced browning and shelf life for mushrooms can be enhanced.

In 2015, the Cpf1 nuclease was characterized from the bacterium *Francisella novicida* (Zetsche et al. 2015; Fonfara et al. 2016). Cpf1 have an advantage over Cas9 as the following: (i) It produces "staggered" cut in double-stranded DNA. (ii) It can be applied to "T-rich" PAM regions. (iii) It requires only one CRISPR RNA for successful targeting. (iv) Cpf1 cleaves 18–23 base pairs downstream to PAM site without disrupting the recognition sequence following repair, and hence Cpf1 enables multiple attempts of DNA cleavage, in turn providing an increased opportunity for the desired genomic editing.

CRISPR/Cas9 has been widely applied in genome editing, and in the last few years, more 5000 articles have been published by using this technique. CRISPR/Cas involves the following simple steps, viz., (i) targeting of a gene and identification of PAM sequence within it; (ii) single gRNA (sgRNA) designing by using various public databases such as CRISPR-P, CRISPR, etc. and its offsite target analysis following its synthesis either commercially or in laboratory; (iii) delivery into plant/host/cell lines either by using simple cloning and transformation of the selected sgRNA into a suitable vector or by direct delivery by particle bombardment; and (iv) screening and validation of mutants (Fig. 24.2).

3 CRISPR for Crop Improvement to Biotic and Abiotic Stresses

The CRISPR/Cas9 has been extensively applied for a point mutation, functional gene analysis, gene modification, and integration of foreign genes for gene pyramiding, gene knockouts, protein delivery to genomic loci, gene expression repression/activation, and epigenome editing in various organisms (Zhang et al. 2017a). Unfortunately, the reports on the use of this magic tool in plant genome engineering are still limited, and ~ 20 crop species so far have been edited for multiple traits (Ricroch et al. 2017). Most of the research articles published so far describe only the utility of CRISPR/Cas9 for the generation of reported gene mutants involved in stress tolerance. However, these studies are limited to some model crop plants. CMGE has been employed for the production of biotic as well as abiotic stress tolerance in some crop plants. Table 24.1 shows a snapshot of the CRISPR/Cas9-based genome-edited crops till date in various crops for biotic and abiotic stress tolerance.

Soon after the discovery of CRISPR/cas9 as a GET (Jinek et al. 2012), three different research groups in the same year simultaneously successfully targeted modification of plants genome by applying CRISPR/Cas9 (Nekrasov et al. 2013; Li et al. 2013; Shan et al. 2013). Following this, a large number of scientists have successfully demonstrated the CRISPR/Cas9 as an efficient genome editing tool in model plants as well as in crops (Belhaj et al. 2013; Feng et al. 2013; Jiang et al. 2013; Mao

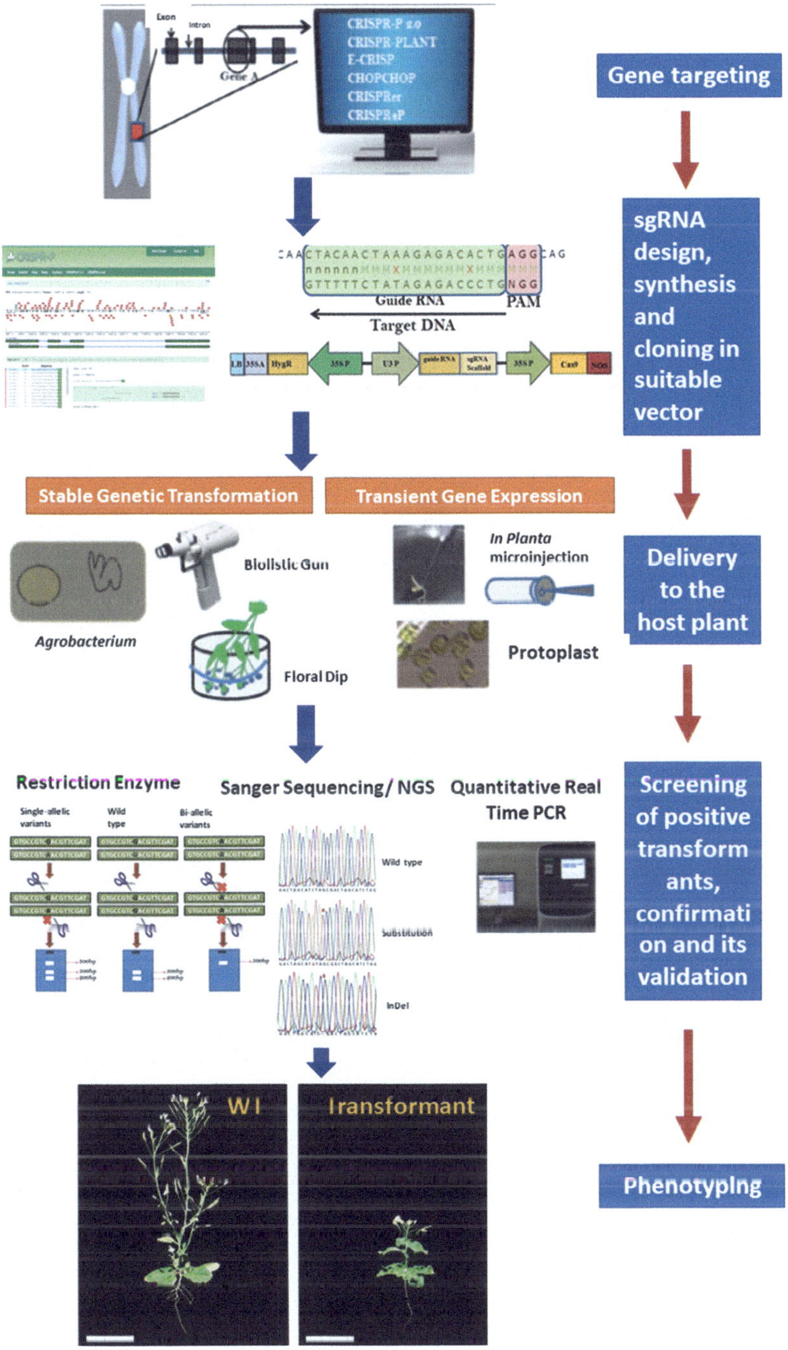

Fig. 24.2 Steps involved in CRISPR/Cas9-mediated genome editing. (Modified from: Jaganathan et al. 2018)

Table 24.1 Inventory of crops edited with CRISPR/Cas9 toward the generation of stress tolerance

Stress	Crop	Gene target	Stress trait	References
Biotic stress	*Arabidopsis thaliana*/ *Nicotiana benthamiana*	Virus dsDNA (A7, B7, and C3 regions)	Resistance to beet severe curly top virus	Ji et al. (2015)
	A. thaliana	eIF(iso)4E	Resistance to Turnip mosaic virus (TuMV)	Pyott et al. (2016)
	N. benthamiana	BeYDV	Resistance to Bean yellow dwarf virus (BeYDV)	Baltes et al. (2015)
	N. benthamiana	ORFs and the IR sequence sDNA of virus	Tomato yellow leaf curl virus (TYLCV) and Merremia mosaic virus (MeMV)	Ali et al. (2015)
	Rice *(Oryza sativa)*	OsERF922 (ethylene-responsive factor)	Blast resistance	Wang et al. (2016a, b)
	Rice *(Oryza sativa)* IR24	OsSWEET13	Bacterial blight disease resistance	Zhou et al. (2015)
	Bread wheat *(Triticum aestivum)*	TaMLO-A1, TaMLO-B1, and TaMLOD1	Powdery mildew resistance	Wang et al. (2014)
	Cucumber (*Cucumis sativus*)	eIF4E (eukaryotic translation initiation factor 4E)	Zucchini yellow mosaic virus (ZYMV), Cucumber vein yellowing virus (CVYV), and Papaya ring spot mosaic virus type-W (PRSV-W)	Chandrasekaran et al. (2016)
	Cassava (*Manihot esculenta*)	eIF4E isoforms nCBP-1 and nCBP-2	Resistance to the Cassava brown streak disease	Gomez et al. (2017, 2019)
	Cocoa beans *(Theobroma cacao)*	TcNPR3, a suppressor of the defense response	Resistance to the *cacao* pathogen *Phytophthora tropicalis*	Fister et al. (2018)
	Cotton (*Gossypium hirsutum*)	CLCuD IR and Rep regions	Resistance to cotton leaf curl disease	Iqbal et al. (2016)

(continued)

Table 24.1 (continued)

Stress	Crop	Gene target	Stress trait	References
	Avocado (*Persea americana*)	PaNPR2 and PaNPR4	*Phytophthora cinnamomi* resistance	Backer et al. (2015)
	Avocado (*Persea americana*)	PAL and LOX	Anthracnose disease resistance	Bill et al. (2017)
	Banana (*Musa acuminata*)	MaATG8s	*Fusarium oxysporum* f. sp. *cubense* (Foc) resistance	Wei et al. (2017a)
	Banana (*Musa acuminata*)	Hrap, Pflp	*Xanthomonas campestris* pv. *musacearum* resistance	Tripathi et al. (2010, 2014), Namukwaya et al. (2012)
	Cassava (*Manihot esculenta*)	RXam1	*Xanthomonas axonopodis* pv. *manihotis* (Xam) strain-specific resistance to XamCIO136	Díaz Tatis et al. (2018)
	Cassava (*Manihot esculenta*)	MeWRKY20-MeATG8a/8f/8h (MeATG8a, MeATG8f, MeATG8h)	Cassava bacterial blight (CBB), caused by *Xanthomonas Axonopodis* pv. *manihotis* (Xam) resistance	Yan et al. (2017)
	Cassava (*Manihot esculenta*)	MeDELLAs	Cassava bacterial blight (CBB) resistance	Li et al. (2018a, b, c)
	Cassava (*Manihot esculenta*)	MebZIPs (MebZIP3 and MebZIP5)	Cassava bacterial blight (CBB) resistance	Li X. et al. (2017b)
	Cassava (*Manihot esculenta*)	MeRAV1 and MeRAV2	Cassava bacterial blight (CBB) resistance	Wei et al. (2017b)
	Coconut (*Cocos nucifera*)	PTI5	Root wilt disease (RWD) resistance	Verma et al. (2017)

(continued)

Table 24.1 (continued)

Stress	Crop	Gene target	Stress trait	References
	Coconut (*Cocos nucifera*)	NBS-LRR-type RGAs	Coconut root (wilt) disease resistance	Rajesh et al. (2015)
	Cotton (*Gossypium hirsutum*)	GhERF-IIb3	Bacterial blight caused by *Xanthomonas citri* pv. *malvacearum* (Xcm) resistance	Cacas et al. (2017)
	Sugarcane (*Saccharum officinarum*)	ScGluA1, ScCAT1, ScChiIV1, and ScChiVI1	Smut (*Sporisorium scitamineum*) resistance	Su et al. (2013, 2014, 2015)
	Sugarcane (*Saccharum officinarum*)	ScGluD2	Smut and abiotic stresses (salt and heavy metal) resistance	Su et al. (2016)
Abiotic stress	Wheat (*Triticum spp.*)	TaDREB2 and TaERF3	Abiotic stress response	Kim et al. (2018)
	Maize (*Zea mays*)	ARGOS8	Increased grain yield under drought stress	Shi et al. (2017)
	Tomato (*Solanum lycopersicum*)	SlMAPK3	Drought tolerance	Wang et al. (2017)
	Arabidopsis	UGT79B2, UGT79B3, MIR169a	Susceptibility to cold, salt, and drought stresses, drought tolerance	Zhao et al. (2016)
	Arabidopsis	OST2 (OPEN STOMATA 2) (AHA1)	Increased stomatal closure in response to abscisic acid (ABA)	Osakabe et al. (2016)
	Rice *(Oryza sativa)*	OsPDS, OsMPK2, OsBADH2	Involved in various abiotic stress tolerance	Shan et al. (2013)
	Rice *(Oryza sativa)*	OsMPK5	Various abiotic stress tolerance and disease resistance	Xie and Yang (2013)

(continued)

Table 24.1 (continued)

Stress	Crop	Gene target	Stress trait	References
	Rice *(Oryza sativa)*	OsMPK2, OsDEP1	Yield under stress	Shan et al. (2014)
	Rice *(Oryza sativa)*	OsPMS3, OsMSH1, OsDERF1, OsMYB5, OsEPSPS	Drought tolerance	Zhang et al. (2014a, b)
	Rice *(Oryza sativa)*	OsHAK-1	Low cesium accumulation	Cordones et al. (2017)
	Rice *(Oryza sativa)*	OsPRX2	Potassium deficiency tolerance	Mao et al. (2017)
	Rice *(Oryza sativa)*	OsAOX1a, OsAOX1b, OsAOX1c, OsBEL	Various abiotic stress tolerance	Xu et al. (2015)
	Banana (*Musa acuminata*)	MaSWEET-4c, MaSWEET-14c, MaSWEET-4d, MaSWEET-1a, MaSWEET-4b, MaSWEET-14b, MaSWEET-14d, and MaSWEET-14h	Foc 4 TR4 and abiotic stresses (cold and salt) resistance	Miao et al. (2017a)
	Banana (*Musa acuminata*)	MaAPS1 and MaAPL3	Abiotic stresses (cold and salt) and *Fusarium oxysporum* resistance	Miao et al. (2017b)
	Cassava (*Manihot esculenta*)	MeKUPs	Abiotic stresses (salt, osmosis, cold, drought) resistance	Ou et al. (2018)
	Cassava (*Manihot esculenta*)	MeMAPKKK	Abiotic stress (drought) resistance	Ye et al. (2017)
	Cotton (*Gossypium hirsutum*)	GhPIN1–3 and GhPIN2	Abiotic stress (drought) resistance	He et al. (2017)
	Cotton (*Gossypium hirsutum*)	GhRDL1	Abiotic stress (drought) resistance	Dass et al. (2017)

(continued)

Table 24.1 (continued)

Stress	Crop	Gene target	Stress trait	References
	Cotton (*Gossypium hirsutum*)	GhPDS, GhCLA1, and GhEF1	Abiotic stress response and chloroplast biogenesis	Gao et al. (2017)
	Cotton (*Gossypium hirsutum*)	Vacuolar H^+-pyrophosphatase (GhVP) and Cloroplastos alterados 1 (GhCLA1)	Abiotic stress response	Chen et al. (2017a, b, c)
	Date palm (*Phoenix dactylifera*)	Pdpcs and Pdmt	Abiotic stress (Cd and Cr) resistance	Chaâbene et al. (2018)
	Date palm (*Phoenix dactylifera*)	Pdpcs and Pdmt	Abiotic stress (metals) resistance	Chaâbene et al. (2017)
	Papaya (*Carica papaya*)	CpDreb2	Abiotic stresses (drought, heat, and cold) resistance	Arroyo-Herrera et al. (2016)
	Papaya (*Carica papaya*)	CpRap2.4a and CpRap2.4b	Abiotic stresses (heat and cold) resistance	Figueroa-Yañez et al. (2016)
	Sugarcane (*Saccharum officinarum*)	ScAPX6	Abscisic acid (ABA), methyl jasmonate (MeJA), and copper (Cu) stress resistance	Liu et al. (2017a, b)
	Sugarcane (*Saccharum officinarum*)	ScNsLTP	Abiotic stresses (drought and chilling) resistance	Chen et al. (2017a, b, c)
	Strawberry (*Fragaria vesca*)	FaTM6 MADS-box gene	Anther and pollen development	Martín-Pizarro et al. (2019)

et al. 2013; Miao et al. 2013; Sugano et al. 2014; Jiang and Wang 2014; Wang et al. 2014; Chen et al. 2017a, b, c; Chaâbene et al. 2018). Diverse variants of Cas9, such as native Cas9, Cas9 nickase, and dCas9 (nuclease-deficient Cas9), have been employed for different purposes. For the first time in 2013, CRISPR/Cas9 mediated targeted genome editing in rice and common wheat genome was done by

(Shan et al. 2013). Another report in the same year demonstrated targeted mutagenesis by the use of CRISPR/Cas9 in model plants such as *Arabidopsis* and *Nicotiana benthamiana* (Li et al. 2013). Furthermore, in the same year, Feng et al. (2013) optimized single sgRNA to target the different target genes individually in *Arabidopsis* and rice. Again, two sites within the same gene by the use of two sgRNAs have been targeted simultaneously in the *Arabidopsis* genome (Mao et al. 2013), which signifies the use of the CRISPR/Cas9 for multiplex GE in the plant. This technology has also been applied for a particular mutation of only one of the three homeo-alleles of mildew resistance locus (MLO-A1) in hexaploid wheat (Wang et al. 2014). Furthermore, the CRISPR has demonstrated that it can offer the fastest production of transgenic crop plants with specific and homozygous targeted gene editing (Zhang et al. 2014a, b).

4 CRISPR and Progress in Genome Editing of Monocot Crops

4.1 *Rice* (Oryza sativa *L.*)

Rice is a major staple crop grown worldwide for food, and CRISPR has been successfully applied for its genome editing to enhance tolerance to stresses (Table 24.1). The first successful application of CRISPR in crop plant (rice) was by using particle bombarded approach of three abiotic stress-responsive genes of rice. viz., *phytoene desaturase* (*OsPDS*), *mitogen-activated protein kinase* (*OsMPK2*), and *betaine aldehyde dehydrogenase* (*OsBADH2*) (Shan et al. 2013). However, by using suitable genome editing vectors for CRISPR/Cas9, Xie and Yang (2013) successfully edited a negative regulator of stress-responsive gene *OsMPK5* gene of rice. Again, Zhang et al. (2014a, b) demonstrated the CMGE-based mutagenesis and heritability of rice genes (*OsPMS3*, *OsEPSPS*, *OsMYB5*, *OsMSH1*, and *OsDERF1*). For the first time, Cas9 fused with cytidine deaminase known as dCas9 programmed for gene editing in rice without induction of DSBs (Shimatani et al. 2017) and in other crops such as maize and wheat (Zong et al. 2017). BE3 (base editor) an advanced version of GETs in combination with ncas9, a D10 cas9 mutation (known as nicked cas9), uracil glycosylase inhibitor (UGI), and cytosine deaminase was used for the first time for editing of rice genes (Li et al. 2017a). Multiplex gene editing in rice as well as model *Arabidopsis* plant by employing one binary vector has also been done by Zhang et al. (2016a, b) and Shen et al. (2017a, b). Furthermore, Biswas et al. (2019) demonstrated an efficient method for the identification of CRISPR/Cas9-generated indels in rice.

Enhanced resistance against the *Magnaporthe oryzae*-induced blast disease was obtained by CRISPR-based targeted mutation of *ethylene-responsive factor22* (*OsERF922*) in rice (Liu et al. 2012). Furthermore, promoter element editing of *OsSWEET13*, which is responsible for *Xanthomonas oryzae* pv. *oryzae*-triggered blight disease in rice, offers resistance to this disease (Zhou et al. 2015). Annexins provide a substantial role in plant development and abiotic stress tolerance in plants and after editing with CRISPR rice *annexin gene* (*OsAnn3*), the edited lines susceptible to cold stress (Shen et al. 2017a, b). It is worthwhile to mention that various agronomically important traits such as yield and stress tolerance are polygenic, and therefore improvement of crops needs a substantial amount of work on the identification of such polygenic genes or quantitative regions (also known as quantitative trait loci – QTL). Moreover, several such identified QTL regions have already been introgressed for the development of elite varieties. However, the generation of introgression lines is a tedious job if these QTLs are closely linked in the elite line because of harmful effects. Shen et al. (2018) have developed a CRISPR-based QTL editing method for the first time that not only functionally dissected the two QTLs, i.e., *grain size3* (*GS3*) and *grain numbers1a* (Gn1a) in rice simultaneously, but also showed that the same QTL may possess high variability and opposing effects of each other.

*4.2 Wheat (*Triticum *spp.)*

Wheat is an important cereal widely cultivated for its seed as a source of staple food. *Triticum* is the most commonly grown common wheat. First successful application of CRISPR/Cas9 for the generation of knock outline of *Triticum aestivum* mildew resistance locus O gene (*TaMLO*), which offers resistance to *Blumeria graminis* f. sp. *tritici* (Btg)-induced powdery mildew disease in wheat, was done independently by Shan et al. (2014) and Wang et al. (2014). Unfortunately, they used T-DNA-based delivery systems and obtained tiny transgenic lines. However, DNA-virus-based amplicons were later identified as an efficient construct delivery method for GET (Gil-Humanes et al. 2017). The application of DNA replicons based on *Geminivirus*, i.e., Wheat dwarf virus (WDV) in wheat, enhanced the expression of CRISPR/Cas9 compared to the ubiquitin reference gene and proposed that it could be a potential tool CMGE in wheat (Gil-Humanes et al. 2017).

Furthermore, Kim et al. (2018) successfully applied CMGE in protoplast wheat for editing of *dehydration-responsive element-binding protein 2* (TaDREB2) and *ethylene-responsive factor 3* (TaERF3). However, off-targeting and transgene integration into genome are two major pitfalls of CRISPR-mediated gene editing in crops. Therefore, an efficient method of editing by using CRISPR/Cas9 ribonucleoproteins (RNPs) reconciled biolistic delivery approach could be applied (Kanchiswamy 2016; Liang et al. 2017). Permanent genome integration of CRISPR/Cas9 leads to stable expression of CRISPR/Cas9. However, the RNP-based biolistic delivery offers a transient expression of CRISPR/cas9, and its rapid degradation following editing reduces off-targeting (Liang et al. 2017). RNP-based gene editing has been successfully applied for the gene editing in bread wheat (Liang et al. 2018). Hence, transgene-free editing methods are not only time-efficient but also eco-friendly to obtain non-GM crops. However, RNP-based delivery method holds certain limitations such as low efficiency and need of a laborious mutant screening process. Zhang et al. (2018a) for the first time performed the editing of upstream open reading frames (ORFs) of abiotic stress-responsive genes, which enables translational control of gene expression in the model plants *Arabidopsis thaliana* and *Lactuca sativa*. This approach could also be applied for genome-free editing in crop plants. Genome multiplexing by CRISPR/Cas has also been demonstrated for model crops for simultaneous editing of various agronomic traits. Recently three genes, viz., *TaGW2* (grain traits negative regulator), *TaMLO* (resistance to powdery mildew), and *TaLpx-1* (lipoxygenase; offers resistance to *Fusarium graminearum*), were targeted by CRISPR multiplexing (Wang et al. 2018a, b, c). It has been reported that this transgenerational gene editing (TGE) not only serves as a source of variation in the CRISPR/Cas9 progenies but is also applicable for polyploid genome multiplexing.

Moreover, CRISPR/Cas9 has been successfully applied for *male sterility1* (*Ms1*) mutation resulting in the generation of complete male sterility in commercial wheat cultivars Fielder and Gladius (Okada et al. 2019). It is worthwhile to mention that the main hindrance in plants gene replacement is the efficient delivery of a donor repair template (DRT, hereinafter) into the nucleus for HDR of DSBs. Hence, the production of RNA templates in vivo for transcript-templated HDR (TT-HDR) may overcome this issue. However, primary transcripts are often processed and transported to the cytosol, rendering them unavailable for HDR in the nucleus. Li et al. (2019) showed that coupling CRISPR-Cpf1 (CRISPR from *Prevotella* and *Francisella 1*) to a CRISPR RNA (crRNA) array flanked with ribozymes, along with a DRT flanked with either ribozymes or crRNA targets, produces primary transcripts, which undergoes self-processing and releases the crRNAs as well as DRT inside the nucleus. Li et al. (2019) first time successfully replaced the rice acetolactate synthase (ALS) gene with a mutated version using a DNA-free ribonucleoprotein complex (having recombinant Cpf1, crRNAs, and DRT transcripts).

4.3 Maize (Zea mays)

Maize, also known as corn, is a chief cereal crop worldwide and domesticated by the indigenous population of Southern Mexico. Phytic acid is a significant constituent of corn seeds, with no nutritional role, and also a source of environmental pollution; three genes, viz., *ZmIPK1A*, *ZmIPK*, and *ZmMRP4*, have been reported to be involved in its biosynthesis (Liang et al. 2014). Editing of these genes in *Z. mays* by using CRISPR reduces the phytic acid content in corn seed (Liang et al. 2014). Likewise, CRISPR-mediated stable editing of *phytoene synthase gene1* (PSY1) involved in carotenoid biosynthesis (Zhu et al. 2016) and *Zmzb7* engaged in the generation of albino phenotype (Feng et al. 2016) was successfully demonstrated in maize plant. CRISPR-mediated simple editing of three transcription factors (*MADS*, *MYBR*, and *AP2*) and multiplexing of *RPL*, *PPR*, and *IncRNA* was successfully demonstrated in maize by Qi et al. (2016). CMGE has also been applied for editing of male sterility-responsive gene *thermosensitive genic male-sterile 5* (*ZmTMS5*) gene maize (Li et al. 2017a). Likewise, the *auxin-regulated gene involved in organ size* (*ARGOS*) genes have been reported as negative regulators of the ethylene-responsive genes. Two allelic variants of *ARGOS8-v1*and *ARGOS8-v2* possessing improved yield under stress than those of wild type have been generated by CRISPR-mediated editing of *ARGOS8* (Shi et al. 2017). Commercially, a researcher from Syngenta, Shujie Dong, has successfully edited the corn genome by using pollen having CRISPR/cas9 (Cohen 2019). Aforesaid studies demonstrated that CRISPR/Cas9 genome editing has revolutionized the genetic editing for the generation of novel varieties with improved tolerance to biotic as well as abiotic stresses. Withstanding above CMGE has also been applied to other monocot crops such as barley (Kapusi et al. 2017), banana (Kaur et al. 2018), date palm (Chaâbene et al. 2017, 2018), coconut (Verma et al. 2017; Rajesh et al. 2015), sugarcane (Su et al. 2013, 2014, 2015, 2016), and others (Table 24.1).

5 CRISPR-Based Genome Editing in Dicot Plants

5.1 Arabidopsis*: As a Model Dicot Plant*

The importance of CRISPR/Cas9 for the efficient gene correction and deletion of large genomic fragments were independently reported at first by two independent researchers in the same year (Feng et al. 2013; Mao et al. 2013). Using CRISPR/Cas9, three genes related to phenology (*gibberellic acid insensitive*, *GAI*; *jasmonate-zim-domain protein 1*, *JAZ1*; and *brassinosteroid insensitive1*, *BRI1*) have been edited successfully in *Arabidopsis* for the first time to obtain a high efficiency of mutation (Feng et al. 2013). However, Mao et al. (2013) successfully edited two genes located at different loci within the genome by CMGE. Again, Feng et al. (2013) for the first time studied the CMGE efficiency, specificity, and heritability by editing of 7 genes localized at 12 different loci across the *Arabidopsis* genome. However, CRISPR/Cas9-based multiplex editing of *transparent testa4* (*TT4*) gene by using a golden gateway and multisite gateway LR recombination in *Arabidopsis thaliana* was done by Lowder et al. (2017). Furthermore, by using multiple sgRNAs, five *Arabidopsis* genes, viz., *AtFLS2* (*flagellin sensitive* 2), *AtPDS3* (*phytoene desaturase*), *AtRACK1* (*receptor for activated c kinase 1; AtRACK1b and 1c*), and *AtCYCD3* (*cyclin D-type* 3), were also successfully edited (Li et al. 2013). Wang et al. (2015) performed tissue-specific editing for the first time by using a tissue-specific promoter and terminator. Fifty different regulatory elements invovled in floral organs development in in Arabidopsis were edited for the first time by Mao et al. (2016). CRISPR/Cas9 was also applied for locus-specific editing of eukaryotic elongation factor in offering resistance against *Turnip mosaic virus* (TuMV) (Pyott et al. 2016), and CRISPRi system has also been established for TuMV resistance in *Arabidopsis* (Zhang et al. 2018b). Similarly, translational editing was also applied in *Arabidopsis* for abiotic stress-responsive genes (Zhang et al. 2018a). Furthermore, a higher frequency of CRISPR-induced mutation was obtained under heat stress in *Arabidopsis* (LeBlanc et al. 2018). Papikian et al. (2019) developed adCas9-SunTag system for the engineering of a gene involved in DNA methylation, located at various loci, and transposable elements.

5.2 *Cotton as a Fiber Crop* (Gossypium *spp.*)

Complex genome makes cotton very difficult for editing by first-generation tools; however, the availability of second-generation editing tool like CRISPR/Cas9 makes possible editing of the complex genome of cotton (Janga et al. 2017). Further, a higher genome editing efficiency of CRISPR/Cas9 in cotton was obtained for the first time by the generation of two gRNAs targeting two genes, viz., *vacuolar HC-pyrophosphatase* (*GhVP*) and *Cloroplastos alterados 1* (*GhCLA1*) genes (Chen et al. 2017a, b, c). Later on, using CRISPR as an efficient editing tool, two homologous genes by one gRNA and multiple genes by single gRNA (hexaploidy nature of genome) were successfully edited in cotton (Gao et al. 2017). By using CRISPR/Cas9 technique, MYB transcription factors (GhMYB25-like A and GhMYB25-like D) of *Gossypium* have been shown to be involved in fiber development (Li et al. 2017c). Successful editing of *Gh14-3-3d* genes in cotton offers resistance to *Verticillium dahliae* (Zhang et al. 2018c). Again Lu et al. (2018) have used an endogenous GhU6 promoter for efficient editing of the genome of the cotton plant and obtained higher mutation efficiency.

5.3 *Legumes: Soybean* (Glycine max)

Soybean was successfully subjected to edit endogenous (*GmFEI2* and *GmSHR*) and exogenous (*bar*) genes for the first time by CRISPR/Cas9 (Cai et al. 2015) using six and one sgRNA, respectively. Likewise, two genomic loci at chromosome 4 (*DD20* and *DD43*) of soybean targeted by using CRISPR/cas9 provided small INDELs (Li et al. 2015). Du et al. (2016) have identified that efficient and simultaneous editing of various homeotic genes can be performed by *Glycine max* ubiquitin promoter (*GmU6-16-1*). Tang et al. (2016) have demonstrated the involvement and specificity of soybean nodulation restriction gene (*Rj4*) in nodulation inhibition by using CRISPR/Cas9. *Phytophthora sojae virulence gene* (*Avr4/6*) mutated by CRISPR leads to the identification of soybean virulence gene locus (Fang and Tyler 2016). Furthermore, a delayed flowering under both short-day and long-day conditions was obtained by CRISPR-based mutagenesis of soybean *flowering time gene 2* (*GmFT2*) (Cai et al. 2018).

5.4 *Vegetable Crops: Tomato (***Solanum lycopersicum** *L.) and Potato (***Solanum tuberosum***)*

The availability of complete genome, efficient transformation methodologies, and functional genomic characterization make tomato (*Solanum lycopersicum* L.) a model plant for research on crops of Solanaceae family. An efficient CRISPR-mediated gene editing of tomato *argonaute* (SlAGO7) was demonstrated by Brooks et al. (2014), and in the same year, *short-root*(*SHR*) gene of tomato has been shown to be involved in regulation of expression of *scarecrow* (*SCR*) transcription factor and root length (Ron et al. 2014). A MADS-Box transcription factor gene, viz., *ripening inhibitor* (RIN), has been demonstrated to be involved in the regulation of tomato fruit ripening by the generation of RIN mutants by CMGE (Ito et al. 2015). Similarly, CMGE of tomato lncRNA1459 and *SlORRM4* (*RNA recognition motif*) genes delays the onset of fruit ripening (Li et al. 2018a, b, c; Yang et al. 2017). Yu et al. (2017) enhanced the shelf life of tomato by CMGE of dominant *ALC* (*ALCOBACA*) gene generated recessive *alcobaca* (*alc*). Furthermore, the CRISPR technique has also been applied for generation of parthenocarpy in tomato fruits by a mutation in the parthenocarpy-related gene, SlIAA9 (Ueta et al. 2017). CMGE of a tomato flowering repressor gene *SP5G* enhances the inflorescence and fruit yield (Soyk et al. 2017). Mitogen-activated protein kinases (MAPKs) have been reported to be involved in the regulation of drought stress-responsive genes and protection of cell membrane under drought stress. However, CMGE-generated *Slmapk3* mutants identified as sensitive to drought stress (Wang et al. 2017).

Impairment of the SlPMR4 encoding a callose synthase protein, an ortholog of *powdery mildew resistant4* (PMR4) in tomato, offers protection against *Oidium neolycopersici* (Koseoglou 2017). Multiplex CRISPR was applied for targeting of five tomato essential genes involved in g-aminobutyric acid (GABA) pathway to enhance the GABA content in leaves and fruits (Li et al. 2018b). Similarly, the CMGE was also applied to alter the *phytoene desaturase* (*SlPDS*) and *phytochrome-interacting factor4* (*SlPIF4*) genes to enhance the tomato nutritional value (Pan et al. 2016). Again, CMGE was also applied for precise segmental reshuffling between the homologous chromosomes by using a visual gene marker coupled with CRISPR in somatic cells of tomato (Hayut et al. 2017). Hence, somatic homologous recombination could be used for shortening of crop generation and allelic replacement. Tomato fruit ripening process is primarily reported to be controlled by three players, viz., ethylene, ripening transcription factors (SBP-CNR and NAC-NOR), and DNA methylation. Gao et al. (2019) employed CMGE to mutate the ripening transcription factors (SBP-CNR and NAC-NOR) and enhance the shelf life of a tomato. Furthermore, the potato is a principal carbohydrate food crop across the globe, and its starch quality is very crucial for its application as a food. CMGE was successfully applied for the development of starchy hexaploid genotype of potato by mutating *StGBSS* (*granule-bound starch synthase*) (Andersson et al. 2017) and *StALS1* (*acetolactate synthase1*) genes (Butler et al. 2016). Kusano et al. (2019) by using the translational editing approach successfully enhanced the expression of *granule-bound starch synthase I* (*GBSSI*) in potato. Likewise, by applying cytidine base editors (CBEs) along with CRISPR, *acetolactate synthase* (*ALS*) gene was edited in potato (Veillet et al. 2019). Moreover, by applying the RNP complex, Andersson et al. (2018) successfully edited the *GBSSI* gene in potato.

5.5 Citrus Fruit Crops

CRISPR/Cas9 has been extensively applied in citrus species to make them resistant to the biotic and abiotic stresses (Jia et al. 2019). Phytoene desaturase genes (*CsPDS* and *CpPDS*) of two different citrus cultivars, i.e., sweet orange (*Citrus sinensis*) and grapefruit (*Citrus paradisi*), have been successfully edited by using CMGE (Jia and Wang 2014; Jia et al. 2017). CRISPR-mediated targeted modification of *lateral organ boundaries1* (*Cslob1*) gene regulatory elements offers improved resistance to common citrus canker disease. *Cslob1* in citrus has been reported as citrus canker-susceptible gene and is known to play a vital role in pathogen growth, and its various alleles possess an *effector-binding element4* (*EBEPthA4*). CRISPR-mediated editing of the *EBEPthA4* and *Cslob1* promoter offers a high degree of resistance to orange citrus canker (Peng et al. 2017). Furthermore, editing of the *Cslob1* alleles has been demonstrated to offer protection against grapefruit canker (Jia et al. 2017). Zhang et al. (2017a, b) showed that the early-stage shoot regeneration in citrus is the only active stage of CMGE and hence in citrus time-specific genome editing was performed. *L-idonate dehydrogenase* (*IdnDH*) gene in grape was first time successfully targeted by application of CRISPR/Cas9 by Ren et al. (2016), and Wang et al. (2016a, b) have successfully advocated the optimal CMGE procedure in grapes. Furthermore, Nakajima et al. (2017) have successfully edited the grape *phytoene desaturase* (*VvPDS*) gene. CRISPR/Cas9 as ribonucleoprotein (RNP) was first time successfully delivered in grape protoplasts to offer protection against the susceptibility gene for powdery mildew *MLO-7* (Malnoy et al. 2016). Furthermore, *WRKY52* editing in *Vitis* has been demonstrated to offer protection against biotic stress responses (Wang et al. 2018a, b, c).

5.6 Other Genome-Edited Dicots

CRISPR/Cas9 has revolutionized the plant biotechnology field by providing plant biologist a magic tool for targeted genetic alterations, and technology has been successfully also applied in other dicot crops. Protein involved in the first step of abscisic acid biosynthesis *Lactuca sativa* is encoded by *9- cis-epoxy-carotenoid dioxygenase4* (*NCED4*), and its editing enhanced the high-temperature seed germination in two different cultivars of lettuce (Bertier et al. 2018). Likewise, *Theobroma cocoa non-expressor of pathogenesis-related* (*TcNPR3*) gene on edition confers an enhanced tolerance against *Phytophthora tropicalis* (Fister et al. 2018). Furthermore, successful CMGE of *flavanone-3-hydroxylase* (*F3H*), has been done, which is an anthocyanin biosynthetic pathway gene used as a visual marker in the carrot to identify edited transformants (Klimek-Chodacka et al. 2018). Droplet digital PCR (ddPCR) have been successfully applied for identification targeted mutations in *Medicago sativa* (Gao et al. 2019). Furthermore, *phytoene desaturase* (*MtPDS*) gene of another legume crop (*Medicago truncatula*) has been successfully edited by CMGE (Meng et al. 2017), and in the same year, Tian et al. (2017) successfully applied CRISPR approach for the generation of *phytoene desaturase* (*ClPDS*)-edited watermelons. Chandrasekaran et al. (2016) developed edited cucumber (*Cucumis sativus*) cultivar resistant to viral infection. Furthermore, CMGE was also applied for enhancing yield, shelf life, and disease resistance in various horticulturally important crops because of the availability of genomics data. CMGE of ornamental plant gene *Ipomoea carotenoid cleavage dioxygenase4* (*Inccd4*) produces a pale yellow flower (Watanabe et al. 2018). Moreover, *symbiosis receptor-like kinase* of *Lotus japonicus* responsible for symbiotic nitrogen fixation has been demonstrated by using CMGE (Wang et al. 2016a, b). CMGE of *rosmarinic acid synthase* (*SmRAS*) gene of *Salvia miltiorrhiza* enhances the accumulation of rosmarinic acid (RA) and lithospermic acid (Zhou et al. 2018). Likewise, Wang et al. (2018c) edited *phytoene desaturase* gene and generated albino kiwifruit, and recently, multiplex editing has been applied for editing of five rapeseed SPL3 homologs in allotetraploid oilseed rape (Li et al. 2018a, b, c). Although CMGE has revolutionized the agriculture, a large number of most important cultivated crops possess polyploid genome. Martín-Pizarro et al. (2019) recently applied CMGE in octoploid and heterozygotic cultivated strawberry (*Fragaria ananassa* cv. Camarosa). Hence, many more results are yet to come about this enigmatic approach, and more complex genomes are yet to be edited in the near future.

6 Practical Problems with CRISPR/Cas9

Shortly after the first reports of CMGE in animal, the plant community enthusiastically started this technique in crops as well as model plants species (Nekrasov et al. 2013; Li et al. 2013; Shan et al. 2013; Belhaj et al. 2013; Feng et al. 2013; Jiang et al. 2013; Mao et al. 2013; Miao et al. 2013; Sugano et al. 2014; Jiang and Wang 2014; Wang et al. 2014; Chen et al. 2017a, b, c; Chaâbene et al. 2018). CMGE technology thus not only has an enormous potential of unlocking the plenty of information for biosynthetic pathways as well as the generation of future crops, but also transgene from CMGE crops is easily removed by backcrossing (Mao et al. 2017). Although CMGE has been widely applied successfully in various crops, its application is not universal for many recalcitrant crops such as *Gossypium*, *Fragaria*, *Sorghum*, *Mangifera*, etc. Similarly, transgene removal from heterozygous and self-incompatible fruit crops such as pineapple is challenging. Therefore, the development of non-transgenic approaches by using ribonucleoprotein complexes or a transient nucleic acids expression system and a robust efficiency of CMGE could be a solution (Zhang et al. 2016a, b; Liang et al. 2017).

7 Concluding Remarks and Future Perspectives

CRISPR/Cas is serving as a magic tool for plant biologists as it combines with several desirable features such as simple, efficient, highly specific, reduced off-target effects, and easy to multiplex; these features make this technique a most powerful tool for plant genome editing. The targeted mutagenesis of various genes offers essential insights into their function as well as their regulatory role in response to multiple stimuli in crops. Furthermore, genes multiplexing by CRISPR provides a clue of metabolic and gene regulatory pathways as well as helpful in the engineering of complex multigenic agronomic traits in plants. Again, the sequence-specific integration of foreign genes by CRISPR makes it a robust tool for gene pyramiding in elite cultivars in a shorter time. CRISPR/Cas could also be utilized for the generation of novel allelic variants for crops breeding programs and production of haploids. Furthermore, the introduction of desired variations (genetic/epigenetic) via the CRISPR/Cas may offer a promising platform to edit useful agronomic traits, such as enhanced stress tolerance and improved yield in crops. Altogether, the CRISPR/Cas opens an exciting opportunity for plant biologists for implementation of novel approaches for crop improvement to feed the growing population. Starting from the model plant, i.e., *Arabidopsis thaliana*, to the hard ones with most of the "impossible" genomes are now edited with the use of CRISPR/Cas9. Hence, CRISPR/Cas9 is a ubiquitous technology for crop improvement programs, and it has already been employed in complicated crops such as hexaploid bread wheat (Zhang et al. 2016a, b; Liang et al. 2017) and one step beyond on the most complicated one strawberry (Martín-Pizarro et al. 2019). From this point of view, it seems that the CMGE can offer to edit the complex genomes with high efficiency. The safe side of CRISPR technology is that it provides non-transgenic plants, which can address ethical concerns. A declaration by the United States Department of Agriculture (USDA) allows the cultivation of CMGE crops for the ever-increasing population. According to an estimate, global CRISPR genome editing market is increasing at an annual growth rate of 10.2% and will reach $5.7 billion by 2023. However, CMGE is still in infancy, and its safety and efficacy in various applications require further studies.

Although CRISPR/Cas9 has established a milestone in plant genome editing, still some problems associated with this technique such as the specific molecular mechanism, influence on chromatin context, the sgRNA length for high efficiency, off-target probability, and efficient delivery methods for plants remain elusive. Furthermore, researches are also required for the evaluation of germline transmission and the mutations heritability mediated through CRISPR/Cas9. In the nutshell, CRISPR is serving as a magic tool for genome editing in plants, in a controlled manner by the scientists. All the enduring efforts and future advancement in CRISPR technology will accelerate the basic as well as applied research and will open the door for plant biologists for the improvement of crops toward the abiotic and biotic stress tolerance. We anticipate the application of CRISPR/Cas in various crops reform agriculture to ensure food security of the ever-increasing population of the world.

Acknowledgments Authors are thankful to the Agriculture Research Organisation (ARO), the Volcani Center, Israel, for Visiting Scientist Fellowship through the Israel Ministry of Agriculture and rural development.

References

Abdelrahman M, Al-Sadi AM, Pour-Aboughadareh A, Burritt DJ, Tran LP (2018) Genome editing using CRISPR/Cas9-targeted mutagenesis: an opportunity for yield improvements of crop plants grown under environmental stresses. Plant Physiol Biochem 131:31–36. https://doi.org/10.1016/j.plaphy.2018.03.012

Ali Z, Abulfaraj A, Idris A, Ali S, Tashkandi M, Mahfouz MM (2015) CRISPR/Cas9-mediated viral interference in plants. Genome Biol 16:238. https://doi.org/10.1186/s13059-015-0799-6

Andersson M, Turesson H, Nicolia A, Falt AS, Samuelsson M, Hofvander P (2017) Efficient targeted multiallelic mutagenesis intetraploid potato (*Solanum tuberosum*) by transient CRISPR-Cas9 expression in protoplasts. Plant Cell Rep 36:117–128. https://doi.org/10.1007/s00299-016-2062-301045

Andersson M, Turessona M, Olssona N, Fälta A, Ohlssona P, Gonzalezb MN, Samuelssond M, Hofvandera P (2018) Genome editing in potato via CRISPR-Cas9 ribonucleoprotein delivery. Physiol Plant 164:378–384

Arroyo-Herrera A, Figueroa-Yanez L, Castano E, Santamari'a J, Pereira-Santana A, Espadas-Alcocer J, Sánchez-Teyer F, Espadas-Gil F, Alcaraz LD, López-Gómez R, Sánchez-Calderón L, Rodríguez-Zapata LC (2016) A novel Dreb2-type gene from *Carica papaya* confers tolerance under abiotic stress. Plant Cell Tissue Org Cult 125:119–133. https://doi.org/10.1007/s11240-015-0934-9

Backer R, Mahomed W, Reeksting BJ, Engelbrecht J, Ibarra-Laclette E, van den Berg N (2015) Phylogenetic and expression analysis of the NPR1-like gene family from *Persea americana* (Mill). Front Plant Sci 6:300. https://doi.org/10.3389/fpls.2015.00300

Baltes NJ, Hummel AW, Konecna E, Cegan R, Bruns AN, Bisaro DM, Vyotas DF (2015) Conferring resistance to geminiviruses with the CRISPR-Cas prokaryotic immune system. Nat Plants 1:15145. https://doi.org/10.1038/nplants.2015.145

Barrangou R (2015) The roles of CRISPR-Cas systems in adaptive immunity and beyond. Curr Opin Immunol 32:36–41. https://doi.org/10.1016/j.coi.2014.12.008

Belhaj K, Chaparro-Garcia A, Kamoun S, Nekrasov V (2013) Plant genome editing made easy: targeted mutagenesis in model and crop plants using the CRISPR/Cas system. Plant Methods 9:39

Bertier LD, Ron M, Huo H, Bradford KJ, Britt AB, Michelmore RW (2018) High-resolution analysis of the efficiency, heritability, and editing outcomes of CRISPR-Cas9 -induced modifications of NCED4 in lettuce (*Lactuca sativa*). G3 8:1513–1521. https://doi.org/10.1534/g3.117.300396

Bhaya D, Davison M, Rodolphe B (2011) CRISPR-Cas systems in bacteria and archaea: versatile small RNAs for adaptive defense and regulation. Annu Rev Genet 45:273–297

Bill M, Korsten L, Remize F, Glowacz M, Sivakumar D (2017) Effect of thyme oil vapours on phenylalanine ammonia-lyase (PAL) and lipoxygenase (LOX) genes expression, and control of anthracnose in "Hass" and "Ryan" avocado fruit. Sci Hortic 224:32–237. https://doi.org/10.1016/j.scienta.2017.06.026

Biswas S, Li R, Yuan Z, Zhang D, Zhao X, Shi J (2019) Development of methods for effective identification of CRISPR/Cas9-induced indels in rice. Plant Cell Rep. https://doi.org/10.1007/s00299-019-02392-3

Brooks C, Nekrasov V, Lippman ZB, Van Eck J (2014) Efficient gene editing in tomato in the first generation using the clustered regularly interspaced short palindromic repeats/CRISPR-associated9 system. Plant Physiol 166:1292–1297. https://doi.org/10.1104/pp.114.247577

Butler NM, Baltes NJ, Voytas DF, Douches DS (2016) Geminivirus-mediated genome editing in potato (*Solanum tuberosum* L.) using sequence-specific nucleases. Front Plant Sci 7:1045. https://doi.org/10.3389/fpls.2016

Cacas JL, Pré M, Pizot M, Cissoko M, Diedhiou I, Jalloul A, Doumas P, Nicole M, Champion A (2017) GhERF-IIb3 regulates the accumulation of jasmonate and leads to enhanced cotton resistance to blight disease. Mol Plant Pathol 18:825–836. https://doi.org/10.1111/mpp.12445

Cai Y, Chen L, Liu X, Sun S, Wu C, Jiang B, Han T, Hou W (2015) CRISPR/Cas9-mediated genome editing in soybean hairy roots. PLoS One 10:e0136064. https://doi.org/10.1371/journal.pone.0136064

Cai Y, Chen L, Liu X, Guo C, Sun S, Wu C, Jiang B, Han T, Hou W (2018) CRISPR/Cas9-mediated targeted mutagenesis of GmFT2a delays flowering time in soybean. Plant Biotechnol J 16:176–185. https://doi.org/10.1111/pbi.12758

Chaâbene Z, Hakim IR, Rorat A, Elleuch A, Mejdoub H, Vandenbulcke F (2017) Copper toxicity and date palm (*Phoenix dactylifera*) seedling tolerance: monitoring of related biomarkers. Environ Toxicol Chem 37:797–806. https://doi.org/10.1002/etc.4007

Chaâbene Z, Rorat A, Hakim I, Bernard F, Douglas GC, Elleuch A, Vandenbulcke F, Mejdoub H (2018) Insight into the expression variation of metal-responsive genes in the seedling of date palm (*Phoenix dactylifera*). Chemosphere 197:123–134. https://doi.org/10.1016/j.chemosphere.2017.12.146

Chandrasekaran J, Brumin M, Wolf D, Leibman D, Klap C, Pearlsman M, Sherman A, Arazi T, Gal-On A (2016) Development of broad virus resistance in non-transgenic cucumber using CRISPR/Cas9 technology. Mol Plant Pathol 17:1140–1153. https://doi.org/10.1111/mpp.12375

Chen X, Lu X, Shu N, Wang S, Wang J, Wang D, Guo L, Ye W (2017a) Targeted mutagenesis in cotton (*Gossypium hirsutum* L.) using the CRISPR/Cas9 system. Sci Rep 7:44304. https://doi.org/10.1038/srep44304

Chen X, Lu X, Shu N, Wang S, Wang J, Wang D, Guo L, Yea W (2017b) Targeted mutagenesis in cotton (*Gossypium hirsutum* L.) using the CISPR/Cas9 system. Sci Rep 7:44304. https://doi.org/10.1038/srep44304

Chen Y, Ma J, Zhang X, Yang Y, Zhou D, Yu Q, Que Y, Xu L, Guo J (2017c) A novel non-specific lipid transfer protein gene from sugarcane (NsLTPs), obviously responded to abiotic stresses and signaling molecules of SA and MeJA. Sugar Technol 19:17–25. https://doi.org/10.1007/s12355-016-0431-4

Cohen J (2019) Corn and other important crops can now be gene edited by pollen carrying CRISPR. Science News. Date 04 March, 2019. www.sciencemag.org/news/2019/03

Cordones MN, Mohamed S, Tanoi K, Natsuko Kobayashi NI, Takagi K, Vernet A, Guiderdoni E, Périn C, Sentenac H, Véry AA (2017) Production of low-Cs^+ rice plants by inactivation of the K^+ transporter OsHAK1 with the CRISPR-Cas system. Plant J 92:43–56. https://doi.org/10.1111/tpj.13632

Dass A, Abdin MZ, Reddy VS, Leelavathi S (2017) Isolation and characterization of the dehydra tion stress-inducible GhRDL1 promoter from the cultivated upland cotton (*Gossypium hirsutum*). J Plant Biochem Biotechnol 26:113–119. https://doi.org/10.1007/s13562-016-0369-3

Díaz Tatis PA, Herrera Corzo M, Ochoa Cabezas JC, Medina Cipagauta A, Prías MA, Verdier V, Chavarriaga Aguirre P, López Carrascal CE (2018) The overexpression of RXam1, a cassava gene coding for an RLK, confers disease resistance to *Xanthomonas axonopodis* pv. *Manihotis*. Planta 247:1031–1042. https://doi.org/10.1007/s00425-018-2863-4

Ding Y, Li H, Chen LL, Xie K (2016) Recent advances in genome editing using CRISPR/Cas9. Front Plant Sci 7:703. PMC4877526

Du H, Zeng X, Zhao M, Cui X, Wang Q, Yang H, Cheng H, Yu D (2016) Efficient targeted mutagenesis in soybean by TALENs and CRISPR/Cas9. J Biotechnol 217:90–97. https://doi.org/10.1016/j.jbiotec.2015.11.005

Fang Y, Tyler BM (2016) Efficient disruption and replacement of an effector gene in the Oomycete *Phytophthora sojae* using CRISPR/Cas9. Mol Plant Pathol 17:127–139. https://doi.org/10.1111/mpp.12318

FAOSTAT (2016) FAOSTAT database. http://faostat3.fao.org/faostat gateway/go/to/download/Q/QC/E. Accessed 8 Feb 2018

Feng Z, Zhang B, Ding W, Liu X, Yang DL, Wei PL, Cao F, Zhu S, Zhang F, Mao Y, Zhu JK (2013) Efficient genome editing in plants using a CRISPR/Cas system. Cell Res 23:1229–1232

Feng C, Yuan J, Wang R, Liu Y, Birchler JA, Han F (2016) Efficient targeted genome modification in maize using CRISPR/Cas9 system. J Genet Genomics 43:37–43. https://doi.org/10.1016/j.jgg.2015.10.002

Figueroa-Yañez L, Pereira-Santana A, Arroyo-Herrera A, Rodriguez-Corona U, Sanchez-Teyer F, Espadas-Alcocer J, Espadas-Gil F, Barredo-Pool F, Castaño E, Rodriguez-Zapata LC (2016) RAP2.4a is transported through the phloem to regulate cold and heat tolerance in papaya tree (*Carica papaya* cv. Maradol): implications for protection against abiotic stress. PLoS ONE 11:0165030. https://doi.org/10.1371/journal.pone.0165030

Fister AS, Landherr L, Maximova SN, Guiltinan MJ (2018) Transient expression of CRISPR/Cas9 machinery targeting TcNPR3 enhances defense response in *Theobroma cacao*. Front Plant Sci 9:268. https://doi.org/10.3389/fpls.2018.00268

Flint-Garcia SA (2018) Genetics and consequences of crop domestication. J Agric Food Chem 61:8267–8276. https://doi.org/10.1021/jf305511d

Fonfara I, Richter H, Bratovič M, Le Rhun A, Charpentier E (2016) The CRISPR-associated DNA-cleaving enzyme Cpf1 also processes precursor CRISPR RNA. Nature 532:517–521. https://doi.org/10.1038/nature17945

Gao W, Long L, Tian X, Xu F, Liu J, Singh PK, Botella JR, Song CP (2017) Genome editing in cotton with the CRISPR/Cas9 system. Front Plant Sci 8:1364. https://doi.org/10.3389/fpls.2017.01364

Gao Y, Zhu N, Zhu X, Wu M, Jiang CZ, Grierson D, Luo Y, Shen W, Zhong S, Fu DQ, Qu G (2019) Diversity and redundancy of the ripening regulatory networks revealed by the fruit-ENCODE and the new CRISPR/Cas9CNR and NOR mutants. Hortic Res 6:39. https://doi.org/10.1038/s41438-019-0122-x

Gil-Humanes J, Wang Y, Liang Z, Shan Q, Ozuna CV, Sanchez-Leon S, Baltes NJ, Starker C, Barro F, Gao C, Voytas DF (2017) High-efficiency gene targeting in hexaploid wheat using DNA replicons and CRISPR/Cas9. Plant J 89:1251–1262. https://doi.org/10.1111/tpj.13446

Gomez MA, Lin ZD, Moll T, Luebbert C, Chauhan RD, Vijayaraghavan A, Renninger K, Beyene G, Taylor NJ, Carrington JC, Staskawicz BJ, Bart R (2017) Simultaneous CRISPR/Cas9-mediated editing of cassava elF4E isoforms nCBP-1 and nCBP-2 confers elevated resistance to cassava brown streak disease. bioRxiv:209874. https://doi.org/10.1101/209874

Gomez MA, Lin ZD, Moll T, Chauhan RD, Hayden L, Renninger K, Beyene G, Taylor NJ, Carrington JC, Staskawicz BJ, Bart RS (2019) Simultaneous CRISPR/Cas9-mediated editing of cassava eIF4E isoforms nCBP-1 and nCBP-2 reduces cassava brown streak disease symptom severity and incidence. Plant Biotechnol J 17:421–434. https://doi.org/10.1111/pbi.12987

Hayut SF, Melamed Bessudo C, Levy AA (2017) Targeted recombination between homologous chromosomes for precise breeding in tomato. Nat Commun 8:15605. https://doi.org/10.1038/ncomms15605

He P, Zhao P, Wang L, Zhang Y, Wang X, Xiao H, Jianing Yu, Xiao G (2017) The PIN gene family in cotton (*Gossypium hirsutum*): genome-wide identification and gene expression analyses during root development and abiotic stress responses. BMC Genomics 18:507. https://doi.org/10.1186/s12864-017-3901-5

Iqbal Z, Sattar MN, Shafiq M (2016) CRISPR/Cas9: a tool to circumscribe cotton leaf curl disease. Front Plant Sci 7:475. https://doi.org/10.3389/fpls.2016.00475

Ishino Y, Shinagawa H, Makino K, Amemura M, Nakata A (1987) Nucleotide sequence of the IAP gene, responsible for alkaline phosphatase isozyme conversion in *Escherichia coli*, and identification of the gene product. J Bacteriol 169:5429–5433. https://doi.org/10.1128/jb.169.12.5429-5433.1987

Ito Y, Nishizawa-Yokoi A, Endo M, Mikami M, Toki S (2015) CRISPR/Cas9-mediated mutagenesis of the RIN locus that regulates tomato fruit ripening. Biochem Biophys Res Commun 467:76–82. https://doi.org/10.1016/j.bbrc.2015.09.117

Jaganathan D, Ramasamy K, Sellamuthu G, Jayabalan S, Venkataraman G (2018) CRISPR for Crop improvement: an updated review. Front Plant Sci 9:985. https://doi.org/10.3389/fpls.2018.00985

Janga MR, Campbell LM, Rathore KS (2017) CRISPR/Cas9-mediated targeted mutagenesis in upland cotton (*Gossypium hirsutum* L.). Plant Mol Biol 94:349–360. https://doi.org/10.1007/s11103-017-0599-3

Jansen R, Embden JDV, Gaastra W, Schouls LM (2002) Identification of genes that are associated with DNA repeats in prokaryotes. Mol Microbiol 43:1565–1575. https://doi.org/10.1046/j.1365-2958.2002.02839.x

Jansson S (2018) Gene-edited plants on the plate: the 'CRISPR cabbage story. Physiol Plant 164:396–405

Ji X, Zhang H, Zhang Y, Wang Y, Gao C (2015) Establishing a CRISPRCas-like immune system conferring DNA virus resistance in plants. Nat Plants 1:15144. https://doi.org/10.1038/nplants.2015.144

Jia H, Wang N (2014) Targeted genome editing of sweet orange usingCas9/sgRNA. PLoS One 9:e93806. https://doi.org/10.1371/journal.pone.0093806

Jia H, Xu J, Orbovic V, Zhang Y, Wang N (2017) Editing citrus genome via SaCas9/sgRNA system. Front Plant Sci 8:2135. https://doi.org/10.3389/fpls.2017.02135

Jia H, Zou X, Orbovic V, Wan N (2019) Genome editing in citrus tree with CRISPR/Cas9. In: Qi Y (ed) Plant genome editing with CRISPR systems, Humana Press, Methods in molecular biology, vol 1917, New York

Jiang W, Zhou H, Bi H, Fromm M, Yang B, Weeks DP (2013) Demonstration of CRISPR/Cas9/sgRNA mediated targeted gene modification in *Arabidopsis*, tobacco, sorghum and rice. Nucleic Acids Res 41:e188. https://doi.org/10.1093/nar/gkt780

Jinek M, Chylinski K, Fonfara I, Hauer M, Doudna JA, Charpentier E (2012) A programmable dual-RNA-guided DNA endonuclease in adaptive bacterial immunity. Science 337:816–821. https://doi.org/10.1126/science.1225829

Kamburova VS, Nikitina EV, Shermatov SE, Buriev ZT, Kumpatla SP, Emani C, Abdurakhmonov EY (2017) Genome editing in plants: an overview of tools and applications. Int J Agron 2017:7315351. https://doi.org/10.1155/2017/7315351

Kapusi E, Corcuera-Gómez M, Melnik S, Stoger E (2017) Heritable genomic fragment deletions and small indels in the putative ENGase gene induced CRISPR/Cas9 in barley. Front Plant Sci 8:540. https://doi.org/10.3389/fpls.2017.00540

Kanchiswamy CN (2016) DNA-free genome editing methods for targeted crop improvement. Plant Cell Rep 35:1469. https://doi.org/10.1007/s00299-016-1982-2

Kaur N, Alok A, Shivani, Kaur N, Pandey P, Awasthi P, Tiwari S (2018) CRISPR/Cas9-mediated efficient editing in phytoene desaturase (PDS) demonstrates precise manipulation in banana cv. Rasthali genome. Funct Integr Genomics 18:89–99. https://doi.org/10.1007/s10142-017-0577-5

Kim D, Alptekin B, Budak H (2018) CRISPR/Cas9 genome editing in wheat. Funct Integr Genomics 18:31–41. https://doi.org/10.1007/s10142-017-0572-x

Klimek-Chodacka M, Oleszkiewicz T, Lowder LG, Qi Y, Baranski R (2018) Efficient CRISPR/Cas9-based genome editing in carrot cells. Plant Cell Rep 37:575–586. https://doi.org/10.1007/s00299-018-2252-2

Koseoglou E (2017) The study of SlPMR4 CRISPR/Cas9- mediated tomato allelic series for resistance against powdery mildew. Master thesis, Wageningen University and Research, Wageningen

Kusano H, Ohnuma M, Mutsuro-Aoki H, Asahi T, Ichinosawa D, Onodera H, Asano K, Noda T, Horie T, Fukumoto K, Kihira M, Teramura H, Yazaki K, Umemoto N, Muranaka T, Shimada H (2019) Establishment of a modified CRISPR/Cas9 system with increased mutagenesis frequency using the translational enhancer dMac3 and multiple guide RNAs in potato. Sci Rep 8.13753. https://doi.org/10.1038/s41598-018-32049-2

LeBlanc C, Zhang F, Mendez J, Lozano Y, Chatpar K, Irish VF, Jacob Y (2018) Increased efficiency of targeted mutagenesis by CRISPR/Cas9 in plants using heat stress. Plant J 93:377–386. https://doi.org/10.1111/tpj.13782

Li J, Norville JE, Aach J, McCormack M, Zhang D, Bush J, Church GM, Sheen J (2013) Multiplex and homologous recombination-mediated genome editing in Arabidopsis and *Nicotiana benthamiana* using guide RNA and Cas9. Nat Biotechnol 31:688–691. https://doi.org/10.1038/nbt.2654

Li Z, Liu ZB, Xing A, Moon BP, Koellhoffer JP, Huang L, Ward RT, Clifton E, Falco SC, Cigan AM (2015) Cas9-Guide RNA directed genome editing in soybean. Plant Physiol 169:960–970. https://doi.org/10.1104/pp.15.00783

Li J, Sun Y, Du J, Zhao Y, Xia L (2017a) Generation of targeted point mutations in rice by a modified CRISPR/Cas9 system. Mol Plant 10:526–529. https://doi.org/10.1016/j.molp.2016.12.001

Li X, Fan S, Hu W, Liu W, Wei Y, He C, Shi H (2017b) Two cassava Basic Leucine Zipper (bZIP) transcription factors (MebZIP3 and MebZIP5) confer disease resistance against cassava bacterial blight. Front Plant Sci 8:2110. https://doi.org/10.3389/fpls.2017.02110

Li J, Zhang H, Si X, Tian Y, Chen K, Liu J, Chen H, Gao C (2017c) Generation of thermosensitive male-sterile maize by targeted knockout of the ZmTMS5 gene. J Genet Genomics 44:465–468. https://doi.org/10.1016/j.jgg.2017.02.002

Li C, Unver T, Zhang B (2017d) A high-efficiency CRISPR/Cas9 system for targeted mutagenesis in Cotton (*Gossypium hirsutum* L.). Sci Rep 7:43902. https://doi.org/10.1038/srep43902

Li X, Liu W, Li B, Liu G, Wei Y, He C, Shi H (2018a) Identification and functional analysis of cassava DELLA proteins in plant disease resistance against cassava bacterial blight. Plant Physiol Biochem 124:70–76. https://doi.org/10.1016/j.plaphy.2017.12.022

Li R, Li R, Li X, Fu D, Zhu B, Tian H, Luo Y, Zhu H (2018b) Multiplexed CRISPR/Cas9-mediated metabolic engineering of gamma-aminobutyric acid levels in *Solanum lycopersicum*. Plant Biotechnol J 16:415–427. https://doi.org/10.1111/pbi.12781

Li C, Hao M, Wang W, Wang H, Chen F, Chu W et al (2018c) An efficient CRISPR/cas9 platform for rapidly generating simultaneous mutagenesis of multiple gene homoeologs in allotetraploid oilseed rape. Front Plant Sci 9:442. https://doi.org/10.3389/fpls.2018.00442

Li S, Li J, He Y, Xu M, Zhang J, Du W, Zhao Y, Xia L (2019) Precise gene replacement in rice by RNA transcript-templated homologous recombination. Nat Biotechnol. https://doi.org/10.1038/s41587-019-0065-7

Liang Z, Zhang K, Chen K, Gao C (2014) Targeted mutagenesis in *Zea mays* using TALENs and the CRISPR/Cas system. J Genet Genomics 41:63–68. https://doi.org/10.1016/j.jgg.2013.12.001

Liang Z, Chen KL, Li TD, Zhang Y, Wang Y, Zhao Q, Liu J, Zhang H, Liu C, Ran Y, Gao C (2017) Efficient DNA-free genome editing of bread wheat using CRISPR/Cas9 ribonucleoprotein complexes. Nat Comm, p 8. https://doi.org/10.1038/ncomms14261

Liang Z, Chen K, Zhang Y, Liu J, Yin K, Qiu JL, Gao C (2018) Genome editing of bread wheat using biolistic delivery of CRISPR/Cas9 *in vitro* transcripts or ribonucleoproteins. Nat Protoc 13:413–430. https://doi.org/10.1038/nprot.2017.145

Liu D, Chen X, Liu J, Ye J, Guo Z (2012) The rice ERF transcription factor OsERF922 negatively regulates resistance to *Magnaporthe oryzae* and salt tolerance. J Exp Bot 63:3899–3912. https://doi.org/10.1093/jxb/ers097

Liu F, Huang N, Wang L, Ling H, Sun T, Ahmad W, Guo J, Xu L, Gao S, Que Y, Su Y (2017a) A novel L-ascorbate Peroxidase 6 gene, ScAPX6, plays an important role in the regulation of response to biotic and abiotic stresses in sugarcane. Front Plant Sci 8:2262. https://doi.org/10.3389/fpls.2017.02262

Liu X, Wu S, Xu J, Sui C, Wei J (2017b) Application of CRISPR/Cas9 in plant biology. Acta Pharm Sin B 7:292–302. https://doi.org/10.1016/j.apsb.2017.01.002

Lowder L, Malzahn A, Qi Y (2017) Rapid construction of multiplexedCRISPR-Cas9 systems for plant genome editing. Methods Mol Biol 1578:291–307. https://doi.org/10.1007/978-1-4939-6859-6_25

Lu L, Guo D, Gao W, Yang W, Hou L, Ma X, Miao Y, Botella JR, Song CP (2018) Optimization of CRISPR/Cas9 genome editing in cotton by improved sgRNA expression. BMC Plant Biol 14:85. https://doi.org/10.1186/s13007-018-0353-0

Ma X, Zhu Q, Chen Y, Liu YG (2016) CRISPR/Cas9 platforms for genome editing in plants: developments and applications. Mol Plant 9:961–974. https://doi.org/10.1016/j.molp.2016.04.009

Makarova KS, Haft DH, Barrangou R, Brouns SJ, Charpentier E, Horvath P, Moineau S, Mojica FJ, Wolf YI, Yakunin AF, Oost VDJ, Koonin EV (2011) Evolution and classification of the CRISPR-Cas systems. Nat Rev Microbiol 9:467–477

Malnoy M, Viola R, Jung MH, Koo OJ, Kim S, Kim JS, Velasco R, Kanchiswamy CN (2016) DNA Free genetically edited grapevine and apple protoplast using CRISPR/Cas9 ribonucleoproteins. Front Plant Sci 7:1904. https://doi.org/10.3389/fpls.2016.01904

Mao Y, Zhang H, Xu N, Zhang B, Gou F, Zhu JK (2013) Application of the CRISPR-Cas system for efficient genome engineering in plants. Mol Plant 6:2008–2011

Mao Y, Zhang Z, Feng Z, Wei P, Zhang H, Botella JR, Zhu JK (2016) Development of germ-line-specific CRISPR-Cas9 systems to improve the production of heritable gene modifications in *Arabidopsis*. Plant Biotechnol J 14:519–532. https://doi.org/10.1111/pbi.12468

Mao Y, Botella JR, Zhu JK (2017) Heritability of targeted gene modifications induced by plant-optimized CRISPR systems. Cell Mol Life Sci 74:1075–1093

Martín-Pizarro C, Triviño JC, Posé D (2019) Functional analysis of the TM6 MADS-box gene in the octoploid strawberry by CRISPR/Cas9-directed mutagenesis. J Exp Bot 70:885–895

Meng Y, Hou Y, Wang H, Ji R, Liu B, Wen J, Niu L, Lin H (2017) Targeted mutagenesis by CRISPR/Cas9 system in the model legume *Medicago truncatula*. Plant Cell Rep 36:371–374. https://doi.org/10.1007/s00299-016-2069-9

Miao J, Guo D, Zhang J, Huang Q, Qin G, Zhang X, Wan J, Gu H, Qu LJ (2013) Targeted mutagenesis in rice using CRISPR-Cas system. Cell Res 23:1233–1236

Miao H, Sun P, Liu Q, Miao Y, Liu J, Zhang K, Hu W, Zhang J, Wang J, Wang Z, Jia C, Xu B, Jinet Z (2017a) Genome-wide analyses of SWEET family proteins reveal involvement in fruit development and abiotic/biotic stress responses in banana. Sci Rep 7:3536. https://doi.org/10.1038/s41598-017-03872-w

Miao H, Sun P, Liu Q, Miao Y, Liu J, Xu B, Jin Z (2017b) The AGPase family proteins in banana: genome-wide identification, phylogeny, and expression analyses reveal their involvement in the development, ripening, and abiotic/biotic stress responses. Int J Mol Sci 18:1581. https://doi.org/10.3390/ijms18081581

Mojica FJ, Diez-Villasenor C, Garcia-Martinez J, Soria E (2005) Intervening sequences of regularly spaced prokaryotic repeats derive from foreign genetic elements. J Mol Evol 60:174–182. https://doi.org/10.1007/s00239-004-0046-3

Nakajima I, Ban Y, Azuma A, Onoue N, Moriguchi T, Yamamoto T, Toki S, Endo M (2017) CRISPR/Cas9-mediated targeted mutagenesis in grape. PLoS One 12:e0177966. https://doi.org/10.1371/journal.pone.0177966

Namukwaya B, Tripathi L, Tripathi JN, Arinaitwe G, Mukasa SB, Tushemereirwe WK (2012) Transgenic banana expressing Pflp gene confers enhanced resistance to *Xanthomonas* wilt disease. Trans Res 4:855–865. https://doi.org/10.1007/s11248-011-9574-y

Nekrasov V, Staskawicz B, Weigel D, Jones JD, Kamoun S (2013) Targeted mutagenesis in the model plant *Nicotiana benthamiana* using Cas9 guided endonuclease. Nat Biotechnol 31:691–693

Okada A, Arndell T, Borisjuk N, Sharma N, Watson-Haigh NS, Tucker EJ, Baumann U, Langridge P, Whitford R (2019) CRISPR/Cas9-mediated knockout of Ms1 enables the rapid generation of male-sterile hexaploid wheat lines for use in hybrid seed production. Plant Biotechnol J. https://doi.org/10.1111/pbi.13106

Oladosu Y, Rafii MY, Abdullah N, Hussin G, Ramli A, Rahim HA, Miah G, Usman M (2016) Principle and application of plant mutagenesis in crop improvement: a review. Biotechnol Biotechnol Equip 30:1–16

Osakabe Y, Watanabe T, Sugano SS, Ueta R, Ishihara R, Shinozaki K, Osakabe K (2016) Optimization of CRISPR/Cas9 genome editing to modify abiotic stress responses in plants. Sci Rep 6:26685. https://doi.org/10.1038/srep26685

Ou W, Mao X, Huang C, Tie W, Yan Y, Ding Z, Wu C, Xia Z, Wang W, Zhou S, Li K, Hu W (2018) Genome-wide identification and expression analysis of the KUP family under abiotic stress in cassava (*Manihot esculenta* Crantz). Front Physiol 9:17. https://doi.org/10.3389/fphys.2018.00017

Pan C, Ye L, Qin L, Liu X, He Y, Wang J, Chen L, Lu G (2016) CRISPR/Cas9-mediated efficient and heritable targeted mutagenesis in tomato plants in the first and later generations. Sci Rep 6:24765. https://doi.org/10.1038/srep24765

Papikian A, Liu W, Gallego-Bartolomé J, Jacobsen SE (2019) Site-specific manipulation of *Arabidopsis* loci using CRISPR-Cas9 Sun Tag systems. Nat Commun 10:729. https://doi.org/10.1038/s41467-019-08736-7

Paszkowski J, Baur M, Bogucki A, Potrykus I (1988) Gene targeting in plants. EMBO J 7(p):4021–4026

Peng A, Chen S, Lei T, Xu L, He Y, Wu L, Yao L, Zou X (2017) Engineering canker resistant plants through CRISPR/Cas9-targeted editing of the susceptibility gene CsLOB1 promoter in citrus. Plant Biotechnol J 15:1509–1519. https://doi.org/10.1111/pbi.12733

Puchta H, Dujon B, Hohn B (1996) Two different but related mechanisms are used in plants for the repair of genomic double-strand breaks by homologous recombination. Proc Natl Acad Sci USA 93:5055–5060. https://doi.org/10.1073/pnas.93.10.5055

Pyott DE, Sheehan E, Molnar A (2016) Engineering of CRISPR/Cas9-mediated potyvirus resistance in transgene-free *Arabidopsis* plants. Mol Plant Pathol 17:1276–1288. https://doi.org/10.1111/mpp.12417

Qi W, Zhu T, Tian Z, Li C, Zhang W, Song R (2016) High-efficiency CRISPR/Cas9 multiplex gene editing using the glycine tRNA-processing system-based strategy in maize. BMC Biotechnol 16:58. https://doi.org/10.1186/s12896-016-0289-2

Rajesh MK, Rachana KE, Naganeeswaran SA, Shafeeq R, Thomas RJ, Shareefa M, Merin B, Anitha K (2015) Identification of expressed resistance gene analogue sequences in coconut leaf transcriptome and their evolutionary analysis. Turk J Agric For 39:489–502. https://doi.org/10.3906/tar-1409-75

Ren C, Liu X, Zhang Z, Wang Y, Duan W, Li S, Liang Z (2016) CRISPR/Cas9-mediated efficient targeted mutagenesis in chardonnay (*Vitis vinifera* L.). Sci Rep 6:32289. https://doi.org/10.1038/srep32289

Richardson CD, Kazane KR, Feng SJ, Zelin E, Bray NL, Schäfer AJ, Floor SN, Corn JE (2018) CRISPR-Cas9 genome editing in human cells occurs via the Fanconi anemia pathway. Nat Genetics 50:1132–1139

Ricroch A, Clairand P, Harwood W (2017) Use of CRISPR systems in plant genome editing: toward new opportunities in agriculture. Emerg Top Life Sci 1:169–182. https://doi.org/10.1042/etls20170085

Ron M, Kajala K, Pauluzzi G, Wang D, Reynoso MA, Zumstein K, Garcha J, Winte S, Masson H, Inagaki S, Federici F, Sinha N, Deal RB, Bailey-Serres J, Brady SM (2014) Hairy root transformation using *Agrobacterium rhizogenes* as a tool for exploring cell type-specific gene expression and function using tomato as a model. Plant Physiol 166:455–469. https://doi.org/10.1104/pp.114.239392

Rousseau C, Gonnet M, Le Romancer M, Nicolas J (2009) CRISPI: a CRISPR interactive database. Bioinformatics 25:33173–33318

Salomon S, Pucht H (1998) Capture of genomic and T-DNA sequences during double-strand break repair in somatic plant cells. EMBO J 17:6086–6095. https://doi.org/10.1093/emboj/17.20.6086

Shan Q, Wang Y, Li J, Zhang Y, Chen K, Liang Z, Zhang K, Liu J, Xi JJ, Qiu JL, Gao C (2013) Targeted genome modification of crop plants using a CRISPR-Cas system. Nat Biotechnol 31:686–688. https://doi.org/10.1038/nbt.2650

Shan Q, Wang Y, Li J, Gao C (2014) Genome editing in rice and wheat using the CRISPR/Cas system. Nat Protoc 9:2395–2410. https://doi.org/10.1038/nprot.2014.157

Shen C, Que Z, Xia Y, Tang N, Li D, He R, Cao M (2017a) Knock out of the annexin gene OsAnn3 via CRISPR/Cas9-mediated genome editing decreased cold tolerance in rice. J Plant Biol 60:539–547. https://doi.org/10.1007/s12374-016-0400-1

Shen L, Hua Y, Fu Y, Li J, Liu Q, Jiao X, Xin G, Wang J, Wang X, Yan C, Wang K (2017b) Rapid generation of genetic diversity by multiplex CRISPR/Cas9 genome editing in rice. Sci China Life Sci 60:506–515. https://doi.org/10.1007/s11427-017-9008-8

Shen L, Wang C, Fu Y, Wang J, Liu Q, Zhang X, Yan C, Qian Q, Wang K (2018) QTL editing confers opposing yield performance in different rice varieties. J Integr Plant Biol 60:89–93. https://doi.org/10.1111/jipb.12501

Shi J, Gao H, Wang H, Lafitte HR, Archibald RL, Yang M, Hakimi SM, Mo H, Habben JE (2017) ARGOS8 variants generated by CRISPR-Cas9 improve maize grain yield under field drought stress conditions. Plant Biotechnol J 15:207–216. https://doi.org/10.1111/pbi.12603

Shimatani Z, Kashojiya S, Takayama M, Terada R, Arazoe T, Ishii H, Teramura H, Yamamoto T, Komatsu H, Miura K, Ezura H, Nishida K, Ariizumi T, Kondoet A (2017) Targeted base editing in rice and tomato using a CRISPR-Cas9 cytidine deaminase fusion. Nat Biotechnol 35:441–443. https://doi.org/10.1038/nbt.3833

Soyk S, Lemmon ZH, Oved M, Fisher J, Liberatore KL, Park SJ, Goren A, Jiang K, Ramos A, Knaap E, Eck EV, Zamir D, Eshed Y, Lippman ZB (2017) Bypassing negative epistasis on yield in tomato imposed by a domestication gene. Cell 169:1142–1155.e12. https://doi.org/10.1016/j.cell.2017.04.032

Stephens J, Barakate A (2017) Gene editing technologies- ZFNs, TALENs, and CRISPR/Cas9. In: Thomas B, Murray BG, Murphyp DJ (eds) Encyclopedia of applied plant sciences, 2nd edn. Academic Press, Cambridge, MA, pp 157–161. https://doi.org/10.1016/B978-0-12-394807-6.00242-2

Su YC, Xu LP, Xue BT, Wu QB, Guo JL, Wu LG, Que Y (2013) Molecular cloning and characterization of two pathogenesis-related *beta*-1,3- glucanase genes ScGluA1 and ScGluD1 from sugarcane infected by *Sporisorium scitamineum*. Plant Cell Rep 32:1503–1519. https://doi.org/10.1007/s00299-013-1463-9

Su Y, Guo J, Ling H, Chen S, Wang S, Xu L, Allan AC, Que Y (2014) Isolation of a novel peroxisomal catalase gene from sugarcane, which is responsive to biotic and abiotic stresses. PLoS ONE 9:84426. https://doi.org/10.1371/journal.pone.0084426

Su Y, Xu L, Wang S, Wang Z, Yang Y, Chen Y, Que Y (2015) Identification, phylogeny, and transcript of chitinase family genes in sugarcane. Sci Rep 5:10708. https://doi.org/10.1038/srep10708

Su Y, Wang Z, Liu F, Li Z, Peng Q, Guo JL, Xu L, Que Y (2016) Isolation and characterization of ScGluD2, a new sugarcane *beta*-1,3-glucanase D family gene induced by *Sporisorium scitamineum*, ABA, H_2O_2, NaCl, and $CdCl_2$ stresses. Front Plant Sci 7:1348. https://doi.org/10.3389/fpls.2016.01348

Sugano SS, Shirakawa M, Takagi J, Matsuda Y, Shimada T, Hara-Nishimura I, Kohchi T (2014) CRISPR/Cas9 mediated targeted mutagenesis in the liverwort *Marchantia polymorpha* L. Plant Cell Physiol 55:475–481

Tang F, Yang S, Liu J, Zhu H (2016) Rj4, a gene controlling nodulation specificity in soybeans, encodes a thaumatin-like protein but not the one previously reported. Plant Physiol 170:26–32. https://doi.org/10.1104/pp.15.01661

Tian S, Jiang L, Gao Q, Zhang J, Zong M, Zhang H, Ren Y, Guo S, Gong G, Liu F, Xu Y (2017) Efficient CRISPR/Cas9-based gene knockout in watermelon. Plant Cell Rep 36:399–406. https://doi.org/10.1007/s00299-016-2089-5

Tripathi L, Mwaka H, Tripathi JN, Tushemereirwe WK (2010) Expression of sweet pepper Hrap gene in banana enhances resistance to *Xanthomonas campestris* pv. musacearum. Mol Plant Pathol 11:721–731. https://doi.org/10.1111/j.1364-3703.2010.00639.x

Tripathi L, Tripathi JN, Kiggundu A, Korie S, Shotkoski F, Tushemereirwe WK (2014) Field trial of *Xanthomonas* wilt disease-resistant bananas in East Africa. Nat Biotechnol 32:868–870. https://doi.org/10.1038/nbt.3007

Ueta R, Abe C, Watanabe T, Sugano SS, Ishihara R, Ezura H, Osakabe Y, Osakabeet K (2017) Rapid breeding of parthenocarpic tomato plants using CRISPR/Cas9. Sci Rep 7:507. https://doi.org/10.1038/s41598-017-00501-4

Veillet F, Perrot L, Chauvin L, Kermarrec MP, Guyon-Debast A, Chauvin JE, Nogué F, Mazier M (2019) Transgene-free genome editing in tomato and potato plants using *Agrobacterium*-mediated delivery of a CRISPR/Cas9 cytidine base editor. Int J Mol Sci 20:402. https://doi.org/10.3390/ijms20020402

Verma SK, Jasrotia RS, Iquebal MA, Jaiswal S, Angadi UB, Rai A, Kumar D (2017) Deciphering genes associated with root wilt disease of coconut and development of its transcriptomic database (CnTDB). Physiol Mol Plant Pathol 100:255–263. https://doi.org/10.1016/j.pmpp.2017.03.011

Waltz E (2018) With a free pass, CRISPR-edited plants reach market in record time. Nat Biotechnol 36:6–7. https://doi.org/10.1038/nbt0118-6b

Wang Y, Cheng X, Shan Q, Zhag Y, Liu J, Gao C, Qiu JL (2014) Simultaneous editing of three homoeoalleles in hexaploid bread wheat confers heritable resistance to powdery mildew. Nat. Biotechnol 32:947–951. https://doi.org/10.1038/nbt.2969

Wang ZP, Xing HL, Dong L, Zhang HY, Han CY, Wang XC, Chen QJ (2015) Egg cell-specific promoter-controlled CRISPR/Cas9 efficiently generates homozygous mutants for multiple target genes in *Arabidopsis* in a single generation. Genome Biol 16:144. https://doi.org/10.1186/s13059-015-0715-0

Wang F, Wang C, Liu P, Lei C, Hao W, Gao Y, Liu YG, Zhao K (2016a) Enhanced rice blast resistance by CRISPR/Cas9-targeted mutagenesis of the ERF transcription factor gene OsERF922. PLoS One 11:e0154027. https://doi.org/10.1371/journal.pone.0154027

Wang Y, Liu X, Ren C, Zhong GY, Yang L, Li S, Liang Z (2016b) Identification of genomic sites for CRISPR/Cas9-based genome editing in the *Vitis vinifera* genome. BMC Plant Biol 16:96. https://doi.org/10.1186/s12870-016-0787-3

Wang L, Chen L, Li R, Zhao R, Yang M, Sheng J, Shen L (2017) Reduced drought tolerance by CRISPR/Cas9-mediated SlMAPK3 mutagenesis in tomato plants. J Agric Food Chem 65:8674–8682. https://doi.org/10.1021/acs.jafc.7b02745

Wang W, Pan Q, He F, Akhunova A, Chao S, Trick H, Akhunov E (2018a) Transgenerational CRISPR-Cas9 activity facilitates multiplex gene editing in allopolyploid wheat. CRISPR J 1:65–74. https://doi.org/10.1089/crispr.2017.0010

Wang X, Tu M, Wang D, Liu J, Li Y, Li Z, Wang Y, Wang X (2018b) CRISPR/Cas9-mediated efficient targeted mutagenesis in grape in the first generation. Plant Biotechnol J 16:844–855. https://doi.org/10.1111/pbi.12832

Wang Z, Wang S, Li D, Zhang Q, Li L, Zhong C, Liu Y, Huang H (2018c) Optimized paired-sgRNA/Cas9 cloning and expression cassette triggers high-efficiency multiplex genome editing in kiwifruit. Plant Biotechnol J. https://doi.org/10.1111/pbi.12884

Watanabe K, Oda-Yamamizo C, Sage-Ono K, Ohmiya A, Ono (2018) Alteration of flower colour in *Ipomoea* nil through CRISPR/Cas9-mediated mutagenesis of carotenoid cleavage dioxygenase 4. Transgenic Res 27:25–38. https://doi.org/10.1007/s11248-017-0051-0

Wei Y, Liu W, Hu W, Liu G, Wu C, Liu W, Shi H (2017a) Genome-wide analysis of autophagy-related genes in banana highlights MaATG8s in cell death and autophagy in immune response to *Fusarium* wilt. Plant Cell Rep 36:1237–1250. https://doi.org/10.1007/s00299-017-2149-5

Wei Y, Chang Y, Zeng H, Liu G, He C, Shi H (2017b) RAV transcription factors are essential for disease resistance against cassava bacterial blight via activation of melatonin biosynthesis genes. J Pineal Res 64:12454. https://doi.org/10.1111/jpi.12454

Wiedenheft B, Zhou K, Jinek M, Coyle SM, Ma W, Doudna JA (2009) Structural basis for DNase activity of a conserved protein implicated in CRISPR-mediated genome defense. Structure 17:904–912

Wiedenheft B, Sternberg SH, Doudna JA (2012) RNA guided genetic silencing systems in bacteria and archaea. Nature 482:331–338

Xie K, Yang Y (2013) RNA-guided genome editing in plants using a CRISPR-Cas system. Mol Plant 6:1975–1983. https://doi.org/10.1093/mp/sst119

Xu H, Xiao T, Chen CH, Li W, Meyer CA, Wu Q, et al. (2015) Sequence determinants of improved CRISPR sgRNA design. Genome Res 25:1147–1157

Yan Y, Wang P, He C, Shi H (2017) MeWRKY20 and its interacting and activating autophagy-related protein 8 (MeATG8) regulate plant disease resistance in cassavaMeWRKY20 and its interacting and activating autophagy related protein 8 (MeATG8) regulate plant disease resistance in cassava. Biochem Biophysic Res Commun 494:20–26. https://doi.org/10.1016/j.bbrc.2017.10.091

Yang Y, Zhu G, Li R, Yan S, Fu D, Zhu B, Tian H, Luo Y, Zhu H (2017) The RNA editing factor SlORRM4 is required for normal fruit ripening in tomato. Plant Physiol 175:690–1702. https://doi.org/10.1104/pp.17.01265

Ye J, Yang H, Shi H, Wei Y, Tie W, Ding Z, Yan Y, Luo Y, Xia Z, Wang W, Peng M, Li K, Zhang H, Hu W (2017) The MAPKKK gene family in cassava: genome-wide identification and expression analysis against drought stress. Sci Rep 7:14939. https://doi.org/10.1038/s41598-017-13988-8

Yu QH, Wang B, Li N, Tang Y, Yang S, Yang T, Xu J, Guo C, Yan P, Wang Q, Asmutola P (2017) CRISPR/Cas9-induced targeted mutagenesis and gene replacement to generate long-shelf life tomato lines. Sci Rep 7:11874. https://doi.org/10.1038/s41598-017-12262-1

Zetsche B, Gootenberg JS, Abudayyeh OO, Slaymaker IM, Makarova KS, Essletzbichler P, Volz SE, Joung J, van der Oost J, Regev A, Koonin EV, Zhang F (2015) Cpf1 is a single RNA-guided endonuclease of a class 2 CRISPR-Cas system. Cell 163:59–771. https://doi.org/10.1016/j.cell.2015.09.038

Zhang F, Wen Y, Guo X (2014a) CRISPR/Cas9 for genome editing: progress, implications and challenges. Hum Mol Gen 23:R40–R46. https://doi.org/10.1093/hmg/ddu125

Zhang H, Zhang J, Wei P, Zhang B, Gou F, Feng Z, Mao Y, Yang L, Zhang H, Xu N, Zhu JK (2014b) The CRISPR/Cas9 system produces specific and homozygous targeted gene editing in rice in one generation. Plant Biotechnol J 12:797–807. https://doi.org/10.1111/pbi.12200

Zhang Y, Liang Z, Zong Y, Wang YP, Liu JX, Chen KL, Qiu JL, Gao CX (2016a) Efficient and transgene-free genome editing in wheat through transient expression of CRISPR/Cas9 DNA or RNA. Nat Commun 7. https://doi.org/10.1038/ncomms12617

Zhang Z, Mao Y, Ha S, Liu W, Botella JR, Zhu JK (2016b) A multiplex CRISPR/Cas9 platform for fast and efficient editing of multiple genes in *Arabidopsis*. Plant Cell Rep 35:1519–1533. https://doi.org/10.1007/s00299-015-1900-z

Zhang H, Zhang J, Lang Z, Botella JR, Zhu JK (2017a) Genome editing-principles and applications for functional genomics research and crop improvement. Critic Rev Plant Sci 36:291–309

Zhang F, LeBlanc C, Irish VF, Jacob Y (2017b) Rapid and efficient CRISPR/Cas9 gene editing in citrus using the YAO promoter. Plant Cell Rep 36:1883–1887. https://doi.org/10.1007/s00299-017-2202-4

Zhang H, Si X, Ji X, Fan R, Liu J, Chen K, Wang D, Gao C (2018a) Genome editing of upstream open reading frames enables translational control in plants. Nat Biotechnol 36:894–898

Zhang T, Zheng Q, Yi X, An H, Zhao Y, Ma S, Zhau G (2018b) Establishing RNA virus resistance in plants by harnessing CRISPR immune system. Plant Biotechnol J 16:1415–1423. https://doi.org/10.1111/pbi.12881

Zhang Z, Ge X, Luo X, Wang P, Fan Q, Hu G, Xiao J, Li F, Wu J (2018c) Simultaneous editing of two copies of Gh14-3-3d confers enhanced transgene-clean plant defense against *Verticillium dahliae* in allotetraploid upland cotton. Front Plant Sci, p 842. https://doi.org/10.3389/fpls.2018.00842

Zhao Y, Zhang C, Liu W, Gao W, Liu C, Song G, Li WX, Mao L, Chen B, Xu Y, Li X, Xie C (2016) An alternative strategy for targeted gene replacement in plants using a dual-sgRNA/Cas9 design. Sci Rep 6:23890. https://doi.org/10.1038/srep23890

Zhou J, Peng Z, Long J, Sosso D, Liu B, Eom JS, Huang S, Liu S, Vera Cruz C, Frommer WB, White FF, Yang B (2015) Gene targeting by the TAL effector PthXo2 reveals cryptic resistance gene for bacterial blight of rice. Plant J 82:632–643. https://doi.org/10.1111/tpj.12838

Zhou Z, Tan H, Li Q, Chen J, Gao S, Wang Y, Chen W, Zhang L (2018) CRISPR/Cas9-mediated efficient targeted mutagenesis of RAS in *Salvia miltiorrhiza*. Phytochemistry 148:63–70. https://doi.org/10.1016/j.phytochem.2018.01.015

Zhu J, Song N, Sun S, Yang W, Zhao H, Song W, Lai J (2016) Efficiency and inheritance of targeted mutagenesis in maize using CRISPR-Cas9. J Genet Genomics 43:25–36. https://doi.org/10.1016/j.jgg.2015.10.006

Zhu J, Chen J, Gao F, Xu C, Wu H, Chen K, Si Z, Yan H, Zhang T (2017) Rapid mapping and cloning of the *virescent-1* gene in cotton by bulked sergeant analysis-next generation sequencing and virus-induced gene silencing strategies. J Exp Bot 68:4125–4135

Zong Y, Wang Y, Li C, Zhang R, Chen K, Ran Y, Qiu JL, Wang D, Gao C (2017) Precise base editing in rice, wheat and maize with a Cas9-cytidine deaminase fusion. Nat Biotechnol 35:438–440. https://doi.org/10.1038/nbt.3811

Chapter 25
Application of Bioinformatics for Crop Stress Response and Mitigation

Anubhab Laha, Priyanka Chakraborty, Chiranjib Banerjee, Anindya Sundar Panja, and Rajib Bandopadhyay

Abstract Due to the present changes in the global environmental scenario, every day several crop plants face different types of stress from abiotic and biotic factors. The cost of overcoming the obstacle of stress has a direct impact on the yield of the plants. The abiotic stresses can range between drought, cold, high temperature, high relative humidity, salinity and heavy metals. The plant pathogens can be a reason of biotic stress. Just like in the field of any applicative studies, the role of bioinformatics is undeniable too in the field of study of stress responses. Stress has a direct effect on the growth and development of the plants, which consequently has an adverse effect on the productivity of the crop plants. This has created a huge interest among the scientific masses. Several experiments are being conducted in order to analyse the response of crop plants to stress. The development of recent computational facilities and new algorithms has ushered the use of the bioinformatics in the scientific study of the stress response. All the stress responses are in fact controlled by different genes of the affected plants. Bioinformatics has helped in the analysis of genome sequence and quantitative trait loci (QTLs), which are believed to play a major role in stress responses. The hypothesis thus generated by bioinformatics can later be confirmed by transgenics and mutant analysis. Bioinformatics has harnessed the knowledge from molecular and physiological experiments and has

A. Laha
Department of Botany, Chandernagore College, Chandernagore, West Bengal, India

Department of Botany (UGC-Centre for Advanced Study), The University of Burdwan, Burdwan, West Bengal, India

P. Chakraborty · R. Bandopadhyay (✉)
Department of Botany (UGC-Centre for Advanced Study), The University of Burdwan, Burdwan, West Bengal, India

C. Banerjee
Department of Environmental Science & Engineering, Indian Institute of Technology (ISM), Dhanbad, Jharkhand, India

A. S. Panja
Post-Graduate Department of Biotechnology and Biochemistry, Oriental Institute of Science and Technology, Burdwan, West Bengal, India

R. Roychowdhury et al. (eds.), *Sustainable Agriculture in the Era of Climate Change*, https://doi.org/10.1007/978-3-030-45669-6_25

applied a multidisciplinary approach to alleviate the effect of stress from plants. The idea of stress management is thus being strengthened by the application of open-access bioinformatics tools.

Keywords Abiotic stress · Biotic stress · Bioinformatics tools · Crop plants · Database · Stress response

1 Introduction: Crop's Response Under Stress Environments

The population of the world is increasing at an exponential rate. Crop production has to be increased at the same rate in order to meet this huge food demand (Godfray et al. 2010). In the meantime, the climate has undergone a disastrous change due to global warming (Chakraborty et al. 2014; Roychowdhury 2014). This has contributed to a huge amount of crop loss due to various kinds of abiotic stresses. The co-occurrence of several abiotic stresses has a lethal effect on plants. Drought is a major phenomenon affecting crop yield throughout the world. This has necessitated the development of stress-resistant crops. But in order to develop stress-resistant crops, one must have a clear idea about the stress response in plants. Modern computational technology and metabolomic approach have provided an insight into the stress response mechanism (Mir et al. 2012). The molecular dissection of stress tolerance will help to develop important strategies for development of stress-resistant cultivars, thereby ensuring the food security of the global population (Roychowdhury et al. 2014).

1.1 Stress

Every day millions of plants are being exposed to several kinds of environmental stress both in the agricultural and natural condition. Stress can be defined as any external factor that has adverse effects on the physical, chemical and biological status of any system, like plant system. The geographical reach and spread of the plants can get easily limited by climatic factors due to the ensuing stress factors. Though it is known to all that abiotic factors can exert stress to plants, it must also be kept in mind that there are lots of biotic factors like insects, weeds and pathogens that can produce stress too (Knight and Knight 2001; Roychowdhury et al. 2013a, b, c; Mamgain et al. 2013; Hasanuzzaman et al. 2015) (Fig. 25.1).

Stress can lead to a huge reduction in the yield of crop plants. So, agriculturists and scientists are greatly concerned with the primary assimilation processes, biomass accumulation processes, survival and growth of the crop plants. Any deviation in these criteria is useful for measuring the amount of stress faced by the plants.

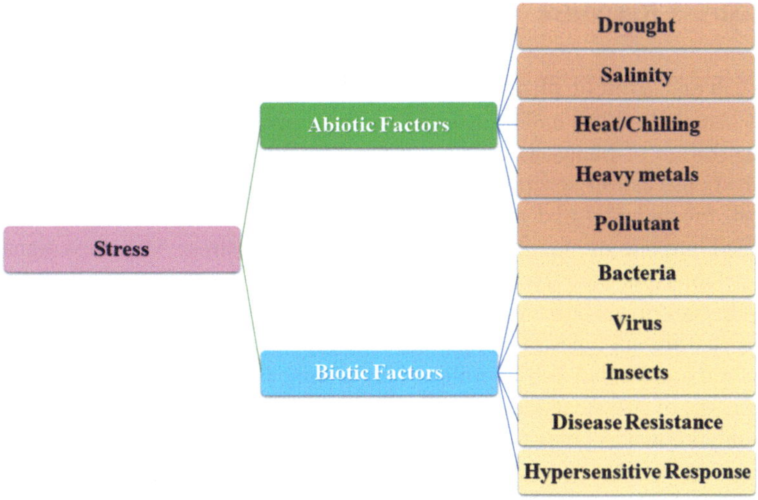

Fig. 25.1 Few abiotic and biotic stress factors

1.2 Stress Tolerance

Different plants have different abilities to cope up with the stress factors. Thus, a stressful environment can have a varied range of effects in a diverse group of plants. It may happen that a calm environment for one plant is quite stressful to the other. For example, *Pisum sativum* exhibits heat stress in temperatures above 30 °C, unlike that of *Glycine max*. Here comes the role of stress tolerance, or, in other words, the ability and fitness of the plants to cope with a stressful environment. The higher the level of fitness, the higher the stress tolerance; consequently, the higher the chance of survival and unaffected growth of plants. Future research studies have to be focused on the development of plants tolerant to a combination of stress conditions (Lobell et al. 2008).

1.3 Acclimatization and Adaptation

A plant can have a genetic predisposition of stress tolerance, which it has acquired through evolution. This can be regarded as an adaptation to stress (Kramer 1980). It is very much different from acclimatization. Acclimatization involves a prior exposure to stress that eventually increases the level of stress tolerance in some plants. Acclimatization can thus help the plant to overcome a stressful environment in the future.

1.4 Stress Response

Stress can take a toll on plants, ranging from deterioration, production of abnormal or subnormal seedlings and decrease in productivity to even death. Adaptation and acclimatization of plants work hand in hand to aid in the survival of plants. The stress response causes several changes at the molecular, cellular, anatomical, physiological, biochemical and morphological levels (Fig. 25.2). All these events are integrated into one single response, i.e. overcoming the stress. The stress signals are better perceived at the molecular level. These lead to the changes of cell division, cell cycle pattern, vacuolization and cell wall structure. Biochemically the plant can synthesize different types of compatible osmoregulatory products like glycine betaine (Ashraf and Foolad 2007), proline, beta-alanine betaine, etc. Often acclimatization to a type of stress can induce tolerance to another type of stress – a phenomenon commonly known as cross-tolerance.

2 Metabolic Changes in Plants Due to Stresses

Stress can inflict noticeable changes in the metabolic pathways of the plant, thereby disrupting its metabolic homeostasis (Shulaev et al. 2008). Generally, the changes can be noticed in three types of chemical compounds – chemicals that help in acclimatization (e.g. osmoprotectants, antioxidants, etc.), the metabolic by-products of stress (e.g. ROS, phenolic compounds, antioxidants, etc.) (Mittler 2002; Mittler et al. 2004) and the molecules involved in signal transduction of acclimatization response due to stress (e.g. plant hormones like salicylic acid) (Mittler 2006).

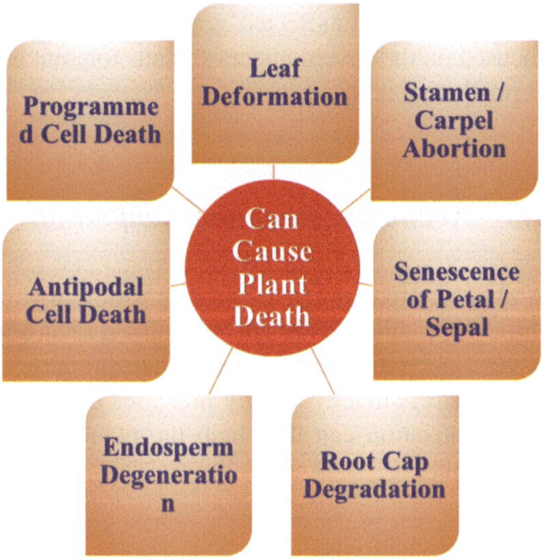

Fig. 25.2 Effect of stress on growth and development of plants

A large number of chemicals are produced in response to different types of stress in plants. In order to combat drought and salinity stress, osmoprotectants like dimethylsulfoniopropionate (Husband et al. 2012), glycine betaine, mannitol, sorbitol, proline, trehalose, sucrose, etc. are produced in plants and which helps the plants to uptake water by reducing the chemical potential. Wax deposition helps to prevent water loss and acts as a barrier to deter pathogens. Plants can withstand chilling stress by increasing the amount of saturated fatty acids in the cell membrane. During oxidative stress, the accumulated reactive oxygen species (ROS) can be scavenged with the help of carotenoids, glutathione, ascorbic acid, etc. (Hasanuzzaman et al. 2013; Roychowdhury et al. 2018, 2019). The signal transduction molecules like salicylic acid, jasmonic acid, etc. help in the generation of stress response and acclimatization. At the end, phytoalexin is produced, lignin is biosynthesized and the phenyl propanoid pathway gets activated. These are some of the changes that take place in the plant body when it gets exposed to stress. Thus, metabolic profiling can help in the study of stress biology in plants. These data can be correlated with the changes at the transcriptome and proteome level. Mutant analysis has to be done for further confirmation of the results.

The studies in metabolomics generally involve three processes – targeted analysis, metabolite profiling and metabolic fingerprinting (Fiehn 2002). The main advantage of targeted analysis over the other two processes is its ability of quantitative analysis. Targeted analysis can also detect very low levels of metabolites in a high-throughput mode. However, it also has some disadvantages, the major one being that the information of the targeted product must be known a priori, in a purified manner (Shulaev 2006).

3 Stress Biology: A Combined Approach

Bioinformatics helps in the integrated analysis of the previously studied response. It can help to study the stress response at the transcription, protein and metabolite level. This combined approach has to be utilized in order to understand the plant stress response. Sobhanian et al. (2010) applied this combined metabolomic and proteomic approach in order to study the effect of salt stress on *Aeluropus lagopoides*, a C_4 halophyte plant. Le Lay et al. (2006) had utilized this combined bioinformatics approach to study the caesium stress response on *Arabidopsis*. Integration of all experimental data is a major problem in the metabolomics study Since different specimens and samples are used for metabolite, protein and transcript analysis, it becomes quite difficult to correlate the data between metabolite, protein and RNA (Weckwerth et al. 2004). In an ideal analytical approach, the same biological sample must be used at all stages. The analysis will be much more refined if the stress response dynamics is also studied with the help of mathematical modelling (Kell 2006). It is otherwise known as the approach of "computational system biology" (Kitano 2002). The measurement of enzyme activity and absolute quantities of metabolites at every step, in a time course experiment, can provide sufficient high-throughput data, which can be utilized in system biology for mathematical modelling (Goel et al. 2006; Martins et al. 2007). Morioka et al. (2007) noted changes in metabolite and gene expression level in lipid metabolism of *Arabidopsis* during sulphur deficiency. This Markov model proved to be efficient in analysis of the time series data profile.

4 Role of Bioinformatics in Analysis of Stress Response

There has been a huge development in the field of bioinformatics during the last decade. This has ushered the creation of various databases and tools which can be utilized for different forms of research applications. There are some general bioinformatics resources that can be exploited at any step of the analysis. A glimpse of the general bioinformatics resources, viz. database, sequence alignment tools, protein structure prediction tools, etc., has been discussed.

4.1 Databases

For the first time, in 1988, the general scientific community got access to the National Center for Biotechnology Information (NCBI) (Smith 2013). This formed the infrastructural base of bioinformatics research analysis. It served the basic purpose of a biological database by storing a large array of data from the field of metabolomics, proteomics, gene expression data, genomics and phylogenetic data (Altman 2004). The nucleotide sequence was destined to be stored in three major databases – DNA Data Bank of Japan (DDBJ), European Molecular Biology Laboratory's European Bioinformatics Institute (EMBL-EBI) and GenBank. The sharing of the information between these three databases are constantly monitored by the International Nucleotide Sequence Database Collaboration (INSDC) unit (Cochrane et al. 2015). Ensembl Genomes (http://ensemblgenomes.org/; latest release 42 in January 2019) deserves a special mention. In addition to genome sequence data of vertebrates, it also stores the genome data from five different kingdoms of life, viz. bacteria, protists, fungi, plants and invertebrate metazoa. It also helps in data mining and visualization of the comparative analysis between different genomes (Kersey et al. 2017) (Table 25.1).

4.1.1 Metabolic Pathway Databases

A large amount of raw data is being generated by various expensive instruments during the metabolomics study. A large number of bioinformatics tools and databases can aid in the analysis of the experimental data of metabolomics.

Table 25.1 Different types of databases

Types of databases	Examples
Genome database (contains raw sequencing data assembly information along with its gene mapping and phylogenetic information)	EMBL (Mcwilliam et al. 2009), Ensembl (Hubbard et al. 2002), GenBank (Clark et al. 2015)
Nucleotide database (information relating to sequencing, annotations and structures of nucleotides like DNA and RNA are stored here)	DNA Data Bank of Japan (Tateno et al. 2002), European Nucleotide Archive (Harrison et al. 2018), Rfam (Griffiths-Jones et al. 2003)
Protein database (it acts as a knowledge centre of different protein sequence and structure along with its functional domain)	InterPro (Mitchell et al. 2018), JIPID (Barker et al. 1992), MIPS (Mewes et al. 2006), Pfam 32.0 (Finn et al. 2015), PIR (Barker et al. 2000), PROSITE (Hulo et al. 2006), RCSB PDB (Rose et al. 2016), SWISS-PROT (Khoury et al. 2011), UniProt (Magrane 2011)
Protein motif database (the biochemical reactions and structural information of the functional protein motifs can be accessed here)	EMOTIF (Huang and Brutlag 2001)
Taxonomic database (the taxonomic description and literature of a particular organism can be accessed through these databases)	NCBI (Federhen 2011), ProClass (Huang et al. 2000)
Phylogenetic database (phylogenetic relationships and trees along with various ribosomal data, viz. 28S rRNA and 16S rRNA, can be studied from these open-access platforms)	RDP (Cole et al. 2013), RDP-II (Cole et al. 2005)

Plant Metabolic Network (PMN) has been formed in June 2008, which is working continuously to integrate all the available plant biochemical pathway information (Schläpfer et al. 2017). It also annotates the plant genomes and curates several kinds of literature relating to plant metabolism. The latest PMN RELEASE 13.0 was announced in July 2018 which confirms the collection of metabolic pathways from 100 different plants (https://www.plantcyc.org/releases). Some of the metabolic pathway databases and tools are given in Table 25.2.

4.2 The Tools of Bioinformatics

The functional and structural relationship of biological molecules can be established by the use of different bioinformatics tools. All of these tools are available on the web; the majority of them includes open-access programs. The basic step involves sequence alignment followed by an analysis of the structure.

Table 25.2 Different types of metabolic pathway database

Databases	Information accessed	Websites	References
AraCyc 15.0	The Arabidopsis Information Resource (TAIR) has developed AraCyc database which contains nearly 610 biochemical pathways occurring in model plant *Arabidopsis thaliana*	https://www.plantcyc.org/databases/aracyc/15.0	Mueller et al. (2003)
BioCyc	Pathway Tools software is utilized by the BioCyc metabolic database containing numerous metabolic pathway data. It curates all the omics data along with the published articles that aids in metabolic modelling	https://biocyc.org/	Karp et al. (2017)
BioPath version 3	It is a database providing access through a web interface to the wall chart of Roche Applied Science "Biochemical Pathways". Computer programs like CORINA and ROTATE help in the 3D structure generation of chemical compounds, and this huge database is mined by the help of C@ROL, a reaction retrieval tool	https://www.mn-am.com/databases/biopath	Reitz et al. (2004)
KaPPA-View 4 KEGG (Kazusa Plant Pathway Viewer)	It accesses all the pathway-related information stored in the Kyoto Encyclopedia of Genes and Genomes (KEGG). It helps to interpret the correlation between genes and metabolites from the metabolome and transcriptome data	http://kpv.kazusa.or.jp/kpv4-kegg/	Sakurai et al. (2010)
KEGG Release 89.0, January 1, 2019	Kyoto Encyclopedia of Genes and Genomes (KEGG) is an informative database correlating the functional information stored in PATHWAY database with its respective genomic information stored in GENES database	https://www.genome.jp/kegg/	Kanehisa and Goto (2000)
MapMan	It is a user-friendly tool used for imposing and displaying large genomic data sets into metabolic pathways	https://mapman.gabipd.org/home	Thimm et al. (2004)
MetaCyc	MetaCyc is an online encyclopaedia of 2666 metabolic pathways from nearly 2960 organisms. All of this experimentally elucidated information was curated into the database	https://metacyc.org/	Caspi et al. (2013)
MetNetDB	The metabolic pathway information of *Arabidopsis* along with its regulation and different interactions are curated in MetNetDB	http://metnetweb.gdcb.iastate.edu/MetNet_db.htm	Li et al. (2008)
PlantCyc 12.0	PlantCyc 12.0 was released in May 2017 as a metabolic pathway database containing experimentally curated information on 1200 metabolic pathways along with their catalytic enzymes and genes	https://www.plantcyc.org/databases/plantcyc/12.0	Schlapfer et al. (2017)

4.2.1 Sequence Alignment Programs

In order to identify the similarity between two sequences of biological origin (DNA, RNA or protein), sequence alignment has to be performed. This is a very important step to identify the evolutionary relationship between the sequences (Baxevanis and Ouellette 2004). Thus, sequence alignment assists in the prediction of the structures and inferring of functions. Generally, two types of sequence alignment are followed for analysis, viz. global alignment and local alignment. The conserved sequences are aligned with the help of local alignment algorithms like Smith-Waterman algorithm (Smith and Waterman 1981). However, the overall optimized end-to-end comparison of whole sequences is performed using the global alignment algorithms like the Needleman-Wunsch algorithm (Needleman and Wunsch 1970). Pairwise sequence alignment is used to find the best possible alignment among two query sequences. It involves three methods such as dot matrix, dynamic programming and word methods. Multiple sequence alignment is used to detect conserved sequences in more than two sequences (Mount David 2001). This helps in the construction of a phylogenetic tree by identifying the conserved sequence across a group of organisms. Some researchers also take the help of FASTA and BLAST to search similar sequences in a database (Pearson 2014) (Table 25.3).

4.2.2 Protein Structure Prediction and Molecular Dynamics Simulation Tools

All the physiological processes in the plants are controlled by the efficient function of the inter- and intracellular proteins. The 3D structure of the protein controls its catalytic function. Any change of structure can alter the function of a protein, thereby disrupting the normal physiological pathway of the plant (Zhu 2016). Abiotic and biotic stress can affect the 3D conformation and function of the proteins (Rodziewicz et al. 2014). So, the study of protein structure can provide insight into

Table 25.3 Different forms of sequence alignment

Type of alignment	Sequence alignment tools (website)
Pairwise sequence alignment	BLAST (https://blast.ncbi.nlm.nih.gov/Blast.cgi), FASTA (https://www.ebi.ac.uk/Tools/sss/fasta/)
Multiple sequence alignment	ClustalW (https://www.genome.jp/tools-bin/clustalw), T-COFFEE (http://tcoffee.crg.cat/), MUSCLE (https://www.ebi.ac.uk/Tools/msa/muscle/), MAFFT (https://mafft.cbrc.jp/alignment/software/), DIALIGN (http://dialign.gobics.de/)
Local alignment	EMBOSS Water (https://www.ebi.ac.uk/Tools/psa/emboss_water/nucleotide.html)
Global alignment	EMBOSS Needle (https://www.ebi.ac.uk/Tools/psa/emboss_needle/)

understanding the stress response of plants (Wang et al. 2017). Methods like fold recognition, homology or template-based modelling and ab initio protein structure prediction can be applied to predict the three-dimensional structure of proteins (Moraes Filho et al. 2017).

The homology or comparative modelling is used to construct an experimental 3D structure of a query protein sequence, using a related protein structure as a template (Aszódi and Taylor 1996). A model of protein having the same folds sequence, like that of known protein structures, can be created by the protein threading or fold recognition methods (Cristobal et al. 2001). When no similar protein is available, then the ab initio method can be applied to fold the protein into random confirmation, following thermodynamic hypothesis (Hardin et al. 2002). Virtual plants can be modelled to express a variety of genetic responses, under an environment of abiotic stress. In order to reduce the expense of genetic manipulation, a transgenic plant carrying a set of quantitative trait loci can be modelled using mathematical modelling and studied beforehand (Tardieu 2003).

Several software have come up with an efficient algorithm to predict the structure of the protein. For example, JNet algorithm and Max/Hom/HSSP algorithm are used by JPred 4 and PHD software, respectively, in order to predict the secondary structure of the proteins. Some programs like the APSSP2 utilize the nearest neighbour and neural network approach for the creation of 3D protein structure. These software not only render an idea about the secondary and tertiary protein structure, but they also provide enough data about the backbone torsion angle, functions, mutations and homolog sequences present in the query protein sequence. This de novo hierarchical approach aids in the probabilistic alignment of protein sequence too (Table 25.4).

After the successful prediction of the protein structure, it becomes necessary to study its effect in the biological system. The interaction of the enzymes (i.e. proteins) with different substrates can be found out by studying their respective molecular domains. The molecular docking tools aid in observing the interaction between ligands and the target proteins. The molecular dynamics simulation can identify the changes in protein structure during stress responses. Molecular dynamics study found out an array of genes to be involved in chromate shock response of metal *Shewanella oneidensis* MR-1, a metal-reducing bacterium (Brown et al. 2006). Bioinformatics tools like SOPMA, MODELLER, PROSA and PROCHECK were used to determine the structure, 3D model and stability of the DREB1A protein of *Oryza sativa* var. IR6. Then, MOE, Hex and UCSF chimera software was used to study the molecular docking phenomenon of DREB1 (Nawaz et al. 2014). Docking programs, viz. AutoDock Vina, FlexX, GOLD (Genetic Optimisation for Ligand Docking) and FRED, were used to identify the metaloninergic inhibitors of N-acetylserotonin O-methyltransferase (ASMT) (Azam and Abbasi 2013). Vamsidhar et al. (2010) also utilized the GOLD docking program for analysis of binding efficacy of extracellular regulated kinase 2(ERK2) with 266 different compounds from 7 different plants, viz. *Allium sativum*, *Calendula officinalis*, *Coriandrum sativum*, *Daucus carota*, *Eucalyptus globulus*, *Murraya koenigii* and

Table 25.4 Different tools for protein structure prediction

Software	Websites	References
APSSP2	http://crdd.osdd.net/raghava/apssp2/	Raghava (2002)
CATH	http://www.cathdb.info/	Dawson et al. (2016)
CABS-fold	http://biocomp.chem.uw.edu.pl/CABSfold/	Blaszczyk et al. (2013)
CPHmodels	http://www.cbs.dtu.dk/services/CPHmodels/	Nielsen et al. (2010)
ESyPred3D	https://www.unamur.be/sciences/biologie/urbm/bioinfo/esypred/	Lambert et al. (2002)
HMMSTR	http://www.bioinfo.rpi.edu/bystrc/hmmstr/server.php	Bystroff et al. (2000)
I-TASSER (Iterative Threading ASSEmbly Refinement)	https://zhanglab.ccmb.med.umich.edu/I-TASSER/	Yang et al. (2015)
JPred 4	http://www.compbio.dundee.ac.uk/jpred4/index.htmlD	Drozdetskiy et al. (2015)
LOMETS (Local Meta-Threading Server)	https://zhanglab.ccmb.med.umich.edu/LOMETS/	Wu and Zhang (2007)
MODELLER	https://salilab.org/modeller/	Webb and Sali (2014)
PEP-FOLD	http://bioserv.rpbs.univ-paris-diderot.fr/services/PEP-FOLD/	Shen et al. (2014)
PHD (Profile network from Heidelberg)	http://www.biopred.net/sekundaerstruktur.html, email to "PredictProtein@EMBL-Heidelberg.DE"	Rost et al. (1994)
Phyre	http://www.sbg.bio.ic.ac.uk/phyre/html/	Kelley and Sternberg (2009)
Phyre2	http://www.sbg.bio.ic.ac.uk/~phyre2/html/page.cgi?id=index	Kelley et al. (2015)
RaptorX	http://raptorx.uchicago.edu/	Källberg et al. (2014)
SWISS-MODEL	https://swissmodel.expasy.org/	Waterhouse et al. (2018)

Lycopersicon esculentum. Stigmasterol exhibited the highest binding affinity to 3ERK in the above-mentioned study.

The molecular interaction and dynamics tools help to study the ligand-receptor model. They help to visualize the molecular docking and virtual screening of protein-protein interaction with lowest intermolecular energy. They try to correlate the information with several previously reported experimental data (Table 25.5). Some software also contains additional tools for torsion analysis and epitope mapping. Complex intermolecular reactions can be found out from a highly interconnected network of protein pathways. The role of a dense cluster of genes in the metabolism pathway can be observed too. The sequences conserved through the evolution help to identify and annotate the protein domains. Mechanical force fields are created at the molecular level in order to simulate the effect of mutations on the structure, stability and function of proteins (Table 25.6). Biologically meaningful analysis of biomolecular simulation at the atomic level can be done easily by multithreading approach in multicore machines.

Table 25.5 Different kinds of molecular interaction tools

Software	Websites	References
AutoDock	http://autodock.scripps.edu/	Goodsell et al. (1996)
AutoDock Vina	http://vina.scripps.edu/	Trott and Olson (2010)
BindingDB	https://www.bindingdb.org/bind/index.jsp	Gilson et al. (2015)
CFinder	http://www.cfinder.org/	Adamcsek et al. (2006)
GEMDOCK	http://gemdock.life.nctu.edu.tw/dock/	Yang and Chen (2004)
Glide	https://www.schrodinger.com/glide/	Friesner et al. (2006)
Graemlin	http://graemlin.stanford.edu/	Flannick et al. (2006)
HADDOCK	https://haddock.science.uu.nl/	Dominguez et al. (2003)
IntAct	https://www.ebi.ac.uk/intact/	Kerrien et al. (2011)
MCODE	http://baderlab.org/Software/MCODE	Bader and Hogue (2003)
MIMO	http://www.picb.ac.cn/ClinicalGenomicNTW/software.html	Di Lena et al. (2013)
MOE 2018.01	https://www.chemcomp.com/MOE2018.htm	Awale and Reymond (2019)
PathBLAST	http://www.pathblast.org/	Kelley et al. (2004)
SMART	http://smart.embl-heidelberg.de/	Letunic and Bork (2017)
STRING	https://string-db.org/	Szklarczyk et al. (2016)

Table 25.6 Simulation tools for molecular dynamics study

Software	Websites	References
Abalone II	http://www.biomolecular-modeling.com/Abalone/	Redhu and Jindal (2013)
Amber	http://ambermd.org/	Salomon-Ferrer et al. (2013)
CHARMM (Chemistry at HARvard Macromolecular Mechanics)	https://www.charmm.org/charmm/?CFID=eb547acf-c5d5-420c-8ef9-9834f4370e52&CFTOKEN=0	Brooks et al. (2009)
FoldX	http://foldxsuite.crg.eu/	Schymkowitz et al. (2005)
GROMACS	http://www.gromacs.org/About_Gromacs	Abraham et al. (2015)
RasMol	http://www.openrasmol.org/	Bernstein (2000)

5 Bioinformatics Application for Stress-Metabolomics Data Analysis

Stress is a complex interaction between the environment and the plants. Plants have to face different forms of stress in both agricultural and natural conditions. The normal functioning of the plants gets affected due to the adverse environmental conditions (temperature extremities, salinity, radiation, etc.). Plant pathogens exert biotic stress in plants. All of these stress factors impose a negative impact on the growth and development of the plant. Many databases have come up which store the gene sequences, annotations and experimentally approved data relating to the plant stress response.

5.1 DroughtDB

Experimentally verified protein function during drought tolerance can be found in DroughtDB (http://pgsb.helmholtz-muenchen.de/droughtdb/). Climate change has resulted in the spread of the devastating scenario of drought throughout the world. The increased photorespiration during drought leads to reactive oxygen species (ROS) accumulation. It can have a negative effect on cell health and may even cause death. A study of 182 publications has led to the identification of 199 molecularly characterized genes from different plant species. The orthologs of the drought stress genes from *Aegilops tauschii*, *Arabidopsis*, *Brachypodium distachyon*, *Hordeum vulgare*, *Oryza sativa*, *Secale cereale*, *Solanum lycopersicum*, *Sorghum bicolour* and *Zea mays* can also be found out from this database. This aids in the evolutionary analysis and study of drought stress response in plants (Alter et al. 2015).

5.2 PhytAMP

The taxonomic, physiochemical and microbiological information of antimicrobial plant peptides are stored in the PhytAMP database (http://phytamp.hammamilab.org/main.php). Plant antimicrobial peptides (AMPs) are cysteine-rich proteins. They help to increase the innate immunity of the plants, thereby increasing the plant defence response against the pathogens. Currently, PhytAMP is a data repository of nearly 271 AMPs (antimicrobial plant peptides) from different plant families like Violaceae, Brassicaceae, Vicieae, Santalaceae, Oryzeae, Amaranthaceae, etc. This database can also be utilized to understand the mechanism of biotic stress response of plants. The peptide sequences were derived from the UniProt and were correlated with the literature using the PubMed. Hidden Markov models-based structure prediction, sequence alignment and homology searching tools are also available in this database. The functional relationship of AMPs helps to exploit the antimicrobial property of plant peptides in agriculture and pharmaceutical industry (Hammami et al. 2008).

5.3 PSRN (Plant Stress RNA-seq Nexus: A Stress-Specific Transcriptome Database in Plant Cells)

Biotic and abiotic stress encompasses various types of infections by pathogen, insects and environmental stress like salinity, drought, chill and heat, etc. Environmental stress can affect the metabolic pathway of the plant, which is reflected in the transcriptome profile of the plants. Profiling of coding transcript and long non-coding RNA (lncRNA) can be done with the help of RNA sequencing (RNA-seq) tool. It is being used extensively to detect the changes in the expression of long non-coding RNA and coding genes that takes place during environmental stress. PSRN (http://syslab5.nchu.edu.tw/PSRN) is a large-scale database of plant RNA-seq that can study the phenotypic changes and large-scale coding and non-coding RNA profiling along with the visualization of differential expression profile. This database encompasses the RNA-seq data from NCBI GEO (Gene Expression Omnibus) and SRA (Sequence Read Archive). Phytozome, PopGenIE and Ensembl Plants provide the reference transcriptome. All these RNA-seq data are correlated with the transcriptomic data, followed by Bowtie sequence alignment and eXpress expression analysis. This helps in the construction of mRNA-lncRNA co-expression network and the web interface of PSRN database (Li et al. 2018). PSRN contains 133 stress-specific subsets constructed from 12 plant species, 937 samples and 26 plant-stress RNA-seq databases. A systemic RNA-seq analysis of the data stored in PSRN can easily identify the differentially expressed (DE) stress-specific transcripts.

5.4 Plant Stress Gene Database

Plants, being sessile in nature, cannot escape the multitude of environmental stress, to which they are exposed every now and then. This has a serious impact on the crop plants and causes a rapid decline in agricultural production. Plant Stress Gene Database (http://ccbb.jnu.ac.in/stressgenes/frontpage.html) is a compendium of different genes involved in plant stress response. The role of 259 genes, from 11 plant species, in stress response has been studied extensively. All of this information along with their paralog and ortholog are stored in the database. Majority of the identified genes have a distinct role in combating oxidative stress and salinity stress response. This data is useful to the biologists to understand the effect of the stress-related genes on the defence system and productivity of crop plants (Prabha et al. 2011).

5.5 Plant Stress Protein Database (PSPDB)

A total of 2064 plant proteins of 134 plant species have been curated into Plant Stress Protein Database (http://www.bioclues.org/pspdb/) from UniProt. This database provides information on both abiotic and biotic stress proteins. Nearly 30 different types of stress proteins have been identified after experimental and functional validation. MySQL (5.0), Perl (ver. 5.8.8) and PHP (5.2.4) have been integrated together for the proper functioning of the database. This database provides rapid citation through the UniProt and several integrated tools like HMMER, NCBI BLAST, NJPLOT and ClustalW. The unknown function of a protein during stress can be interpreted with the help of this database. Thus, a suitable protein candidate can be chosen for further wet lab experiments (Kumar et al. 2014).

5.6 *PlantPReS v3.0 Database*

PlantPReS Database (http://www.proteome.ir/) comprises of 37,930 entries from nearly 534 research publications with data on more than 10,600 unique proteins. This database is mostly focused on providing verified literature information on stress-related proteins. A customized BLAST (NCBI version 2.2.28+) tool is available to look for a homologous sequence in this database. Nearly 20 different protein identifiers can be employed simultaneously for homology searching. All of this multifactorial analysis helps in the proper characterization of the plant stress response (Mousavi et al. 2016).

5.7 *PRGdb 3.0*

Every plant is susceptible to attacks by a large variety of pathogen like bacteria, virus, insects, etc. The pathogen attack leads to the biotic stress in crop plants. In order to have an idea about the biotic stress response, one has to understand the defence mechanism imposed by plants through the R genes (i.e. plant disease resistance genes). The R genes play a key role in the identification of pathogenic proteins. The PRGdb (http://prgdb.org/prgdb/) is a comprehensive database of pathogen recognition genes (PRGs). Information about nearly 268 plant species is present in this database. Of them, 99 species are present with complete annotated PRGs. The updated PRGdb 3.0 sports a collection of 177,072 annotated PRGs. DRAGO 2 tool is being used extensively for automated annotation in this database. The disease resistance mechanism of plants can be analysed with the help of this database (Osuna-Cruz et al. 2017).

5.8 *QlicRice*

QlicRice (http://cabgrid.res.in/nabg/qlicrice.html) is a web database of abiotic stress-responsive QTLs (quantitative trait loci) and their corresponding gene annotations in rice. The user can perform data mining of gene location, associated traits and ESTs too. QTL is the location in the DNA that can affect a phenotypic trait, measurable on a linear quantitative scale. Since QTL helps in the correlation of genotypic information with that of the phenotype, QTL analysis helps to understand the abiotic stress response in plants. The data on 53 traits represented by more than 970 abiotic stress-responsive QTLs have been stored in the QlicRice database.

This interactive platform helps to search for different abiotic stress-related QTLs and explore their role in the biochemical pathway along with their regulations. All the genetic information related to one specific rice variety, viz. *Oryza sativa* ssp. *japonica* cv. *Nipponbare*, can be accessed here too (Smita et al. 2011).

5.9 Rice Stress-Responsive Transcription Factor Database (SRTFDB)

Rice is a staple food for a large section of the world's population and has created a lot of interest in stress response studies among scientists. Productivity gets highly affected due to salinity and drought. In order to increase the yield, one must overcome the obstacle of abiotic stress. Rice SRTFDB (http://www.nipgr.res.in/RiceSRTFDB.html) has been created to understand the underlying mechanism of the stress response. It curates all the information on rice transcription factors' expressions and functions during abiotic stress. The transcription of many genes is controlled by the binding of cis-regulatory elements and the transcription factors. All the transcriptions factors are classified into related families and help in easy discovery from the database. Rice SRTFDB provides this information on transcription factors and ToS17 mutant. The mutation of ToS17, a long terminal repeat retrotransposon, helps to study the gene function by disrupting the gene sequence. This database has curated its information from a comprehensive study of 2478 nonredundant rice transcription factors and 456 Affymetrix GeneChip rice genome arrays data. This unique database provides all the knowledge required to understand the stress regulatory mechanism in rice (Priya and Jain 2013).

5.10 STIFDB V2.0 (Stress-Responsive Transcription Factor Database)

In order to develop a stress-adaptive or stress-tolerant plant species, one must have a clear understanding of the molecular regulatory pathways involved in stress management. The understanding of the intercellular and intracellular changes will render an outlook of the physiological changes in the stress response. STIFDB2 (http://caps.ncbs.res.in/stifdb2/) has included the genomes of *Oryza sativa* subsp. *japonica* and *Oryza sativa* subsp. *indica* in addition to the stress-responsive genes of *Oryza sativa* and *Arabidopsis thaliana*. HMM-based model study has identified ten families of transcription factors, specifically involved in abiotic and biotic stress response, along with their binding sites at the promoter region. More than 3000 stress-responsive genes have been described here in addition to seven new stress signals, viz. aluminium, iron, heat, bacterial blight, wounding, UV-B and osmotic stress. Ortholog inclusion can give an evolutionary outlook to the analysis done via STIFB2. Data extrapolation helps to identify the interactions between upregulated genes and the transcription factors, during stress signals (Naika et al. 2013).

6 Generation Challenge Programme: A Practical Step to Provide Food Security

A time-bound 10-year (2004–2013) Generation Challenge Programme (http://www.generationcp.org/) was undertaken in order to provide better food security to the world. It used the prevailing genetic diversity along with the knowledge of stress tolerance and advanced crop science in order to increase the productivity of the crop plants. This programme achieved a huge success and came to an end in December 2014. This programme performed the clustering of orthologous and paralogous genes across many crop plants, in order to identify the stress-responsive genes. Phylogenomic inference helped to construct the phylogenetic trees, and Dayhoff method was applied for comparative analysis of the stress responsive gene sequences. The genes involved in abiotic stress, especially drought stress, were compiled in the *plant stress-responsive gene catalogue* of the Generation Challenge Programme (Wanchana et al. 2007).

7 Conclusion

Climate change has imposed drought and salinity stress upon the plants, thereby making the developing countries vulnerable to problems of food scarcity. Lipiec et al. (2013) have found out that nearly 50% of agricultural crop yield is lost as a result of abiotic stress, especially drought. The functional interpretation of different stress-responsive genes can elucidate the stress response pathways and show the path to achieve stress tolerance (Kumar et al. 2018). The decrease in arable lands due to urban encroachment and soil degradation is inhibitory to increase crop productivity (Godfray et al. 2010). The photosynthetic efficiency is also affected by abiotic stress (Foyer and Shigeoka 2011).

Any change in the plant proteome due to stress response can be noted easily with the help of bioinformatics. The in silico analysis of the proteins can be done with the tools. The change in protein structure due to stress and its subsequent effects along the metabolic pathway is easily observed during atomic simulations. In fact, these tools can greatly reduce the time and cost involved in this type of stress response analysis at a high-end laboratory facility. This pool of knowledge can be used to develop different stress-tolerant variety effective in combating the adverse environmental condition.

8 Future Trends

In the coming years, humanity is going to face a threat to their food security. This can only be overcome by intensifying our concern with stress-tolerant plant development. Both the new-age and conventional research approach has to be applied in this regard. CRISP/Cas9 gene editing technology can showcase an innovative approach in transgenic crop development (Voytas and Gao 2014). But, at first, one must gather a basic understanding of the plant stress response, with the help of various simulation tools offered by the bioinformatics.

A huge ocean of bioinformatics tools is easily available on the web. Most of the software can be accessed freely by anyone. This wealth of information and bioinformatics tools is enough to make one confused about which tool to use for efficient stress response analysis. An integrated database containing the list of tools along with their shortcomings will surely help one biologist to clear up his confusions. The predictive nature of the bioinformatics tools is to be praised indeed, but they need to be explanatory too. It will make the analysis easy to understand. The system biology approach, integrating the genomics, transcriptomics, proteomics and metabolomics data all at the same time, is a powerful method to establish the effects of the stress in plants. This repertoire of information will aid in the production of next-generation transgenic crop plant. However, field testing has to be done with the transgenic varieties, before they are being released for cultivation. The path towards the study of the biotic and abiotic stress response can thus be made easy for the budding researchers. It is vital for the development of next-generation crops, an essential step towards global food security.

Acknowledgements Authors are thankful to UGC-Center of Advanced Study, Department of Botany, the University of Burdwan, for pursuing research activities. AL is thankful to Principal of Chandernagore College for allowing for research work.

References

Abraham MJ, Murtola T, Schulz R, Páll S, Smith JC, Hess B, Lindahl E (2015) GROMACS: high-performance molecular simulations through multi-level parallelism from laptops to supercomputers. SoftwareX 1:19–25. https://doi.org/10.1016/j.softx.2015.06.001

Adamcsek B, Palla G, Farkas IJ, Derényi I, Vicsek T (2006) CFinder: locating cliques and overlapping modules in biological networks. Bioinformatics 22(8):1021–1023. https://doi.org/10.1093/bioinformatics/btl039

Alter S, Bader KC, Spannagl M, Wang Y, Bauer E, Schön CC, Mayer KF (2015) DroughtDB: an expert-curated compilation of plant drought stress genes and their homologs in nine species. Database 2015:1–7. https://doi.org/10.1093/database/bav046

Altman RB (2004) Building successful biological databases. Brief Bioinform 5(1):4–5. https://doi.org/10.1093/bib/5.1.4

Ashraf MF, Foolad MR (2007) Roles of glycine betaine and proline in improving plant abiotic stress resistance. Environ Exp Bot 59(2):206–216. https://doi.org/10.1016/j.envexpbot.2005.12.006

Aszódi A, Taylor WR (1996) Homology modelling by distance geometry. Fold Des 1(5):325–334. https://doi.org/10.1016/S1359-0278(96)00048-X

Awale M, Reymond JL (2019) Web-based tools for Polypharmacology prediction. In: Ziegler S, Waldmann H (eds) Systems chemical biology, Methods molecular biology, vol 1888. Humana Press, New York, pp 255–272. https://doi.org/10.1007/978-1-4939-8891-4_15

Azam SS, Abbasi SW (2013) Molecular docking studies for the identification of novel melatoninergic inhibitors for acetylserotonin-O-methyltransferase using different docking routines. Theor Biol Med Model 10(1):63. https://doi.org/10.1186/1742-4682-10-63

Bader GD, Hogue CW (2003) An automated method for finding molecular complexes in large protein interaction networks. BMC Bioinforma 4(1):2. https://doi.org/10.1186/1471-2105-4-2

Barker WC, George DG, Mewes HW, Tsugita A (1992) The PIR-international protein sequence database. Nucleic Acids Res 20(Suppl):2023–2026

Barker WC, Garavelli JS, Huang H, McGarvey PB, Orcutt BC, Srinivasarao GY, Xiao C, Yeh LL, Ledley RS, Janda JF, Pfeiffer F, Mewes HW, Tsugita A, Wu C (2000) The protein information resource (PIR). Nucleic Acids Res 28(1):41–44. https://doi.org/10.1093/nar/28.1.41

Baxevanis AD, Ouellette BFF (2004) Bioinformatics: a practical guide to the analysis of genes and proteins, vol 43. Wiley, Hoboken

Bernstein HJ (2000) Recent changes to RasMol, recombining the variants. Trends Biochem Sci 25(9):453–455. https://doi.org/10.1016/S0968-0004(00)01606-6

Blaszczyk M, Jamroz M, Kmiecik S, Kolinski A (2013) CABS-fold: server for the de novo and consensus-based prediction of protein structure. Nucleic Acids Res 41(W1):W406–W411. https://doi.org/10.1093/nar/gkt462

Brooks BR, Brooks CL III, Mackerell AD Jr, Nilsson L, Petrella RJ, Roux B, Won Y, Archontis G, Bartels C, Boresch S, Caflisch A, Caves L, Cui Q, Dinner AR, Feig M, Fischer S, Gao J, Hodoscek M, Im W, Kuczera K, Lazaridis T, Ma J, Ovchinnikov V, Paci E, Pastor RW, Post CB, Pu JZ, Schaefer M, Tidor B, Venable RM, Woodcock HL, Wu X, Yang W, York DM, Karplus M (2009) CHARMM: the biomolecular simulation program. J Comput Chem 30(10):1545–1614. https://doi.org/10.1002/jcc.21287

Brown SD, Thompson MR, VerBerkmoes NC, Chourey K, Shah M, Zhou J, Hettich RL, Thompson DK (2006) Molecular dynamics of the *Shewanella oneidensis* response to chromate stress. Mol Cell Proteomics 5(6):1054–1071. https://doi.org/10.1074/mcp.M500394-MCP200

Bystroff C, Thorsson V, Baker D (2000) HMMSTR: a hidden Markov model for local sequence-structure correlations in proteins. J Mol Biol 301(1):173–190. https://doi.org/10.1006/jmbi.2000.3837

Caspi R, Altman T, Billington R, Dreher K, Foerster H, Fulcher CA, Holland TA, Keseler IM, Kothari A, Kubo A, Krummenacker M, Latendresse M, Mueller LA, Ong Q, Paley S, Subhraveti P, Weaver DS, Weeransinghe D, Zhang P, Karp PD (2013) The MetaCyc database of metabolic pathways and enzymes and the BioCyc collection of pathway/genome databases. Nucleic Acids Res 42(D1).D459–D471. https://doi.org/10.1093/nar/gkt1103

Chakraborty S, Pattanayak A, Mandal S, Das M, Roychowdhury R (2014) An overview of climate change: causes, trends and implications. In: Roychowdhury R (ed) Crop improvement in the era of climate change. IK International Publishing House, New Delhi, pp 1–29

Clark K, Karsch-Mizrachi I, Lipman DJ, Ostell J, Sayers EW (2015) GenBank. Nucleic Acids Res 44(D1):D67–D72. https://doi.org/10.1093/nar/gkv1276

Cochrane G, Karsch-Mizrachi I, Takagi T, International Nucleotide Sequence Database Collaboration (2015) The international nucleotide sequence database collaboration. Nucleic Acids Res 44(D1):D48–D50. https://doi.org/10.1093/nar/gkv1323

Cole JR, Chai B, Farris RJ, Wang Q, Kulam SA, McGarrell DM, Garrity GM, Tiedje JM (2005) The ribosomal database project (RDP-II): sequences and tools for high-throughput rRNA analysis. Nucleic Acids Res 33(Suppl 1):D294–D296. https://doi.org/10.1093/nar/gki038

Cole JR, Wang Q, Fish JA, Chai B, McGarrell DM, Sun Y, Brown CT, Porras-Alfaro A, Kuske CR, Tiedje JM (2013) Ribosomal database project: data and tools for high throughput rRNA analysis. Nucleic Acids Res 42(D1):D633–D642. https://doi.org/10.1093/nar/gkt1244

Cristobal S, Zemla A, Fischer D, Rychlewski L, Elofsson A (2001) A study of quality measures for protein threading models. BMC Bioinforma 2(1):5. https://doi.org/10.1186/1471-2105-2-5

Dawson NL, Lewis TE, Das S, Lees JG, Lee D, Ashford P, Orengo CA, Sillitoe I (2016) CATH: an expanded resource to predict protein function through structure and sequence. Nucleic Acids Res 45(D1):D289–D295. https://doi.org/10.1093/nar/gkw1098

Di Lena P, Wu G, Martelli PL, Casadio R, Nardini C (2013) MIMO: an efficient tool for molecular interaction maps overlap. BMC Bioinforma 14(1):159. https://doi.org/10.1186/1471-2105-14-159

Dominguez C, Boelens R, Bonvin AMJJ (2003) HADDOCK: a protein– protein docking approach based on biochemical or biophysical information. J Am Chem Soc 125(7):1731–1737. https://doi.org/10.1021/ja026939x

Drozdetskiy A, Cole C, Procter J, Barton GJ (2015) JPred4: a protein secondary structure prediction server. Nucleic Acids Res 43(W1):W389–W394. https://doi.org/10.1093/nar/gkv332

Federhen S (2011) The NCBI taxonomy database. Nucleic Acids Res 40(D1):D136–D143. https://doi.org/10.1093/nar/gkr1178

Fiehn O (2002) Metabolomics—the link between genotypes and phenotypes. In: Town C (ed) Functional genomics, Plant molecular biology, vol 48. Springer, Dordrecht, pp 155–171. https://doi.org/10.1007/978-94-010-0448-0_11

Finn RD, Coggill P, Eberhardt RY, Eddy SR, Mistry J, Mitchell AL, Potter SC, Punta M, Qureshi M, Sangrador-Vegas A, Salazar GA, Tate J, Bateman A (2015) The Pfam protein families database: towards a more sustainable future. Nucleic Acids Res 44(D1):D279–D285. https://doi.org/10.1093/nar/gkv1344

Flannick J, Novak A, Srinivasan BS, McAdams HH, Batzoglou S (2006) Graemlin: general and robust alignment of multiple large interaction networks. Genome Res 16(9):1169–1181. https://doi.org/10.1101/gr.5235706

Foyer CH, Shigeoka S (2011) Understanding oxidative stress and antioxidant functions to enhance photosynthesis. Plant Physiol 155(1):93–100. https://doi.org/10.1104/pp.110.166181

Friesner RA, Murphy RB, Repasky MP, Frye LL, Greenwood JR, Halgren TA, Sanschagrin PC, Mainz DT (2006) Extra precision glide: docking and scoring incorporating a model of hydrophobic enclosure for protein–ligand complexes. J Med Chem 49(21):6177–6196. https://doi.org/10.1021/jm051256o

Gilson MK, Liu T, Baitaluk M, Nicola G, Hwang L, Chong J (2015) BindingDB in 2015: a public database for medicinal chemistry, computational chemistry and systems pharmacology. Nucleic Acids Res 44(D1):D1045–D1053. https://doi.org/10.1093/nar/gkv1072

Godfray HCJ, Beddington JR, Crute IR, Haddad L, Lawrence D, Muir JF, Pretty J, Robinson S, Thomas SM, Toulmin C (2010) Food security: the challenge of feeding 9 billion people. Science 327(5967):812–818. https://doi.org/10.1126/science.1185383

Goel G, Chou IC, Voit EO (2006) Biological systems modeling and analysis: a biomolecular technique of the twenty-first century. J Biomol Tech 17(4):252–269

Goodsell DS, Morris GM, Olson AJ (1996) Automated docking of flexible ligands: applications of AutoDock. J Mol Recognit 9(1):1–5. https://doi.org/10.1002/(SICI)1099-1352(199601)9:1<1::AID-JMR241>3.0.CO;2-6

Griffiths-Jones S, Bateman A, Marshall M, Khanna A, Eddy SR (2003) Rfam: an RNA family database. Nucleic Acids Res 31(1):439–441. https://doi.org/10.1093/nar/gkg006

Hammami R, Ben Hamida J, Vergoten G, Fliss I (2008) PhytAMP: a database dedicated to antimicrobial plant peptides. Nucleic Acids Res 37(Suppl 1):D963–D968. https://doi.org/10.1093/nar/gkn655

Hardin C, Pogorelov TV, Luthey-Schulten Z (2002) *Ab initio* protein structure prediction. Curr Opin Struct Biol 12(2):176–181. https://doi.org/10.1016/S0959-440X(02)00306-8

Harrison PW, Alako B, Amid C, Cerdeño-Tárraga A, Cleland I, Holt S, Hussein A, Jayathilaka S, Kay S, Keane T, Leinonen R, Liu X, Martínez-Villacorta J, Milano A, Pakseresht N, Rajan J, Reddy K, Richards E, Rosello M, Silvester N, Smirnov AL, Vijayraja S, Cochrane G (2018)

The European nucleotide archive in 2018. Nucleic Acids Res 47(D1):D84–D88. https://doi.org/10.1093/nar/gky1078

Hasanuzzaman M, Nahar K, Alam M, Roychowdhury R, Fujita M (2013) Physiological, biochemical, and molecular mechanisms of heat stress tolerance in plants. Int J Mol Sci 14(5):9643–9684

Hasanuzzaman M, Roychowdhury R, Karmakar J, Dey N, Nahar K, Fujita M (2015) Recent advances in biotechnology and genomic approaches for abiotic stress tolerance in crop plants. In: Devarajan T, Jeyabalan S (eds) Genomics and proteomics: concepts, technologies and applications. Apple Academic Press, Mistwell Crescent, Oakville, Canada, pp 333–366

Huang JY, Brutlag DL (2001) The EMOTIF database. Nucleic Acids Res 29(1):202–204. https://doi.org/10.1093/nar/29.1.202

Huang H, Xiao C, Wu CH (2000) ProClass protein family database. Nucleic Acids Res 28(1):273–276. https://doi.org/10.1093/nar/28.1.273

Hubbard T, Barker D, Birney E, Cameron G, Chen Y, Clark L, Cox T, Cuff J, Curwen V, Down T, Durbin R, Eyras E, Gilbert J, Hammond M, Huminiecki L, Kasprzyk A, Lehvaslaiho H, Lijnzaad P, Melsopp C, Mongin E, Pettett R, Pocock M, Potter S, Rust A, Schmidt E, Searle S, Slater G, Smith J, Spooner W, Stabenau A, Stalker J, Stupka E, Ureta-Vidal A, Vastrik I, Clamp M (2002) The Ensembl genome database project. Nucleic Acids Res 30(1):38–41

Hulo N, Bairoch A, Bulliard V, Cerutti L, De Castro E, Langendijk-Genevaux PS, Pagni M, Sigrist CJA (2006) The PROSITE database. Nucleic Acids Res 34(suppl 1):D227–D230. https://doi.org/10.1093/nar/gkj063

Husband JD, Kiene RP, Sherman TD (2012) Oxidation of dimethylsulfoniopropionate (DMSP) in response to oxidative stress in *Spartina alterniflora* and protection of a non-DMSP producing grass by exogenous DMSP+ acrylate. Environ Exp Bot 79:44–48. https://doi.org/10.1016/j.envexpbot.2012.01.006

Källberg M, Margaryan G, Wang S, Ma J, Xu J (2014) RaptorX server: a resource for template-based protein structure modeling. In: Kihara D (ed) Protein structure prediction, Methods molecular biology, vol 1137. Humana Press, New York, pp 17–27. https://doi.org/10.1007/978-1-4939-0366-5_2

Kanehisa M, Goto S (2000) KEGG: Kyoto encyclopedia of genes and genomes. Nucleic Acids Res 28(1):27–30. https://doi.org/10.1093/nar/28.1.27

Karp PD, Billington R, Caspi R, Fulcher CA, Latendresse M, Kothari A, Ingrid M Keseler IM, Krummenacker M, Midford PE, Ong Q, Ong WK, Paley SM, Subhraveti (2017) The BioCyc collection of microbial genomes and metabolic pathways. Brief Bioinform bbx085. https://doi.org/10.1093/bib/bbx085

Kell DB (2006) Metabolomics, modelling and machine learning in systems biology–towards an understanding of the languages of cells. FEBS J 273(5):873–894. https://doi.org/10.1111/j.1742-4658.2006.05136.x

Kelley LA, Sternberg MJ (2009) Protein structure prediction on the web: a case study using the Phyre server. Nat Protoc 4(3):363–371. https://doi.org/10.1038/nprot.2009.2

Kelley BP, Yuan B, Lewitter F, Sharan R, Stockwell BR, Ideker T (2004) PathBLAST: a tool for alignment of protein interaction networks. Nucleic Acids Res 32(suppl 2):W83–W88. https://doi.org/10.1093/nar/gkh411

Kelley LA, Mezulis S, Yates CM, Wass MN, Sternberg MJ (2015) The Phyre2 web portal for protein modeling, prediction and analysis. Nat Protoc 10(6):845. https://doi.org/10.1038/nprot.2015.053

Kerrien S, Aranda B, Breuza L, Bridge A, Broackes-Carter F, Chen C, Duesbury M, Dumousseau M, Feuermann M, Hinz U, Jandrasits C, Jimenez RC, Khadake J, Mahadevan U, Masson P, Pedruzzi I, Pfeiffenberger E, Porras P, Raghunath A, Roechert B, Orchard S, Hermjakob H (2011) The IntAct molecular interaction database in 2012. Nucleic Acids Res 40(D1):D841–D846. https://doi.org/10.1093/nar/gkr1088

Kersey PJ, Allen JE, Allot A, Barba M, Boddu S, Bolt BJ, Carvalho-Silva D, Christensen M, Davis P, Grabmueller C, Kumar N, Liu Z, Maurel T, Moore B, McDowall MD, Maheswari U, Naamati G, Newman V, Ong CK, Paulini M, Pedro H, Perry E, Russell M, Sparrow H, Tapanari

E, Taylor K, Vullo A, Williams G, Zadissia A, Olson A, Stein J, Wei S, Tello-Ruiz M, Ware D, Luciani A, Potter S, Finn RD, Urban M, Hammond-Kosack KE, Bolser DM, Silva ND, Howe KL, Langridge N, Maslen G, Staines DM, Yates A (2017) Ensembl genomes 2018: an integrated omics infrastructure for non-vertebrate species. Nucleic Acids Res 46(D1):D802–D808. https://doi.org/10.1093/nar/gkx1011

Khoury GA, Baliban RC, Floudas CA (2011) Proteome-wide post-translational modification statistics: frequency analysis and curation of the Swiss-Prot database. Sci Rep 1:90. https://doi.org/10.1038/srep00090

Kitano H (2002) Computational systems biology. Nature 420(6912):206–210. https://doi.org/10.1038/nature01254

Knight H, Knight MR (2001) Abiotic stress signalling pathways: specificity and cross-talk. Trends Plant Sci 6(6):262–267. https://doi.org/10.1016/S1360-1385(01)01946-X

Kramer PJ (1980) Drought, stress, and the origin of adaptations 7–20

Kumar SA, Kumari PH, Sundararajan VS, Suravajhala P, Kanagasabai R, Kishor PK (2014) PSPDB: plant stress protein database. Plant Mol Biol Report 32(4):940–942. https://doi.org/10.1007/s11105-014-0698-0

Kumar S, Sachdeva S, Bhat KV, Vats S (2018) Plant responses to drought stress: physiological, biochemical and molecular basis. In: Vats S. (eds) Biotic and abiotic stress tolerance in plants. Springer, Singapore, pp1–25. doi:https://doi.org/10.1007/978-981-10-9029-5_1

Lambert C, Leonard N, De Bolle X, Depiereux E (2002) ESyPred3D: prediction of proteins 3D structures. Bioinformatics 18(9):1250–1256. https://doi.org/10.1093/bioinformatics/18.9.1250

Le Lay P, Isaure MP, Sarry JE, Kuhn L, Fayard B, Le Bail JL, Bastien O, Garin J, Roby C, Bourguignon J (2006) Metabolomic, proteomic and biophysical analyses of *Arabidopsis thaliana* cells exposed to a caesium stress. Influence of potassium supply. Biochimie 88(11):1533–1547. https://doi.org/10.1016/j.biochi.2006.03.013

Letunic I, Bork P (2017) 20 years of the SMART protein domain annotation resource. Nucleic Acids Res 46(D1):D493–D496. https://doi.org/10.1093/nar/gkx922

Li J, Babka HL, Miller L, Ransom N, Xia T, Wurtele ES (2008) Data integration for biological network databases: MetNetDB labeled graph model and graph matching algorithm. Graduate theses and dissertations, Iowa State University 10921, pp 102–107

Li JR, Liu CC, Sun CH, Chen YT (2018) Plant stress RNA-seq Nexus: a stress-specific transcriptome database in plant cells. BMC Genomics 19(1):966. https://doi.org/10.1186/s12864-018-5367-5

Lipiec J, Doussan C, Nosalewicz A, Kondracka K (2013) Effect of drought and heat stresses on plant growth and yield: a review. Int Agrophys 27(4):463–477. https://doi.org/10.2478/intag-2013-0017

Lobell DB, Burke MB, Tebaldi C, Mastrandrea MD, Falcon WP, Naylor RL (2008) Prioritizing climate change adaptation needs for food security in 2030. Science 319(5863):607–610. https://doi.org/10.1126/science.1152339

Magrane M (2011) UniProt knowledgebase: a hub of integrated protein data database. https://doi.org/10.1093/database/bar009

Mamgain A, Roychowdhury R, Tah J (2013) *Alternaria* pathogenicity and its strategic controls. Res J Biol 1:1–9

Mcwilliam H, Valentin F, Goujon M, Li W, Narayanasamy M, Martin J, Miyar T, Lopez R (2009) Web services at the European bioinformatics Institute-2009. Nucleic Acids Res 37(suppl 2):W6–W10. https://doi.org/10.1093/nar/gkp302

Mewes HW, Frishman D, Mayer KF, Münsterkötter M, Noubibou O, Pagel P, Rattei T, Oesterheld M, Ruepp A, Stümpflen V (2006) MIPS: analysis and annotation of proteins from whole genomes in 2005. Nucleic Acids Res 34(suppl 1):D169–D172. https://doi.org/10.1093/nar/gkj148

Mir RR, Zaman-Allah M, Sreenivasulu N, Trethowan R, Varshney RK (2012) Integrated genomics, physiology and breeding approaches for improving drought tolerance in crops. Theor Appl Genet 125(4):625–645. https://doi.org/10.1007/s00122-012-1904-9

Mitchell AL, Attwood TK, Babbitt PC, Blum M, Bork P, Bridge A, Brown SD, Chang H, El-Gebali S, Fraser MI, Gough J, Haft DR, Huang H, Letunic I, Lopez R, Luciani A, Madeira F, Marchler-Bauer A, Mi H, Natale DA, Necci M, Nuka G, Orengo C, Pandurangan AP, Paysan-Lafosse T, Pesseat S, Potter SC, Qureshi MA, Rawlings ND, Redaschi N, Richardson LJ, Rivoire C, Salazar GA, Sangrador-Vegas A, Sigrist CJA, Sillitoe I, Sutton GG, Thanki N, Thomas PD, Tosatto SCE, Yong S, Finn RD (2018) InterPro in 2019: improving coverage, classification and access to protein sequence annotations. Nucleic Acids Res 47(D1):D351–D360. https://doi.org/10.1093/nar/gky1100

Mittler R (2002) Oxidative stress, antioxidants and stress tolerance. Trends Plant Sci 7(9):405–410. https://doi.org/10.1016/S1360-1385(02)02312-9

Mittler R (2006) Abiotic stress, the field environment and stress combination. Trends Plant Sci 11(1):15–19. https://doi.org/10.1016/j.tplants.2005.11.002

Mittler R, Vanderauwera S, Gollery M, Van Breusegem F (2004) Reactive oxygen gene network of plants. Trends Plant Sci 9(10):490–498. https://doi.org/10.1016/j.tplants.2004.08.009

Moraes Filho RM, Menezes AF, Martins LSS (2017) In silico modeling and characterization of phytoparasitic nematodes translationally-controlled tumor proteins. Genet Mol Res 16(3). https://doi.org/10.4238/gmr16039800

Morioka R, Kanaya S, Hirai MY, Yano M, Ogasawara N, Saito K (2007) Predicting state transitions in the transcriptome and metabolome using a linear dynamical system model. BMC Bioinforma 8(1):343. https://doi.org/10.1186/1471-2105-8-343

Mount David W (2001) Bioinformatics–sequence and genome analysis. CSHL, New York, pp 75–85

Mousavi SA, Pouya FM, Ghaffari MR, Mirzaei M, Ghaffari A, Alikhani M, Ghareyazie M, Komatsu S, Haynes PA, Salekdeh GH (2016) PlantPReS: a database for plant proteome response to stress. J Proteome 143:69–72. https://doi.org/10.1016/j.jprot.2016.03.009

Mueller LA, Zhang P, Rhee SY (2003) AraCyc: a biochemical pathway database for *Arabidopsis*. Plant Physiol 132(2):453–460. https://doi.org/10.1104/pp.102.017236

Naika M, Shameer K, Mathew OK, Gowda R, Sowdhamini R (2013) STIFDB2: an updated version of plant stress-responsive transcription factor database with additional stress signals, stress-responsive transcription factor binding sites and stress-responsive genes in *Arabidopsis* and rice. Plant Cell Physiol 54(2):e8. https://doi.org/10.1093/pcp/pcs185

Nawaz M, Iqbal N, Idrees S, Ullah I (2014) DREB1A from *Oryza sativa* var. IR6: homology modelling and molecular docking. Turk J Bot 38(6):1095–1102. https://doi.org/10.3906/bot-1403-45

Needleman SB, Wunsch CD (1970) A general method applicable to the search for similarities in the amino acid sequence of two proteins. J Mol Biol 48(3):443–453. https://doi.org/10.1016/0022-2836(70)90057-4

Nielsen M, Lundegaard C, Lund O, Petersen TN (2010) CPHmodels-3.0 – remote homology modeling using structure guided sequence profiles. Nucleic Acids Res 38(suppl 2):W576–W581. https://doi.org/10.1093/nar/gkq535

Osuna-Cruz CM, Paytuvi-Gallart A, Di Donato A, Sundesha V, Andolfo G, Aiese Cigliano R, Sanseverino W, Ercolano MR (2017) PRGdb 3.0: a comprehensive platform for prediction and analysis of plant disease resistance genes. Nucleic Acids Res 46(D1).D1197–D1201. https://doi.org/10.1093/nar/gkx1119

Pearson WR (2014) BLAST and FASTA similarity searching for multiple sequence alignment. In: Russell D (ed) Multiple sequence alignment methods, Methods molecular biology, vol 1079. Humana Press, Totowa, pp 75–101. https://doi.org/10.1007/978-1-62703-646-7_5

Prabha R, Ghosh I, Singh DP (2011) Plant stress gene database: a collection of plant genes responding to stress condition. ARPN J Sci Technol 1(1):28–31

Priya P, Jain M (2013) RiceSRTFDB: a database of rice transcription factors containing comprehensive expression, cis-regulatory element and mutant information to facilitate gene function analysis. Database 2013:bat027. https://doi.org/10.1093/database/bat027

Raghava GPS (2002) APSSP2: a combination method for protein secondary structure prediction based on neural network and example based learning. CASP5 A-132

Redhu S, Jindal A (2013) Molecular modelling: a new scaffold for drug design. Int J Pharm Pharm Sci 5(Suppl 1):5–8

Reitz M, Sacher O, Tarkhov A, Trümbach D, Gasteiger J (2004) Enabling the exploration of biochemical pathways. Org Biomol Chem 2(22):3226–3237. https://doi.org/10.1039/B410949J

Rodziewicz P, Swarcewicz B, Chmielewska K, Wojakowska A, Stobiecki M (2014) Influence of abiotic stresses on plant proteome and metabolome changes. Acta Physiol Plant 36(1):1–19. https://doi.org/10.1007/s11738-013-1402-y

Rose PW, Prlić A, Altunkaya A, Bi C, Bradley AR, Christie CH, Costanzo LD, Duarte JM, Dutta S, Feng Z, Green RK, Goodsell DS, Hudson B, Kalro T, Lowe R, Peisach E, Randle C, Rose AS, Shao C, Tao Y, Valasatava Y, Voigt M, Westbrook JD, Woo J, Yang H, Young JY, Zardecki C, Berman HM, Burley SK (2016) The RCSB protein data bank: integrative view of protein, gene and 3D structural information. Nucleic Acids Res 45(D1):D271–D281. https://doi.org/10.1093/nar/gkw1000

Rost B, Sander C, Schneider R (1994) PHD-an automatic mail server for protein secondary structure prediction. Bioinformatics 10(1):53–60. https://doi.org/10.1093/bioinformatics/10.1.53

Roychowdhury R (2014) Crop improvement in the era of climate change. IK International Publishing House, New Delhi, p 496

Roychowdhury R, Karmakar J, Karmakar J, Adak MK, Dey N (2013a) Physio-biochemical and microsatellite based profiling of lowland rice (*Oryza sativa* L.) landraces for osmotic stress tolerance. Am J Plant Sci 4(12):52

Roychowdhury R, Abdel Gawwad MR, Banerjee U, Bishnu S, Tah J (2013b) Status, trends and prospects of organic farming in India: a review. J Plant Biol Res 2:38–48

Roychowdhury R, Banherjee U, Slofkova S, Tah J (2013c) Organic farming for crop improvement and sustainable agriculture in the era of climate change. OnLine J Biol Sci 13(2):50–65

Roychowdhury R, Taoutaou A, Hakeem KR, Gawwad MR, Tah J (2014) Molecular marker-assisted technologies for crop improvement. In: Roychowdhury R (ed) Crop improvement in the era of climate change. IK International Publishing House, New Delhi, pp 241–258

Roychowdhury R, Khan MH, Choudhury S (2018) Arsenic in rice: an overview on stress implications, tolerance and mitigation strategies. In: Hasanuzzaman M, Nahar K, Fujita M (eds) Plants under metal and metalloid stress. Springer, Singapore, pp 401–415

Roychowdhury R, Khan MH, Choudhury S (2019) Physiological and molecular responses for metalloid stress in rice – a comprehensive overview. In: Hasanuzzaman M, Fujita M, Nahar K, Biswas J (eds) Advances in rice research for abiotic stress tolerance. Woodhead Publishing/Elsevier, New York, pp 341–369

Sakurai N, Ara T, Ogata Y, Sano R, Ohno T, Sugiyama K, Hiruta A, Yamazaki K, Yano K, Aoki K, Aharoni A, Hamada K, Yokoyama K, Kawamura S, Otsuka H, Tokimatsu T, Kanehisa M, Suzuki H, Saito K, Shibata D (2010) KaPPA-View4: a metabolic pathway database for representation and analysis of correlation networks of gene co-expression and metabolite co-accumulation and omics data. Nucleic Acids Res 39(suppl 1):D677–D684. https://doi.org/10.1093/nar/gkq989

Salomon-Ferrer R, Case DA, Walker RC (2013) An overview of the Amber biomolecular simulation package. Wiley Interdiscip Rev Comput Mol Sci 3(2):198–210. https://doi.org/10.1002/wcms.1121

Schlapfer P, Zhang P, Wang C, Kim T, Banf M, Chae L, Dreher K, Chavali AK, Nilo-Poyanco R, Bernard T, Kahn D, Rhee SY (2017) Genome-wide prediction of metabolic enzymes, pathways and gene clusters in plants. Plant Physiol 173(4):2041–2059. https://doi.org/10.1104/pp.16.01942

Schymkowitz J, Borg J, Stricher F, Nys R, Rousseau F, Serrano L (2005) The FoldX web server: an online force field. Nucleic Acids Res 33(suppl_2):W382–W388. https://doi.org/10.1093/nar/gki387

Shen Y, Maupetit J, Derreumaux P, Tufféry P (2014) Improved PEP-FOLD approach for peptide and miniprotein structure prediction. J Chem Theory Comput 10(10):4745–4758. https://doi.org/10.1021/ct500592m

Shulaev V (2006) Metabolomics technology and bioinformatics. Brief Bioinform 7(2):128–139. https://doi.org/10.1093/bib/bbl012

Shulaev V, Cortes D, Miller G, Mittler R (2008) Metabolomics for plant stress response. Physiol Plant 132(2):199–208. https://doi.org/10.1111/j.1399-3054.2007.01025.x

Smita S, Lenka SK, Katiyar A, Jaiswal P, Preece J, Bansal KC (2011) QlicRice: a web interface for abiotic stress responsive QTL and loci interaction channels in rice. Database 2011:bar037. https://doi.org/10.1093/database/bar037

Smith K (2013) A brief history of NCBI's formation and growth. The NCBI Handbook [Internet], 2nd edn. National Centre for Biotechnology Information (US), Bethesda

Smith TF, Waterman MS (1981) Identification of common molecular subsequence. Mol Biol 147:195–197. https://doi.org/10.1016/0022-2836(81)90087-5

Sobhanian H, Motamed N, Jazii FR, Nakamura T, Komatsu S (2010) Salt stress induced differential proteome and metabolome response in the shoots of *Aeluropus lagopoides* (Poaceae), a halophyte C4 plant. J Proteome Res 9(6):2882–2897. https://doi.org/10.1021/pr900974k

Szklarczyk D, Morris JH, Cook H, Kuhn M, Wyder S, Simonovic M, Santos A, Doncheva NT, Roth A, Bork P, Jensen LJ, von Mering C (2016) The STRING database in 2017: quality-controlled protein–protein association networks, made broadly accessible. Nucleic Acids Res 45(D1):D362–D368. https://doi.org/10.1093/nar/gkw937

Tardieu F (2003) Virtual plants: modelling as a tool for the genomics of tolerance to water deficit. Trends Plant Sci 8(1):9–14. https://doi.org/10.1016/S1360-1385(02)00008-0

Tateno Y, Imanishi T, Miyazaki S, Fukami-Kobayashi K, Saitou N, Sugawara H, Gojobori T (2002) DNA data Bank of Japan (DDBJ) for genome scale research in life science. Nucleic Acids Res 30(1):27–30. https://doi.org/10.1093/nar/30.1.27

Thimm O, Bläsing O, Gibon Y, Nagel A, Meyer S, Krüger P, Selbig J, Müller LA, Rhee SY, Stitt M (2004) MAPMAN: a user-driven tool to display genomics data sets onto diagrams of metabolic pathways and other biological processes. Plant J 37(6):914–939. https://doi.org/10.1111/j.1365-313X.2004.02016.x

Trott O, Olson AJ (2010) AutoDock Vina: improving the speed and accuracy of docking with a new scoring function, efficient optimization, and multithreading. J Comput Chem 31(2):455–461. https://doi.org/10.1002/jcc.21334

Vamsidhar E, Swamy GV, Chitti S, Babu PA, Venkatasatyanarayana G, Raju AD (2010) Screening and docking studies of 266 compounds from 7 plant sources as antihypertensive agents. J Comput Sci Syst Biol 3(1):016–020. https://doi.org/10.4172/jcsb.1000050

Voytas DF, Gao C (2014) Precision genome engineering and agriculture: opportunities and regulatory challenges. PLoS Biol 12(6):e1001877. https://doi.org/10.1371/journal.pbio.1001877

Wanchana S, Thongjuea S, Ulat VJ, Anacleto M, Mauleon R, Conte M, Rouard M, Ruiz M, Krishnamurthy N, Sjolander K, van Hintum T, Bruskiewich RM (2007) The generation challenge Programme comparative plant stress-responsive gene catalogue. Nucleic Acids Res 36(suppl 1):D943–D946. https://doi.org/10.1093/nar/gkm798

Wang L, Guo Z, Zhang Y, Wang Y, Yang G, Yang L, Wang R, Xie Z (2017) Characterization of LhSorP5CS, a gene catalyzing proline synthesis in oriental hybrid lily Sorbonne: molecular modelling and expression analysis. Bot Stud 58(1):10. https://doi.org/10.1186/s40529-017-0163-0

Waterhouse A, Bertoni M, Bienert S, Studer G, Tauriello G, Gumienny R, Heer FT, de Beer TAP, Rempfer C, Bordoli L, Lepore R, Schwede T (2018) SWISS-MODEL: homology modelling of protein structures and complexes. Nucleic Acids Res 46(W1):W296–W303. https://doi.org/10.1093/nar/gky427

Webb B, Sali A (2014) Comparative protein structure modeling using MODELLER. Curr Protoc Bioinforma 47(1):5–6. https://doi.org/10.1002/0471250953.bi0506s47

Weckwerth W, Wenzel K, Fiehn O (2004) Process for the integrated extraction, identification and quantification of metabolites, proteins and RNA to reveal their co-regulation in biochemical networks. Proteomics 4(1):78–83. https://doi.org/10.1002/pmic.200200500

Wu S, Zhang Y (2007) LOMETS: a local meta-threading-server for protein structure prediction. Nucleic Acids Res 35(10):3375–3382. https://doi.org/10.1093/nar/gkm251

Yang JM, Chen CC (2004) GEMDOCK: a generic evolutionary method for molecular docking. Proteins Struct Funct Bioinf 55(2):288–304. https://doi.org/10.1002/prot.20035

Yang J, Yan R, Roy A, Xu D, Poisson J, Zhang Y (2015) The I-TASSER suite: protein structure and function prediction. Nat Methods 12(1):7–8. https://doi.org/10.1038/nmeth.3213

Zhu JK (2016) Abiotic stress signaling and responses in plants. Cell 167(2):313–324. https://doi.org/10.1016/j.cell.2016.08.029

Chapter 26
Nanobiotechnological Applications for Crop Improvement

Rachna Gupta and Parth Malik

Abstract Biotechnological solutions have revamped unpredictable agricultural vulnerabilities affecting the crop yield via numerous intricately woven mechanisms. Increasing population pressure and unregulated climate changes have been the major causes of mounting pressure on natural conditions. The depleting forest cover alongside deteriorating soil texture is the pivotal factor responsible for diminishing crop productivity and yield, forcing the pressure-ridden farmer to employ non-scientific curative measures with an expectation of improving crop harvest. Similarly, continuous planting of the same crop on a particular land area with unregulated pesticide and insecticide usage has depleted the nutrient content that has culminated into large areas of barren land. Biotechnological remedies offer valuable solutions to these multitude of risks, through practices such as crop rotation and newer automated irrigated methods. The emergence of nanotechnology principles has further consolidated the controlling grip of such biotechnological remedies. Recent studies have enabled numerous remedial recourses in this regard, with probe regulated pesticide distribution ensuring the prevention of arbitrary fertilizer loading on the soil. Prior testing of a pesticide formulation can provide adequate knowledge of its distribution potential, paving way for its uniform distribution across the entire vegetation area. Such interventions have not only reduced the chemical burden on soil but also reduced the usage of synthetic chemicals as fertilizers. On a similar basis, the feeding of nanotechnology-based foods to cattle has improved the manure quality and texture whereby betterment in existing crop yields is being exercised through simplistic environment-friendly procedures. With such insights, the present article sheds light on nanotechnology-based solutions to improve the agricultural output and quality.

Keywords Agricultural vulnerabilities · Biotechnology · Nanotechnology · Pesticide distribution · Pesticide formulation

R. Gupta
Department of Biotechnology, Visva-Bharati, Santiniketan, Bolpur, West Bengal, India

P. Malik (✉)
School of Nano Sciences, Central University of Gujarat, Gandhinagar, Gujarat, India

R. Roychowdhury et al. (eds.), *Sustainable Agriculture in the Era of Climate Change*, https://doi.org/10.1007/978-3-030-45669-6_26

1 Introduction

Agricultural activities and their financial contributions have always been at the central stage of the economic balance sheets amongst the different economies of the world. The dependence on agriculture is more significant for the rural population, where agricultural income is sometimes the sheer source of livelihood (Kydd 2002; Sharma 2015). With increasing population, the pressure on available resources, like functional cultivable area and the use of quality seeds, and the functional extent of fertilizers and pesticides have been increasing (Snodgrass and Wallace 1980; Mathur 1999). Consequently, a circumstantial necessity has made it mandatory to introduce technological breakthroughs for obtaining increasing outputs from the available land area. In the last 50 years, a number of technologies have been successfully integrated to uplift agricultural outputs from the economical as well as yield point of view (Kole et al. 1999; Keating et al. 2010; Mahadevan 2003). Though biotechnologies have always remained the forefront in this regard, a number of other sustainable measures have enabled a much better arrest of unpredictable scenarios. A major hurdle of conventional agricultural improvement measures is the controlling extent of their integrated application, such as the use of fertilizers and pesticides (including weedicides and insecticides) and regulating the dosage frequency at varying intervals of crop growth (Roychowdhury et al. 2013a, b). Though theoretical understanding of such probes creates reasonable reliability, the practical implementation remains a challenge as there are various factors which govern the accessible expressive mode of such measures. For example, the mere availability of a good pesticide (having broad functioning performance) is not sufficient since farmers with a lack of scientific understanding regarding its usage most likely encounter the menace of chemical clogging (Gupta et al. 1984; Forget 1993). This will not only result in economic loss to a farmer but will also mount the future uncertainties as the land area may become non-functional and viable for a long time. In such scenarios, it becomes dire essential to deliver the stimulators as and when required, in the needful proportions, and ensuring that a homogeneous distribution on the land area is available. To address such concerns, we need to have sharper tools with dual attributes of sensing and homogeneous delivery (Perlatti et al. 2013; Collins et al. 1973; Garrido-Herrera et al. 2009).

Apart from the above-mentioned aspects, a number of conventional techniques bring about the improvement through chemical actions whereby the deterioration risk of natural soil quality becomes high. So technologies need to be more bio-friendly and eco-friendly which would enable the exercised remedial measures to manifest as habitual tendencies of the concerned land area. Such cautions would enhance the longevity of implemented solutions and will be a boost to naturally enhance the response, rather than only till the external monitoring. The emergence of nanoscale technologies has revolutionized the implementation of the multiplying tapping controls, substantially attributed to their precise control and regulating extents (Fig. 26.1). Amongst the several incentives, the foremost is the sensing advance where the levels of even till 10^{-12} units are now being

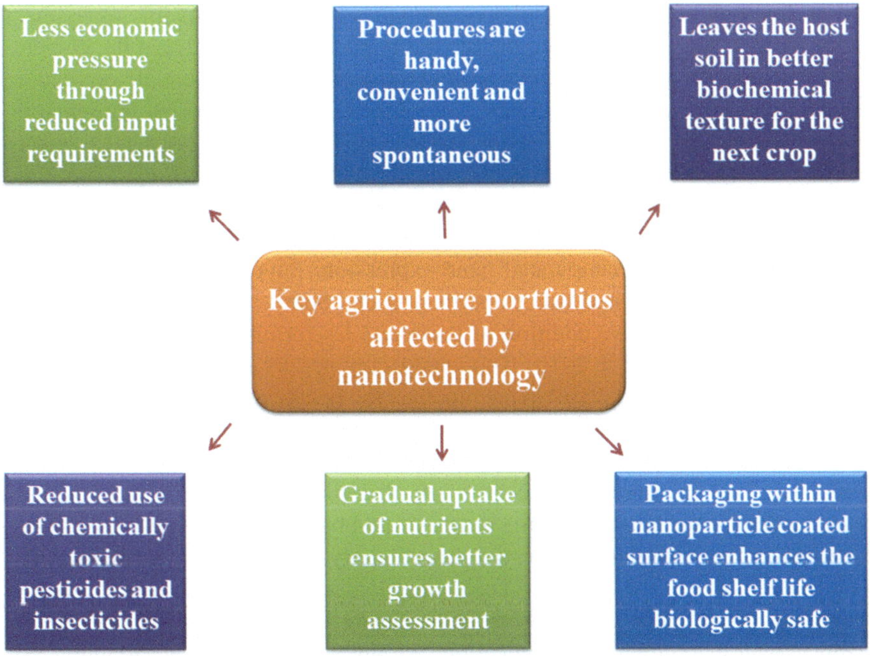

Fig. 26.1 The key agricultural activities that could be revolutionized using nanotechnological tools and techniques. Reduced material usage, quick detection of crop deteriorating pathogens and growth monitoring agents, with gradual fertilizer and pesticide delivery, are some of the crucial breakthroughs

estimated (Devreese 2007; Patolsky and Lieber 2005). Undoubtedly, the larger surface area of nanomaterials, particularly the various compatible nanoparticles (NPs) (prepared using non-chemical routes), makes it feasible to deliver simultaneous probes in the form of eco-friendly microbial species. For instance, the scenario where NPs and genetically modified organisms (suited to peculiar soil chemistry) are delivered together or feeding the crops with residual water streams whose chemical load is tolerable along with variable population of bioactive NPs could be the breakthrough solutions to augment the soil chemistry of a particular region (Oliver 2014).

Numerous NPs prepared from microbes are being used as catalysts to restore and increase the organic content in a particular soil area, which makes soil fit for all crops. The particular NPs used for this purpose include Au and Ag, owing to their least chemical reactivity along with the robust synthesis methods, requiring minimal chemical tediousness. Nanoscale probes have enabled contaminant detection even at the lowest possible levels so that the land area sustainability is not lost. Advanced measures of remote sensing (through geographical information systems [GIS]) have augmented the predictive assessments to select the suitability of a land area with respect to a specific crop. These advances are being used by developed economies to multiply their agricultural outputs with higher accuracy and in much lower times than conventional techniques. Similarly, the integration of nanotechnology to molecular biology techniques such as seed quality assessment and genetic testing has increased the reliability standards and reduced the vulnerabilities in the implemented solutions. Despite enormous potential, continuing research on agricultural nanotechnology over the past decade has left the scientific community amidst several tricky uncertainties. The clear-cut thrust areas where nanotechnology made its mark in the agricultural sector are increasing global food security and unpredictable climatic dilemmas. Till date, application of nanotechnology in agriculture has focused on reducing the use of deleterious chemicals, enhancing the yield through manipulated gene expression, improving the irrigation via homogeneous fertilizer and water distribution and better measures of soil quality improvement through live addition of nanoparticles forming genetically modified microbes (Maruyama et al. 2016). Though the commercial application of any major nanotechnology solution is still in the pipeline, the results of nanoscale control have been fabulous. This has not only reduced the intake of raw material but has also provided amicable solutions with reproducible and predictable benefits, in much lesser time, not to forget the incentive of allocating manpower to more serious issues, whereby sustainable income sources could be strengthened.

A wide range of agricultural products are already in the market and are booming well, particularly in the food and healthcare domain (Table 26.1) (the Project on Emerging Nanotechnologies (webpage); Huang 2012; Qureshi et al. 2012; Nanotechnology products and applications. Product: Guard IN Fresh. Honolulu, HI; Nanowerk; Nanotechnology products and applications. Product: NanoCeram-PAC. Honolulu, HI: Nanowerk; Miller and Senjen 2008; Hamad et al. 2018; Kalia and Parshad 2015). With such encouragements and multiple expectations, this chapter sheds light on the application of nanobiotechnological measures to the current agriculture techniques to improve the process as well as product quality.

2 Fascinations Behind the Expectations from Nanobiotechnology

Agricultural quality monitoring has been substantially dominated by biotechnological solutions till date, encompassing improvement in the seed quality, betterment in irrigation techniques and keeping the harvested crop in proper storage conditions before it is sent to the market. A lot has been made possible through controlled gene expression and exercising desired enzyme activity, which are the initial stages of any agricultural crop development process. The nanomaterials prove to have value-added benefits in this context, since their restricted dimensions (usually <500 nm) enable their robust accommodation so that larger functionalities could be integrated compared to conventional measures. Complementary to this size quality is the manifestation of quantum confinement effects, owing to which the salient material properties, ranging from optical, mechanical, electronic, chemical and electromechanical, are significantly altered. The foremost reason for such size dependent material

Table 26.1 An overview of nanotechnology-conceptualized commercial agriculture and food crop plants. The products offer numerous advantages, ranging from lesser input needed to controlling the delivery of the loaded active component

Product name	Marketed under	Specialty/purpose	Reference
Canola active oil	Shemen Industries, Tel Aviv, Israel	Comprises additive "nanodrops" optimized to transport vitamins, minerals and phytochemicals through the digestive system and urea	Project on Emerging Nanotechnologies (webpage), 2013
Nano tea	Qinhuangdao Taiji Ring Nano-Products Co., Ltd., Hebei, People's Republic of China	Capable of releasing all tea essences, boosting the adsorption (adsorbing viruses, free radicals, cholesterol and blood fat) and annihilation of viruses through penetration	Huang (2012)
Nanoceuticals slim shake chocolate	RBC Life Sciences Inc., Irving, TX, USA	Comprises cocoa-infused "nanoclusters" to enhance taste and health benefits of cocoa without adding extra sugar	Qureshi et al. (2012)
Guard IN fresh	Fayetteville, AR, USA	Delays ripening of perishable foods and floral products through scavenging the ethylene gas	Nanotechnology products and applications. Product: Guard IN fresh, Honolulu, HI; Nanowerk, 2014
TopScreen DS13	TopChim, Wommelgem, Belgium	Recyclable water-based coating comprising of biopolymer facilitated monodispersed NPs to replace hard to recycle wax-based coatings and reduce any negative impact on food packaging biodegradability	Nanotechnology products and applications. Product: NanoCeram-PAC, Honolulu, HI; Nanowerk, 2014
NanoCeram-PAC	The Aquarian Environmental Group Pty Ltd., Sydney (Australia)	Substantial external surface area facilitates rapid absorption of soluble contaminants developing undesired taste and odour	Miller and Senjen (2008)
Food contact material	Nanosilver baby mug comprising of Ag NPs	Ag NPs confer better protection against bacterial contamination	Hamad et al. (2018) and Kalia and Parshad (2015)
Food packaging material	Durethan KU 2-2601 plastic wrapping, Bayer	Silica NPs embedded in a polymer-based nanocomposite prevent oxidative degradation of the product	Hamad et al. (2018) and Kalia and Parshad (2015)
Nutritional drink	Oat chocolate nutritional drink mix, toddler health, SunActive Fe	300 nm Fe particles ensure an improved bioavailability	Hamad et al. 2018 and Kalia and Parshad (2015)

properties is the closer spacing of discrete molecular energy levels as the dimensions progressively approach the nanoscale. Typically, this leaves us with the ultimate picture of having an exposed larger surface area, where multiple probes could be integrated. Such attributes confer multifunctional abilities to nanomaterials, making way for nanoparticle(s) conjugated with a dye, serving its native function as well as performing sensibly. So the foremost advantage of nanotechnology is the reduced material usage and the ability to perform multiple jobs with more precision than conventional materials. The next argument for preferential nanomaterial usage is the better integration of nanotechnology with several biotechnological practices being associated with agricultural output monitoring. For example, the use of enzyme-based sensors could be enhanced in its accuracy through the incorporation of metallic NPs, which would increase the precision limit and therefore make the detection more accurate. A detailed look on the diversity of differently abled nano-biosensors can be had in one of our earlier contributions (Malik et al. 2013). Similarly, fitting of nanoscale controllers along with dosage monitoring mechanisms could be exercised in conventional water and pesticide sprinklers, since the crops do not require the same input of water and pesticides in their entire growth process. Low toxicities of NPs alongside their robust synthesis routes from plants and even microbial species are the specialties for their robust implementation. For a detailed look of biological attributes of the plant and microbial synthesised NPs, readers are yet again advised to consult our earlier contributions (Malik et al. 2014; Gupta et al. 2019; Roy 2017). With such insights, the section ahead focuses on the recent progress in getting enhanced agricultural output through integrating applications of nanomaterials and nanodevices.

3 Recent Progress in Nanotechnology-Based Diversified Agricultural Betterments'

After the green revolution, there has not been any major technological intervention to uplift the agricultural outputs and the income generated thereof. The aftermaths of the green revolution left the soils in most of the well-cultivable locations across the globe overloaded with residual pesticide and chemical effects. Conventional methods of agricultural improvement still rely much on the increased use of fertilizers and pesticides, leaving the soil exposed to the chemical vulnerability which drastically affects its organic content. This undesired chemical loading of soil not only affects the existing crop quality but also leads to augmented risks in aquatic environments, via eutrophication (Kalwasińska et al. 2011; Lew et al. 2013). Several studies discuss the potential risks of different metal- and metal oxide-based NPs, amongst which Ag, CuO and CeO_2 do pose a significant concern with regard to their polluting tendencies. The higher toxicity of Ag than Au seems to be due to its higher reactivity, which also imparts its substantially stronger antifungal attributes. Interestingly, CuO NPs have been recurrently proposed as toxic, affecting crop germination and growth from as low as 0.1 to 600 mg/L concentrations (Atha et al. 2012; Nair et al. 2014; Moon et al. 2014; Corral-Diaz et al. 2014; Saha and Dutta 2017). Such indications do pose a concern for NP usage but exact generalization about all NP kinds is relatively impractical. This is so as several other factors, like soil composition, geography, microbial content, organic content, water retention ability, pH and aggregation-promoting/aggregation-inhibiting response towards NPs, remain critical for inducing the toxic response of NPs. Thereby, it is impulsively desired to regulate the requirement driven fertilizer and pesticide delivery so that the concurrent damages could be arrested before acquiring irreversible extent.

3.1 Nanotechnology-Enabled Pesticide and Fertilizer Delivery

The emergence of nanoscale formulations having pesticides and fertilizers in dissolved form has emerged as a reliable solution to this problem. The nanoemulsions enable uniform distribution of pesticide and fertilizer droplets, owing to which much lower concentration of chemicals goes to the soil being treated. Nanodispersions carrying nano-encapsulated pesticides enable their controlled release at the desired sites, alongside preventing the premature degradation and reactivity of carrier molecules. This would enable a much lesser amount of pesticides being used (which eases the economic pressure) and also ensure its effective usage at the desired rate, thereby causing little damage to the soil texture (Chen et al. 2011). A systematic understanding of nanofertilizers is provided by Kah et al., categorized into three broad categories, namely, (a) nanomaterials comprising of macronutrients, (b) nanomaterials comprising of micronutrients and (c) nanomaterials acting as carriers of macronutrients (Kah 2015). Unlike the first two categories, the third category does not use nanomaterials as nutrients, but as additives. The added nanomaterials could either be NPs or any other nanostructured material, where popular examples of three categories are hydroxyapatite, layered double hydroxides intercalated with phosphate ions and ZnO NPs (Koilraj and Kannan 2010, Novillo et al. 2014, Ittekhar et al. 2018). The products in the third category generally comprise a wide variety of materials that are like nutrient-loaded zeolites or some materials that are principally not considered as nanoscale material (biochar) (Servin et al. 2017). Several carrier systems using nanomaterials have been developed that facilitate effective delivery of active ingredients having insecticidal, fungicidal or herbicidal properties. Materials as diverse as silica NPs, carbon nanotubes, graphene oxide, solid-lipid NPs and polymers are being increasingly applied for nanoscale control in the delivery of pesticides. Amongst the inorganic nanomaterials, Cu NPs are widely used for their antifungal applications (Pyrzynska 2011; Rastogi et al. 2019; Beltrán-Partida et al. 2019).

3.2 NPs for Faster and Robust Sensing

The second potential area having witnessed the distinction of using nanomaterials is the application of NPs and nanosystems for monitoring the seed quality (such as the preharvest measure) as well as monitoring the standards of subsequent agricultural practices. The use of ultrasensitive NPs, such as Au, Ag, silica, Fe_3O_4 and several others, has enhanced the accuracy of sensing significantly compared to the conventional enzyme-controlled measures. There are two potential advantages of using these NPs: first is that they can be made using a variety of methods and in different sizes and geometry and second is that the quantum confinement effect in these entities makes them capable of detecting even a little attenuation in their SPR frequencies, owing to which even the minute-level changes in the sensed environments could be detected. The different shapes and geometries of these NPs are extremely handy incentives in some practical scale-up operations. For example, nanorods have an elongated surface, so they have the ability to detect pathogens or harmful stimulus over a large volume. Similarly, NPs of Au can be conjugated with quantum dots that detect foreign substances (even at picometer levels) through variations in their fluorescence properties. Likewise, where there is a risk of magnetically responsive contaminations, the magnetic NPs could be used. The key advantages with respect to conventional sensing methods are the detection limit (which is usually highly sensitive) and the efficacy of detection since higher chemical reactivity and larger surface areas enhance the binding activities of detection agents.

The distinct performance of nanomaterial conferred sensing could be viewed in a 2006 study, proposing the utility of acidic polystyrene microparticle-based fluorescent sensor for detection of *Staphylococcus aureus* enterotoxin B (SEB), a highly thermoresistant toxin, contaminating drinking water and milk samples. The sensing system comprised of polystyrene microparticles labelled with fluorescein isothiocyanate (FITC) and anti-SEB. Following SEB binding, the variations in fluorescence pattern of FITC were noted as estimators of SEB concentration, which was noted as low as 0.125 ng/mL in drinking water and 0.5 ng/mL in milk. So if microparticles can detect such limits of pathogens, extending the probe with NPs would replicate the detection limit to still lower limits (Medina 2006). Some other breakthrough sensing systems developed using Au, Ag NPs, CNTs, and quantum dots, operational either alone or in combination mode, are listed in Table 26.2 (Schofield et al. 2007; Chien et al. 2008; Radoi et al. 2008; Pathak et al. 2001; Taton et al. 2000; Jean et al. 2010; Liu et al. 2014; Han et al. 2015; Wang et al. 2008; Zhang et al. 2010; Li et al. 2015; Singh et al. 2012). These studies speak volumes about the extraordinary precision level of nanomaterials through their distinguished surface properties, thereby making the diagnosis not only faster but more accurate.

Table 26.2 Sensing distinctions attained by different nanomaterials, highlighting precision and detectable damage modes on reducing limits

NP type	Special attribute	Major application accomplished	Reference
Gold	Robust surface functionalization, detection of little quantities via spectrophotometric changes	Ten-minute detection of cholera toxin through GM1 ganglioside ECM terminal moiety recognition, Shiga-like toxin via globotriose conjugation and toxin B subunit interaction	Schofield et al. (2007) and Chien et al. (2008)
Iron oxide	Modulation of shape and magnetic response via controlling polymer addition time, variation of temperature and using specified capping agents	Used to detect and quantify aflatoxin from milk samples, following conjugation with membrane antibodies	Radoi et al. (2008)
QDs	Size-dependent optical excitation and relaxation, highly bright and extremely photostable	In vitro and in vivo detection of cancers, detection of Y-chromosome in fixed human sperm cells, detection of genetic diseases through low-target DNA concentrations	Pathak et al. (2001) and Taton et al. (2000)
Silica NPs	Core-shell structure enables higher stability, accuracy and sensitivity, ability of being used in combination with a range of other NPs	Detection of melamine when used in combination with Ag nanospheres, fluorescent detection of Cu NPs in tap water in combination with carbon dots	Jean et al. (2010) and Liu et al. (2014)
CNTs	Elongated sensing possibilities through a rod-shaped structure, tuneable geometry, extraordinary strength and electromechanical responses	Electrochemical detection of bisphenol A, organophosphorus pesticides, hydrazine and nitrides, calorimetric detection of melamine, ochratoxin A, mercury and silver	Han et al. (2015), Wang et al. (2008), Zhang et al. (2010), and Li et al. (2015)
ZnO NPs	Robust functionality, flexible design and surface modulation, range of synthesis options	Large-scale sensing of nitrogen dioxide, ability to sense free radicals through its native antioxidant traits	Singh et al. (2012)

3.3 *Nanochips, Nanoarrays and Nano-Barcoding*

The terminologies "chips, arrays and barcodes" are basically the molecular platforms for sensing a biomolecule, be it DNA, a live organism as a contaminant or any other. The prefix nano, with each methodology, infers the use of nanomaterials that range from a multitude of NPs to quantum dots and hybrid nanomaterials comprising of NPs immobilized on thin and flexible supports. The whole story revolves around the higher surface area of nanomaterials, which harbours the potential of storing greater information compared to conventional materials. The lower dimensions of nanomaterials with their faster electromechanical response (triggered by their constricted energy levels) not only enhance the detection level but also occupy much lesser space alongside enabling flexible handling and control. With nanoscale dimensions, multiple analysis could be done at the same time as compared to the conventional array and chip technology. Biochips are devices which have cross-linked networks on their surface that could be immobilized with sensing probes or curative elements for any internal injury. Though studies with nanochip development are presently restricted to animal models, they have established a reliable method of gene replacement using an electric field to deliver specific genes to the tissues underlying the skin layer (Kricka 2000). The replication of this technology to the seeds of crop plants could enable the seeds with desired gene expression thereby minimizing the productivity uncertainty and could even be used to test new and more efficient stress-tolerant species. Similarly, barcodes are the assembly of parallel devices containing information about any product (anything in the grocery shop or medical dispensary). The use of nanomaterials, having high aspect ratios, could revolutionize this technology by occupying less space and containing much higher information. Storing information on a nanomaterial will obviously provide the benefit of having greater information within a small space. The functioning of these materials is critically affected by the encoding stimulus, which varies according to the intended purpose. The stimulus used for coding varies from being fluorescently sensitive, optically active, magnetically active or even heat sensitive. Each code, therefore, could be tracked only in a specific manner, thereby providing exclusive information storage for systematic record maintenance. Tagging such specific labels to seeds having unique attributes in any of the growth features could facilitate a predictable study of combining differing features (Shikha et al. 2017; Valentini et al. 2017). For example, stress-tolerant species could be tagged using antibodies while species tolerant to salinity could be selected with a marker delivering more salt. In this way, the selection of seeds with desired features could be fastened and crop productivity could be improved, irrespective of the geographical conditions. Though such conceptualizations are in the research phase, they have proved effective on lab-scale samples. To replicate these performances to the agricultural fields, several factors need to be optimized, the foremost of which is an unpredictable climate. In this context, modelling techniques exercising controls on the varying parameters could be helpful. Trials for commercialization are in progress but limited to developed countries only.

3.4 Sustainable Use of Agrochemicals Using Nanoplatforms

Apart from pesticides and fertilizers, a number of other chemicals are also needed to regulate crop growth at different stages. These chemicals could be the growth factors, microbial proteins, materials enriching the organic content of a particular soil, detoxifying bacteria and other microbes. Although the impacts of adding such materials using nanocarriers have still not been commercialized, on the contrary the studies suggest that with nanoscale devices and control features, 20–30% gain in crop productivity could be attained. A substantial concern in this regard arises from the concern that whether this gain is considerable with respect to the expenditure incurred in implementing the nanocarrier-mediated delivery of these chemicals. Studies focused on this assessment reveal a gloomy picture, with multiple trials facing obstructions due to the scarcity of public funding (Mukhopadhyay 2014; Report of the National Nanotechnology Initiative Workshop; Arlington, VA, USA 2009). Furthermore, though results on a small scale have been promising, establishing a clear picture from the viewpoint of a large agricultural field has proved to be rather impractical. This is because the crops grown in a field are always a witness of so many uncontrolled and unpreventable stresses and the conditions in the laboratory are significantly different. So attaining the nanoscale benefits of nanocarrier-delivered agrochemicals involves the optimization of several factors owing to which results are not yet being promisingly replicated. Oftentimes, the modified protocols could not be tested due to the ethical concerns being imposed on the food crops, enlightening the societal concern.

4 Useful Nanomaterials with Distinguished Mechanisms

Although reports on any major improvement in crop yield and quality of harvest are barely minimal, the laboratory-scale or research-level efforts frequently demonstrate nearly similar nanomaterials for obtaining improved crop response. The nanomaterials finding favour for such applications include inert NPs (substantially, Au and Ag), carbon nanotubes, quantum dots, nanorods and nanosuspensions. All these nanosystems are characterized by unique structural and functional attributes, making them suitable for faster and robust sensing, ability to be diversely functionalized and their inherent flexible nature (soft boundaries of nanosuspensions). The text ahead describes the unique features of several NPs and their integrated systems, describing the differences and key benefits of their singular and integrated application (Fig. 26.2).

4.1 Au and Ag NPs

NPs are primarily entities having sizes less than 100 nm. This size limit is not clearly defined and standardized, and therefore till 500 nm, several nanoscale properties are exhibited in variable extents (Babick et al. 2016). The Au NPs are one of the most widely used nanomaterials for faster and more efficient sensing because of three major reasons. The first is the availability of a wide variety of synthesis methods, utilizing low energy requirements from the external end. Preparation of Au NPs using plants, microbes and sodium citrate (burst method) could be traced in several eminent publications (Menon et al. 2017; Kimling et al. 2006). Such robust synthesis methodologies make the preparation of Au NPs an eco-friendly and inexpensive

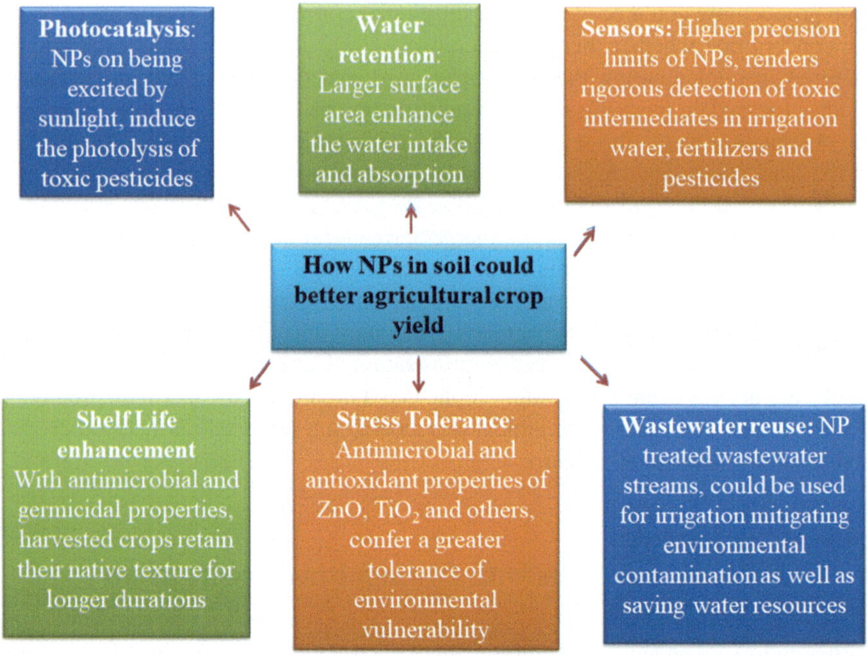

Fig. 26.2 The specific advantages of supplementing NPs to agriculturally lucid soils, projecting the usefulness of metal and metal oxide NP robust responses

process, which is the prime reason for their increasing multiple applications. The second reason for preferential Au NP usage is the availability of different shapes and geometries, which remains significantly important as properties at nanoscale remain implicit functions of their size and shape. At the same time, the elongated rod shape is capable of some exclusive functions which the spherical shape cannot perform. For sensing purposes, generally spherical or rod shapes are employed, which exhibit a characteristic plasmon resonance peak due to the coherent existence of free electrons in the conduction band. Interaction or binding with any stimulus at or adjacent to the surface results in changes in the SPR peak of the native nanomaterial and a proportionate change in any of the size-dependent characteristic properties. The third reason for increasing preference for Au NPs is the low reactivity of Au, making the development of conjugated assays easier, simpler and easily controllable (Zhao et al. 2008). Similar to Au, Ag NPs can also be prepared in various shapes and geometries; through changes in a precursor to reducing agent stoichiometry, the triangular, spherical and rod shapes are well known. The capability of being available in so many diverse shapes with easier and greener synthesis methods is the reason for their highly precise sensing applications in that a stimulus <1 nm could be easily detected. These particles are not only fitted to function as improved sensing agents but these can also be fed directly to the soil, where the nutrient replenishment via degradation of texturally and compositionally complex chemical substances can be fastened. Strategies with combined delivery of genetically modified microorganisms and these NPs are in active consideration, since many bacterial species can metabolize these NPs, and thereafter, the enzymatic controls in their body could be highly potent biocatalysts for enhancing soil fertility (Pallavi et al. 2016).

4.2 Carbon Nanotubes (CNTs)

One of the well-understood nanostructures, CNTs, prevails in cylindrical morphology, with two distinctly characterized forms, known as single-walled and multi-walled. Multiwalled nanotubes (MWNTs) are more common and readily prepared but single-walled nanotubes (SWNTs) are isolated from MWNTs following purification. The advantage with these nanostructures is that their cylindrical and elongated structure allows for multiple complementary binding sites, making the sensing quicker and more robust. Additionally, their compatibility with the carbon skeleton (an inherent constituent of these materials) makes them feasible for multiple functionalizations, so there is always an incentive of preparing need-based sensing probes. Recently, many studies have demonstrated a preferential uptake of these nanostructures at varying time intervals of the identification of damage or a troubled metabolic abnormality (Kobayashi et al. 2017; Liu et al. 2013). Since cylindrical nanostructures can stay longer within the physiological boundaries, so a number of modifications are being rapidly pursued to prolong their physiological existence. Such controls have facilitated the delivery of drugs in a time-phased manner, allowing for systemic regulation of disease cure analysis. Apart from their elongated structure, the arrangement of carbon atoms in these structures accomplishes them a conducting and semiconducting behaviour, allowing for simultaneous sensing possibility. The only concern behind using these nanomaterials is the restriction of their vulnerable toxic responses, whereby the risk of cross-reactivity could reach an uncontrollable extent. With recent advances and understanding of functionalization, better structural and morphological controls are being exercised to attune the biocompatibility of these materials. For multidisciplinary biological advances of CNTs, readers are suggested to have a look at numerous resourceful literature contributions (Kumar et al. 2017; Bekyarova et al. 2005; Schnorr and Swager 2011).

4.3 Quantum Dots (QDs)

After NPs and CNTs, QDs are the next nanomaterials which are theoretically well understood. These nanomaterials restrict their constituent free electrons and constituent atoms in all three dimensions, which makes it possible to perform multiple analyses at the same time, thereby saving overall resources of materials. A range of compositions is documented for the diversely studied QDs, each working on its own specific fluorescence sensitivity (Rosi and Mirkin 2005). For example, the

CdSe-based QDs are not preferred for delivering drugs due to Cd and Se toxicity, whereas QDs made up of only Si or P are more readily used for drug delivery. The three-dimensional quantum confinements (of charge carriers) in these nanomaterials allow a higher size control, paving the way for even slight changes in the fluorescence intensities. Once a specific toxin or harmful species binds to the QD-based sensing probe, its accurate estimation is facilitated through changes in the fluorescence intensity. Nowadays, faster diagnosis applications are being conceptualized via tagging fluorescent active biocompatible dyes with QDs and using the entire assembly as sensing probe. Higher photosensitivity of QDs along with the smaller dimensions enables their channelling into plant roots, where seed germination and nutrient absorption processes could be regulated via photocatalytic attributes (Bakalova et al. 2004; Das et al. 2015). The only concern regarding the QD use relates to their unpredictable metabolic responses, which could result in the enhancement of oxidative stress via generation of free radicals (Liu et al. 2011). Therefore, regulatory cautions evaluating proper monitoring of working efficacy external to the cellular environment should be practised to minimize unpredictable vulnerabilities.

4.4 Nanoemulsions and Nanosuspensions

As also detailed in the section of improved pesticide and fertilizer delivery, nanoemulsions are fluidic systems that allow controlled expression of a particular compound or species, through its sustained interactions. These systems are ideal carriers for a progressively controlled expression of organic compounds which are, otherwise, water-insoluble, remain poorly absorbed and could result in undesired chemical toxicities. With potential substitutes of surfactant-like molecules, the nanoemulsions prepared using non-ionic surfactants are highly suitable for drug and nutrient delivery to remote locations. The advantage of using non-ionic surfactants is that these systems do not require pH optimization before practical implementation and the toxicity of these systems is much lower, owing to the low chemical reactivity of non-ionic surfactants than their ionic counterparts. Nowadays, amino acids, protein-derived secondary biomolecules and plant metabolites have ably replaced the use of ionic surfactants (Mcclements et al. 2007; McClements 2004). The inclusion of such materials has not only reduced the toxicity associated with conventional surfactants but also reduced the energy required to make the nanoemulsions. Enhanced expression of pesticides, fungicides or any other complex fertilizer could be engineered through increased kinetic stability of nanoemulsions, allowing enhanced Brownian motions and interactions mediated through multiple binding sites. The implicit advantage of using such systems is the effective delivery of intended compounds at very low concentrations compared to conventional systems, which is highly instrumental to control the toxic responses. An alternative terminology of nanoemulsions is nanosuspension or microemulsion, differing on the basis of particle sizes and the use of external energy (for intended thermodynamic stability), with all other functional activities remaining the same. It is obvious that the smaller the particle size or the higher the distribution, the greater will be the nullification of gravitational effects or coalescence-related turbulence. So the smaller size of emulsions (in nanoemulsions) allows for greater enhancement in the dispersed phase chemical expression. Some nanoemulsion models optimized to deliver pesticides and nutrients in crop plants are listed in Table 26.3, where pH and stoichiometric composition of constituent phases play a critical role in attaining the homogeneous expression of the dispersed phase (Jiang et al. 2011; Kumar et al. 2004; Yang et al. 2009; Casanova et al. 2005; Wang et al. 2007; Wilson et al. US Patent Number 2011/0052654 A1; Latheef et al. 1993; Arthur 1999; Takei et al. 2008).

4.5 ZnO, Fe_3O_4 and TiO_2 NPs

Apart from the above-mentioned nanomaterials, ZnO, TiO_2 and Fe_3O_4 NPs have been well studied for their antioxidant and antimicrobial attributes, which jointly provide optimum growth conditions to various crop plants. For example, deficiency of Zn is the most common micronutrient adversity affecting the crop yield in alkaline soils. Soils that are too alkaline (due to aggravated $CaCO_3$ levels) often hinder the right Zn availability to crop plants due to the interference caused by high pH and $CaCO_3$-initiated Zn absorption and precipitation. In such circumstances, the use of ZnO NPs can help increase Zn bioavailability to crop plants, as compared to micron- or millimetre sized Zn particles in conventional Zn fertilizers (ZnO and $ZnSO_4$). A number of leaf extracts have been used to synthesize ZnO NPs, providing remarkable control on size limits through adjusting the precursor-reducing agent stoichiometries. For example, *Moringa oleifera* leaf extract has been used to obtain 16–20-nm ZnO NPs that have further provided antibacterial and antifungal responses towards *Staphylococcus aureus*, *Bacillus subtilis*, *Pseudomonas aeruginosa*, *Proteus mirabilis*, *Escherichia coli* and *Candida albicans* and *Candida tropicalis* species. Maximum activity was noted against *Staphylococcus aureus* (Elumalai et al. 2015). Similarly, *Parthenium* leaf extracts have been used to prepare spherical and hexagonal ZnO NPs of up to 32- and 86-nm sizes, which subsequently showed varying antifungal responses towards *Aspergillus flavus* and *Aspergillus niger* (Rajiv et al. 2013). Spherical ZnO NPs, sized between 23 and 57 nm using zinc acetate and sodium hydroxide, have been evaluated against *Escherichia coli*, *Pseudomonas aeruginosa* (ATCC 15442), *Staphylococcus aureus* (ATCC 6538) and *Bacillus thuringiensis* (ATCC 10792), where all, except *Pseudomonas aeruginosa*, showed resistance to ZnO NPs.

Like ZnO, TiO_2 is a widely used photocatalyst, where Ti is known to catalyze the production of carbohydrates that increases the rate of photosynthesis and growth (Owolade et al. 2008; Khodakovskaya and Lahiani 2014; Chen et al. 2014). The photocatalytic attributes of TiO_2 have aided in its pesticide degradation and also towards plant protection as TiO_2 does not form any toxic and dangerous compounds

Table 26.3 Commercialized nanoemulsion- and nanoparticle-based pesticide/herbicide/insecticide delivery systems. The option of varying constitutional stoichiometries offers considerable advantages to control textural properties of encapsulated compounds, to minimize their adverse reactivity

Compound delivered	Key regulatory activity	Major advance notified	Reference
Nanoemulsions			
Glyphosate (herbicide)	Broad-spectrum systemic herbicide and crop desiccant	Reduced soil and water pollution risks, better control for specific crop type	Jiang et al. (2011)
Imidacloprid (insecticide)	Controls sucking insects, termites, selective soil insects and fleas on pets	Delivery through nanoemulsions has enhanced applicability range through reduced cytotoxicity	Kumar et al. (2004)
Polyethylene glycol (PEG) NPs loaded with garlic oil via melt dispersion method (insecticide)	Antibacterial and antibiotic characteristics	>80% encapsulation efficacy over a 5-month period, slow and controlled release of oil, effective for stored products	Yang et al. (2009)
Nicotine carboxylate (insecticide)	Modulation of oxidative balance	Monomodal size distribution, increased bioactivity with decrease in the fatty acid chain length	Casanova et al. (2005)
β-Cypermethrin (broad-spectrum insecticide)	Neurodegeneration, reproductive failure, dermal and ophthalmologic toxicity	30 nm droplets using poly(oxyethylene) lauryl ether and methyl decanoate, increased bioefficacy and dissolution	Wang et al. (2007)
Nanoparticles			
Diazinon (organophosphate insecticide)	Inhibition of neurotransmitter activity	Poly (n-alkyl acrylate)-stabilized temperature-sensitive microcapsules; encapsulated formulation exhibited nearly 90% insect mortality over 8 weeks	Wilson et al. US patent number 2011/0052654 A1
Sulprofos (insecticide)	Inhibition of neurotransmitter activity	Ethyl cellulose-stabilized formulations showing good results against eggs and larvae of the tobacco budworm *Heliothis virescens* in cotton plants	Latheef et al. (1993)
Cyfluthrin (insecticide and pesticide)	Induces muscle weakness, shortness of breath, headache, nausea and seizures	Controlled release and long-term action towards rice weevil *Sitophilus oryzae* over an 8-month period	Arthur (1999)
Acetamiprid (an alkaline and high-temperature-sensitive insecticide)	Nicotinic activity against acetylcholine receptors	Diffusion-controlled 2–20-μm-sized microcapsules with improved thermal degradation, controlled liberation for up to 10 weeks	Takei et al. (2008)

(Pelaez et al. 2012). TiO_2 NPs prepared using *Psidium guajava* leaf extract (32.58 nm in size) have been studied against *Aeromonas hydrophila* (MTCC-1739), *Proteus mirabilis* (MTCC-442), *Escherichia coli* (MTCC-1677), *Staphylococcus aureus* (MTCC-3160) and *Pseudomonas aeruginosa* (MTCC-4030), where highest activities were noted against *Staphylococcus aureus* and *Escherichia coli*, at 20 μg/mL. The synthesized NPs showed higher antioxidant activity than ascorbic acid and also superseded the antibacterial activity of tetracycline (Santhoshkumar et al. 2014). Besides ZnO and TiO_2, delivery of iron oxide NPs is also being investigated to tackle iron deficiency in high pH and calcareous soils. Several studies document the effect of spraying iron oxide NPs on wheat growth, yield and quality. Improvements in spike and grain weight and biological and grain yield alongside grain protein content were noted altogether (Bakhtiari et al. 2015). Enhanced chlorophyll contents were noted in the subapical soybean leaves under the greenhouse test and hydroponic conditions, upon being subjected to a low concentration of superparamagnetic iron oxide NPs, with Fe NPs compensating the Fe deficiency via reduction of chlorotic symptoms (Ghafariyan et al. 2013). Fe NPs (at 500 mg/L) affected 47% increase in the number of pods in black-eyed peas, 7% increase in the seed weight and 10% enhancement in the chlorophyll content of leaves, where iron salt application proved less effective as compared to the Fe NPs alone (Delfani et al. 2014). Fe NPs bettered the beneficial effect of Mg NPs (frequently used as nanofertilizer) on black-eyed peas (Hoagland and Arnon 1950).

Thus, the different kinds of NPs show variable efficacy in improving the crop yield through bettering the responses of various growth control features, such as stress tolerance, optimized use of fertilizers and pesticides, providing metals as essential macro- or micronutrients and minimizing the unpredictability as well as a financial burden on the farmers.

5 Effect of Soil Type on the NP-Regulated Nutrient Regulation

Having a sound knowledge of soil chemistry is the foremost knowledge base to assure rightful return from the crop harvest since all crops are obviously not likely to grow equally well on the same kind of soil. Furthermore, even all NPs do not have a similar kind of working actions in the same soil variety which makes it important to study which kind of NPs are suitable for which kind of soil. Furthermore, the heterogeneous composition of soil characterized by its varied physicochemical properties such as pH, texture, organic content, water retention and others can alter the NP interacting behaviour once they are within the soil. Altogether, there are three main kinds of soil well known with respect to their compositional diversity and texture, namely, sandy, clayey and loamy. The particles in sandy soil are larger in size and have considerable spaces in between so that there is proper air circulation but little water retention. As a result, sandy soil cannot retain much water and is fit for the growth of only those plants that can survive water scarcity. Contrary to this, clayey soil has comparatively smaller-sized particles which are closely placed with each other, thereby resulting in higher water retention but poor aeration. Loamy soil is an intermediate within these two categories and has the right kind of particle size and interparticle separation. Comparing the suitability of NP activity in different soil types, the above differences reveal a better NP working efficacy in sandy soil, substantially because of the proper particle space. The only factor which needs to be assured is that the chemical composition of NPs should aid in optimal water retention and absorption. Soil microflora also occupies a central role in optimum nutrient absorption and distribution, such as in the case of rhizobium species occupying the root nodules in leguminous crops. Unless and until the rhizobium population is not in desired proportions, the nitrogen fixation extent of the host soil is inadequate, affecting the total protein content of agricultural soil. So the chosen NP concentrations and fed dosages should not be detrimental to the soil microbial population; otherwise, many useful functions of the soil could be inadvertently lost. Table 26.4 summarizes the concentration-dependent potential threats of various metal and metal oxide NPs in different soil varieties (Colman et al. 2013; Asadishad et al. 2017; Shah et al. 2014; Ge et al. 2012; Kumar et al. 2012; McGee et al. 2017).

Critical roles are depicted in hampering enzymatic activities and microbial population that have driven responses for the proper growth of crop plants. Several studies suggest exposure to fullerenes does not alter the structure and function of soil microflora, while nanoscale ZnO and TiO_2 particles have growth-inhibiting effects on the bacteria living inside the soil (Tan et al. 2018; Ge et al. 2011). To assure no loss in water retention capabilities of the host soil, natural zeolites are rapidly being used as potential alternatives to improve soil quality alongside the impacts of chemical and organic fertilizers (Najafi-Ghiri 2014). Nanozeolites facilitate slow release

Table 26.4 Soil-type-dependent toxicity of popularly used NP expressions (neat as well as oxidized forms)

Type of NP	Peculiar growth-dampening effect	Type of soil	Reference
Ag (0.14 mg•kg^{-1})	Attenuated bacterial activity through suppressed enzymatic activities	Sandy and sandy loam	Colman et al. (2013)
Au (0.1–100 mg•kg^{-1})	Significant effect on soil microflora and nutrient cycling	Sandy	Asadishad et al. (2017)
Fe, Ag and Co (550 mg pot)	No significant effect on microbial population (in collective mode) but individual activity affects bacterial activities	Acidic topsoil	Shah et al. (2014)
TiO_2 (20 g•kg^{-1})	Decrement and modification of bacterial diversity	Sandy clay loam	Ge et al. (2012)
Mixture of Cu, Ag and Si	Reduced C and N biomass along with modification of microbial community structure	Sandy peat Arctic soil	Kumar et al. (2012)
Ag, SiO_2 and Al_2O_3	Reduced dehydrogenase and urease activity, bacterial and archaeal amoA gene abundance	Pastureland soil	McGee et al. (2017)

of water sources and therefore increase the water-holding capacity of a soil (Manikandan and Subramanian 2014). Zeolites and nanozeolites improve the soil physical characteristics, such as water conduction, infiltration and ventilation, made possible by their porous and capillary properties. It is due to this reason that zeolite action keeps a particular soil uniformly aerated and that leads to zeolites being frequently termed as natural wetting agents, regulating the water conduction in plants (Szerment et al. 2014; Ghazavi 2015). Similarly, Si NPs are being used as relievers of the heavy metal toxicity risk in different soils as well as salinity stress and dehydration (Abdel-Haliem et al. 2017; Jullok et al. 2016). Most of the studies focusing on the effect of NP activities in the soil voice about their toxicity concerns although the exact mechanisms of the long-term manifestations are yet to be understood. In this context, it is of paramount importance to know about the peculiar source of NP entry to the soil, as there are numerous kinds of NPs which naturally prevail in every soil due to persistent environmental activities. A second major consideration is whether the NPs existing in any soil type are biologically or physically or chemically prepared. Studies predict that biologically prepared NPs are more compatible to crop plants since surface coatings (for aggregation prevention) are biocompatible in nature and most of the used biological sources (primarily leaf extracts or microbial population in subcultured fractions) are water soluble. Contrary to this, the NPs prepared through physical or chemical methods and are being used in electronic semiconducting applications or rather faster mechanical purposes bear chemically complex coatings on their surface. These coatings are not easily degraded and often result in agglomeration owing to the exposure towards host soil carrying several different kinds of materials with different reactivities. Increasing percentages of such nanomaterials often cause undesirable interference in seed germination, root and shoot growth and photosynthesis rate through diminishing chlorophyll concentrations. The risks present a gloomy picture beyond doubt because the effect at hand is practical of many significant proportions. Crops as common as onion, spinach, coriander, wheat, rice, soybean, mung bean, radish, lettuce, barley and cucumber have been affected in multifold undesirable extents (Shaw and Hossain 2013; Frazier et al. 2014; Hong et al. 2015; Yang et al. 2015; Rajput et al. 2018; Priester et al. 2017). In this reference, Josko et al. have reported the negative effect of high NP concentrations on dehydrogenase activity, while Janvier et al. and Suresh et al. reported their detrimental effect on the self-cleaning facilitated nutrient balancing, which considerably affects the plant nutrition and soil fertility (Josko et al. 2014; Janvier et al. 2007).

Typical factors ascertaining the harmful influences of NPs disposed to soil include their concentration, soil type and enzymatic activity. Soil properties, such as pH, chemical texture, structure and relative organic content, affect the microbial content of the soil as well as the ability of pollutant species to express toxic effects on the microorganisms (Fierer and Jackson 2006; Simonin and Richaume 2015). Not all soils are similarly affected by the NP exposure since there are some intentional preparations, such as supplementing the soil with digested and fly ash, which would reduce the pollutant bioavailability. Similarly, Calvario et al. in 2014 showed the role of particle size distribution and organic matter content (in the soil) as the critical factors affecting the microbial populations of host soil (Calvarro et al. 2014). Exposure to NPs is not always detrimental to the microbial and biological efficacy of the soil as revealed by the results of the studies on some intentionally prepared soils. Biochar, utilizing charcoal as a nutrient supplement to the soil, as soil amendment expresses minimal toxic effects of CeO_2 NPs on the plants grown, although NP- and biochar-amended soil interactions are not entirely studied (Servin et al. 2017).

6 Future Insights with Respect to Sustainable Progress Monitoring

Sustainable development is aimed at the optimum extraction of all potential benefits so that the advantage of one positive is not compromised or compensated by the negative of another aspect. Rather, the positive aspects of different contributing factors are expressed with reasonable control avoiding even a meagre negative expression of any of the variables. For instance, activities of NPs other than in the soil are finely regulated since a robust knowledge and database of potential alternatives is available for the NPs suitable in pesticide and fertilizer delivery or in the quantification of seed quality inspection, pathogen detection and related activities. Contrary to all such requirements (where the usage of NPs and nanomaterials is in our own hands and offers optional implementation, in sync with the intended purpose), the scenario where NP exposure to agricultural crops is accidental presents a differentiating aspect of NP activities. A careful look at the literature suggests repeated referrals to modelling several features of NP parameters so that the optimum combination of more than one attribute is attained (Raies and Bajic 2016; Valerio 2009; Deeb and Goodarzi 2012). For example, using an NP or nanocomposite system to deliver pesticides will ultimately deliver the nanomaterials to the soil after which it is well known that exposure of such nanomaterials to the soil could affect the soil texture and fertility through more than one way. In such circumstances, the choice needs to be rationally made so that the chosen nanosystem is moderately interactive which also aids in the controlled extent of pesticide delivery at the targeted site.

Making such a choice would simultaneously reassure or reduce any untoward chemical risk to the host soil so that the microbes existing in the soil are not hampered or affected wrongly. As studies till date mostly refer to critical detrimental effects on crop plant, there is a need to uptake scale-up models for studying the long-term NP effects (in terms of exposure dosages and duration). The risks with irregular NP toxic menaces are substantially due to their unceremoniously higher reactivity which could be a source of formation of heterogeneous and chemically more complex forms. Thus, modelling the nature of surface passivation materials with respect to the composition of a particular soil type in such cases could be the remedial measure. The next level of control could be exercised via collection of databases listing the harmful effects of NPs on different growth-regulating properties of crop plants. These insights could be useful to select the rightful NP concentrations and choose only those NPs which cause a lower negative influence on one factor and comparatively better effect on another. In vitro modelling of such permutations using computational techniques or docking could be a crucial link to choose optimal combinations. Physically and chemically prepared NPs generally possess chemically constituted aggregation-preventing coatings on their surfaces, which are hydrophobic in nature. Such artefacts are likely to complicate the degradation of NPs and ultimately increase their toxicity via an enhancement in oxidative stress (Manke et al. 2013; Saddick et al. 2017; Brown et al. 2004). The occurrence of such threats in several animals in the course of their drug delivery trials is proof of such vulnerable metabolic fates. The enzyme diversity of microbes is already a familiar part of several symbiotic associations between microbes and plants. Owing to a differential enzyme expression in microbial species alongwith the higher chemical reactivity of NPs and the dependence of enzyme activity on the working temperature, NPs having significantly distinct specificity (such as magnetic sensitivity, fluorescence sensitivity, antibiotic sensitivity, stress sensitivity or any other) could be utilized as need-based sensing agents. These entities have been used to monitor the stress levels of desalinated soils in terms of heavy metal pollution and another toxic constituent. NPs along with microbes serve as dual stress mitigating agents since microbes metabolize these harmful materials due to their greater stress tolerance levels and NPs do so by virtue of their higher surface areas. Furthermore, there could be a possibility that microbes metabolize these toxic species and synthesize them as NPs through enzymes of their metabolic machinery. In such interactions, it could be possible that modification in microbial genes could provide us with enhanced fluorescence or chemical responses towards any sensing material. This dual strategy could be a vital breakthrough as having more than one kind of NPs in the soil base could aid not only in detoxification but also in better nutrient and water absorption. The repercussions of such potential remedies are expressed via reduced water and more nutrient supply. Microbes with abilities to survive in < 0 and $> 100\ ^{\circ}C$ temperatures are enormous powerhouses to enhance the nanoscale specialities of NPs and other nanomaterials.

7 Conclusions

The application of nanoscale miniaturization tools has minimized agricultural uncertainties along with improvement in the qualitative and quantitative features of crop plants. The advantage of nanodevices towards exceptionally low-valued detection levels has monitored the seed quality and growth potential with more reliability. Faster and accurate sensing using NPs and integrated nanoprobes has improved disease detection and gene administration in obtaining wholesomely new and more stress-tolerant varieties. Though studies involving NP administration to crop plants project some serious inconclusive flaws, a proper selection of NP with reference to the soil concerned could minimize toxicity and crop quality deterioration issues.

The availability of nanocarriers with tuneable design features has already simplified the need-based targeted delivery of pesticides and fertilizers, thereby reducing the chemical hazards to soil and aquatic habitats. The optimization of NP usage is swiftly on the way of being better controlled, to detect the pathogen level, to enhance the photosynthesis rate through photocatalytic attributes and to detoxify the toxic soil ingredients via judicious utilization of solar energy. With a rapid interest in interdisciplinary studies, expectations to achieve better control on agricultural crop yields and their quality standards are significantly on a high.

References

Abdel-Haliem MEF, Hegazy HS, Hassan NS et al (2017) Effect of silica ions and nano silica on rice plants under salinity stress. Ecol Eng 99:282–289

Arthur FH (1999) Evaluation of an encapsulated formulation of cyfluthrin to control Sitophilus oryzae (L.) on stored wheat. J Stored Prod Res 35(2):159–166

Asadishad B, Chahal S, Cianciarelli V et al (2017) Effect of gold nanoparticles on extracellular nutrient-cycling enzyme activity and bacterial community in soil slurries: role of nanoparticle size and surface coating. Environ Sci Nano 4:907–918

Atha DH, Wang H, Petersen EJ et al (2012) Copper oxide nanoparticle mediated DNA damage in terrestrial plant models. Environ Sci Technol 46(3):1819–1827

Babick F, Mielke J, Wohlleben W et al (2016) How reliably can a material be classified as a nanomaterial? Available particle-sizing techniques at work. J Nanopart Res 18:158

Bakalova R, Ohba H, Zhelev Z et al (2004) Quantum dots as photosensitizers? Nat Biotechnol 22(11):1360–1361

Bakhtiari M, Moaveni P, Sani B (2015) The effect of iron nanoparticles spraying time and concentration on wheat. Biol Forum Int J 7(1):679–683

Bekyarova E, Ni Y, Malarkey EB et al (2005) Applications of carbon nanotubes in biotechnology and biomedicine. J Biomed Nanotechnol 1(1):3–17

Beltrán-Partida E, Valdez-Salas B, Valdez-Salas E et al (2019) Synthesis, characterization, and in situ antifungal and cytotoxicity evaluation of ascorbic acid-capped copper nanoparticles. J Nanomater Article ID 5287632

Brown DM, Donaldson K, Borm PJ et al (2004) Calcium and ROS-mediated activation of transcription factors and TNF-alpha cytokine gene expression in macrophages exposed to ultrafine particles. Am J Physiol Lung Cell Mol Physiol 286(2):344–353

Calvarro LM, de Santiago-Martín A, Gomez JQ et al (2014) Biological activity in metal contaminated calcareous agricultural soils: the role of the organic matter composition and the particle size distribution. Environ Sci Pollut Res 21(9):6176–6187

Casanova H, Araque P, Ortiz C (2005) Nicotine carboxylate insecticide emulsions: effect of the fatty acid chain length. J Agric Food Chem 53:9949–9953

Chen J, Wang W, Xu Y, Zhang X (2011) Slow-release formulation of a new biological pesticide, pyoluteorin, with mesoporous silica. J Agric Food Chem 59(1):307–311

Chen H, Seiber JN, Hotze M (2014) ACS select on nanotechnology in food and agriculture: a perspective on implications and applications. J Agric Food Chem 62(6):1209–1212

Chien YY, Jan MD, Adak AK et al (2008) Globotriose-functionalized gold nanoparticles as multivalent probes for Shiga-like toxin. Chembiochem 9:1100–1109

Collins RL, Doglia S, Mazak RA et al (1973) Controlled release of herbicides theory. Weed Sci 21:1–5

Colman BP, Arnaout CL, Anciaux S et al (2013) Low concentrations of silver nanoparticles in biosolids cause adverse ecosystem responses under realistic field scenario. PLoS One 8:57189

Corral-Diaz B, Peralta-Videa JR, Alvarez-Parrilla E et al (2014) Cerium oxide nanoparticles alter the antioxidant capacity but do not impact tuber ionome in *Raphanus sativus* (L). Plant Physiol Biochem 84:277–285

Das S, Wolfson BP, Tetard L et al (2015) Effect of N-acetyl cysteine coated CdS:Mn/ZnS quantum dots on seed germination and seedling growth of snow pea (*Pisum sativum* L.): imaging and spectroscopic studies. Environ Sci Nano 2(2):203–212

Deeb O, Goodarzi M (2012) In silico quantitative structure toxicity relationship of chemical compounds: some case studies. Curr Drug Saf 7(4):289–297

Delfani M, Firouzabadi MB, Farrokhi N et al (2014) Some physiological responses of black-eyed pea to iron and magnesium nanofertilizers. Commun Soil Sci Plant Anal 45(4):530–540

Devreese JT (2007) Importance of nanosensors: Feynman's vision and the birth of nanotechnology. Mater Res Soc Symp Proc. Materials Research Society 0952-F01-01

Elumalai K, Velmurugan S, Ravi S et al (2015) Green synthesis of zinc oxide nanoparticles using *Moringa oleifera* leaf extract and evaluation of its antimicrobial activity. Spectrochim Acta Mol Biomol Spectrosc 143:158–164

Fierer N, Jackson RB (2006) The diversity and biogeography of soil bacterial communities. PNAS 103(3):626–631

Forget G (1993) Balancing the need for pesticides with the risk to human health. In: Forget G, Goodman T, de Villiers A (eds) Impact of pesticide use on health in developing countries. IDRC, Ottawa, p 2

Frazier TP, Burklew CE, Zhang B (2014) Titanium dioxide nanoparticles affect the growth and microRNA expression of tobacco (*Nicotiana tabacum*). Funct Integr Genomics 14(1):75–83

Garrido-Herrera FJ, Daza-Fernandez I, Gonzalez-Pradas E et al (2009) Lignin-based formulations to prevent pesticides pollution. J Hazard Mater 168(1):220–225

Ge YG, Schimel JP, Holden PA (2011) Evidence for negative effects of TiO_2 and ZnO nanoparticles on soil bacterial communities. Environ Sci Technol 45(4):1659–1664

Ge YG, Schimel JP, Holden PA (2012) Identification of soil bacteria susceptible to TiO_2 and ZnO nanoparticles. Appl Environ Microbiol 78:6749–6758

Ghafariyan MH, Malakouti MJ, Dadpour MR et al (2013) Effects of magnetite nanoparticles on soybean chlorophyll. Environ Sci Technol 47(18):10645–10652

Ghazavi R (2015) The application effects of natural zeolite on soil runoff, soil drainage and some chemical soil properties in arid land área. Int J Innov Appl Stud 13(1):172–177

Gupta SK, Jani JP, Saiyed HN et al (1984) Health hazards in pesticide formulators exposed to a combination of pesticides. Indian J Med Res 79:666–672

Gupta R, Malik P, Das N, Singh M (2019) Antioxidant and physicochemical study of *Psidium guajava* prepared zinc oxide nanoparticles. J Mol Liq 275:749–767

Hamad AF, Han JH, Kim BC et al (2018) The intertwine of nanotechnology with the food industry. Saudi J Biol Sci 25(1):27–30

Han J, Li F, Jiang L et al (2015) Electrochemical determination of bisphenol a using a polyacrylamide-multiwalled carbon nanotube-modified glassy carbon electrode. Anal Methods 7(19):8220–8226

Hoagland DR, Arnon DI (1950) The water-culture method for growing plants without soil. Calif Agric Exp Station Circ 347:1–32

Hong J, Rico CM, Zhao L et al (2015) Toxic effects of copper-based nanoparticles or compounds to lettuce (*Lactuca sativa*) and alfalfa (*Medicago sativa*). Environ Sci Process Impact 17(1):177–185

Huang Q (2012) Nanotechnology in the food, beverage and nutraceutical industries. Woodhead Publishing Limited, Cambridge

Iftekhar S, Küçük ME, Srivastava V et al (2018) Application of zinc-aluminium layered double hydroxides for adsorptive removal of phosphate and sulfate: equilibrium, kinetic and thermodynamic. Chemosphere 209:470–479

Janvier C, Villeneuve F, Alabouvette C et al (2007) Soil health through soil disease suppression: which strategy from descriptors to indicators? Soil Biol Biochem 39:1–23

Jean RD, Chiu KC, Chen TH et al (2010) Functionalized silica nanoparticles by nanometallic ag decoration for optical sensing of organic molecule. J Phys Chem C 114(37):15633–15639
Jiang LC, Basri M, Omar D et al (2011) Physicochemical characterization of non-ionic surfactants in oil-in-water (o/w) nanoemulsions for new pesticide formulations. Int J Appl Sci Technol 1:131–142
Josko I, Oleszczuk P, Futa B (2014) The effect of inorganic nanoparticles (ZnO, Cr_2O_3, CuO and Ni) and their bulk counterparts on enzyme activities in different soils. Geoderma 232-234:528–537
Jullok N, Van Hooghten R, Luis P, Volodin A, Van Haesendonck C, Vermant J, Van der BB (2016) J Clean Prod 112:4879–4889
Kah M (2015) Nanopesticides and nanofertilizers: emerging contaminants or opportunities for risk mitigation? Front Chem 3:64
Kalia A, Parshad VR (2015) Novel trends to revolutionize preservation and packaging of fruits/ fruit products: microbiological and nanotechnological perspectives. Crit Rev Food Sci Nutr 55(2):159–182
Kalwasińska A, Kęsy J, Wik I et al (2011) Neustonic versus epiphytic bacteria of eutrophic lake and their biodegradation ability on deltamethrin. Biodegradation 22:699–707
Keating BA, Carberry PS, Bindraban PS et al (2010) Eco-efficient agriculture: concepts, challenges and opportunities. Crop Sci 50:S109–S119 [Symposia]
Khodakovskaya MV, Lahiani MH (2014) Nanoparticles and plants: from toxicity to activation of growth. In: Sahu SC, Casciano DA (eds) Handbook of nanotoxicology, nanomedicine and stem cell use in toxicology. Wiley Sciences, pp 121–130
Kimling J, Maier M, Okenve B, Kotaidis V, Ballot H, Plech A (2006) Turkevich method for gold nanoparticle synthesis revisited. J Phys Chem B 110(32):15700–15707
Kobayashi N, Izumi H, Morimoto Y (2017) Review of toxicity studies of carbon nanotubes. J Occup Health 59(5):394–407
Koilraj P, Kannan S (2010) Phosphate uptake behaviour of ZnAlZr ternary layered hydroxides through surface precipitation. J Colloid Interface Sci 341:289–297
Kole RK, Banerjee H, Bhattacharyya A et al (1999) Phototransformation of some pesticides. J Indian Chem Soc 76(11–12):595–600
Kricka LJ (2000) Microchips, microarrays, biochips and nanochips – personal laboratories for the 21st century. EJIFCC 12(4):105–108
Kumar J, Singh G, Walia S et al (2004) Controlled release formulations of Imidacloprid: water and soil release kinetics. Pestic Res J 16:13–17
Kumar N, Shah V, Walker VK (2012) Influence of a nanoparticle mixture on an arctic soil community. Environ Toxicol Chem 31:131–135
Kumar S, Rani R, Dilbaghi N et al (2017) Carbon nanotubes: a novel material for multifaceted applications in human healthcare. Chem Soc Rev 46(1):158–196
Kydd J (2002) Agriculture and rural livelihoods: is globalisation opening or blocking paths out of rural poverty? Agricultural research and extension network paper no.121, ISBN 085003 5856
Latheef MA, Dailey OD Jr, Franz E (1993) Efficacy of polymeric controlled release formulations of sulprofos against tobacco budworm, Heliothis virescens (Lepidopetea: Noctuidae) on cotton. In: Berger PD, Devisetty BN, Hall FR (eds) Pesticide formulations and applications systems: 13th volume, ASTM STP 1183. American Society for Testing and Materials, Philadelphia, pp 300–311
Lew S, Lew SM, Biedunkiewicz A, Szarek J (2013) Impact of pesticide contamination on aquatic microorganism populations in the littoral zone. Arch Environ Contam Toxicol 64:399–409
Li Y, Xu J, Sun C (2015) Chemical sensors and biosensors for the detection of melamine. RSC Adv 5:1125–1147
Liu W, Zhang S, Wang L et al (2011) CdSe quantum dot (QD)-induced morphological and functional impairments to liver in mice. PLoS One 6:24406
Liu Y, Zhao Y, Sun B, Chen C (2013) Understanding the toxicity of carbon nanotubes. Acc Chem Res 46(3):702–713

Liu X, Zhang N, Bing T et al (2014) Carbon dots based dual-emission silica nanoparticles as a ratiometric nanosensor for Cu^{2+}. Anal Chem 86(5):2289–2296

Mahadevan R (2003) Productivity growth in Indian agriculture: the role of globalization and economic reform. Asia-Pac Dev J 10(2):57–72

Malik P, Katyal V, Malik V et al (2013) Nanobiosensors: concepts and variations. ISRN Nanomater Article ID 327435

Malik P, Shankar R, Malik V et al (2014) Green chemistry based benign routes for nanoparticle synthesis. J Nanopart Article ID 302429

Manikandan A, Subramanian KS (2014) Fabrication and characterization of nanoporous zeolite based N fertilizer. Afr J Agric Res 9(2):276–284

Manke A, Wang L, Rojanasakul Y (2013) Mechanisms of nanoparticle-induced oxidative stress and toxicity. Biomed Res Int:942916

Maruyama CR, Guilger M, Pascoli M et al (2016) Nanoparticles based on chitosan as carriers for the combined herbicides Imazapic and Imazapyr. Sci Rep 6:19768

Mathur SC (1999) Future of Indian pesticides industry in next millennium. Pestic Inf 24:9–23

McClements DJ (2004) Protein-stabilized emulsions. Curr Opin Colloid Interface Sci 9(5):305–313

McClements DJ, Decker EA, Weiss J (2007) Emulsion-based delivery systems for lipophilic bioactive components. J Food Sci 72(8):R109–R124

McGee CF, Storey S, Clipson N et al (2017) Soil microbial community responses to contamination with silver, aluminium oxide and silicon dioxide nanoparticles. Ecotoxicology 26(3):449–458

Medina MB (2006) Development of a fluorescent latex microparticle immunoassay for the detection of *Staphylococcal Enterotoxin* B (SEB). J Agric Food Chem 54(14):4937–4942

Menon S, Rajeshkumar S, Kumar V (2017) A review on biogenic synthesis of gold nanoparticles, characterization, and its applications. Res Effic Technol 3(4):516–527

Miller G, Senjen R (2008) Out of the laboratory and on to our plates nanotechnology in food and agriculture. Friends of the Earth

Moon YS, Park ES, Kim TO, Lee HS, Lee SE (2014) SELDI-TOF MS-based discovery of a biomarker in *Cucumis sativus* seeds exposed to CuO nanoparticles. Environ Toxicol Phar 38:922–931

Mukhopadhyay SS (2014) Nanotechnology in agriculture: prospects and constraints. Nanotechnol Sci Appl 7:63–71

Nair PMG, Kim SH, Chung IM (2014) Copper oxide nanoparticle toxicity in mung bean (*Vigna radiata* L.) seedlings: physiological and molecular level responses of in vitro grown plants. Acta Physiol Plant 36:2947–2958

Najafi-Ghiri M (2014) Effects of zeolite and vermicompost applications on potassium release from calcareous soils. Soil Water Res 9(1):31–37

Nanowerk (Webpage). Nanotechnology products and applications. Product: Guard IN Fresh. Honolulu, HI; Nanowerk; 2014. Available from: http://www.nanowerk.com/products/product. Accessed 19 Apr 2014

Nanowerk (Webpage). Nanotechnology products and applications. Product: NanoCeram-PAC. Honolulu, HI: Nanowerk; 2014. Available from: http://www.nanowerk.com/products/product. Accessed 19 Apr 2014

Novillo C, Guaya D, Avendano AAP et al (2014) Evaluation of phosphate removal capacity of Mg/Al layered double hydroxides from aqueous solutions. Fuel 138:72–79

Oliver MJ (2014) Why we need GMO crops in agriculture. Mo Med 111(6):492–507

Owolade O, Ogunleti D, Adenekan M (2008) Titanium dioxide effects disease development and yield of edible cowpea. Electron J Environ Agric Food Chem 7:2942–2947

Pallavi MCM, Srivastava R et al (2016) Impact assessment of silver nanoparticles on plant growth and soil bacterial diversity. 3 Biotech 6:254

Pathak S, Choi SK, Arnheim N et al (2001) Hydroxylated quantum dots as luminescent probes for in situ hybridization. J Am Chem Soc 123(17):4103–4104

Patolsky F, Lieber CM (2005) Nanowire nanosensors. Mater Today 8(4):20–28

Pelaez M, Nolan NT, Pillai SC et al (2012) A review on the visible light active titanium dioxide photocatalysts for environmental applications. Appl Catal B Environ 125(2012):331–349
Perlatti B, Bergo BPLS, da Silva MFGF, Fernandes JB, Forim MR (2013) (Chapter-20) Polymeric nanoparticle-based insecticides: a controlled release purpose for agrochemicals in insecticides – development of safer and more effective technologies, ed. Trdan S. INTECH Publishers
Priester JH, Moritz SC, Espinosa K et al (2017) Damage assessment for soybean cultivated in soil with either CeO_2 or ZnO manufactured nanomaterials. Sci Total Environ 579:1756–1768
Pyrzynska K (2011) Carbon nanotubes as sorbents in the analysis of pesticides. Chemosphere 83(11):1407–1413
Qureshi MA, Karthikeyan S, Punita K et al (2012) Application of nanotechnology in food and dairy processing: an overview. Pak J Food Sci 22(1):23–31
Radoi A, Targa M, Prieto-Simon B et al (2008) Enzyme-Linked Immunosorbent Assay (ELISA) based on superparamagnetic nanoparticles for aflatoxin M1 detection. Talanta 77(1):138–143
Raies AB, Bajic VB (2016) In silico toxicology: computational methods for the prediction of chemical toxicity. Wiley Interdiscip Rev Comput Mol Sci 6(2):147–172
Rajiv P, Rajeshwari S, Venckatesh R (2013) Bio-fabrication of zinc oxide nanoparticles using leaf extract of *Parthenium hysterophorus* L. and its size-dependent antifungal activity against plant fungal pathogens. Spectrochim Acta Mol Biomol Spectrosc 112:384–387
Rajput VD, Minkina T, Sushkova S et al (2018) Effect of nanoparticles on crops and soil microbial communities. J Soils Sediments 18(6):2179–2187
Rastogi A, Tripathi DK, Yadav S et al (2019) Application of silicon nanoparticles in agriculture. 3 Biotech 9(3):90
Rosi NL, Mirkin CA (2005) Nanostructures in biodiagnostics. Chem Rev 105(4):1547–1562
Roy A (2017) Synthesis of silver nanoparticles from medicinal plants and its biological applications: a review. Res Rev Biosci 12(4):138
Roychowdhury R, Abdel Gawwad MR, Banerjee U, Bishnu S, Tah J (2013a) Status, trends and prospects of organic farming in India: a review. J Plant Biol Res 2:38–48
Roychowdhury R, Banherjee U, Slofkova S, Tah J (2013b) Organic farming for crop improvement and sustainable agriculture in the era of climate change. OnLine J Biol Sci 13(2):50–65
Saddick S, Afifi M, Zinada OAA (2017) Effect of zinc nanoparticles on oxidative stress-related genes and antioxidant enzymes activity in the brain of *Oreochromis niloticus* and *Tilapia zillii*. Saudi J Biol Sci 24(7):1672–1678
Saha N, Dutta GS (2017) Low-dose toxicity of biogenic silver nanoparticles fabricated by *Swertia chirata* on root tips and flower buds of *Allium cepa*. J Hazard Mater 330:18–28
Santhoshkumar T, Rahuman AA, Jayaseelan C et al (2014) Green synthesis of titanium dioxide nanoparticles using *Psidium guajava* extract and its antibacterial and antioxidant properties. Asian Pac J Trop Med 7:968–976
Schnorr JM, Swager TM (2011) Emerging applications of carbon nanotubes. Chem Mater 23(3):646–657
Schofield CL, Field RA, Russell DA (2007) Glyconanoparticles for the colorimetric detection of cholera toxin. Anal Chem 79(4):1356–1361
Servin AD, De la Torre-Roche R, Castillo-Michel H et al (2017) Exposure of agricultural crops to nanoparticle CeO2 in biochar-amended soil. Plant Physiol Biochem 110:147–157
Shah V, Collins D, Walker VK et al (2014) The impact of engineered cobalt, iron, nickel and silver nanoparticles on soil bacterial diversity under field conditions. Environ Res Lett 9:024001
Sharma AK (2015) National council of applied economic research, Annual Report 2015–16
Shaw AK, Hossain Z (2013) Impact of nano-CuO stress on rice (*Oryza sativa* L.) seedlings. Chemosphere 93(6):906–915
Shikha S, Salafi T, Cheng J et al (2017) Versatile design and synthesis of nano-barcodes. Chem Soc Rev 46(22):7054–7093
Simonin M, Richaume A (2015) Impact of engineered nanoparticles on the activity, abundance, and diversity of soil microbial communities: a review. Environ Sci Pollut Res 22(18):13710–13723

Singh O, Kohli N, Singh RC (2012) Effect of pH on the morphology and gas sensing properties of ZnO nanostructures. Sensors Actuators B Chem 166:438–443

Snodgrass MM, Wallace LT (1980) Agricultural economics and resource management. Prentice Hall, Inc, New Delhi, pp 79–80

Szerment J, Ambrozewich-Nita A, Kedziora K et al (2014) Use of zeolite in agriculture and environmental protection. A short review. UDC 666(96):691.5

Takei T, Yoshida M, Hatate Y et al (2008) Preparation of polylactide/poly(ε-caprolactone) microspheres enclosing acetamiprid and evaluation of release behaviour. Polym Bull 61:391–397

Tan W, Peralta-Videa JR, Gardea-Torresdey JL (2018) Interaction of titanium dioxide nanoparticles with soil components and plants: current knowledge and future research needs-a critical review. Environ Sci Nano 5(2):257–278

Taton TA, Mirkin CA, Letsinger RL (2000) Scanometric DNA array detection with nanoparticle probes. Science 289(5485):1757–1760

The Nanoscale Science, Engineering, and Technology Subcommittee of the Committee on Technology of the National Science and Technology Council Nanotechnology-Enabled Sensing: Report of the National Nanotechnology Initiative Workshop; Arlington, VA, USA (2009)

The Project on Emerging Nanotechnologies (Webpage). Consumer products inventory. Washington, DC; 2013. Available from: http://www.nanotechproject.org/cpi. Accessed 19 Apr 2014

Valentini P, Galimberti A, Mezzasalma V et al (2017) DNA barcoding meets nanotechnology: development of a universal colorimetric test for food authentication. Angew Chem Int Ed 56:8094–8098

Valerio LG Jr (2009) In silico toxicology for the pharmaceutical sciences. Toxicol Appl Pharmacol 241(3):356–370

Wang L, Li X, Zhang G et al (2007) Oil-in-water nanoemulsions for pesticide formulations. J Colloid Interface Sci 314(1):230–235

Wang J, Timchalk C, Lin Y (2008) Carbon nanotube-based electrochemical sensor for assay of salivary cholinesterase enzyme activity: an exposure biomarker of organophosphate pesticides and nerve agents. Environ Sci Technol 42(7):2688–2693

Wilson SL, Boucher RE Jr, Inventors, Dow Agrosciences LLC., assignee. US Patent Number 2011/0052654 A1. 2010 Mar 1

Yang FL, Li XG, Zhu F et al (2009) Structural characterization of nanoparticles loaded with garlic essential oil and their insecticidal activity against *Tribolium castaneum* (Herbst) (Coleoptera: Tenebrionidae). J Agric and Food Chem 57(21):10156–10162

Yang Z, Chen J, Dou R et al (2015) Assessment of the phytotoxicity of metal oxide nanoparticles on two crop plants, maize (*Zea mays* L.) and rice (*Oryza sativa* L.). Int J Environ Res Public Health 12(12):15100–15109

Zhang L, Li T, Li B et al (2010) Carbon nanotube-DNA hybrid fluorescent sensor for sensitive and selective detection of mercury(ii) ion. Chem Commun 46(9):1476–1478

Zhao W, Ali MM, Aguirre SD et al (2008) Paper-based bioassays using gold nanoparticle colorimetric probes. Anal Chem 80(22):8431–8437

Chapter 27
Agrobiodiversity and Advances in the Development of Millets in Changing Environment

D. A. Patil

Abstract Selective utilization of crops and their forms of germplasm, of late, have threatened the world's agrobiodiversity. This trend along with forces of expansion of commercial agriculture, market links, unsustainable uses, modification of landscape, and largely changing climate led to the rapid erosion of food sources. These, in turn, affected the nutritional security of people particularly in poverty-driven human societies and those people suffering from "hidden hunger." One such neglected or underutilized group of grain crops is millets. Actually, these are ancient crops but orphaned or forgotten for some obvious reasons. However, they are regaining the fame as "nutritious" or "superfood grains" because of great flavor, taste, nutritional profile, high antioxidants, gluten-free, evidence-based health benefits, vital trace elements, etc. The present author attempts to review the entire realm of development of millets particularly when climate change is putting long strides the world over. Both biotic and abiotic impacts are thought over while evaluating millets in the present all-pervasive examination of problems. This communication helps divulge agrobiodiversity of major and minor millets worldwide in various forms of germplasm. To cope with agrarian crisis, the present researches especially in developing countries elsewhere have been carried out to redeem the situation. Agrobiodiversity of millets generally available and attempts to save and conserve them for human welfare are reviewed to unearth pros and cons of development of millets to date. More efforts in collecting germplasm; conserving, evaluating, and utilizing with value addition; and promoting cultivation, besides awareness about benefits from millet consumption under climate change scenario, are required for sustainable millet farming.

Keywords Agrobiodiversity · Millets · Germplasm · Climate change

D. A. Patil (✉)
Post-Graduate Department of Botany, S.S.V.P. Sanstha's L. K. Dr. P. R. Ghogrey Science College, Dhule, Maharashtra, India

R. Roychowdhury et al. (eds.), *Sustainable Agriculture in the Era of Climate Change*, https://doi.org/10.1007/978-3-030-45669-6_27

1 Introduction

Agriculture is the major land use across the globe. It is also a major economic, social, and cultural activity. Agrobiodiversity focuses mainly on that portion of the biodiversity that has undergone selection and modification over millennia by human civilization to serve human needs better (Subramanian and Thirumeni 2007). In general, agrobiodiversity refers to the variety and variability of plants or other organisms on Earth that are important to food and all other agricultural production systems along with the management systems and traditional practices used by farmers and farming communities (Rana 2002). Agrobiodiversity has a great role in sustaining and strengthening the food and nutritional security and health of humanity. One side of a coin represents a national emblem, while the other side gives its value. Biodiversity represents the former side of a coin, while the latter side expresses agrobiodiversity, which is most valued for humankind. So, it is very important to improve the crops under ongoing climatic changes (Hasanuzzaman et al. 2013, 2015; Roychowdhury 2014; Roychowdhury and Tah 2013; Roychowdhury et al. 2013a, b, c, 2014, 2018, 2019; Chakraborty et al. 2014; Anumalla et al. 2015).

We are living in an age that is witnessing an unprecedented agrarian crisis. In the modern period, just 150 plant species out of nearly 5000 plant species useful nutritionally are significant to meet our requirement of food. Of these, again just 20–30 plant species are mostly emphasized (Bermejo and Lean 1994). This trend culminated ultimately concentrating only these few crops. Their intensive agriculture resulted in diets containing inadequate nutrient-rich components and genetic erosion. Green evolution also aggravated this trend of neglecting local nutritionally useful resources. Many neglected or underutilized crop species are still surviving in some pockets. These are generally thought to be poor man's diet. One such group of crop species is "millets." They serve the dual purpose of meeting food (nutrition) for humankind and feed or fodder security for domestic animals and birds. Nevertheless, they are often classified as underutilized or orphan crops because of the paucity of scientific attention. They are available even in modern period as landraces, folk varieties, farmers' varieties, local varieties, or wild relatives of cultivated crops. There is a dire necessity to emphasize them from nutritional and conservation point of view. Millets stand sixth in most important grain crops of the world and sustain more than one-third of the world's population (Changmei and Dorothy 2014). A word about taxonomic limits especially about barnyard millets appears necessary. The identities of species complex have remained confused in the past and still need clarifications. I, therefore, have treated them here as they originally appeared in literature published.

2 Millets

Small-seeded edible grains yielded by grasses are collectively called "millets." These do not constitute a taxonomic entity; however, they project as a functional or agronomic one. Mostly, they are consumed whole, rich in fat, and generally gluten-free. They are being cultivated since ancient times. However, none of the millet species have been able to compete with cereals in world commerce. Because of fast rate of acculturation worldwide and movements like the "Green Revolution," they have been largely threatened.

2.1 Distribution

In the arid and semiarid tropical regions of Asia and Africa, millets constitute an important source of food for mankind, apart from feed and fodder for domestic animals. These regions generally represent developing nations. Millets are conveniently grouped as major millets, e.g., sorghum and pearl millet, and minor (small) millets, e.g., finger millet, foxtail millet, proso millet, kodo millet, little millet, teff millet, and barnyard millet. This grouping is obviously based on the size of grains and plants. Major millets are tall and drought resistant. Minor millets obviously belong to grasses and have short slender culms and small grains and can survive under severe drought. Their worldwide distribution is under Table 27.1.

2.2 Morphological Characteristics

They are typically annuals attaining maturity within 3–4 months. They range from 30 to 130 cm in height, except pearl millet and sorghum which grow 1.3–3.00 m tall. Florets are born in spikes, racemes, or panicles with dense clusters of small florets. Seeds (called grains) are enclosed usually by hulls, except pearl millet and sorghum seeds which are usually ovoid and white and sometimes otherwise. The agro-morphological characteristics, however, differ from species to species of millets' w.r.t. plant height, number of tillers, days to flowering, and maturity and grain yield per plant. Morphological or qualitative characteristics include leaf (blade, color, sheath, pubescence), flag leaf (angle, ligules, and auricle shape), culm (culm angle, node color, internode color, and pubescence), panicles (inflorescence) type and their shattering, stigma (number and color), stamen (number and color), grain/seed (shape, grain awning, lemma and palea pubescence, seed coat color, and grain color), and root (type of root system).

Table 27.1 Worldwide distribution of commonly cultivated millets

Sl. No.	Common Name	Scientific Name	Countries/Areas of Cultivation
1	Sorghum	*Sorghum bicolor* (L.) Moench.	Africa, India, Pakistan, USA, Nepal.
2	Pearl millet	*Pennisetum americanum* L.	Africa, India, Pakistan, Arabian Peninsula
3	Finger millet	*Eleusine coracana* (L.) Gaertn.	India, Nepal, China, Myanmar, Sri Lanka, Uganda, Kenya, Eritrea, Sudan, Zambia, Zimbabwe, Malawi, Rwanda, Burundi, Madagascar.
4	Foxtail millet	*Setaria italica* (L.) P. Beauv.	Asia, Europe, North Africa, North America, Australia
5	Proso millet	*Panicum miliaceum* L.	Asia, Africa, Australia, Europe, North America
6	Little millet	*Panicum sumatrense* Roth ex Roem. and Schult.	India, Sri Lanka, Pakistan, Myanmar, China, Malaysia, Caucasus
7	Kodo millet	*Paspalum scrobiculatum* L.	India
8	Teff	*Eragrostis tef* (Zucc.) Trotter	Ethiopia
9	Barnyard millet (Indian+ Japanese)	*Echinochloa crus-galli* (L.) Beauv. and E. Colona (L.) Link.	India, China, Nepal, Japan, Korea

2.3 Agricultural vis-à-vis Agronomic Attributes

Millets are resilient to the extreme climatic and soil conditions prevalent in the semiarid and arid regions of Asia and Africa. They are grown under inadequate moisture and poor soil fertility, which are poorly suited to the major crops of the world (Bermejo and Lean 1994; Baker 2003). Millets also possess a C_4 photosynthesis system (Brutnell et al. 2010; Warner and Edwards 1988). They have prevalent photorespiration and so utilize the scarce moisture present in such regions. In spite agronomic, nutritional and health-related benefits, millets are low yielding in comparison to major cereals such as wheat, rice, etc. Their considerably low productivity is related to the challenging environment under which they are widely cultivated. Millets generally benefit agronomically as they are drought- and heat-tolerant to biotic stresses such as early maturity, antifungal, pest-tolerant, biotic stress-tolerant, and salt-tolerant, and even few millets are rain-fed and survive in marginal lands and environments. Agronomic or quantitative characters include (i) number of days of seedling emergence; (ii) at vegetative stage, days from emergence to panicle (inflorescence) initiation, number of leaves produced on the main culm from planting to panicle initiation, number of tillers produced from planting to panicle initiation, average panicle height, and final plant height; and (iii) at reproductive stage, ripening stage. Apart from these, shoots and roots are considered also for the following features: (a) shoots – number of days from blooming to ripening, fresh weight of aboveground biomass, fresh weight of panicle per plant, number of grains per panicle, dry weight 1000 grains per plant, and number of seeds per 1 gm per plant and (b) roots – root fresh weight, dry root weight, number of roots per plant, average length of roots per plant, and diameter of the biggest root per plant.

3 Status of Germplasm – Diversity of Millets

Diversity of organisms, where plant or animals, is not only expressed by its taxonomic categories like families, genera, or species but also now by its landraces, folk varieties, cultivars, farmers, or local varieties. Some of these are but natural and others released experimentally. All these are glimpsed in the following to know forms of germplasm especially of millets worldwide (Table 27.2).

4 Advances in Development of Millets in the Perspective of Climate Changes

4.1 Sorghum

Tongcheng et al. (2016) investigated potential impacts of climate change on grain sorghum productivity using the CERES-Sorghum model in the Decision Support System for Agrotechnology Transfer 44.S. The model was first calibrated in 1998 for a sorghum cultivar grown in a free-air CO_2 enrichment experiment at the University of Arizona (USA). This model was validated later in 1999. The simulated grain yield, growth, and soil water of sorghum for both years were found in statistical agreement with the corresponding measurement, respectively. The simulated and measured yields both did not respond to elevated CO_2; however, they both were found sensitive to water supply. This validated model was then implemented in western North America during 2080–2100 to simulate possible effects of climate change on sorghum grain yield and water use efficiency. The projected CO_2 fertilizer effect on grain yield was dominated by the adverse effect of projected temperature increases. Temperature, therefore, seems to be a dominant driver of the global climate change having an effect on future sorghum productivity. The CERES-Sorghum model provided a valuable preview of sorghum crop response to potential climate change forcing factors including CO_2, temperature and precipitation. It exhibited its capability to simulate the impact of global climate change on sorghum production.

Gupta et al. (2014) studied the impact of rainfall and temperature in India during the period of 1966–1999. They inferred that higher rainfall means higher yield. Similarly, the higher the average temperature, the lower the yield. These results also, in their opinion, corroborate with agronomic studies. They opined that

Table 27.2 Different types of millets and their number of germplasm, country of collections

Sl. No.	Form and No. of Germplasm	Country/Region	Reference
Sorghum [*Sorghum bicolor* (L.) Moench. (Great Millet)]			
1.	01 Accession	Arunachal Pradesh, India	Pal et al. (2011)
2.	25 Cultivars	Khandesh, Maharashtra, India	Khairnar et al. (2016a, b)
3.	18 Accessions of landraces	Andhra Pradesh, Odisha, India	Elangovan et al. (2017)
4.	36,771 Accessions	Different countries	Upadhyaya et al. (2006)
5.	7366 Germplasm collections	India	Sharma and Brahmi (2006)
6.	15 Accessions	Nepal	Ghimire et al. (2017)
7.	47 Landraces	Zimbabwe	Claid and Ereck (2008)
8.	02 Landraces	India and Nepal	Reghupathy et al. (2016)
9.	57 Varieties released	16 Countries	Dalton and Zereyeus (2013)
10.	25 Local varieties	Khandesh (M.S.) India	Khairnar et al. (2016a, b)
Pearl millet (*Pennisetum americanum* Schum.)			
1.	12 Accessions of landraces	Andhra Pradesh, Odisha, India	Elangovan et al. (2017)
2.	21,563 Accessions	Different countries	Upadhyaya et al. (2006)
3.	3100 Germplasm collections	India	Sharma and Brahmi (2006)
4.	169 Landraces	Rajasthan, India	Yadav (2008)
5.	180 Landraces	India	Khairwal (2007)
6.	225 Accessions (landraces or local varieties)	Sudan	Bashir et al. (2014)
7.	(i) 20,800 Accessions (ii) 750 Wild relatives	28 Different countries	Upadhyaya et al. (2007a, b)
8.	81 Accessions from 78 landraces	28 West and Central African countries	Sattler et al. (2018)
9.	21,594 Accessions including 750 accessions of 24 species of genus *Pennisetum*	51 Countries	Manga (2015)
10.	(i) 22 Cultivars (ii) 36 Landraces	Different states, India	Chowdari et al. (1998)
11.	27 Landraces	Lake Chad Basin	Naino Jika et al. (2017)
Finger millet [*Eleusine coracana* (L.) Gaertn.]			
1.	58 Accessions	Ethiopia	Tesfaye and Mengistu (2017)

(continued)

Table 27.2 (continued)

Sl. No.	Form and No. of Germplasm	Country/Region	Reference
2.	27 Accessions	Arunachal Pradesh, India	Pal et al. (2011)
3.	64 Landraces	Maharashtra, India	Kazi and Auti (2017)
4.	02 Cultivars	Khandesh (M.S.), India	Khairnar et al. (2016a, b)
5.	24 Accession of landraces	Andhra Pradesh, Odisha, India	Elangovan et al. (2017)
6.	5949 Accessions	Different countries	Upadhyaya et al. (2007a, b)
7.	(i) Local variety 01	Kenya	Amos (2016)
	(ii) Local landraces 16		
8.	(i) 850 Accessions	Nepal	Ghimire et al. (2017)
	(ii) 05 Improved varieties		
	(iii) 38 Landraces		
9.	07 Landraces	India and Nepal	Reghupathy et al. (2016)
10.	100 Accessions	Uganda	Owere et al. (2015)
11.	24 Improved varieties	Worldwide	Puranik et al. (2017)
12.	39 Varieties	South Asia	Anonymous (2014)
13.	10 Varieties released	Uganda	Wanyera (2007)
14.	03 Varieties released	Kenya	Oduori and Kanyenii (2007)
15.	282 Landraces	Tanzania	Kisandu et al. (2007)
16.	5949 Accessions	12 Asian and African Countries ICRSSAT, India	Upadhyaya et al. (2007a, b)
17.	37 Landraces	Bhutan	Anonymous (2008)
18.	9522 Accessions	India (NBPGR)	Dwivedi et al. (2012)
19.	6804 Accessions	India (ICRISAT)	Dwivedi et al. (2012)
20.	6257 Accessions	India (AICMMP)	Dwivedi et al. (2012)
21.	32,875 Accessions	Kenya (KARI)	Dwivedi et al. (2012)
22.	2156 Accessions	Ethiopia (IBC)	Dwivedi et al. (2012)
23.	1231 Accessions	Uganda (SAARI)	Dwivedi et al. (2012)
24.	1037 Accessions	Lusaka, Zambia	Dwivedi et al. (2012)
25.	869 Accessions	Nepal (CPBBD)	Dwivedi et al. (2012)
26.	702 Accessions	USA (NCGRP)	Dwivedi et al. (2012)
27.	06 Local varieties	(Tamil Nadu) India	Padulosi et al. (2009)
Foxtail millet (Italian or German millet) [*Setaria italica* (L.) P.Beauv] (Syn. *Panicum italicum* L.)			
1.	02 Accessions	Arunachal Pradesh, India	Pal et al. (2011)
2.	01 Cultivars	Khandesh (M.S.), India	Khairnar et al. (2016a, b)

(continued)

Table 27.2 (continued)

Sl. No.	Form and No. of Germplasm	Country/Region	Reference
3.	741 Accessions	Tamil Nadu, India	Nirmalkumari and Vetriventham (2010)
4.	16 Accession of landraces	Andhra Pradesh, Odisha, India	Elangovan et al. (2017)
5.	1535 Accessions	Different countries	Upadhyaya et al. (2006)
6.	55 Accessions	Nepal	Ghimire et al. (2017)
7.	26,000 Accessions (90% of it landraces)	Beijing, China	Doust et al. (2009)
8.	250 Landraces	China	Wang et al. (2012)
9.	223 Accessions	India	Chander et al. (2017)
10.	04 Landraces	India and Nepal	Reghupathy et al. (2016)
11.	27 Landraces	Nepal	Yadav et al. (2018)
12.	324 Landraces	Taiwan	Lin et al. (2012)
13.	348 Improved varieties	Different provinces, China	Jia et al. (2015)
14.	36 Landraces	Bhutan	Anonymous (2008)
15.	26,670 Accessions	China	Wang et al. (2012)
16.	4330 Accessions	India (NBPGR)	Dwivedi et al. (12)
17.	3500 Accessions	France	Dwivedi et al. (12)
18.	2512 Accessions	(Bangalore) India	Dwivedi et al. (12)
19	1000 Accessions	USA	Dwivedi et al. (12)
20.	850 Accessions	France	Dwivedi et al. (12)
21.	712 Accessions	Kenya	Dwivedi et al. (12)
22.	350 Accessions	Mexico	Dwivedi et al. (12)
23.	07 Local varieties	(Tamil Nadu) India	Padulosi et al. (2009)
Proso millet (common millet or Vari) (*Panicum miliaceum* L.)			
1.	01 Cultivar	Khandesh (M.S.), India	Khairnar et al. (2016b)
2.	835 Accessions	Different countries	Upadhyaya et al. (2006)
3.	835 Accessions	Nepal	Ghimire et al. (2017)
4.	(i) 88 Accessions	China	Liu et al. (2016)
	(ii) 56 Cultivated varieties		
	(iii) 32 Landraces		
5.	849 Accessions	India (ICRISAY)	Singh and Upadhyaya (2016)
6.	8778 Accessions	Russian Federation	Habiyaremye et al. (2017)
7.	6517 Accessions	China	Habiyaremye et al. (2017)
8.	5022 Accessions	Ukraine	Habiyaremye et al. (2017)

(continued)

Table 27.2 (continued)

Sl. No.	Form and No. of Germplasm	Country/Region	Reference
9.	1432 Accessions	USA	Habiyaremye et al. (2017)
10.	842 Accessions	India	Habiyaremye et al. (2017)
11.	721 Accessions	Poland	Habiyaremye et al. (2017)
12.	400 Accessions	Mexico	Habiyaremye et al. (2017)
13.	302 Accessions	Japan	Habiyaremye et al. (2017)
14.	8778 Accessions	Russian Federation	Dwivedi et al. (2012)
15.	6517 Accessions	China	Dwivedi et al. (2012)
16.	3976 Accessions	Ukraine	Dwivedi et al. (2012)
17.	1046 Accessions	Ukraine	Dwivedi et al. (2012)
18.	721 Accessions	Poland	Dwivedi et al. (2012)
19	713 Accessions	USA	Dwivedi et al. (2012)
20.	400 Aaccessions	Mexico	Dwivedi et al. (2012)
Little millet (Sawa) (*Panicum sumatrense* Roth ex Roem. and Schult.)			
1.	20 Cultivars	Madhya Pradesh, India	Kumar et al. (2017a, b)
2.	03 Cultivars	Khandesh (M.S.), India	Khairnar et al. (2016a, b)
3.	08 Accessions of landraces	Andhra Pradesh, Odisha, India	Elangovan et al. (2017)
4.	462 Accessions	Different Countries	Upadhyaya et al. (2006)
5.	105 Accessions	Coimbatore, Tamil Nadu, India	Selvi et al. (2014)
6.	07 Landraces	India and Nepal	Reghupathy et al. (2016)
7.	(i) 20 Cultivars (ii) 100 Landraces	Madhya Pradesh, India	Kumar et al. (2017a, b)
8.	15 Varieties	South Asia	Anonymous (2014)
9.	10 Varieties released	India	Kumar (2005)
10.	473 Accessions	India (ICRISAT)	Singh and Upadhyaya (2016)
11.	544 Accessions	India	Dwivedi et al. (2012)
12.	08 Local varieties	(Tamil Nadu) India	Padulosi et al. (2009)
13.	460 Accessions	(ICRISAT) India	Upadhyaya et al. (2014)
14.	109 Accessions	(Tamil Nadu) India	Nirmalkumari et al. (2010)
Kodo millet (Kodra) (*Paspalum scrobiculatum* L.)			

(continued)

Table 27.2 (continued)

Sl. No.	Form and No. of Germplasm	Country/Region	Reference
1.	01 Cultivar	Khandesh (M.S), India	Khairnar et al. (2016a, b)
2.	02 Accessions of landraces	Andhra Pradesh, Odisha, India	Elangovan et al. (2017)
3.	656 Accessions	Different countries	Upadhyaya et al. (2006)
4.	05 Landraces	India and Nepal	Reghupathy et al. (2016)
5.	07 Varieties	South Asia	Anonymous (2014)
6.	665 Accessions	India (SCRIBT)	Singh and Upadhyaya (2016)
7.	2170 Accessions	India (Delhi)	Dwivedi et al. (2012)
8.	1111 Accessions	India (Banglore)	Dwivedi et al. (2012)
9.	656 Accessions	India (Patancheru)	Upadhyaya et al. (2014)
10.	01 Local variety	India (Tamil Nadu)	Padulosi et al. (2009)
Teff [*Eragrostis tef* (Zucc.) Trotter]			
1.	3842 Accessions	(PGRC/E) Ethiopia	Seyfu (1997)
2.	35 Cultivars	Ethiopia	Ebba (1975)
3.	60 Genotypes	Ethiopia	Abraha (2016)
4.	36 Genotypes	Ethiopia	Jefar (2015)
5.	15 Genotypes (10 accessions + 5 varieties)	Ethiopia	Asfaw and Danno (2011)
6.	10 Varieties	Ethiopia	Abate et al. (2013)
7.	42 Varieties	Ethiopia	Marga (2018)
8.	10 Genotypes	Ethiopia	Worku et al. (2018)
Indian barnyard millet (*Echinochloa frumentacea* Link.) [Syn. *Echinochloa colona* var. *frumentacea* (Link.) Ridl.]			
1.	01 Cultivar	Khandesh (M.S.), India	Khairnar et al. (2016a, b)
2.	743 Accessions	Different Countries	Upadhyaya et al. (2006)
3.	02 Accessions	Nepal	Ghimire et al. (2017)
4.	05 Landraces	India and Nepal	Reghupathy et al. (2016)
5.	11 Variety	South Asia	Anonymous (2014)
Japanese barnyard millet (*Echinochloa utilis* Ohwi *et* Yabuno) [Syn. *E. esculenta* (A. Braun) H.Scholz.]			
1.	120 Cultivars	Japan	Yabuno (1987)
2.	01 Accession	Florida, USA	Dvorakova et al. (2015)
3.	02 Varieties	Japan	Anonymous (2005)
Barnyard millet [*Echinochloa colona* (L.) Link.]			
1.	65 Accessions	ICRISAT, India	Wallace (2015)

(continued)

Table 27.2 (continued)

Sl. No.	Form and No. of Germplasm	Country/Region	Reference
Barnyard millet [*Echinochloa crus-galli* (L.) Beauv.]			
1.	30 Accessions	ICRISAT, India	Wallace (2015)

sorghum is better adapted to dry and cool conditions. Irrigation has a positive and significant effect on yield.

Olatoye et al. (2018) investigated signatures of clonal adaptation in sorghum to the precipitation gradient in West Africa using a panel (n = 607) of sorghum accessions from diverse agroclimatic zones of Nigeria. They observed significant correlations between common garden phenotypes of three putative climate-adaptive traits (flowering time, plant height, and panicle length) and climatic variables. They characterized the panel at >400,000 single-nucleotide polymorphisms (SNPs) using genotyping by sequencing (GBS). Redundancy analysis indicated that a small of SNP variation can be explained by climate (1%), space (1%), and climate collinear with space (3%). Discriminant analysis of principal component identified three genetic groups that are distributed differently along the precipitation gradient. Genome-wise association studies were conducted with phenotype and three climatic variables of overall enrichment of associations near a priori candidate genes implicated in flowering time, height, and inflorescence architecture in cereals, but several significant associations were found near a priori candidates including photoperiodic flowering regulators SbCN12 and Ma6. These findings together suggested that a small (3%) but significant proportion of nucleotide variation in Nigerian sorghum landraces reflects clinal adaptation along the West African precipitation gradient. Maccarthy and Vlek (2012) evaluated the potential impact of climate change on sorghum grain yield under different crop residue and nutrient management system in a small holder farming system. They used Agricultural Production System Simulator (APSIM) in this scenario analysis. They employed two crop residue management types (crop residue retention in soil and crop residue removal) and fertilizer management (no fertilization and application of 41.30 kg ha^{-1}). The impact of crop residue management on grain yield was lower under climate change weather conditions. This can be attributed, in their opinion, to higher soil moisture stress, which also contributed to lower rate of soil carbon decomposition in the topsoil. They noted instability (interannual standard deviation) in grain yield, which was higher under climate change (0.13–0.21) weather conditions than under historical (0.04–0.11) weather conditions. This was reflected in a higher change in yield and thereby rendering sorghum production under rain-fed agriculture riskier.

Dossou-Aminon et al. (2014) examined farmers' perceptions and adaptation strategies to mitigate impact of climate change on sorghum production and diversity in Northeastern Benin. They interviewed 300 sorghum farmers from 15 villages. They found that farmers in these villages were able to recognize that temperatures were increased and several fluctuations were observed in the rainfall pattern. Perceptions of farmers about climate scenarios involved low productivity, soil poverty, increase in damage by insects, and sorghum varietal diversity loss. Sorghum farmers developed strategies to face these impacts due to climate change such as (i) sowing of drought-resistant varieties, (ii) utilization of fertilizers, (iii) resowing, and (iv) rotation and/association of sorghum production with leguminous crops. These authors, therefore, also made suggestions like organization of information campaigns and institutional strengthening including farmers, agricultural extension agencies, NGOs, decision-makers, and public investment programs. They further called attention of breeders and policy-makers to create an enabling environment to lend support to farmers' adaptation to climate change.

4.2 Pearl Millet

A fair realization of the impacts of climate on crop productivity is a fundamental requirement to enhance climate resilience in crop varieties through breeding or for adapting current varieties more resilient to climate-induced stress through management options employing different strategies to respond the contrary impacts of climate change on crop economic part.

Pearl millet is stiff, climate-smart grain crop, idyllic for environments prone to stresses (drought and heat). It is a water-saving, drought-tolerant, and climate change-compliant crop. It germinates well under optimum temperature (25–30 °C) but can be also planted under cool soil conditions, before the soil temperature reaches 23 °C. Tillering from primary tillers at all stages of apical development is every 40–45 °C days. The optimal rainfall prerequisite of pearl millet differs between 300 and 350 mm and can be also cultivated in low annual rainfall receiving (<300 mm) areas. Due to deep root penetration, pearl millet performs reasonably in unpredicted weather conditions. It quickly reacts to good production options such as planting time, planting density, inter-/intra-row spacing, nitrogen application, and irrigation, besides high growth rate, large leaf area index, and high radiation use efficiency (Ullah et al. 2016). Ullah et al. (2016) adapted and dilated various strategies in respect of climate change such as (i) improving resource use efficiency (e.g., water use efficiency, nitrogen use efficiency, radiation use efficiency), (ii) production options under changing climate (e.g., adjusting planting time intra-row species), (iii) crop modeling to avoid wasting time as a strategy for further implementation in the fields, and (iv) briefing farmers/cultivars regarding fundamentals of climate information inclusive of all possible mitigation and adaptation strategies.

Singh et al. (2017) used the modified CSM-CERES-Pearl Millet model to evaluate the genetic traits of pearl millet for adaptation to climate change at selected sites in India and West Africa. In higher rainfall environments of Aurangabad and Bijapur, the potential yield gains with the 10% longer maturity cultivar are observed to the extent of 47% as compared to the baseline cultivar (Sharda) under current climate and climate change by midcentury. Although the yields increased with 10% longer maturity at other sites, viz., Hisar, Jaipur, Jodhpur, and Bikaner, they were nonsignificant statistically in both present and future climates. In their opinion, the baseline cultivar life cycle duration (ICMK-356) remained the highest yielding at the four locations under both climate regimes, possibly due to better fit to the rainfall patterns of those sites, which minimized the benefit of the longer maturity types. Likewise, at Sadore and Cinzana (West Africa) under base climate, the yields were higher for the baseline cultivar (CIVT) as compared to type yields with the 10% shorter or longer maturity cultivar. The baseline cultivar went the highest yield at Sadore under climate change. However, at Cinzana, a 4% increase in yield was simulated with a 10% longer maturity cultivar, which was, in their opinion, statistically nonsignificant. They reached to a conclusion that the cultivars that are of longer maturity in current climate will generally be more suitable as the warmer climate typically shortens the life cycle and longer maturity cultivars will compensate for these conditions and produce higher yields than the default baseline cultivars. They further stated that identification of a proper cultivar according to the length of growing period is the best way to tackle climate change impacts because sufficient genetic diversity exists in pearl millet maturity groups. This would minimize drought and heat stress during the crop life cycle, and the available seasonal resources would be fully utilized. This investigation will help the breeders to assess new promising traits of pearl millet for adapting to climate change.

Renolds et al. (2016) discussed a need for an integrated and coordinate approach worldwide to maintain productivity under climate change. They focused sorghum and pearl millet in both millets, besides other cereals. In their opinion, the global condition of agricultural research will greatly improve ability to develop crops and cropping systems which will be more resilient in the face of climate change. They proposed a better-coordinating and fairly standardized way to crop research in domains such as (i) characterization of target agroecosystems, (ii) standardized experimental environments, (iii) phenotyping platform, and (iv) comparative biology. This approach will boost, according to them, the cost-effectiveness and facilitate genetic gains of these crops.

Haussmann (2012) experimented for breeding strategies for adaption of pearl millet to climatic variability and changes in West Africa. They proposed a wide range of crop improvement options for enhancing adaptation to climate variability. These include the choice of type of cultivar (degree of heterozygosity and heterogeneity); direct selection in multiple environments, including farmer participatory testing; indirect selection for individual adaptation traits using conventional or genomic selection methods; a dynamic gene pool management approach; and selection for responsiveness or compatibility to improved crop and soil management techniques. Seed systems, in their opinion, need to be strengthened so that they effectively provide access to new varieties and a diverse range of varieties that respond to farmer's current and evolving needs, including adaptation to variable and changing climatic conditions. They made observations on traits such as photoperiod sensitivity, plastic tillering, very early maturity, and flooding tolerance.

Gupta et al. (2014) studied the impact of climate change, especially the effects of rainfall and temperature, on pearl millet in India during 1966–1999. They concluded that the greater the rainfall, the higher the yield. Irrigation has a positive and considerable effect on the yield. They noted, however, that the average temperature is highly insignificant. The sign, in their opinion, is positive, suggesting the possible hardiness of pearl millet to increasing temperatures. They further stated that these results corroborate with agronomic researches which indicate that the pearl millet is resistant to drought and it is also thought more efficient in the utilization of soil moisture. It shows a higher degree of heat tolerance than sorghum.

Johannes (2015) observed trends of pearl millet yields under climate variability conditions in the Oshana region of Namibia. Climate change impacts have been felt in Namibia over the past years. It was noted that in 2009–2011, pearl millet yields were significantly reduced due to severe floods. The rainfall data showed a change in rainfall intensity with shorter rainfall seasons and late arrival on the rainfall. Pearl millet is a rain-fed, dry crop type and does not grow well in waterlogged soils, contributing the low yields under flood conditions. The author suggested farmers to start sowing on the every 25 December every season. He also advised farmers to make use, at least, of both pearl millet available cultivars through intercropping, one with long duration to reach maturity and the other one with short duration to reach maturity depending on the climate situation. He further suggested having alternative crop varieties like cassava and rice as a possible solution to climate change adaptation. This alternative, in his opinion, will enhance diversification of pearl millet cultivars within a crop field to adapt to climate scenarios. A method of variety types of pearl millet cultivars with high degrees of heterozygosity and genetic heterogeneity for adaptation traits will help to achieve better individual and population buffering capacity in pearl millet. He opined that crop improvement exclusively cannot produce miracles. Therefore, the development of new improved and climate-proof cultivars must go simultaneously with sustainable soil fertility management and water conservation and drainage techniques.

4.3 Finger Millet

Onyango (2016) highlighted the positive attributes of finger millet based on agricultural research and development reports in Kenya, especially when the frequency and increased intensity of extreme climatic events have become additional challenges for global agriculture. The author pointed out a need for focusing on sound nutritional and medicinal values of finger millet to the residents since consumption patterns of finger millet are very specific and continue to remain region-specific. Their population in the broader range is essential. He further emphasized to prepare ready-to-use or ready-to-cook products, which would help in increasing its consumption among non-millet consumers and address the problem of food insecurity. He clearly earmarked four broad areas of adaptation of crop production systems as climate change: (i) new crop introduction and phasing out of previous ones, (ii) development of new varieties of existing crops, (iii) evolution of crop management practices, and (iv) dealing with climate uncertainty through the provision of information. These adaptions, in his opinion, will involve many trade-offs and possibly some synergies at different scales, requiring decisions to be made. He specifically emphasized a need for characterization of finger millet varieties to support farmers' decision-making.

Shibario et al. (2016) carried out an investigation on finger millet production in lower eastern Kenya in view of constraints and climate change. Eastern Kenya is characterized by aridity and semiaridity region. Because of drought-resistant nature, high nutritional content, and the ability to produce with few inputs, availability of finger millet is one of the crops to combat food insecurity. These authors used Logit model to determine the effect of education, land size, age, and gender on finger millet production. The imports revealed constraints like lack of seeds, pests and diseases, overdependence on maize, and climate change. The author saw potential for production in aspects of climate change, extension services, nutritional content, and marketability. In their opinion, education and land size had a positive effect on finger millet production, while gender had a negative effect. Based on the Logit model results, extension service providers should lay emphasis on farmer's age, education, land size, and gender when deciphering a target group for finger millet dissemination. Respondents thought finger millet to have a lot of potential in dealing with effects of climate change due to its drought-resistant nature and ability to provide good yield with low rainfall.

Masood and Azam Ali (2007) grew two landraces of finger millet, viz., TZA-01 and T2M-01, in glasshouses under two moisture regimes (fully irrigated and after a drought) to investigate the effects of environmental stress on the growth, SPAD measurement, radiation use efficiency, and yield. They imposed the drought treatment at 28 DAS beyond what was applied to the drought treatment. Growth and development were monitored between 21 DAS and 105 DAS. They showed a clear subjection of the two finger millet landraces to a progressively severe treatment of soil moisture stress. They recorded reduction in the growth and development of both the landraces in this study between 35 DAS and 105 DAS. The similarity in the moisture content at the time of imposing drought treatment (i.e., at 28 DAS) indicated the uniformity of the condition in the glasshouses with respect to the soil moisture content. These authors reached to a conclusion that drought has a significant influence on the vegetative and reproductive growth of finger millet. Besides the water use, other parameters were dependent on soil moisture content. This knowledge will be, in their opinion, useful to strengthen the research activities.

4.4 Foxtail Millet

Ning Na et al. (2017) studied correlation between grain quality of foxtail millet and environmental factors. Variations in climate and ecological resource lead to variations in grain quality. Quantification of effects of environmental conditions is critical for the large-scale promotion of high quality. These authors analyzed the said correlations during the growing season (May to September) using multivariate statistical analysis under different ecological/climate conditions at five locations in China. They inferred that the difference in grain quality across different locations was mainly affected by altitude, followed by precipitation, diurnal temperature range, latitude, sunshine hours, and ≥20 °C accumulated temperature. The precipitation of July and diurnal temperature range of July to September had the greatest effect on grain quality. Precipitation and ≥20 °C accumulated temperature showed a significant negative effect. Thus, the effect of environmental factors on grain quality of foxtail millet is the result of a combination of factors. According to these authors, the regression equation proposed in this study can be used to predict and forecast grain quality of foxtail millet.

Yang et al. (2016) investigated water use efficiency (WUE) of foxtail millet in relation to climate charge in China, particularly in arid and semiarid northwest regions. They examined the impact of climate change on WUE and considered yield, soil water content, rainfall, and temperature data at three experimental sites between 1978 and 2007 at Pingliang, Yulin, and Huhebot. The accumulated temperature increased by 502, 541, and 857 °C, respectively, at these sites. These temperatures have a significant climate-warming trend during the period of study. Temperature warming and decreasing precipitation, according to these authors, caused severe droughts in the said region. These resulted in high WUE to adapt arid environment during the said period. Strategies such as improving crop distribution, increasing plant areas of foxtail millet with high WUE, and the adoption of drought-resistant farming techniques are advised by the authors to mitigate the influences of global warming on crop production, especially in arid and semiarid regions.

4.5 Proso Millet

Zhang et al. (2012) studied leaf senescence and antioxidant enzymes in three cultivars of proso millet (*Panicum miliaceum* L.) after anthesis. They investigated the changes in chlorophyll content, antioxidant enzymes (SOD, CAT, POD), MAD, and superoxide anion during seed filling to maturity with the primary goal of using these indices in the selection of drought-resistant varieties. The cultivar 'Ningmi 13' was noted for slower degradation ratio of chlorophyll content, higher activity of SOD and CAT, lesser POD, and smaller accumulation of MDA and superoxide anion, resulting to delayed leaf senescence and prolonged leaf functional period. The longer functional leaf period and higher SOD activities can be, therefore, used as indices for selection of drought-tolerant genotypes (cf. Dai et al. 2011).

Lin et al. (2006b) used a forward subtracted cDNA library constructed from normally watered leaves and leaves dehydrated after drought to investigate the genes induced by drought in *Panicum miliaceum* (proso millet). They employed a suppressive subtraction hybridization technique to construct the cDNA library and 60 positive clones identified and sequenced. Out of a total of 60 sequences, only 32 EST were observed highly homologous to known plant sequences manifested in a response to abiotic or biotic stress. Furthermore, 28 ESTs are homologous to known proteins involved in signal transduction, transcription, and protein processing. AFLP markers were generated from this cDNA library (Lin et al. 2006a) to analyze genes differentially expressed in seedlings watered normally, those subjected to drought, and seedlings rehydrated after drought. Twelve fragments were amplified from the leaf samples under drought and rehydration regimes. Using the same cDNA library, Lin et al. (2008) carried their study further. Their results showed that its expression declined under drought, increased after rehydration, and then settled to normal levels 6 h after rehydration. Thus proso millet utilizes moisture efficiently.

Nielsen and Vigil (2017) collected proso millet water use data and yield data from 1995 to 2016 as a part of ongoing long-term alternative crop rotation experiment conducted at the USDA-ARS Central Great Plains Research Station under dry land conditions. The objectives of this experiment were to determine the water-limited yield relationship for proso millet and to identify environmental factors that cause yields to be lower than predicted by the water-limited yield relationship. They used stepwise linear regression analysis to determine important environmental factors influencing yield. The water-limited yield relationship had a slope on growing season water use, plant-available soil water at planting, precipitation received from 12 to 18 August, number of days in July and August with maximum temperature greater than 36 °C, daily average wind run, and maximum wind gust during the week before swathing explained 88% of yield variability. The regression parameters suggested that plant breeding efforts should be directed toward improving shattering resistance and heat tolerance and that cropping systems management should be directed toward crop sequencing and no-till production methods. This improves precipitation storage efficiency during the non-crop period prior to millet planting and increases available soil water at millet planting.

4.6 Little Millet

Matsuura et al. (2012) using PBC tubes filled with sandy soil investigated the effect of moisture before and after flowering in little millet. A significant yield reduction was recorded, as compared to the well-watered plants, when the drought treatment was implemented at early developmental stage, i.e., before flowering. Terminal drought condition occurring from the flowering stage to the harvesting phase of little millet contributed to a significant yield loss.

4.7 Kodo Millet

Kumar et al. (2017c) carried out experimental analysis, looking for climate-resilient potential, among 43 advanced breeding lines of kodo millet (*Paspalum scrobiculatum*) to interpret the stress tolerance mechanism and homogenize crop improvement parameters of widespread economic domestication of the hilltop. They observed maximum canopy length in genotype BK31 (74 cm) followed by BK6 (73 cm), BK48 (68 cm), BK2, and BK23 (64 cm each). They noted 50% flowering by 64–75

DAS. In contrast, very early blooming genotype (PCGK18, 50 DAS; PGCK 8 and 19, 50 DAS; PGCK 13, 59 DAS) exhibited comparative lower yield owing to exceedingly short vegetative phase. Among early maturing accessions, viz., PGCK 18 (81 DAS); PGCK 8, PCK 16, and BK60 (94 DAS); and others in similar category suffered from yield penalty. Hence, genotypes should be bread for 100–105 DAS to optimize yield potential. They concluded that optimal vegetative growth is essential for grain yield physiology and yield increases in accordance with total biomass following normal distributional curve. Genotype BK 48 with 70.50 cm plant height turned to reproductive phase by 76 DAS, accomplished crop cycle by 111 DAS, and produced significantly higher biomass and maximum yield. The plant vegetative organs, during primary development phase, are well protected by vegetative tissues, and unless the stress is semilethal or lethal, the reproductive cells and or structures respond to unfavorable conditions indirectly, as mediated by the vegetative plant organs. The author concluded that the success of reproduction as well as the realization of yield potential of a given genotype, however, are dependent not only on the stress sensitivity of the reproductive and grain-filling stages but on overall plant growth and development. They further opined that resourceful photosynthesis system and stem reserve accumulation throughout the vegetative development segment have a definite function in the formation of reproductive organs and thus may directly affect final yield.

4.8 Teff

The production areas of teff range from the cool highlands to the dry lowlands that are generally associated with moisture deficit during critical stages of plant development. Studies are conducted to investigate the effect of moisture deficit on the performance of teff plants (Degu et al. 2008; Mengistu 2009; Ginhot and Farrant 2011). These studies showed that there is genetic variability among the genotypes investigated suggesting that the teff gene pool harbors moisture stress-tolerant genotypes that could be examined through efficient tools such as molecular markers.

Teff is also investigated for the anticipated changes in the climate and expansion of farmlands in the rift valley areas. Asfaw and Dano (2011) showed presence of broad intraspecific variability among the ten teff accessions studied for salinity tolerance. One accession also showed presence of genetic variability for tolerance to soil acidity and aluminum toxicity in selected genotypes (Abate et al. 2013).

Lodging is the major constraint to yield increases in teff. It is so weak in nature and cannot endure several internal and external factors like wind and rains. Recently, several semidwarf and lodging-tolerant candidate lines have been developed (Marga 2018; Worku et al. 2018). Zhu et al. (2012) also experimented on a similar line and succeeded to improve yield by semidwarfing mutants. Jost et al. (2015) also stated that productivity in teff is extremely low mainly due to susceptibility to lodging. This feature, viz., lodging in his opinion, is aggravated by wind, rain, or application of nitrogen fertilizer. These anthers, therefore, developed semidwarf lodging-tolerant teff line, called "Kegne."

Salinization of soil is a major factor to limit crop production particularly in arid and semiarid regions of the world. Teff is cultivated in such areas. Salt stress is known to perturb a multitude of physiological processes. It exerts its undesirable effects through inhabitation and ionic toxicity (Norean and Ashraf 2008; Munns et al. 2006). Increased salinity causes a significant reduction in germination percentage, germination rate, and root and shoot length vis-à-vis weight (Jamil et al. 2006). To deal with this problem, in case of teff, Asfaw and Dann screened 15 lowland teff genotypes. They divulged the presence of broad intraspecific genetic variation in teff accessions and varieties for salt tolerance.

Araya (2015) assessed impacts of climate change on teff (*Eragrostis tef*) productivity in Debre Zeit area of Ethiopia in three different periods using different models and pathway. They recorded median yields which increased and decreased by up to 10% and 38% for early and late sowing, respectively. Increase in yield was observed mainly due to early sowing and efficient use of rainwater over the growing period, relatively conductive early seedling establishment and better synchronization of the crop growing cycle with the rainy period. On the contrary, they noted late sowing resulting in significant yield reduction because of poor synchronization of the rainy period with the growing cycle of the crop, especially exposure to the long dry period after the reproductive phase. They concluded that rainfall distribution and amount have the greatest impact on teff yield under future time, and, therefore, early sowing can be on adaptive strategy for teff under future climate.

Felix (2018) investigated the impact of climate change on teff production in southeast Tigray (Ethiopia). He employed farm characteristics and socioeconomical settings in a low-income developing country. He used the Ricardian model to evaluate data gained. Out of 14 predictor variables fitted in the model, 6 variables, viz., climate factors, adaptation strategies, production factors, weather and climate information, socioeconomic factors, and agroecology, were found to have significant influence on net revenues. Increase or decrease in temperatures affects teff revenues. Thus, climate factors and adaptation to climate change seem, according to the author, to be strong determinants in influencing teff revenues.

Scientist evaluated climate change impacts on crop productivity of teff using the Geographic Information Systems (GIS). They estimated the effects of altered environments on teff's productivity. They recorded a nonlinear relationship between suitability indices, the output of spatial analysis, and teff yield data collected from varied ecological zones. They also conducted a socioeconomic survey to understand the agricultural activities in the study area. Their results indicated that crop yield varied significantly as a function of climatic variation, and the model is applicable at different levels into consideration of spatial variability of climate.

4.9 Barnyard Millet

Trivedi et al. (2017) assessed barnyard millet (*Echinochloa frumentacea*) diversity in the central Himalayan region for environmental stress tolerance. They noted significant variability in days to 50% flowering, days to 80% maturity, and 1000 seed weight and yield potential of the germplasm. These traits are considered crucial for tailoring new varieties for different agroclimatic conditions. Variations in biochemical traits such as lipid peroxidation, nmol malondialdehyde formed, total glutathione, and total ascorbate content indicated the potential of collected germplasm for abiotic stress tolerance. They identified trait-specific populations that could be useful in crop improvement programs and climate-resilient agriculture.

5 Efforts Dealt with Climate Change and Millets: General Considerations

In semiarid and arid environments where millets are the dominant crops, drought or inadequate moisture is the major abiotic stress affecting productivity. Ajitkumar and Pannerselvam (2014), in the case of pearl millet, showed that drought impacts include growth, yield, membrane integrity, pigment, osmatic adjustment, water relations, and photosynthetic activity. Drought is also usually manifestation of a shortage or absence of rainfall causing drought a loss in rain-fed agriculture. For example, the decline in the level of rainfall during severe drought years in Ethiopia was accompanied by serious reductions in rain-fed agricultural outputs.

Matssura et al. (2012) earmarked the effects of moisture deficit before and after flowering on four millets, viz., proso millet, little millet, foxtail millet, and wild millet [*Setaria glauca* (L.) Beauv.]. Compared to the well-watered plants, a considerable yield decrease was obtained in all of them when the drought treatment was implemented at the early developmental stage before flowering. Nevertheless, terminal drought, which occurs from the flowering stage to the harvesting of the crop, culminated into a considerable yield loss only in proso and little millets, while the effect on foxtail and wild millets was negligible.

The annual rainfall in Niger is about 200 mm. Winkel et al. (1997) investigated the impact of water deficit at three stages of pearl millet development. These three stages were (i) prior to flowering, (ii) at flowering, and (iii) at the end of flowering. The authors concluded that the grain yield of pearl millet was severely reduced when moisture was limited prior to and at the flowering stage but not at the end of flowering. Terminal drought in which irrigation was terminated from the flowering until crop maturity was severe. It resulted in 60% yield loss (Bidinger et al. 1987).

Drought is defined as a temporary reduction in moisture availability in which the amount of available water is significantly below normal for a specified period. Drought can be meteorological, hydrological, or agricultural. Agricultural drought occurs when there is not enough soil moisture to meet the needs of a particular crop at a particular time. The plants or crops have to cope with drought using strategies, viz., drought escape, drought avoidance and drought tolerance, or even drought recovery. Traits associated with drought escape are rapid growth, early flowering, high leaf nitrogen level, and high photosynthetic capacity (Kooyers 2015). Pearl millet matches its phenology to the mean distribution of the rainfall where precipitation in limited and erratic (Sivakumar 1992). The development of main panicle, in case of pearl millet, coincides with an increasing period of rain and thus reduces the risks associated with drought occurring prior to or at the beginning of flowering. Drought avoidance mechanisms generally reduce water loss through transpiration or maintain water uptake during drought period (Fang and Xiong 2015; Kooyers 2015). Traits associated with drought tolerance are increased osmoprotectants and osmotic adjustment (Blum 2005; Kooyers 2015). Desiccation-tolerant or resurrection plants, e.g., wild *Eragrostis nindensis*, stabilize their cells or membranes at desiccated stage (Vander Wilingen et al. 2004).

Initiating to promote millet production is one way to conserve germplasm and to utilize their full potential. For example, the Government of India launched some programs to promote millet farming: (i) Initiative for Nutritional Security through Intensive Millet Promotion (INSIMP) as a part of Rashtriya Krishi Vikas Yojana (RKVY) and (ii) Rainfed Area Development Programme (RADP) as a part of RKVY. Some Indian state governments also aid in improving both production and consumption of millets, for example, (i) state of Kerala, its Agriculture Department implemented mega millet cultivation drive in the backward region with active involvement of the local community; (ii) state of Odisha, took measures to improveme a millet mission in 2016 to fillip farming of millets and also provided market linkage to millet farmers; (iii) state of Maharashtra, announced subsidies for millet; and (iv) state of Karnataka, selling of finger millet in the south India and sorghum in north India through networks (Behera 2017).

Millets exhibit several morphological, molecular, biochemical, and physiological attributes. These confer better tolerance to environmental stresses than other crops such as cereals. First, the short life cycle of millets helps in escaping from stress since they need only 12–14 weeks to complete their seed-to-seed life cycle. Nevertheless, the prevalence of stress conditions and their impacts are overcome by some traits, e.g., small leaf areas, short stature, thicker cell walls, and the capability to form dense root system (Li and Brutnell 2011). Millets have enhanced photosynthetic rates at worm conditions and confer immediate water use efficiency and nitrogen use efficiency (Li and Brutnell 2011). Also, increase in biochemical activities, e.g., enhanced levels of antioxidants, reactive oxygen species and their scavenging enzymes, activities by catalase and superoxide, and synthesis of osmolytes and other stress-related proteins, is known in response to the abiotic stresses in case of foxtail millet (Lata et al. 2011), little millet (Ajithkumar and Pannerselvam 2014), and teff (Smirnoff and Colombe 1988).

The use of molecular biomarkers, sequence information, creation of mapping populations, and mutants has led to the development and release of high-yielding varieties of millets (Joel et al. 2005; Brink 2006). Newly developed hybrids are resistant to diseases and have increased per hectare production as compared to their parent varieties (Joel et al. 2005; ICAR 2017). Millets have vast natural diversity (Tables 27.1 and 27.2), and the release of new hybrids increases this variation by multifold.

The production of edible millets was limited in the past due to lack of suitable machinery and traditional methods like pounding, winnowing, etc. used for the decortication of millet grains. These methods were labor-intensive. Recently, millet-specific threshers, decorticators, and polishers have been fabricated or designed. These have eased the postharvest operations of millets. These paved the way of utilization of millets in the developments of food and products and to check loss of millet production because of climate change and thereby benefit the millet farmers.

6 Miscellaneous Constraints

Various socioeconomics have restricted uses of millet consumption and hence contributed to a loss of a cultivated diversity: (i) millets typically labor-intensive, manual postharvest processing, grain threshing and milling (Rengalakshmi 2005), (ii) low yield as a result of the lack of scientific attention (Plaza-Wuthrick and Tadele 2012), (iii) family farm-level diversity heavily affected by community access to seed which again limited by current rural seed system (Nagarajan et al. 2007), (iv) agricultural policies in different nations having negative impact on cultivation of and research of small millets, and (v) displacement of production partially or completely in many areas by mainstream cereals.

7 Remedial Measures

A target of 70% more food production by 2050 has been set by the World Summit on Food Security (Tester and Langridge 2010). Additional difficulties will be caused by climate change as many regions are becoming drier. The small millets have the potential to meet these challenges because of (i) their drought tolerance and ability to grow under low-input conditions and (ii) health-promoting traits valued by mankind. Our priority, therefore, should be

(I) Exploitation of diversity within seed banks: The small millets possess considerable morphological and genetic sequence variation, which can be used by the breeders to generate improved varieties. The seed bank infrastructure and associated reporting in the scientific literature and in online databases should be more accessible for breeders. Improved funding, coordination, communication, and sharing of genetic resources are necessary to face various problems.

(II) Genes from wild relatives: The wild relations of small millets can serve as donors of useful genes for crop improvement. The wild germplasm generally does not find place in gene/seed/germplasm banks. Only traditional ones are favored. They are also treated as weeds or avoided being invasively mature in some cases.

(III) Weightage for traditional knowledge: Small millets are usually cultivated in remote regions of the world. Diversity is rich in these inaccessible pockets in the hands of some ethnic or aboriginal communities. These human societies have also their own system of classification of landraces or cultivars. Their wisdom can be profitably used using modern technology and made accessible for the world community at large.

(IV) Molecular and genetic knowledge: Several small millets are still lacking molecular and genetic markers, and linkage maps are available for breeders for the crop improvement programs. Sequencing of genome is again a rarity for this group of crop species.

(V) Molecular mechanism: Millet diversity has largely remained untapped at the level of molecular mechanism. Study on this line is very limited. An understanding of the molecular mechanisms underlying various traits can lead to agronomic improvement. A special drive is a need of hour on this line.

(VI) Dominance of few crops: Modern agriculture is dominated by few crop species which has obviously marginalized some indigenous crop species. To redeem this situation, traditional landraces or cultivars or species of small millets should be continued for cultivation. Awareness about millet production because of nutritional content, health benefits, and low inputs should be familiarized to modern communities.

(VII) Support for water-intensive crops should be phased out.

(VIII) Efforts should be pursued to improve agronomic practices of farmers to understand the strategic choice of crop associations and rotations within the production systems.

8 Promises from Minor Millets

Climate change scenario calls for urgent and strategic interventions toward adaptive agricultural measures. A great ally to that end is represented by the genetic resources of minor millets:

(i) Minor millets are fairly adapted to enhance resilience of local production systems and strengthen food and nutrition security, particularly among the poor.

(ii) They have a wide genetic adaptation and are able to grow successfully in diverse soils, varying rainfall regimes, diverse photoperiods, and in marginal, arid, and mountainous terrains wherein major cereals have less possibility of success.

(iii) They have the potential to thrive with low inputs and can withstand severe edaphoclimatic stresses.

(iv) These qualities are also combined with excellent nutritional values and opportunities for strengthening income generation through value addition.
(v) Millets are generally thermophilic and xerophilic. They are hardy crops with short growth periods.
(vi) Millets have an excellent nutritional profile and are usually a non-glutinous food. This renders them easily digestible and nonallergic foods. They are safer for consumption especially for diabetic patients having a low glycemic index.
(vii) Millets have powerful root systems and are able to penetrate down easily to a great depth of soil to extract water and minerals.
(viii) They resist against drought. Also, they are a good element to diversity crop rotations.
(ix) Millets possess a C_4 photosynthesis system; hence they prevent photorespiration and, as a consequence, efficiently utilize the scarce moisture present in the semiarid regions. Being C_4 plants, they are able to close stomatal openings for longer periods and thereby reduce moisture loss significantly through the foliar surfaces.
(x) They have been proved to release fewer greenhouse gases and hence are beneficial in reducing the contributions of the agro-fed sector to global warming.
(xi) The nutrient requirement for millets is minimal, and a few millet varieties can be grown in soils with low fertility, e.g., sandy loam and slightly acid soils.
(xii) Also, millets are mostly pest-free due to their strong disease resistance traits. They thus benefit from reducing the use of pesticides and consequent soil pollution caused by pesticide use.
(xiii) Millets have the potential to reduce the carbon footprint as it has the least global warming potential.

9 Epilogue

This review summarizes influences of climate change on the millets and their diversity and benefits for well-being of humans, apart from human attempts to modify them. Millets have an excellent nutritional profile and are also important for health benefits. Being C_4 plants, they efficiently use water in semiarid or arid regions of the world. Moreover, they have sufficient potential to thrive with low inputs and are genetically diverse with fair number of forms of germplasm. Still, they are called orphan crops, underutilized or minor or small crops, etc. This is so because they often sustain the poverty-driven populations and hence remained neglected scientifically. Live forms of germplasm of millets are available in rather inaccessible pockets in arid or semiarid areas. They are also preserved in gene banks or seed banks. But both sources have not received desired attention for welfare of mankind. The modern and conventional crop improvement techniques have not been employed sufficiently. Moreover, environmental changes and disasters are putting long strides and severely affecting economy especially in arid and semiarid regions of the world.

Millets certainly provide alternative climate smart crops since they adapted better to changing climates. At this backdrop, concrete efforts of donors, policy-makers, agronomists, breeders, and NGOs are dire necessities to enhance the productivity of millets.

Millets as climate change-compliant crops score highly over other grain crops. Furthermore, millet production is helpful to mitigate climate change since it emits fewer greenhouse gases than other grain crops and has less environmental impacts. They have climate-resilient features. Millets also hold great promise for food security and nutrition when agricultural costs are ever-increasing. All these features accentuate millets as crops of choice for the world population. Even though they are ancient crops of primitive human societies, they have been proved to be nutritionally excellent, and hence, of late, they are being accepted as sources of "superfood." Millets also provide the best option to the farmers for achieving the triple objectives of farming, viz., sustainability, profitability, and adaptability. Of course, there is still a need for increasing awareness among the populace by the scientific community and urban elites worldwide for increased consumption of millets. The challenge to feed the ever-growing population with healthy balanced diets and the threats faced by agricultural crops due to changing climate highlight the immediate requirement to exploit the beneficial attributes of millets. Of late, postharvest operation of millets has been eased due to newly designed machinery. A ray of hope to combat food and nutrition security is indiscernible but needs further substantial coordination on all grounds.

References

Abate E et al (2013) Quantitative responses of tef [Eragrostis tef (Zucc.) Trotter] and weeping love grass [Eragrostis curvula (Schrad.) Nees] varieties to acid soil. Aust J Crop Sci 7:1854–1860

Abraha MT (2016) Assessment of the genetic relationship of tef (Eragrostis tef) genotypes using SSR markers. S Afr J Bot 105:106–110

Ajithkumar IP, Pannerselvam R (2014) ROS scavenging system, osmotic maintenance, pigment and growth status of *Panicum sumatrense* Roth under drought stress. Cell Biochem Biophys 68:587–595

Amos OO (2016) Finger millet: food security crop in the arid and semi-arid lands (ASALs) of Kenya. World Environ 6(2):62–70

Anonymous (2005) Guide to growing summer grain and forages in the South Coast region, Western Australia. Miscellaneous Publication 20/104, Andrea Hills, Sally-Anne Penny, GRDC Project DAW 722 https://researchlibrary.agric.wa.Gov.au/viewcontent.Cgi?article=1035 ontext=misc...

Anonymous (2008) Plant genetic resources of Bhutan Vol. I: Field crops. National Biodiversity Centre, Ministry of Agriculture, Serbithang, Thimphu, Bhutan

Anonymous (2014) Report on: Characterization of land races and local varieties of finger millet and other small millets (Annexure 2.1). DHAN Foundation, India, pp.1–65

Anumalla M, Roychowdhury R, Geda CK, Mazid M, Rathoure AK (2015) Utilization of plant genetic resources and diversity analysis tools for sustainable crop improvement with special emphasis on rice. Int J Adv Res 3(3):1155–1175

Araya A (2015) Assessing impacts of climate change on tef (Eragrostis tef) productivity in Debrezeit area, Ethiopia. Int J Agric Sci Res 4(3):39–48
Asfaw KG, Dano FI (2011) Effects of salinity on yield and yield components of tef [*Eragrostis tef* (Zucc.)Trotter] accessions and varieties. Cur Res J Biol Sci 3:289–299
Baker RD (2003) Millet production. Guide A-414 2003: cooperative Extension Service. Las Cruces, USA, College of Agriculture and Home Economics, New Mexico University
Bashir EMA et al (2014) Characterization of Sudanese Pearl millet germplasm for agro-morphological traits and grain nutritional values. Plant Gen Reso Char Util 12(1):35–37
Behera MK (2017) Assessment of the state of millets farming in India. MOJ ECO Environ 2(1):16–20
Bermejo JEH, Leon J (eds) (1994) Neglected Crops: 1452 from a different perspective. Plant Production and Protection Series No 26 FAO, Rome
Bidinger FR et al (1987) Assessment of drought resistance in pearl millet [*Pennisetum americanum* (L.)] Leeke. 2 estimation of genotypes response to stress. Aust J Agric Res 38(1):49–59
Blum A (2005) Drought resistance, water-use efficiency and yield potential-are they compatible, dissonant or mutually exclusive? Aust J Agri Res 56(11):1159–1168
Brink M (2006) Setaria italica (L.) P. Beauv. Record from Protabase. In: Brink M, Belay G (eds) PROTA (Plant Resources of Tropical Africa/Resources-vegetables de I'Afrique tropicale). Wageningen, Netherlands
Brutnell TP et al (2010) Setaria viridis: A model for C_4 photosynthesis. Plant Cell 22(8):2537–2544
Chakraborty S, Pattanayak A, Mandal S, Das M, Roychowdhury R (2014) An overview of climate change: causes, trends and implications. In: Roychowdhury R (ed) Crop improvement in the era of climate change. IK International Publishing House, New Delhi, pp 1–29
Chander S et al (2017) Analysis of spatial distribution of genetic diversity and validation of Indian foxtail millet core collection. Physiol Mol Biol Plants 23(3):663–673
Changmei S, Dorothy J (2014) Millet-the frugal grain. Int J Sci Res Rev 3:75–90
Chowdari KV et al (1998) Genotype identification and assessment of genetic relationships in pearl millet [*Pannisetum glaucum* (L.) R.Br.] using microsatellites and RAPDs. Theo Appl Genet 97:154–162
Claid M, Ereck C (2008) Morphological variation of Sorghum landrace accessions on-farm in semi-arid areas of Zimbabwe. Int J Bot 4(4):376–382
Dai H et al (2011) Leaf senescence and reactive oxygen species metabolism of broomcorn millet (*Panicum miliceum* L.) under drought condition. Aust J Crop Sci 5:1655–1660
Dalton TJ, Zereyeus YA (2013) Economic impact assessment of Sorghum, millet and other grains CRSP: Sorghum and Millet germplasm development research. INTSORMIL Scientific Publications, 20
Degu HD et al (2008) Drought tolerance of Eragrostis tef and development of roots. Int J Plant Sci 169:768–775
Dossou-Aminon I et al (2014) Farmers' perceptions and adaptation strategies to mitigate impact of climate change scenario on sorghum production and diversity in North eastern of Benin. Int J Curr Microbiol App Sci 3(10):496–509
Doust AN et al (2009) Foxtail millet-A sequence-driven grass model system. Plant Physiol 149:137–141
Dvorakova Z et al (2015) Comparative analysis of genetic diversity of 8 millet genera revealed by ISSR markers. Emirates J Food Agri 27(8):617–628
Dwivedi S et al (2012) Millets: genetic and genomic resources. In: Janick J (ed) Plant breeding reviews. Wiley, pp 247–374
Ebba T (1975) Tef cultivars: morphology and classification. University College of Agriculture, Dire Dawa
Elangovan M et al (2017) Collection and characterization of millet genetic resources from Andhra Pradesh and Odisha. Int J Agric Sci 9(4):4650–4658
Fang VJ, Xiong LZ (2015) General mechanisms of drought response of their application in drought resistance improvement in plants. Cell Mol Life Sci 72(4):673–689

Felix MT (2018) The impact of climate change on teff production in southeast Tigray, Ethiopia. J Agric Econ Rural Devel 4(1):389–396

Ghimire KH et al (2017) Diveresity and utilization status of millets genetic resources in Nepal. In: Proceedings of 2nd National Workshop on CUAPGR (22–23 May) Dhulikhet. NAGRC, FDD, DaO and MoAD, Kathmandu, Nepal, pp 215–225

Ginhot ZG, Farrant JM (2011) Physiological response of selected Eragrostis species to water-deficit stress. African J Biotechnol 10:10405–10417

Gupta S et al (2014) Impact of climatic change on the Indian economy: evidence from food grain yields. Clim Chang Eco 5(2):1–29

Habiyaremye C, Matanguihan JB, et al (2017) Proso Millet (*Panicum miliaceum* L.) and its Potential for Cultivation in the Pacific Northwest, U.S.: A Review. Front Plant Sci. https://doi.org/10.3389/fpls.2016.01961

Hasanuzzaman M, Nahar K, Alam M, Roychowdhury R, Fujita M (2013) Physiological, biochemical, and molecular mechanisms of heat stress tolerance in plants. Int J Mol Sci 14(5):9643–9684

Hasanuzzaman M, Roychowdhury R, Karmakar J, Dey N, Nahar K, Fujita M (2015) Recent advances in biotechnology and genomic approaches for abiotic stress tolerance in crop plants. In: Devarajan T, Jeyabalan S (eds) Genomics and Proteomics: concepts, technologies and applications. Apple Academic Press, Canada, pp 333–366

Haussmann BIG (2012) Breeding strateging for adaptation of pearl millet and sorghum to climate variability of change in West Africa. Journal of Agronomy And Crop Sci:1–37. https://doi.org/10.1111/j.1439-037x2012.00526x

ICAR-Indian Institute of Millets Research (2017) Millets annual report 2016–17. http://millets.res.in/annual report/ar16-17pdf

Jamil M et al (2006) Effect of salt (NaCl) stress on germination and early seedling growth of four vegetable species. J Central Eur Agric 7(2):272–282

Jefar H (2015) Grain yield variation and association of major traits in brown-seeded genotypes of tef [*Eragrostis tef* (Zucc.) Trotter]. Agri Food Sec 4:7–9

Jia G et al (2015) Microsatellite variations of elite Setaria varieties released during last six decades in China. PLoS One 10(5):1–15

Joel A et al (2005) A high-yielding finger millet variety CO (Ra) 14. Madras Agric J 92.375–380

Johannes A (2015) Trends of pearl millet (*Pennisetum glaucum*) yields under climate variability conditions in Oshana region, Namibia. Int J Ecol Ecosolut 2(4):49–62

Jost M et al (2015) Semi-dwarfism and lodging tolerance in tef (*Eragrostis tef*) is linked to a mutation in the a-tubulin 1 gene. J Exp Bot 66(3):933–944

Kazi T, Auti SG (2017) Screening of higher mineral containing finger millet landraces from Maharashtra. Int J Food Sci Nutr 2(3):21–25

Khairnar SB et al (2016a) Diversity and utilities of Sorghum in tribal areas of Khandesh region (Maharashtra) India. Int J Recent Sci Res 7(12):14774–14778. http://www.recentscientific.com

Khairnar SB et al (2016b) Agrobiodiversity of minor millets in tribal tehsils of Khandesh (Maharshtra: India). Scholars's World (Special Issue) pp 100–105. www.scholarsworld.net

Khairwal IS et al (2007) Evaluation of identification of promising pearl millet germplasm for grain and fodder traits. J SAT Agric Res 5(1):1–6

Kisandu DB et al (2007) Germplasm collection and evaluation of finger millet in Tanzania. challenges and opportunities for improved production. In: Mgonja MA, Leene JM, Emanyasa E, Srinivasprasad S (eds) Proceedings of the first International Finger Millet Stakeholder Workshop, 13–14 September 2005. ICRISAT, Nairobi, pp 23–33

Kooyers NJ (2015) The evolution of drought escape and avoidance in natural herbaceous populations. Plant Sci 234:155–162

Kumar A (2005) Studies on grain smut of little millet (*Panicum sumatrense* Roth ex Roemer and Schultes) caused by *Maclpinomyces sharmae* K. Vanky. Thesis for Master of Science, Jawaharlal Nehru Krishi Vishwa Vidyalaya, Jabalpur (M.P.), India

Kumar A et al (2017a) Agro-morphological characters of little millet (*Panicum sumatrense*) associated with grain smut incidence. J Entomol Zool Stud 5(5):356–359

Kumar A et al (2017b) Screening of Little Millet germplasm against smut caused by *Macalpinomyces sharmae*. Int J Curr Microbiol App Sci 6(4):2187–2193

Kumar P et al (2017c) Resourceful photosynthesis system and stem reserve accumulation plays decisive role in grain yields of Kodo millet (*Paspalum scrobiculatum*). Int J Pure Bio Sci 5(2):420–426

Lata C et al (2011) Association of an SNP in a novel DREB2-like gener SiDREB with stress tolerance in foxtail millet [*Setaria italica* (L)]. J Expt Bot:62, 3387–3401

Li P, Brutnell TP (2011) Setaria viridis and *Setaria italica*, model genetic system for the panicoid grasses. J Expt Bot 65:5415–5427

Lin F et al (2006a) Analysis on the responsive genes during drought and rehydration in broomcorn millet (*Panicum miliaceum* L.) by means of cDNA-AFLP. Agric Res Arid Areas 3:19

Lin F et al (2006b) Isolation and analysis of genes induced by rehydration after serous drought in broomcorn millet (*Panicum milliaceum* L.) by SSH. Chin J Agric Biotechnol 3:227–242

Lin F et al (2008) Cloning of broomcorn millet thionine synthetase gene from broomcorn millet (*Panicu milliaceum* L.) and its expression during drought and re-watering. Acta Agron Sin 34:777–782

Lin H-S et al (2012) Genetic diversity in the foxtail millet (*Setaria italica*) germplasm as determined by agronomic traits and microsatellite markers. Aus J Crop Sci (AJSC) 6(2):342–349

Liu M et al (2016) Genetic diversity and population structure of Broomcorn millet (*Panicum miliaceum* L.) cultivars and landraces in China based on microsatellite markers. Int J Mol Sci 17:1–18

Maccarthy DS, Vlek PLG (2012) Impact of climate change on Sorghum production under different nutrient and crop residue management in semi-arid region of Ghana: a modeling perspective. Afr Crop Sci J 20(S2):243–259

Manga VK (2015) Diversity in pearl millet [*Pennisetum glucu*m (L.) R.Br.] and its management. Indian J Plant Sci 4(1):38–51

Marga M (2018) Progress, achievements and challenges to tef breeding in Ethiopia. J Agri Sci Food Res 9(1):1–8

Masood M, Azam Ali SN (2007) Effects of environmental stress on growth, radiation use and yield of finger millet (*Eleusine coracana*). Pak J Bot 39(2):463–474

Matsuura A et al (2012) Effect of pre-and post-heading water deficit on growth and grain yield of four millets. Plant Prod Sci 15(4):323–331

Mengistu DK (2009) The influence of soil water deficit imposed during various developmental phases on physiological processes of tef (*Eragrostis tef*). Agric Ecocyst Environ 132:283–289

Munns R et al (2006) Approaches to increasing the salt tolerance of wheat and other cereals. J Exp Bot 57:1025

Nagarajan L et al (2007) Determinants of millet diversity at the household-farm and village-community levels in the drylands of India: the role of local seed systems. Agric Econ 36:157–167

Naino Jika AK et al (2017) Unexpected Pattern of Pearl millet genetic diversity among ethno-linguistic groups in the Lake Chad Basin. Heredity 118:491–502

Nielsen DC, Vigil MF (2017) Water use and environmental parameters influence proso millet yield. Field Crop Res 212:34–44

Ning Na et al (2017) Correlation between grain quality of foxtail millet [*Setaria italica* (L.) P.Beauv.] and environmental factors of multivariate statistical analysis. Chilean Journal of Agricultural Research 77(4):303–310

Nirmalkumari A, Vetriventham M (2010) Characterization of foxtail millet germplasm collections for yield contributing traits. Elect J Plant Breed 1(2):140–147

Nirmalkumari A et al (2010) Morphological characterization and evaluation of little millet (*Panicum sumatrense* are Roen and Schultz.) germplasm. Elect J Plant Breed 1(2):148–155

Norean S, Ashraf M (2008) Alleviation of advance effects of salt stress on sunflower (Helianthus annus L.) by exogenous application of salicylic acid: growth and photosynthesis. Pak J Bot 40(4):1657–1663

Oduori C, Kanyenii B (2007) Finger millet in Kenya: importance, advances in RandD, Challenges and opportunities for improved production and profitability. In: Mgonja MA, Leene JM, Emanyasa E, Srinivasprasad S (eds) Proceedings of the first International Finger Millet Stakeholder Workshop, 13–14 Sepetember 2005. ICRISAT, Nairobi, pp 10–22

Olatoye MO et al (2018) Genomic signatures of adaptation to a precipitation gradient in Nigerian sorghum. G3 Genes/Genomes/Genetics 8:3269–3281

Onyango AO (2016) Finger-Millet: food security crop in the arid and semi-arid lands (ASALs) of Kenya. World Environ 6(2):62–70

Owere L et al (2015) Variability and trait relationships among finger millet accessions in Uganda. Uganda J Agri Sci 16(2):161–176

Padulosi S et al (2009) Food security and climate change: role of plant genetic resources of minor millets. Indian J Plant Gonct Resour 22(1):1–16

Pal AK et al (2011) Collection of multi-crop germplasm from lower and upper Subansiri district of Arunachal Pradesh, India and their range of diversity. Indian J Hill Farm 24(1):1–6

Plaza-Wuthrick S, Tadele Z (2012) Milet improvement through regeneration and transformation. Biotechnol Mol Bil Reve 7:48–61

Puranik S et al (2017) Harnessing finger millet to combat deficiency in humans: challenges and prospects. Front Plant Sci 8:1–16

Rana R (2002) Agro-ecosystem: an assessment. Indian J Pl Ganet Res 15(1):1–16

Reghupathy S et al (2016) DNA record of some traditional small millet landraces in India and Nepal. Biotech 6:133–152

Rengalakshmi R (2005) Folk biological classification of minor millet species in Kolli hills, India. BioOne 25:59–70

Renolds MP et al (2016) An integrated approach to maintaining cereal productivity under climate change. Glob Food Sec 8:9–18

Roychowdhury R (2014) Crop improvement in the era of climate change. IK International Publishing House, New Delhi, p 496

Roychowdhury R, Tah J (2013) Mutagenesis – a potential approach for crop improvement. In: Hakeem KR, Ahmad P, Ozturk M (eds) Crop improvement – new approaches and modern techniques. Springer, Boston, MA, pp 149–187

Roychowdhury R, Karmakar J, Karmakar J, Adak MK, Dey N (2013a) Physio-biochemical and microsatellite based profiling of lowland rice (*Oryza sativa* L.) landraces for osmotic stress tolerance. American J Plant Sci 16(12):52

Roychowdhury R, Abdel Gawwad MR, Banerjee U, Bishnu S, Tah J (2013b) Status, trends and prospects of organic farming in India: a review. J Plant Biol Res 2:38–48

Roychowdhury R, Banherjee U, Slofkova S, Tah J (2013c) Organic farming for crop improvement and sustainable agriculture in the era of climate change. OnLine J Biol Sci 13(2):50–65

Roychowdhury R, Taoutaou A, Hakeem KR, Gawwad MR, Tah J (2014) Molecular marker-assisted technologies for crop improvement. In: Roychowdhury R (ed) Crop improvement in the era of climate change. IK International Publishing House, New Delhi, pp 241–258

Roychowdhury R, Khan MH, Choudhury S (2018) Arsenic in rice: an overview on stress implications, tolerance and mitigation strategies. In: Hasanuzzaman M, Nahar K, Fujita M (eds) Plants under metal and metalloid stress. Springer, Singapore, pp 401–415

Roychowdhury R, Khan MH, Choudhury S (2019) Physiological and molecular responses for metalloid stress in rice – a comprehensive overview. In: Hasanuzzaman M, Fujita M, Nahar K, Biswas J (eds) Advances in rice research for abiotic stress tolerance. Woodhead Publishing, Elsevier, pp 341–369

Sattler FT et al (2018) Characterization of West and Central African accessions from a pearl millet reference collection for agro-morphological traits and Striga resistance. Plan Gene Res Char Util 16(3):260–272

Selvi VM et al (2014) Genetics and interrelationships of yield traits for enhancing productivity of little millet. Elect J Plant Breed 5(1):82–86

Seyfu K (1997) Tef. Eragrostis tef (Zucc.) Trotter. Promoting the conservation and use of underutilized and neglected crops. 12. Institute of Plant Genetics and Crop Plant research. Gatersleben/IPGRI, Rome, Italy pp.1–52
Sharma SK, Brahmi P (2006) Gene Bank Curators (Chapter 14). PGRFA, India, pp 182–196
Shibario SI et al (2016) Finger millet (Eleusine coracana) production in Lower Eastern, Kenya: status, constraints and opportunities. British J Appl Sci Technol 14(66):1–10
Singh M, Upadhyaya HD (2016) Genetic and genomic resources for grain cereals improvement. Academic Press, New York, pp xiii–xix
Singh P et al (2017) An assessment of yield gains under climate change due to genetic modification of pearl millet. Sci Total Environ 601-602:1226–1237
Sivakumar MVK (1992) Empirical-analysis of dry spells for agricultural applications in West Africa. J Clim 5(5):532–539
Smirnoff N, Colombe SV (1988) Drought influences the activity of enzymes of the chloroplast hydrogen-peroxide scavenging system. J Expt Bot 39:1097–1108
Subramanian M, Thirumeni S (2007) Genetic diversity in rice and conservation of germplasm. In: Kannaiyan S, Gopalan A (eds) Crop genetic resources and conservation, vol I. Associated Publishing Co., New Delhi, pp 1–17
Tesfaye K, Mengistu S (2017) Phenotypic characterization of Ethiopian finger millet accessions [Eleusine coracana (L.) Gaertn.] for their agronomically important traits. Acta Univ Sapientiae Agric Environ 9:107–118
Tester M, Langridge P (2010) Breeding technologies to increase crop production in a changing world. Science 327:818–822
Tongcheng F et al (2016) Simulation of climate change impacts on grain sorghum production grown under free air CO_2 enrichment. Int Agrophys 30:311–322
Trivedi AK et al (2017) Evaluation of barnyard millet diversity in central Himalayan region for environmental stress tolerance. J Agric Sci 15(10):1497–1507
Ullah A et al (2016) Recognizing production options for pearl in Pakistan under changing climate scenarios. J Integr Agric 15:1–13
Upadhyaya HD et al (2006) Current status of genetic resources conservation, characterization and utilization. The Rajendra S Paroda Genebank. ICRISAT, Patancheru
Upadhyaya HD et al (2007a) Genetic resources diversity of finger millet-a global perspective. In: Mgonja MA, Leene JM, Emanyasa E, Srinivasprasad S (eds) Proceedings of the first International Finger Millet Stakeholder Workshop, 13–14 September 2005. ICRISAT, Nairobi, pp 90–101
Upadhyaya HD et al (2007b) Pearl millet germplasm of ICRISAT genebank: status of impact. JSAT Agric Res 3:1–5
Upadhyaya H et al (2014) Forming core collection in barnyard, Kodo and little millets using morphoagronomic descriptors. Crop Sci 54:1–10
Vander Wilingen C et al (2004) Mechanical stabilization of desiccated vegetative tissues of the resurrection grass Eragrostis nindensis: does a TIP 3; 1 and/or compartmentalization of subcellular components and metabolites play a role? J Exp Bot 55(397):651–661
Wallace JG (2015) The genetic makeup of a global Barnyard millet germplasm collection. Plant Genome 8(1):1–7
Wang C et al (2012) Genetic diversity and population structure of Chinese foxtail millet [Setaria italica (L.) Beauv.] landraces. G3 Geres/Genomes/Genetics 2:769–777
Wanyera NMW (2007) Finger milet [Eleusine coracana (L.) Gaertn] in Uganda. In: Mgonja MA, Leene JM, Emanyasa E, Srinivasprasad S (eds) Proceedings of the first international finger millet stakeholder workshop, 13–14 September 2005. ICRISAT, Nairobi, pp 1–9
Warner DA, Edward GE (1988) C_4 photosynthesis and leaf anatomy in diploid and autotetraploid Pennisetum americanum (Pearl Millet). Plant Sci 56(1):85–92
Winkel T et al (1997) Effect of the timing of water deficit on growth, phenology and yield of pearl millet [Pennisetum glaucum (L.) R.Br.] grown in Sahelian conditions. J Exp Bot 48(310):1001–1009

Worku K et al (2018) Tef (Eragrostis tet) variety 'Tesfa'. Ethiop J Agric Sci 28(2):107–112
Yabuno T (1987) Japanese barnyard millet (Echinochloa utilis, Poaceae) in Japan. Econ Bot 41(4):484–493
Yadav OP (2008) Collection, characterization and evaluation of genetic diversity in Pearl millet and landraces from arid and semi-arid regions of Rajasthan. Ann Arid Zone 47(1):33–39
Yadav RK et al (2018) Diversity and on-farm evaluation. Cogent Food Agric 4:1–17
Yang Q et al (2016) Water use efficiency of foxtail millet (Panicum italicum L.) under climate change conditions in northwest region of China. Agrociencia 50:665–676
Zhang P et al (2012) Leaf senescence and activities of antioxidant enzymes in different broomcorn millet (Panicum miliaceum L.) cultivars under simulated drought condition. J Food Agric Environ 10:438–444
Zhu Q et al (2012) High-throughout discovery of mutations in tef semi-dwarfing genes by next-generation sequencing analysis. Genetics 192:819–829

Index

R. Roychowdhury et al. (eds.), *Sustainable Agriculture in the Era of Climate Change*, https://doi.org/10.1007/978-3-030-45669-6

B

O

The manufacturer's authorised representative in the EU is Springer
Nature Customer Service Centre GmbH, Europaplatz 3, 69115 Heidelberg,
Germany. If you have any concerns regarding our products, please
contact ProductSafety@springernature.com

Printed and bound by CPI Group (UK) Ltd, Croydon, CR0 4YY
30/09/2026
02225697-0018